Physics of Thermal Therapy

Physics of Thermal Therapy
Fundamentals and Clinical Applications

Edited by

Eduardo G. Moros

CRC Press
Taylor & Francis Group
Boca Raton London New York

CRC Press is an imprint of the
Taylor & Francis Group, an **informa** business

A TAYLOR & FRANCIS BOOK

First published 2013 by CRC Press

Published 2019 by CRC Press
Taylor & Francis Group
6000 Broken Sound Parkway NW, Suite 300
Boca Raton, FL 33487-2742

First issued in paperback 2020

© 2013 by Taylor & Francis Group, LLC
CRC Press is an imprint of the Taylor & Francis Group, an informa business

No claim to original U.S. Government works

ISBN 13 : 978-0-367-57663-9 (pbk)
ISBN 13 : 978-1-4398-4890-6 (hbk)

Library of Congress Cataloging-in-Publication Data

Physics of thermal therapy : fundamentals and clinical applications / edited by Eduardo Moros.
    p. ; cm. -- (Imaging in medical diagnosis and therapy)
   Includes bibliographical references and index.
   ISBN 978-1-4398-4890-6 (hardcover : alk. paper)
   I. Moros, Eduardo. II. Series: Imaging in medical diagnosis and therapy.
   [DNLM: 1. Hot Temperature--therapeutic use. 2. Nanostructures--therapeutic use. 3. Neoplasms--therapy. 4. Ultrasonic Therapy--methods. WB 469]

616.99'407543--dc23

2012033768

Visit the Taylor & Francis Web site at
http://www.taylorandfrancis.com

and the CRC Press Web site at
http://www.crcpress.com

To Kimberly,

your love and noble character strengthen me,

and

to our wonderful sons, Jonas and Ezra

Contents

PART I: Foundations of Thermal Therapy Physics

PART II: Clinical Thermal Therapy Systems

PART III: Physical Aspects of Emerging Technology for Thermal Therapy

Series Preface

Advances in the science and technology of medical imaging and radiation therapy are more profound and rapid than ever before, since their inception over a century ago. Further, the disciplines are increasingly cross-linked as imaging methods become more widely used to plan, guide, monitor, and assess treatments in radiation therapy. Today the technologies of medical imaging and radiation therapy are so complex and so computer driven that it is difficult for the persons (physicians and technologists) responsible for their clinical use to know exactly what is happening at the point of care, when a patient is being examined or treated. The professionals best equipped to understand the technologies and their applications are medical physicists, and these individuals are assuming greater responsibilities in the clinical arena to ensure that what is intended for the patient is actually delivered in a safe and effective manner.

The growing responsibilities of medical physicists in the clinical arenas of medical imaging and radiation therapy are not without their challenges, however. Most medical physicists are knowledgeable in either radiation therapy or medical imaging and expert in one or a small number of areas within their discipline. They sustain their expertise in these areas by reading scientific articles and attending scientific meetings. In contrast, their responsibilities increasingly extend beyond their specific areas of expertise. To meet these responsibilities, medical physicists periodically must refresh their knowledge of advances in medical imaging or radiation therapy, and they must be prepared to function at the intersection of these two fields. How to accomplish these objectives is a challenge.

At the 2007 annual meeting in Minneapolis of the American Association of Physicists in Medicine, this challenge was the topic of conversation during a lunch hosted by Taylor & Francis Publishers and involving a group of senior medical physicists (Arthur L. Boyer, Joseph O. Deasy, C.-M. Charlie Ma, Todd A. Pawlicki, Ervin B. Podgorsak, Elke Reitzel, Anthony B. Wolbarst, and Ellen D. Yorke). The conclusion of this discussion was that a book series should be launched under the Taylor & Francis banner, with each volume in the series addressing a rapidly advancing area of medical imaging or radiation therapy of importance to medical physicists. The aim would be for each volume to provide medical physicists with the information needed to understand technologies driving a rapid advance and their applications to safe and effective delivery of patient care.

Each volume in the series is edited by one or more individuals with recognized expertise in the technological area encompassed by the book. The editors are responsible for selecting the authors of individual chapters and ensuring that the chapters are comprehensive and intelligible to someone without such expertise. The enthusiasm of the volume editors and chapter authors has been gratifying and reinforces the conclusion of the Minneapolis luncheon that this series of books addresses a major need of medical physicists.

Imaging in Medical Diagnosis and Therapy would not have been possible without the encouragement and support of the series manager, Luna Han of Taylor & Francis Publishers. The editors and authors and, most of all, I are indebted to her steady guidance of the entire project.

William Hendee
Series Editor
Rochester, Minnesota

Preface

The field of thermal therapy has been growing tenaciously in the last few decades. The application of heat to living tissues, from mild hyperthermia to high temperature thermal ablation, produces a host of well-documented genetic, cellular, and physiological responses that are being intensely researched for medical applications, in particular for the treatment of solid cancerous tumors using image guidance. The controlled application of thermal energy (heat) to living tissues has proven to be a most challenging feat, and thus it has recruited expertise from multiple disciplines leading to the development of a great number of sophisticated preclinical and clinical devices and treatment techniques. Among the multiple disciplines involved, physics plays a fundamental role because controlled heating demands knowledge of acoustics, electromagnetics, thermodynamics, heat transfer, fluid mechanics, numerical modeling, imaging, and many other topics traditionally under the umbrella of physics. This book attempts to capture this highly multidisciplinary field! Therefore, it is not surprising that when I was offered the honor of editing a book on the physics of thermal therapy, I was faced with trepidation. After 25 years of research in thermal therapy physics and engineering and radiation oncologic physics, I was keenly aware of the vastness of the field and my humbling ignorance. Even worse, the rapid growth of the field makes it impossible, in my opinion, to do it justice in one tome. Consequently, tough decisions had to be made in choosing the content of the book, and these were necessarily biased by my experience and the kindness of the contributing authors.

The book is divided into three parts. Part I covers the fundamental physics of thermal therapy. Since thermal therapies imply a source of energy and the means for the controlled delivery of energy, Part I includes chapters on bio-heat transfer, thermal dose, thermometry, electromagnetic and acoustic energy sources, and numerical modeling. This part of the book, although not exhaustive, can be thought of as an essential requirement for any person seriously seeking to learn thermal therapy physics.

Part II offers an overview of clinical systems (or those expected to be clinical in the near future) covering internally and externally applied electromagnetic and acoustic energy sources. Despite the large number of devices and techniques presented, these must be regarded as a sample of the current clinical state of the art. A future book on the same topic may have a similar Part I while the contents of Part II would be significantly different, as clinical technology experiences advances based on clinical practice and new needs.

The last section of the book, Part III, is composed of chapters describing the physical aspects of an emerging thermal therapy technology. The spectrum is wide, from new concepts relatively far from clinical application, such as thermochemical ablation, through technologies at various stages in the translational continuum, such as nanoparticle-based heating and heat-augmented liposomal drug delivery, to high-intensity-focused ultrasound interventions that are presently being investigated clinically. Imaging plays a crucial role in thermal therapy, and many of the newer approaches are completely dependent on image guidance *during* treatment administration. Therefore, Part III also covers both conventional as well as emerging imaging technologies and tools for image-guided therapies.

Although there are published books covering the physics and technology of hyperthermia, therapeutic ultrasound, radiofrequency ablation, and other related topics, to my knowledge this is the first book with the title *Physics of Thermal Therapy*. For this I have to thank Dr. William Hendee, a medical physicist *par excellence* and the series editor, who had the original idea for the book. In regard to the target audience, the book has been written for physicists, engineers, scientists, and clinicians. It will also be useful to graduate students, residents, and technologists.

Finally, I must confess that it is extremely difficult to remain modest about the list of outstanding contributors. A well-established expert, at times in collaboration with his/her colleagues, graduate student(s), or postdoctoral fellow(s), has authored each chapter. I am profoundly grateful to all for the time and effort they invested in preparing their chapters. I would also like to thank Luna Han and Amy Blalock from Taylor & Francis for their patience, assistance, and guidance during the entire process leading to this book.

MATLAB® is a trademark of The MathWorks, Inc. and is used with permission. The MathWorks does not warrant the accuracy of the text or exercises in this book. This book's use or discussion of MATLAB® software or related products does not constitute endorsement or sponsorship by The MathWorks of a particular pedagogical approach or particular use of the MATLAB® software.

MATLAB® is a registered trademark of The MathWorks, Inc. For product information, please contact:

The MathWorks, Inc.
3 Apple Hill Drive
Natick, MA, 01760-2098 USA
Tel: 508-647-7000
Fax: 508-647-7001
E-mail: info@mathworks.com
Web: www.mathworks.com

Editor

Eduardo G. Moros earned a PhD in mechanical engineering from the University of Arizona, Tucson, in 1990. His graduate studies were performed at the radiation oncology department in the field of scanned focused ultrasound hyperthermia for cancer therapy. After a year as a research associate at the University of Wisconsin, Madison in the human oncology department, he joined the Mallinckrodt Institute of Radiology at Washington University School of Medicine, St. Louis, Missouri, where he was the chief of hyperthermia physics (1991–2005) and the head of the research physics section (2001–2005). He was promoted to associate professor with tenure in 1999 and to professor in 2005. In August 2005, Dr. Moros joined the University of Arkansas for Medical Sciences as the director of the division of radiation physics and informatics. Currently, he is the chief of medical physics for the departments of radiation oncology and diagnostic imaging at the H. Lee Moffitt Cancer Center and Research Institute in Tampa, Florida.

Dr. Moros served as president of the Society for Thermal Medicine (2004–2005), as associate editor for the journal *Medical Physics* (2000–2007) and the *International Journal of Hyperthermia* (2006–2009), and was a permanent member of the NIH Radiation Therapeutics and Biology Study Section (2002–2005). He is an associate editor of the *Journal of Clinical Applied Medical Physics* and the *Journal of Radiation Research*. He is an active member of several scientific and professional societies, such as the American Association for Physicists in Medicine, the American Society for Therapeutic Radiology and Oncology, the Bioelectromagnetics Society, the Radiation Research Society, the Society for Thermal Medicine, and the International Society for Therapeutic Ultrasound. Dr. Moros holds a certificate from the American Board of Radiology in therapeutic radiologic physics.

Dr. Moros's strength has been to collaborate with scientists and clinicians in the application of physics and engineering to facilitate biomedical research and promote translational studies. He has published more than one hundred peer-reviewed articles and has been a principal investigator/coinvestigator on multiple research grants from the National Institutes of Health, other federal agencies, and industry. He was a recipient of an NIH Challenge Grant in Health and Science Research (RC1) in 2009.

Contributors

R. Martin Arthur
Department of Electrical and Systems Engineering
Washington University in St. Louis
St. Louis, Missouri

Filip Banovac
Department of Radiology
Georgetown University Medical Center
Washington, DC

John C. Bischof
Department of Mechanical Engineering
Department of Biomedical Engineering
Department of Urologic Surgery
University of Minnesota
Minneapolis, Minnesota

Chris Brace
Department of Radiology
Department of Biomedical Engineering
University of Wisconsin, Madison
Madison, Wisconsin

Victoria Bull
Division of Radiotherapy and Imaging
Institute of Cancer Research
Sutton, Surrey, United Kingdom

Lili Chen
Department of Radiation Oncology
Fox Chase Cancer Center
Philadelphia, Pennsylvania

Kevin Cleary
The Sheikh Zayed Institute for Pediatric Surgical Innovation
Children's National Medical Center
Washington, DC

Erik N. K. Cressman
Department of Radiology
University of Minnesota Medical Center
Minneapolis, Minnesota

Mark W. Dewhirst
Department of Radiation Oncology
Duke University Medical Center
Durham, North Carolina

Chris J. Diederich
Department of Radiation Oncology
University of California, San Francisco
San Francisco, California

Kenneth R. Diller
Biomedical Engineering Department
University of Texas
Austin, Texas

Michael L. Etheridge
Department of Mechanical Engineering
Department of Biomedical Engineering
University of Minnesota
Minneapolis, Minnesota

Dieter Haemmerich
Department of Pediatrics
Medical University of South Carolina
Charleston, South Carolina

Jeffrey W. Hand
King's College London
London, United Kingdom

Kullervo Hynynen
Sunnybrook Health Sciences Centre
Toronto, Ontario, Canada

Andreas Jordan
Department of Radiology
Charité-University Medicine
Berlin, Germany

Niels Kuster
Foundation for Research on Information Technologies in
    Society (IT'IS)
and
Swiss Federal Institute of Technology (ETHZ)
Zurich, Switzerland

Faqi Li
College of Biomedical Engineering
Chongqing Medical University
Chongqing, China

Robert J. McGough
Department of Electrical and Computer Engineering
Michigan State University
East Lansing, Michigan

Eduardo G. Moros
H. Lee Moffitt Cancer Center and Research Institute
Tampa, Florida

Esra Neufeld
Foundation for Research on Information Technologies in
    Society (IT'IS)
and
Swiss Federal Institute of Technology (ETHZ)
Zurich, Switzerland

Meaghan A. O'Reilly
Sunnybrook Health Sciences Centre
Toronto, Ontario, Canada

Maarten M. Paulides
Department of Radiation Oncology
Erasmus MC Daniel den Hoed Cancer Center
Rotterdam, The Netherlands

John A. Pearce
Department of Electrical and Computer Engineering
University of Texas at Austin
Austin, Texas

Zhenpeng Qin
Department of Mechanical Engineering
University of Minnesota
Minneapolis, Minnesota

R. Jason Stafford
Department of Imaging Physics
University of Texas MD Anderson Cancer Center
Houston, Texas

Brian A. Taylor
Department of Radiological Sciences
St. Jude Children's Research Hospital
Memphis, Tennessee

Gail R. ter Haar
Division of Radiotherapy and Imaging
Institute of Cancer Research
Sutton, Surrey, United Kingdom

Gerard C. van Rhoon
Department of Radiation Oncology
Erasmus MC Daniel den Hoed Cancer Center
Rotterdam, The Netherlands

Emmanuel Wilson
The Sheikh Zayed Institute for Pediatric Surgical
    Innovation
Children's National Medical Center
Washington, DC

Feng Wu
Institute of Ultrasonic Engineering in
    Medicine
Chongqing Medical University
Chongqing, China

and

Nuffield Department of Surgical Sciences
University of Oxford
Oxford, United Kingdom

I

Foundations of Thermal Therapy Physics

Fundamentals of Bioheat Transfer

Kenneth R. Diller
University of Texas

1.1 Introduction

The science of heat transfer deals with the movement of thermal energy across a defined space under the action of a temperature gradient. Accordingly, a foundational consideration in understanding a heat transfer process is that it must obey the law of conservation of energy, or the first law of *thermodynamics*. Likewise, the process must also obey the second law of thermodynamics, which, for most practical applications, means that heat will flow only from a region of higher temperature to one of lower temperature. We make direct and repeated use of thermodynamics in the study of heat transfer phenomena, although thermodynamics does not embody the tools to tell us the details of how heat flows across a spatial temperature gradient.

A more complete analysis of heat transfer depends on further information about the *mechanisms* by which energy is driven from a higher to a lower temperature. Long experience has shown us that there are three primary mechanisms of action: *conduction*, *convection*, and *radiation*. The study of heat transfer involves developing a quantitative representation for each of the mechanisms that can be applied in the context of the conservation of energy in order to reach an overall description of how the movement of heat by all of the relevant mechanisms influences changes in the thermal state of a system.

Biological systems have special features beyond inanimate systems that must be incorporated in the expressions for the heat transfer mechanisms. Many of these features result in effects that cause mathematical nonlinearities and render the analytical description of bioheat transfer more complex than

more routine problems. For that reason, you will find numerical methods applied for the solution of many bioheat transfer problems, including a large number in this book. The objective of this chapter is to provide a simple introduction of bioheat transfer principles without attempting to delve deeply into the details of the very large number of specific applications that exist. The following chapters will provide this particular analysis where appropriate.

1.2 Heat Transfer Principles

In this section we will review the general principles of heat transfer analysis without reference to the special characteristics of biological tissues that influence heat transfer and the energy balance. These matters will be addressed in the next section. Here we will first consider the energy balance as it applies to all types of heat transfer processes and then each of the three heat transport mechanisms.

1.2.1 Thermodynamics and the Energy Balance

The starting point for understanding the movement of heat within a material is to consider an energy balance for the system of interest. When an appropriate system has been identified in conjunction with a heat transfer process, an energy balance shows that the rate at which the internal energy storage within the system changes is equal to the summation of all energy interactions the system experiences with its environment. This relationship is expressed as the first law of thermodynamics, the conservation of energy:

$$\frac{dE}{dt} = \sum \dot{Q} - \dot{W} + \sum \dot{m}\left(h_{in} - h_{out}\right) + \dot{Q}_{gen} \qquad (1.1)$$

where E is the energy of the system; $\Sigma\dot{Q}$ is the sum of all heat flows, taken as positive into the system; $\dot{W}$ is the rate at which work is performed on the environment; $\Sigma\dot{m}(h_{in} - h_{out})$ is the sum of all mass flows crossing the system boundary, with each having a defined enthalpy, h, as it enters or leaves the system; and $\dot{Q}_{gen}$ is the rate at which energy generation and dissipation occur on the interior of the system. These terms are illustrated in Figure 1.1, depicting how the energy interactions with the environment affect the system energy. In this case the system is represented on a macroscopic scale, but there are alternative situations in which it is of advantage to define the boundary as having microscopic differential scale dimensions.

For the special case of a steady state process, all properties of the system are constant in time, including the energy, and the time derivative on the left side of Equation 1.1 is zero. For these conditions, the net effects of all boundary interactions are balanced.

Each term in Equation 1.1 may be expressed in terms of a specific constitutive relation, which describes the particular energy flow as a function of the system temperature, difference between the system and environmental temperatures, and/or spatial temperature gradients associated with the process as well as many thermal properties of the system and environment. When the constitutive relations are substituted for the individual terms in the conservation of energy (Equation 1.1), the result is a partial differential equation that can be solved for the temperature within the system during a heat transfer process as a function of position and time. There are well-known solutions for many of the classical problems of heat transfer (Carslaw and Jaeger 1959), but numerous biomedical problems involve nonlinearities that require a numerical solution method.

Development of the specific equations for the various constitutive relations constitutes a major component of heat transfer analysis. We will review these relations briefly in the following sections. The one constitutive equation we will discuss here is that for system energy storage.

Although there are a large number of energy storage mechanisms in various materials, those that are likely to be most relevant to processes encountered in biomedical applications include: *mechanical*, related to velocity (kinetic), relative position in the gravity field (potential), and elastic stress; *sensible*, related to a change in temperature; and *latent*, related to a change in phase or molecular reconfiguration such as denaturation. Thus,

$$E = KE + PE + SE + U + L \qquad (1.2)$$

where substitution of a constitutive relation for each term yields

$$KE = \tfrac{1}{2}mV^2,\, PE = mgz,\ SE = \tfrac{1}{2}\kappa x^2,\, U = mc_pT,\, L = m\Lambda \quad (1.3)$$

with properties defined as: KE is kinetic energy; V is velocity; PE is potential energy; g is the acceleration of gravity; z is position along the gravity field; SE is the elastic energy; κ is the spring constant; x is the elastic deformation; U is the internal energy; c_p is the specific heat; T is the temperature; L is the latent energy; and Λ is the latent heat. The most commonly encountered mode of energy storage is via temperature change.

1.2.2 Conduction Heat Transfer

Energy can be transmitted through materials via conduction under the action of an internal temperature gradient. Conduction occurs in all phases of material: solid, liquid, and gas, although the effectiveness of the different phases in transmitting thermal energy can vary dramatically as a function of the freedom of their molecules to interact with nearest neighbors. The conductivity and temperature of a material are key parameters used to describe the process by which a material may be engaged in heat conduction.

The fundamental constitutive expression that describes the conduction of heat is called Fourier's law:

$$\dot{Q}_{cond} = -kA\frac{dT}{dr} \qquad (1.4)$$

where r is a coordinate along which a temperature gradient exists, and A is the area normal to the gradient and the cross section through which the heat flows. The negative sign accounts for the fact that heat must flow along a negative gradient from a higher to a lower temperature. This phenomenon is described by the second law of thermodynamics and is illustrated in Figure 1.2.

For a process in which the only mechanism of heat transfer is via conduction, a microscopic system may be defined as shown in Figure 1.3. Equation 1.4 may be applied to the conservation of energy (Equation 1.1) to obtain a partial differential equation for the temporal and spatial variations in temperature. A microscopic system of dimensions dx, dy, and dz is defined in the interior of the tissue as shown. The various properties and boundary flows illustrated represent the individual terms

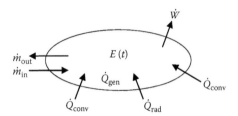

FIGURE 1.1 A thermodynamic system that interacts with its environment across its boundary by flows of heat, mass, and work that contribute to altering the stored internal energy.

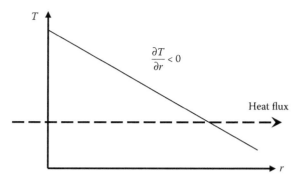

FIGURE 1.2 A positive flow of heat along a coordinate occurs by application of a negative gradient in temperature along the direction of flow.

to be accounted for in applying conservation of energy to this elemental volume.

The time rate of change of energy stored in the elemental system is described as

$$\dot{E}_{st} = \rho c_p \frac{dT}{dt} dx \cdot dy \cdot dz. \tag{1.5}$$

The individual conduction exchanges across the system boundary are written in terms of the Fourier law:

$$\dot{Q}_x = -k \cdot dy \cdot dz \frac{\partial T}{\partial x}$$

$$\dot{Q}_y = -k \cdot dx \cdot dz \frac{\partial T}{\partial y}. \tag{1.6}$$

$$\dot{Q}_z = -k \cdot dx \cdot dy \frac{\partial T}{\partial z}$$

The differentials in the conduction terms on opposing boundary faces are approximated via a Taylor series expansion with all higher order terms dropped:

$$\dot{Q}_x - \dot{Q}_{x+dx} = -\frac{\partial \dot{Q}_x}{\partial x} dx$$

$$\dot{Q}_y - \dot{Q}_{y+dy} = -\frac{\partial \dot{Q}_x}{\partial y} dy. \tag{1.7}$$

$$\dot{Q}_z - \dot{Q}_{z+dzx} = -\frac{\partial \dot{Q}_x}{\partial z} dz$$

When Fourier's law is substituted for the heat flows, the boundary interactions are written in terms of the temperature gradients

$$\dot{Q}_x - \dot{Q}_{x+dx} = \frac{\partial}{\partial x}\left(\frac{\partial T}{\partial x}\right) dx \cdot dy \cdot dz$$

$$\dot{Q}_y - \dot{Q}_{y+dy} = \frac{\partial}{\partial y}\left(\frac{\partial T}{\partial y}\right) dx \cdot dy \cdot dz. \tag{1.8}$$

$$\dot{Q}_z - \dot{Q}_{z+dzx} = \frac{\partial}{\partial z}\left(\frac{\partial T}{\partial z}\right) dx \cdot dy \cdot dz$$

The constitutive Equations 1.5 and 1.8 may be substituted in the conservation of energy Equation 1.1 for the limited boundary interactions assumed in this analysis, noting that each resulting term contains the system volume, $dx \cdot dy \cdot dz$, which can be divided out:

$$\rho c_p \frac{\partial T}{\partial t} = \frac{\partial}{\partial x}\left(k\frac{\partial T}{\partial x}\right) + \frac{\partial}{\partial y}\left(k\frac{\partial T}{\partial y}\right) + \frac{\partial}{\partial x}\left(k\frac{\partial T}{\partial z}\right). \tag{1.9}$$

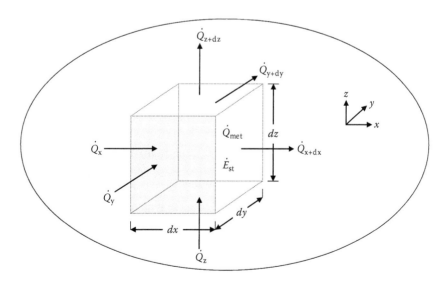

FIGURE 1.3. A small interior elemental system for analysis of heat conduction consisting of differential lengths dx, dy, and dz in Cartesian coordinates as identified within a larger overall system.

This expression is known as *Fourier's equation*, and it has units of W/m³. Although Equation 1.9 was derived in Cartesian coordinates, it can be generalized to be applicable for alternate coordinate systems:

$$\rho c_p \frac{\partial T}{\partial t} = \nabla \cdot (k \nabla T). \tag{1.10}$$

The foregoing equation may be divided by the product ρc_p to isolate the temperature term on the left side. The resulting thermal property is the thermal diffusivity, $\alpha = k/\rho c$:

$$\frac{\partial T}{\partial t} = \nabla \cdot (\alpha \nabla T). \tag{1.11}$$

Applications involving therapeutic hyperthermia generally involve the deposition of a temporally and spatially distributed internal energy source to elevate the temperature within a target tissue. In this case, the energy generation term must be included in the conservation of energy equation, resulting in

$$\frac{\partial T}{\partial t} = \nabla \cdot (\alpha \nabla T) + \frac{\dot{Q}_{gen}}{\rho c_p}. \tag{1.12}$$

The complete solution of Equation 1.11 requires the specification of one (initial) boundary condition in time and two spatial boundary conditions for each coordinate dimension along which the temperature may vary independently. These boundary conditions are used to evaluate the constants of integration that result from solution of the partial differential equation. They are determined according to: (a) the geometric shape of the system, including whether there is a composite structure with component volumes having distinct material properties; (b) what the temperature field interior to the system is like at the beginning of the process; (c) the geometry of imposed heat transfer interactions with the environment, such as radiation and/or convection; and (d) how these environmental interactions may change over time. As an aggregate, these four types of conditions dictate the form and complexity of the mathematical solution to Equation 1.11, and there are many different outcomes that may be encountered. Mathematical methods for solving this equation have been available for many decades, and some of the most comprehensive and still useful texts are true classics in the field (Morse and Feshback 1953; Carslaw and Jaeger 1959).

The temporal boundary condition is generally defined in terms of a known temperature distribution within the system at a specific time, usually at the beginning of a process of interest. However, definition of the spatial boundary conditions is not so straightforward. There are three primary classes of spatial boundary conditions that are encountered most frequently. The thermal interaction with the environment at the physical boundary of the system may be described in terms of a defined temperature, heat flux, or convective process. The energy source applied to create a hyperthermia state in tissue nearly always results in a geometrically complex internal temperature field imposed onto the system of analysis. The source can be viewed as a type of internal boundary condition. The solution of the Fourier equation issues in an understanding of the spatial and temporal variations in temperature, $T(x,y,z,t)$, which can then be applied to predict the therapeutic outcome of a procedure. This analysis is covered in Chapter 2, this book. The solution for the temperature field in tissue may also be incorporated into feedback control algorithms to achieve specific therapeutic outcomes.

Several classes of boundary conditions will be discussed to illustrate how different environmental interactions influence the flavor of the solution for the temperature field. We will first consider semi-infinite geometries for which there is an exposed surface of the tissue and an elevated temperature develops over space and time in the interior. The overall tissue dimensions are assumed to be large enough so that the effects of the free surface on the opposing side of the body are not encountered. This geometry simplifies to a one-dimensional Cartesian coordinate system, which we will represent in the coordinate x. The three classes of boundary conditions we will consider for semi-infinite geometry are: (a) constant temperature, (b) convection, and (c) specified heat flux.

1.2.2.1 Semi-Infinite Geometry—Constant Surface Temperature: $T(0,t) = T_s$

A temperature T_s is assumed to be applied instantaneously to the surface of a solid and then to be held constant for the duration of the process. The solution for this problem is the Gaussian error function, $erf\varphi$, where

$$erf\varphi = \frac{2}{\sqrt{\pi}} \int_0^\varphi \exp(-\xi^2)d\xi.$$

For a uniform initial temperature, T_i, throughout the material, the solution is expressed as a dimensionless ratio as

$$\frac{T(x,t) - T_s}{T_i - T_s} = erf\left(\frac{x}{2\sqrt{\alpha t}}\right). \tag{1.13}$$

1.2.2.2 Semi-Infinite Geometry—Convection: $-k\frac{\partial T(0,t)}{\partial t} = h[T_\infty - T(0,t)]$

Here, the symbol h is the convective heat transfer coefficient (in other contexts it may be used for specific enthalpy (Equation 1.1) or for the Planck constant (Equation 1.71)), which is a function of the boundary interaction between a solid substrate and the environmental fluid that is at a temperature T_∞. Convective heat transfer analysis is focused primarily on determining the value for h to be applied as the boundary condition for a conduction process within a solid immersed in a fluid environment. The solution for the internal temperature field is

$$\frac{T(x,t)-T_s}{T_i-T_s} = erfc\left(\frac{x}{2\sqrt{\alpha t}}\right) - \left[\exp\left(\frac{hx}{k}+\frac{h^2\alpha t}{k^2}\right)\right]$$

$$\times\left[erfc\left(\frac{x}{2\sqrt{\alpha t}}+\frac{h\sqrt{\alpha t}}{kw}\right)\right] \qquad (1.14)$$

where *erfc*φ is the complementary error function defined as *erfc*φ = 1 − *erf*φ.

1.2.2.3 Semi-Infinite Geometry—Defined Surface Heat Flux: $-k\frac{\partial T(0,t)}{\partial t} = \frac{\dot{Q}_s}{A} = \dot{q}_s$

A heat flow per unit area of the surface is assumed to be applied instantaneously and then maintained continuously for the duration of the process. Typical causes of this boundary condition are an external noncontact energy source that is in communication with the surface of a solid via electromagnetic radiation. The solution of this problem is

$$T(x,t)-T_i = \frac{2\dot{q}_s\sqrt{\alpha t/\pi}}{k}\exp\left(\frac{-x^2}{4\alpha t}\right) - \frac{\dot{q}_s x}{k}erfc\left(\frac{x}{2\sqrt{\alpha t}}\right) \qquad (1.15)$$

1.2.2.4 Finite Dimensioned System with Geometric and Thermal Symmetry

Another boundary condition encountered frequently occurs when a finite-sized solid is exposed to a new convective environment in a stepwise manner. If the system and process both exhibit geometric and thermal symmetry, an explicit mathematical solution exists for one-dimensional Cartesian, cylindrical, and spherical coordinates in the form of an infinite series. As will become apparent, it is advantageous to write the problem statement and solution in terms of dimensionless variables.

The temperature is scaled to the environmental value as $\theta = T - T_\infty$ and is normalized to the initial value:

$$\theta^* = \frac{\theta}{\theta_i} = \frac{T-T_\infty}{T_i-T_\infty}. \qquad (1.16)$$

Likewise, the independent variables for position and time are normalized to the size and thermal time constant of the system,

$$x^* = \frac{x}{L} \qquad (1.17)$$

where *L* is the half width of the system along the primary thermal diffusion vector,

$$t^* = Fo = \frac{\alpha t}{L^2} \qquad (1.18)$$

where *Fo* is called the Fourier number, representing a dimensionless time. It is the ratio of the actual process time compared to the thermal diffusion time constant for the system.

The Fourier equation (Equation 1.11) in one dimension can be written in terms of these dimensionless variables as

$$\frac{\partial\theta^*}{\partial Fo} = \frac{\partial^2\theta^*}{\partial x^{*2}} \qquad (1.19)$$

for which the initial and boundary conditions are written as

$$\theta^*(x^*,0) = 1 \qquad (1.20)$$

$$\left.\frac{\partial\theta^*}{\partial x^*}\right)_{x^*=0} = 0 \qquad (1.21)$$

which is a result of thermal and geometric symmetry,

$$-\left.\frac{\partial\theta^*}{\partial x^*}\right)_{x^*=1} = Bi\theta^*(1,Fo) \qquad (1.22)$$

where *Bi* is defined as the Biot number, which represents the ratio of thermal resistances by condition on the interior of the solid and by convection at the surface interface with a fluid environment:

$$Bi = \frac{hL}{k} = \frac{1/kA}{1/hA}. \qquad (1.23)$$

The solution for this problem is in the form of an infinite series,

$$\theta^* = \sum_{n=1}^{\infty} C_n e^{-\zeta_n^2 Fo}\cos(\zeta_n x^*) \qquad (1.24)$$

where C_n satisfies for each value of *n*,

$$C_n = \frac{4\sin\zeta_n}{2\zeta_n+\sin(2\zeta_n)} \qquad (1.25)$$

and the eigenvalues ζ_n are defined as the positive roots of the transcendental equation

$$\zeta_n\tan\zeta_n = Bi. \qquad (1.26)$$

Thus, there are unique values of C_n and ζ_n for each value of *Bi*. The first six roots of this expression have been compiled as a function of discrete values for *Bi* between 0 and ∞, and are available widely (Carslaw and Jaeger 1959).

Likewise, a fully analogous analysis can be applied for systems modeled in cylindrical and spherical coordinates. For cylindrical geometry, the dimensionless temperature is given by

$$\theta^* = \sum_{n=1}^{\infty} C_n e^{-\zeta_n^2 Fo} J_0\left(\zeta_n r^*\right) \qquad (1.27)$$

where $Fo = \alpha t / R^2$. C_n satisfies for each value of n

$$C_n = \frac{2 J_1(\zeta_n)}{\zeta_n J_0^2(\zeta_n) + J_1^2(\zeta_n)} \qquad (1.28)$$

and the eigenvalues ζ_n are defined as the positive roots of the transcendental equation

$$\zeta_n \frac{J_1(\zeta_n)}{J_0(\zeta_n)} = Bi \qquad (1.29)$$

where $Bi = hR/k$.

For a spherical geometry, the dimensionless temperature is given by

$$\theta^* = \sum_{n=1}^{\infty} C_n e^{-\zeta_n^2 Fo} \frac{1}{\zeta_n r^*} \sin\left(\zeta_n r^*\right) \qquad (1.30)$$

where $Fo = \alpha t / R^2$. C_n satisfies for each value of n

$$C_n = \frac{4\left[\sin(\zeta_n) - \zeta_n \cos(\zeta_n)\right]}{2\zeta_n + \sin(2\zeta_n)} \qquad (1.31)$$

and the eigenvalues ζ_n are defined as the positive roots of the transcendental equation

$$1 - \zeta_n \cot \zeta_n = Bi \qquad (1.32)$$

where $Bi = hR/k$.

Although the exact solution takes the form of an infinite series, for many problems it is adequate to use only a limited number of terms and still maintain an acceptable level of accuracy. If the analysis can be restricted to portions of the process following the initial transient for which $Fo > 0.2$, then only the first term is required. The closer the analysis must approach the process beginning, the more terms must be included in the calculation. In these cases the exact solution still can be computed in a relatively straightforward manner (Diller 1990a, b), although the detail that must be included increases with each additional term. Unfortunately, in many classes of biomedical processes, information concerning the initial transient behavior is of greatest interest, and it is not possible to use the single term approximation.

The following two sections present brief descriptions for how the convective heat transfer coefficient, h, and the radiation heat flux incident on a surface, $\dot{q}_s$, can be computed to provide quantitative boundary conditions for conduction problems as may be needed.

1.2.3 Convection Heat Transfer

Convective boundary conditions occur when a solid substrate is in contact with a fluid at a different temperature. The fluid may be in either the liquid or vapor phase. The convective process involves relative motion between the fluid and the substrate. The magnitude of the heat exchange is described in terms of Newton's law of cooling, for which the relevant constitutive property of the system is the convective heat transfer coefficient, $h(W/m^2K)$. The primary objective of convection analysis is to determine the value of the convective coefficient, h, to apply in Newton's law of cooling, which describes the convective flow at the surface, $\dot{Q}_s$, in terms of h, the interface area, A, between the fluid and solid, and the substrate surface and bulk fluid temperatures, (T_s) and (T_∞):

$$\dot{Q}_s = hA(T_s - T_\infty). \qquad (1.33)$$

There are four distinguishing characteristics of convective flow that determine the nature and intensity of a convection heat transfer process. It is necessary to evaluate each of these characteristics to calculate the value for the convective heat transfer coefficient, h. These characteristics and the various options they may take are:

1. The source of relative motion between the fluid and solid, resulting in forced (pressure driven) or free (buoyancy driven) convection.
2. The geometry and shape of the boundary layer region of the fluid in which convection occurs, producing internal or external flow. In addition, for free convection the orientation of the fluid/solid interface in the gravitational field is important.
3. The boundary layer flow domain, being laminar or turbulent.
4. The chemical composition and thermodynamic state of the fluid in the boundary layer that dictate numerical values for the constitutive properties relevant to the convective process.

The influence of each of the four principal characteristics must be evaluated individually and collectively, and the value determined for h may vary over many orders of magnitude depending on the combined effects of the characteristics. Table 1.1 presents the range of typical values for h for various combinations of the characteristics as most commonly encountered.

The relative motion between a fluid and solid may be caused by differing kinds of energy sources. Perhaps most obviously, an external force can be applied to the fluid or solid to produce the motion (which is termed *forced convection*). This force is most

TABLE 1.1 Ranges of Typical Values for h as Encountered for Various Combinations of Convective Transport Process Characteristics

Process Characteristics	Range of h (W/m²·K)
Free Convection	
Vapors	3–25
Liquids	20–1000
Forced Convection—Internal/External	
Vapors	10–500
Liquids	100–15,000
Phase Change	
Between liquid and vapor	5,000–100,000

frequently a mechanical force to move the solid or to impose a pressure gradient on the fluid. However, in the absence of an external motivational force, the heat transfer process itself will cause relative motion. Owing to the constitutive properties of fluids, the existence of a temperature gradient produces a concomitant density gradient. When the fluid is in a force field such as gravity or centrifugation, the density gradient causes internal motion within the fluid by buoyancy effects as the less dense fluid rises and the more dense fluid falls under the action of the force field. This phenomenon is called *free convection* since no external energy source is applied to cause the motion directly. Any time there is a temperature gradient in a fluid, there is the potential for having free convection heat transfer. As can be anticipated, the fluid flow patterns for forced and free convection

are very different, and therefore forced and free convection produce quite disparate heat transfer effects. Also, analysis of the fluid flow characteristics in forced and free convection is unique because of the differing patterns of motion. Usually the magnitude of forced convection effects is much larger than for free convection, as indicated in Table 1.1. Thus, although the potential for free convection will be present whenever a temperature field exists in a fluid, if there is also an imposed forced source of fluid motion, the free convection effects usually will be masked since they are much smaller, and they can be neglected.

The convection process consists of the sum of two separate effects. First, when there is a temperature gradient in a fluid, heat conduction will occur consistent with the thermal conductivity of the chemical species and its thermodynamic state. The conduction effect can be very large as in a liquid metal or very small as in a low density vapor. The conduction occurs via microscopic scale interactions among atoms and molecules, with no net translation of mass. Second, there will be transport of energy associated with the bulk movement of a flowing fluid. The component due to only bulk motion is referred to as *advection*. Convection involves a net aggregate motion of the fluid, thereby carrying the energy of the molecules from one location to another. These two effects are additive and superimposed.

A fundamental aspect of convection heat transfer is that the processes involve both velocity and temperature boundary layers in the fluid adjacent to a solid interface. Illustrations of these boundary layers are shown in Figure 1.4. The velocity boundary

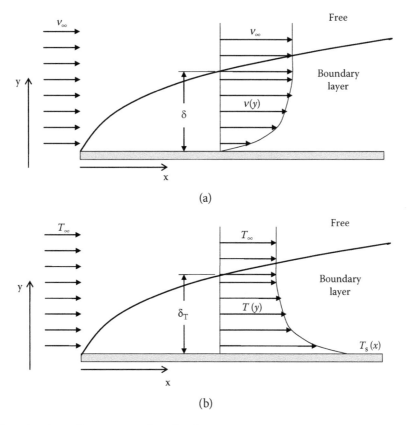

FIGURE 1.4 (a) Velocity boundary layer; (b) temperature boundary layer.

layer defines the region wherein viscous drag causes a velocity gradient as the interface is approached. The region outside the boundary layer where the viscous properties do not affect the flow pattern is called the inviscid free stream. The fluid velocity outside the boundary layer is designated by v_∞ (m/s), and the boundary layer thickness by δ(m), which increases with distance along the interface from the point of initial contact between the fluid and solid. In this case, it is assumed that the interface is planar and the fluid flow is parallel to the interface. In like manner, a thermal boundary layer develops in the flowing fluid as heat transfer occurs between a solid substrate and fluid that are at dissimilar temperatures.

The velocity and temperature boundary layers have similar features. Both define a layer in the fluid adjacent to a solid in which a property gradient exists. The temperature boundary layer develops because there is a temperature difference between the fluid in the free stream T_∞ and the solid surface T_s. A temperature gradient exists between the free stream and the surface, with the maximum value at the surface, and which diminishes to zero at the outer limit of the boundary layer at the free stream. The temperature gradient at the surface defines the thermal boundary condition for conduction in the solid substrate. The boundary condition can be written by applying conservation of energy at the interface. Since the interface has no thickness, it has no mass and is therefore incapable of energy storage. Thus, the conductive inflow is equal to the convective outflow as illustrated in Figure 1.5.

An important feature of the convection interface is that there is continuity of both temperature and heat flow at the surface, the latter of which is expressed in Equation 1.34:

$$-k_f \frac{dT}{dy}\bigg|_{y=0} = h(T_s - T_\infty).$$ (1.34)

The magnitude of convection heat transfer is directly dependent on the size and flow characteristics within the boundary layer. As a general rule, thicker boundary layers result in a larger resistance to heat transfer and, thus, a smaller value for h. The result is that there can be local variations in convective transport over different regions of an interface as a function of the local boundary layer characteristics. In some cases it is necessary to determine these local variations, requiring more detailed calculations. Often it is sufficient to use a single average value over the entire interface, thereby simplifying the analysis. The averaged heat transfer coefficient is denoted by $\bar{h}_L$, where the subscript L defines the convective interface dimension over which the averaging occurs.

Values for the convective heat transfer coefficient appropriate to a given physical system are usually calculated from correlation equations written in terms of dimensionless groups of system properties. The most commonly applied dimensionless groups are defined as follows.

The *Nusselt number* is a dimensionless expression for the convective heat transfer coefficient defined in Equation 1.35. It can be written in terms of local or averaged (over an entire interface surface) values:

$$Nu_x = \frac{hx}{k_f} \quad \text{or} \quad \bar{Nu}_L = \frac{\bar{h}_L L}{k_f}.$$ (1.35)

An important distinction to note is that k_f is the thermal conductivity of the fluid, not of the solid substrate. The Nusselt number describes the ratio of total convection effects to pure thermal conduction in the fluid. In principle, its value should always be greater than 1.0 since convection is a combination of both conduction plus advection in the fluid.

Nu represents the ratio of the temperature gradient in the fluid at the interface with the solid to an overall reference temperature gradient based on a physical dimension of the system, L. This dimension has a different meaning, depending on whether the flow geometry is internal or external. For an internal flow in which the boundary layer occupies the volume normal to the interface, the dimension represents the cross-sectional size of the flow passage, such as the diameter, D. For an external flow configuration in which the size of the boundary layer can grow normal to the interface with no physical restriction, the relevant dimension is the distance along the interface from the leading edge at which the fluid initially encounters the solid substrate, such as the length, L. It is important to identify

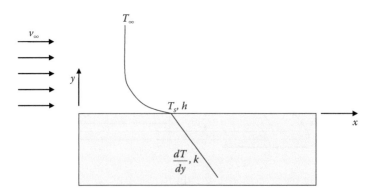

FIGURE 1.5 Convective boundary condition at the surface of a conducting tissue.

the flow geometry properly in order to use the Nusselt number to calculate a value for the convective heat transfer coefficient h using Equation 1.35.

The Nusselt number is generally determined for a particular convection process as a function of the interface geometry, flow properties of the fluid at the interface, and thermodynamic state of the fluid. These properties are in turn represented by dimensionless ratios defined as the Reynolds, Prandtl, and Grashoff numbers, as described below.

The *Reynolds number* is defined by the dimensionless ratio

$$\text{Re}_L = \frac{\rho v L}{\mu} \tag{1.36}$$

where all of the constitutive properties refer to the fluid: ρ is the density (kg/m^3), v is a representative flow velocity (m/s), and μ is the viscosity ($N \cdot s/m^2$). An appropriate physical dimension of the interface is indicated by L. The Reynolds number is a primary property applied to describe forced convection processes. It describes the ratio of the inertial and viscous forces associated with the fluid flow. For low Re values the flow is dominated by the viscous resistance resulting in a laminar boundary layer in which the movement of fluid is highly ordered. High values of Re have a much larger inertial component, which produces a turbulent boundary layer. The magnitude of convective heat transfer is strongly influenced by whether the boundary layer is laminar or turbulent. The geometry of the interface and boundary layer also plays an important role in the convection process. Accordingly, the Reynolds number can be written in terms of either an effective diameter, D, for an internal flow geometry, or an interface length, L, for an external flow geometry. The transition between the laminar and turbulent regimes is defined in terms of a threshold value for Re, and is very different for internal and external flow geometries. Thus, the *transition values* for Re are given as

$$\text{Re}_{trans,int} = \frac{\rho \upsilon D}{\mu} = 2300 \tag{1.37}$$

and

$$\text{Re}_{trans,ext} = \frac{\rho \upsilon L}{\mu} = 5 \times 10^5. \tag{1.38}$$

The *Prandtl number* is defined by the dimensionless ratio

$$\text{Pr} = \frac{c_p \mu}{k} = \frac{\nu}{\alpha} \tag{1.39}$$

where all of the constitutive properties refer to the fluid. The symbol ν is the kinematic viscosity (m^2/s), which is the ratio of the dynamic viscosity and the density. The Prandtl number describes the ratio of momentum diffusivity to thermal

diffusivity. It represents a measure of the relative effectiveness of diffusive momentum and heat transport in the velocity and thermal boundary layers. It provides an indication of the relative thickness of these two boundary layers in a convective system. As a general guideline for the broad range of Prandtl number values that may be encountered: for vapors, $Pr_v \approx \delta/\delta_T \approx 1$; for liquid hydrocarbons such as oils, $Pr_{hc} \approx \delta/\delta_T \gg 1$; and for liquid metals, $Pr_{lm} \approx \delta/\delta_T \ll 1$.

For forced convection processes the Nusselt number is calculated from a relation of the form

$$Nu_x = f(x, \text{Re}_x, \text{Pr}) \quad \text{or} \quad \bar{Nu}_L = f(\text{Re}_L, \text{Pr}). \tag{1.40}$$

In most cases these relations are based on an empirical fit of the equation to experimental data.

The *Grashof number* is defined by the dimensionless ratio

$$Gr_L = \frac{g\beta(T_s - T_\infty)L^3}{\nu^2} \tag{1.41}$$

where g is the acceleration of gravity (1/s^2), and β is the volumetric thermal expansion coefficient of the fluid (1/K). β is given by the relationship

$$\beta = -\frac{1}{\rho}\left(\frac{\partial \rho}{\partial T}\right)_p \tag{1.42}$$

which, for an ideal gas becomes

$$\beta = \frac{1}{T} \tag{1.43}$$

where the temperature is given in absolute units (K). The Grashof number is a primary property applied to describe free convection processes. It describes the ratio of the buoyant and viscous forces associated with the fluid flow and is the equivalent of the Reynolds number for free convection heat transfer. Accordingly, for free convection processes the Nusselt number is calculated from a relation of the form

$$Nu_x = f(x, Gr_x, \text{Pr}) \quad \text{or} \quad \bar{Nu}_L = f(Gr_L, \text{Pr}). \tag{1.44}$$

There are many correlation equations of the type given in Equations 1.40 and 1.44 for forced and free convection, respectively. Comprehensive compendia of these relations are available in dedicated books (Bejan 2004; Kays, Crawford, and Weigand 2004; Incroprera et al. 2007). Only the most generally used relations are presented in this chapter, with a twofold purpose: to illustrate the format of the correlations for various convective domains and to provide a basic set of correlations that can be applied to the solution of many frequently encountered convection problems.

1.2.3.1 Interior Forced Convection Correlations

For the sake of simplicity, assume the flow to be through a circular conduit of diameter, D, and length, L. The conduit length is assumed to be greater than the entrance region at the inlet over which the boundary layers on opposing surfaces grow until they meet at the centerline. Downstream of this point the entire volume of the conduit is filled with boundary layer flow and is termed fully developed. Fluid properties are evaluated at a mean temperature, T_m, which is an integrated average value for fluid flowing in the boundary layer through the conduit. T_m depends on the velocity and temperature profiles within the flowing fluid, which are quite different for laminar and turbulent boundary layers. In addition, T_m will change along the conduit from the inlet to the outlet as heat is exchanged between the fluid and the wall. Overall, the temperature at which the properties are evaluated should reflect the average value for all of the fluid contained in the conduit at any given time. If v_m is the mean flow velocity over the cross-sectional area, A_c, of a conduit, then the mass flow rate, $\dot{m}$, is given by

$$\dot{m} = \rho A_c v_m. \tag{1.45}$$

The net convective heat exchange between the fluid and conduit over the entire length equals the change in enthalpy of the fluid between the inlet and outlet:

$$\dot{Q} = H_{out} - H_{in} = \dot{m}(h_{out} - h_{in}) = \dot{m}c_p(T_{m,out} - T_{m,in}). \tag{1.46}$$

At any cross section along the length of the conduit the rate of energy flow associated with movement of the fluid (which is the advection rate) is obtained by integrating across the boundary layer:

$$\dot{m}c_p T_m = \int_c \rho v c_p T \, dA_c. \tag{1.47}$$

The velocity change with radius over the cross-sectional area in the above integral is substantially different for laminar and turbulent boundary layers. Eliminating the mass flow rate between Equations 1.45 and 1.47 yields an expression for the mean temperature over a circular cross-sectional area of outer radius r_o, for constant density and specific heat:

$$T_m = \frac{2}{v_m r_o^2} \int_0^{r_o} v(r)T(r)r\,dr. \tag{1.48}$$

The functions $v(r)$ and $T(r)$ are determined by the profiles of the velocity and temperature boundary layers specific to the flow conditions of interest. They provide a basis for determining the mean temperature for defining the state at which the fluid properties in the following convection correlation relations are evaluated. The following are some of the most commonly applied convection correlation equations with the conditions noted for which they are valid.

Conditions of validity: fully developed, laminar, uniform temperature of wall surface, T_s.

$$Nu_D = 3.66 \tag{1.49}$$

Conditions of validity: fully developed, laminar, uniform heat flux at the wall surface, $\dot{q}_s$.

$$Nu_D = 4.36 \tag{1.50}$$

Conditions of validity: fully developed, turbulent, $Re_D \geq 10^4$, $L/D \geq 10$, $0.6 \leq Pr \leq 160$, $T_s > T_m$.

$$Nu_D = 0.23 Re^{0.8} Pr^{0.4} \tag{1.51}$$

Conditions of validity: fully developed, turbulent, $Re_D \geq 10^4$, $L/D \geq 10$, $0.6 \leq Pr \leq 160$, $T_s < T_m$.

$$Nu_D = 0.23 Re^{0.8} Pr^{0.3} \tag{1.52}$$

Conditions of validity: fully developed, turbulent, $3 \times 10^3 \leq Re_D \leq 5 \times 10^6$, $L/D \geq 10$, $0.5 \leq Pr \leq 2000$.

$$Nu_D = \frac{1}{8(0.79\ln Re_D - 1.64)^2} \frac{(Re_D - 1000)Pr}{1 + \frac{12.7(Pr^{2/3}-1)}{8^{1/2}(0.79\ln Re_D - 1.64)}} \tag{1.53}$$

As illustrated by Equations 1.51, 1.52, and 1.53, in some cases alternative correlation relations are available to calculate a value for h under the same conditions.

1.2.3.2 Exterior Forced Convection Correlations

The properties of the fluid are determined for a state defined by the temperature T_f where

$$T_f = \frac{T_s + T_\infty}{2} \tag{1.54}$$

which is the average of the wall and free stream fluid temperatures. The length of the fluid/substrate interface is L. The following are some of the most commonly applied convection correlation equations with the conditions noted for which they are valid.

Conditions of validity: local convection in laminar region for flow over a flat plate, $0.6 \leq Pr$.

$$Nu_x = 0.332 Re_x^{0.5} Pr^{0.33} \tag{1.55}$$

Conditions of validity: convection averaged across laminar region, L, for flow over a flat plate, $0.6 \leq \text{Pr}$.

$$\overline{Nu}_L = 0.664 \, \text{Re}_L^{0.5} \, \text{Pr}^{0.33} \tag{1.56}$$

Conditions of validity: local convection in turbulent region for flow over a flat plate, $\text{Re}_x \leq 10^8$, $0.6 \leq \text{Pr} \leq 60$.

$$Nu_x = 0.296 \, \text{Re}_x^{0.8} \, \text{Pr}^{0.33} \tag{1.57}$$

Conditions of validity: convection averaged across the combined laminar and turbulent regions of total length, L, for flow over a flat plate, $0.6 \leq \text{Pr}$.

$$\overline{Nu}_L = (0.037 \, \text{Re}_L^{0.8} - 871) \text{Pr}^{0.33} \tag{1.58}$$

Conditions of validity: convection averaged across the entire surface around a cylinder of diameter, D, in perpendicular flow, $0.4 \leq \text{Re}_D \leq 4 \times 10^5$, $0.7 \leq \text{Pr}$.

$$\overline{Nu}_D = C \, \text{Re}_D^n \, \text{Pr}^{0.33} \tag{1.59}$$

where the values of C and n are functions of Re_D as given below:

Re_D	C	n
0.4–4	0.989	0.330
4–40	0.911	0.385
40–4,000	0.683	0.466
4,000–40,000	0.193	0.618
40,000–400,000	0.027	0.805

Conditions of validity: convection averaged across the entire surface around a sphere of diameter, D, properties based on T_∞, $3.5 \leq \text{Re}_D \leq 7.6 \times 10^4$, $0.71 \leq \text{Pr} \leq 380$.

$$\overline{Nu}_D = 2 + (0.4 \, \text{Re}_D^{0.5} + 0.06 \, \text{Re}_D^{0.67}) \text{Pr}^{0.4} \left(\frac{\mu}{\mu_s} \right)^{0.25} \tag{1.60}$$

1.2.3.3 Free Convection Correlations

Since free convection processes are driven by buoyant effects, determination of the relevant correlation relations to determine the convection coefficient must start with analysis of the shape and orientation of the fluid/solid interface. This effect is illustrated with the free convection boundary layer adjacent to a vertical cooled flat plate shown in Figure 1.6. Note that the flow velocity is zero at both the inner and outer extremes of the boundary layer, although the gradient is finite at the solid interface owing to viscous drag of the fluid. The environment is assumed to be quiescent so that there is no viscous shearing action at the outer region of the boundary layer. The following

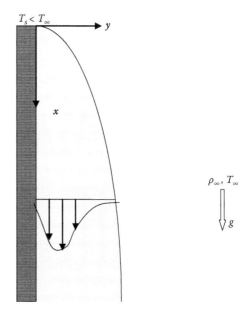

FIGURE 1.6 Growth of a free convection boundary layer on a vertical cooled plate.

are some of the most commonly applied free convection correlation relations. Many make use of a dimensionless constant, the *Rayleigh number*, $Ra = Gr \cdot Pr$.

Conditions of validity: free convection averaged over a vertical plate of length L, including both the laminar and turbulent flow regions over the entire range of Ra.

$$\overline{Nu}_L = \left\{ 0.825 + \frac{0.387 \, Ra_L^{1/6}}{\left[1 + (0.492 / \text{Pr})^{9/16} \right]^{8/27}} \right\}^2 \tag{1.61}$$

Conditions of validity: free convection averaged over a vertical plate of length L for laminar flow defined by $10^4 \leq Ra_L \leq 10^9$.

$$\overline{Nu}_L = 0.59 \, Ra_L^{1/4} \tag{1.62}$$

Conditions of validity: free convection averaged over a vertical plate of length L for turbulent flow defined by $10^9 \leq Ra_L \leq 10^{13}$.

$$\overline{Nu}_L = 0.1 \, Ra_L^{1/3} \tag{1.63}$$

Conditions of validity: free convection averaged over the upper surface of a heated plate or lower surface of a cooled plate having a dimension L; $10^4 \leq Ra_L \leq 10^7$.

$$\overline{Nu}_L = 0.5 \, Ra_L^{1/4} \tag{1.64}$$

Conditions of validity: free convection averaged over the upper surface of a heated plate or lower surface of a cooled plate having a dimension L; $10^7 \leq Ra_L \leq 10^{11}$.

$$\overline{Nu}_L = 0.15 \, Ra_L^{1/3} \tag{1.65}$$

Conditions of validity: free convection averaged over the lower surface of a heated plate or upper surface of a cooled plate having a dimension L; $10^5 \leq Ra_L \leq 10^{10}$.

$$\bar{Nu}_L = 0.27\,Pa_L^{1/4} \tag{1.66}$$

Conditions of validity: free convection averaged over the entire circumferential surface of a horizontal cylinder having an isothermal surface and a diameter D; $Ra_D \leq 10^{12}$.

$$\bar{Nu}_D = \left\{ 0.6 + \frac{0.387\,Pr_D^{1/6}}{[1+(0.559/pr)^{9/16}]^{8/27}} \right\}^2 \tag{1.67}$$

Conditions of validity: alternatively, free convection averaged over the entire circumferential surface of a horizontal cylinder having an isothermal surface and a diameter D.

$$\bar{Nu}_D = CRa_D^n \tag{1.68}$$

where the values of C and n are functions of Ra_D as given in the table below:

Ra_D	C	n
10^{-10}–10^{-2}	0.675	0.058
10^{-2}–10^{2}	1.02	0.148
10^{2}–10^{4}	0.85	1.88
10^{4}–10^{7}	0.48	0.25
10^{7}–10^{12}	0.125	0.333

Conditions of validity: free convection averaged over the entire circumferential surface of a sphere having an isothermal surface and a diameter D; $Ra_D \leq 10^{11}$; $Pr \geq 0.7$.

$$\bar{Nu}_D = 2 + \frac{0.589\,Ra_D^{1/4}}{[1+(0.469/Pr)^{9/16}]^{4/9}} \tag{1.69}$$

1.2.4 Radiation Heat Transfer

Thermal radiation is primarily a surface phenomenon as it interacts with a conducting medium (except in transparent or translucent fluids, which will be considered at the end of this discussion), and it is to be distinguished from laser irradiation, which comes from a different type of source. Thermal radiation is important in many types of heating, cooling, and drying processes. In the outdoor environment, solar thermal radiation can have a significant influence on the overall heat load on the skin.

Thermal radiation occurs via the propagation of electromagnetic waves. It does not require the presence of a transmitting material as do conduction and convection. Therefore, thermal radiation can proceed in the absence of matter, such as in the radiation of heat from the sun to earth. All materials are continuously emitting thermal radiation from their surfaces as a function of their temperature and radiative constitutive properties. All surfaces also are continuously receiving thermal energy from their environment. The balance between radiation lost and gained defines the net radiation heat transfer for a body. The wavelengths of thermal radiation extend across a spectrum from about 0.1 μm to 100 μm, embracing the entire visible spectrum. It is for this reason that some thermal radiation can be observed by the human eye, depending on the temperature and properties of the emitting surface.

The foregoing observations indicate that there are three properties of a body (i.e., a system) and its environment that govern the rate of radiation heat transfer: (1) the *surface temperature*, (2) the *surface radiation properties*, and (3) the *geometric sizes, shapes, and configurations* of the body surface in relation to the aggregate surfaces in the environment. Each of these three effects can be quantified and expressed in equations used to calculate the magnitude of radiation heat transfer. The objective of this presentation is to introduce and discuss how each of these three factors influences radiation processes and to show how they can be grouped into a single approach to analysis.

1.2.4.1 Temperature Effects

The first property to consider is temperature. The relationship between the temperature of a perfect radiating (black) surface and the rate at which thermal radiation is emitted is known as the *Stefan-Boltzmann law*:

$$E_b = \sigma T^4 \tag{1.70}$$

where E_b is the blackbody emissive power [W/m²], and σ is the Stefan-Boltzmann constant, which has the numerical value

$$\sigma = 5.678 \times 10^{-8}\ W/_{m^2 \cdot K^4}.$$

Note that the temperature must be expressed in absolute units (K). E_b is the rate at which energy is emitted diffusely (without directional bias) from a surface at temperature T(K) having perfect radiation properties. It is the summation of radiation emitted at all wavelengths from a surface. A perfect radiating surface is termed black and is characterized by emitting the maximum possible radiation at any given temperature. The blackbody monochromatic (at a single wavelength, λ) emissive power is calculated from the Planck distribution as

$$E_{\lambda,b}(\lambda,T) = \frac{2\pi hc_o^2}{\lambda^2 \left[\exp\left(\dfrac{hc_o}{\lambda kT}\right) - 1 \right]} \left[W/m^2 \cdot \mu m \right] \tag{1.71}$$

where $h = 6.636 \times 10^{-34}$ [J·s] is the Planck constant, $k = 1.381 \times 10^{-23}$ (J/K) is the Boltzmann constant, and $c_o = 2.998 \times 10^8$ (m/s) is the speed of light in vacuum. The Planck distribution can be plotted showing $E_{\lambda,b}$ as a function of Λ for specific constant values of absolute temperature, T. The result is the nest of spectral emissive power curves in Figure 1.7. Note that for each

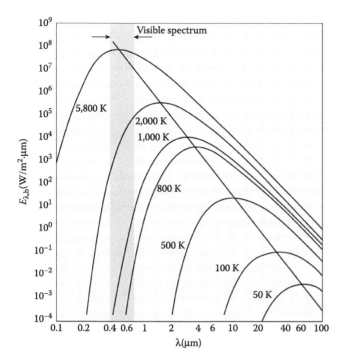

FIGURE 1.7 Spectral blackbody emissive power as a function of surface temperature and wavelength.

temperature there is an intermediate wavelength for which $E_{\lambda,T}$ has a maximum value, and this maximum increases monotonically in magnitude and occurs at shorter wavelengths with increasing temperature. Wien's displacement law, Equation 1.72, describes the relationship between the absolute temperature and the wavelength at which maximum emission occurs:

$$\lambda_{max}T = 2898[\mu m \cdot k]. \tag{1.72}$$

Equation 1.71 is integrated over the entire emission spectrum to obtain the expression for the total emitted radiation, Equation 1.70.

$$E_b(T) = \int_0^\infty \frac{2\pi hc_o^2}{\lambda^5 \left[\exp\left(\dfrac{hc_o}{\lambda kT}\right) - 1\right]} d\lambda = \sigma T^4 \tag{1.73}$$

Equation 1.73 represents the area under an isothermal curve in Figure 1.7 depicting the maximum amount of energy that can be emitted from a surface at a specified temperature. This set of equations provides the basis for quantifying the temperature effect on thermal radiation. It applies to idealized, black surfaces.

1.2.4.2 Surface Effects

Next we will consider the effect of real, rather than idealized, surface properties on thermal radiation exchange. Real surfaces emit less than blackbody radiation at a given temperature. The ratio of real to black radiation levels defines a property called the

emissivity, ε. In general, radiation properties are functions of the radiation wavelength and for many practical systems can change significantly over the thermal spectrum. Thus,

$$\varepsilon_\lambda(\lambda,T) = \frac{E_\lambda(\lambda,T)}{E_{\lambda,b}(\lambda,T)}. \tag{1.74}$$

An idealized real surface has radiant properties that are wavelength independent and is termed a gray surface. For these conditions,

$$\varepsilon(T) = \frac{E(T)}{E_b(T)} = \frac{E(T)}{\sigma T^4}. \tag{1.75}$$

A gray surface has the effect of decreasing the magnitude of the curves in Figure 1.7 by a constant factor over all wavelengths.

In addition to emission, surfaces are continually bombarded by thermal radiation from their environments. The net radiant flux at a surface is the difference between the energies received and lost. As with emission, the surface radiation properties play an important role in determining the amount of energy absorbed by a surface. A black surface absorbs all incident radiation, whereas real surfaces absorb only a fraction that is less than one. The total radiant flux onto a surface from all sources is called the *radiosity* and is denoted by the symbol $G[W/m^2]$. In general, the incident radiation will be composed of many wavelengths, denoted by G_λ. A surface can have three modes of response to incident radiation: the radiation may be absorbed, reflected, and/or transmitted. The fractions of incident radiation that undergo each of these responses are determined by three dimensionless properties: the coefficients of absorption, α, reflection, ρ, and transmission, τ. Conservation of energy applied at a surface dictates that the relationship among these three properties must be

$$\alpha + \rho + \tau = 1. \tag{1.76}$$

Figure 1.8 illustrates these phenomena for radiation incident onto a surface that is translucent, allowing some of the radiation to pass through. All three of the properties are wavelength dependent.

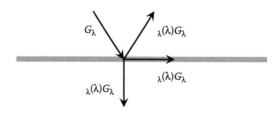

FIGURE 1.8 Absorption, reflection, and transmission phenomena for a surface irradiated with a multi-wavelength incident radiation, G_λ.

The three properties are defined according to the fraction of radiosity that is absorbed, reflected, and transmitted:

$$\alpha_\lambda (\lambda) = \frac{G_{\lambda,abs}(\lambda)}{G_\lambda (\lambda)} \tag{1.77}$$

$$\rho_\lambda (\lambda) = \frac{G_{\lambda,ref}(\lambda)}{G_\lambda (\lambda)} \tag{1.78}$$

$$\tau_\lambda (\lambda) = \frac{G_{\lambda,tr}(\lambda)}{G_\lambda (\lambda)}. \tag{1.79}$$

It is well known that there is a very strong spectral (wavelength) dependence of these properties. For example, the greenhouse effect occurs because glass has a high transmissivity (τ) at relatively short wavelengths in the visible spectrum that are characteristic of the solar flux. However, the transmissivity is very small in the infrared spectrum in which terrestrial emission occurs. Therefore, heat from the sun readily passes through glass and is absorbed by interior objects. In contrast, radiant energy emitted by these interior objects is reflected back to the source. The net result is a warming of the interior of a system that has a glass surface exposed to the sun. The lens of a camera designed to image terrestrial sources of thermal radiation, predominantly in the infrared spectrum, must be fabricated from a material that is transparent at those wavelengths. It is important to verify whether the spectral dependence of material surface properties is important for specific applications involving thermal radiation.

An additional important surface property relationship is defined by *Kirchhoff's law*, which applies for a surface that is in thermal equilibrium with its environment. Most thermal radiation analyses are performed for processes that are steady state. For the surface of a body n having a surface area A_n, at steady state the radiation gained and lost is balanced so that the net exchange is zero. To illustrate, we may consider a large isothermal enclosure at a temperature, T_s, containing numerous small bodies, each having unique properties and temperature. See Figure 1.9.

Since the surface areas of the interior bodies are very small in comparison to the enclosure area, their influence on the radiation field is negligible. Also, the radiosity to the interior bodies is a combination of emission and reflection from the enclosure surface. The net effect is that the enclosure acts as a blackbody cavity regardless of its surface properties. Therefore, the radiosity within the enclosure is expressed as

$$G = E_b (T_s). \tag{1.80}$$

For thermal equilibrium within the cavity, the temperatures of all surfaces must be equal. $T_{n-1} = T_n = T.. = T_s$. A steady state energy balance between absorbed and emitted radiation on one of the interior bodies yields

$$\alpha_n G A_n - E_n (T_s) \cdot (A_n) = 0. \tag{1.81}$$

The term for radiosity may be eliminated between the two foregoing equations:

$$E_b (T_s) = \frac{E_n (T_s)}{\alpha_n}. \tag{1.82}$$

This relationship holds for all of the interior bodies. Comparison of Equations 1.75 and 1.82 shows that the emissivity and absorptivity are equal:

$$\alpha = \varepsilon, \quad or \quad \alpha_\lambda = \varepsilon_\lambda. \tag{1.83}$$

The general statement of this relationship is that for a gray surface, the emissivity and absorptivity are equal and independent of spectral conditions.

1.2.4.3 Surface Geometry Effects

The third factor influencing thermal radiation transfer is the geometric sizes, shapes, and configurations of body surfaces in relation to the aggregate surfaces in the environment. This effect is quantified in terms of a property called the shape factor, which is solely a function of the geometry of a system and its environment. By definition, the shape factor is determined for multiple bodies, and it is related to the size, shape, separation, and orientation of the bodies. The shape factor $F_{m \to n}$ is defined between two surfaces, m and n, as the fraction of energy that leaves surface m that is incident onto the surface n. It is very important to note that the shape factor is directional. The shape factor from body m to n is probably not equal to that from body n to m.

Values for shape factors have been compiled for a broad range of combinations of size, shape, separation, and orientation and are available as figures, tables, and equations (Howell 1982; Siegel and Howell 2002). There are a number of simple relations that govern shape factors and that are highly useful in working many types of problems. One is called the reciprocity relation:

$$A_n F_{n \to m} = A_m F_{m \to n}. \tag{1.84}$$

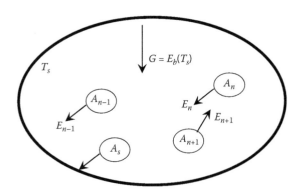

FIGURE 1.9 Steady state thermal radiation within a large isothermal enclosure containing multiple small bodies.

This equation is applied for calculating the value of a second shape factor between two bodies if the first is already known.

A second relation is called the summation rule, which says the sum of shape factors for the complete environment of a body equals 1.0:

$$\sum_{k=1}^{n} F_{m \to k} = 1.0. \tag{1.85}$$

The summation rule accounts for the entire environment for an object.

A limiting case is shown in Figure 1.9 in which $A_s \gg A_n$. For this geometry, the reciprocity relation dictates that $F_{s \to n}$ be vanishingly small since only a very small fraction of the radiation leaving the large surface s will be incident onto the small surface n. The summation rule then shows that effectively $F_{n \to s} = 1$.

The third geometric relationship states that the shape factors for a surface to each component of its environment are additive. If a surface n is divided into l components, then

$$F_{m \to (n)} = \sum_{j=1}^{l} F_{m \to j} \tag{1.86}$$

where

$$A_n = \sum_{j=1}^{l} A_j.$$

Equation 1.86 is often useful for calculating the shape factor for complex geometries that can be subdivided into an assembly of more simple shapes.

There exist comprehensive compendia of data for determination of a wide array of shape factors (Howell, 1982). The reader is directed to such sources for detailed information. Application of geometric data to radiation problems is very straightforward.

Evaluation of the temperature, surface property, and geometry effects can be combined to calculate the magnitude of radiation exchange among a system of surfaces. The simplest approach is to represent the radiation process in terms of an equivalent electrical network. For this purpose, two special properties are used: the irradiation, G, which is the total radiation incident onto a surface per unit time and area, and the radiosity, J, which is the total radiation that leaves a surface per unit time and area. Also, for the present time it is assumed that all surfaces are opaque (no radiation is transmitted), and the radiation process is at steady state. Thus, there is no energy storage within any components of the radiating portion of the system.

The radiosity can be written as the sum of radiation emitted and reflected from a surface, which is expressed as

$$J = \varepsilon E_b + \rho G. \tag{1.87}$$

The net energy exchanged by a surface is the difference between the radiosity and the irradiation. For a gray surface with $\alpha = \varepsilon$, and therefore $\rho = 1 - \varepsilon$,

$$\frac{\dot{Q}}{A} = \dot{q} = J - G = \varepsilon E_b + (1 - \varepsilon) G - G \tag{1.88}$$

which can also be written as

$$\dot{Q} = \frac{E_b - J}{(1 - \varepsilon)/\varepsilon A}. \tag{1.89}$$

This relationship of a flow across a resistance between a potential difference is shown in Figure 1.10. The format of Equation 1.89 is in terms of a flow that equals a difference in potential divided by a resistance. In this case, the equation represents the drop in potential from a black to a gray surface associated with a finite surface radiation resistance. The equation can be represented graphically in terms of a steady state resistance. This resistance applies at every surface within a radiating system that has non-black radiation properties. Note that for a black surface for which $\varepsilon = 1$, the resistance goes to zero.

A second type of radiation resistance is due to the geometric shape factors among multiple radiating bodies. The apparent radiation potential of a surface is the radiosity. For the exchange of radiation between two surfaces A_1 and A_2, the net energy flow equals the sum of the flows in both directions. The radiation leaving surface 1 that is incident on surface 2 is

$$J_1 A_1 F_{1 \to 2}$$

and in like manner, the radiation from 2 to 1 is

$$J_2 A_2 F_{2 \to 1}.$$

The net interchange between surfaces 1 and 2 is then the sum of these two flows

$$\dot{Q}_{1 \to 2} = J_1 A_1 F_{1 \to 2} - J_2 A_2 F_{2 \to 1} = \frac{J_1 - J_2}{1/A_1 F_{1 \to 2}}. \tag{1.90}$$

This process can also be modeled via an electrical network as shown in Figure 1.11.

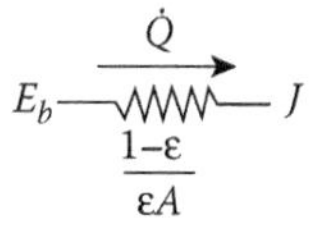

FIGURE 1.10 Electrical resistance model for the drop in radiation potential due to a gray surface defined by the property ε.

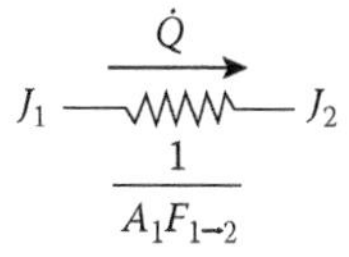

FIGURE 1.11 Electrical resistance model for the radiation exchange between two surfaces with a shape factor $F_{1 \to 2}$.

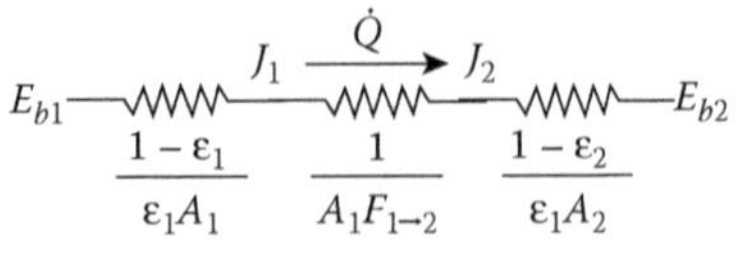

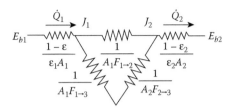

$$\dot{Q}_{1\rightarrow2} = \frac{E_{b1} - E_{b2}}{\frac{1-\varepsilon_1}{\varepsilon_1 A_1} + \frac{1}{A_1 F_{1\rightarrow2}} + \frac{1-\varepsilon_2}{\varepsilon_2 A_2}} = \frac{\sigma\left(T_1^4 - T_2^4\right)}{\frac{1-\varepsilon_1}{\varepsilon_1 A_1} + \frac{1}{A_1 F_{1\rightarrow2}} + \frac{1-\varepsilon_2}{\varepsilon_2 A_2}}. \quad (1.91)$$

FIGURE 1.12 Electrical resistance model for the radiation exchange between two surfaces with a shape factor $F_{1\rightarrow2}$.

These two types of resistance elements can be applied to model the steady state interactions among systems of radiating bodies. The simplest example is of two opaque bodies that exchange radiation only with each other. This problem is characterized by the network shown in Figure 1.12.

This network can be solved to determine the radiation heat flow in terms of the temperatures (T_1, T_2), surface properties (ε_1, ε_2), and system geometry (A_1, A_2, $F_{1\rightarrow2}$):

Note that the second term in this equation has a linear differential in the driving potential, whereas the third term has a fourth power differential. A major advantage of the electrical circuit analogy is that a radiation problem can be expressed as a simple linear network, as compared to the thermal formulation in which temperature must be raised to the fourth power.

Given the network modeling tools, it becomes straightforward to describe radiation exchange among the components of an n-bodied system. A three-bodied system can be used to illustrate this analysis as shown in Figure 1.13.

In this system each of the three bodies experiences a unique radiant heat flow. For the special case in which one surface, such as 3, is perfectly insulated, meaning that all incident radiation is reradiated, then the diagram is simplified to a combined series/parallel exchange between surfaces 1 and 2 as seen in Figure 1.14.

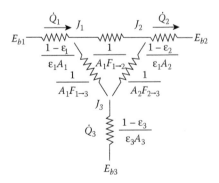

FIGURE 1.13 Electrical resistance model for the radiation exchange among the surfaces of a three-bodied system.

FIGURE 1.14 Electrical resistance model for the radiation exchange among the surfaces of a three-bodied system in which surface 3 is perfectly insulated.

The radiation heat flow for this system is written as a function of the system properties, with the shape factor reciprocity relation applied for $A_1 F_{1\rightarrow2} = A_2 F_{2\rightarrow1}$, as

$$Q_{1\rightarrow2} = \frac{\sigma A_1 \left(T_1^4 - T_2^4\right)}{\left(\frac{1}{\varepsilon_1}-1\right) + \frac{A_1 + A_2 - 2A_1 F_{1\rightarrow2}}{A_2 - A_1(F_{1\rightarrow2})^2} + \frac{A_1}{A_2}\left(\frac{1}{\varepsilon_2}-1\right)}. \quad (1.92)$$

A final case to be discussed addresses the effect of an absorbing and transmitting medium included in a radiating system. Glasses and gases are examples of this type of media. If the medium is nonreflective, then

$$\alpha_m + \tau_m = \varepsilon_m + \tau_m = 1. \quad (1.93)$$

Radiation that is absorbed by the medium is transmitted and then emitted to its environment. For a system consisting of two surfaces 1 and 2 that see only each other, plus an intervening medium m, the net energy leaving surface 1 that is transmitted through the medium and arrives at surface 2 is

$$J_1 A_1 F_{1\rightarrow2}\tau_m.$$

Likewise, the energy flow in the opposite direction is

$$J_2 A_2 F_{2\rightarrow1}\tau_m.$$

The net interchange between surfaces 1 and 2 via transmission through the medium is then the sum of these two flows.

$$Q_{1\rightarrow2} = J_1 \tau_m A_1 F_{1\rightarrow2} - J_2 \tau_m F_{2\rightarrow1} = \frac{J_1 - J_2}{1/A_1 F_{1\rightarrow2}(1-\varepsilon_m)} \quad (1.94)$$

The effect of the medium can be represented by a radiation network element in Figure 1.15.

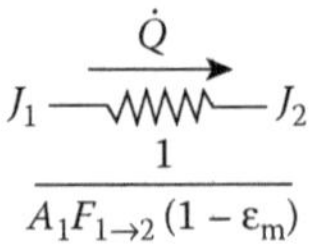

FIGURE 1.15 Electrical resistance model for the effect of an absorbing and transmitting medium on the thermal radiation between two surfaces.

This circuit element can be included in a radiation network model as appropriate to represent the effect of an interstitial medium between radiating surfaces.

Note that in this analysis of radiation, all of the equivalent electrical networks contain only resistors, and specifically there are no capacitors. The explicit interpretation of this arrangement is that all of the radiation processes considered are at steady state such that no energy storage occurs. For our present analysis, the mass of all radiating bodies has been neglected. Since many thermal radiation processes are approximated as surface phenomena, this is a reasonable assumption. Under conditions that demand more comprehensive and sophisticated analysis, this assumption may have to be relaxed, which leads to a significant increase in the complexity of the thermal radiation analysis.

1.3 Special Features of Heat Transfer in Biomedical Systems

Living tissues present a special set of complications for solving heat transfer problems. Among the often-encountered issues are: composite materials structures, anisotropic properties, complex geometric shapes not amenable to convenient mathematical description, nonhomogeneously distributed internal energy generation, constitutive properties that may change dramatically with temperature, nonlinear feedback control (such as for thermoregulatory function), and a diffuse and complex internal circulation of blood that has a significant effect on the body's thermal state and energy distribution via convective heat transfer. Plus, therapeutic, diagnostic, and prophylactic procedures that are energy based, such as hyperthermia protocols, frequently introduce intricate formulations for energy deposition in tissue as a function of time and position. This latter topic is addressed in detail in other chapters throughout this text and will not be discussed here except to note that the various energy sources applied to create a state of hyperthermia are embodied into the $\dot{Q}_{gen}$ term in the conservation of energy equation (Equation 1.1). Neither does space allow us to discuss all of the unique features of heat transfer in living tissues as listed before. There are many more comprehensive analyses and presentations of bioheat transfer to which the reader is directed (Charney 1992; Diller 1992; Diller et al. 2005; Roemer 1990; Roselli and Diller 2011). Here we will discuss only two aspects of bioheat transfer that are of greatest relevance to the design and application of therapeutic hypothermia protocols: the influence of convective flow of blood through blood vessels and the thermal properties of living tissues, including local blood perfusion rates.

1.3.1 Blood Perfusion Effects

Bioheat transfer processes in living tissues are often influenced by blood perfusion through the vascular network. When there is a significant difference between the temperature of blood and the tissue through which it flows, convective heat transport will occur, altering the temperatures of both the blood and the tissue. Perfusion-based heat transfer interaction is critical to a number of physiological processes such as thermoregulation and inflammation. The blood/tissue thermal interaction is a function of several parameters including the rate of perfusion and the vascular anatomy, which vary widely among the different tissues, organs of the body, and pathology. Diller et al. (2005) contains an extensive compilation of perfusion rate data for many tissues and organs and for many species.

The rate of perfusion of blood through different tissues and organs varies over the time course of a normal day's activities, depending on factors such as physical activity, physiological stimulus, circadian cycle, and environmental conditions. Further, many disease processes are characterized by alterations in blood perfusion, and some therapeutic interventions result in either an increase or decrease in blood flow in a target tissue. For these reasons, it is very useful in a clinical context to know what the absolute level of blood perfusion is within a given tissue. There are numerous techniques that have been developed for this purpose over the past several decades. In some of these techniques, the coupling between vascular perfusion and local tissue temperature is applied to advantage to assess the flow through local vessels via inverse solution of equations that model the thermal interaction between perfused blood and the surrounding tissue.

Pennes (Pennes 1948; Wissler 1998) published the seminal work on developing a quantitative basis for describing the thermal interaction between tissue and perfused blood. His work consisted of a series of experiments to measure temperature distribution as a function of radial position in the forearms of nine human subjects. A butt-junction thermocouple was passed completely through the arm via a needle inserted as a temporary guideway, with the two leads exiting on opposite sides of the arm. The subjects were unanesthetized so as to avoid the effects of anesthesia on blood perfusion. Following a period of normalization, the thermocouple was scanned transversely across the mediolateral axis to measure the temperature as a function of radial position within the interior of the arm. The environment in the experimental suite was kept thermally neutral during the experiments. Pennes's data showed a temperature differential of three to four degrees between the skin and the interior of the arm, which he attributed to the effects of metabolic heat generation and heat transfer with arterial blood perfused through the microvasculature.

Pennes proposed a model to describe the effects of metabolism and blood perfusion on the energy balance within tissue. These two effects were incorporated into the standard thermal diffusion equation, which is written in its simplified form as

$$\rho c \frac{\partial T}{\partial t} = \nabla \cdot k \nabla T + (\rho c)_b \omega_b (T_b - T) + \dot{Q}_{met} \qquad (1.95)$$

where ω_b is perfusion of blood through the microvasculature in units of volume of blood flowing per unit time per unit of tissue volume, and the subscript b denotes a property of blood.

Equation 1.95 is written to describe the thermal effects of blood flow through a local region of tissue having a temperature T. It contains the familiar energy storage and conduction terms, plus terms to account for convection with perfused blood and metabolic heat generation. The middle term on the right side of the equation corresponds to the enthalpy term in Equation 1.1 that accounts for the effects of mass entering and leaving a system influencing the stored energy. The temperature of entering perfused blood into a tissue region is that of the arterial supply, T_a, and the leaving temperature is T because the relatively small volume of flowing blood completely equilibrates with the surrounding tissue via the very large surface area to volume ratio of the microvasculature through which it flows. Indeed, the level of the vasculature at which thermal equilibration is achieved between perfused blood and the surrounding tissue has been a topic of considerable interest and importance for many applications of bioheat transfer (Chato 1980; Chen and Holmes 1980; Shrivastava and Roemer 2006). There is little doubt that blood comes to the temperature of the tissue through which it is flowing within the arteriolar network long before the capillaries are reached (in stark contrast to mass transport, which is focused in the capillaries).

The Pennes model contains no specific information about the morphology of the vasculature through which the blood flows. The somewhat simple assumption is that the fraction of blood flowing through a tissue that is diverted through the microvasculature comes to thermal equilibration with the local tissue as it passes to the venous return vessels. A major advantage of the Pennes model is that the added term to account for perfusion heat transfer is linear in temperature, which facilitates the solution of Equation 1.95. Since the publication of this work, the Pennes model has been adapted by many researchers for the analysis of a variety of bioheat transfer phenomena. These applications vary in physiological complexity from a simple homogeneous volume of tissue to thermal regulation of the entire human body. As more scientists have evaluated the Pennes model for application in specific physiological systems, it has become increasingly clear that some of the assumptions foundational to the model are not valid for some vascular geometries that vary greatly among the various tissues and organs of the body (Charney 1992).

Given that the validity of the Pennes model has been questioned for many applications, Wissler (1998) has revisited and reanalyzed Pennes's original data. Given the hindsight of five decades of advances in bioheat transfer plus greatly improved computational tools and better constitutive property data, Wissler's analysis pointed out flaws in Pennes's work that had not been appreciated previously. However, he also showed that much of the criticism that has been directed toward the Pennes model is not justified, in that his improved computations with the model demonstrated a good standard of agreement with the experimental data. Thus, Wissler's conclusion is that "those who base their theoretical calculations on the Pennes model can be somewhat more confident that their starting equations are valid" (Wissler, 1998).

Another important issue relating to convective heat transfer between blood and tissue is the regulation of local perfusion rate. The flow of blood to specific regions of the body, organs, and tissues is a function of many variables. These include maintenance of thermogenesis, thermoregulatory function, type and level of physical activity, existence of a febrile state, and others. The regulation and distribution of blood flow involves a complex, nonlinear feedback process that involves a combination of local and central control inputs. Many existing models include these various inputs on a summative basis. In contrast, Wissler (2008) has recently introduced a multiplicative model that provides an improved simulation of physiological performance. In summary, quantitative analysis of the effects of blood perfusion on the internal temperature distribution in living tissue remains a topic of active research after a half century of study.

1.3.2 Thermal Properties of Living Tissues

Compilation of tables of the thermal properties of tissues has lagged behind that of properties for inanimate materials. One of the major challenges faced in measuring tissue properties is the fact that inserting a measurement probe into a live specimen will alter its state by causing trauma and modifying local blood perfusion. Also, there are large variations among different individuals, and physical access to internal organs is difficult. Nonetheless, there are increasing broad compilations of tissue thermal properties such as that prepared by Ken Holmes in Diller et al. (2005).

Thermal probe techniques are used frequently to determine the thermal conductivity and the thermal diffusivity of biomaterials (Balasubramaniam and Bowman 1977; Chato 1968; Valvano et al. 1984). Common to these techniques is the use of a thermistor bead either as a heat source or a temperature sensor. Various thermal diffusion probe techniques (Valvano 1992) have been developed from Chato's first practical use of the thermal probe (Chato 1968). Physically, for all of these techniques, heat is introduced to the tissue at a specific location and is dissipated by conduction through the tissue and by convection with blood perfusion.

Thermal probes are constructed by placing a miniature thermistor at the tip of a plastic catheter. The volume of tissue over which the measurement occurs depends on the surface area of the thermistor. Electrical power is delivered simultaneously to a spherical thermistor positioned invasively within the tissue of interest. The tissue is assumed to be homogeneous within the mL surrounding the probe. The electrical power and the resulting temperature rise are measured by a microcomputer-based instrument. When the tissue is perfused by blood, the thermistor heat is removed both by conduction and by heat transfer due to blood flow near the probe. *In vivo*, the instrument measures effective thermal properties that are the combination of conductive and convective heat transfer. Thermal properties are derived from temperature and power measurements using equations that describe heat transfer in the integrated probe/tissue system.

The measurement technique consists of inserting the thermistor probe of nominal radius a into a target and allowing an initial period of thermal equilibration. Then, a current is supplied to the thermistor with a control to maintain a constant temperature rise of the probe above the initial baseline, $\bar{T} = T - T_o$. The Pennes equation is applied in spherical coordinates to model the transient temperatures in the probe and in the surrounding tissue:

$$\rho_p c_p \frac{\partial \bar{T}_p}{\partial t} = k_p \frac{\partial}{\partial r}\left(r^2 \frac{\partial \bar{T}_p}{\partial r}\right) + \frac{A + Bt^{-\frac{1}{2}}}{\frac{4}{3}\pi a^2} \quad \text{for } r < a \quad (1.96)$$

$$\rho_t c_t \frac{\partial \bar{T}_t}{\partial t} = k_t \frac{\partial}{\partial r}\left(r^2 \frac{\partial \bar{T}_t}{\partial r}\right) - \omega_b c_b (T_b - T_t) \quad \text{for } r < a \quad (1.97)$$

where the subscripts p and t refer to the probe and tissue. The constants A and B are characteristics of the heating regime applied to the thermistor. The initial power input to the thermistor to maintain $\bar{T}_p$ at a constant value is maximal, followed by a decline to a steady state value in which there is an equilibrium between the heating of the thermistor and the rate of loss by conduction in the tissue and convection to perfused blood. The solution of Equations 1.96 and 1.97 is complex, and for details the reader may reference Valvano (1992) and Diller et al. (2005).

Currently, there is no method to quantify simultaneously the major three parameters: the intrinsic tissue thermal conductivity, k_m, the tissue thermal diffusivity, α_m, and perfusion, ω. Either the knowledge of k_m is required prior to the perfusion measurement, or even when k_m is measured in the presence of perfusion, the thermal diffusivity cannot be measured.

Acknowledgments

This chapter was prepared with support from NSF Grant No. CBET 0966998 and the Robert and Prudie Leibrock Professorship in Engineering.

References

Balasubramaniam, T.A. and Bowman, H.F. 1977. Thermal conductivity and thermal diffusivity of biomaterials: A simultaneous measurement technique, *J. Biomech. Engr.* 99:148–154.

Bejan, A. 2004. *Convection heat transfer*, Hoboken, NJ, Wiley.

Carslaw, H.S. and Jaeger, J.C. 1959. *Conduction of heat in solids*, 2nd ed., London, Oxford University Press.

Charney, C.K. 1992. Mathematical models of bioheat transfer, *Adv. Heat Trans.* 22:19–155.

Chato, J.C. 1968. A method for the measurement of thermal properties of biologic materials, in *Symposium on thermal problems in biotechnology*, ed. J.C. Chato, 16–25, New York, ASME.

Chato, J.C. 1980. Heat transfer in blood vessels, *J. Biomech. Engr.* 102:110–118.

Chen, M.M. and Holmes, K.R. 1980. Microvascular contributions in tissue heat transfer, *Ann. N.Y. Acad. Sci.* 335:137–143.

Diller, K.R. 1990a. A simple procedure for determining the spatial distribution of cooling rates within a specimen during cryopreservation I. analysis, *Proc. Inst. Mech. Engrs, J. Engr. in Med.* 204:179–187.

Diller, K.R. 1990b. Coefficients for solution of the analytical freezing equation in the range of states for rapid solidification of biological systems, *Proc. Inst. Mech. Engrs, J. Engr. in Med.* 204:199–202.

Diller, K.R. 1992. Modeling of bioheat transfer processes at high and low temperatures, *Adv. Heat Trans.* 22:157–357.

Diller K.R., Valvano, J.W., and Pearce, J.A. 2005. Bioheat transfer, in *The CRC handbook of mechanical engineering, 2nd ed.*, ed. F. Kreith and Y. Goswami, 4-282–4-361, Boca Raton, CRC Press.

Howell, J.R. 1982. *Catalogue of radiation configuration factors*, New York, McGraw Hill.

Incroprera, F.P., DeWitt, D.P., Bergman, T.L. et al. 2007. *Fundamentals of heat and mass transfer*, 6th ed., Hoboken, NJ, Wiley.

Kays, W.M., Crawford, M.F., and Weigand, B. 2004. *Convective heat and mass transfer*, 4th ed., New York, McGraw Hill.

Morse, P.M. and Feshback. H. 1953. *Methods of theoretical physics, Parts I and II*, New York, McGraw Hill.

Pennes, H.H. 1948. Analysis of tissue and arterial blood temperatures in the resting forearm. *J. Appl. Physiol.* 1:92–122 (republished on 50th anniversary in 1998. *J. Appl. Physiol.* 85:5–34.)

Roemer, R.B. 1990. Thermal dosimetry, in *Thermal dosimetry and treatment planning*, ed. M. Gautherie, 119–208, Berlin, Springer.

Roselli, R.J. and Diller, K.R. 2011. *Biotransport principles and applications*, New York, Springer.

Shrivastava, D. and Roemer, R.B. 2006. Readdressing the issue of thermally significant blood vessels using a countercurrent vessel network, *J. Biomech. Engr.* 128:210–216.

Siegel, R. and Howell, J.R. 2002. *Thermal radiation heat transfer*, 4th ed. New York, Taylor & Francis.

Valvano, J.W. 1992. Temperature measurement, *Adv. Heat Trans.* 22:359–436.

Valvano, J.W., Allen, J.T., and Bowman, H.F. 1984. The simultaneous measurement of thermal conductivity, thermal diffusivity, and perfusion in small volume of tissue, *J. Biomech. Engr.* 106:198–191.

Wissler, E.H. 1998. Pennes' 1948 paper revisited. *J. Appl. Physiol.* 85:35–41.

Wissler, E.H. 2008. A quantitative assessment of skin blood flow in humans. *Eur. J. Appl. Physiol.* 104:145–157.

2

Thermal Dose Models: Irreversible Alterations in Tissues

John A. Pearce
University of Texas at Austin

2.1 Introduction

Irreversible thermal alterations in tissues result from denaturation of native state tissue enzymes, proteins, and related structures, such as cell membranes, micro tubules, and the like. Mathematical description of these processes, while somewhat cumbersome, yields valuable insight into experimentally observed tissue changes and is a worthwhile undertaking. Predictions of temperature alone are just not adequate since the time of exposure is a defining parameter: hour-long hyperthermia treatments at moderate temperatures exhibit different thermal alteration processes than minutes-long treatments at higher temperatures in ablation procedures. This chapter reviews thermal damage processes in tissues, their physical-chemical thermodynamic governing relations, and useful approximations that have been used to describe them. Specific modeling approaches that have proven useful are also discussed.

We begin with the classical Arrhenius formulation, the Theory of Absolute Reaction Rates, which is designed to predict the "yield" of a reaction (i.e., the fraction of product formed). A discussion of thermal dose units, cumulative equivalent minutes (CEM) as typically used in hyperthermia studies, which derive from Relative Reaction Rates—more properly, relative reaction times—follows. Thermal dose units effectively normalize different thermal histories to a common basis for comparison. The typical assessment criterion in hyperthermia work is to regard a CEM value in excess of a defined threshold number of minutes as evidence of an effective treatment, or an estimate of a treatment physical boundary. CEM values do not, *per se*, provide a prediction of an outcome in the tissues of interest.

The underlying physical principles of the two descriptions are the same; however, their physical significance is quite different. In thermal damage analyses Arrhenius calculations can be used to estimate the volume fraction of damaged tissue constituent, or the probability of observing a particular discrete damage event in an ensemble of experiments.[1] In contrast, thermal dose units constitute a normalizing method for thermal histories: a particular thermal exposure is scaled into an equivalent exposure time at the reference temperature, typically 43°C. It is therefore fair to describe thermal dose units as comparative rather than predictive.

2.2 Irreversible Thermal Alterations in Tissues

More thorough reviews of thermally induced tissue alterations may be found in either of the book chapters[2,3] and in two recent papers by Thomsen[4] and by Godwin and Coad.[5] The brief overview given here will serve to introduce the wide spectrum and complexity of the intricate cascade of thermally induced pathology, both the acutely observable immediate effects and delayed effects. A short summary of representative healing responses is also included. It should be appreciated that many tissue thermal responses, such as hemorrhage and the healing response, are only observable *in vivo* in perfused tissues and will not be observed *in vitro*.

2.2.1 Low Temperature Tissue Effects

Cells and tissues can survive moderate heating for limited times depending on their functional architecture and internal adaptive mechanisms. Protective mechanisms include, for example, the evolution of heat shock proteins—shepherd molecules that assist damaged proteins to refold into their native state conformation. The development of heat shock proteins protects the cell from moderate thermal damage, and makes the cell somewhat robust to repeated heat treatment. Other nonlethal changes are secondary to accelerated metabolism or reversible inactivation of enzymes, or the like. Governing thermal mechanisms in cells and cell cultures are often different than those observed in organs and tissue structures—the tissue structure support mechanisms somewhat ameliorate thermal alterations, and often a substantial volume of an organ can be thermally deactivated and it still may be able to perform to required levels.

Low temperature heating first results in cell swelling, which is largely reversible. The lowest temperature irreversible alterations are apoptosis and necrosis.[6] Apoptosis, or programmed cell death (PCD), is a form of cellular suicide in which the cell dies in response to either an intrinsic or extrinsic trigger.[7] Intrinsic triggers include the presence of a virus, or tissue differentiation during embryonic development; extrinsic triggers include heat, radiation, and circulating cytokines, such as tumor necrosis factor (TNF). Apoptotic processes comprise complex biochemical cascades that shut down cell metabolic functions (the mitochondria) and particular enzymes.[4] Markers of apoptosis include blebbing of cell membranes (the formation of irregular buds in the margin), nuclear shrinking and pyknosis, and chromatin condensation. The improved TUNEL assay (Terminal deoxynucleotidyl transferase dUTP and Nick End Labeling) is now an accepted method for identifying the DNA fragments that result from the last stage of an apoptotic cascade.[8] Apoptotic processes may continue to occur over several days after heating.

Necrosis is cell death due to traumatic injury from external agents—chemicals, poisons, and of course, heat and radiation. Markers include membrane disruption, mitochondrial degeneration, or other morphological changes. Heat-induced apoptosis

and necrosis occur over essentially similar heating times and temperature ranges and are often observed together. In fact, one of the early criticisms of the TUNEL assay work was that the original study included some necrosis mislabeled as apoptosis.[8] More recent TUNEL techniques have successfully addressed this issue. Apoptosis and necrosis are both relatively low temperature "slow" thermal processes and are not usually observed after short-term higher temperature heating. Triggering an apoptotic cascade is one hypothesis for the clinical effectiveness of tumor hyperthermia therapy.[7] Bhowmick et al.[9] report Arrhenius coefficients for combined apoptosis and necrosis in human prostate (see Table 2.1).

Moderate heating also induces alterations in tissue microvasculature. Moderate heating increases the inter-endothelial cell gaps in capillaries, leading to edema—extravasation of plasma proteins. Green et al.[10] and others[11,12] have studied the migration of high molecular weight fluorescent-tagged molecules in the dorsal skin flap model. Fluorescent dye and vital stain techniques are also very useful in cell damage studies, and provide quantitatively measurable markers of thermal damage.[13–17] At slightly higher temperatures the endothelial cells die and blood in the vessels coagulates; hemolysis of red blood cells and the clotting cascade result. Vascular disruption is one of the primary differences between low temperature thermal injury in cells and those in tissues and organs; propagation of cell death downstream from vascular damage may occur for many days after the initial thermal insult, occurring even in unheated tissue volumes as a result of disruptions in local perfusion. Similar thermal damage occurs in tubular endothelial cells, for example, in breast, pancreas, and the testes.[4]

2.2.2 Higher Temperature Tissue Effects

Higher temperature heating ranges from "heat fixation" of tissues—deactivation all life processes without significant alteration of morphology—up to significant structural disruption resulting in amorphous masses. For this discussion, we limit the consideration to temperatures below the boiling point, at which steam evolution dominates other thermodynamic forces.

2.2.2.1 Blanching and Heat Fixation

Heat-fixed tissue is morphologically similar to viable tissue, but the complete absence of reperfusion (in contrast to a cryotherapy lesion) and metabolic processes give it a blanched appearance grossly; consequently, it is often referred to as "white coagulation." At higher temperatures the cellular and tissue structure is visibly disrupted. In successful ablation lesions the central zone next to the applicator frequently exhibits heat fixation and substantial coagulative necrosis and denaturation of the cellular and extracellular matrix proteins that resists standard wound healing.[5] The central lesion is surrounded by coagulative necrosis that undergoes classical wound healing. The outermost lesion site (*in vivo*) is a transition zone of intermixed processes in which substantial extravasation of red blood cells (hemorrhage)

TABLE 2.1 Collected Arrhenius Parameters for Example Thermal Damage Process

Process	Process Parameters			Notes
	A (s^{-1})	E_a (J mole^{-1})	T_{crit} (°C)	
Cell Death				
Sapareto[32]	2.84×10^{99}	6.18×10^5	51.4	CHO Cells, $T > 43$°C
Beckham[34]	6.9×10^{116}	7.3×10^5	53.2	Murine w/o Hsp70
	3.7×10^{157}	9.8×10^5	51.7	Murine w/Hsp70
Bhowmick[9]	7.78×10^{22}	1.61×10^5	94.2	H. Prostate apoptosis/necrosis
Bhowmick[44]	1.66×10^{91}	5.68×10^5	52.1	AT-1 Cells < 50°C
	173.5	1.97×10^4	186	AT-1 Cells > 50°C
Borrelli[35]	2.984×10^{80}	5.064×10^5	55.5	BhK Cells
He[40]	4.362×10^{43}	2.875×10^5	71.0	SN12 cells, suspended
	3.153×10^{47}	3.149×10^5	73.1	SN12 cells, attached
Erythrocytes				
Lepock[37]	7.6×10^{66}	4.55×10^5	82.2	Hemoglobin denaturation
Przybylska[36]	a1.08×10^{44}	2.908×10^5	71.8	Hemolysis, Normal RBCs
	a3.7×10^{43}	2.88×10^5	72.1	Hemolysis, Down syndrome RBCs
Skin Burns				
Henriques[22]	3.1×10^{98}	6.28×10^5	59.9	Not recommended
Diller[45]	8.82×10^{94}	6.03×10^5	58.6	$T \leq 53$°C (same data)
	1.297×10^{31}	2.04×10^5	69.3	$T > 53$°C
Weaver[41]	2.19×10^{124}	7.82×10^5	55.4	$T \leq 50$°C
	1.82×10^{51}	3.27×10^5	60.1	$T > 50$°C
Brown[39]	1.98×10^{106}	6.67×10^5	54.6	Microvessels in muscle
Retinal Damage				
Welch[46]	3.1×10^{99}	6.28×10^5	56.6	Whitening
Collagen Changes				
Maitland[47]	1.77×10^{56}	3.676×10^5	68.2	Rat tail birefringence
Pearce[48]	1.61×10^{45}	3.06×10^5	80.4	Rat skin birefringence
Aksan[49]	1.136×10^{86}	5.623×10^5	68.1	Rabbit patellar tendon
Miles[50]	6.658×10^{79}	5.21×10^5	67.8	Rat tail tendon, DSC
Chen[51–53]	1.46×10^{66}	4.428×10^5	76.4	Shrinkage, rat tail tendon
Muscle				
Jacques[54]	2.94×10^{39}	2.596×10^5	70.4	Myocardium whitening
Liver				
Jacques[55]	5.51×10^{41}	2.769×10^5	73.4	Whitening, pig liver

a Value for A is estimated from Wright's line.

usually accompanies cell death. The transition zone contains tissues that may regenerate postheating.[5]

2.2.2.2 Structural Protein Denaturation

Collagen is the ubiquitous basic structural protein for almost all tissues, an especially prominent feature of blood vessels, connective tissues such as tendons and ligaments, and bone. There are at least 13 various forms of the collagen molecule.[18] Heating collagen results in unraveling of the rope-like molecular and macromolecular structure that is the source of its strength and resilience. At higher temperatures collagen shrinks in length and swells in diameter in an approximately isovolumic process as the macromolecules unwind. As the temperature increases, the collagen assumes an amorphous glassy, or hyalinized, state. The completely denatured collagen is resorbed during healing and eventually replaced by scar tissue.[4,5]

Elastin is a thermally robust structural tissue protein responsible for the elasticity of large arteries and other mechanical energy-storing structures, such as the *ligamentum nuchae* in the necks of grazing animals. Elastin is thermally stable until temperatures exceed the boiling point. Tissue fusion processes rely on the complete denaturation of collagen and also elastin, if present.

2.2.3 Healing Processes

The two major healing pathways are: (1) a "repair in place" mechanism, wherein the tissue is rebuilt into its original composition, architectural structure, and function with little to no observable scarring, and (2) replacement of the damaged tissue by fibroblastic scar with concomitant loss of function.[5] In certain cases the damaged tissues may be replaced with metaplastic structures that include nonfibroblastic scarring or tissues of a type not native to the damaged region.

Healing processes begin with the autolytic enzymatic breakdown of injured cells—heat-fixed tissues do not undergo these changes and do not participate in healing. Heat-fixed tissues maintain their *in situ* appearance for the first several days to a week while the coagulative necrotic tissues fade out and assume a ghost-like appearance. After about a week, heat-fixed tissues gradually fade to the appearance of coagulative necrosis and become brittle. The adjacent zone of coagulative necrosis is eventually phagocytized and replaced by scar tissue, and a layer of histocytes forms a barrier around the heat-fixed tissue as a foreign body reaction.[5] The isolated foreign body can remain stable for extended times.

2.3 Physical Chemical Models: Arrhenius Formulation

Arrhenius models of irreversible thermal alterations in tissues were first introduced in a seminal series of papers by Henriques and Moritz in 1947.[19–22] In their work skin burns were created in pig skin by flowing water over the surface at constant temperature, and the relative damage was calculated in the form of a dimensionless parameter, Ω:

$$\Omega(\tau) = \int_0^\tau A\, e^{\left[\frac{-E_a}{R T(t)}\right]} dt \qquad (2.1)$$

where τ = the duration of the exposure (s), A = the frequency factor (s⁻¹), E_a = the activation energy (J/mole), R = the universal gas constant, 8.3143 (J mole⁻¹ K⁻¹), T = temperature (K), and t the time (s). Ω was calculated based on the temperature and exposure time of the flowing water and represents the thermal history

at the skin surface. A surface Ω value of 0.53 corresponded to first-degree, 1.0 to second-degree, and 10^4 to third-degree burns by their definition. We will look more closely at this particular process in a later section.

The Arrhenius model is based on the kinetic model of relative reaction rates that the Swedish scientist Svante Arrhenius first formulated and described in 1889,[23] which was extended to the theory of absolute reaction rates in later studies. That original work was published in German, and is well described by Johnson et al.[24] in their first chapter. Henriques and Moritz, unfortunately, did not fully employ the rate process description as a prediction of the extent of damage, and left the result in terms of Ω alone, evaluated at the surface of the skin.

Historically, the kinetic model was thoroughly studied in the context of protein denaturation *in vitro*—a dominant process in tissue thermal damage. The review by Eyring and Stearn in 1939[25] (see Table 2.2) effectively summarizes much of the early work. Perhaps the first rigorous application of Arrhenius analysis to the skin burn problem was by Fugitt in 1955.[26] Successful models of thermal radiation (i.e., "flash") skin burns were a matter of great interest in the decades immediately following the initial development of nuclear weapons. In this discussion the underlying formulation for the Arrhenius model is briefly reviewed from the physical chemistry point of view, and cast into a form that can readily be compared to quantitative measures of thermal alteration in tissues, such as histologic section, relative fluorescence intensity, relative birefringence intensity, or vital stain assay.

2.3.1 Chemical Thermodynamics Basis for the Arrhenius Model

For a first-order unimolecular reaction, the rate of disappearance of the reactant is described by a Bernoulli differential equation:

$$\frac{dC}{dt} = -kC \qquad (2.2)$$

where C = the concentration of reactant, and k = the velocity of the reaction (s⁻¹). Physically, the reactant is described as surmounting an energy barrier to form an "activated" complex, C^*, in quasi-equilibrium with the normal inactivated state. The

TABLE 2.2 Arrhenius Parameters for Selected Proteins and Enzymes from Erying and Stearn

| | Process Parameters | | | |
Protein/Enzyme	A (s⁻¹)	E_a (J mole⁻¹)	@ T (°C)	ΔS^* (J mole⁻¹ K⁻¹)
Insulin	1.177×10^{18}	1.49×10^5	80	99.63
Egg albumin	7.51×10^{81}	5.53×10^5	65	1321.5
Hemoglobin	170×10^{46}	3.165×10^5	60.5	639.2
Pancreatic proteinase	5.074×10^{21}	1.585×10^5	50	170.0
Pancreatic lipase	5.503×10^{27}	1.899×10^5	50	285.5
Hydrated invertase	1.931×10^{53}	3.615×10^5	55	774.4

Source: Eyring H, Stearn AE, *Chemical Reviews*, 24, 1939.

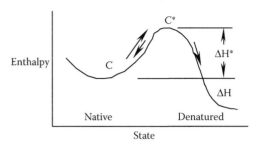

FIGURE 2.1 Activation energy barrier, ΔH^*, between native and denatured states. Activated complex can either relax back to the native state or progress to denatured at the net forward velocity, k.

activated complex may either relax to the inactivated native state or may proceed to a denatured state, as depicted in Figure 2.1, at overall denaturation forward velocity, k.

In the figure the enthalpy represents the total energy; the enthalpy of activation, ΔH^*, should not be confused with the sensible reaction heat, ΔH. The reaction velocity, k, is in turn given by the Gibb's free energy of activation, ΔG^*:

$$k = e^{\left[\frac{-G^*}{RT}\right]}. \tag{2.3}$$

The activation enthalpy, ΔH^*, includes both the Gibb's free energy of activation and the entropy of activation, ΔS^*:

$$H^* = G^* + T S^*. \tag{2.4}$$

Consequently, in terms of the activation entropy and enthalpy, the Eyring-Polyani equation gives:

$$k = \left(\frac{RT}{Nh_p}\right) e^{\left[\frac{S^*}{R}+1\right]} e^{\left[\frac{-H^*}{RT}\right]} \cong \left(\frac{RT}{Nh_p}\right) e^{\left[\frac{S^*}{R}\right]} e^{\left[\frac{-E_a}{RT}\right]} \tag{2.5}$$

where N = Avogadro's number (6.023×10^{23} mole^{-1}), and h_p = Planck's constant (6.625×10^{-34} J s mole^{-1}). Here the activation enthalpy has been truncated to the activation energy, E_a, as an approximation—in fact, $\Delta H^* = E_a - iRT$, where i is the order of the reaction. In practice iRT is small compared to E_a for a first-order damage process—i.e., the 1 in Equation 2.5 is small compared to $\Delta S^*/R$. Note also that the fraction preceding the entropy exponential varies little over the few K temperature range that typifies the active region of a damage thermal history. Consequently, the entire preexponential factor is usually treated as a constant, A, as in Equation 2.1. The physical significance of A is that it is an indication of the collision frequency between reactants, or some analogous quantity in a unimolecular reaction.

The solution of the governing differential equation (Equation 2.2) is determined by an appropriate integrating factor, and given by

$$C(\tau) = C(0)e^{\left\{-\int_0^\tau k\,dt\right\}}$$

so that

$$\ln\left\{\frac{C(0)}{C(\tau)}\right\} = \int_0^\tau A\, e^{\left[\frac{-E_a}{RT(t)}\right]} dt = \Omega(\tau) \tag{2.6}$$

and the physical significance of the damage parameter, Ω, as used in Equation 2.1 is now easily seen. Equation 2.6 constitutes the theory of absolute reaction rates. Using this approach, the remaining undamaged tissue constituent, surviving cell fraction, or probability of thermal damage is

$$C(\tau) = 100e^{-\Omega}\ (\%) \quad \text{or, the probability is} \quad P(\%) = 100[1 - e^{-\Omega}]. \tag{2.7}$$

2.3.2 Arrhenius Process Functional Behavior and Determination of Process Coefficients

The functional behavior of Equations 2.1 and 2.6 warrants some discussion at this point. Additionally, the standard methods to determine damage process coefficients are worthy of review. There are three possibilities: (1) the damage process can be measured in real time as it develops, such as fluorescence intensity, in which case determining the appropriate coefficients, A and E_a, amounts to a curve-fitting exercise; (2) the result can only be determined *a posteriori*, as by histologic assay, but the experiments are at constant temperature, or nearly so; or (3) the result can only be determined *a posteriori*, but the experiments are plainly transient, for example, as in a laser heating experiment. This section treats these issues in turn.

2.3.2.1 Functional Behavior

The damage parameter, Ω, makes a convenient measure of the process progress; $\Omega = 1$ is a convenient calculational reference point and is often referred to as the "threshold." However, it should be borne in mind that by the time $\Omega = 1$ the process is 63.2% complete—hardly the "threshold" for the process. Using $\Omega = 1$ as the reference point, the "threshold temperature" in a constant temperature exposure, T_{TH} (K), is given by:

$$T_{TH} = \frac{E_a}{R[\ln\{A\} + \ln\{\tau\}]}. \tag{2.8}$$

For a different reference damage level, Ω_{ref}, the threshold constant temperature is:

$$T_{TH} = \frac{E_a}{R[\ln\{A\} + \ln\{\tau\} - \ln\{\Omega_{ref}\}]}. \tag{2.9}$$

Finally, an effective characterizing parameter for a higher temperature damage process is its "critical temperature," T_{crit} (K), the temperature at which the rate of damage accumulation, $d\Omega/dt = 1$ (s^{-1}):

$$T_{crit} = \frac{E_a}{R\ln\{A\}}. \qquad (2.10)$$

Functionally, the remaining undamaged tissue constituent, $C(\tau)$, decreases sigmoidally with time at constant temperature, and also decreases sigmoidally with temperature for a constant exposure time. Several calculations follow in a later section that illustrate this behavior. Also, the rate of damage accumulation, $d\Omega/dt$, is vanishingly small at low temperature and increases precipitously above T_{crit}. As a usual consequence, it is far too tempting to describe thermal damage as occurring at a specific temperature; however, as in Equation 2.9, the time of exposure is important as well and must be reported when discussing results. For example, a hypothetical damage process for which $E_a = 5 \times 10^5$ (J mole^{-1}) and $A = 7.07 \times 10^{78}$ (s^{-1}) has $T_{crit} = 58.1°C$. For an exposure typical of hyperthermia treatment times, 1 hour ($\tau = 3600$ s), and 90% thermal damage probability ($\Omega = 2.303$), and from Equation 2.9, $T_{TH} = 45.2°C$. For an exposure typical of an ablation procedure (e.g., $\tau = 60$ s) the threshold temperature increases to $52.2°C$; and if the exposure is a 1 ms laser pulse, $T_{TH} = 72.8°C$.

There is a logical trap to be wary of in universally applying the Arrhenius formulation. Mathematically, thermal damage can occur at extremely low rates at normal body temperatures. For example, using the Diller et al. skin burn coefficients[27] from Table 2.1—$A = 8.82 \times 10^{94}$ (s^{-1}), $E_a = 6.03 \times 10^5$ (J mole^{-1})—at $37°C$ $d\Omega/dt = 2.47 \times 10^{-9}$ (s^{-1}). We would expect 63.2% thermal damage in about 4.04×10^6 (s), or 7.69 years. This is, of course, nonsense. It is difficult, if not impossible, to accept any thermal damage prediction at normal body temperatures. Of course, skin cells are continually replaced at a much higher rate than that by normal attrition (originating in the granulating layer) due to multiple damage processes (dehydration, ultra violet exposure, abrasion, and so forth); and resting exposed skin temperature is less than $37°C$, ranging from $30°C$ to $35°C$ depending on covering, latitude, and season of the year. Nevertheless, it is important not to calculate any thermal damage process below the lowest temperature at which the process has been observed. For the classical skin burn example the traditionally accepted lower temperature limit is $45°C$, at which temperature $\Omega = 1$ at 11,309 s, or 188 hours from those coefficients.

2.3.2.2 Determining Process Coefficients in Constant Temperature Experiments

The process coefficients, A and E_a, can only be determined from theoretical calculations in the simplest of reactions in gas phase at low pressure. All processes of practical interest must be studied experimentally. We do have predictive limits to work within, however. Eyring and Stearn[25] point out that the Gibb's free energy of activation, ΔG^*, varies only over a relatively narrow range: they list 21 values for the denaturation of hydrated enzymes from literature that range from a low of $\Delta G^* = 91.7$ (kJ mole^{-1}) for pepsin at $25°C$—that is, $\Delta H^* = 232.7$ (kJ mole^{-1}) and $\Delta S^* = 474.3$ (J mole^{-1} K^{-1}), with $A = 3.7 \times 10^{37}$—to a high of $\Delta G^* = 107.6$ (kJ mole^{-1}) for invertase, an enzyme that catalyzes the hydrolysis of sucrose, at $55°C$—that is, $\Delta H^* = 361.5$ (kJ mole^{-1}), and $\Delta S^* = 774$ (J mole^{-1} K^{-1}), with $A = 1.84 \times 10^{53}$. The consequence is that ΔH^* and ΔS^* are approximately linearly related, as derived by Miles[28] using a "polymer in a box" construct to describe collagen denaturation. In an extremely insightful recent article, Wright[29] plotted a large number of published values for tissue damage process coefficients on $\ln\{A\}$–E_a axes with the result that:

$$E_a = 2.61 \times 10^3 \ln\{A\} + 2.62 \times 10^4 \qquad (2.11a)$$

or

$$\ln\{A\} = 3.832 \times 10^{-4} E_a - 10.042 \qquad (2.11b)$$

with a very high degree of correlation—and E_a is in units of J mole^{-1}, as in the preceding discussion. Figure 2.2 compares Equation 2.11b to the data in Tables 2.1 and 2.2.

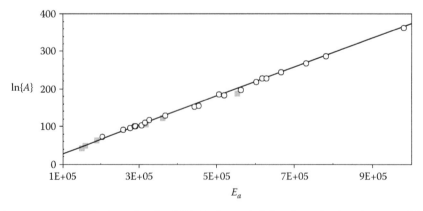

FIGURE 2.2 Plot of Arrhenius parameters for tissues listed in Table 2.1 (open circles) and proteins from Erying and Stearn[25] listed in Table 2.2 (solid squares) compared to Wright's Line, Equations 2.11a,b. A subset of the data included in Table 2.1 was used by Wright to determine the line, along with many additional sources.

The classical method to determine damage process coefficients is to expose the tissue to constant temperatures for measured times and determine Ω by some quantitative measure, such as cell survival fraction or vital stain/fluorescence intensity. Looking at Equation 2.1 and rearranging, the logarithm of the result follows:

$$\ln\{\tau\} - \ln\{\Omega\} = \frac{1}{T}\left(\frac{E_a}{R}\right) - \ln\{A\} = \ln\{\tau_{eq}\} \qquad (2.12)$$

where $\tau_{eq} = \tau/\Omega$, the time of an equivalent exposure at T for which $\Omega = 1$. An ensemble of such experiments is plotted on Arrhenius axes—$\ln\{\tau_{eq}\}$ on the ordinate and $(1/T)$ on the abscissa—for which a least-squares fit has slope = E_a/R and intercept = $\ln\{A\}$. The determination is very sensitive to small uncertainties in temperature owing to the hyperbolic dependence in the exponent; consequently, extremely careful attention must be given to accurate temperature measurements, above all other considerations. In all of these experiments, the linear regression line effectively holds the short end of a very long "stick," as it were, and the uncertainty in the intercept, A, is huge as a result.

2.3.2.3 Transient Experiments

The case of transient experiments in which only a single assay can be determined at the conclusion of the experiment is an especially thorny one. Unfolding Equation 2.1 to a suitable form for calculation from the experiment results is intractable. There is an approach that can be used in this case, however.[1,3] Each experiment has a measured transient thermal history, $T_i(t)$, and corresponding level of damage, Ω_i, for which E_a and A are unknown. For each experiment in an ensemble, a selected segment of the $\ln\{A\} - E_a$ plane can be scanned for values that yield the measured Ω_i from the transient history, $T_i(t)$. Define a "cost" function that is minimized when A and E_a give the correct integral, such as:

$$Cost = \left| \ln \left\{ \frac{\int_0^\tau A\, e^{\left[\frac{-E_a}{R\, T_i(t)}\right]} dt}{\Omega_i} \right\} \right|. \qquad (2.13)$$

The locus of points in the $\ln\{A\} - E_a$ plane that yield the required Ω_i value lies along a straight line. Rearranging Equation 2.12 one obtains:

$$\ln\{\tau_{eq}\} = \frac{1}{T_{eq}}\left(\frac{E_a}{R}\right) - \ln\{A\}. \qquad (2.14)$$

The slope of the solution line for the A–E_a pairs gives T_{eq}, and the intercept τ_{eq}; the temperature and exposure time, respectively, for an equivalent constant temperature exposure for which $\Omega = 1$. It is not difficult to estimate the likely range of $\ln\{A\}$ values that will match a selected search range for E_a from Equations 2.11a,b above. The ensemble of equivalent constant temperature points for the experiment series is then plotted on standard Arrhenius axes to determine A and E_a for the process under study.

2.3.3 Example Thermal Damage Processes

Thermal damage has been measured in a very large number of processes over the past four decades. A few representative processes have been selected for illustrative purposes in this section. More complete reviews of this literature may be found in several recent publications.[1,6,29,30] The process parameters for the examples as described may be found in Table 2.1.

2.3.3.1 Cell Survival Studies

In the classical experiments of the thermal sensitivity of Chinese hamster ovary cells by Sapareto et al.[31,32] "survival" is indicated by the ability of the cells to continue to form colonies. The surviving fraction at constant temperature was measured *vs* time analogously to Equation 2.6. The slope of the survival curve, D_0 corresponds to $1/k$:

$$\left.\frac{S}{N_0}\right|_t = e^{-\frac{t-t_0}{D_0(T)}} = e^{-k(t-t_0)} = \frac{C(\tau)}{C(0)} \qquad (2.15)$$

where N_0 cells are counted at time t_0 and S of them survive at time t. The Arrhenius plot of D_0 *vs* T has an obvious break point at $T = 43°C$ in its data, and this temperature is used as the reference temperature in hyperthermia work. A break point in an Arrhenius plot indicates that different thermal processes are dominant above and below the break point temperature.

For the Chinese hamster ovary cells above 43°C, the Arrhenius coefficients work out to: $A = 2.84 \times 10^{99}$ (s⁻¹) and $E_a = 6.18 \times 10^5$ (J mole⁻¹), with a critical temperature of 51.4°C. Figure 2.3 plots the predicted surviving fraction *vs* time at 44°C for these cells; the response is characteristically sigmoidal, as has been described. The curve shifts to the left as the temperature increases.

In cell survival studies involving longer term heating at lower temperatures it is often more instructive to use the time at 43°C to reach $\Omega = 1$ as a comparative parameter, rather than T_{crit}:

$$\tau_{43} = \exp\left\{\frac{E_a}{2.629 \times 10^3} - \ln\{A\}\right\} \qquad (2.16)$$

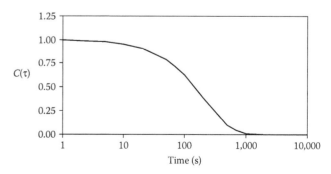

FIGURE 2.3 Cell survival fraction for asynchronous Chinese hamster ovary cells[31,32] at 44°C.

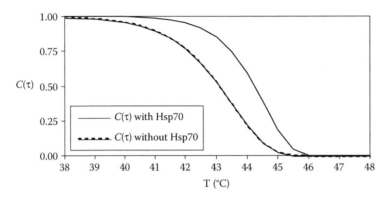

FIGURE 2.4 The protective effects of Hsp70 in murine embryonic blast cells may be seen in a plot of the surviving fraction *vs* temperature for exposures of 60 minutes, also a sigmoidal relationship.

where the constant in the denominator is RT at 43°C. For the Chinese hamster ovary cells, $\tau_{43} = 447$ minutes. Beckham et al.[33,34] studied the protective role of the heat shock protein Hsp70 in murine embryonic fibroblasts using bioluminescent imaging. The thermal sensitivity was measured in intact cells and those with Hsp70 production blocked: cells were preheated, and then heated again 4 hours later for the measurement.

With Hsp70 production intact the Arrhenius parameters for reheated cells are: $A = 3.7 \times 10^{157}$ (s^{-1}) and $E_a = 9.8 \times 10^5$ (J mole^{-1}), $\tau_{43} = 22.1 \times 10^3$ s (368 minutes). In the Hsp70 deficient cells the reheating damage coefficients are: $A = 6.9 \times 10^{116}$ (s^{-1}) and $E_a = 7.3 \times 10^5$ (J mole^{-1}), $\tau_{43} = 5.87 \times 10^3$ s (97.9 minutes). Hsp70 provides nearly a 4:1 improvement in thermotolerance in these cells by this measure—as in Figure 2.4. Interestingly, the T_{crit} values are reversed in the two experiment groups: 51.7°C and 53.2°C, respectively.

Borelli et al. determined thermal damage coefficients for Bhk cells *in vitro*.[35]

2.3.3.2 Membrane Disruption

Cell membranes experience a mostly a nonproteinaceous damage process, although breakdown of intrinsic and extrinsic membrane proteins is equally likely to occur in parallel with loss of integrity in the bi-lipid layer. Przybylska et al.[36] compared hemolysis rates in normal and Down syndrome patient RBCs, but did not report values for A or ΔS^*—the estimates for A given in Table 2.1 were derived from Wright's line, Equations 2.11a,b. Lepock et al.[37] measured thermal denaturation of hemoglobin.

A comparison of the two sets of parameters at a high fever temperature of 42.8°C (109°F) is shown in Figure 2.5, along with the Down syndrome results. Despite the slight difference in rates (i.e., E_a) between Down syndrome and normal patients reported by Przybylska et al., there is no detectable difference in the predicted hemolysis at this temperature. Note also that the Lepock et al. coefficients predict that the hemoglobin would be robust against denaturation for at least 100 hours (more than 4 days) at this temperature. Practical clinical experience suggests that the Przybylska et al. coefficients do provide a reasonable estimate of the onset of hemolysis under these conditions.

2.3.3.3 Skin Burns

Skin burns make an interesting example because, in addition to being the classical regime for thermal damage studies, the dominant process in skin burns is disruption in the vasculature, mostly the capillary microvasculature. Diller et al.[38] provide

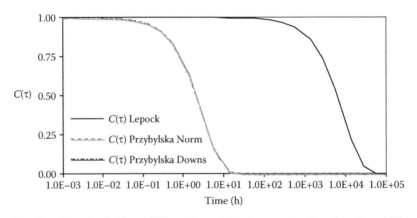

FIGURE 2.5 Prediction of hemolysis from Przybylska et al.[36] and hemoglobin denaturation from Lepock et al.[37]

an excellent description of skin burn physical and physiological processes. In a related study Brown et al.[39] determined process parameters for normal muscle microvasculature that agree very well in model work with the published skin burn values. That study also illuminated the thermal hypersensitivity of KHT fibrosarcoma microvasculature.

In their original skin burn experiments, Henriques and Mortiz calculated only Ω values, defining $\Omega = 0.53$ as corresponding to superficial irreversible erythema (first-degree), $\Omega = 1.0$ to transepidermal necrosis (partial thickness, or second-degree), and $\Omega = 10^4$ to complete involvement of the dermis (full-thickness, or third-degree). The reported Ω values refer to the thermal history at the skin surface in their work, as previously mentioned. The value of Ω at the damage propagation front is much lower, however. Numerical model studies[40] suggest that a threshold of $C(\tau) = 80$ to 85% (i.e., 15% to 20% predicted damage) provides a reasonable estimate of the depth of skin burn under both the Diller-Pearce coefficients[40] and the Weaver-Stoll coefficients.[41]

A word of caution: the Arrhenius coefficients originally reported by Henriques and Moritz[22] have since been widely used but actually do not fit their own data very well—Diller et al.[27] reanalyzed their original data and developed the values listed in Table 2.1, and these are recommended in their stead. Weaver and Stoll[41] studied thermal radiation skin burns with similar results. In both sets of data there is a breakpoint, 53°C in the Henriques and Moritz data and 50°C in Weaver and Stoll's measurements.

2.3.3.4 Recent Models for Ablation

In two recent publications, Breen et al.[42] and Chen and Saidel[43] have suggested a different calculational approach that they claim (without demonstration) has advantages over the more traditional Arrhenius models in predicting ablation results. Their calculation preserves much of the form of Equation 2.6 but relies on elaborate curve fitting independent of consideration of the processes listed in Table 2.1. In brief, the damage is calculated from:[42]

$$W(\tau) = 1 - C(\tau) = 1 - \exp\left[-\int_0^\tau \beta\, T(t)\, dt\right] \qquad (2.17)$$

where the integration kernel is given by:

$$\beta\, T(t) = \begin{cases} 0, & T(t) < T_C \\ A[T(t) - T_C]^N & T(t) \geq T_C \end{cases} \qquad (2.18)$$

with: T_C = the "critical temperature" (°C), and a particular value for W, W_{crit}, indicates "cell death" and is held by the authors to combine several cell-damage processes. Their original notation has been slightly modified to coincide more closely with that used in this chapter. The fit parameters A and N are determined by experiment—and this A is not the frequency factor from Equation 2.1. It remains to be seen whether this model gives predictions significantly different from the classical Arrhenius approach.

To approach this question, we take as a calculational framework a constant temperature exposure of 2 minutes duration, such as might be observed in an ablation procedure. For illustration we select mid-range parameters from[42]: $A = 0.0067$, $N = 1.0672$, $T_C = 45.65°C$ with $W_{crit} = 0.6782$—i.e., $C(\tau) = 0.3218$. Three of the comparison processes in Figure 2.6 agree remarkably well with the Equation 2.17 prediction—surprisingly, the BhK cells and murine fibroblasts (with Hsp70 intact) are virtually identical for this exposure time and agree extremely closely with the calculation of Equation 2.17.

Using the 32% remaining undamaged cell criterion (i.e., $C(\tau) = 0.322$) for comparison, this occurs just slightly above 47.5°C for the two cell lines and just above 47°C in the Equation 2.17 calculation. The two thermal damage models are indistinguishable in this example for all practical purposes. It is interesting to note that CHO cells appear to be much more thermally sensitive, the muscle microvasculature is an acceptably close

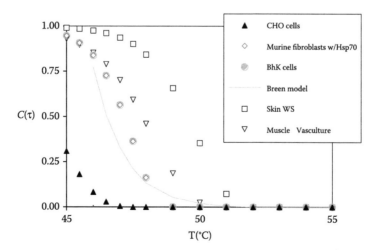

FIGURE 2.6 Comparison of the damage prediction from the formulation of Breen et al.[42] with representative processes from Table 2.1 for 2-minute exposures.

match, and the skin burn coefficients from Weaver and Stoll predict substantially less thermal damage.

Also, the shape of the Breen et al. model prediction is noticeably different from the Arrhenius processes, as would be expected from their respective functional forms. If the Breen function is used to calculate the equivalent exposures for $\Omega = 1$ at temperatures between 47°C and 60°C, the resulting plot on Arrhenius axes is reasonably fit by a parabola—plainly not first-order reaction behavior:

$$\ln\{\tau\} = 1.562 \times 10^8 \left(\frac{1}{T}\right)^2 - 9.384 \times 10^5 \left(\frac{1}{T}\right) + 1.412 \times 10^3 (r^2 = 0.987). \quad (2.19)$$

The Arrhenius model is based on well-established fundamental physical principles, so it is difficult to see any real advantage to the reformulation in the style advocated by Breen et al.

2.4 Comparative Measures for Thermal Histories: Thermal Dose Concept

For approximately the past three decades the standard method for assessing hyperthermia treatment effectiveness has been the thermal dose unit. The concept of thermal dose units, in the form of cumulative equivalent minutes (CEM) of exposure at 43°C, dates at least from the pioneering work of Sapareto et al. in 1978[32] and is also very well described in their chapter in the subsequent 1982 book.[31] Thermal dose units derive from the observation that above 43°C for many cell lines studied a similar level of damage resulted in approximately half the time when the temperature increased by 1°C.[31,32,56] A CEM value in excess of an accepted threshold is considered indicative of a successful treatment. The applied thermal dose, CEM, is calculated from:

$$\text{CEM}_{43} = \sum_{i=1}^{N} [R_{\text{CEM}}]^{(43-T_i)} t_i = \int_0^\tau [R_{\text{CEM}}]^{(43-T)} dt \quad (2.20)$$

where the exposure is either in discrete intervals, t_i, or continuous, and $R_{\text{CEM}} = 0.5$ is typically used above 43°C (and $R_{\text{CEM}} = 0.25$ below 43°C). The authors suggest the use of a normalized time of exposure in order to provide a means for quantizing hyperthermia treatments, CEM, or cumulative equivalent minutes at 43°C. In that way, an arbitrary transient temperature history can be directly compared to a constant temperature exposure at 43°C.

2.4.1 Foundation of Thermal Dose Concept

Thermal dose units may be seen to derive directly from Arrhenius's original description, as described in the summary of his pioneering work in the first chapter of Johnson et al.[24]

Briefly, Arrhenius measured the rate of hydrolysis of sucrose in the presence of various acids in his original experiments. The observations indicated that the temperature dependence of the reaction rate was too great to be described by either the temperature effect on the kinetic energy of the molecules or on the dissociation of the acids. The measured reaction velocities at temperatures T_1 and T_2 (K) were related by:

$$\frac{k|_{T_2}}{k|_{T_1}} = e^{\frac{B(T_2-T_1)}{T_2 T_1}} = \left[e^{\frac{B}{T_2 T_1}} \right]^{(T_2-T_1)} \quad (2.21)$$

where Arrhenius's original notation has been revised to that used in this chapter for clarity, and B is an experimentally determined constant. The reason for the particular regrouping in the right-hand expression will become clear in the following discussion.

R_{CEM} in Equation 2.20 is the ratio of exposure times required to result in the same survival for a 1°C rise in temperature (note that $\tau_2/\tau_1 = k_1/k_2$, an inverse relationship). That is, for $T_2 = T_1 + 1$ (K) and $\Omega_2 = \Omega_1$, then:[30]

$$R_{\text{CEM}} = \frac{\tau_2}{\tau_1} = \frac{A \, e^{\left[\frac{-E_a}{R T_1}\right]}}{A \, e^{\left[\frac{-E_a}{R T_2}\right]}} = e^{\frac{-E_a}{R T_1 (T_1+1)}} = e^{\frac{-B}{T_1 T_0}}. \quad (2.22)$$

The relationship to Arrhenius's original work is now apparent, and the value of B is seen to be E_a/R. At 44°C R_{CEM} for the Chinese hamster ovary cells is 0.479 (see Table 2.3), very close to 0.5. For uniform computation of CEM values the representative value for R_{CEM} has historically been taken to be 0.5 above 43°C and 0.25 below 43°C to match the data for the several cell types originally listed.[31]

2.4.2 Example Process Calculations

Values for R_{CEM} are given in Table 2.3, computed at $T_1 = 317.16$ K = 44°C for the thermal damage processes previously discussed.

Note that no information about the process temperature offset, A, is contained in R_{CEM}; consequently, by default all damage processes are referred to the 43°C reference point in this measure. R_{CEM} depends more strongly on the activation energy, E_a, than it does on the temperature. At 43°C $R_{\text{CEM}} = \exp(1.199 \times 10^{-6} E_a)$, while at 50°C $R_{\text{CEM}} = \exp(1.148 \times 10^{-6} E_a)$, only very slightly different.

2.5 Applications in Thermal Models

Implementation of Equation 2.6 (the Arrhenius model) or Equation 2.20 (the CEM calculation) in a thermal model is fairly straightforward, but may require some care. For example, the value of Ω increases very rapidly above threshold temperatures estimated from Equation 2.8, and may quickly saturate the computer arithmetic if not limited. Values of Ω above about 10—i.e., $C(\tau) \leq 4.5 \times 10^{-6}$—yield no new information since the damage process is plainly saturated;

TABLE 2.3 R_{CEM} for Collected Representative Arrhenius Kinetic Coefficients

Process	A (s^{-1})	E_a (J mole^{-1})	R_{CEM} 44°C	Notes
Cell Death				
Sapareto	2.84×10^{99}	6.18×10^{5}	0.479	CHO Cells
Beckham	6.9×10^{116}	7.3×10^{5}	0.419	without Hsp70
	3.7×10^{157}	9.8×10^{5}	0.311	with Hsp70
Bhowmick	7.78×10^{22}	1.61×10^{5}	0.825	H. Prostate Apoptosis
Bhowmick	1.66×10^{91}	5.68×10^{5}	0.508	AT-1 Cells < 50°C
	173.5	1.97×10^{4}	0.977	AT-1 Cells > 50°C
Borrelli	2.984×10^{80}	5.064×10^{5}	0.547	BhK Cells
He	4.362×10^{43}	2.875×10^{5}	0.710	SN12 cells, suspended
	3.153×10^{47}	3.149×10^{5}	0.687	SN12 cells, attached
Erythrocytes				
Lepock	7.6×10^{66}	4.55×10^{5}	0.581	Hemoglobin denaturation
Przybylska	[a]1.08×10^{44}	2.908×10^{5}	0.707	Hemolysis Normal
	[a]3.7×10^{43}	2.88×10^{5}	0.709	Hemol. Down's Syndrome
Skin Burns				
Henriques	3.1×10^{98}	6.28×10^{5}		<u>Not</u> Recommended
Diller	8.82×10^{94}	6.03×10^{5}	0.487	$T \le 53$°C (same data)
	1.297×10^{31}	2.04×10^{5}		$T > 53$°C
Weaver	2.19×10^{124}	7.82×10^{5}	0.394	$T \le 50$°C
$T > 50$°C				
	1.82×10^{51}	3.27×10^{5}		
Brown	1.98×10^{106}	6.67×10^{5}	0.452	Microvessels
Retinal Damage				
Welch	3.1×10^{99}	6.28×10^{5}	0.473	Whitening

[a] Value of A estimated from Wright's line (Equation 2.11b).

consequently, it is sensible to limit Ω and $d\Omega/dt$ calculations to that neighborhood in numerical model work. CEM calculations do not have this limitation. Both Arrhenius and CEM estimates may be easily included in commercial numerical modeling software that permits superposition of additional model modes on the model space. All that is necessary is the superposition of one or more additional model modes (i.e., one mode for each active damage process calculation) that can accomplish integration by time:

$$\frac{\partial y}{\partial t} = f \qquad (2.23)$$

where y is either Ω or CEM and f, the forcing function, is either from Equation 2.6 or 2.20.

2.5.1 Apoptosis/Necrosis Example Prediction

The low value of E_a in the apoptosis/necrosis coefficients, 1.61×10^5 (J mole^{-1}),[57] indicates that it is a slow process; it develops over long times at lower temperatures (in combination with the low value for A), and is likely not to be observed in higher

temperature, shorter exposures since it will be overwhelmed by faster processes that develop at the higher temperatures.

To illustrate this behavior, Figure 2.7 compares the apoptosis/necrosis coefficients to a high energy process, the murine fibroblasts (with Hsp 70 production intact, $E_a = 9.8 \times 10^5$), at 2 hours (Figure 2.7a) and 2 minutes (Figure 2.7b) of exposure. The long exposure shows more damage at lower temperatures for the apoptosis coefficients, and the short exposure is plainly dominated by the murine fibroblast damage process at the lower temperatures.

2.5.2 Alterations in Structural Proteins: Muscle and Collagen Thermal Damage Examples

Muscle and collagen are birefringent in their native state—that is, able to rotate the polarization angle of a polarized light beam. Their birefringence properties arise from ordered arrays that act as a sort of "quarter-wave transformer" because of the specific dimensions of the proteins. In muscle cells it is the organized structure of the actin-myosin array in the sarcomere. In collagen it is the regimented ordering of the rope-like macromolecular twists. In both cases thermal alterations disrupt the arrays and

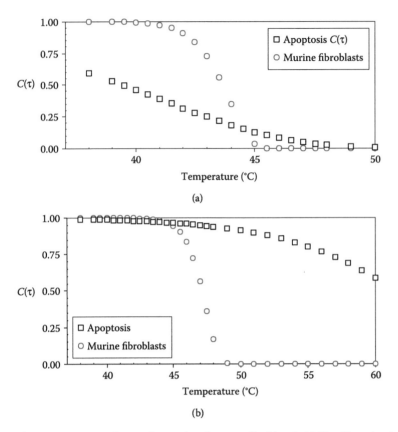

(a)

(b)

FIGURE 2.7 (a) Comparison of apoptosis/necrosis damage (squares) and murine fibroblast (with Hsp 70 production, circles) Arrhenius predictions *vs* temperature for a 2-hour exposure (i.e., reheating in the case of the murine fibroblasts); (b) analogous plot to Figure 2.7a for 2 minutes of exposure.

birefringence is lost—at higher temperatures in collagen than in muscle—and birefringence loss is specific to thermal damage. Relative birefringence intensity constitutes a quantitative histologic marker, and is a useful indicator of thermal damage. In the case of muscle, it is a certain indicator of anatomical disruption leading to irreversible electrophysiologic inactivation, although this probably actually occurs at lower temperatures when the transmembrane charge distribution collapses. In muscle the surface appearance of "whitening" corresponds approximately to the boundary of birefringence loss, and representative Arrhenius coefficients are given in Table 2.1.[1] There are also Arrhenius coefficients for birefringence loss in collagen in the table.

Collagen shrinkage is in routine clinical use for surgical procedures ranging from vessel fusion and correction of nasal septal defects to cosmetic procedures. In the late 1990s, Chen et al.[51-53] presented a useful Arrhenius-based model that predicts collagen shrinkage and ultimate jellification. Their formulation also includes the effects of applied stresses, which somewhat stabilize the collagen molecules. Adapting the method for use in numerical models of shrinkage processes is described by Pearce.[17] Briefly, the collagen exhibits a slow shrinkage phase (indicated by τ_1 in Figure 2.8) followed by a rapid shrinkage phase up to a

maximum of approximately $\xi = 60\%$ in length, indicated by τ_2 in Figure 2.8.

The authors were able to collapse their entire experimental data set into a single normalized equivalent exposure time, ν, in terms of τ_2, where:

$$\nu = \ln\left\{\frac{t}{\tau_2}\right\}. \qquad (2.24)$$

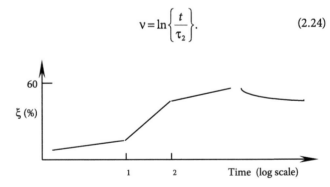

FIGURE 2.8 Sketch of collagen shrinkage model as derived from experimental data by Chen et al.[17] and used by permission. The slow shrinkage phase up to τ_1 is followed by rapid shrinkage up to τ_2 to a maximum of about 60% in length, after which the collagen jellifies as the temperature continues to increase. Collagen shrinkage is followed by relaxation during cooling.

In a numerical model the equivalent increment in the τ_2 time of exposure is Arrhenius-based:

$$\tau_2 = e^{[\alpha + \beta P + M/T]} \tag{2.25}$$

where $\alpha = -152.35 = -\ln\{A\}$; $\beta = 0.0109$ (kPa^{-1}); $P = $ applied stress (kPa); and $M = 53,256$ (K) $= E_a/R$. The signs are reversed in this expression because τ_2 appears in the denominator. Increments in t/τ_2 (i.e., dt/τ_2) are calculated at the local temperature as the model progresses. At the end of heating the accumulated value for t/τ_2 is converted to the shrinkage in two normalizing steps:

$$f(\nu) = \frac{e^{a(\nu - \nu_m)}}{1 + e^{a(\nu - \nu_m)}} \tag{2.26a}$$

where $a = 2.48 \pm 0.438$, and $\nu_m = \ln\{\tau_1/\tau_2\} = -0.77 \pm 0.26$, and then:

$$\xi = (1 - f(\nu))[a_0 + a_1 \nu] + f(\nu)[b_0 + b_1 \nu] \tag{2.26b}$$

where $a_0 = 1.80 \pm 2.25$; $a_1 = 0.983 \pm 0.937$; $b_0 = 42.4 \pm 2.94$; and $b_1 = 3.17 \pm 0.47$ (all in %). Values of ξ above about 60% indicate jelled collagen, which in the case of tissue fusion is the desired result.

2.5.3 Example Multiple Damage Numerical Model

Multiple damage processes may be studied in parallel to reveal underlying trends and analyze planned heating protocols. Here an example numerical model is used to illustrate the relative transient development of several of the damage processes listed in Table 2.1. A short CO_2 laser activation of 20 s followed by 10 s of cooling was implemented in a 101 × 51 node axisymmetric finite difference method (i.e., finite control volume) grid. The model space included equilibrium boiling at 1 atmosphere pressure—it also includes temperature- and water-dependent optical and thermal properties, but in this instance there was not sufficient water vaporization to warrant recalculations. The CO_2 wavelength (10.6 μm) is dominantly absorbed in water with an absorption coefficient of $\mu_a = 792$ (cm^{-1}), and in the numerical model it was assumed that 80% of the total absorption was in tissue water (water = 50% concentration by mass) and 20% in residual (i.e., dry) tissue constituent proteins—as a result the effective tissue absorption coefficient was 492 (cm^{-1}). The 7 mm diameter Gaussian profile beam had a total power of 0.5 W, resulting in a center fluence rate of 2.6 (W cm^{-2})—i.e., a maximum adiabatic heating rate of 460 (°C s^{-1}). The tissue was 2 mm thick with convection and thermal radiation boundary conditions on both surfaces. The numerical model results are shown in Figure 2.9.

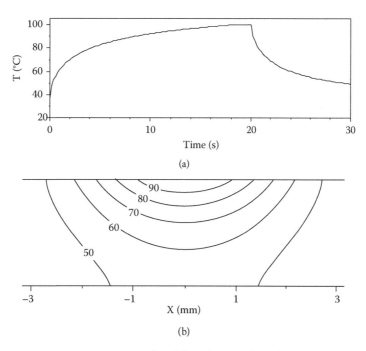

(a)

(b)

FIGURE 2.9 Laser beam heating axisymmetric FDM numerical model result. A CO_2 laser ($\lambda = 10.6$ μm) at total power of 0.5 W with Gaussian beam profile, 2σ diameter = 7 mm for 20 s on tissue 2 mm thick with 50% mass fraction water. The effective laser absorption coefficient was $\mu_a = 492$ (cm^{-1}); and the beam center surface fluence rate 2.59 (W cm^{-2}). (a) Beam center surface temperature history; 100°C reached at $t = 18$ s. (b) Spatial distribution of temperature at the end of heating, 20 s. (c) Apoptosis/necrosis 10% and 90% damage contours. (d) Chinese hamster ovary cell 10–90% contours. (e) BhK cell 10–90% contours. (f) Microvascular damage 10–90% contours. (g) Cardiac muscle whitening 10–90% contours. (h) Skin burn coefficient (Diller [27]) 10–90% contours. (i) Collagen birefringence loss 10–90% contours. (j) Collagen shrinkage 10–60% contours.

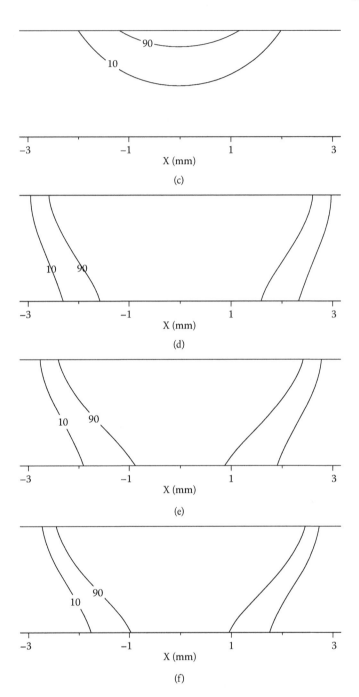

FIGURE 2.9 (*Continued*) Laser beam heating axisymmetric FDM numerical model result. A CO_2 laser ($\lambda = 10.6\ \mu m$) at total power of 0.5 W with Gaussian beam profile, 2σ diameter = 7 mm for 20 s on tissue 2 mm thick with 50% mass fraction water. The effective laser absorption coefficient was $\mu_a = 492\ (cm^{-1})$; and the beam center surface fluence rate 2.59 (W cm^{-2}). (a) Beam center surface temperature history; 100°C reached at $t = 18$ s. (b) Spatial distribution of temperature at the end of heating, 20 s. (c) Apoptosis/necrosis 10% and 90% damage contours. (d) Chinese hamster ovary cell 10–90% contours. (e) BhK cell 10–90% contours. (f) Microvascular damage 10–90% contours. (g) Cardiac muscle whitening 10–90% contours. (h) Skin burn coefficient (Diller[27]) 10–90% contours. (i) Collagen birefringence loss 10–90% contours. (j) Collagen shrinkage 10–60% contours.

Note that the top center transient temperature follows a square-root-of-time dependence very closely (Figure 2.9a) and equilibrium boiling at 100°C commences at about 18 s. In Figure 2.9b, substantial heating occurs throughout the thickness, and out to a radius of about 3 mm.

Eight Arrhenius integral damage processes (Equation 2.6) acting in parallel have been included for illustration (Figures 2.9c–j) in the form of the probability of damage, P (%), as in Equation 2.7. In Figure 2.9c, the model predicts that apoptosis/necrosis damage is not likely be observed in this short heating

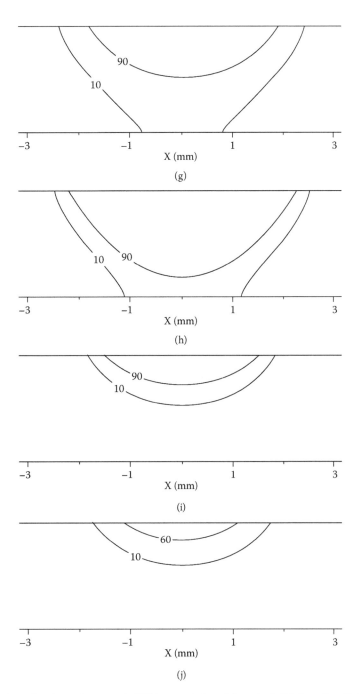

FIGURE 2.9 (*Continued*) Laser beam heating axisymmetric FDM numerical model result. A CO_2 laser ($\lambda = 10.6\,\mu m$) at total power of 0.5 W with Gaussian beam profile, 2σ diameter = 7 mm for 20 s on tissue 2 mm thick with 50% mass fraction water. The effective laser absorption coefficient was $\mu_a = 492$ (cm^{-1}); and the beam center surface fluence rate 2.59 (W cm^{-2}). (a) Beam center surface temperature history; 100°C reached at $t = 18$ s. (b) Spatial distribution of temperature at the end of heating, 20 s. (c) Apoptosis/necrosis 10% and 90% damage contours. (d) Chinese hamster ovary cell 10–90% contours. (e) BhK cell 10–90% contours. (f) Microvascular damage 10–90% contours. (g) Cardiac muscle whitening 10–90% contours. (h) Skin burn coefficient (Diller [27]) 10–90% contours. (i) Collagen birefringence loss 10–90% contours. (j) Collagen shrinkage 10–60% contours.

time lesion, since the 10% and 90% boundaries fall well inside the boundaries of all of the other damage processes, extending only to a depth of 1 mm and radius of 1.5 mm, in keeping with its slowly evolving nature. In contrast, it is highly likely that Chinese hamster ovary and BhK cells will be killed out to a radius of about 2.5 mm over the full thickness. Similarly, cardiac muscle whitening, gross microvascular damage, and skin burns are equally likely to be observed over about the same tissue volume. Significant damage to tissue collagen is a much more severe sort of thermal damage, and is predicted to have dimensions similar to the apoptosis/necrosis volume; consequently, collagen effects should swamp out all other histologically observable processes in this volume. It is interesting that we could also reasonably expect to observe collagen jellification near the tissue surface.

2.6 Summary

Quantitative predictions of thermal damage in tissues yield unique insights into governing processes, much more so than predictions of temperature fields alone. An Arrhenius integral approach yields more insight than the thermal dose—that is, "cumulative equivalent minutes at 43°C"—that is most often used in hyperthermia treatment since CEM values only provide comparisons to the reference point. Thermal dose units do accurately compare diverse thermal histories, and have been used for many years for that purpose. At the same time, however, R_{CEM} = 0.5 is quite appropriate for several important reference cell lines, as demonstrated by Sapareto et al,[31] Overgaard et al,[56] and others—it's just not appropriate for all thermal damage processes, as can easily be seen in Table 2.1, and doesn't seem to provide useful results at high temperatures.

The number of processes for which Arrhenius parameters are available is actually quite large, much larger than just those that have been included in this chapter. In fact, any process for which either R_{CEM} or the rate constant, E_a, is known can be converted into an Arrhenius model by application of Wright's line, Equations 2.11a,b, if needed.

It is a worthwhile exercise to apply the absolute reaction rate formulation to assess and compare thermal treatments, especially for the case of shorter heating times at higher temperatures, such as are typical of ablation procedures.

References

1. Pearce JA. Models for Thermal Damage in Tissues: Processes and Applications. *Critical Reviews in Biomedical Engineering.* 2010; 38(1): 1–20.
2. Pearce JA, Thomsen S. Ch. 17: Rate Process Analysis of Thermal Damage. In: Welch AJ, vanGemert MJC, eds. *Optical-Thermal Response of Laser-Irradiated Tissue.* New York: Plenum Press; 1995. pp. 561–606.
3. Thomsen S, Pearce JA. Ch. 13: Thermal Damage and Rate Processes in Biologic Tissue. In: Welch AJ, van Gemert MJC, eds. *Optical-Thermal Response of Laser-Irradiated Tissue.* 2nd ed. Dordrecht (The Netherlands): Springer-Verlag; 2011.
4. Thomsen S. Targeted Thermal Injury: Mechanisms of Cell and Tissue Death. *Energy-based Treatment of Tissue and Assessment V;* 2009; San Jose, CA. Proceedings of SPIE, Bellingham, WA; pp. 718102-1–15.
5. Godwin BL, Coad JE. Healing Responses Following Cryothermic and Hyperthermic Tissue Ablation. *Energy-based Treatment of Tissue and Assessment V;* 2009; San Jose, CA, Proceedings of SPIE, Bellingham, WA; p. 718103-1–9.
6. He X, Bischof JC. Quantification of Temperature and Injury Response in Thermal Therapy and Cryosurgery. *Critical Reviews in Biomedical Engineering.* 2003; 31(5 & 6): 355–421.
7. Wust P, Hildebrandt B, Sreenivasa G, Rau B, Gellermann J, Riess H et al. Hyperthermia in Combined Treatment of Cancer. *The Lancet Oncology.* [Review]. 2002; 3: 487–97.
8. Lim C-U, Zhang Y, Fox MF. Cell Cycle Dependent Apoptosis and Cell Cycle Blocks Induced by Hyperthermia in HL-60 Cells. *International Journal of Hyperthermia.* 2006; 22(1): 77–91.
9. Bhowmick P, Coad JE, Bhowmick S, Pryor JL, Larson T et al. *In Vitro* Assessment of the Efficacy of Thermal Therapy in Human Benign Prostatic Hyperplasia. *International Journal of Hyperthermia.* 2004; 20(4): 412–39.
10. Green DM, Diller KR. Measurement of Burn-induced Leakage of Macromolecules in Living Tissue. *Journal of Biomechanical Engineering.* 1978; 100(3): 153–8.
11. Aggarwal SJ, Shah SJ, Diller KR, Baxter CR. Fluorescence Digital Microscopy of Interstitial Macromolecular Diffusion in Burn Injury. *Computers in Biology and Medicine.* 1989; 19(4): 245–61.
12. Taormina M, Diller KR, Baxter CR. Burn Induced Alteration of Vasoactivity in the Cutaneous Microcirculation. *Heat and Mass Transfer in the Microcirculation of Thermally Significant Vessels;* 1986; Anaheim, CA. American Society of Mechanical Engineers, Heat Transfer Division; pp. 81–5.
13. He X, Bischof JC. The Kinetics of Thermal Injury in Human Renal Carcinoma Cells. *Annals of Biomedical Engineering.* 2005; 33(4): 502–10.
14. Gourgouliatos ZF, Welch AJ, Diller KR. Microscopic Instrumentation and Analysis of Laser-tissue Interaction in a Skin Flap Model. *Journal of Biomechanical Engineering.* 1991; 113(3): 301–7.
15. Hoffmann NE, Bischof JC. Cryosurgery of Normal and Tumor Tissue in the Dorsal Skin Flap Chamber: Part I—Thermal Response. *Journal of Biomechanical Engineering.* 2001; 123(4): 301–9.
16. Hoffmann NE, Bischof JC. Cryosurgery of Normal and Tumor Tissue in the Dorsal Skin Flap Chamber: Part II—Injury Response. *Journal of Biomechanical Engineering.* 2001; 123(4): 310–6.
17. Pearce JA. Corneal Reshaping by Radio Frequency Current: Numerical Model Studies. *Thermal Treatment of Tissue: Energy Delivery and Assessment;* Proceedings of SPIE—The International Society for Optical Engineering; 2001, pp. 109–18.
18. Nimmi ME. Ch. 1: Molecular Structures and the Functions of Collagen. In *Collagen: Vol. 1, Biochemistry.* Boca Raton, FL: CRC Press; 1988, pp. 1–78.

19. Moritz AR. Studies of Thermal Injury III. The Pathology and Pathogenesis of Cutaneous Burns: An Experimental Study. *American Journal of Pathology*. 1947; 23: 915–34.

20. Moritz AR, Henriques FC. Studies in Thermal Injury II: The Relative Importance of Time and Surface Temperature in the Causation of Cutaneous Burns. *American Journal of Pathology*. 1947; 23: 695–720.

21. Henriques FC, Moritz AR. Studies of Thermal Injury in the Conduction of Heat to and Through Skin and the Temperatures Attained Therein: A Theoretical and Experimental Investigation. *American Journal of Pathology*. 1947; 23: 531–49.

22. Henriques FC. Studies of Thermal Injury V: The Predictability and Significance of Thermally Induced Rate Processes Leading to Irreversible Epidermal Injury. *Archives of Pathology*. 1947; 43: 489–502.

23. Arrhenius S. Uber die Reaktionsgeschwindigkeit bei der Inversion von Rohrzucker durch Sauren. *Zeitschrift für Physikalische Chemie*. 1889; 4: 226–48.

24. Johnson FH, Eyring H, Stover BJ. *The Theory of Rate Processes in Biology and Medicine*. New York: John Wiley & Sons; 1974.

25. Eyring H, Stearn AE. The Application of the Theory of Absolute Reaction Rates to Proteins. *Chemical Reviews*. 1939; 24: 253–70.

26. Fugitt CH. A Rate Process Theory of Thermal Injury. In: Division WE, editor. Washington, D.C.: Armed Forces Special Weapons Project; 1955.

27. Diller KR, Klutke GA. Accuracy Analysis of the Henriques Model for Predicting Thermal Burn Injury. *Advances in Bioheat and Mass Transfer*; 1993. ASME, Heat Transfer Division; 1993, pp. 117–23.

28. Miles CA, Ghelashvili M. Polymer-in-a-Box Mechanism for the Thermal Stabilization of Collagen Molecules in Fibers. *Biophysics Journal*. 1999; 76: 3243–52.

29. Wright NT. On a Relationship Between the Arrhenius Parameters from Thermal Damage Studies. *Journal of Biomechanical Engineering*. 2001; 125(2): 300–4.

30. He X, Bhowmick S, Bischof JC. Thermal Therapy in Urologic Systems: A Comparison of Arrhenius and Thermal Isoeffective Dose Models in Predicting Hyperthermic Injury. *Journal of Biomechanical Engineering*. 2009; 131.

31. Sapareto SA. Ch. 1: The Biology of Hyperthermia In Vitro. In: Nussbaum GH, ed. *Physical Aspects of Hyperthermia*. New York: Am. Inst. Phys.; 1982.

32. Sapareto SA, Hopwood LE, Dewey WC, Raju MR, Gray JW. Effects of Hyperthermia on Survival and Progression of Chinese Hamster Ovary Cells. *Cancer Research*. 1978; 38(2): 393–400.

33. Beckham JT, Mackanos MA, Crooke C, Takahashi T, O'Connell-Rodwell C, Contag CH et al. Assessment of Cellular Response to Thermal Laser Injury Through Bioluminescence Imaging of Heat Shock Protein 70. *Photochemistry and Photobiology*. 2004; 79(1): 76–85.

34. Beckham JT, Wilmink GJ, Mackanos MA, Takahashi K, Contag CH, Takahashi T et al. Role of HSP70 in Cellular Thermotolerance. *Lasers in Surgery and Medicine*. 2008; 40: 704–15.

35. Borrelli MJ, Thompson LL, Cain CA, Dewey WC. Time-temperature Analysis of Cell Killing of BhK Cells Heated at Temperatures in the Range of 43.5°C to 57.0°C. *International Journal of Radiation Oncology and Biological Physics*. 1990; 19: 389–99.

36. Przybylska M, Bryszewska M, Kedziora J. Thermosensitivity of Red Blood Cells from Down's Syndrome Individuals. *Bioelectrochemistry*. 2000; 52(2): 239–49.

37. Lepock JR, Frey HE, Bayne H, Markus J. Relationship of Hyperthermia-induced Hemolysis of Human Erythrocytes to the Thermal Denaturation of Membrane Proteins. *Biochimie Biophysical Acta*. 1989; 980: 191–201.

38. Diller KR. Analysis of Skin Burns. In: Shitzer A, Eberhart RC, editors. *Heat Transfer in Medicine and Biology: Analysis and Applications*. New York: Plenum Press; 1984.

39. Brown SL, Hunt JW, Hill RP. Differential Thermal Sensitivity of Tumour and Normal Tissue Microvascular Response During Hyperthermia. *International Journal of Hyperthermia*. 1992; 8: 501–4.

40. Pearce JA. Relationship Between Arrhenius Models of Thermal Damage and the CEM 43 Thermal Dose. *Energy-Based Treatment of Tissue and Assessment V*; 2009; San Jose, CA. Proceedings of SPIE, Bellingham, WA; p. 718104-1–15.

41. Weaver JA, Stoll AM. Mathematical Model of Skin Exposed to Thermal Radiation. *Aerospace Medicine*. 1967; 40(1): 24–30.

42. Breen MS, Breen M, Butts K, Chen L, Saidel GM, Wilson DL. MRI-guided Thermal Ablation Therapy: Model and Parameter Estimates to Predict Cell Death from MR Thermometry Images. *Annals of Biomedical Engineering*. 2007; 35(8): 1391–403.

43. Chen X, Saidel GM. Modeling of Laser Coagulation of Tissue with MRI Temperature Monitoring. *Journal of Biomechanical Engineering*. 2010; 135(6): 064503-1–4.

44. Bhowmick S, Swanlund DJ, Bischof JC. Supraphysiological Thermal Injury in Dunning AT-1 Prostate Tumor Cells. *Journal of Biomechanical Engineering*. 2000; 122(1): 51–9.

45. Diller KR, Valvano JW, Pearce JA. Bioheat Transfer. In: Kreith F, editor. *CRC Handbook of Thermal Engineering*. Boca Raton: CRC Press; 2000, pp. 114–215.

46. Welch AJ, Polhamus GD. Measurement and Prediction of Thermal Injury in the Retina of Rhesus Monkey. *IEEE Transactions on Biomedical Engineering*. 1984; 31: 633–44.

47. Maitland DJ, Walsh JT, Jr. Quantitative Measurements of Linear Birefringence During Heating of Native Collagen. *Lasers in Surgery and Medicine*. 1997; 20: 310–8.

48. Pearce JA, Thomsen SLMD, Vijverberg H, McMurray TJ. Kinetics for Birefringence Changes in Thermally Coagulated Rat Skin Collagen. Society of Photo-Optical Instrumentation Engineers, Bellingham, WA; 1993, pp. 180–6.

49. Aksan A, McGrath JJ, Nielubowicz DSJ. Thermal Damage Prediction for Collagenous Tissues Part I: A Clinically Relevant Numerical Simulation Incorporating Heating Rate Dependent Denaturation. *Journal of Biomechanical Engineering*. 2005; 127(1): 85–97.

50. Miles CA. Kinetics of Collagen Denaturation in Mammalian Lens Capsules Studied by Differential Scanning Calorimetry. *International Journal of Biology and Macrobiology*. 1993; 15: 265–71.

51. Chen SS, Wright NT, Humphrey JD. Heat-induced Changes in the Mechanics of a Collagenous Tissue: isothermal Free Shrinkage. *Journal of Biomechanical Engineering*. 1997; 119(4): 372–8.

52. Chen SS, Wright NT, Humphrey JD. Heat-induced Changes in the Mechanics of a Collagenous Tissue: Isothermal, Isotonic Shrinkage. *Journal of Biomechanical Engineering*. 1998; 120: 382–8.

53. Chen SS, Wright NT, Humphrey JD. Phenomenological Evolution Equations for Heat-induced Shrinkage of a Collagenous Tissue. *IEEE Transactions on Biomedical Engineering*. 1998; 45: 1234–40.

54. Jacques SL, Gaeeni MO. Thermally Induced Changes in Optical Properties of Heart. *IEEE Eng Med Biol Mag*. 1989; 11: 1199–200.

55. Jacques SL, Newman C, He XY. Thermal Coagulation of Tissues: Liver Studies Indicate a Distribution of Rate Parameters Not a Single Rate Parameter Describes the Coagulation Process. Proc Winter Annual Meeting; 1991; Atlanta, GA, USA. American Society of Mechanical Engineers, Heat Transfer Division, HTD; 1991. pp. 71–3.

56. Overgaard K, Overgaard J. Investigations on the Possibility of a Thermic Tumour Therapy. I. Shortwave Treatment of a Transplanted Isologous Mouse Mammary Carcinoma. *European Journal of Cancer*. 1972; 8(1): 65–78.

57. Bhowmick S, Swanlund DJ, Bischof JC. *In Vitro* Thermal Therapy of AT-1 Dunning Prostate Tumours. *International Journal of Hyperthermia*. 2004; 20(1): 73–92.

Practical Clinical Thermometry

R. Jason Stafford
University of Texas MD
Anderson Cancer Center

Brian A. Taylor
St. Jude Children's
Research Hospital

3.1 Introduction

The goal of thermal therapy is to alter tissue temperature in a targeted region over time for the purpose of inducing a desired biological response (Goldberg et al. 2000; Dewhirst et al. 2005; Stauffer 2005; Hurwitz 2010). The target temperature may be only a slight deviation from physiological temperature for a prolonged period of time (hypo- or hyperthermia) or more extreme deviations for shorter periods of time focused on tissue coagulation (cryo- or thermal ablation). Regardless of the treatment modality or anatomical location, the majority of these therapies are designed to conformally deliver the thermal therapy to a target tissue volume with minimal impact on intervening or surrounding tissues. In particular, during therapies engaged in heating the tissue, isoeffects are rarely predicted by a simple temperature threshold and are more generally a complex function of time and temperature, which depend on the rate at which the transition occurs. Endpoints, such as tissue destruction, have been modeled using Arrehnius rate processes and derivatives, such as equivalent minutes spent at 43°C. These models relate the biological response to the cumulative temperature history of the tissue over time, thereby relying on accurate temperature measurements in the tissues of concern.

One of the key challenges associated with safely and effectively implementing these therapies is spatiotemporal control of the induced temperature changes. This challenge is usually addressed by a combination of therapy planning and monitoring, the specific implementation of which is often tailored to the treatment modality and site of therapy. Generally, treatment planning is the use of models and simulation of the delivery of energy and resulting heating to aid in optimizing the logistical approach to therapy delivery, such as location of applicators and applied power. In most cases, patient-specific imaging is incorporated to better incorporate patient-specific anatomy and characterize the target tissue. However, while an excellent tool for optimizing the approach to therapy delivery, in most cases the complexities of modeling and simulating the heat deposition in tissue do not accurately predict outcomes or fully assure patient safety. Therefore, in order to increase the efficacy of these procedures as well as enhance the safety aspects of delivery, feedback of the therapy is often necessary for many of these procedures. While indirect measurements of heating, such as estimating the specific absorption rate (SAR), can be used, direct observation of temperature provides the most vital information with regard to evaluating delivery in most cases. Therefore, many therapies employ some combination of imaging, modeling, and thermometry to aid in planning and monitoring of thermal therapies.

Thermometry, particularly for heat-mediated regional therapies of deep-seated lesions or tissue that cannot be directly visualized, can play an important role in improving treatment safety and efficacy. This is done by providing direct feedback of the temperature in the target tissue, nearby critical structures, and, in cases of minimally invasive treatment modalities, of the probe itself. In terms of treatment safety, the role of thermometry is to provide feedback, which can help control the therapy and avoid destructive temperatures from being reached during the course of treatment, such as excessive heating in the region of the treatment probe, nearby normal tissue, or adjacent critical structures, the damage of which may result in complications. Examples of effectively incorporating thermometry for safety include monitoring for areas of elevated temperature ("hot spots") inside and outside the target volume during hyperthermia or thermal ablation. Excessive heating outside the target volume can result in damage to normal tissue. Near invasive probes, monitoring maximum temperature in the target tissue aids in avoiding catastrophic tissue damage, such as vaporization or charring.

Often, the choice to monitor everywhere is not feasible, so choosing the locations of thermometry must weigh in to planning the procedure. Hot spots could occur during hyperthermia due to interaction of the electromagnetic or ultrasound field with unexpected impedance mismatches that lead to heating at boundaries. During high-intensity focused ultrasound (HIFU) ablations, cumulative heating in areas slightly distal or proximal to the focus may result in tissue damage. During interstitial laser ablation, allowing the tissue temperature adjacent to the probes to exceed 100°C may lead to vaporization or charring, which can damage the treatment probe. Effective use of thermometry for enhancing procedure safety includes assessing the areas at risk and using temperature feedback to help minimize these risks.

In terms of treatment efficacy, the role of thermometry is to provide feedback to aid in making better decisions with respect to predicting, or controlling, the treatment outcome. This could involve evaluating how much of the target tissue exceeds a threshold temperature, which may be linked to activation of a specific bioeffect or release of a drug. However, since damage to tissue from heat is a cumulative effect (Sapareto et al. 1984; Dewhirst et al. 2003), often, the most effective strategy is to take the temporal temperature history ("exposure") into account and formulate a metric that helps make decisions about the likelihood of tissue destruction or a particular bioeffect. For these purposes, thermometry is combined with a biological model of damage, such as an Arrhenius rate process or an effective exposure that can be compared to known empirical outcomes, such as cumulative minutes spent at 43°C (Sapareto et al. 1984; Dewey 1994; McDannold et al. 2000; Dewhirst et al. 2003).

In practical terms, there are really two general approaches to obtaining temperature measurements in the body during thermal therapy delivery, invasively and noninvasively. Invasive techniques generally rely on interstitial, or intracavitary, placement of thermometers, while noninvasive techniques may rely on strategic superficial placement of thermometers in combination with modeling, or on temperature-sensitive imaging techniques. Regardless of technique employed, it must be amenable to achieving the goals of the therapeutic intervention in order to be of value. Consideration must be made in the trade-offs associated with the level of invasiveness and associated risks, interactions between the thermometry system and therapy or imaging modalities, spatial and temporal sampling of points, and the uncertainty in temperature estimates.

3.2 Invasive Thermometry

Small measurement devices can be inserted directly into the patient. This minimally invasive approach to thermometry provides temperature feedback in real-time finite points within or near the treatment volume. In general, the more points desired, the more invasive the procedure becomes, as well as more time consuming and difficult if the probes must be accurately placed at depth in tissue, particularly moving organs.

Thermocouples utilize a heat-induced potential difference from a junction of two different metals to measure temperature and have the advantage of being small, accurate, stable, and cost efficient. Known weaknesses include possible interactions between the device or its sheath and sources of heating, such as ultrasound (Hynynen et al. 1983) or radiofrequency fields (Gammampila et al. 1981; Chakraborty et al. 1982), and thermal

smearing due to the conductivity of probe leads (Fessenden et al. 1984). Measurements that do not perturb ultrasound fields may be made using small, unsheathed leads. Alternatively, thermistors provide accurate measurements using temperature sensitive changes in resistance, and are generally less sensitive to electromagnetic field interactions (Bowman 1976; Hjertaker et al. 2005). More recently, fluorescent fiberoptic (fluoroptic) thermometry technology has entered wide use. Fluoroptic thermometry uses minimally invasive fiberoptic applicators with phosphor sensors (magnesium fluorogermanate activated with tetravalent manganese) in the tip, which have a temperature dependent fluorescent decay. Self-heating limits use in near infrared (NIR) laser applications unless these artifacts are controlled by capping of the fluoroptic probe, particularly when close to the laser (Reid et al. 2001; Davidson et al. 2005).

Thermal ablation techniques utilizing invasive radiofrequency electrode systems have incorporated thermistors or thermocouples to monitor the tip of the probe from exceeding 100°C. Additionally, interstitial placement of thermocouples (Buy et al. 2009), thermistors (Diehn et al. 2003), or fluoroptic fibers (Wingo et al. 2008) may be used to monitor temperature of nearby critical structures to help mitigate complications. For hyperthermia applications, applicators may be placed in superficial locations on the skin or intraluminally (i.e., transurethral, transvaginal, endorectal, etc.) (Wust et al. 2006). These locations may be used to monitor safety and coupled with modeling to help estimate spatiotemporal heating deeper in the tissue. Probes placed directly in the target tissue itself can be used as an aid to inferring dose delivered to the target for aiding in efficacy.

Despite real-time feedback and accuracy better than 0.5°C, from a monitoring perspective invasive probes are extremely limited by their finite sampling capability. Additionally, invasive probes are not only clinically difficult to accurately place within the patient but they increase the invasiveness of the procedure. This tends to increase the time and complexity associated with planning as well as increase the potential for complications associated with the procedure. Because of this, a substantial amount of interest has been spent on development of noninvasive thermometry methods.

3.3 Noninvasive Thermometry

Noninvasive thermometry for thermal therapies has been investigated using a variety of approaches, including infrared thermography, microwave tomography (Meaney et al. 2003), impedance tomography (Amasha et al. 1988; Blad et al. 1992), CT Hounsfield unit changes (Fallone et al. 1982; Bruners et al. 2010), pulsed-echo ultrasound using estimation of the temperature-dependent echo shifts via speckle tracking (Maass Moreno et al. 1996) or temperature-dependent spectral shifts (Seip et al. 1995), photoacoustic imaging, and temperature-sensitive magnetic resonance imaging (MRI). Of these techniques, MRI is the only technique currently evolved to the point of clinical use.

3.3.1 Magnetic Resonance Temperature Imaging

MRI is a noninvasive 3D-imaging modality that does not use ionizing radiation and provides a plethora of contrast mechanisms with which to image both anatomy and function, making it an attractive modality for image-guided thermal therapy in that it aids in the planning, targeting, monitoring/control, and verification of treatment delivery (Figure 3.1). Of all these qualities, it is the ability to noninvasively monitor temperature changes in the body either qualitatively or quantitatively that makes MRI a particularly attractive modality for enhancing the safety and efficacy of thermal therapies through thermometry feedback. Therefore, despite the challenges associated with performing thermal therapy procedures in the magnetic resonance (MR) environment, MRI is quickly emerging as a modality of choice for guiding many of these therapies, and many vendors are working hard to develop MR compatible therapy delivery equipment.

Because the nuclear magnetic resonance (NMR) phenomenon is by its very nature a thermodynamic process exchanging energy with its chemical environment, MR is exquisitely sensitive to the microscopic chemical and physical state in which the spin system resides. So, it should come as no surprise that many measureable MR imaging parameters exhibit various degrees of temperature sensitivity. Under most circumstances, the temperature sensitivity of MR parameters have always tended to be regarded as more of a nuisance than a useful technique. Physicists need to pay close attention to the temperature of their phantoms and the temperature sensitivity of their phantom materials lest unexpected anomalies arise during the assignment of peak locations in spectroscopy or unexplained signal changes in quality assurance phantoms arise. However, the desire to leverage the temperature sensitivity of MRI as a noninvasive modality for monitoring temperature changes during delivery of thermal therapies such as hyperthermia, and later, thermal ablation, sparked an interest in the development of MR temperature imaging (MRTI), also commonly referred to as MR thermometry.

MRI is an attractive platform for noninvasive temperature measurement as there are several temperature-sensitive parameters that remain linear within the biological range of tissue measurement. Additionally, MRI is a nonionizing, noninvasive modality capable of making dynamic volume measurements to map these changes out over a volume. This ability to estimate temperature in three-dimensional space via a variety of mechanisms and correlate these measurements with high-resolution anatomical MR images made MRTI-driven approaches highly desirable when a noninvasive and highly sensitive ($\sigma_T \leq 1°C$) approach to temperature measurement was required of an MR-compatible object and MR-compatible heating device.

Despite the availability of multiple temperature-sensitive parameters, it is important to note that not all are equally appropriate for measuring temperature in all materials and under all circumstances. The technique to be utilized for noninvasive estimation of temperature should be carefully evaluated and chosen so as to be congruent with the needs of the therapy, and a thorough review of the literature conducted to assess potential measurement biases and errors.

For instance, delivery of minimally invasive thermal ablative therapy is generally characterized by a rapid, local delivery of high temperatures ($\geq 54°C$) for the purpose of coagulating tissue. This demands a technique that has relatively high spatiotemporal resolution, but not necessarily extensive volume coverage, and can often accommodate more uncertainty in the temperature ($\sigma_T \leq 5°C$) while still providing extremely useful information for monitoring outcomes and safety (upper temperature limits). However, as the exposure times increase and the spatial gradient of temperature becomes lower, accuracy and precision at lower temperatures become increasingly important, as does volume coverage.

For instance, when guiding hyperthermia, a clinical thermal therapy procedure in which tissue temperature is raised several degrees (usually >42°C) for long, continuous periods of time (>20 min), the tolerance for temperature uncertainty is much tighter ($\sigma_T \leq 1°C$) across a very large volume. This volume often encompasses areas not amenable to MR temperature imaging as well as potentially moving organs. MR temperature-imaging techniques for such slow heating methods with low spatial

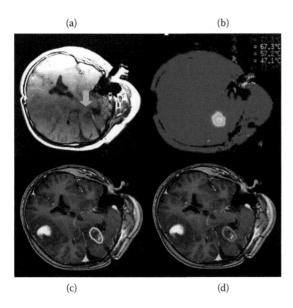

(a)   (b)

(c)   (d)

FIGURE 3.1 (See color insert.) MRI guidance of laser ablation of brain lesion. In addition to planning the procedure, MRI is useful for targeting the volume of interest and verifying correct location (green arrow) of devices for therapy delivery (a). MR temperature imaging provides a spatiotemporal map of the temperature that can be used to help control delivery for safety and efficacy (b). The temperature history can be integrated with biological models that predict damage (orange contour) as shown in (c), which may be used as a surrogate for predicting the treatment endpoints during the course of therapy as opposed to more time-consuming posttreatment verification imaging, such as contrast-enhanced imaging, which demonstrates the perfusion deficit left by therapy and enhancing ring of edema (d).

temperature gradients often sacrifice spatiotemporal resolution for volume coverage, increased temperature sensitivity, and artifact reduction techniques.

3.3.2 Temperature Sensitivity of Several Intrinsic MR Parameters

There are numerous temperature-sensitive MR parameters intrinsic to tissue aqueous solutions, as well as some exogenous agents, that may be exploited for *in vivo* temperature measurements (Rieke et al. 2008; Ludemann et al. 2010). There are actually too many to fully review here, so attention will focus on techniques that tend to impact clinical temperature monitoring. The primary parameters to be discussed here include the molecular diffusion constant of water (D), water proton density (PD), spin-lattice (T_1) and spin-spin (T_2) relaxation times, magnetization transfer contrast, and the water proton resonance frequency (PRF) shift.

The molecular diffusion coefficient, D, is used to describe the thermal Brownian motion of molecules. The general relationship of D to temperature can be represented by an Arrhenius rate process,

$$D \approx e^{-E_a(D)/kT} \qquad (3.1)$$

where k is Boltzmann's constant, T is the absolute temperature in Kelvin, and $E_a(D)$ is the activation energy of the diffusing substance, such as water, which has a self-diffusion coefficient measured by MR to be approximately 2.3×10^{-5} cm²/sec at 25°C (Carr et al. 1954) with diffusion coefficients for other solvents of interest in MR being compiled by various investigators or various temperature ranges (Holz et al. 2000). Being the primary *in vivo* signal for MR imaging, the temperature dependence of the apparent diffusion coefficient of water has been researched extensively for temperature imaging.

Differentiation of Equation 3.1 results in an expression for the temperature dependence of the diffusion coefficient:

$$\frac{dD}{dT} = \frac{E_a(D)}{kT^2} e^{-E_a(D)/kT}. \qquad (3.2)$$

Here we see we need a minimum of two diffusion measurements, D_{ref} (D at reference temperature, T_{ref}) and D, and knowledge of the basal temperature (T_{ref}) in order to estimate the temperature change, which is given for a two-point measurement by:

$$T = \frac{kT_{ref}^2}{E_a(D)}\left(\frac{D - D_{ref}}{D_{ref}}\right) \qquad (3.3)$$

where we have assumed that changes in D and T are relatively small and that $E_a(D)$ does not change with temperature. Deviations of these conditions may happen for temperatures over 40°C (Simpson et al. 1958).

The apparent diffusion coefficient of water was first investigated in context of MR temperature imaging in the employment of hyperthermia studies, where low temperature changes and low resolution imaging would not be problematic (Le Bihan et al. 1989). The theoretical temperature sensitivity of the apparent diffusion coefficient is actually relatively high across tissues and solvents (2%/°C) and is relatively insensitive to magnetic field strength, making it consistently one of the more sensitive MRTI techniques by comparison (de Senneville et al. 2005; Rieke et al. 2008). However, the diffusion of water depends heavily on tissue type and the microenvironment since restricted diffusion (i.e., muscle, white matter, etc.) and microperfusion changes can impact accurate measurement. Techniques to limit the impact of such confounding effects on the diffusion coefficient have been proposed.

Measurement of the apparent diffusion coefficient is accomplished via a pulsed gradient technique (Stejskal et al. 1965). Generally, diffusion acquisitions are signal-to-noise ratio (SNR) limited, and spatial resolution is relatively low in order to speed up the acquisition and increase SNR. Because the measurement is looking at microscopic motion, the measurements can be highly sensitive to motion. Line-scan (Morvan et al. 1993) and single-shot echo-planar imaging (EPI) (Bleier et al. 1991) are useful techniques for reducing imaging time and motion artifacts and are likely the best options on modern scanners with high-performance, eddy current–corrected gradient subsystems.

Like most techniques to be discussed, heating that invokes a strong physiological response from tissue, such as edema, perfusion changes, tissue coagulation, and so forth, is another source of artifact in diffusion-based MRTI. These events can be difficult, if not impossible, at times, to isolate from temperature changes (Moseley et al. 1990). Additionally, in cases where adipose tissue is not completely suppressed during diffusion measurements, use of the diffusion coefficient remains difficult due to restricted diffusion of the lipids and varying temperature sensitivity of lipid diffusion compared to that of soft tissues (Rieke et al. 2008). Therefore, lipid suppression is recommended when diffusion measurements are used to estimate temperature changes in mixed water-lipid tissue environments.

The water proton density (PD) varies approximately linearly with the equilibrium magnetization, M_0, determined by the Boltzmann distribution

$$\text{PD} \propto M_0 = \frac{N\gamma^2\hbar^2 I(I+1)B_0}{3\mu_0 kT} = \chi_0 B_0 \qquad (3.4)$$

where N is the number of spins, γ is the gyromagnetic ratio (42.58 MHz/T for hydrogen protons), $\hbar$ is Planck's constant, I is the quantum number of the spin system (1/2 for hydrogen protons), B_0 is the magnetic flux density, μ_0 is the permeability of free space, k is the Boltzmann constant, T is the temperature (in Kelvin), and χ_0 is the susceptibility. Note χ_0 and T have an inverse relationship (Curie's law), which relates changes in susceptibility to the temperature in PD-weighted images. PD temperature sensitivity goes inversely with sample temperature and is relatively low in sensitivity with ranges around $-0.30 \pm 0.01\%/°C$ reported from 37°C to 80°C (Johnson 1974) so that very high SNR is required to limit uncertainty in the temperature estimates. To reduce the impact of the temperature-dependent T_1, very long repetition times are required, which means spatiotemporal resolution is often sacrificed (Chen et al. 2006). When lower TR values are used, it may be difficult to isolate changes in PD from other relaxation times (Gultekin et al. 2005). Because of these limitations, and the uncertainty in measurement, like diffusion, PD is rarely used for clinical MRTI.

The temperature dependence of relaxation parameters was predicted early on in NMR theory (Bloembergen et al. 1947). The temperature-dependence of T_1 has been well documented since early NMR studies related it to the correlation time and hence diffusion (Bloembergen et al. 1948). T_1 was one of the earliest and most aggressively pursued parameters considered for MRTI measurement (Lewa et al. 1980; Parker et al. 1983; Matsumoto et al. 1992; Matsumoto et al. 1994; Hynynen et al. 2000). In tissues, T_1 depends primarily on dipolar interactions during the temperature-dependent rotational and translational motion of molecules. As with diffusion, T_1 can be described by the following first-order Arrhenius rate process where

$$T_1(T) \propto e^{-E_a(T_1)/kT} \tag{3.5}$$

and $E_a(T_1)$ is the activation energy of the relaxation. This activation energy is tied to dipole-dipole interaction and therefore is extremely sensitive to the microenvironment. High temperatures that can effect changes in the tissue state, such as denaturation and coagulation, can cause changes in the activation energy that affect temperature measurements (Peller et al. 2002).

The heavy tissue dependence of the T_1 temperature dependence has been extensively documented. The sensitivity has been measured in several tissues with different values including 0.97%/°C in adipose tissue (Hynynen et al. 2000) and 1–2%/°C in soft tissue (Lewa et al. 1980; Parker et al. 1983; Matsumoto et al. 1992; Cline et al. 1994; Matsumoto et al. 1994). It is important to note that it is one of the only techniques conducive to measuring temperature in adipose tissue.

Quantitative mapping of T_1 requires a long imaging time, so often, T_1 temperature dependence is related to the change in signal from T_1-weighted spin echo or gradient echo images. Generally, the signal, S, can be described as

$$S(T) \propto M_0(T)\sin\theta \frac{1-e^{-TR/T_1(T)}}{1-\cos\theta e^{-TR/T_1(T)}} \tag{3.6}$$

where θ is the flip angle. The signal decreases with increasing temperature due to increasing T_1 and decreasing M_0. By taking into account the small nonlinear temperature dependence of M_0, the change in S as a function of temperature can be described as

$$\frac{dS}{dT} = \alpha \frac{dS}{dT_1} - \frac{S}{T} \tag{3.7}$$

which takes into account both the decrease in M_0 and the increase in temperature or vice versa (α is the temperature sensitivity). Note that

$$T_1 = T_{1,ref} + \alpha(T - T_{ref}) \tag{3.8}$$

where *ref* is the baseline for T_1 and temperature, T. Therefore, by taking Equations 3.7 and 3.9 and by neglecting the small T_2 (or T_2^* for gradient echo acquisitions) temperature dependence and weighting in T_1-W images, the change in signal in T_1-W images as a function of temperature can be described as

$$T = T - T_{ref} = -\frac{S-S_{ref}}{S_{ref}} \frac{1}{\frac{\alpha TR(1-\cos\theta)E}{(1-E)(1-E\cos\theta)T_{1,ref}^2} + \frac{1}{T_{ref}}} \tag{3.9}$$

where $E = e^{-TR/(T_{1,ref}+\alpha(T-T_{ref}))}$.

For many of these techniques, the signal changes remain approximately linear up to about 54°C, where irreversible changes in tissue tend to lead to an altered temperature sensitivity coefficient. While this may seem unsatisfactory for monitoring tissue temperature for safety, it may be a useful adjuvant measurement for determining tissue damage.

The spin-spin relaxation time (T_2) has a similar relationship with temperature as T_1, but with a different activation energy and break points (Lewa et al. 1990). In observing the signal from a T_2-W image with increasing temperature, there are large sigmoidal decreases that can even remain during cooling of the tissue (Graham et al. 1998). This is an important characteristic of irreversible tissue damage that is useful to define the outcome of a thermal treatment. Therefore, although T_2 may not be a good quantitative means to measure temperature changes, the irreversible changes with thermal damage means it can play a role in treatment verification of high-temperature thermal ablations.

Magnetization transfer contrast (MTC) techniques use spectrally selective radiofrequency (RF) pulses to saturate protons bound to macromolecules, which are normally not visible in MR due to very short T_2 or T_2^* values. During the RF pulse, the bound, saturated protons either enter the primary pool of water protons or transfer the magnetization to the primary pool (Wolff et al. 1989). These exchanges are temperature dependent. For example, MTC signals were determined to be tissue dependent, with some tissues showing relatively no change in signal, such as in adipose and brain tissues. In tissues where a change in signal was seen with temperature, effects on MT-W images were nonlinear with either increasing signal (as in the muscle, heart, prostate, liver) or decreasing signal (blood) (Young et al. 1994;

Graham et al. 1999). These variances in temperature sensitivities do not make MT a readily reliable method for MRTI.

3.3.3 Temperature Dependence of the Water Proton Resonance Frequency (PRF)

Temperature sensitivity of the water PRF was first reported by a group led by Pople in 1958 (Schneider 1958) and later by Hindman (1966). This dependence is due to the relatively weak hydrogen bonding between hydrogen protons and oxygen nuclei of water. When temperature increases, the increased kinetic energy of the water protons results in a longer hydrogen bond with other water molecules and a shorter covalent bond between the hydrogen and parent oxygen nuclei. This results in the proton lying in closer proximity to the electron cloud of oxygen, thereby changing the proton chemical shift, σ (Figure 3.2). The PRF (f) can therefore be expressed as a function of the chemical shift where

$$f = \gamma B_0 (1 - \sigma), \tag{3.10}$$

γ is the gyromagnetic ratio (42.58 MHz/Tesla for hydrogen atoms), and B_0 is the applied magnetic flux density. The shielding constant, σ, theoretically is related to the magnetic field associated with the electronic structure around the nucleus. Usually empirical approaches are used to describe the shielding constant since theoretical calculation requires extensive knowledge of the electron density in the ground and excited states of the molecule as well as the excitation energies. Generally, this chemical shift caused by the shielding constant is the contribution of the electrons surrounding the nucleus as well as other surrounding molecules around the nucleus (Haacke et al. 1999).

Measuring the PRF shift as a function of temperature or measuring the distances between two spectral peaks to measure temperature has long been used in NMR in the field of analytical chemistry. When compared to other MRTI-based parameters, the temperature sensitivity coefficient of the water PRF is relatively the same for different tissues and is close to pure water, which has a temperature sensitivity coefficient (TSC) of –0.010 ppm/°C (Hindman 1966; Figure 3.3). Many studies have been performed both *ex vivo* and *in vivo* to measure the temperature sensitivity of the PRF. McDannold presented a thorough review of the TSC of the PRF using both complex phase-difference (CPD) and chemical shift imaging (CSI) techniques. Most TSC values in tissues have been found to be between –0.009 and –0.010 ppm/°C (McDannold 2005). However, some *in vivo* studies have measured values above and below this range. The difficulty of accurately measuring the temperature sensitivity coefficient *in vivo* remains a challenge that needs to be addressed. In practical implementation, many investigators measure tissue sensitivity *ex vivo* or use literature-obtained values. These values tend to be near –0.01 ppm/°C for a wide variety of tissue.

A source of error in the PRF technique arises when there is a shift in the local magnetic field unrelated to the temperature-dependent chemical shift. One instance in which this can happen is a temperature-dependent change in the bulk magnetic susceptibility (De Poorter 1995; Peters et al. 1999), which can also be affected by the orientation of the heat source in relation to the magnetic field. Additionally, the temperature-dependent electrical conductivity of tissue can induce temperature-dependent phase errors when large amounts of tissue are heated (Peters et al. 2000). While this effect is unlikely to be insignificant for ablation of small volumes of tissue, the impact may become significant when large volumes of tissue are heated, such as with hyperthermia. In these cases, a multi-echo sequence is recommended to offset this shift.

In fatty tissue, De Poorter et al. measured an overall TSC of –0.0097 ppm/°C with a susceptibility constant of –0.0013 ppm/°C, giving a corrected field shift of –0.0088 ppm/°C (De Poorter 1995). In a separate study by a different investigator using lipid (bulk methylene) as an internal reference, the TSC of the difference was measured and found to be around –0.00852 ppm/°C (Kuroda 2005). A study using a line

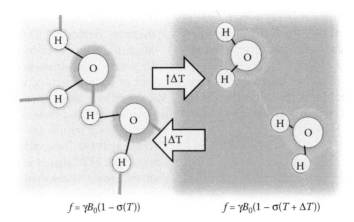

$$f = \gamma B_0 (1 - \sigma(T)) \qquad\qquad f = \gamma B_0 (1 - \sigma(T + \Delta T))$$

FIGURE 3.2 The proton resonance frequency phenomenon. As temperature increases, hydrogen bonds lengthen (gray) and covalent bonds (black) pull the proton (H) closer to the parent oxygen (O). The proton experiences a downfield shift in its Larmor resonance frequency (f) due to increased shielding ($\Delta\sigma$) from the electron cloud.

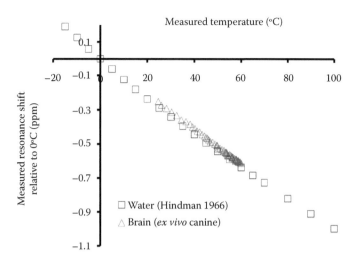

FIGURE 3.3 Calibration of the proton resonance frequency shift. The measured proton resonance frequency shift versus temperature is plotted for the experiments of Hindman for water along with *ex vivo* normal canine brain heated using a 980-nm cooled laser-applicator at 1.5T with measurements from a fluoroptic probe and fast chemical shift imaging sequence (Taylor 2011). The temperature sensitivity coefficient is taken from the slope of the fitted line, which in this case is -0.0103 ± 0.00007 ppm/°C for the canine brain ($R^2 = 0.998$) and $-0.0103 + 0.00016$ ppm/°C for water ($R^2 = 0.996$).

scan echo-planar spectroscopic imaging (LSEPSI) technique on a mayonnaise and lemon juice phantom also found a similar TSC when using bulk methylene as an internal reference (McDannold et al. 2001). More recently, a rapid CSI technique using a multi-gradient echo showed a similar temperature sensitivity of −0.0088 ppm/°C in a similar fat-water phantom (Taylor et al. 2008). This is an important finding in that it suggests that if lipid is present in the tissue, the lipid susceptibility effect corrections are necessary in order to give more accurate measurements. Basic suppression of lipid or selective excitation of water for PRF-based temperature measurements will not suppress the effects from lipid since water will still experience susceptibility from the lipid. Correcting for the susceptibility by using lipid as a reference provides higher accuracy and should be considered when monitoring therapies in fatty tissues that require high accuracy.

It is also important to note that the PRF sensitivity does not change when tissues coagulate, which is in contrast to what is seen in the majority of other parameters (Peters et al. 1998; Kuroda 2005). This is important with respect to safety aspects as the ability to measure temperature after damage has occurred is necessary to assure temperatures do not reach excessive levels, particularly near high-temperature ablation applicators where tissue charring can occur.

There is also growing interest in trying to use the PRF, and particularly CSI, for absolute temperature estimations. Currently, MRTI methods are used to measure relative temperature changes, not the actual tissue temperature. Absolute temperature estimation is an established method in NMR experiments. For instance, in a sample tube, the shift between ethyl (-CH₂-CH₃) or methyl (-CH₃) protons to the hydroxyl (-OH) protons is commonly used in the NMR setting. For MR, the goal is to quantify water, lipid, and/or metabolite peak locations and the relative differences between

them to make absolute temperature measurements (Figure 3.4). There are studies where N-acetylaspartate and water were used to measure temperature in the brain with low spatial and temporal resolution (Cady et al. 1995; Corbett et al. 1995). However, there are many factors that affect the distribution of the metabolite,

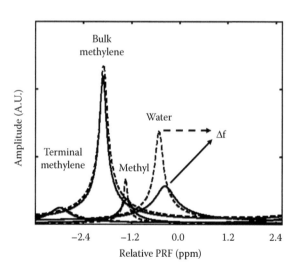

FIGURE 3.4 Lipid-water spectrum. The chemical shift spectrum of laser irradiated bone marrow *ex vivo* as observed during fast chemical shift imaging during heating. The spectrum is aliased (i.e., lipid is shown to the left of water) owing to spectral undersampling by the sequence. As temperature increases, the water peak shifts from its original position (dashed line) to a new position (solid line). The temperature difference can by calculated from the frequency difference (Δf) and knowledge of the temperature sensitivity coefficient. Covalently bonded lipid protons do not shift significantly with heating and thus may serve as an internal reference to correct for background shifts in the field.

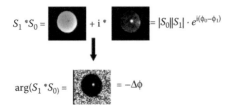

$$S_1 {}^*S_0 = \quad + i * \quad = |S_0||S_1| \cdot e^{i(\phi_0 - \phi_1)}$$

$$\arg(S_1 {}^*S_{0)} = \quad = -\Delta\phi$$

FIGURE 3.5 Complex-phase difference image of a focused ultrasound beam in phantom. The complex phase-difference is calculated by taking the real and imaging components of the quadrature-detected image and multiplying the reference baseline image (S_0) by the complex conjugate of the subsequent image (S_1). By taking the argument over all pixels, a map of the $\Delta\phi$ is formed. Note that uniformly distributed phase occurs in regions with low or no signal. This is often masked by SNR-based thresholding. Additionally, care must be taken to make sure the phase is unwrapped in the temporal dimension before converting to temperature (Equation 3.11).

lipid, and water chemical shifts including intravoxel temperature variations, pH, magnetic ion concentration, blood-oxygen level dependent (BOLD) effect, J-coupling effects, and magnetic susceptibility effects (Kuroda 2005). According to a recent review of MRTI, the "feasibility of absolute temperature imaging has not yet been established; further detailed investigations will be required for that" (Kuroda 2005).

Complex phase-difference (CPD) techniques utilizing fast gradient echo imaging (Ishihara et al. 1995) have been the preferred method to estimate the PRF shift indirectly. To estimate the temperature change in a voxel, the difference between the current phase image, Φ, and the reference image, Φ_{ref}, can be related to the TSC described as (Figure 3.5)

$$T = \frac{\Phi - \Phi_{ref}}{2\pi \cdot \alpha \cdot \gamma \cdot B_0 \cdot TE} \tag{3.11}$$

where α is the TSC (ppm/°C) and TE is the echo-time (ms).

In CPD techniques with sufficient SNR (≥ 5), the uncertainty in the CPD image, $\sigma_{\Delta\phi}$, can be expressed as (Figure 3.6a):

$$\sigma_\phi \cong \frac{\sigma}{A}\sqrt{2} = \frac{\sqrt{2}}{SNR_A} \tag{3.12}$$

where A is the magnitude signal, σ is the noise of the magnitude signal (which is assumed to be approximately Gaussian distributed), and SNR_A is the signal-to-noise ratio of the magnitude image (Conturo et al. 1990). A contrast-to-noise ratio in the phase difference image, $CNR_{\Delta\phi}$, can then be defined by (Figure 3.6b):

$$CNR_\phi \propto \quad \cdot SNR_A \propto TE \cdot e^{-TE/T2^*} \cdot \sin(\theta) \cdot \frac{1 - e^{-TR/T1}}{1 - e^{-TR/T1} \cdot \cos(\theta)} \tag{3.13}$$

assuming a spoiled gradient-echo acquisition. Note that this is essentially the product of the TE (which is proportional to $\Delta\phi$)

and the SNR. The dependence of TE on the phase difference constrains the $CNR_{\Delta\phi}$ to be optimal at $TE = T_2^*$, which can be shown by differentiation of Equation 3.13 (Conturo et al. 1990).

A major advantage of the CPD technique for PRF-based thermometry is its high spatial and temporal resolution (de Senneville et al. 2007). Typically, fast temperature imaging can be achieved with GRE acquisitions (Ishihara et al. 1995) or EPI (Stafford et al. 2004). As stated before, the optimal TE for CNR in CPD is at the T_2^* of the tissue. As a result, standard GRE acquisitions have a relatively long TR, which lowers spatial and temporal resolution. If higher resolutions are needed, echo shifts are applied with $TR < TE$ at an expense of SNR (de Zwart et al. 1999). Alternatively, the use of EPI allows for longer TEs without sacrificing resolution or SNR. Techniques using balanced steady-state free-precession (bSSFP) have also been studied for PRF-based temperature mapping. This is accomplished by measuring a linear fit along several TEs acquired. However, the phase behavior was found to be highly nonlinear with this technique, so simple phase to frequency mapping is not feasible (Mulkern et al. 1998).

Although CPD techniques are advantageous when high spatial and temporal resolution are required, there are several well-known limitations of this technique. Intravoxel lipid contamination (Kuroda et al. 1997; de Zwart et al. 1999), inter- and intra-scan motion (Hynynen et al. 2001), tissue susceptibility changes (De Poorter 1995), and magnetic field drift (De Poorter 1994) all cause artifacts that are commonly encountered with CPD techniques.

It is important to remember that the PRF shift is a function of temperature primarily due to changes in the hydrogen bonds between water molecules. These hydrogen bonds are absent for protons in lipid molecules, which are covalently bonded. Therefore, the temperature sensitivity of lipid is a stronger function of the macroscopic susceptibility. The presence of lipid can alter the phase in CPD acquisitions, which, in turn, can affect temperature estimates. A common practice to address this is to simply suppress the lipid signal (Kuroda et al. 1997; de Zwart et al. 1999). Although lipid signal is suppressed, it does not correct susceptibility effects that lipid can have on the nearby water molecules. As stated previously, the effect of a mixed water and lipid environment cannot be handled completely with suppression because lipid is still physically present in the tissue and can have an effect on temperature estimation in the voxel via susceptibility effects.

As with most imaging techniques, intra-scan and inter-scan motion is another common issue that must be addressed with CPD approaches (Ludemann et al. 2010; Roujol et al. 2010). Intra-scan motion is motion during the acquisition that results in view to view k-space errors seen as image blurring and ghosting. To address this issue, imaging times can be decreased to reduce intra-scan motion, often at the expense of SNR or resolution.

Inter-scan motion, from organ motion or development of edema during therapy delivery, is a more difficult problem to mitigate since images are often subtracted to obtain temperature estimates (Figure 3.7). In the case of the PRF, motion not

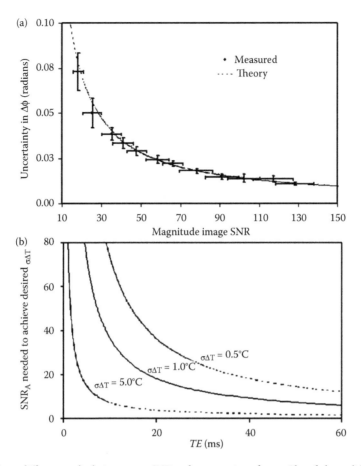

FIGURE 3.6 Uncertainty in phase-difference calculation versus SNR and propagation of error. Plot of phase-difference noise (s_{Df}) versus magnitude image signal-to-noise (SNR_A) (a). The dashed line is the theoretical calculation based on the probability density function for phase-difference noise. Data points are from measured data taken from an 8-interleaved MRTI iGE-spiral sequence (*TE* varied from 10 ms–65 ms, *TR* = 375 ms, *FA* = 60°, 5 imaging planes updated every 3 seconds). Vertical error bars indicate the standard error of the mean of the measured s_{Df} and the horizontal error bars indicate the uncertainty of the mean of the measured SNR_A. The magnitude image SNR necessary to achieve a particular temperature uncertainty (s_{DT}) is plotted versus the echo-time of the gradient-echo sequence (b). Note that significantly higher SNR is needed at shorter echo-times to keep the uncertainty in the measurement small.

only affects the spatial registration between scans, but can also nonlinearly affect the background magnetic field, leading to errors in temperature estimation. Multiple strategies have been developed in an effort to overcome artifacts from inter-scan motion. For respiratory motion, the simplest approach is to use a breath-hold technique. However, in many cases this may not be a feasible or desirable solution. Gating with mechanical respiration has been successful, but this, of course, is invasive (Morikawa et al. 2004). Gating with free-breathing has been used, but it is challenging when there is irregular breathing (Lepetit-Coiffe et al. 2006). Additionally, the use of navigator echoes has been studied for triggering the temperature acquisition during reproducible phases of the respiratory cycle. Another easy-to-implement approach is gathering of multiple baselines followed by retrospective identification of the most appropriate reference image. All these techniques have problems with nonperiodic motion. On the other hand, "referenceless" thermometry attempts to robustly approximate and remove the slowly varying background magnetic field effects.

The background phase in the treated area is estimated by fitting a polynomial from unheated regions and extrapolating into the region of interest under the assumption of slowly varying phase (Rieke et al. 2004). This technique can be very powerful for smaller regions of heating located away from boundaries that result in steep magnetic field gradients. However, when large regions of tissue are heated, such as with hyperthermia, the slowly varying changes in the magnetic field from motion are difficult to separate from the effects of slowly varying changes due to heating.

Studies involving motion correction schemes to date have had limited success. According to a recent review of MRTI, "motion is the most prevalent problem for temperature monitoring with PRF phase mapping (CPD) and the main reason that has impeded its acceptance for clinical applications in areas that are subject to motion" (Rieke et al. 2008). Hybrid techniques, model-based "referenceless" techniques, and multi-parametric acquisitions are all being considered as solutions to the motion problem.

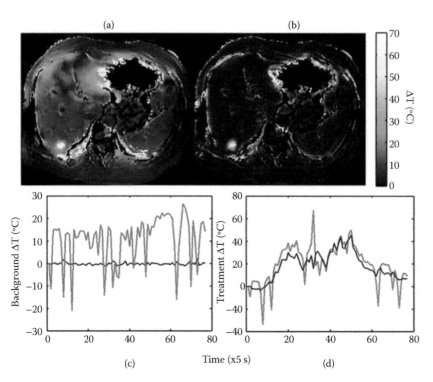

FIGURE 3.7 Motion errors in PRF MRTI of liver. Respiratory motion can result in shifts of the background magnetic field, which decreases the conspicuity of the edges of the temperature changes in the lesion (a). Corrections that estimate the background magnetic field outside the treatment area as well as extrapolating into the treatment area can minimize the errors and often produce usable temperature images (b). Suppression of the background errors outside the treatment area can aid in visualization of the heating (c), while corrections inside the treatment area can minimize the large errors associated with large movements (d).

As mentioned before, macroscopic susceptibility changes in tissue can also affect the accuracy of temperature estimates with the CPD technique (De Poorter 1995). Susceptibility does change with temperature, but when compared to the PRF as a function of temperature, this effect is four to five times smaller in magnitude. It has been approximated as being linear with 0.0026 ppm/°C for pure water and 0.0016 ppm/°C in muscle (De Poorter 1995). It has been found though that the temperature dependence of the susceptibility is tissue dependent (Young et al. 1996). There is also a temperature dependence on the orientation and geometry of the heating source (Peters et al. 1999). One possible method to correct for susceptibility effects is to measure the PRF response to temperature in a chemical specimen that is not sensitive to temperature changes. Lipid is such a specimen. It is expected that lipid will experience susceptibility effects in the voxel. Therefore, using lipid, or any covalently bonded proton, as an internal reference to correct for susceptibility is a possible means to improve temperature measurements (Kuroda 2005).

System magnetic field drift during the therapy is yet another effect that must be considered when the PRF is used for MRTI. Temporal drift of the local magnetic field due to increased gradient duty cycles and eddy currents can change the PRF (El-Sharkawy et al. 2006). These changes are most prominent on older systems, or when high-speed imaging sequences are used. Simple corrections are often possible by using a reference phase

in unheated nearby tissue or a phantom at a fixed temperature (De Poorter 1994). More elaborate corrections are also possible using multiple reference phantoms to estimate the linear shifts across the image (De Poorter et al. 1995). A method that calculates the apparent diffusion coefficient (ADC) and PRF has also been proposed to correct for field drift (Das et al. 2005). As with susceptibility, lipid or another internal covalently bonded proton reference can be used to aid in correcting for field drifts (Kuroda 2005).

An alternative to the indirect CPD method is a direct method of measuring PRF shift via chemical shift imaging (CSI) techniques (Kuroda 2005). CSI has several advantages over CPD techniques. One is that the water proton resonant frequency is measured separately from other resonances whereas CPD measures the mixture of frequency components in each voxel, resulting in a variable response and relieving the problem of intravoxel lipid contamination seen in CPD. Also, the lipid signal can be used as an internal reference to account for field drifts, susceptibility, and motion since it is relatively insensitive to temperature (Kuroda 2005). Therefore, CSI would be particularly useful in areas with high lipid content, such as bone marrow, breast, and head and neck lesions (McDannold et al. 2001). In addition, it may be useful in patients with fatty-liver for minimally-invasive treatments of liver lesions (Hussain et al. 2009). It is also important to note that areas with little or no lipid can also benefit from CSI-based temperature imaging since measuring the water PRF directly can potentially result in higher sensitivity given appropriate SNR with adequate spectral analysis techniques (McDannold et al. 2001; Taylor et al. 2008). The main disadvantages for CSI methods have been low spatial and temporal resolution. Therefore, researchers have worked to develop CSI techniques that increase the sensitivity and spatiotemporal resolution while maintaining the ability to reduce artifacts seen in CPD.

Magnetic resonance spectroscopic imaging (MRSI) was proposed for MRTI before the CPD technique was introduced (Kuroda et al. 1996). However, the acquisition time needed for the desired spectral resolution (with or without suppression) was deemed impracticable for MRTI. For example, MRSI measurements using a small 32×32 matrix with a spectral bandwidth of 10 ppm at 1.5T and a spectral resolution of 0.01 ppm (1°C) required 26 minutes per acquisition.

One approach to overcoming some of the time limitations of the MRSI approach is to use echo planar spectroscopic imaging (EPSI) techniques. These techniques have been proposed for temperature monitoring, and prior studies have demonstrated that temperature changes can successfully be monitored, lipid can be used as an internal reference when present, and artifacts due to motion can be reduced (Kuroda et al. 2000). Although this method was demonstrated to be considerably faster than conventional MRSI, initial EPSI acquisitions still required 3 minutes to acquire a low spatial resolution temperature image (5×5 mm^2) on a 3.0T system. This is because the technique used 4 interleaved shots of 16 echoes (5.2 ms echo-spacing, total of 64 echoes) to achieve an effective echo-spacing of 1.3 ms to avoid aliasing of the lipid signal due to a narrow spectral bandwidth. Data interleaving resulted in a spectral bandwidth of 769 Hz (12.04 ppm) with a spectral resolution of 12.0 Hz (0.188 ppm). In addition, spectra from the EPSI technique can be degraded by instability in the magnetic field and motion during the phase-encoding process of the acquisition (Kuroda 2005).

McDannold et al. applied a line scan EPSI (LSEPSI) technique (Mulkern et al. 1997) to address limitations to previous CSI-based MRTI techniques (McDannold et al. 2001; McDannold et al. 2007). LSEPSI is a combination of EPSI with a voxel-selective technique for column scanning (Mulkern et al. 1997). This method substantially improved the spatial and temporal resolution. Specifically, in an *in vivo* breast scan at 1.5T, a 32×32 cm^2 FOV was acquired in 4.2 seconds with a spatial resolution of $5 \times 5 \times 5$ mm^3.

More recently, model-based approaches of multi-gradient echo acquisitions have been investigated to encode the chemical shift of multiple chemical species such as water, methyl, and methylene (Taylor et al. 2008; Li et al. 2009; Taylor et al. 2009; Sprinkhuizen et al. 2010). These techniques hold potential to overcome artifacts commonly seen with CPD techniques, such as lipid contamination, to provide highly accurate and precise thermometry at sufficient resolutions for even rapid thermal ablations. One of the disadvantages is the increased computation time compared to CPD techniques. To address this, parallel implementation on a stand-alone or portable workstation, which can be easily adapted to the MR scanner, is now possible. For example, implementation of the algorithms on a single graphics processing unit (GPU) can, at best, achieve a speed-up factor of 125 (i.e., reduce the processing time by a factor of 125).

3.4 Summary

Invasive thermometry techniques are limited in their ability to sample and complicate thermal therapy procedures. Techniques utilizing invasive probes have the benefit of providing accurate, real-time measurements in a specified location of interest, but also the obvious limitations of being time consuming to localize, increasing the risk of complications, and, depending on the procedure, may not be able to provide the feedback needed to ensure the proper level of safety and efficacy desired. Bioheat models can be used for limited predictions of temperature beyond the probes but still do not provide accurate enough feedback for many thermal therapies.

MRTI provides a means of noninvasive thermometry that has facilitated the ability to perform some thermal therapy procedures that would otherwise not be feasible, as well as potentially increasing the safety and efficacy of procedures that can be performed in those environments. Current techniques for MRTI have been incorporated into thermal ablation and hyperthermia protocols at numerous sites, and commercial products for performing thermometry are beginning to get FDA clearance from MR and third-party vendors. The next generation of these techniques will have basic motion and field drift correction capability built into the systems. As with any thermometry technique, implementation of quality control procedures for evaluating the accuracy of MRTI in a relevant phantom with the heating modality to be used clinically is critical. Additionally, using this same equipment, verification of the temperature-sensitivity coefficient should be considered using *ex vivo* samples of relevant tissue when possible.

Despite more than a decade of progress, there are numerous challenges and hurdles that remain for robust implementation of the PRF technique. Attempts to mitigate PRF errors arising from tissue motion have been addressed from both the acquisition and postprocessing side, but these techniques are still not fully robust on their own. Additionally, susceptibility and signal loss from equipment and heating can complicate monitoring near interstitially placed applicators. Potential solutions to these problems are currently under investigation and may include incorporation of multi-parametric data, such as diffusion (Das et al. 2005) or T_1 (Ong et al. 2003; Taylor et al. 2009), to complement the PRF measurements. Additionally, the incorporation of real-time modeling and simulation utilize MRTI feedback to provide more optimal temperature estimates and extrapolations into regions with errors or lacking signal. In all cases, the techniques being investigated are making use of the steady advances seen in high-performance computing in order to apply more rigorous correction schemes in real time. This is a phenomenon observable throughout medical imaging.

Lastly, noninvasive temperature imaging has emerged as a powerful tool that is amenable to thermal therapy delivery in many cases. Use of MRI for thermometry is costly and requires MR-compatible instrumentation to be effective. In addition to research being conducted to mitigate current MRTI limitations as well as to develop compatible therapy instrumentation, other alternatives, such as ultrasound temperature imaging, are desperately needed for those procedures in which MRI is contraindicated or presents a logistical or financial barrier to implementation.

References

Amasha, H. M., A. P. Anderson, J. Conway, and D. C. Barber 1988. "Quantitative assessment of impedance tomography for temperature measurements in microwave hyperthermia." *Clin Phys Physiol Meas* 9 Suppl A: 49–53.

Blad, B., B. Persson, and K. Lindstrom 1992. "Quantitative assessment of impedance tomography for temperature measurements in hyperthermia." *Int J Hyperthermia* 8(1): 33–43.

Bleier, A. R., F. A. Jolesz, M. S. Cohen, R. M. Weisskoff, J. J. Dalcanton, N. Higuchi et al. 1991. "Real-time magnetic resonance imaging of laser heat deposition in tissue." *Magn Reson Med* 21(1): 132–7.

Bloembergen, N., E. M. Purcrll and R. V. Pound 1947. "Relaxation effects in nuclear magnetic resonance absorption." *Nature Phys Rev* 73: 679.

Bloembergen, N. P., E.M. Pound, R.V. 1948. "Relaxation effects in nuclear magnetic resonance absorption." *Phys Rev* 73: 679–712.

Bowman, R. R. 1976. "Probe for measuring temperature in radio-frequency-heated material." *IEEE Transactions on Microwave Theory and Techniques* 24(1): 43–45.

Bruners, P., E. Levit, T. Penzkofer, P. Isfort, C. Ocklenburg, B. Schmidt et al. 2010. "Multi-slice computed tomography: A tool for non-invasive temperature measurement?" *Int J Hyperthermia* 26(4): 359–65.

Buy, X., C. H. Tok, D. Szwarc, G. Bierry, and A. Gangi 2009. "Thermal protection during percutaneous thermal ablation procedures: Interest of carbon dioxide dissection and temperature monitoring." *Cardiovasc Intervent Radiol* 32(3): 529–34.

Cady, E. B., P. C. D'Souza, J. Penrice and A. Lorek 1995. "The estimation of local brain temperature by *in vivo* 1H magnetic resonance spectroscopy." *Magn Reson Med* 33(6): 862–7.

Carr, H. Y. and E. M. Purcell 1954. "Effects of diffusion on free precession in nuclear magnetic resonance experiments." *Physical Review* 94(3): 630.

Chakraborty, D. P. and I. A. Brezovich 1982. "Error sources affecting thermocouple thermometry in RF electromagnetic fields." *J Microw Power* 17(1): 17–28.

Chen, J., B. L. Daniel, and K. B. Pauly 2006. "Investigation of proton density for measuring tissue temperature." *J Magn Reson Imaging* 23(3): 430–4.

Cline, H. E., K. Hynynen, C. J. Hardy, R. D. Watkins, J. F. Schenck, and F. A. Jolesz. 1994. "MR temperature mapping of focused ultrasound surgery." *Magn Reson Med* 31(6): 628–36.

Conturo, T. E. and G. D. Smith 1990. "Signal-to-noise in phase angle reconstruction: Dynamic range extension using phase reference offsets." *Magn Reson Med* 15(3): 420–37.

Corbett, R. J., A. R. Laptook, G. Tollefsbol, and B. Kim 1995. "Validation of a noninvasive method to measure brain temperature *in vivo* using 1H NMR spectroscopy." *J Neurochem* 64(3): 1224–30.

Das, S. K., J. Macfall, R. McCauley, O. Craciunescu, M. W. Dewhirst, and T. V. Samulski. 2005. "Improved magnetic resonance thermal imaging by combining proton resonance frequency shift (PRFS) and apparent diffusion coefficient (ADC) data." *Int J Hyperthermia* 21(7): 657–67.

Davidson, S., I. Vitkin, M. Sherar, and W. Whelan 2005. "Characterization of measurement artefacts in fluoroptic temperature sensors: Implications for laser thermal therapy at 810 nm." *Lasers in Surgery and Medicine* 36(4): 297–306.

De Poorter, J. 1995. "Noninvasive MRI thermometry with the proton resonance frequency method: Study of susceptibility effects." *Magn Reson Med* 34(3): 359–67.

De Poorter, J., C. De Wagter, Y. De Deene, C. Thomsen, F. Stahlberg, and E. Achten. 1995. "Noninvasive MRI thermometry with the proton resonance frequency (PRF) method: *In vivo* results in human muscle." *Magn Reson Med* 33(1): 74–81.

De Poorter, J. D. W., C. De Deene, Y. Thomsen, C. Stahlberg, and E. Achten. 1994. "The proton-resonance-frequency-shift method compared with molecular diffusion for quantitative measurement of two-dimensional time-dependent temperature distribution in a phantom." *J Magn Reson B* 103(3): 234.

de Senneville, B. D., C. Mougenot, B. Quesson, I. Dragonu, N. Grenier, and C. T. W. Moonen. 2007. "MR thermometry for monitoring tumor ablation." *Eur Radiol* 17(9): 2401–10.

de Senneville, B. D., B. Quesson, and C. T. Moonen 2005. "Magnetic resonance temperature imaging." *Int J Hyperthermia* 21(6): 515–31.

de Zwart, J. A., F. C. Vimeux, C. Delalande, P. Canioni, and C. T. Moonen 1999. "Fast lipid-suppressed MR temperature mapping with echo-shifted gradient-echo imaging and spectral-spatial excitation." *Magn Reson Med* 42(1): 53–9.

Dewey, W. C. 1994. "Arrhenius relationships from the molecule and cell to the clinic." *Int J Hyperthermia* 10(4): 457–83.

Dewhirst, M. W., B. L. Viglianti, M. Lora-Michiels, M. Hanson, and P. J. Hoopes 2003. "Basic principles of thermal dosimetry and thermal thresholds for tissue damage from hyperthermia." *Int J Hyperthermia* 19(3): 267–294.

Dewhirst, M. W., Z. Vujaskovic, E. Jones, and D. Thrall 2005. "Re-setting the biologic rationale for thermal therapy." *Int J Hyperthermia* 21(8): 779–90.

Diehn, F., Z. Neeman, J. Hvizda, and B. Wood 2003. "Remote thermometry to avoid complications in radiofrequency ablation." *Journal of Vascular and Interventional Radiology: JVIR* 14(12): 1569.

El-Sharkawy, A. M., M. Schar, P. A. Bottomley, and E. Atalar 2006. "Monitoring and correcting spatio-temporal variations of the MR scanner's static magnetic field." *Magma* 19(5): 223–36.

Fallone, B., P. Moran, and E. Podgorsak 1982. "Noninvasive thermometry with a clinical x-ray CT scanner." *Medical Physics* 9: 715.

Fessenden, P., E. R. Lee, and T. V. Samulski 1984. "Direct temperature measurement." *Cancer Res* 44(10 Suppl): 4799s–4804s.

Gammampila, K., P. B. Dunscombe, B. M. Southcott, and A. J. Stacey 1981. "Thermocouple thermometry in microwave fields." *Clin Phys Physiol Meas* 2(4): 285–92.

Goldberg, S. N., G. S. Gazelle, and P. R. Mueller 2000. "Thermal ablation therapy for focal malignancy: A unified approach to underlying principles, techniques, and diagnostic imaging guidance." *AJR Am J Roentgenol* 174(2): 323–31.

Graham, S. J., M. J. Bronskill, and R. M. Henkelman 1998. "Time and temperature dependence of MR parameters during thermal coagulation of *ex vivo* rabbit muscle." *Magn Reson Med* 39(2): 198–203.

Graham, S. J., G. J. Stanisz, A. Kecojevic, M. J. Bronskill, and R. M. Henkelman 1999. "Analysis of changes in MR properties of tissues after heat treatment." *Magn Reson Med* 42(6): 1061–71.

Gultekin, D. H. and J. C. Gore 2005. "Temperature dependence of nuclear magnetization and relaxation." *J Magn Reson* 172(1): 133–41.

Haacke, E. M., R. W. Brown, M. R. Thompson, and R. Venkatesan 1999. *Magnetic resonance imaging: Physical principles and sequence design.* New York City, NY, John Wiley & Sons, Inc.

Hindman, J. C. 1966. "Proton resonance shift of water in the gas and liquid states." *Journal of Chemical Physics* 44(12): 4582–92.

Hjertaker, B. T., T. Froystein, and B. C. Schem 2005. "A thermometry system for quality assurance and documentation of whole body hyperthermia procedures." *Int J Hyperthermia* 21(1): 45–55.

Holz, M., S. R. Heil, and A. Sacco 2000. "Temperature-dependent self-diffusion coefficients of water and six selected molecular liquids for calibration in accurate 1 H NMR PFG measurements." *Physical Chemistry Chemical Physics* 2(20): 4740–42.

Hurwitz, M. D. 2010. "Today's thermal therapy: Not your father's hyperthermia: Challenges and opportunities in application of hyperthermia for the 21st century cancer patient." *Am J Clin Oncol* 33(1): 96–100.

Hussain, K. and H. B. El-Serag 2009. "Epidemiology, screening, diagnosis and treatment of hepatocellular carcinoma." *Minerva Gastroenterol Dietol* 55(2): 123–38.

Hynynen, K., C. J. Martin, D. J. Watmough, and J. R. Mallard 1983. "Errors in temperature measurement by thermocouple probes during ultrasound induced hyperthermia." *Br J Radiol* 56(672): 969–70.

Hynynen, K., N. McDannold, R. V. Mulkern, and F. A. Jolesz 2000. "Temperature monitoring in fat with MRI." *Magn Reson Med* 43(6): 901–4.

Hynynen, K., O. Pomeroy, D. N. Smith, P. E. Huber, N. J. McDannold, J. Kettenbach et al. 2001. "MR imaging-guided focused ultrasound surgery of fibroadenomas in the breast: A feasibility study." *Radiology* 219(1): 176–85.

Ishihara, Y., A. Calderon, H. Watanabe, K. Okamoto, Y. Suzuki, K. Kuroda et al. 1995. "A precise and fast temperature mapping using water proton chemical shift." *Magn Reson Med* 34(6): 814–23.

Johnson, F. E. and H. Stover, B. 1974. *Theory of rate processes in biology and medicine.* New York, NY, John Wiley & Sons.

Kuroda, K. 2005. "Non-invasive MR thermography using the water proton chemical shift." *Int J Hyperthermia* 21(6): 547–60.

Kuroda, K., R. V. Mulkern, K. Oshio, L. P. Panych, T. Nakai, T. Moriya et al. 2000. "Temperature mapping using the water proton chemical shift: Self-referenced method with echo-planar spectroscopic imaging." *Magn Reson Med* 43(2): 220–5.

Kuroda, K., K. Oshio, A. H. Chung, K. Hynynen, and F. A. Jolesz 1997. "Temperature mapping using the water proton chemical shift: A chemical shift selective phase mapping method." *Magn Reson Med* 38(5): 845–51.

Kuroda, K., Y. Suzuki, Y. Ishihara, and K. Okamoto 1996. "Temperature mapping using water proton chemical shift obtained with 3D-MRSI: Feasibility in vivo." *Magn Reson Med* 35(1): 20–9.

Le Bihan, D., J. Delannoy, and R. L. Levin 1989. "Temperature mapping with MR imaging of molecular diffusion: Application to hyperthermia." *Radiology* 171(3): 853.

Lepetit-Coiffe, M., B. Quesson, O. Seror, E. Dumont, B. Le Bail, C. T. Moonen et al. 2006. "Real-time monitoring of radiofrequency ablation of rabbit liver by respiratory-gated quantitative temperature MRI." *J Magn Reson Imaging* 24(1): 152–9.

Lewa, C. J. and M. Lewa 1990. "Temperature dependence of 1H NMR relaxation time, T2, for intact and neoplastic plant tissues* 1." *Journal of Magnetic Resonance (1969)* 89(2): 219–26.

Lewa, C. J. and Z. Majewska 1980. "Temperature relationships of proton spin-lattice relaxation time T1 in biological tissues." *Bull Cancer* 67(5): 525–30.

Li, C., X. Pan, K. Ying, Q. Zhang, J. An, D. Weng et al. 2009. "An internal reference model-based PRF temperature mapping method with Cramer-Rao lower bound noise performance analysis." *Magn Reson Med* 62(5): 1251–60.

Ludemann, L., W. Wlodarczyk, J. Nadobny, M. Weihrauch, J. Gellermann, and P. Wust. 2010. "Non-invasive magnetic resonance thermography during regional hyperthermia." *Int J Hyperthermia* 26(3): 273–82.

Maass Moreno, R. and C. Damianou 1996. "Noninvasive temperature estimation in tissue via ultrasound echo shifts. Part I. Analytical model." *Journal of the Acoustical Society of America* 100: 2514.

Matsumoto, R., K. Oshio, and F. A. Jolesz 1992. "Monitoring of laser and freezing-induced ablation in the liver with T1-weighted MR imaging." *J Magn Reson Imaging* 2(5): 555–62.

Matsumoto, R., R. V. Mulkern, S. G. Hushek, and F. A. Jolesz 1994. "Tissue temperature monitoring for thermal interventional therapy: Comparison of T1-weighted MR sequences." *J Magn Reson Imaging* 4(1): 65–70.

McDannold, N. 2005. "Quantitative MRI-based temperature mapping based on the proton resonant frequency shift: Review of validation studies." *Int J Hyperthermia* 21(6): 533–46.

McDannold, N., A. S. Barnes, F. J. Rybicki, K. Oshio, N. K. Chen, K. Hynynen et al. 2007. "Temperature mapping considerations in the breast with line scan echo planar spectroscopic imaging." *Magn Reson Med* 58(6): 1117–23.

McDannold, N., K. Hynynen, K. Oshio, and R. V. Mulkern 2001. "Temperature monitoring with line scan echo planar spectroscopic imaging." *Med Phys* 28(3): 346–55.

McDannold, N. J., R. L. King, F. A. Jolesz, and K. H. Hynynen 2000. "Usefulness of MR imaging-derived thermometry and dosimetry in determining the threshold for tissue damage induced by thermal surgery in rabbits." *Radiology* 216(2): 517–23.

Meaney, P. M., K. D. Paulsen, M. W. Fanning, D. Li, and Q. Fang 2003. "Image accuracy improvements in microwave tomographic thermometry: Phantom experience." *Int J Hyperthermia* 19(5): 534–50.

Morikawa, S., T. Inubushi, Y. Kurumi, S. Naka, K. Sato, K. Demura et al. 2004. "Feasibility of respiratory triggering for MR-guided microwave ablation of liver tumors under general anesthesia." *Cardiovasc Intervent Radiol* 27(4): 370–3.

Morvan, D., A. Leroy-Willig, A. Malgouyres, C. A. Cuenod, P. Jehenson, and A. Syrota. 1993. "Simultaneous temperature and regional blood volume measurements in human muscle using an MRI fast diffusion technique." *Magn Reson Med* 29(3): 371–7.

Moseley, M. E., Y. Cohen, J. Mintorovitch, L. Chileuitt, H. Shimizu, J. Kucharczyk et al. 1990. "Early detection of regional cerebral ischemia in cats: Comparison of diffusion- and T2-weighted MRI and spectroscopy." *Magn Reson Med* 14(2): 330–46.

Mulkern, R. V., A. H. Chung, F. A. Jolesz, and K. Hynynen 1997. "Temperature monitoring of ultrasonically heated muscle with RARE chemical shift imaging." *Med Phys* 24(12): 1899–906.

Mulkern, R. V., L. P. Panych, N. J. McDannold, F. A. Jolesz, and K. Hynynen 1998. "Tissue temperature monitoring with multiple gradient-echo imaging sequences." *J Magn Reson Imaging* 8(2): 493–502.

Ong, J. T., J. A. d'Arcy, D. J. Collins, I. H. Rivens, G. R. ter Haar, and M. O. Leach. 2003. "Sliding window dual gradient echo (SW-dGRE): T1 and proton resonance frequency (PRF) calibration for temperature imaging in polyacrylamide gel." *Phys Med Biol* 48(13): 1917–31.

Parker, D. L., V. Smith, P. Sheldon, L. E. Crooks, and L. Fussell 1983. "Temperature distribution measurements in two-dimensional NMR imaging." *Med Phys* 10(3): 321–5.

Peller, M., H. M. Reinl, A. Weigel, M. Meininger, R. D. Issels, and M. Reiser. 2002. "T1 relaxation time at 0.2 Tesla for monitoring regional hyperthermia: Feasibility study in muscle and adipose tissue." *Magn Reson Med* 47(6): 1194–201.

Peters, R. D. and R. M. Henkelman 2000. "Proton-resonance frequency shift MR thermometry is affected by changes in the electrical conductivity of tissue." *Magn Reson Med* 43(1): 62–71.

Peters, R. D., R. S. Hinks, and R. M. Henkelman 1998. "Ex vivo tissue-type independence in proton-resonance frequency shift MR thermometry." *Magn Reson Med* 40(3): 454–9.

Peters, R. D., R. S. Hinks, and R. M. Henkelman 1999. "Heat-source orientation and geometry dependence in proton-resonance frequency shift magnetic resonance thermometry." *Magn Reson Med* 41(5): 909–18.

Reid, A., M. Gertner, and M. Sherar 2001. "Temperature measurement artefacts of thermocouples and fluoroptic probes during laser irradiation at 810 nm." *Physics in Medicine and Biology* 46: N149.

Rieke, V. and K. Butts Pauly 2008. "MR thermometry." *J Magn Reson Imaging* 27(2): 376–90.

Rieke, V., K. K. Vigen, G. Sommer, B. L. Daniel, J. M. Pauly, and K. Butts. 2004. "Referenceless PRF shift thermometry." *Magn Reson Med* 51(6): 1223–31.

Roujol, S., M. Ries, B. Quesson, C. Moonen, and B. Denis de Senneville 2010. "Real-time MR-thermometry and dosimetry for interventional guidance on abdominal organs." *Magn Reson Med* 63(4): 1080–7.

Sapareto, S. A. and W. C. Dewey 1984. "Thermal dose determination in cancer therapy." *Int J Radiat Oncol Biol Phys* 10(6): 787–800.

Schneider, W. G., Bernstein, H.J., Pople, J.A. 1958. "Proton magnetic resonance chemical shift of free (gaseous) and associated (liquid) hydride molecules." *J. Chem. Phys.* 284: 601.

Seip, R. and E. Ebbini 1995. "Noninvasive estimation of tissue temperature response to heating fields using diagnostic ultrasound." *Biomedical Engineering, IEEE Transactions on* 42(8): 828–39.

Simpson, J. H. and H. Y. Carr 1958. "Diffusion and nuclear spin relaxation in water." *Physical Review* 111(5): 1201–2.

Sprinkhuizen, S. M., C. J. Bakker, and L. W. Bartels 2010. "Absolute MR thermometry using time-domain analysis of multi-gradient-echo magnitude images." *Magn Reson Med* 64(1): 239–48.

Stafford, R. J., R. E. Price, C. J. Diederich, M. Kangasniemi, L. E. Olsson, and J. D. Hazle. 2004. "Interleaved echo-planar imaging for fast multiplanar magnetic resonance temperature imaging of ultrasound thermal ablation therapy." *J Magn Reson Imaging* 20(4): 706–14.

Stauffer, P. R. 2005. "Evolving technology for thermal therapy of cancer." *Int J Hyperthermia* 21(8): 731–44.

Stejskal, E. O. and J. E. Tanner 1965. "Spin diffusion measurements: Spin echoes in the presence of a time-dependent field gradient." *The Journal of Chemical Physics* 42(1): 288.

Taylor, B. A., K. P. Hwang, A. M. Elliott, A. Shetty, J. D. Hazle, and R. J. Stafford. 2008. "Dynamic chemical shift imaging for image-guided thermal therapy: Analysis of feasibility and potential." *Med Phys* 35(2): 793–803.

Taylor, B. A., K. P. Hwang, J. D. Hazle, and R. J. Stafford 2009. "Autoregressive moving average modeling for spectral parameter estimation from a multigradient echo chemical shift acquisition." *Med Phys* 36(3): 753–64.

Wingo, M. S. and R. J. Leveillee 2008. "Central and deep renal tumors can be effectively ablated: Radiofrequency ablation outcomes with fiberoptic peripheral temperature monitoring." *Journal of Endourology* 22(6): 1261–7.

Wolff, S. D. and R. S. Balaban 1989. "Magnetization transfer contrast (MTC) and tissue water proton relaxation in vivo." *Magn Reson Med* 10(1): 135–44.

Wust, P., C. Cho, B. Hildebrandt, and J. Gellermann 2006. "Thermal monitoring: Invasive, minimal-invasive and non-invasive approaches." *Int J Hyperthermia* 22(3): 255–62.

Young, I. R., J. V. Hajnal, I. G. Roberts, J. X. Ling, R. J. Hill-Cottingham, A. Oatridge et al. 1996. "An evaluation of the effects of susceptibility changes on the water chemical shift method of temperature measurement in human peripheral muscle." *Magn Reson Med* 36(3): 366–74.

Young, I. R., J. W. Hand, A. Oatridge, and M. V. Prior 1994. "Modeling and observation of temperature changes *in vivo* using MRI." *Magn Reson Med* 32(3): 358–69.

Physics of Electromagnetic Energy Sources

Jeffrey W. Hand
King's College London

4.1 Introduction

The use of electromagnetic energy to cause heating and/or ablation of biological tissue is widespread as evidenced in other chapters in this book. To understand and optimize the use of such energy sources it is necessary to consider some fundamental aspects of the electromagnetic spectrum (see Figure 4.1). We shall be interested in nonionizing electromagnetic fields for which the photon energy, given by the product of the frequency and Planck's constant ($= 6.626 \times 10^{-34}$ Js), is insufficient to cause ionization. In particular we shall discuss aspects of interactions between the body and electromagnetic fields such as microwaves (MW) and radiofrequency (RF) fields. The term RF is often used in the biological effects and medical applications literature to cover the ranges from 3 kHz to 300 GHz, respectively. The frequency range from 300 MHz to 300 GHz is also referred to as the "microwave range." The consensus of scientific opinion is that interactions between such fields and the human body are thermal, and although there have been claims for other mechanisms of interaction, the plausibility of the various nonthermal mechanisms that have been proposed is very low (ICNIRP, 2009).

4.2 Static Electric and Magnetic Fields

Electric and magnetic fields are produced by electric charges and their motion. Electric charge may be positive or negative. The force F_e between two charged spherical bodies whose radii are small compared to the distance between them, r, and which are remote from other dielectric media is

$$F_e = \frac{1}{4\pi\varepsilon}\,\frac{q_1 q_2}{r^2}\, r \qquad (4.1)$$

where q_1 and q_2 are the charges and ε is the permittivity of the medium in which they are located. In free space $\varepsilon = \varepsilon_0 = 8.854 \times 10^{-12}$ F m^{-1}. **Bold** typeface indicates a vector quantity. When the charge q_2 is distributed over a region of space the force exerted on q_1 is the vector sum of the forces due to all the elemental charges dq_n that make up q_2 and so

$$F_e = \frac{1}{4\pi\varepsilon}\sum_n \frac{q_1 dq_n}{r_n^2}\, r_n \qquad (4.2)$$

where r_n is the distance between q_1 and dq_n and r_n is the unit vector along the line joining them. From Equation 4.2, the force per unit charge at the location of q_1 is

$$E = \frac{F_e}{q_1} = \frac{1}{4\pi\varepsilon}\sum_n \frac{dq_n}{r_n^2}\, r_n. \qquad (4.3)$$

E is known as the electric field. If the charge distribution can be expressed in terms of charge density $\rho(r)$ within a volume V then

$$E = \frac{1}{4\pi\varepsilon}\int_V \frac{\rho(r)r}{r^2}\, dV \qquad (4.4)$$

where dV is an elemental volume located at position r.

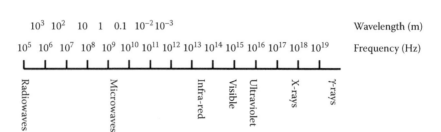

FIGURE 4.1 The electromagnetic spectrum.

The electric field flux Φ_e through a surface is given by

$$\Phi_e = \int_S E.dS \qquad (4.5)$$

where dS is an elemental surface area and dS is the vector of magnitude dS directed along the outward normal to the elemental area. For a single charge, from Equation 4.1, the right-hand side of Equation 4.5 can be written

$$\int_S E.dS = \int_S \frac{1}{4\pi\varepsilon}\frac{q}{r^2}r.dS = \frac{1}{4\pi\varepsilon}\int_S d\Omega \qquad (4.6)$$

where $d\Omega$ is the solid angle subtended at q by dS. Since the integral of the solid angle over a closed surface is 4π:

$$\oint_S E.dS = \frac{q}{\varepsilon} \qquad (4.7)$$

and when the charge can be represented by a continuous charge density $\rho(r)$:

$$\oint_S E.dS = \frac{1}{\varepsilon}\int_V \rho dV. \qquad (4.8)$$

The total electric field flux passing out through a closed surface is equal to the total charge within the surface divided by ε. This is the integral form of Gauss's law. Using Equation 4.7 and the divergence theorem:

$$\oint_S E.dS = \int_V \nabla.E\, dV \qquad (4.9)$$

Equation 4.8 may be rewritten as

$$\int_V \left(\nabla.E - \frac{\rho}{\varepsilon}\right)dV = 0 \quad \text{or} \quad \nabla.E = \frac{\rho}{\varepsilon} \qquad (4.10)$$

which is the differential form of the Gauss law and states that the electric field flux out per unit volume from a point is proportional to the charge density at that point.

In the case of the magnetic field, moving charges need to be considered. The magnetic flux density B is defined through

$$F_m = q\left(v_q \times B\right) \qquad (4.11)$$

where v_q is the velocity of charge q. The vector product in Equation 4.11 shows that the force F_m is normal to both v and B.

The total magnetic flux density B_P at a point P due to a complete current circuit is given by

$$B_P = \frac{\mu_0}{4\pi} I \int_l \frac{dl_l \times r}{r^2} \qquad (4.12)$$

where μ_0 is the permeability of free space ($= 4\pi \times 10^{-7}$ H m^{-1}), I is the current in the circuit, dl is an elemental length of wire, r is the distance between dl, and point P and r is the unit vector directed from dl to P. This relationship is known as the Biot-Savart law.

4.3 Time-Varying Electric and Magnetic Fields

Time-varying electric and magnetic fields are described by the set of equations known as Maxwell's equations. In integral form, the equations describing fields in free space are

$$\oint_S E.dS = \frac{1}{\varepsilon}\int_V \rho dV \qquad (4.13a)$$

$$\oint_l E.dl = -\int_S \frac{\partial B}{\partial t}.dS \qquad (4.13b)$$

$$\oint_S B.dS = 0 \qquad (4.13c)$$

$$\oint_l B.dl = \mu_0 \int_S J.dS + \varepsilon_0\mu_0 \int_S \frac{\partial E}{\partial t}.dS \qquad (4.13d)$$

where J is the current density (A m^{-2}).

Equation 4.13b states that the electromotive force (emf) around a closed loop is equal to the rate of change of the magnetic flux cutting the loop. This is known as Faraday's law. Equation 4.13c indicates that the total magnetic field flux through a closed surface is zero and so magnetic field lines always form closed loops. Equation 4.13d shows the dependence of the magnetic field on steady current (the first term on the right-hand side) and on the displacement current, which is related to the rate of change of electric field (the second term), which ensures conservation of charge. Since the total current passing out through any closed surface must be equal to the total change in charge within the surface,

$$\int_S \mathbf{J}.d\mathbf{S} = -\int_V \frac{\partial \rho}{\partial t} \, dV. \tag{4.14}$$

The differential forms of Maxwell's equations are

$$.\mathbf{E} = \frac{\rho}{\varepsilon_0} \tag{4.15a}$$

$$\times \mathbf{E} = -\frac{\partial \mathbf{B}}{\partial t} \tag{4.15b}$$

$$\nabla.\mathbf{B} = 0 \tag{4.15c}$$

$$\times \mathbf{B} = \mu_0 \left(\mathbf{J} + \varepsilon_0 \frac{\partial \mathbf{E}}{\partial t} \right). \tag{4.15d}$$

Equation 4.15a indicates that charge density is a source of electric field and that electric field lines begin and end on charges. Equation 4.15c shows that magnetic field lines are always closed loops. From Equations 4.15b and 4.15d it can be seen that time-varying electric and magnetic fields are intimately related but that static electric and magnetic fields can be considered separately since in this case time derivatives are zero. However, even when the time derivatives are not zero but remain small, the approximation that electric and magnetic fields are independent remains valid. This is known as a quasi-static approximation and in practical terms requires that the dimensions of the problem are small compared to the wavelength.

The $\mathbf{E}$ and $\mathbf{B}$ fields also satisfy the wave equation:

$$^2\mathbf{E} - \varepsilon_0 \mu_0 \frac{\partial^2 \mathbf{E}}{\partial t^2} = 0 \tag{4.16}$$

$$^2\mathbf{B} - \varepsilon_0 \mu_0 \frac{\partial^2 \mathbf{B}}{\partial t^2} = 0 . \tag{4.17}$$

The simplest solution is to assume a sinusoidal wave of a single frequency:

$$\mathbf{E}(\mathbf{r},t) = Re\left\{ \mathbf{E}(\mathbf{r})e^{j\omega t} \right\} \quad \text{or} \quad \mathbf{B}(\mathbf{r},t) = Re\left\{ \mathbf{B}(\mathbf{r})e^{j\omega t} \right\} \tag{4.18}$$

and in the case of a plane travelling wave:

$$\mathbf{E}(\mathbf{r}) = E_0 e^{-j\mathbf{k}.\mathbf{r}} \quad \text{or} \quad \mathbf{B}(\mathbf{r}) = B_0 e^{-j\mathbf{k}.\mathbf{r}} \tag{4.19}$$

where E_0, B_0 are amplitudes and $\mathbf{k}$ is the wave vector that is related to the angular frequency through

$$|\mathbf{k}| = \frac{\omega}{c} = \frac{2\pi}{\lambda}. \tag{4.20}$$

c is the wave velocity and is determined by

$$c = \frac{1}{\sqrt{\mu_0 \varepsilon_0}}. \tag{4.21}$$

4.4 Interaction of Electric and Magnetic Fields with Tissues

Interactions between electric and magnetic fields and media in general can be explained in terms of conduction currents, dielectric polarization, and magnetization. Tissues are essentially nonmagnetic, and so only the first two mechanisms will be outlined here. The electric field exerts a force on charges and results in a drift of ions that is imposed on their random thermal motion. This gives rise to a conduction current $\mathbf{J}_c = \sigma \mathbf{E}$ where σ (S m^{-1}) is the electrical conductivity of the tissue. The electric field also interacts with polar molecules within tissues causing small displacements and reorientation from their equilibrium positions in the absence of the field. The polarization density $\mathbf{P}$ is related to the $\mathbf{E}$ field through the electric susceptibility χ:

$$\mathbf{P} = \varepsilon_0 \chi \mathbf{E}. \tag{4.22}$$

χ accounts for both types of polarization and is related to the relative permittivity ε_r through

$$\chi = \varepsilon_r - 1. \tag{4.23}$$

The permittivity of a medium ε indicates the ease with which it is polarized:

$$\varepsilon = \varepsilon_r \varepsilon_0 = (1 + \chi)\varepsilon_0. \tag{4.24}$$

A quantity known as the complex permittivity $\varepsilon*$ is used to account for both drift of conduction charges and polarization:

$$\varepsilon^* = (\varepsilon' - j\varepsilon'') \tag{4.25}$$

The real part of this complex quantity ε' $(= \varepsilon_r)$ accounts for polarization and is a measure of the energy stored in the medium. The imaginary part ε'' accounts for conduction and is a measure of energy loss from the E-field to the medium. It can be expressed in terms of the conductivity σ and angular frequency ω as

$$\varepsilon'' = \frac{\sigma}{\omega \varepsilon_0}. \tag{4.26}$$

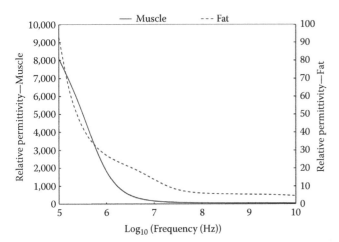

FIGURE 4.2 Frequency dependence of relative permittivity of typical high- and low-water-content tissues (muscle and fat, respectively).

Both ε' and σ are dependent upon the frequency of the field and electrolyte content of the tissue as shown in Figures 4.2 and 4.3, respectively. At low frequencies (lower than approximately 100 MHz) the cell membranes present a high impedance and essentially restrict the flow of current to extracellular regions within the tissue. In this case the capacitance and permittivity of the tissue are relatively high since the cell membranes can be charged and discharged during each cycle of the applied field. Most tissues can be classed as either being of low or high water content. Fat and bone lie in the first group, while muscle and brain are typical of the second group. Large variations in ε' and σ of tissues of low water can be caused by relatively small variations in free water and bound water close to biological macromolecules.

In all types of tissue, the impedance of the membranes decreases as the frequency of the field increases, giving rise to an increase in the conductivity. There is also insufficient time to charge and discharge the cell membranes as the frequency increases, and this leads to the decrease in ε' observed. In the case of fields of frequency greater than approximately 100 MHz,

the impedance of the membranes is so small that cells are essentially short-circuited, and ε' and σ are then determined by the water, salt, and protein content of the tissue. As the frequency increases further to around 1 GHz there is a small decrease in ε' due to the interaction of the field with large polar molecules. The rapid increase in σ above about 1 GHz and decrease in ε' above 3 GHz are due to the polar properties of water within the tissue. Values of ε' and σ for several tissue types at 0.1, 1, 127, 100, 433, 915, and 2450 MHz are listed in Tables 4.1 and 4.2. These are based on measurements and theoretical models described by Gabriel (1996) and Gabriel et al. (1996a,b,c) and were computed online from http://niremf.ifac.cnr.it/tissprop/.

The question of uncertainty regarding values of ε' and σ has been addressed by Gabriel and Peyman (2006). They suggested that random variations from repeat measurements were the major contribution to uncertainty in the case of biological tissues and reported total uncertainties ranging from 0.8% to 7.1% for ε and 1.3% to 10.6% for σ regarding measurements made on porcine gray and white matter, cornea, long bone, cartilage, liver, and fat over the frequency range 50 MHz to 20 GHz. Many dielectric data are based on *ex vivo* measurements of animal tissues. Stuchly et al. (1982) compared *in vivo* measurements over the range 100 MHz to 10 GHz made on several tissues in cats and rats and found relatively small differences between species. However, O'Rourke et al. (2007) measured dielectric properties of *in vivo* and *ex vivo* human liver tissues between 500 MHz and 20 GHz. Measurements indicated that properties of normal liver tissue *in vivo* were 16% and 43% higher than those *ex vivo* at 0.915 and 2.45 GHz. Lazebnik et al. (2007) carried out a large-scale multi-institutional study in which the dielectric properties of normal breast tissue samples obtained from reduction surgeries were measured over the frequency range 500 MHz to 20 GHz. Results showed that the dielectric properties of breast tissue are primarily determined by the adipose content of each tissue sample and that the dielectric properties of some types of normal breast tissues are much higher than previously reported. They also found no statistically significant difference between the within-patient and between-patient variability in ε' and σ.

4.5 Propagation of Electromagnetic Fields in Tissues

To describe fields within lossy dielectric media such as tissues, the permittivity ε in Equations 4.15a to 4.15d is replaced by the complex permittivity ε^*. The plane travelling wave solution analogous to Equation 4.19 becomes

$$\boldsymbol{E}(r) = E_0 e^{-\gamma \cdot r} \quad \text{or} \quad \boldsymbol{B}(r) = B_0 e^{-\gamma \cdot r} \tag{4.27}$$

with a complex propagation constant γ (m^{-1}) given by

$$\gamma = j\omega\mu_0\varepsilon_0\left(\varepsilon' - j\frac{\sigma}{\omega\varepsilon_0}\right). \tag{4.28}$$

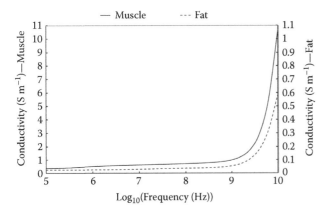

FIGURE 4.3 Frequency dependence of conductivity of typical high- and low-water-content tissues (muscle and fat, respectively).

TABLE 4.1   Relative Permittivity of Tissues versus Frequency

	Frequency (MHz)							
	0.1	0.5	1	27	100	433	915	2450
Bladder	1231	535	343	31.5	22.7	19.6	18.9	18.0
Blood	5120	4189	3026	127	76.8	63.8	61.3	58.3
Blood Vessel	930	312	218	88.9	59.8	46.7	44.7	42.5
Bone Cancellous	472	308	249	42.0	27.6	22.3	20.8	18.5
Bone Cortical	228	175	145	21.8	15.3	13.1	12.4	11.4
Bone Marrow	111	49.0	39.8	10.5	6.5	5.7	5.5	5.3
Brain Gray Matter	3222	1187	860	164	80.1	56.8	52.7	48.9
Brain White Matter	2108	712	480	107	56.8	41.7	38.8	36.2
Breast Fat	70.6	30.7	23.7	6.3	5.7	5.5	5.4	5.1
Cartilage	2572	1939	1391	90.4	55.8	45.1	42.6	38.8
Cerebellum	3515	1475	1141	221	89.8	55.1	49.3	44.8
Cerebrospinal Fluid	109	109	109	106	88.9	70.6	68.6	66.2
Cervix	1751	614	448	98.3	60.3	51.4	49.7	47.6
Colon	3722	2370	1679	141	81.8	62.0	57.9	53.9
Cornea	10567	4637	2878	129	76.0	58.8	55.2	51.6
Duodenum	2861	2065	1678	119	77.9	67.2	65.0	62.2
Dura	326	264	253	122	60.5	46.4	44.4	42.0
Esophagus	2861	2065	1678	119	77.9	67.2	65.0	62.2
Eye Sclera	4745	3252	2178	105	67.9	57.4	55.2	52.6
Fat	92.9	34.6	27.2	8.5	6.1	5.6	5.5	5.3
Gallbladder	107	101	100	96.4	79.0	60.9	59.1	57.6
Gallbladder Bile	120	120	120	117	95.0	72.4	70.2	68.4
Heart	9846	3265	1967	159	90.8	65.3	59.8	54.8
Kidney	7652	3443	2251	188	98.1	65.5	58.6	52.7
Lens	2068	1502	1227	84.2	55.1	48.0	46.6	44.6
Liver	7499	2770	1536	120	69.0	50.7	46.8	43.0
Lung Deflated	5145	1884	1171	104	67.1	54.2	51.4	48.4
Lung Inflated	2581	1025	733	57.8	31.6	23.6	22.0	20.5
Lymph	3301	2140	1433	94.1	68.8	61.3	59.7	57.2
Muscle	8089	3647	1836	95.9	66.0	56.9	55.0	52.7
Ovary	1942	873	678	170	87.2	56.7	50.4	44.7
Pancreas	3301	2140	1433	94.1	68.8	61.3	59.7	57.2
Prostate	5717	4002	2683	120	75.6	63.0	60.5	57.6
Retina	4745	3252	2178	105	67.9	57.4	55.2	52.6
Skin Dry	1119	1062	991	166	72.9	46.1	41.3	38.0
Skin Wet	15357	3610	1833	115	66.0	49.4	46.0	42.9
Small Intestine	13847	8594	5676	203	96.5	65.3	59.4	54.4
Spinal Cord	5133	1488	926	82.1	47.3	35.0	32.5	30.1
Spleen	4222	2789	2290	189	90.7	62.5	57.1	52.4
Stomach	2861	2065	1678	119	77.9	67.2	65.0	62.2
Tendon	472	201	160	78.3	53.9	47.1	45.8	43.1
Testis	5717	4002	2683	120	75.6	63.0	60.5	57.6
Thyroid	3301	2140	1433	94.1	68.8	61.3	59.7	57.2
Tongue	4746	3252	2178	105	67.9	57.4	55.2	52.6
Tooth	228	175	145	21.8	15.3	13.1	12.4	11.4
Trachea	3735	1158	775	80.7	53.0	43.9	42.0	39.7
Uterus	3411	1489	1168	144	78.0	64.0	61.1	57.8
Vitreous Humor	98	91.4	84	69.3	69.1	69.0	68.9	68.2

Sources: http://niremf.ifac.cnr.it/tissprop/; Gabriel, C, *Compilation of the dielectric properties of body tissues at RF and microwave frequencies*, Report N.AL/OE-TR- 1996-0037, Occupational and Environmental Health Directorate, Radiofrequency Radiation Division, Brooks Air Force Base, Texas, 1996; Gabriel, C., Gabriel, S., and Corthout, E., *Phys. Med. Biol.* 41, 1996a; Gabriel, S., Lau, R.W., and Gabriel, C., *Phys. Med. Biol.* 41, 1996b; Gabriel, S., Lau, R.W., and Gabriel, C., *Phys. Med. Biol.* 41, 1996c.

TABLE 4.2 Conductivity (S m⁻¹) of Tissues versus Frequency

	Frequency (MHz)							
	0.1	0.5	1	27	100	433	915	2450
Bladder	0.219	0.228	0.236	0.276	0.294	0.330	0.385	0.685
Blood	0.703	0.748	0.822	1.158	1.233	1.361	1.545	2.545
Blood Vessel	0.319	0.324	0.327	0.375	0.462	0.569	0.701	1.435
Bone Cancellous	0.084	0.087	0.090	0.142	0.173	0.241	0.344	0.805
Bone Cortical	0.021	0.022	0.024	0.052	0.064	0.094	0.145	0.394
Bone Marrow	0.003	0.004	0.004	0.017	0.028	0.030	0.041	0.095
Brain Gray Matter	0.134	0.152	0.163	0.412	0.559	0.751	0.949	1.808
Brain White Matter	0.082	0.095	0.102	0.225	0.324	0.454	0.595	1.215
Breast Fat	0.025	0.025	0.026	0.029	0.030	0.035	0.050	0.137
Cartilage	0.179	0.201	0.233	0.413	0.475	0.598	0.789	1.756
Cerebellum	0.154	0.172	0.185	0.568	0.790	1.048	1.270	2.101
CSF	2.000	2.000	2.000	2.015	2.114	2.260	2.419	3.458
Cervix	0.548	0.557	0.562	0.686	0.744	0.830	0.964	1.726
Colon	0.248	0.278	0.314	0.566	0.680	0.873	1.087	2.038
Cornea	0.499	0.577	0.656	0.938	1.0369	1.206	1.401	2.295
Duodenum	0.536	0.554	0.584	0.838	0.900	1.013	1.193	2.211
Dura	0.502	0.503	0.504	0.629	0.737	0.835	0.966	1.669
Esophagus	0.536	0.554	0.584	0.838	0.900	1.013	1.193	2.211
Eye Sclera	0.519	0.562	0.619	0.844	0.905	1.014	1.173	2.033
Fat	0.024	0.025	0.025	0.033	0.036	0.042	0.051	0.105
Gallbladder	0.900	0.900	0.900	0.916	1.014	1.145	1.261	2.059
Gallbladder Bile	1.400	1.400	1.4000	1.419	1.541	1.704	1.844	2.801
Heart	0.215	0.281	0.328	0.588	0.733	0.983	1.238	2.256
Kidney	0.171	0.228	0.278	0.624	0.811	1.116	1.401	2.430
Lens	0.340	0.353	0.375	0.559	0.600	0.675	0.798	1.504
Liver	0.085	0.148	0.187	0.382	0.487	0.668	0.861	1.686
Lung Deflated	0.272	0.307	0.334	0.484	0.559	0.695	0.864	1.683
Lung Inflated	0.107	0.123	0.136	0.259	0.306	0.380	0.459	0.804
Lymph	0.537	0.566	0.603	0.751	0.794	0.886	1.044	1.968
Muscle	0.362	0.446	0.503	0.654	0.708	0.805	0.948	1.739
Ovary	0.339	0.350	0.358	0.568	0.751	1.032	1.299	2.264
Pancreas	0.537	0.566	0.603	0.751	0.794	0.886	1.044	1.968
Prostate	0.439	0.491	0.562	0.838	0.911	1.038	1.216	2.168
Retina	0.519	0.562	0.619	0.844	0.905	1.014	1.173	2.033
Skin Dry	0.0005	0.004	0.013	0.328	0.491	0.702	0.872	1.464
Skin Wet	0.066	0.178	0.221	0.427	0.523	0.681	0.850	1.592
Small Intestine	0.594	0.715	0.865	1.476	1.655	1.921	2.173	3.173
Spinal Cord	0.081	0.111	0.130	0.268	0.338	0.456	0.578	1.089
Spleen	0.122	0.147	0.182	0.637	0.802	1.043	1.280	2.238
Stomach	0.536	0.554	0.584	0.838	0.900	1.013	1.193	2.211
Tendon	0.389	0.391	0.392	0.439	0.490	0.568	0.724	1.685
Testis	0.439	0.491	0.562	0.838	0.911	1.038	1.216	2.168
Thyroid	0.537	0.566	0.603	0.751	0.794	0.886	1.044	1.968
Tongue	0.288	0.331	0.388	0.613	0.674	0.783	0.942	1.803
Tooth	0.021	0.022	0.024	0.052	0.064	0.094	0.145	0.394
Trachea	0.338	0.359	0.373	0.495	0.548	0.644	0.776	1.449
Uterus	0.531	0.550	0.564	0.847	0.942	1.087	1.276	2.247
Vitreous Humor	1.500	1.500	1.5007	1.502	1.504	1.534	1.641	2.478

Sources: http://niremf.ifac.cnr.it/tissprop/; Gabriel, C, *Compilation of the dielectric properties of body tissues at RF and microwave frequencies*, Report N.AL/OE-TR- 1996-0037, Occupational and Environmental Health Directorate, Radiofrequency Radiation Division, Brooks Air Force Base, Texas, 1996; Gabriel, C., Gabriel, S., and Corthout, E., *Phys. Med. Biol.* 41, 1996a; Gabriel, S., Lau, R.W., and Gabriel, C., *Phys. Med. Biol.* 41, 1996b; Gabriel, S., Lau, R.W., and Gabriel, C., *Phys. Med. Biol.* 41, 1996c.

The real and imaginary parts of γ, β, and α, respectively, are

$$\alpha = \omega \sqrt{\frac{\mu_0 \varepsilon_0 \varepsilon'}{2} \left[\sqrt{1 + \left(\frac{\sigma}{\omega \varepsilon_0 \varepsilon'}\right)^2} - 1 \right]} \qquad (4.29)$$

$$\beta = \omega \sqrt{\frac{\mu_0 \varepsilon_0 \varepsilon'}{2} \left[\sqrt{1 + \left(\frac{\sigma}{\omega \varepsilon_0 \varepsilon'}\right)^2} + 1 \right]} \qquad (4.30)$$

where α (Np m^{-1}) is the attenuation coefficient and β (radian m^{-1}) is the phase constant. In the engineering literature the penetration depth d is usually taken to be the distance into a medium over which the electric field is reduced by a factor e. For plane waves, d is determined using

$$d = \frac{1}{\omega} \sqrt{\frac{2c}{\sqrt{\epsilon'^2 + \epsilon''^2} - \epsilon'}}. \qquad (4.31)$$

The wavelength in the lossy medium is given by

$$\lambda = \frac{2\pi}{\beta}. \qquad (4.32)$$

Since the dielectric properties of tissues are dispersive, then the wavelength and attenuation are dependent upon frequency. Figures 4.4 and 4.5, respectively, show these dependencies for muscle and fat.

When investigating the energy transmitted via the electromagnetic field and/or deposited in tissues, it is useful to begin by considering Poynting's theorem. This is a statement of the conservation of energy in a volume contained within a closed surface and can be expressed as

$$\frac{\partial}{\partial t} \int_V \left(\varepsilon \boldsymbol{E}.\boldsymbol{E} + \mu_0 \boldsymbol{H}.\boldsymbol{H} + w_{cp} \right) dV + \int_S (\boldsymbol{E} \times \boldsymbol{H}).d\boldsymbol{S} \qquad (4.33)$$

The volume integral in Equation 4.33 contains terms representing the instantaneous energy stored in the electric field, in the magnetic field, and that possessed by charged particles, and gives the instantaneous total energy within the enclosed volume. The vector term $\boldsymbol{E} \times \boldsymbol{H}$ is known as the Poynting vector and represents power density (W m^{-2}). The surface integral gives the instantaneous total power passing out through the closed surface. Thus, Equation 4.33 shows that the time rate of change of the total energy inside V is equal to the total power passing out through the surface S. For sinusoidal fields, the time-averaged energy stored in the electric and magnetic fields is zero, and so, from Equation 4.33, it follows that

$$-\int_S \langle \boldsymbol{P} \rangle.d\boldsymbol{S} = \int_V \langle P_{cp} \rangle dV \qquad (4.34)$$

where the parentheses represent time-averaged values, $-P = -E \times H$ is the total average power passing into the volume through the

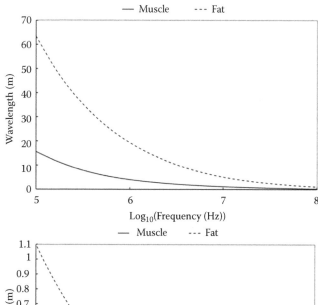

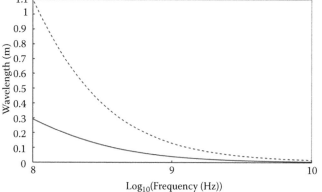

FIGURE 4.4 Frequency dependence of wavelength in muscle and fat tissues. Upper: 100 kHz to 100 MHz. Lower: 100 MHz to 10 GHz.

surface S, and P_{cp} is the total average power transferred to the charged particles within the volume. For steady-state sinusoidal fields,

$$\langle P_{cp} \rangle = \sigma \frac{|E_0|^2}{2}. \qquad (4.35)$$

The specific absorption rate (SAR) is defined as the time derivative of the incremental energy (dW) absorbed by or dissipated in an incremental mass (dm) contained in a volume element (dV) of a medium of density ρ:

$$SAR = \frac{d}{dt}\left(\frac{dW}{dm}\right) = \frac{d}{dt}\left(\frac{dW}{\rho dV}\right) \qquad (4.36)$$

and has units W kg^{-1}. From Equations 4.34 and 4.35 it follows that the whole body averaged SAR, SAR$_{wb}$, is

$$SAR_{wb} = \frac{\int_V \langle P_{cp} \rangle dV}{M} \qquad (4.37)$$

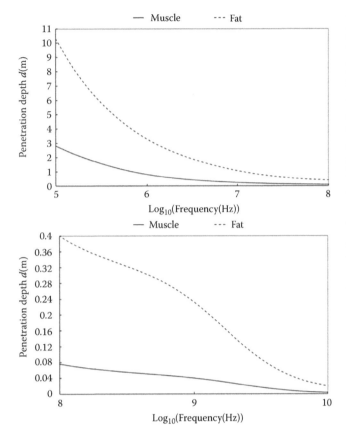

FIGURE 4.5 Frequency dependence of plane wave penetration depth in muscle and fat tissues. Upper: 100 kHz to 100 MHz. Lower: 100 MHz to 10 GHz.

where M is the total mass of the body. The local SAR at a given location is

$$\text{SAR}_{\text{local}} = \frac{\sigma |E_0|^2}{2\rho}. \tag{4.38}$$

If the penetration depth d is taken to be the distance into a medium over which the electric field is reduced by a factor e, the corresponding reduction in the SAR is by a factor e^2 (i.e., by a factor of 7.39 or to approximately 13.5%). In the clinically related literature, a measure of penetration that is of more practical use in thermal therapy is $d_{1/2}$, the distance over which SAR is reduced by a factor of 2. The determination of penetration in practice requires further comment, since in most cases tissues are located in the near field of the electromagnetic source. For example, if SAR is measured at two depths, z_1 and z_2, beneath a source, then

$$\text{SAR at } z_2 / \text{SAR at } z_1 = \exp\left(2(z_2 - z_1)/d\right). \tag{4.39}$$

If z_1 and z_2 are approximately equal to or greater than the wavelength in tissue, then for a constant $(z_1 - z_2)$, this ratio is independent of the actual depth. Under these conditions, the determination of d is clearly defined. However, in most thermal

therapy applications tissues are closer to the source (certainly less than a wavelength away), or even in contact with it. Under these conditions, a parameter often referred to as the effective penetration depth, d_{eff}, is used, which is not only less than d but is also dependent upon the distance from the device. Figure 4.6 shows the dependence of d_{eff} upon frequency and the size of source, assuming that the source has a square aperture and irradiates muscle-like tissue. For relatively large sources ($a \sim 20$ cm), d_{eff} is dependent upon frequency and approaches the plane wave penetration d. However, d_{eff} generally decreases as the size of the aperture decreases. For relatively small apertures ($a \sim 4$ cm), there is little dependence of d_{eff} on frequency.

The human body is comprised of many tissue types, each with differing dielectric properties (see Tables 4.1 and 4.2). Electromagnetic fields must satisfy Maxwell's equations in each tissue type and at boundaries between tissues. The boundary conditions that must be met at an interface between two tissues, tissue A and tissue B, are:

$$\left(E_A - E_B\right) \times n = 0 \tag{4.40a}$$

$$\left(\varepsilon_A E_A - \varepsilon_B E_B\right) \times n = \rho \tag{4.40b}$$

$$\left(H_A - H_B\right) \times n = J \tag{4.40c}$$

$$\mu_0 \left(H_A - H_B\right) \times n = 0 \tag{4.40d}$$

where n is a unit vector normal to the interface. In Equation 4.40d it is assumed that the permeability of each tissue is equal to that of free space μ_0. When there are no surface charges or currents, ρ and J are zero, and it follows from Equations 4.40a to 4.40d that the components of E and H parallel to the interface are continuous:

$$E_{A\parallel} = E_{B\parallel} \quad \text{and} \quad H_{A\parallel} = H_{B\parallel} \tag{4.41}$$

while the components perpendicular to the interface satisfy

$$\varepsilon_A E_{A\perp} = \varepsilon_B E_{B\perp} \quad \text{and} \quad H_{A\perp} = H_{B\perp}. \tag{4.42}$$

4.6 Principles of Electromagnetic Heating Techniques

A general term for the production of heat in body tissues is diathermy. In the literature medical diathermy refers to the moderate heating of tissues used to treat sports injuries and in other forms of rehabilitation where the induced temperature rise remains below the patient's pain threshold. Surgical diathermy refers to more extreme heating such as that used to cut tissue or to produce coagulation. More specific thermal therapies include hyperthermia and thermal ablation. Heating techniques employ frequencies from those in the RF range through to microwaves. Sources may use predominantly electric or magnetic fields and may be located externally to the body or can be implanted within the tissues to be heated.

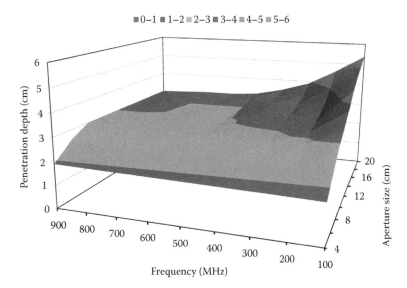

■0–1 ■1–2 ■2–3 ■3–4 ■4–5 ■5–6

FIGURE 4.6 **(See color insert.)** Dependence of effective penetration depth d_{eff} into muscle-like tissue associated with square aperture sources upon frequency f and aperture size a. The surface $d_{eff}(f, a)$ is colored according to 1 cm increments in d_{eff}.

4.7 Invasive Heating Techniques

4.7.1 RF Interstitial Hyperthermia

RF techniques for interstitial hyperthermia rely on Joule (I^2R) heating due to current flow between pairs of electrodes. A frequency in the range 500 kHz to 1 MHz is usually chosen for most RF interstitial hyperthermia techniques (referred to as localized current field [LCF] methods), satisfying the need to avoid direct stimulation of nerve and muscle fibers and for conduction currents to dominate displacement currents (i.e., $\sigma \gg \omega\, \varepsilon'\, \varepsilon_o$).

Figure 4.7 shows a simple model of a bipolar technique in which a pair of RF electrodes, each 1 mm in diameter, is implanted to a depth of 5 cm and with center to center spacing of 10 mm in a muscle-like phantom. The SAR falls off rapidly with distance from the electrodes, although with a ratio of center to center spacing/electrode diameter of 10, the regions around each electrode in which SAR is greater than –3 dB of the peak SAR are contiguous. For larger separations (e.g., a ratio of center to center spacing/electrode diameter of 15), this is no longer the case, and heating of the intervening tissue is essentially dependent upon thermal conduction and tissue blood flow. Since the volume of tissue that can be heated by a single pair of electrodes is relatively small, an array of parallel electrodes is often implanted into the tissue. Experience has shown that the diameter of the electrodes should be 1–1.6 mm and that the interelectrode spacing should not be greater than about 15 mm. It is important that the electrodes are parallel since converging electrodes will lead to excessive heating in those regions where the interelectrode spacing is reduced. If one of the needle electrodes is replaced by a larger electrode at the skin surface, then a monopolar technique can be used. In this case the SAR distribution around the needle remains localized while the larger surface area of the return electrode results in a relatively low current density in that region and therefore relatively low SAR. Although rigid needle electrodes were used originally, flexible electrodes became more common (Kapp et al. 1988). To control the resulting elevated temperature distribution within the targeted volume, longitudinally segmented electrodes (Prionas et al. 1988), hollow electrodes cooled by water flowing through the lumen (Zhu and Gandhi 1988), and means of multiplexing RF power to pairs of electrodes (Leybovich et al. 2000) have been introduced. An associated method operating at 27 MHz, a frequency sufficiently high for the use of capacitive coupling to the tissues rather than direct contact, has also been described (van der Koijk et al. 1997, Crezee et al. 1999, van Vulpen et al. 2002).

4.7.2 RF Ablation (RFA)

Techniques that lead to rapid heating and consequent tissue destruction (thermal ablation) have been developed for treating a range of conditions. Such minimally invasive, image-guided therapy can provide effective local treatment and is also used adjunctively with conventional surgery, systemic chemotherapy, and radiation.

Cardiac catheter RFA involves the delivery of energy from a catheter inserted into the heart through a vein to endocardial tissue responsible for causing changes in the normal heart rhythm. Both monopolar and bipolar techniques have been used to cause desiccation of the targeted tissue. Control of the temperature at the catheter tip is important in determining lesion size and also in achieving reproducibility by preventing tissue boiling, formation of blood coagulum at the electrode surface, and subsequent impedance changes at the catheter/tissue interface (Hindricks et al. 1989, Jackman et al. 1988). Lesions produced by a typical RF catheter with, say, a 4 mm diameter tip, are small (Huang 1991,

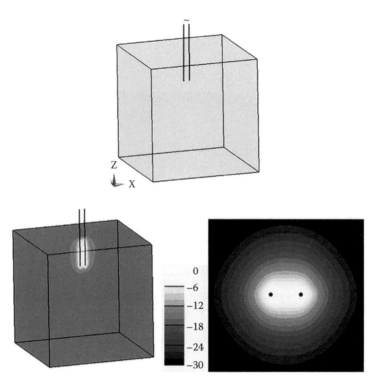

FIGURE 4.7 Invasive needle electrodes. Upper: Two electrodes, diameter 1 mm, are implanted 5 cm into a muscle-like phantom with center-to-center spacing of 10 mm. An ELF source is used to apply a voltage between the electrodes. Lower left: SAR distribution in the plane $y = 0$ that intersects the electrodes (in dB relative to the peak SAR). Lower right: SAR distribution in the z plane at a depth of 26 mm into the phantom. Both SAR distributions are shown between 0 and –30 dB.

Blouin and Marcus 1989). Attempts at producing larger lesions have included the use of larger electrodes (Tsai et al. 1999) and saline-irrigated and cooled electrode tips (Demazumder et al. 2001).

Another common application of RFA is in the treatment of liver tumors. Patients with colorectal liver metastases often present with unresectable disease, and RFA is among several minimally invasive treatments under investigation as an alternative or complementary tool in the management of some of these patients. RFA can be delivered percutaneously, laparoscopically, or through open surgical approaches. Typically, Joule heating caused by a 450–500 kHz current results in coagulative necrosis at a temperature between 50°C and 100°C. RFA procedures involve the insertion of a needle electrode with a noninsulated tip and an insulated needle shaft into the tumor. The method can be either monopolar or bipolar. The dimensions of the volume of ablated tissue are determined largely by the magnitude of the RF current, the length and diameter the electrode tip, and the time for which RF energy is applied (Goldberg et al. 1995). In the event that the tissue temperature at the electrode becomes excessive, carbonization occurs around the electrode, which results in a sharp rise in tissue impedance and interruption of the RF current. This effectively limits the volume of tissue that can be ablated. To avoid this, RFA systems incorporate techniques that monitor and control the temperature at the electrode and/or

monitor the impedance, keeping it below an acceptable value. Volumes up to approximately 20 mm in diameter can be ablated with a conventional single needle electrode, but ablation of larger tissue volumes is possible with recent technical improvements (Goldberg and Gazelle 2001). A cooled-tip electrode avoids charring of tissue immediately around the electrode by cooling the internal chamber of the needle via cold saline infusion, thus allowing the use of a higher power than the conventional needle. Other devices capable of ablating larger volumes use multiple-prong (clustered) electrodes or an expandable electrode with multiple retractable J hooks to create overlapping ablation fields (Livraghi et al. 2000). Pulsing of the RF current and the use of saline are other ways of increasing the ablated volume. Technical aspects of the various approaches are described in Haemmerich et al. (2001, 2002).

4.7.3 Invasive Microwave Techniques

The use of either a small implantable microwave antenna or an array of such devices has the potential for producing larger lesions than is the case with RF electrodes, although the volume remains small for a single antenna. Antenna designs have included monopoles (Nevels et al. 1998), helical structures (Wonnell et al. 1992), and other structures (Lin 1999, Shetty et al. 1996, Liem et al. 1996). Since microwave antennas deliver energy to the targeted tissue despite the presence of coagulum, temperature control of the ablation electrode is important to avoid thromboembolic risks (van der Brink et al. 2000). Reviews of microwave invasive microwave techniques include those by Rhim (2004) and Ryan et al. (2010).

A common antenna design is based around an insulated resonant dipole. In this case the choice of frequency becomes a compromise between energy deposition at increased radial distance from the antenna and the length of the antenna. Antennas designed for use at 915 MHz are of a practical length (Trembly 1985). The use of helical antennas provides greater flexibility in choosing the length of the antenna, and these have been customized for various applications (Stauffer 1988; Sherar et al. 2001; Reeves et al. 2005, 2008).

4.8 External Heating Techniques

4.8.1 Capacitive and Inductive Techniques

RF heating techniques can be classified into two groups: those that use the electric field primarily and the energy is coupled to the tissues through capacitance, and those that use the magnetic field primarily to couple energy to the tissue inductively. A basic capacitive method involves applying power to two electrodes that may be placed either one on each side or both on the same side of the body part to be treated. The electrodes are often small metal plates mounted in cushion-like enclosures, but may also be made of a flexible material such as wire mesh so that they may conform to the shape of the region of interest. The tissue temperature is raised predominantly through Joule (I^2R) heating by conduction currents. An example is shown in Figure 4.8 in which two plate electrodes are positioned around a plane, layered fat-muscle-fat phantom, offset from the fat layers by a water bolus, although other means of spacing the electrodes from tissues may be used. This is necessary to avoid exposing the tissue to the very high electric field close to the electrodes. Relatively high values of SAR are restricted to the region immediately below the small electrode; higher values are located in the fat-like layer because the electric field is essentially perpendicular to the interface between fat and muscle layers and therefore greater in the low permittivity fat than the high permittivity muscle in order to satisfy boundary conditions. The ability to heat at depth is determined by the size of the small electrode.

Inductive coupling of RF energy to tissues may be achieved by placing RF current carrying coils near to or around the body part to be treated. The RF magnetic field produced induces electric fields and circulating currents in the body tissues, which give rise to Joule heating. Figure 4.9 illustrates the method. A three-turn coil is modeled by three concentric circular current-carrying loops (in this case the radii are 10, 8, and 6 cm) located

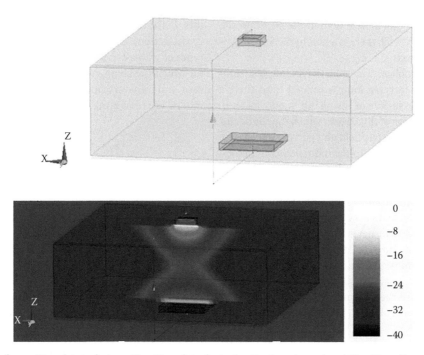

FIGURE 4.8 Example of capacitive plate technique. Top: Two plate electrodes, 5 × 5 cm (upper) and 15 × 15 cm (lower), each placed on a water bolus 2 cm thick, are placed around a phantom consisting of a muscle-like central region 80 × 80 × 26 cm with fat-like layers (80 × 80 × 1 cm) placed on the top and bottom surfaces. The electrodes are connected to a current source and tissue properties at 30 MHz are assumed. Bottom: SAR distribution (in dB relative to the peak SAR) in plane × = 0.

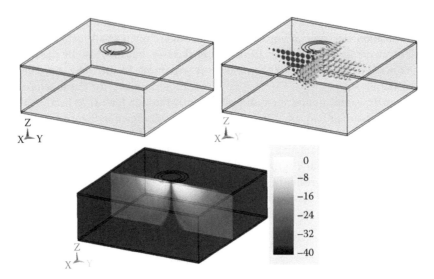

FIGURE 4.9 (**See color insert.**) Top left: Three-turn induction coil modeled as three concentric circular loops (radii 10, 8, and 6 cm) placed 3 cm above a plane-layered fat-muscle-fat phantom. Each loop is driven by a current source at 30 MHz. Top right: Vector plot of current density *J* in planes x = 0 and y = 0. Bottom: The resulting SAR distribution (in dB relative to the maximum SAR) in the x = 0 plane.

parallel to (i.e., in the x-y plane) and 3 cm above the upper surface of a planar fat-muscle phantom. The frequency is 30 MHz. Unlike the case of capacitive electrodes, the maximum SAR is located in the muscle region rather than in the fat layer because the induced electric field and currents are parallel to the phantom surface and the interface between the fat and muscle layers. The boundary conditions require that the E-field is continuous across the fat/muscle interface and so the induced current is larger in the more-conducting muscle than in the fat. However, due to the symmetry of the coil, there is a null in the SAR distribution along the central axis.

To overcome this problem, an inductively coupled technique in which current-carrying loop is rotated (into the x-z plane) with respect to the surface of the phantom may be used. Figure 4.10 shows a variation of this idea in which the conductor is U-shaped. The resulting electric field and current density are parallel to the tissue interfaces, as in the induction coil method above, but the circular symmetry leading to the axial null in SAR is avoided. The SAR distribution is relatively uniform and for practical purposes is constrained within the footprint of the device. In practice, the current-carrying conductors form part of a resonant circuit and such devices are therefore tuned to a narrow-frequency band. However, this type of device, often referred to as "current sheet applicators," can be designed for operation at RF and microwave frequencies. Further theoretical and practical data have been discussed by Morita and Bach Andersen (1982), Johnson et al. (1987), Gopal et al. (1992), and Prior et al. (1995).

4.8.2 Aperture Sources

When the wavelength of the electromagnetic field is comparable to the dimensions of the source, the efficiency of energy transfer through radiation is increased. Several types of sources based on hollow cylindrical waveguides of various cross sections or on microstrip have been developed. When waves propagate between metallic walls or dielectric surfaces, the electric and magnetic fields must be satisfy boundary conditions. In the case of perfectly conducting metallic walls, these conditions are that tangential components of the electric field and normal components of the magnetic field are zero at the surface of the conductors. In the case of a hollow metallic waveguide with rectangular cross section, solution of Maxwell's equations shows that two types of waves can be supported: transverse electric (TE) waves for which the component of electric field along the guide (E_z) is zero, and transverse magnetic waves for which H_z is zero. For TE modes, the remaining components are:

$$E_x = E_{xmn}\cos\left(\frac{m\pi}{a}x\right)\sin\left(\frac{n\pi}{b}y\right) \qquad (4.43a)$$

$$E_y = E_{ymn}\sin\left(\frac{m\pi}{a}x\right)\cos\left(\frac{n\pi}{b}y\right) \qquad (4.43b)$$

$$H_x = H_{xmn}\sin\left(\frac{m\pi}{a}x\right)\cos\left(\frac{n\pi}{b}y\right) \qquad (4.43c)$$

$$H_y = H_{xmn}\cos\left(\frac{m\pi}{a}x\right)\sin\left(\frac{n\pi}{b}y\right). \qquad (4.43d)$$

In the case of a perfect (lossless) dielectric within the waveguide, the propagation constant (along the waveguide) $\gamma_{m,n}$ is

$$\gamma_{m.n} = \sqrt{\left(\frac{m\pi}{a}\right)^2 + \left(\frac{n\pi}{b}\right)^2 - \omega^2\mu\epsilon} \qquad (4.44)$$

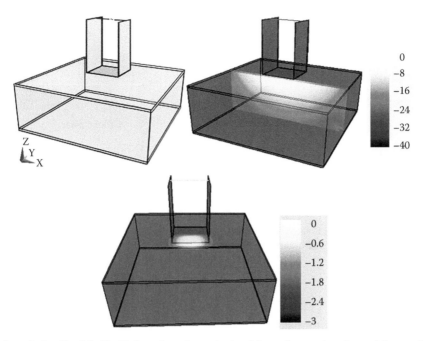

FIGURE 4.10 Current sheet device. Top left: The U-shaped conductor is placed 3 cm above a plane-layered fat-muscle-fat model and driven by a 30 MHz current source. Top right: The resulting SAR distribution (in dB relative to the maximum SAR) in the $y = 0$ plane. Bottom: SAR in the uppermost plane of the muscle-like medium (in dB relative to the maximum in that plane) showing a relatively uniform SAR distribution that is essentially constrained within the footprint of the conductor parallel to the upper surface of the phantom.

where m and n are integers and describe different modes of propagation. If $\left(\frac{m\pi}{a}\right)^2 + \left(\frac{n\pi}{b}\right)^2 > \omega^2\mu\epsilon$ then $\gamma_{m,n}$ is real and there is no significant propagation since the fields for these evanescent modes decay rapidly along the guide. If $\left(\frac{m\pi}{a}\right)^2 + \left(\frac{n\pi}{b}\right)^2 < \omega^2\mu\epsilon$, then $\gamma_{m,n}$ is imaginary and the fields propagate along the waveguide. The transition occurs when $\left(\frac{m\pi}{a}\right)^2 + \left(\frac{n\pi}{b}\right)^2 = \omega^2\mu\epsilon$. The frequency $f_{c_{m,n}}$ that satisfies this condition is known as the cut-off frequency and is given by

$$f_{c_{m,n}} = \frac{1}{2\pi\sqrt{\mu\epsilon}} \sqrt{\left(\frac{m\pi}{a}\right)^2 + \left(\frac{n\pi}{b}\right)^2}. \qquad (4.45)$$

The mode with the lowest cut-off frequency is known as the dominant mode, and the dimensions of the waveguide are often chosen so that only this mode can propagate. For a waveguide of rectangular cross section, the TE_{10} mode is the dominant one and Figure 4.11 shows the dependence of E_y as a function of distance across the wide dimension of the rectangular cross-section waveguide for this mode. A parameter that has been proved useful in assessing the effectiveness of hyperthermia devices is the effective field size (EFS), defined as the area that is enclosed within the contour at 50% maximum SAR at a depth of 1 cm in a flat homogeneous phantom (Hand et al. 1989). If an aperture source supporting a TE_{10} mode is placed directly onto a lossy medium, since $E_y \sin\left(\frac{x}{a}\right)$, the SAR distribution immediately beneath the aperture will be nonuniform varying approximately as $\sin^2\left(\frac{x}{a}\right)$ and consequently the EFS is small compared to the footprint of the aperture. One of the challenges

in designing aperture sources is to achieve a large effective field size. Parameters that may be considered include extending the end of the waveguide into a horn structure, dielectric loading of the waveguide, and the use of waveguides of differing cross sections.

An example of a superficial hyperthermia applicator that produces a larger EFS than a simple waveguide applicator is the lucite cone applicator (LCA), a water-filled rectangular waveguide terminated by a modified horn antenna (van Rhoon et al. 1998; Samaras et al. 2000). The waveguide operates in the TE_{10} mode at 433 MHz, and the metallic walls of the horn section that are parallel to the E-field are replaced by polymethyl methacrylate walls. A polyvinyl chloride cone located centrally in the aperture is also inserted into the applicator as shown in Figure 4.12. In use, a water bolus is placed between the aperture and the tissue to be heated. Laboratory testing shows that the nonmetallic walls and an appropriate choice

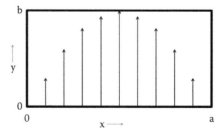

FIGURE 4.11 Electric field $E_y(x)$ for TE_{10} mode in rectangular waveguide of cross-sectional dimensions $a \times b$.

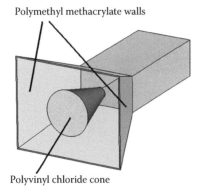

FIGURE 4.12 Schematic drawing of lucite cone applicator (van Rhoon et al. 1998). The side walls of the horn section consist of polymethyl methacrylate sheet, and a PVC cone is located in the horn section at the center of the applicator's aperture.

of dimensions for the PVC cone lead to a threefold increase in EFS compared to a conventional metallic walled-horn applicator of the same dimensions (van Rhoon et al., 1998), and this advantage has also been demonstrated in clinical studies (Rietveld et al. 1999).

Although waveguide-based applicators are generally very useful for delivering superficial hyperthermia, their bulk sometimes makes access to the tumor site difficult. A means of overcoming this problem is to use an applicator based on a microstrip transmission line. A microstrip transmission line consists of a conducting strip separated from a ground plane by a thin dielectric substrate. The line does not support a true transverse electromagnetic (TEM) wave but a set of discrete hybrid modes with non-zero E_z and H_z components (the z direction being along the transmission line). However, the longitudinal components are small, and so the lowest order mode resembles a TEM wave at low frequency. This mode is often known as a quasi-TEM wave. Methods for determining the propagation characteristics of microstrip lines are discussed by James et al. (1985).

A simple microstrip antenna may consist of a rectangular metallic patch with dimensions equal to approximately half the propagation wavelength (Figure 4.13). However, the characteristics of such an antenna are sensitive to the dielectric properties of the media close to it, and so practical designs must be such as to minimize the dependence of performance on changes in loading. In practice the provision of a water bolus approximately 3 cm thick should result in a resonant frequency that does not vary significantly with load and that avoids exposing tissue to the relatively large normal electric field components present at the edges of the patch (Underwood et al. 1992).

In addition to devices based on a rectangular patch, microstrip applicators of different geometry have also been investigated. For example, Archimedean spiral antennas provide a circularly polarized electric field and a wider bandwidth than rectangular

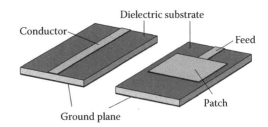

FIGURE 4.13 Schematic view of a microstrip transmission line (left) and a microstrip rectangular patch antenna (right). In both cases a dielectric substrate (thickness t and relative permittivity ε) separates a ground plane and appropriately shaped conductors.

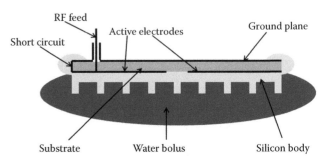

FIGURE 4.14 Schematic diagram showing cross section of a contact flexible microstrip applicator. (After Gelvich, E.A., and Mazokhin, V.N. *IEEE Trans. Biomed. Eng.* 49, 2002.)

or circular patches (Samulski et al. 1990; Jacobsen et al. 2000, 2005).

A type of device that offers flexibility and is lightweight is the contact flexible microstrip applicator (CFMA; Gelvich and Mazokhin 2002). As indicated in Figure 4.14, the CFMA is based on a microstrip structure consisting of two coplanar active electrodes and a ground plane incorporated in a semirigid silicon frame. In use, a thin silicon bolus is filled with circulating distilled water to improve impedance matching as well as to cool superficial tissues. Collapse of the bolus is prevented by the presence of silicon lugs up to 5 mm high.

4.8.3 Arrays of Applicators

The use of a single-heating device when heating superficial tissues is limited in most cases since the distribution of absorbed power cannot be changed in a predictable manner, and only the power delivered to the tissues and the position of the device relative to the patient can be adjusted during treatment. The use of an array of devices in which the power delivered to individual devices can be controlled should offer an improvement. Furthermore, the array of devices can be driven either incoherently or coherently. In the former case the power fed to each device may be controlled, and the resultant SAR distributions from each device are simply added. In the latter case, the E-fields produced by each device are added, and the total E-field produced depends upon both the amplitude and relative phase with which each device is driven;

the resulting SAR distribution is determined by the square of the total E-field. An advantage of coherent operation is that the power deposition may be steered by varying the phases of the devices in the array. For planar or approximately planar arrays, coherent operation often results in more central power deposition while incoherent operation tends to produce more peripheral power deposition and a larger effective field size.

Arrays designed to provide superficial hyperthermia over large surface areas such as the chest wall include a device consisting of 25 microstrip spiral array elements driven at 915 MHz (Lee et al. 1992) and a conformal microwave array consisting of dual concentric conductor applicators (Stauffer et al. 2010). Dual concentric conductor elements consist of multi-fed, quarter wavelength square slot apertures (Rossetto et al. 2000). These devices achieve large effective field sizes through choice of elements and individual control of each element in the array.

The use of nonplanar, for example, cylindrical, coherent arrays offer marked advantages when heating tissues deep in the body. Figure 4.15 shows schematic representations of several methods that have been reported in attempts to achieve deep heating within the body. The use of capacitive coupling produces an E-field that is essentially perpendicular to the skin beneath electrodes' footprints and often results in high SAR in superficial adipose tissue (D'Ambrosio and Dughiero 2007; Kroeze et al. 2003), while a concentric coil encircling the body produces an E-field that is essentially parallel to the skin but because of rotational symmetry is zero on its central axis. In contrast, a cylindrical annular array produces an E-field that is essentially parallel to the skin but radiates from each source, and if the sources are coherent, constructive interference of the fields can occur in deep tissues. In practice, annular arrays have been constructed using, for example, waveguide antennas (van Dijk et al. 1990; Kok et al. 2010), or dipole antennas (Turner et al. 1989) coupled

to the patient by means of a bolus of demineralized water. Early systems operated at a frequency in the range 70–110 MHz and provided limited two-dimensional control of the SAR distribution. Subsequent clinical experience and numerical simulations of annular-phased arrays led to improved designs based on multiple rings of antennas and capable of steering the resulting SAR distribution axially within the patient. Wust et al. (1996) simulated the performance of an array of up to 24 antennas arranged in three rings of eight antennas and driven in various combinations at 90 MHz and predicted that increasing the number of individually controlled (particularly in terms of relative phase) antennas led to improved tumor heating. In another simulation study in which SAR averaged over the volume of targeted tissue was compared with the volume averaged SAR outside the targeted tissue according to three objective functions, Paulsen et al. (1999) investigated the effects of varying the number and distribution of sources, their operating frequency, and the position of the target tissue. Results suggested that improved heating of the pelvic region could be achieved by increasing the operating frequency to around 150 or 200 MHz depending upon the number of antennas, increasing the number of antennas placed circumferentially around the patient, and increasing the number of rings of antennas axially. Seebass et al. (2001) also studied optimal frequency and antenna arrangement for regional hyperthermia and used both power- and temperature-based objective functions to assess performance. Their results were in qualitative agreement with those of Paulsen et al. (1999) and suggested that the optimal system consisted of three rings of sources, each with six or eight channels per ring, and an operating frequency of around 150 MHz. Subsequent experience of simulation of regional hyperthermia arrays and the importance of treatment planning are discussed by Wust and Weihrauch (2009). A common feature of most annular array systems is the incorporation

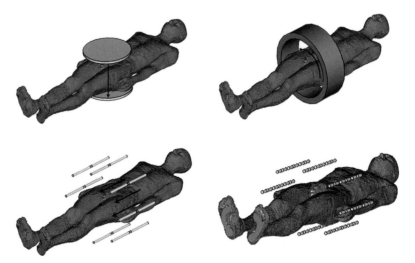

FIGURE 4.15 Examples of techniques reported for deep heating within the body. Top left: Capacitive coupling of E-field. Top right: Inductive coupling using a concentric coil. Lower left: An array of dipole antennas distributed around the body on a circular cylinder. Lower right: 24 dipoles arranged in three rings of eight antennas distributed around the body on a cylinder of elliptical cross section.

of a closed water bolus. Alternative approaches using an open water bolus have been described by De Leeuw and Lagendijk (1987) and Kroeze et al. (2001).

Another clinical application of phased arrays is the induction of hyperthermia in tumors in the head and neck. Paulides et al. (2005) investigated the optimal configuration of applicators, operating frequency, and target volume within this region using simulations of SAR in an anatomically realistic model. They found that the optimal number of antennas was 16 but that the performance of a less complex configuration of 6 to 8 antennas was also close to optimum. The optimal frequency was dependent upon target size; for target volumes in the range of approximately $4 \times 4 \times 4$ cm^3 to $5 \times 5 \times 5$ cm^3, it lay in the range 400–600 MHz. A clinical device, the HYPERcollar applicator system consisting of 12 patch antennas operated at 433 MHz, was subsequently developed (Paulides et al. 2007, Paulides et al. 2010).

References

Blouin, L.T., and Marcus, F.I. 1989. The effect of electrode design on the efficiency of delivery of radiofrequency energy to cardiac tissue in vitro. *Pacing Clin. Electrophysiol.* 12: 136–43.

Crezee, J. 1999. Spatial steering with quadruple electrodes in 27 MHz capacitively coupled interstitial hyperthermia. *Int. J. Hyperthermia* 15: 145–56.

D'Ambrosio, V., and Dughiero, F. 2007. Numerical model for RF capacitive regional deep hyperthermia in pelvic tumors. *Med. Biol. Eng. Comput.* 45: 459–66.

De Leeuw, A.A.C., and Lagendijk, J.J.W. 1987. Design of a deep-body hyperthermia system based on the "Coaxial TEM" applicator. *Int. J. Hyperthermia* 3: 413–21.

Demazumder, D., Mirotznik, M.S., and Schwartzman, D. 2001. Biophysics of radiofrequency ablation using an irrigated electrode. *J. Interv. Card. Electrophysiol.* 5: 377–89.

Gabriel, C. 1996. *Compilation of the dielectric properties of body tissues at RF and microwave frequencies*, Report N.AL/OE-TR- 1996-0037, Occupational and Environmental Health Directorate, Radiofrequency Radiation Division, Brooks Air Force Base, Texas.

Gabriel, C., Gabriel, S., and Corthout, E. 1996a. The dielectric properties of biological tissues: I. Literature survey. *Phys. Med. Biol.* 41: 2231–49.

Gabriel, S., Lau, R.W., and Gabriel, C. 1996b. The dielectric properties of biological tissues: II. Measurements in the frequency range 10 Hz to 20 GHz. *Phys. Med. Biol.* 41: 2251–69.

Gabriel, S., Lau, R.W., and Gabriel, C. 1996c. The dielectric properties of biological tissues: III. Parametric models for the dielectric spectrum of tissues. *Phys. Med. Biol.* 41: 2271–93.

Gabriel, C., and Peyman, A. 2006. Dielectric measurement: Error analysis and assessment of uncertainty. *Phys. Med. Biol.* 51: 6033–46.

Gelvich, E.A., and Mazokhin, V.N. 2002. Contact flexible microstrip applicators (CFMA) in a range from microwaves up to short waves. *IEEE Trans. Biomed. Eng.* 49: 1015–23.

Goldberg, S.N., Gazelle, G.S., Dawson, S.L., Rittman, S.L., Mueller, P.R., and Rosenthal, D.I. 1995. Tissue ablation with radiofrequency: Effect of probe size, gauge, duration, and temperature on lesion volume. *Acad. Radiol.* 2: 399–404.

Goldberg, S.N., and Gazelle, G.S. 2001. Radiofrequency tissue ablation: Physical principles and techniques for increasing coagulation necrosis. *Hepatogastroenterology* 48: 359–67.

Gopal, M.K., Hand, J.W., Lumori, M.L.D., Alkairi, S., Paulsen, K.D., and Cetas, T.C. 1992. Current sheet applicators for superficial hyperthermia of chestwall lesions. *Int. J. Hyperthermia*, 8: 227–40.

Haemmerich, D., Staelin, S.T., Tungjitkusolmun, S., Lee, F.T., Mahvi, D.M., and Webster, J.G. 2001. Hepatic bipolar radiofrequency ablation between separated multiprong electrodes. *IEEE Trans. Biomed. Eng.* 48: 1145–52.

Haemmerich, D., Tungjitkusolmun, S., Staelin, S.T., Lee, F.T., Mahvi, D.M., and Webster, J.G. 2002. Finite-element analysis of hepatic multiple probe radio-frequency ablation. *IEEE Trans. Biomed. Eng.* 49: 836–42.

Hand, J.W., Lagendijk, J.J.W., Andersen, J.B., and Bolomey, J.C. 1989. Quality assurance guidelines for ESHO protocols. *Int. J. Hyperthermia* 5: 421–8.

Hindricks, G., Haverkamp, W., Gülker, H. et al. 1989. Radiofrequency coagulation of ventricular myocardium: Improved prediction of lesion size by monitoring catheter tip temperature. *Eur. Heart J.* 10: 972–84.

Huang, S.K.S. 1991. Advances in applications of radiofrequency current to catheter ablation therapy. *Pacing Clin. Electrophysiol.* 14: 28–41.

ICNIRP. 2009. ICNIRP statement on the "guidelines for limiting exposure to time-varying electric, magnetic, and electromagnetic fields (up to 300 GHz)." *Health Phys.* 97: 257–8.

Jackman, W.M., Kuck, K-H., Naccarelli, G.V., Carmen, L., and Pitha, J. 1988. Radiofrequency current directed across the mitral anulus with a bipolar epicardial-endocardial catheter electrode configuration in dogs. *Circulation* 78: 1288–98.

Jacobsen, S., Stauffer, P. R., and Neuman, D. G. 2000. Dual-mode antenna design for microwave heating and noninvasive thermometry of superficial disease. *IEEE Trans. Biomed. Eng.* 47: 1500–09.

Jacobsen, S., Rolfsnes, H.O., and Stauffer, P.R. 2005. Characteristics of microstrip muscle-loaded single-arm Archimedean spiral antennas as investigated by FDTD numerical computations. *IEEE Trans. Biomed. Eng.* 52: 321–30.

James, J.R., Hall, P.S., and Wood, C. 1985. *Microstrip antenna theory and design*. London: Institution of Engineering and Technology.

Johnson, R.H., Preece, A.W., Hand, J.W., and James, J.R. 1987. A new type of lightweight low-frequency electromagnetic hyperthermia applicator. *IEEE Trans. Microw. Theory Tech.* 35: 1317–21.

Kapp, D.S., Fessenden, P., Samulski, T.V. et al. 1988. Stanford University institutional report. Phase 1 evaluation of equipment for hyperthermic treatment of cancer. *Int. J. Hyperthermia* 4: 75–115.

Kok, H. P., de Greef, M., Wiersma, J., Bel, A., and Crezee, J. 2010. The impact of the waveguide aperture size of the 3D 70 MHz AMC-8 locoregional hyperthermia system on tumour coverage. *Phys. Med. Biol.* 55: 4899–916.

Kroeze, H., Van de Kamer, J. B., De Leeuw, A. A. C., and Lagendijk, J. J. W. 2001. Regional hyperthermia applicator design using FDTD modelling. *Phys. Med. Biol.* 46: 1919–35.

Kroeze, H., van de Kamer, J. B., de Leeuw, A. A. C., Kikuchi, M., and Lagendijk, J. J. W. 2003. Treatment planning for capacitive regional hyperthermia. *Int. J. Hyperthermia* 9: 58–73.

Lazebnik, M., McCartney, L., Popovic, D. et al. 2007. A large-scale study of the ultrawideband microwave dielectric properties of normal breast tissue obtained from reduction surgeries. *Phys. Med. Biol.* 52: 2637–56.

Lee, E.R., Wilsey, T.R., Tarczy-Homoch, P. et al. 1992. Body conformable 915 MHz microstrip array applicators for large surface area hyperthermia. *IEEE Trans. Biomed. Eng.* 39: 470–83.

Leybovich, L.B., Dogan, N., and Sethi, A. 2000. A modified technique for RF-LCF interstitial hyperthermia. *Int. J. Hyperthermia* 16: 405–13.

Liem, L.B., Mead, R.H., Shenasa, M., and Kernoff, R. 1996. *In vitro* and *in vivo* results of transcatheter microwave ablation using forward-firing tip antenna design. *Pacing Clin. Electrophysiol.* 19: 2004–8.

Lin, J.C. 1999. Catheter microwave ablation therapy for cardiac arrhythmias. *Bioelectromagnetics* 20(Suppl 4): 120–32.

Livraghi, T., Goldberg, S.N., Lazzaroni, S. et al. 2000. Hepatocellular carcinoma: Radio-frequency ablation of medium and large lesions. *Radiology* 214: 761–68.

Morita, N., and J. Bach Andersen, J. 1982. Near-field absorption in a circular cylinder from electric and magnetic line sources. *Bioelectromagnetics* 3: 253–74.

Nevels, R.D., Arndt, G.D., Raffoul, G.W., Carl, J.R., and Pacifico, A. 1998. Microwave catheter design. *IEEE Trans. Biomed. Eng.* 45: 885–90.

O'Rourke, A.P.O., Lazebnik, M., Bertram, J.M. et al. 2007. Dielectric properties of human normal, malignant and cirrhotic liver tissue: *In vivo* and *ex vivo* measurements from 0.5 to 20 GHz using a precision open-ended coaxial probe. *Phys. Med. Biol.* 52: 4707–19.

Paulides, M.M., Bakker, J.F., Chavannes, N., and Van Rhoon, G.C. 2007. A patch antenna design for application in a phased-array head and neck hyperthermia applicator. *IEEE Trans. Biomed. Eng.* 54: 2057–63.

Paulides, M.M., Bakker, J.F., Linthorst, M. et al. 2010. The clinical feasibility of deep hyperthermia treatment in the head and neck: New challenges for positioning and temperature measurement. *Phys. Med. Biol.* 55: 2465–80.

Paulides, M.M., Vossen, S.H.J.A., Zwamborn, A.P.M., and Van Rhoon, G.C. 2005. Theoretical investigation into the feasibility to deposit RF energy centrally in the head-and-neck region. *Int. J. Radiation Oncology Biol. Phys.* 63: 634–42.

Paulsen, K.D., Geimer, S., Tang, J., and Boyse, W.E. 1999. Optimization of pelvic heating rate distributions with electromagnetic phased arrays. *Int. J. Hyperthermia* 15: 157–86.

Prionas, S.D., Fessenden, P., Kapp, D.S., Goffinet, D.R., and Hahn, G.M. 1988. Interstitial electrodes allowing longitudinal control of SAR distributions. In *Hyperthermic oncology*, Vol. 2 eds. T. Sugahara and M. Saito, 707–10. London: Taylor and Francis.

Prior, M.V., Lumori, M.L.D., Hand, J.W., Lamaitre, G., Schneider, C.J., and van Dijk J.D.P. 1995. The use of a current sheet applicator array for superficial hyperthermia: Incoherent versus coherent operation. *IEEE Trans. Biomed. Eng.* 42: 694–8.

Reeves, J. W., Meeson, S., and Birch, M. J. 2005. Effect of insertion depth on helical antenna performance in a muscle equivalent phantom. *Phys. Med. Biol.* 50: 2955–65.

Reeves, J.W., Birch, M.J., and Hand, J.W. 2008. Comparison of simulated and experimental results from helical antennas within a muscle-equivalent phantom. *Phys. Med. Biol.* 53: 3057–70.

Rhim, H. 2004. Review of Asian experience of thermal ablation techniques and clinical practices. *Int. J. Hyperthermia* 20: 699–712.

Rietveld, P.J.M., van Putten, W.L.J., van der Zee, J., and van Rhoon, G.C. 1999. Comparison of the clinical effectiveness of the 433 MHz lucite cone applicator with that of a conventional waveguide applicator in applications of superficial hyperthermia. *Int. J. Radiat. Oncol. Biol. Phys.* 43: 681–7.

Rossetto, F., Diederich, C.J., and Stauffer, P.R. 2000. Thermal and SAR characterization of multielement dual concentric conductor microwave applicators for hyperthermia, a theoretical investigation. *Med. Phys.* 27: 745–53.

Ryan, T.P., Turner, P.F., and Hamilton, B. 2010. Interstitial microwave transition from hyperthermia to ablation: Historical perspectives and current trends in thermal therapy. *Int. J. Hyperthermia* 26: 415–33.

Samaras, T., Rietveld, P.J.M., and van Rhoon, G.C. 2000. Effectiveness of FDTD in predicting SAR distributions from the lucite cone applicator. *IEEE Trans. Microwave Theory Tech.* 48: 2059–63.

Samulski, T.V., Fessenden, P., Lee, E.R., Kapp, D.S., Tanabe, E., and McEuen, A. 1990. Spiral microstrip hyperthermia applicators: Technical design and clinical performance. *Int. J. Radiation Oncol. Biol. Phys.* 18: 233–42.

Seebass M., Beck R., Gellermann J., Nadobny J., and Wust, P. 2001. Electromagnetic phased arrays for regional hyperthermia optimal frequency and antenna arrangement. *Int. J. Hyperthermia* 17: 321–36.

Sherar, M.D., Gladman, A.S., Davidson, S.R., Trachtenberg, J., and Gertner, M.R. 2001. Helical antenna arrays for interstitial microwave thermal therapy for prostate cancer: Tissue phantom testing and simulations for treatment. *Phys. Med. Biol.* 46:1905–18.

Shetty, S., Ishii,T.K., Krum, D.P. et al. 1996. Microwave applicator design for cardiac tissue ablations. *J. Microw. Power Electromagn. Energy*, 31: 59–66.

Stauffer, P.R. 1998. Implantable microwave antennas for thermal therapy. *Proc. SPIE* 3249: 38–49.

Stauffer, P.R., Maccarini, P., Arunachalam, K. et al. 2010. Conformal microwave array (CMA) applicators for hyperthermia of diffuse chest wall recurrence. *Int. J. Hyperthermia* 26: 686–98.

Stuchly, M.A., Kraszewski, A., Stuchly, S.S., and Smith, A.M. 1982. Dielectric properties of animal tissues *in vivo* at radio and microwave frequencies: Comparison between species. *Phys. Med. Biol.* 27: 927–36.

Trembly, B.S. 1985. The effects of driving frequency and antenna length on power deposition within a microwave antenna array used for hyperthermia. *IEEE Trans. Biomed. Eng.* 32: 152–7.

Tsai, C-F., Tai, C-T., Yu, W-C. et al. 1999. Is 8-mm more effective than 4-mm tip electrode catheter for ablation of typical atrial flutter? *Circulation* 100: 768–71.

Turner, P.F., Tumeh, A., and Schaefermeyer, T. 1989. BSD-2000 approach for deep local and regional hyperthermia: Physics and technology. *Strahlenther. Onkol.* 165: 738–41.

Underwood, H.R., Peterson, A.F., and Magin, R.L. 1992. Electric-field distribution near rectangular microstrip radiators for hyperthermia heating: Theory versus experiment in water. *IEEE Trans. Biomed. Eng.* 39: 146–53.

van der Brink, B.A., Gilbride, C., Aronovitz, M.J. et al. 2000. Safety and efficacy of a steerable temperature monitoring microwave catheter system for ventricular myocardial ablation. *J. Cardiovasc. Electrophysiol.* 11: 305–10.

van der Koijk, J.F., deBree, J., Crezee, J., and Lagendijk, J.J.W. 1997. Numerical analysis of capacitively coupled electrodes for interstitial hyperthermia. *Int. J. Hyperthermia* 13: 607–19.

van Dijk, J.D.P., Schneider, C.J., van Os, R.M., Blank, L.E., and Gonzalez, D.G. 1990. Results of deep body hyperthermia with large waveguide radiators. *Adv. Exp. Med. Biol.* 267: 315–9.

van Rhoon, G.C., Rietveld, P.J.M., and van der Zee, J. 1998. A 433 MHz lucite cone waveguide applicator for superficial hyperthermia. *Int. J. Hyperthermia* 14: 13–27.

van Vulpen, M., Raaymakers, B.W., Lagendijk, J.J.W. et al. 2002. Three-dimensional controlled interstitial hyperthermia combined with radiotherapy for locally advanced prostate carcinoma—A feasibility study. *Int. J. Radiation Oncology Biol. Phys.* 53: 116–26.

Wonnell, T., Stauffer, P., and Langberg, T. 1992. Evaluation of microwave and radiofrequency catheter ablation in a myocardium-equivalent phantom model. *IEEE Trans. Biomed. Eng.* 39: 1086–95.

Wust, P., Seebass, M., Nadobny, J., Deuflhard, P., Mönich, G., and Felix, R. 1996. Simulation studies promote technological development of radiofrequency phased array hyperthermia. *Int. J. Hyperthermia* 12: 477–94.

Wust, P., and Weihrauch, M. 2009. Hyperthermia classic commentary: "Simulation studies promote technological development of radiofrequency phased array hyperthermia" by Peter Wust et al., *International Journal of Hyperthermia* 1996; 12:477–94. *Int. J. Hyperthermia*, 25: 529–32.

Zhu X-L., and Gandhi, O.P. 1988. Design of RF needle applicators for optimum SAR distributions in irregularly shaped tumors. *IEEE Trans. Biomed. Eng.* 35: 382–8.

The Physics of Ultrasound Energy Sources

Victoria Bull
Institute of Cancer Research

Gail R. ter Haar
Institute of Cancer Research

5.1 Introduction

Many people think of medical ultrasound as being used solely for clinical diagnosis, and they are unaware of its therapeutic potential unless they are sportsmen who have received ultrasound treatment for soft tissue injuries. In fact, the use of therapeutic ultrasound predates its imaging applications. As ultrasound travels through tissues, a number of physical mechanisms take place. These include thermal effects arising from energy absorption and mechanical effects that are caused by the passage of an acoustic pressure wave through tissue. Both these classes of mechanism can be harnessed for therapeutic benefit. A range of biological responses may be sought from the bioeffects achieved. For physiotherapy, where the goal is functional modification for therapeutic reasons, the changes may be reversible or irreversible. In cancer therapy, ultrasound may be used to induce temperatures in the hyperthermic range (43–50°C) for use in conjunction with radio or chemotherapy or may be used on its own to create thermally ablated volumes when the temperature is raised rapidly above 55°C, or, as in histotripsy and drug delivery, to induce cavitation. As will be discussed in the following section, the difference between these regimes lies principally in the amount of power deposited in the tissue and the mode of delivery.

5.2 Ultrasound Transduction

The common feature of most ultrasound transducers is that they are piezoelectric devices. Application of an electric field leads to a change in the thickness of these materials, whereas an incident pressure pulse leads to an imbalance of electric charge and, thus, voltage generation (inverse piezoelectric effect; Ballato 1995, Silk 1984). Naturally occurring piezoelectric materials that are used to generate ultrasound include quartz and lithium niobate. Ferroelectric materials with a microcrystalline structure that can be rendered permanently piezoelectric by the application of a strong electric field are also used. Quartz crystals were used in the early days of medical ultrasound, but as these have narrow resonant peaks they have largely been replaced by ferroelectric ceramics such as lead zirconate titanate (PZT) that have a wider bandwidth. For therapy applications, the low loss material PZT4 is commonly used, whereas PZT5, which has higher sensitivity, is chosen for imaging (O'Donnell et al. 1981, Foster et al. 1991, Ballato 1995). Piezoceramic probes have been used both as therapy ultrasound sources and as hydrophones for detecting and mapping ultrasound fields. Polyvinyledene (di-)fluoride (PVDF) is a synthetic polymer that becomes piezoelectric when polarized using high electric fields at high temperature (Ohigashi 1976, Lewin and Schafer 1988). While PVDF is a poor transmitter of ultrasound, it can be formed into thin membranes that can be used as pressure sensors that disturb an ultrasound field only minimally.

5.2.1 Piezocomposite Transducers

While PZT materials can be used to provide reliable single-element transducers, their use for phased array transducers involves cutting grooves into the ceramic, leaving residual material of the required shape and size. While such transducers are usually highly efficient and capable of operating at high power, they are often fragile, and difficult to manufacture as large area arrays. Their construction is liable to cause significant lateral vibration, which sets up parasitic waves, leading to undesirable hotspots, and they are, by nature, narrow band devices.

In order to circumvent these problems, the use of piezocomposite materials has been explored. These are made by interleaving precut shaped pillars with passive polymer or epoxy host matrix compound (Smith 1989, Berriet and Fleury 2007, Chapelon et al. 2000). This gives considerable flexibility in shape, size, and array geometry. The infill has a low glass transition temperature. The 1-3 structure used in most piezocomposite transducers reduces lateral modes and enhances thickness mode vibration. There are a number of other advantages of this material. The embedding matrix structure can be shaped into a bowl, thus facilitating the production of focused beams. The backing has a low impedance, and the composite material has low electrical and mechanical losses, thus reducing transducer heating. The composite has a high Curie temperature.

5.2.2 CMUTs

Capacitive micromachined ultrasound transducer arrays (CMUTs) rely on microelectromechanical systems (MEMS) technology. These arrays are formed with many adjacent cells arranged in parallel. Originally developed for industrial use in air, their adaptation to use in liquids has led to potential medical applications, initially for imaging but more recently for therapies such as high intensity focused ultrasound (HIFU; Haller and Khuri-Yakub 1994, Khuri-Yakub and Oralkan 2011, Wong et al. 2008). A single CMUT cell is a capacitor formed from a silicon substrate. The principle is shown in Figure 5.1.

A thin membrane is placed over a submicron vacuum gap. A metal layer is placed on top of the membrane to provide one electrode. The second electrode is provided by the conductive substrate that forms the base of the vacuum gap. For high-power applications, such as HIFU, the gap is lined with an insulating layer to prevent short circuiting during operation. On the application of voltage, the top membrane is attracted to the substrate by an electrostatic force. A mechanical restoring force resists this movement. When high-frequency sinusoidal voltage is applied, an ultrasound pressure wave is produced. Conversely, the application of pressure to the surface at a constant bias voltage leads to a capacitance change and a current whose amplitude depends on the frequency, the bias voltage, and the capacitance. An electric field in excess of 10^8 V cm^{-1} must be maintained across the gap. Pressure amplitudes of ~1 MPa peak to peak have been achieved at the transducer surface.

CMUTs provide a number of advantages over more conventional transducer technologies. They have a wide bandwidth and are not subject to self-heating. CMUT arrays are low cost and lend themselves to miniaturization. Basic microlithography techniques allow considerable flexibility in array design, and they have been used, for example, in catheter-based devices.

5.2.3 Transducer Geometries

A variety of transducer geometries have been used for therapy ultrasound, depending on the therapeutic application. Plane discs are used when energy deposition starting from the skin surface is required, as, for example, in some physiotherapy treatments. Where energy deposition is required more locally such as selectively, at depth, focused beams are required. Here, focusing may be achieved in a number of ways. Plane transducers may be combined with appropriate lenses, single piezoelectric elements may be shaped to form spherical bowls, or multielement-phased arrays may be driven to provide electronic focusing and beam steering. Figure 5.2 shows a multielement spherical bowl transducer with multiple connections that allow each element to be driven independently.

For therapy purposes, the ultrasound transducer element is usually held at the rim and has an air backing. This provides the greatest efficiency and a high Q. Imaging elements, on the other hand, have a block of backing material designed to damp out any resonances and provide a more uniform frequency response in reception mode, and to allow short pulse transmission. The amount of energy transmitted into the medium in front of the transducer (the coupling medium) depends on the relative acoustic impedance of the transducer front face and the medium. Acoustic impedance Z is the product of the density and the speed of sound ($Z = \rho c$) (see Section 5.3).

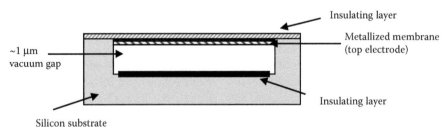

FIGURE 5.1   A CMUT cell.

FIGURE 5.2 A spherical bowl, multi-element focused ultrasound transducer. The central aperture allows for the insertion of an imaging device.

5.2.4 Plane Transducers

The majority of plane transducers are formed from circular discs. It is usually assumed that the plane face of these sources responds linearly to an applied alternating voltage and vibrates as a whole with uniform amplitude and phase. The field from such a transducer is commonly described as being composed of two regions: the near field (Fresnel zone) and the far field (Fraunhofer zone). Interference effects in the near field result in complex pressure distributions both on axis and off axis. Theoretical prediction of the near field is complex. The intensity on axis can be approximated to

$$\frac{I_z}{I_0} = \sin^2\left[\frac{\pi}{\lambda}(a^2 + z^2 - z)\right] \tag{5.1}$$

where z is the axial distance and I_z is the intensity at point z. I_z oscillates in amplitude between I_0 and zero. The intensity varies smoothly on axis in the far field, with the intensity on axis

decreasing exponentially with distance. The directivity function in the far field is given by

$$D_S = \frac{2J_1(2\pi a \sin(\frac{\theta}{\lambda}))}{2\pi a \sin(\frac{\theta}{\lambda})} \tag{5.2}$$

where J_1 is the first order Bessel function, θ is the angle that the line joining the point of interest to the center of the transducer makes with the axis, and λ is the acoustic wavelength in the propagation medium. For most therapeutic ultrasound applications involving plane transducers, the tissues being targeted lie within the near field, and so uniform ultrasound exposure is most easily achieved by moving the transducer.

5.2.5 Focused Fields

5.2.5.1 Focused Bowl Transducers

The simplest way to achieve a focused field is to use a single-element spherical curved shell of piezoelectric or piezoceramic material. It is possible to calculate the field from such a transducer geometry (O'Neill 1949, Penttinen and Luukala 1976). Figure 5.3 illustrates the pressure distribution within a focused ultrasound field, both along the sound axis (a) and in a transverse plane through the focal peak (b).

The ultrasound focus lies on the central axis, near the center of curvature of the bowl. As $2\pi h/\lambda$ increases (where h is the depth of the bowl, as shown in Figure 5.4), the focal point moves closer to the geometric center of curvature. The gain factor (G, the ratio of the intensity at the center of curvature to the average intensity at the transducer surface, I_C/I_0) is given by

$$\frac{I_C}{I_0} = \left(\frac{2\pi h}{\lambda}\right)^2. \tag{5.3}$$

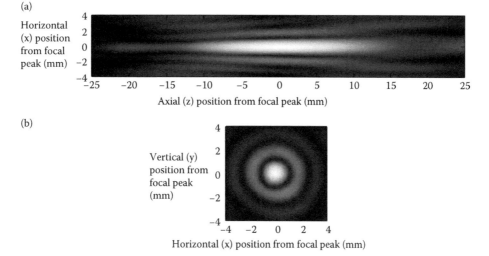

(a)
Horizontal (x) position from focal peak (mm)

Axial (z) position from focal peak (mm)

(b)
Vertical (y) position from focal peak (mm)

Horizontal (x) position from focal peak (mm)

FIGURE 5.3 Pressure distribution close to the focus of a typical focused ultrasound field in the axial (a) and transverse (b) planes.

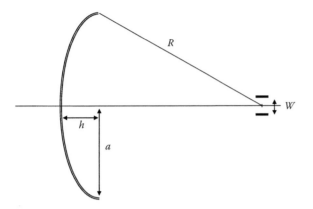

FIGURE 5.4 Geometric parameters of a spherical bowl transducer: h = bowl depth, a = bowl radius, R = radius of curvature, and W = width of focal region.

The width of the focal region, W, is given by

$$W = 1.22 \frac{R\lambda}{a} \tag{5.4}$$

where R is the radius of curvature, and a is the bowl radius. In a nonattenuating medium, diffraction theory shows that 84% of the total energy passes through the focal region (Kossoff et al. 1979). The variation of gain and focal width with transducer radius at different frequencies, for a transducer of radius of curvature 5 cm, is shown in Figure 5.5.

5.2.5.2 Lenses

Converging beams may also be produced by fronting a plane transducer with a concave lens of velocity of sound greater than water, such as polystyrene or Perspex (Fry and Dunn 1962). A plano-concave geometry is the easiest to construct. The focal length of such a lens is given by

$$F = \frac{R}{(1-n)} \tag{5.5}$$

where n is its refractive index.

A limitation of the use of lenses is that the sound is absorbed by the lens material, and the acoustic mismatch between the lens and the medium in which it sits (often water), leads to reflections at the interfaces. Maximum transmission is obtained when the transducer and lens are separated by a quarter wavelength plate of a material that provides a good acoustic match to both.

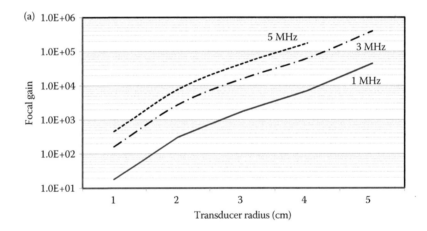

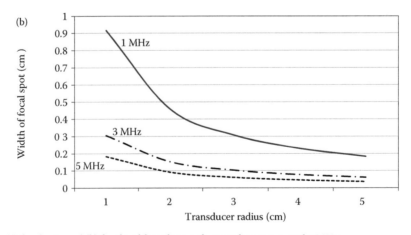

FIGURE 5.5 Variation in (a) focal gain and (b) focal width with transducer radius at 1, 3, and 5 MHz.

5.2.5.3 Phased Arrays

In practice, focused beams are most commonly created using phased arrays. These are usually formed with many individual elements mounted on the surface of a spherical bowl. Piezocomposite materials in particular lend themselves to this type of transducer. Beam focusing and movement is achieved by judicious choice of phasing of the drive signal to each element. These have the advantage that the heating pattern can be changed electronically, where appropriate, in response to feedback from monitoring of, for example, temperature or stiffness in the tissue target. Electronic control of heating has the advantage that the device/patient interface may be easier to provide, and the maximum use of available acoustic windows can be made. Where a single focal spot is desired, changes in phasing of the element may allow an increased heated volume compared to that of a similar single-element transducer. Phased arrays can also be driven to provide multiple foci, which can then be scanned to enlarge the heated volume. This approach is sometimes referred to as multifocus scanning and was first proposed by Ebbini and Cain (1989, 1991). The degree of flexibility in potential field patterns is largely dependent on the number of elements available, and their disposition on the surface of the transducer head. A number of element geometries have been used, depending on the application, with the largest arrays being used for HIFU treatments of the brain (Sun and Hynynen 1998, Clement et al. 2000, Pernot et al. 2005, Clement & Hynynen 2002, Pernot et al. 2003, Hynynen et al. 2004, 2006, Aubry et al. 2007). An added advantage of the phased array approach is the ability to use some of the elements in dual imaging/therapy mode (Bouchoux et al. 2008).

Electronic adjustment of the phasing of each element of concentric annular arrays allows rapid movement of the focal region along the beam axis (Huu and Hartmann 1982, Ibbini and Cain 1990). This is demonstrated in Figure 5.6.

The idea of using several closely spaced focal spots to obtain a larger homogeneous temperature distribution by taking advantage of thermal conductivity has been widely explored. Fan and Hynynen (1995, 1996) modeled the fields from a spherically curved transducer with a 4 × 4 array of square elements. They showed that four focal points at a separation of 2–4 wavelengths could be created. A maximum necrosed volume up to 16 times that of a similar single-element transducer could be obtained.

Subsequently, designs using more elements have been considered (for example, Daum and Hynynen 1999). Sasaki et al. (2003) have described the use of a four-element array mounted in a truncated spherical bowl that has been driven to produce a split focus that gave a 3–4 times larger focal volume than obtained with a single focus exposure.

A problem that arises with these regularly spaced phased-array designs is the presence of grating lobes and other secondary maxima. Goss et al. (1996) showed that sparse, randomly distributed arrays reduce the magnitude of these. Lu et al. (2005) described a generic algorithm for optimization of phase and amplitude distributions for the reduction of grating lobes. There have been a number of other studies of apodization and the use of subsets of elements (Dupenloup et al. 1996; Gavrilov et al. 1997, 2000; Hand and Gavrilov 2000, 2002; Filonenko et al. 2004). In a publication in 2009, Hand et al. (2009) described the performance of a 254 circular element, 1 MHz random array, capable of producing five simultaneous foci. Melodelima et al. (2009) have described a multielement toroidal transducer. This geometry allows the rapid induction of large volumes of tissue within 7 cm of the transducer, and thus offers the possibility of rapid ablation of liver tumors during liver surgery. In the search for a larger ablated volume per unit time Chen et al. (2011) have proposed a cylindrical section single-element transducer design that is capable of generating a line focus. This device is known as the "SonoKnife."

The discussion to this point has focused on extracorporeal transducers. There is also considerable interest in interstitial heating sources. Here miniaturized sources are mounted on the end of catheters or endoscopes. In this geometry, imaging and therapy elements are placed at the end of the probe. The main thrust of this has been in the development of ultrasound based transrectal devices for thermal based therapies of benign and malignant prostate disease (Diederich et al. 2004a,b; Lafon et al. 2004; Ross et al. 2004, 2005; Chopra et al. 2005; Kinsey et al. 2008). Miniaturized image/ablate probes for insertion directly into the tumor of interest have also been proposed (Makin et al. 2005; Mast et al. 2011; Lafon et al. 2002, 2007).

5.3 Acoustic Field Propagation

During its passage through tissue, the energy contained in an ultrasound beam reduces, that is, it is attenuated. The energy reaching a given point in tissue is determined in part by how much is scattered out of the main beam by tissue structures in the beam path and by absorption. This energy loss is characterized for a specific tissue by its attenuation coefficient, which is the sum of the absorption coefficient and the component due to scattering as described below. Energy is also lost by specular reflection and refraction of the beam. For a plane pressure wave propagating in the x-direction, the pressure can be written as

$$P(x,t) = P_0 e^{-\alpha x} e^{i\omega(t - x/c)} \tag{5.6}$$

where α is the amplitude attenuation coefficient, ω is the angular frequency, P_0 is the pressure amplitude at the position $x = 0$, and

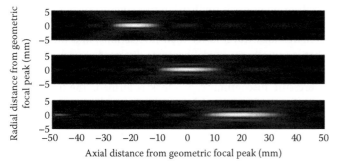

FIGURE 5.6 Field plots for an unsteered transducer (middle) and the same transducer steered backward (top) and forward (bottom) along the acoustic axis.

c is the sound velocity. If a wave of intensity I is incident on an inhomogeneity in tissue, the total power scattered or absorbed is proportional to the intensity. The constant of proportionality is known as the scattering or absorption cross section (σ_s or σ_a). From these definitions we see that, for a beam of power W, and cross-sectional area A, the power scattered by the ith inhomogeneity is $\frac{\sigma_{si}W}{A}$, and that absorbed by the jth is $\frac{\sigma_{aj}W}{A}$. If there are n_i scattering and n_j absorbing inhomogeneities per unit volume, then the power scattered per unit volume is $\sum_i \frac{n_i\sigma_{si}W}{A}$ and that absorbed per unit volume is $\sum_j \frac{n_j\sigma_{aj}W}{A}$. In the absence of multiple scattering, the absorption cross section per unit volume, μ_a, is given by $\mu_a = \sum_j n_j\sigma_{aj}$, and for scattering the cross section per unit volume μ_s is $\sum_i n_i\sigma_{si}$.

For a slice of tissue Δx thick, the total power scattered, ΔW_s, is given by $W_s = \mu_s W \ x$ and the total absorbed power, ΔW_a, is $\mu_a W \ x$. The total loss of power, ΔW, can then be written as:

$$W = \left(\mu_s + \mu_a\right)W \ x. \tag{5.7}$$

Integrating Equation 5.2 for a target of finite thickness, we get

$$W = W_0 e^{-(\mu_a + \mu_s)x} \tag{5.8}$$

where W_0 is the power at position $x = 0$.
This may be rewritten as

$$W = W_0 e^{-\mu x} \tag{5.9}$$

where μ is the attenuation coefficient (also referred to as the intensity coefficient) and $\mu = \mu_a + \mu_s$. An amplitude coefficient, α, can also be defined where $\mu = 2\alpha$ and α is the sum of scattering and absorption coefficients ($\alpha = \alpha_a + \alpha_s$). In soft tissues, absorption represents between 60% and 80% of the attenuation. From Equation 5.9:

$$\mu = -\frac{1}{x}\ln\left(\frac{W}{W_0}\right) \tag{5.10}$$

and so the units of μ are cm^{-1}, or nepers cm^{-1}. In these units μ is numerically twice α. If $\frac{W}{W_0}$ is expressed in decibels,

$$\mu = -\frac{10}{x}\log_{10}\left(\frac{W}{W_0}\right) \tag{5.11}$$

$$\alpha = -\frac{20}{x}\log_{10}\left(\frac{P}{P_0}\right) \tag{5.12}$$

and the units become dB cm^{-1}. Here μ and α are numerically equal. The conversion between the two units is μ dB cm^{-1} = 4.343μ neper cm^{-1}, α dB cm^{-1} = 8.686α neper cm^{-1}.

The choice of exposure frequency to obtain a specific intensity level in a volume in which tissue heating is sought is a compromise between the desire to keep the frequency low in order to minimize attenuation in the tissues overlying the target and the wish to maximize energy absorption within this region. Hill (1994) showed that for HIFU exposures, the frequency of choice is one that gives an overall acoustic attenuation in the path of approximately 10 dB.

5.3.1 Reflection and Refraction

When an ultrasound beam meets an interface between two regions of different acoustic impedance at normal incidence, the fraction of energy reflected is described by the reflection coefficient R given by

$$R = \left(\frac{Z_2 - Z_1}{Z_2 + Z_1}\right)^2 \tag{5.13}$$

where Z_1 is the impedance of the first medium and Z_2 is that of the second medium. The proportion of transmitted energy, T, is then described by

$$T = 1 - R = \frac{4Z_1Z_2}{\left(Z_2 + Z_1\right)^2}. \tag{5.14}$$

When the beam hits an interface at an angle that is not normal to the surface, it changes propagation direction in the second medium. If the beam angle to the normal in medium 1 is θ_1, and that in medium 2 is θ_2, then these angles are related by Snell's law:

$$\frac{\sin\theta_1}{c_1} = \frac{\sin\theta_2}{c_2} \tag{5.15}$$

where c_1 and c_2 are the sound velocities in the two media. A critical angle θ_c is reached for which θ_2 is 90°, that is, the beam does not enter medium 2, but travels parallel to the interface. When $\theta_1 \geq \theta_c$, no sound energy enters the second medium. θ_c is given by

$$\theta_c = \sin^{-1}\frac{c_1}{c_2}. \tag{5.16}$$

5.3.2 Non-Linear Propagation

When particle oscillations are small, as, for example, might be expected in the very low intensity beams used in bone fracture healing, the acoustic propagation is linear. However, for most medical applications (both diagnostic and therapeutic) propagation is nonlinear. Since the speed of sound depends on density, compressions travel faster than rarefactions, and thus a wave that is initially sinusoidal becomes distorted as each compression "catches up" with the preceding rarefaction. In the limit, a pressure discontinuity, or shock is formed. The linear wave equation can be expressed as:

$$P - P_0 = A\left(\frac{\rho - \rho_0}{\rho_0}\right) \tag{5.17}$$

where $A = \rho_0\left[\frac{\partial p}{\partial \rho}\right] = \rho_0 c_0^2$, and ρ, ρ_0 are the instantaneous and equilibrium densities.

This assumes that the particle velocity is very small ($\ll c$) and that the relationship between pressure and density is linear. The linear approximation assumes that there is no heat transfer during the wave propagation. A more accurate description is given by the Taylor expansion:

$$P - P_0 = A\left(\frac{\rho - \rho_0}{\rho_0}\right) + \frac{B}{2}\left(\frac{\rho - \rho_0}{\rho_0}\right)^2 + \frac{C}{3}\left(\frac{\rho - \rho_0}{\rho_0}\right)^3 + \cdots \quad (5.18)$$

$$B = \rho_0^2\left[\frac{\partial^2 p}{\partial \rho^2}\right], \quad C = \rho_0^3\left[\frac{\partial^3 p}{\partial \rho^3}\right]\cdots$$

Commonly, only the first and second order terms are considered.

It can be shown that a point on a wave with particle velocity u will travel with velocity $c + \beta u$ where $\beta = \frac{2B}{A} \cdot \frac{B}{A}$ is the nonlinearity parameter. For a plane wave, a shock parameter σ is defined, where $\sigma = \beta \varepsilon kx$ and $\varepsilon = \frac{u}{c}$, $k = \frac{2\pi}{\lambda}$, and x is the distance traveled. A full shock has developed when $\sigma = \frac{\pi}{2}$, and $\sigma = 1$ indicates the start of shock front formation. In water, for which $\beta = 3.5$, the shock distance at a frequency of 3.5 MHz is 43 mm when $P_0 = 1$ MPa.

The distorted wave has a markedly different frequency content than a pure sine wave, which has only one frequency component. The distorted wave is the sum of the harmonics of the fundamental frequency. The amplitude of each successive harmonic is less than the one preceding it. Thus the fundamental, f_0, is of greater amplitude than the second harmonic, f_2, and so on. This is demonstrated by Figure 5.7. In attenuating media, the higher frequencies are attenuated more rapidly than the fundamental, and thus, as the wave propagates, the degree of nonlinearity decreases. The phenomenon of acoustic saturation occurs when increasing the pressure amplitude at the source does not increase the energy arriving at some distance from the transducer since all the extra energy is contained in higher harmonics that are absorbed in the tissue path.

5.4 Interactions of Ultrasound with Tissue

As described, the resulting effects of a propagating pressure wave on tissue can be both thermal and mechanical in origin. This section deals for the most part with the heating of tissue through absorption of acoustic energy, however, mechanical effects can also have a significant bearing on the overall therapeutic outcome and are therefore also addressed.

Section 5.3 describes the propagation of ultrasound under "free field" conditions, such as occur in a large water bath, and we have seen that attenuation occurs due to a combination of absorption and scatter. These effects become more prominent when an ultrasound wave travels through a medium such as human tissue. As described previously, the speed of propagation of ultrasound is a pressure-dependent parameter, resulting in changes in the nature of a propagating wave. More specifically, the progressive distortion of the wave form with distance traveled at high pressure amplitudes leads to the generation of harmonic frequency components (see Equation 5.18 and Figure 5.4). The levels of attenuation, scattering, and absorption that occur during this process are dependent on the specific properties of the medium traversed as well as the acoustic field parameters. For a plane wave propagating in a homogeneous medium, the total transmitted power experiences an exponential decay, dependent on the attenuation coefficient, which includes absorption and scatter. Measurement of these components in isolation is difficult, however, values for many tissue types have been reported in the literature (Duck 1990, Goss et al. 1979).

5.4.1 Thermal Effects

For the purposes of this text, the most important effect of the interaction of ultrasound with tissue is the deposition of heat in a given volume. This is dependent on the intensity of the ultrasound reaching the region of interest and how much of this is

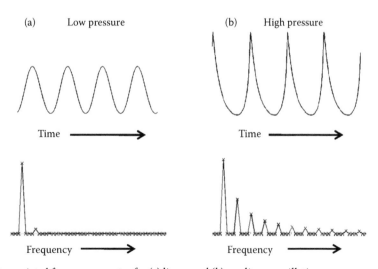

FIGURE 5.7 Waveforms and associated frequency spectra for (a) linear and (b) nonlinear oscillations.

converted into heat. The former is a combined effect of the initial wave amplitude and of how much attenuation has occurred to that point in the path of the beam. Scattering of the beam within the body is dependent on the tissue structure and on the incident speed of sound (Hynynen 1986). Absorption depends on the ultrasound frequency and the intensity, which is a function of pressure amplitude (p), density (ρ), and acoustic impedance (Z). Equation 5.19 describes the instantaneous intensity (I) approximation for a plane wave case, although spatial and temporal averaging is required to relate this quantity to the energy absorbed in bulk within a volume.

$$I = \frac{p^2}{Z}. \qquad (5.19)$$

Different intensity definitions are described in Section 5.5.2. Although the precise mechanisms for the transfer of acoustic energy from motion to heat are not fully understood, the dominant processes can be described broadly in the following way: as an ultrasound pressure field propagates, it causes motion of particles during the compressional and rarefactional phases. In the presence of a low-frequency wave causing gradual pressure changes, tissues will respond in phase, with molecules moving reversibly from regions of high to low density. Increasing the frequency leads to more rapid density fluctuations, and the tissue response becomes out of phase with the incident pressure changes. The motion of molecules and tissue inhomogeneities relative to their surroundings results in viscous heating. Subsequent dissipation of this heat causes an overall bulk temperature rise in the exposed tissue (Bamber 1986).

Figure 5.8 shows the shape of a typical temperature profile with time at different incident spatial peak, temporal average intensities in an absorbing medium during HIFU heating. An extensive study by Goss et al. has shown that absorption and attenuation coefficients of many tissue types exhibit the same relationship with frequency, but that their magnitudes

may be very different. This implies that scatter plays a large part in the loss of ultrasound power with distance, with quoted absorption coefficient to attenuation coefficient ratios spanning a range of 0.22–0.43 at 1 MHz in different tissue types (Goss et al. 1979). It should also be noted that absorption increases with frequency, and thus nonlinearities in the acoustic field increase heating due to the higher harmonic content, as described in Section 5.3.

5.4.1.1 Thermal Index

One way to approach the issue of thermal effects of ultrasound on tissues is through a quantity known as the thermal index (TI). TI as a concept was developed by the National Council of Radiation Protection and Measurements (Duck 2008) and relates primarily to the safety of diagnostic levels of ultrasound. Its definition is based on thermal models of ultrasound fields in tissue. It is defined as the ratio of the total acoustic power to the power required to heat tissue by 1°C (AIUM and NEMA 1992). Three different models exist for the thermal index, depending on the structures through which the beam is propagating. In its simplest form, the soft tissue thermal index, TIS, can be calculated from Equation 5.20:

$$TIS = \frac{f_c W_{01}}{210} \qquad (5.20)$$

where f_c is the center working frequency of the transducer, and W_{01} is the power as measured through a 1-cm aperture. Two more complex indices are also defined: the TIB, used when there is an interface with bone somewhere close to the focus of the beam, for example, when imaging a fetus; and the TIC, used when bone lies close to the surface of the transducer, such as during cranial ultrasound. Detailed discussion of these is outside the scope of this text, however measurements relating to the above quantities are introduced in Section 5.5. Although the TI is not used routinely when discussing therapeutic ultrasound

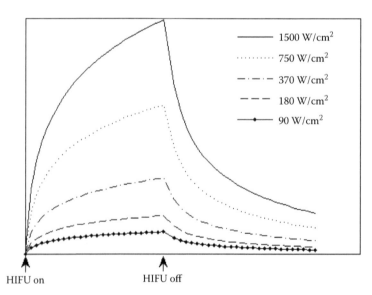

FIGURE 5.8 Temperature as a function of time at the peak of a focused ultrasound field, and its variation with peak intensity.

fields, acoustic power and intensity are important quantities for physiotherapy, hyperthermia, and HIFU applications.

5.4.1.2 Heating Regimes Using Ultrasound Sources

There are three main categories of thermal therapy that use ultrasound as their energy source. Broadly speaking, physiotherapy is intended to induce a beneficial response in cells using temperature rises of less than 5°C. Hyperthermia raises the temperature of tissues by a few degrees, reaching 42–46°C for many minutes, with the aim of causing cell death, and often inducing a synergistic effect with radio- or chemotherapy. Weakly focused transducers are used in hyperthermia to treat volumes of the order of a cubic centimeter in a single exposure over many minutes. In contrast, HIFU uses highly focused transducers to induce rapid cell death in small regions of tissue, fractions of a cubic centimeter in size, relying on mechanical or electronic movement of the focal region to treat larger volumes. This effect, in which tissues undergo protein denaturation (coagulative necrosis) once temperatures in excess of 56°C have been maintained for 1–2 seconds (ter Haar 1986a), is known as thermal ablation. Each of these regimes is described in more detail later in this section. Some typical values of frequency, required power, and treatment depths for each of these regimes are given in Table 5.1.

5.4.1.3 Thermal Dose

The extent of damage to tissues caused by ultrasound depends on a number of factors, again related to the acoustic field parameters and to the exposed tissue properties. For a given tissue type, the overall effect is dependent on a combination of temperature elevation and duration of heating. A single metric, referred to as thermal dose, quantifies these effects using an integration of temperature over time, and the quantity is expressed in terms of cumulative equivalent minutes at 43°C (CEM_{43}). This is calculated from knowledge of temperature history using

$$CEM_{43} = \int_{t=0}^{t=final} R^{43-T} \partial t \qquad (5.21)$$

where the constant R depends on properties of the tissue. Commonly used values are 0.25 for $T \leq 43$°C and 0.5 for $T > 43$°C (Sapareto and Dewey 1984), and the commonly accepted threshold for protein denaturation is 240 CEM_{43} (Sapareto and Dewey 1984, Damianou 1994), although there is some dependence on tissue type.

In order to illustrate the effects of different temperature histories on tissue in such a way that is relevant to ultrasound-based

thermal therapies, it is necessary to divide the discussion into the three regimes described previously, as the overall nature and desired outcome depends strongly on the treatment parameters. Additional nonthermal effects resulting from interactions of ultrasound with tissues are addressed in Section 5.4.2.

5.4.1.4 Physiotherapy

When providing relief from muscle pain or promoting wound healing, the desired effect of heating is not to damage tissues but to induce a therapeutic response. Small temperature rises (~1°C) induced by plane wave transducers have been shown to produce numerous physiological effects. Cellular reactions to temperature elevation may lead to vasodilation, promotion of blood flow, muscle cell activation that may promote relaxation, and tissue regeneration that may be aided by promoting DNA synthesis, and thus cell proliferation, in vivo. Both pulsed and continuous wave (CW) beams are used, depending on the desired level of heating and mechanical effects. Studies have used a vast range of endpoints, such as wound healing (Freitas 2010), muscle healing (Chan 2010), bone healing (Malizos 2006, Guerino 2008), and pain relief and increased mobility of joints (Ozgonenel 2009). It is thought that acoustic streaming plays an important role in these processes (Dyson and Pond 1973), and this mechanical phenomenon is addressed in Section 5.4.2. Despite its common use in physiotherapy clinics, there is still some debate as to whether the effects produced by ultrasound are above and beyond that which can be achieved by physical manipulation and exercise (Marks et al. 2000, Saunders 2003).

5.4.1.5 Hyperthermia

Plane wave transducers are commonly used in hyperthermia to provide bulk heating, but employ higher output powers than those used in physiotherapy. These are designed to give higher spatial and temporal average intensities (see Table 5.1). Many studies of cells in culture and of different tissue types have been performed, and have indicated a range of different effects (ter Haar 1986b). These include microscopic effects such as increased cell membrane permeability and a lowering of the pH level within the cytoplasm, as well as bulk effects such as ischemia and lack of perfusion. Although it is not clear which is the dominant effect of the treatment on a cellular level, it has been shown that the physical state of the cell membrane is critically important. For example, changes in its structure leading to greater fluidity may result from increased temperature. In turn, the permeability of the membrane increases, causing chemical imbalance within the cell. Membrane fluidity is affected by other factors such as cholesterol and the presence of unsaturated fatty acids, however, there is also clear evidence correlating it with temperature rise and with cell death (Anghileri 1986). Also associated with an increase in temperature to hyperthermic levels is the inhibition of DNA synthesis and a loss of ability to repair damaged DNA. This is thought to result from changes in ion concentration in the cell due to membrane permeability alteration, which illustrates that although the cellular effects of heat are many and complex, they may be heavily linked, with one process leading

TABLE 5.1 Typical Ultrasound Parameters and Treatment Depths Associated with Different Thermal Therapies

	Frequency (MHz)	Intensity (W/cm²)	Treatment Depth (cm)
Physiotherapy	0.75–5	<1 SATA	≤5
Hyperthermia	0.5–5	<10 SPTA	<12
HIFU	1–3	>1500 SPTA	<20

on to the next, eventually resulting in programmed cell death (Wong and Dewey 1986). Clinically, hyperthermic treatments have been shown to be effective both alone and in combination with chemo- and radiotherapy, particularly for hypoxic tumors that may have a significant resistance to radiation therapy (Scott 1986). Additionally, the effect of reducing the ability to repair subcellular damage, as described before, suggests that there is a synergistic advantage in using hyperthermia alongside radiation therapy. Finally, hyperthermia has been shown to provide better tumor selectivity than alternative treatment types (Marchal 1992). Its success has improved significantly in recent years due to developments in temperature imaging for treatment guidance through real time feedback.

5.4.1.6 Focused Ultrasound Therapy

In contrast to hyperthermia, the aim of HIFU is to heat cells rapidly to temperatures capable of causing coagulative necrosis in a short time. Necrosis describes premature cell death, whereby the tissue is damaged sufficiently rapidly that the dead tissue remains in situ. This is in contrast to programmed cell suicide, or apoptosis, where signals to neighboring cells allow macrophages to absorb and recycle the waste from dead cells. Coagulative necrosis caused by heating is also referred to as thermal ablation, which, when performed using focused ultrasound beams, leads to a small (typically of the order of a centimeter in length and a millimeter in width at 1 MHz) ellipsoidal volume of dead tissue in the focal region of the acoustic field. The need to ablate whole tumors with this technique can result in lengthy treatment times, as the focal region is moved so as to cover the complete target volume. Similarly to hyperthermia, cell death is thought to arise from the denaturation of cell proteins, causing loss of communication with their environment. However, cell rupture and structural damage can occur due to mechanical effects such as cavitation, as described in the following discussion.

5.4.2 Mechanical Effects

The effects described in this section are not directly related to the thermal mechanisms of the interactions described previously, but occur as a by-product of ultrasound interactions in tissues, which may be undesirable in some circumstances and beneficial in others.

5.4.2.1 Acoustic Cavitation

As described, sound propagates as a longitudinal pressure wave, creating areas of compression (positive pressure) and rarefaction (negative pressure) in the medium it traverses. Negative pressures can cause gas or liquid vapor to be drawn out of solution, creating bubbles, or cavities, in a process known as acoustic cavitation. The resultant effect depends on properties of the field, and on the peak negative pressures achieved at that point in the medium.

Cavitation bubbles will be driven to oscillate by a propagating acoustic field. Stable cavitation describes the activities of bubbles oscillating slowly enough that gas can diffuse across the membrane during compression and rarefaction, in a process referred to as rectified diffusion. Such stable oscillations can continue over many cycles, exerting shear stresses on the surrounding cells. However, in the presence of higher peak negative pressures and more rapid changes, oscillating bubbles no longer undergo rectified diffusion. This leads to inertial, or collapse cavitation. At the peak of rarefaction, a bubble will have its greatest surface area, and more gas may enter the bubble than can be transferred back to the medium during the following rapid compression. Subsequent collapse of the bubble to a fraction of its original size can cause both direct and indirect tissue damage, due to a number of effects. Shear forces, much greater than those associated with stable cavitation, may firstly cause direct mechanical damage, resulting in cell rupture. The large concentration of energy at the center of the collapsing bubble also results in extremely high temperatures, thereby denaturing cells and causing harmful free radicals to be produced. A process known as sonoluminescence may then occur, whereby photons are emitted due to recombination of these free radicals with surrounding atoms, which may lead to additional subcellular damage (Duck 2008). Theoretically, the dissipated power (W) emitted by a bubble of equilibrium radius R_0, which is then converted to heat, is given by

$$W = \left(\frac{4}{3} \pi R_0{}^3 \right) 2\alpha I \qquad (5.22)$$

where I is the acoustic intensity incident on the bubble, and α is the attenuation of the surrounding medium (Coussios 2007). Following collapse, a single bubble can disintegrate into many smaller bubbles, which may dissolve back into solution or act to nucleate further cavitation activity.

A considerable amount of research in HIFU and hyperthermia relates to cavitation, and in particular its potential to increase the amount of energy deposited as heat, and therefore increase the volume of coagulative necrosis achieved in a given time. Stable cavitation bubbles undergoing nonlinear oscillations act as emitters of ultrasound energy at the fundamental frequency of the incident ultrasound beam, as well as harmonic, subharmonic (particularly the half harmonic), and ultraharmonic (multiples of the half harmonic) frequencies. Inertial or collapse cavitation events result in the emission of a broadband noise. These emissions can be detected by listening with a piezoelectric device situated outside the exposed material, a technique known as passive cavitation detection. The energy carried by the acoustic emissions can be deposited locally, with preferential absorption of harmonics at higher frequencies as described in Section 5.4.1, leading to further heating. This additional heat transfer has been found, in some cases, to increase the volume ablated by a single burst of HIFU, or increase the depth of treatment, thus giving the potential to reduce the total treatment time for a given target volume or allow for a greater variety of treatment sites (Holt and Roy 2001, Melodelima et al. 2004, Liu et al. 2006). Alternatively it may allow a lower output power to be used to achieve the desired temperatures within the body, thus reducing the possibility of side effects.

Due to the unpredictable nature of bubble activity in inhomogeneous media, the presence of cavitation in tissues during

HIFU exposure has been shown to result in nonuniform energy deposition due to interactions of bubbles with the incoming sound beam (Bailey et al. 2001). Without full knowledge of the distribution of bubbles and their interaction with the acoustic field, particular areas of tumors may be over- or undertreated through cavitation-enhanced heating, or scattering of the field out of the treatment area. Recent developments in cavitation detection allow bubble populations in vivo to be spatially mapped by listening to the acoustic emissions with an array made up of many piezoelectric crystals. This technique is known as passive cavitation mapping and has the potential for use as a monitoring tool during HIFU treatments (Jensen et al. 2011). Through knowledge of specific bubble locations, researchers may be able to utilize the additional heating they provide in order to enhance thermal ablation.

5.4.2.2 Streaming

Although ultrasound transducers produce an oscillatory pressure field centered about the ambient pressure of the medium, some momentum is transferred as a force, primarily along the direction of propagation. In a generalized sense, the term *streaming* when related to ultrasonic fields refers to the motion of fluid occurring due to this so-called radiation force. The resulting effects on cells from streaming in tissues depend on the magnitude of the force and the proximity of the cells to streaming currents. Motion of fluid close to cell membranes can cause breakages in protein bonds and can damage cells by creating holes in surface membranes, either reversibly or irreversibly, through shear stress (Wu 2001, 2002; Collis et al. 2010). However, as well as causing direct mechanical damage to the closest cells, streaming has been identified as an important contribution to the healing effect of acoustic fields of the levels used in physiotherapy (Dyson and Pond 1973).

In the same way that incident ultrasound fields cause streaming in tissues, the oscillation of cavitation bubbles causes extracellular fluid in the vicinity to be pushed around, forming small, localized currents. This is known as cavitation microstreaming. Not only may this cause direct mechanical damage as described earlier, but it has also been implicated as a mechanism for drug delivery, whereby the cell wall permeability can be increased to allow a targeted agent to enter, in a process termed *sonoporation* (Coussios and Roy 2008, Collis et al. 2010).

5.4.2.3 Mechanical Index

The concept of the mechanical index (MI) is similar in nature to the thermal index described in Section 5.4.1, as it is a safety index used to quantify the potential to cause damage. Similarly to the TI, it is primarily relevant to diagnostic systems, where these effects are to be avoided. In essence, the MI is a measure of the likelihood of cavitation occurring due to the effects of a particular acoustic field. Its form is:

$$MI = \frac{p_{r.3}}{\sqrt{f_c}} \qquad (5.23)$$

where $p_{r.3}$ is the peak rarefactional pressure derated by 0.3 dB cm^{-1} MHz^{-1} (to account for the attenuation of the field by soft tissue) and f_c is the center working frequency of the transducer. The explanation for this relationship is that the higher the negative pressure, the more tension the dissolved gas will be under in order to pull it out of solution, and the lower the frequency, the longer the time period over which the medium will sit below a certain negative pressure level, resulting in greater bubble growth and/or a greater number being produced.

5.4.2.4 Acoustic Dose

We have seen how defining a thermal dose (Section 5.4.1) can relate temperature history to biological effect. A second definition relating to ultrasound exposure has been proposed more recently, namely the acoustic dose, which aims to overcome the limitations of thermal dose such as its lack of ability to relate to different tissue or cell types, and the fact that it ignores additional mechanical effects and their contribution to biological effects. Acoustic dose is defined as the energy deposited per unit mass in a medium within which an ultrasonic wave propagates (Duck 2009). Associated with this is the acoustic dose rate, or acoustic dose per unit time.

One aspect of this that deserves particular discussion is that relating to damage resulting from cavitation activity. For many years there has been discussion of a so-called inertial cavitation dose, pertaining to the amount of broadband noise generated by these events. Chen et al. (2003) have defined a cavitation dose as the cumulative root-mean-square of the detected broadband noise, in the frequency domain, arising from inertial cavitation during ultrasound exposure. Although this appears at first to be a clear definition, the ability to detect this broadband noise is dependent not only on the properties of the medium being exposed but on the response of the detection device. As such, the chosen broadband frequency range for measurement, and thus the specific value quoted for inertial cavitation dose, depends heavily on the experimental arrangement and any signal processing performed. Taking into account the induced temperature rise, radiation force effects, and the presence of cavitation, acoustic dose is a more thorough but complex idea than thermal dose, and has not yet come into widespread use due to the amount of knowledge required of acoustic field parameters and tissue properties in spatially and temporally varying scenarios (Duck 2011).

5.5 Characterization and Calibration

5.5.1 Introduction

The delivery of a successful treatment relies upon some knowledge of the energy deposited in a given region of tissue. It is therefore important to know in detail the distribution of the acoustic field produced by a given transducer and how important parameters vary with the input power. Acoustic fields are commonly described in terms of three main parameters: pressure, intensity, and power. This section addresses the key definitions and measurement techniques of these parameters and introduces the concept of quality assurance (QA), a system by which the user can be confident of the consistency in performance of his or her equipment over time.

5.5.2 Field Parameters and Their Definitions

5.5.2.1 Pressure

Both positive and negative acoustic pressures occur during the oscillation of an acoustic wave. In plane wave linear oscillation, the peak positive and negative values are of equal amplitude. However, nonlinear propagation causes a change in the shape of the pressure waveform as described in Section 5.3, resulting in a greater change in peak positive pressure than in the peak negative. As such, both are important parameters in the description of an acoustic field. Negative pressures are responsible for cavitation activity, and the relationship between negative and positive pressures can help to indicate the potential heating effects through the concept of intensity, as described in the following sections.

5.5.2.2 Intensity

For a plane wave, the instantaneous intensity at any given point is described by Equation 5.19 (Section 5.4). This can be used, for example, to determine the spatial peak, temporal peak intensity within the field from the spatial peak pressure. In order to relate this to the biological effects of real treatments, both spatial and temporal averaging are used. Figure 5.9 helps to illustrate the different definitions of intensity. Although most commonly used in relation to the safety of diagnostic ultrasound systems, they are equally applicable to the output of therapeutic devices.

Spatial peak, pulse average intensity (I_{SPPA}) describes the average intensity value over a single pulse, defined as the ratio of the pulse intensity integral to the pulse duration, at the point in the field at which the value is a maximum. If the pulse were exactly

sinusoidal with a rectangular modulation, this would be given by Equation 5.24:

$$I_{SPPA} = \frac{p^2}{2Z} \qquad (5.24)$$

where p and Z are defined in Equation 5.19. The factor of 2 in the denominator comes from the averaging of a sine-squared wave. However, in a realistic situation, the frequency and modulation are not so well defined. Instead, the pulse intensity integral is calculated and averaged over the pulse duration. Standards published by the American Institute of Ultrasound in Medicine (AIUM 1992) state that the pulse duration is 1.25 times the time interval between the 10% and the 90% levels of the pulse intensity integral.

Where pulsed beams are used, the intensity values defined thus far assume the sound wave is on continuously, and thus overestimate the effect on tissue. To account for this, the spatial peak, temporal average intensity (I_{SPTA}) is often quoted. This is defined as

$$I_{SPTA} = I_{SPPA}\frac{PD}{PRP} \qquad (5.25)$$

where PD is the pulse duration, and PRP is the pulse repetition period. This gives the average intensity over the time between the start of one pulse and the start of the next, to give a better impression of the overall effect on tissue over time. Finally, physiotherapy systems will often be defined by a spatial average, temporal average intensity (I_{SATA}) as they are commonly applying heat to a larger area. This is calculated simply by averaging I_{SPTA} over the effective radiating area of the emitting device.

5.5.2.3 Power

As an ultrasonic field propagates, it exerts a radiation force (F_{rad}) on the traversed medium through which it passes, as described previously. The power in the beam is closely related to this quantity and is expressed by

$$W = F_{rad}c \qquad (5.26)$$

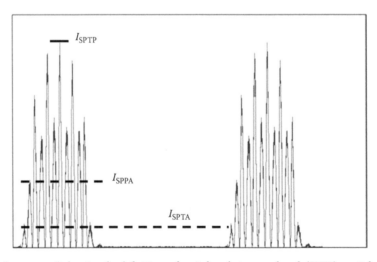

FIGURE 5.9 Intensity pulses ($\propto$ pressure^2) showing the definitions of spatial peak, temporal peak (SPTP), spatial peak, pulse average (SPPA), and spatial peak, temporal average (SPTA) intensity.

where c is the speed of sound in the medium. It is also described by the integral of temporal average intensity over the beam area.

5.5.3 Measurement Methods and Quality Assurance

With all medical ultrasound equipment, both diagnostic and therapeutic, there is a requirement for a complete understanding of the transducer output, not only for the manufacturers but also the end user in order to ensure its consistency and the safety of patients. Two different approaches to ultrasound device characterization and acoustic measurement are possible. The first requires a full understanding of the entire ultrasound field. This involves lengthy measurements over large areas and computationally intensive calculations and is commonly the domain of manufacturers and physicists. The second is based on the needs of the end user to make a regular and relatively quick measurement to check consistency. Both fall under the umbrella of quality assurance (QA). In the following, a brief discussion of acoustic measurement equipment and techniques is given, with reference to the appropriate standards for further information.

5.5.3.1 Hydrophone

Hydrophones are the key devices used in ultrasound measurement. Their mode of action is based on the inverse piezoelectric effect. The most commonly used hydrophones take the form of membranes or needles, shown in Figures 5.10a and b, respectively.

Each has its own advantages. A needle hydrophone has a small active element at its very tip, which is oriented to point toward the ultrasound transducer. Small needle diameters are desirable to avoid excessive disturbance of the acoustic field during measurement; however, reducing the element size results in a lowered sensitivity of the hydrophone. A membrane hydrophone is typically comprised of a piezoelectric PVDF membrane with an active element in the center, stretched over a circular frame and polarized to allow the incident compressions and rarefactions to register as a change in voltage across the element. Membranes are typically positioned in the plane perpendicular to the direction of ultrasound propagation and are large compared to the size of the field, so that the outer frame securing the membrane does not significantly disturb the acoustic field. The membrane is sufficiently thin to be essentially transparent to the incident ultrasound beam. Membrane hydrophones cannot be implanted into tissue, whereas this is possible with the needle geometry. More recently, fiber-optic systems have been developed (Koch 1996, Lim et al. 1999, Takahashi et al. 2000, Beard and Mills 1997, Morris et al. 2009), typically with high sensitivity and narrow (~μm) sensitive tips, allowing high spatial resolution field plots to be acquired. Due to their small geometry, these sensors also lend themselves well to making pressure measurements in situ, through insertion into a medium. This has been taken further into an in vivo situation (Coleman et al. 1998), thus removing the inaccuracies introduced by correcting for the attenuation of sound as it passes through tissue. This process of derating is discussed later.

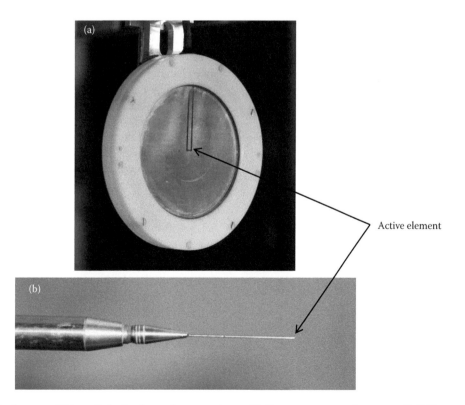

Active element

FIGURE 5.10 (a) Membrane and (b) needle hydrophones for measuring oscillating pressure waves in an acoustic field.

5.5.3.2 Pressure

The voltage detected by a hydrophone is directly related to the pressure at that point in the acoustic field, and so, in order to determine the acoustic pressure accurately, calibration is required. In the United Kingdom, these are performed at the National Physical Laboratory (Teddington, UK) using a "standard" ultrasound source, and a calibration certificate stating the sensitivity (V/Pa) over a range of frequencies for a given hydrophone is issued. A pressure field can then be mapped by scanning the hydrophone through the field and analyzing detected pulses over a region of interest. The peak positive, negative, or root-mean-squared (RMS) pressure is then the maximum of these three quantities within the field.

5.5.3.3 Intensity and Power

The I_{SPTA} of a given therapeutic field is calculable from the peak pressure as measured by a hydrophone at the focus, using Equation 5.24. For full and accurate calculation of temporal average intensities, the pulses can be analyzed as described in Section 5.5.2. Spatial averaging can be performed by plotting a two-dimensional transaxial beam profile and averaging the temporal average intensity values over the beam area. Integration over the beam area then provides a measure of output power.

Alternative methods commonly used for quick measurement of HIFU and physiotherapy transducer output involve calculation of spatially and temporally averaged intensities from measurements of radiation force or power. A number of instruments exist to measure radiation force or power, depending on the device under test. HIFU transducer output is typically measured using a radiation force balance, of which various designs have been developed depending on the application.

FIGURE 5.11 A radiation force balance for measurement of ultrasound transducer output.

The example shown in Figure 5.11 comprises a vertical rod suspended from a horizontal bar with adjustable counterbalance. At the lower end of the rod is a flat plate, onto which the focus of a HIFU device can be directed under free field conditions within a tank of water. This plate is typically at 45 degrees to the direction of propagation of the acoustic field to avoid the generation of standing waves between it and the transducer through multiple reflections of the beam. The force on the plate causes the rod to tilt. Moveable weights on the horizontal arm allow the user to counteract this force by increasing the opposing moment on the rod. By measuring the distance the weight is moved to bring the rod back to a vertical position during the exposure, the force on the plate resulting from the pressure of the acoustic field can be calculated. A number of alternative devices have been developed, based on concepts that are less user dependent and appropriate for unfocused or weakly focused transducers, such as are used for physiotherapy. For example, it is possible to measure the effective weight of an enclosed absorbing oil bath using digital scales when an acoustic field is incident from above (Sutton et al. 2003). This is directly related to the incident radiation force, and therefore the output power, of the transducer. Another example uses an electromagnet to oppose the force of an acoustic field on a sensitive balanced cone of metal. A current is induced in a coil of wire to oppose the motion of the cone, and the resulting voltage is measured and calibrated for acoustic power (Perkins 1989). A solid state device has also been developed, which relies on rapid calculation of the rate of change of temperature of an absorbing material (Zeqiri and Barrie 2008). From physiotherapy transducer power measurements, intensity is calculated by dividing over the effective radiating area (ERA) of the transducer. One disadvantage of many of the earlier devices for power measurement is their high sensitivity, making them inappropriate for measuring the very high powers produced by HIFU devices.

More recent advances have seen the development of a buoyancy phantom as a replacement for the radiation force balance. Developed by the National Physical Laboratory (NPL), this system allows the measurement of both the incident radiation from HIFU transducers on a buoyant container of castor oil, shown in Figure 5.12, and the resulting increase in buoyancy due to thermal expansion of the oil through heating by the beam. Using this type of system, two independent measures that relate to the output power of high-power HIFU transducers can be made (Shaw 2008).

5.5.3.4 Rapid Visualization of Ultrasound Fields

Although the previous measurements are highly relevant to clinical systems, they are likely only to be performed periodically as they may require long set-up or data-acquisition times, which limits their practicality. There is strong justification for developing methods of rapid visualization of the shape or pattern of the acoustic field emitted by a device, which can be used on a day-by-day basis as a qualitative check prior to conducting experiments or clinical treatments. Schlieren imaging, for example, uses either a point light source and high-speed camera or a pulsed laser and conventional CCD camera to photograph a two-dimensional representation of the pressure distribution throughout an ultrasound

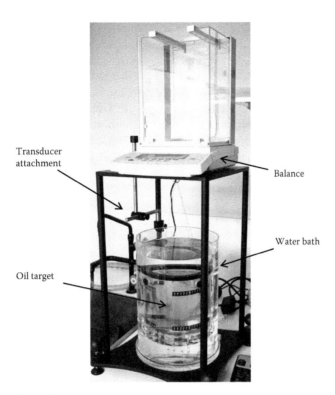

FIGURE 5.12 Buoyancy type system for making ultrasound output measurements.

field (Kudo et al. 2009). Sending light through the ultrasound field onto a screen or detector produces a form of shadow, which reveals the overall shape of the field. Scanning laser vibrometry can be used in a similar way (Zipser and Franke 2004). An alternative approach is to look at the thermal effect of an acoustic field on a known medium, as described in the following section.

5.5.3.5 Thermal Calibration

A relatively quick and simple method of assessing transducer output is to look at the temperature rise produced in a well-characterized medium. Absorbing devices have been produced that rely on thermochromic materials that change color or hue with temperature. These display a reversible change for temperatures appropriate for physiotherapy ultrasound, and can be quantified by photographing the effects and calibrating the color change through image processing (Martin and Fernandez 1997, Shaw et al. 1999). More recently, a pyroelectric device based on an absorbing piezoelectric material with a voltage output relating to temperature rise in the material has been created, as described previously in relation to power measurement (Zeqiri and Barrie 2008). Although these give a measure that relates to the temporal average intensity of an ultrasound field, it is important to make peak pressure and intensity measurements, particularly of HIFU fields, in order to fully understand the potential bioeffects.

5.5.3.6 Derating

Calibration measurements for HIFU devices are typically made under free field conditions, for example, in a large water tank where any field perturbations are minimized. As QA data are used to monitor the consistency of the output of transducers, the main important consideration is that measurements made over time are comparable. However, when relating quantities such as pressure and intensity to safety indices, it is necessary to account for attenuation of the acoustic field as it passes through the body. This is known as derating. The nominal quoted value for the attenuation of ultrasound by soft tissues in the AIUM/NEMA standard (AIUM and NEMA 1992) is 0.3 dB cm^{-1} MHz^{-1}. A known pressure or intensity at a point in water can be derated to provide the corresponding value at a particular depth in the body, provided the frequency of the ultrasound beam is known.

It is well documented that different tissue types have different ultrasound attenuation coefficients, as discussed in Section 5.4. Although having a single accepted value for the approximate attenuation coefficient in the body is useful for giving an estimate of the peak intensities and pressures incident on tissues, this one value is not representative of all scenarios, and inaccurate assumptions have the potential to affect treatment outcomes. It is therefore important to interpret such values with caution and make measurements of attenuation coefficients of exposed media or of in situ pressures where practicable.

5.5.3.7 Safety Indices

Thermal and mechanical indices have been defined and discussed briefly in Section 5.4. Equation 5.20 shows how the TIS can be calculated if the output power has been measured as described in the previous paragraphs, and if the center frequency, f_c, of the acoustic wave is known. The latter can be found using a Fourier transform of an acoustic pulse to analyze the fundamental frequency peak. More details of this and the more complex measurements for TIB and TIC are given in AIUM (1992). Using Equation 5.21, the MI can easily be calculated. Pressures measured at increasing distance from the transducer are first derated, as the peak of the rarefactional pressure in a medium may differ slightly in position from the free field peak. The new peak value is then entered into the equation along with the center frequency of the acoustic wave, calculated as described previously. These values may be of interest to users making detailed output measurements or comparing aspects of safety between devices, however, they are not used for routine QA.

References

American Institute of Ultrasound in Medicine. 1992. Acoustic output measurement and labeling standard for diagnostic ultrasound equipment. American Institute of Ultrasound in Medicine (AIUM).

American Institute of Ultrasound in Medicine and National Electrical Manufacturers Association (AIUM and NEMA). 1992. Standard for real-time display of thermal and mechanical acoustic output indices on diagnostic ultrasound equipment. AIUM/NEMA.

Anghileri, L. J. 1986. Role of Tumour Cell Membrane in Hyperthermia. In: *Hyperthermia in Cancer Treatment: Volume 1*, eds. L. J. Anghileri and J. Robert, 1–36. Boca Raton: CRC Press.

Aubry, J. F., Pernot, M., Tanter, M., Montaldo, G. and Fink, M. 2007. Réseaux de transducteurs ultrasonores: Nouvelles avancées thérapeitiques. *J. Radiol.* 88:1801–1809.

Bailey, M. R., Couret, L. N., Sapozhnikov, O. A., Khokhlova, V. A., ter Haar, G., Vaezy, S. et al. 2001. Use of overpressure to assess the role of bubbles in focused ultrasound lesion shape in vitro. *Ultrasound Med. Biol.* 27:695–708.

Ballato, A. 1995. Piezoelectricity: Old effects, new thrusts. *IEEE Trans. Ultrason. Ferroelectr. Freq. Control* 42: 916–926.

Bamber, J. C. 1986. Attenuation and absorption. In: *Physical Principles of Medical Ultrasonics*, eds. C. R. Hill, J. C. Bamber, and G. R. ter Haar, 93–166. Chichester: John Wiley & Sons.

Beard, P. C. and Mills, T. N. 1997. A miniature optical fibre ultrasonic hydrophone using a Fabry Perot polymer film interferometer. *Electronics Letters* 33:801–803.

Berriet, R. and Fleury, G. 2007. *IEEE Ultrasonics Symposium*, New York: Institute of Electrical and Electronics Engineers.

Bouchoux, G., Lafon, C., Berriet, R., Chapelon, J. Y., Fleury, G., and Cathignol, D. 2008. Dual-mode ultrasound transducer for image-guided interstitial thermal therapy. *Ultrasound Med. Biol.* 34:607–616.

Chan, Y. S., Hsu, K. Y., Kuo, C. H., Lee, S. D., Chen, S. C., Chen, W. J. et al. 2010. Using low-intensity pulsed ultrasound to improve muscle healing after laceration injury: An in vitro and in vivo study. *Ultrasound Med. Biol.* 36:743–751.

Chapelon, J.-Y., Cathiqnol, D., Cain, C., Ebbini, E., Kluiwstra, J-U., Sapozhnikov, O. A. et al. 2000. New piezoelectric transducers for therapeutic ultrasound. *Ultrasound Med. Biol.* 26:153–159.

Chen, D., Xia, R., Chen, X., Shafirstein, G., Corry, P. M., Griffin, R. J. et al. 2011. SonoKnife: Feasibility of a line-focused ultrasound device for thermal ablation therapy. *Med. Phys.* 38:4372–4385.

Chen, W. S., Brayman, A. A., Matula, T. J., and Crum, L. A. 2003. Inertial cavitation dose and hemolysis produced in vitro with or without Optison. *Ultrasound Med. Biol.* 29:725–737.

Chopra, R., Burtnyk, M., Haider, M. A., and Bronskill, M. J. 2005. Method for MRI-guided conformal thermal therapy of prostate with planar transurethral ultrasound heating applicators. *Physics Med. Biol.* 50:4957–4975.

Clement, G. T., Sun, J., Giesecke, T., and Hynynen, K. 2000. A hemisphere array for noninvasive ultrasound brain therapy and surgery. *Phys. Med. Biol.* 45: 3707–3719.

Clement, G. and Hynynen, K. 2002. A non-invasive method for focusing ultrasound through the human skull. *Phys. Med. Biol.* 47:1219–1236.

Coleman, A. J., Draquioti, E., Tiptaf, R., Shotri, N., and Saunders, J. E. 1998. Acoustic performance and clinical use of a fibre-optic hydrophone. *Ultrasound Med. Biol.* 24: 143–151.

Collis, J., Manasseh, R., Liovic, P., Tho, P., Ooi, A., Petkovic-Duran, K. et al. 2010. Cavitation microstreaming and stress fields created by microbubbles. *Ultrasonics* 50:273–279.

Coussios, C. C., Farny, C. H., Haar, G. T., and Roy, R. A. 2007. Role of acoustic cavitation in the delivery and monitoring of cancer treatment by high-intensity focused ultrasound (HIFU). *Int. J. Hyperthermia* 23:105–120.

Coussios, C. C. and Roy, R. A. 2008. Application of acoustics and cavitation to noninvasive therapy and drug delivery. *Annual Review of Fluid Mechanics* 40:395–420.

Damianou, C. and Hynynen, K. 1994. The effect of various physical parameters on the size and shape of necrosed tissue volume during ultrasound surgery. *Journal of the Acoustical Society of America* 95:1641–1649.

Daum, D. R. and Hynynen, K. 1999. A 256-element ultrasonic phased array system for the treatment of large volumes of deep seated tissue. *IEEE Trans. Ultrason. Ferroelectr. Freq. Control* 46:1254–1268.

Diederich, C. J., Nau, W. H., Ross, A. B., Tyreus, P. D., Butts, K., Reike, V. et al. 2004a. Catheter-based ultrasound applicators for selective thermal ablation: Progress towards MRI-guided applications in prostate. *Int. J. Hyperthermia* 20:739–756.

Diederich, C. J., Stafford, R. J., Nau, W. H., Burdette, E. C., Price, R. E., and Hazle, J. D. 2004b. Transurethral ultrasound applicators with directional heating patterns for prostate thermal therapy: In vivo evaluation using magnetic resonance thermometry. *Med. Phys.* 31:405–413.

Duck, F. A. 1990. *Physical Properties of Tissue: A Comprehensive Reference Book*. London: Academic Press Limited.

Duck, F. A. 2008. Hazards, risks and safety of diagnostic ultrasound. *Medical Engineering and Physics* 30:1338–1348.

Duck, F. A. 2009. Acoustic dose and acoustic dose-rate. *Ultrasound Med. Biol.* 35:1679–1685.

Duck, F. A. 2011. A new definition for acoustic dose. *Journal of Physics: Conference Series* 279.

Dupenloup, F., Chapelon, J. Y., Cathignol, D. J., and Sapozhnikov, O. A. 1996. Reduction of the grating lobes of annular arrays used in focused ultrasound surgery. *IEEE Trans. Ultrason. Ferroelectr. Freq. Control* 43:991–998.

Dyson, M. and Pond, J. B. 1973. Biological effects of therapeutic ultrasound. *Rheumatology and Rehabilitation* 12:209–213.

Ebbini, E. S. and Cain, C. A. 1989. Multiple-focus ultrasound phased array pattern synthesis: Optimal driving signal distributions for hyperthermia. *IEEE Trans. Ultrason. Ferroelectr. Freq. Control* 36:540–548.

Ebbini, E. S. and Cain, C. A. 1991. A spherical-section ultrasound phased-array applicator for deep localized hyperthermia. *IEEE Trans. Biomed. Eng.* 38:634–643.

Fan, X. and Hynynen, K. 1995. Control of the necrosed tissue volume during noninvasive ultrasound surgery using a 16 element phased array. *Med. Phys.* 22:297–308.

Fan, X. and Hynynen, K. 1996. Ultrasound surgery using multiple sonications—Treatment time considerations. *Ultrasound Med. Biol.* 22: 471–482.

Filonenko, E. A., Gavrilov, L. R., Khokhlova, V. A., and Hand, J. W. 2004. Heating of biological tissues by two-dimensional phased arrays with random and regular element distributions. *Acoust. Phys.* 50:222–231.

Foster, F. S., Ryan, L. K., and Turnbull, D. H. 1991. Characterization of lead zirconate titanate ceramics for use in high frequency (20–80 MHz) transducers. *IEEE Trans. Ultrason. Ferroelectr. Freq. Control* 38:446–453.

Freitas, T. P., Gomes, M., Fraga, D. B., Freitas, L. S., Rezin, G. T., Santos, P. M. et al. 2010. Effect of therapeutic pulsed ultrasound on lipoperoxidation and fibrogenesis in an animal model of wound healing. *Journal of Surgical Research* 161: 168–171.

Fry, W. J. and Dunn, F. 1962. Ultrasound: Analysis and experimental methods in biological research. In *Physical Techniques in Biological Research Vol. 4*, ed. W. L. Nastuk, 261–394. New York: Academic Press.

Gavrilov, L. R., Hand, J. W., Abel, P., and Cain, C. A. 1997. A method of reducing grating lobes associated with an ultrasound linear phased array intended for transrectal thermotherapy. *IEEE Trans. Ultrason. Ferroelectr. Freq. Control* 44:1010–1017.

Gavrilov, L. R., Hand, J. W., and Yushina, I. G. 2000. Two-dimensional phased arrays for application in surgery: Scanning by several focuses. *Acoust. Phys.* 46:551–558.

Goss, S. A., Frizzell, L. A., and Dunn, F. 1979. Ultrasonic absorption and attenuation in mammalian tissues. *Ultrasound Med. Biol.* 5:181–186.

Goss, S. A., Frizzell, L. A., Kouzmanoff, J. T., Barich, J. M., and Yang, J. M. 1996. Sparse random ultrasound phased array for focal surgery. *IEEE Trans. Ultrason. Ferroel. Freq. Control.* 43:1111–1121.

Guerino, M. R., Santi, F. P., Silveira, R. F., and Luciano, E. 2008. Influence of ultrasound and physical activity on bone healing. *Ultrasound Med. Biol.* 34:1408–1413.

Haller, M. I. and Khuri-Yakub, B. 1994. A surface micromachined electrostatic ultrasonic air transducer. *Proceedings of the IEEE Ultrasonics Symposium*, 1241–1244.

Hand, J. W. and Gavrilov, L. R. August 23, 2000. Ultrasound transducer array. GB patent GB2347043A.

Hand, J. W. and Gavrilov, L. R. December 3, 2002. Arrays of quasi-randomly distributed ultrasound transducers. US patent 6488630.

Hand, J. W., Shaw, A., Sadhoo, N., Rajagopal, S., Dickinson, R. J., and Gavrilov, L. R. 2009. Initial testing of a prototype phased array device for delivery of high intensity focused ultrasound (HIFU). *Phys. Med. Biol.* 54:5675–5693.

Hill, C. R. 1994. Optimum acoustic frequency for focused ultrasound surgery. *Ultrasound Med. Biol.* 20:271–277.

Holt, R. G. and Roy, R. A. 2001. Measurements of bubble-enhanced heating from focused, MHz-frequency ultrasound in a tissue-mimicking material. *Ultrasound Med. Biol.* 27:1399–1412.

Huu, I. and Hartmann, P. 1982. Deep and local heating by an ultrasound phased array applicator. *Proceedings of the 1982 Ultrasonics Symposium*, 735–738.

Hynynen, K. 1986. Generation of ultrasonic fields and the acoustic properties of tissues. In: *Hyperthermia*, eds. D. J. Watmough and W. M. Ross, 76–98. Glasgow: Blackie & Son Ltd.

Hynynen, K., Clement, G. T., McDannold, N., Vykhodtseva, N., King, R., White, P. J. et al. 2004. 500-element ultrasound phased array system for noninvasive focal surgery of the brain: A preliminary rabbit study with ex vivo human skulls. *Magn. Reson. Med.* 52:100–107.

Hynynen, K., McDannold, N., Clement, G., Jolesz, F. A., Zadicario, E., Killiany, R. et al. 2006. Pre-clinical testing of a phased array ultrasound system for MRI-guided noninvasive surgery of the brain—A primate study. *Eur. J. Radiol.* 59: 149–156.

Ibbini, M. S. and Cain, C. A. 1990. The concentric-ring array for ultrasound hyperthermia: Combined mechanical and electrical scanning. *Int. J. Hyperthermia* 6:401–419.

Jensen, C. R., Ritchie, R. W., Gyongy, M., Collin, J. R., Leslie, T., and Coussios, C. 2011. Spatiotemporal monitoring of High-Intensity Focused Ultrasound therapy with passive acoustic mapping. *Radiology* 262:252–261.

Khuri-Yakub, P. and Oralkan, O. 2011. Capacitive micromachined ultrasonic transducers for medical imaging and therapy. *J. Micromech. Microeng.* 21:54004–54015.

Kinsey, A. M., Diederich, C. J., Rieke, V., Nau, W. H., Pauly, K. B., Bouley, D. et al. 2008. Transurethral ultrasound applicators with dynamic multi-sector control for prostate thermal therapy: In vivo evaluation under MR guidance. *Med. Phys.* 35:2081–2093.

Koch, C. 1996. Coated fiber-optic hydrophone for ultrasonic measurement. *Ultrasonics* 34:687–689.

Kossoff, G. 1979. Analysis of focusing action of spherically curved transducers. *Ultrasound in Med & Biol.* 5:359–365.

Kudo, N., Suzuki, R., and Yamamoto, K. 2009. A novel method for visualization of acoustic fields of medical ultrasound equipment and its application for tomographic reconstruction of a 3D ultrasound field. *Ultrasound Med. Biol.* 35:S235.

Lafon, C., De, L., Theillere, Y., Prat, F., Chapelon, J. Y., and Cathignol, D. 2002. Optimizing the shape of ultrasound transducers for interstitial thermal ablation. *Med. Phys.* 29: 290–297.

Lafon, C., Koszek, L., Chesnais, S., Theillere, Y., and Cathignol, D. 2004. Feasibility of a transurethral ultrasound applicator for coagulation in prostate. *Ultrasound Med. Biol.* 30:113–122.

Lafon, C., Melodelima, D., Salomir, R., and Chapelon, J. Y. 2007. Interstitial devices for minimally invasive thermal ablation by high-intensity ultrasound. *Int. J. Hyperthermia* 23:153–163.

Lewin, P. A. and Schafer, M. E. 1988. Wide-band piezoelectric polymer acoustic sources. *IEEE Trans Ultrason Ferroelectr. Freq. Control* 35:175–184.

Lim, T. K., Zhou, Y., Lin, Y., Yip, Y. M., and Lam, Y. L. 1999. Fiber optic acoustic hydrophone with double Mach–Zehnder interferometers for optical path length compensation. *Optics Communications* 159: 301–308.

Liu, H. L., Chen, W. S., Chen, J. S., Shih, T. C., Chen, Y. Y., and Lin, W. L. 2006. Cavitation-enhanced ultrasound thermal therapy by combined low- and high-frequency ultrasound exposure. *Ultrasound Med. Biol.* 32:759–767.

Lu, M., Wan, M., Xu, F., Wang, F., and Zhong, H. 2005. Focused beam control for ultrasound surgery with spherical-section phased array: Sound field calculation and genetic optimization algorithm. *IEEE Transactions on Ultrasonics, Ferroelectrics and Frequency Control* 52:1270–1290.

Makin, I. R. S., Mast, T. D., Faidi, W., Runk, M. M., Barthe, P. G., and Slayton, M. H. 2005. Miniaturized ultrasound arrays for interstitial ablation and imaging. *Ultrasound Med. Biol.* 31:1539–1550.

Malizos, K. N., Hantes, M. E., Protopappas, V., Papachristos, A. et al. 2006. Low-intensity pulsed ultrasound for bone healing: An overview. *Injury* 37:S56–S62.

Marchal, C. 1992. Clinical experience with ultrasound thermotherapy. *Ultrasonics* 30:139–141.

Marks, R., Ghanagaraja, S., and Ghassemi, M. 2000. Ultrasound for osteo-arthritis of the knee: A systematic review. *Physiotherapy* 86:452–463.

Martin, K. and Fernandez, R. 1997. A thermal beam-shape phantom for ultrasound physiotherapy transducers. *Ultrasound Med. Biol.* 23:1267–1274.

Mast, T. D., Barthe, P. G., Makin, I. R., Slayton, M. H., Karunakaran, C. P., Burgess, M. T. et al. 2011. Treatment of rabbit liver cancer in vivo using miniaturised image-ablate ultrasound arrays. *Ultrasound Med. Biol.* 37:1609–1621.

Melodelima, D., Chapelon, J. Y., Theillere, Y., and Cathignol, D. 2004. Combination of thermal and cavitation effects to generate deep lesions with an endocavitary applicator using a plane transducer: ex vivo studies. *Ultrasound Med. Biol* 30:103–111.

Melodelima, D., N'Djin, W. A., Parmentier, H., Chesnais, S., Rivoire, M., and Chapelon, J. Y. 2009. Thermal ablation by high intensity ultrasound using a toroid transducer increases the coagulated volume. *Ultrasound Med. Biol.* 35:425–435.

Mohseni-Bandpei, M. A., Critchley, J., Staunton, T., and Richardson, B. 2006. A prospective randomised controlled trial of spinal manipulation and ultrasound in the treatment of chronic low back pain. *Physiotherapy* 92:34–42.

Morris, P., Hurrell, A., Shaw, A., Zhang, E., and Beard, P. 2009. A Fabry–Pérot fiber-optic ultrasonic hydrophone for the simultaneous measurement of temperature and acoustic pressure. *Journal of the Acoustical Society of America* 125:3611–3622.

O'Donnell, M., Busse, L. J., and Miller, J. G. 1981. Piezoelectric transducers. In *Methods of Experimental Physics, Vol. 19: Ultrasonics,* ed. P. D. Edmonds, 29–65. New York: Academic Press.

Ohigashi, H. 1976. Electromechanical properties of polarized polyvinyledine fluoride films as studied by the piezoelectric resonance method. *J. Appl. Phys.* 47:949–955.

O'Neill, H. T. 1949. Theory of focusing radiators. *J. Acoust. Soc. Am.* 21:516–526.

Ozgonenel, L., Aytekin, E., and Durmusoglu, G. 2009. A double-blind trial of clinical effects of therapeutic ultrasound in knee osteoarthritis. *Ultrasound Med. Biol.* 35:44–49.

Penttinen, A. and Luukkala, M. 1976. The impulse response and pressure nearfield of a curved ultrasonic radiator. *J. Phys. D: Appl. Phys.* 9:1547–1557.

Perkins, M. A. 1989. A versatile force balance for ultrasound power measurement. *Phys. Med. Biol.* 34:1645.

Pernot, M., Aubry, J. F., Tanter, M., Thomas, J. L., and Fink, M. 2003. High power transcranial beam steering for ultrasonic brain therapy. *Phys. Med. Biol.* 48:2577–2589.

Pernot, M., Aubry, J-F., Tanter, M., Boch, A. L., Kujas, M., and Fink, M. 2005. Adaptive focusing for ultrasonic transcranial brain therapy: First in vivo investigation on 22 sheep. *AIP Conference Proceedings* 754:174–177.

Ross, A. B., Diederich, C. J., Nau, W. H., Gill, H., Bouley, D. M., Daniel, B. et al. 2004. Highly directional transurethral ultrasound applicators with rotational control for MRI-guided prostatic thermal therapy. *Phys. Med. Biol.* 49:189–204.

Ross, A. B., Diederich, C. J., Nau, W. H., Rieke, V., Butts, R. K., Sommer, G. et al. 2005. Curvilinear transurethral ultrasound applicator for selective prostate thermal therapy. *Med. Phys.* 32:1555–1565.

Sapareto, S. A. and Dewey, W. C. 1984. Thermal dose determination in cancer therapy. *International Journal of Radiation Oncology Biology Physics* 10:787–800.

Sasaki, K., Azuma, T., Kawabata, K-I., Shimoda, M., Kokue, E-I., and Umemura, S-I. 2003. Effect of split-focus approach on producing larger coagulation in swine liver. *Ultrasound Med. Biol.* 29:591–599.

Saunders, L. 2003. Laser versus ultrasound in the treatment of supraspinatus tendinosis: Randomised controlled trial. *Physiotherapy* 89:365–373.

Scott, R. S. 1986. Hyperthermia in combination with radiotherapy. In: *Hyperthermia in Cancer Treatment: Volume 3*, eds. L. J. Anghileri and J. Robert, 133–168. Boca Raton: CRC Press.

Shaw, A., Pay, N. M., Preston, R. C., and Bond, A. D. 1999. Proposed standard thermal test object for medical ultrasound. *Ultrasound Med. Biol.* 25:121–132.

Shaw, A., Pay, N. M., Preston, R. C., and Bond, A. D. 2008. A buoyancy method for the measurement of total ultrasound power generated by HIFU transducers. *Ultrasound Med. Biol.* 34:1327–1342.

Silk, M. G. 1984. *Ultrasonic Transducers for Nondestructive Testing.* Bristol: Adam Hilger.

Smith, W. A. 1989. The role of piezocomposites in ultrasonic transducers. *IEEE Ultrasonics Symposium* 755–766.

Sun, J. and Hynynen, K. 1998. The potential of transskull ultrasound therapy and surgery using the maximum available skull surface area. *J. Acoust. Soc. Am.* 104: 2519–2527.

Sutton, Y., Shaw, A., and Zeqiri, B. 2003. Measurement of ultrasonic power using an acoustically absorbing well. *Ultrasound Med. Biol.* 29:1507–1513.

Takahashi, N., Yoshimura, K., Takahashi, S., and Imamura, K. 2000. Development of an optical fiber hydrophone with fiber Bragg grating. *Ultrasonics* 38: 581–585.

ter Haar, G. R. 1986a. Therapeutic and surgical applications. In: *Physical Principles of Medical Ultrasonics*, eds. C. R. Hill, J. C. Bamber, and G. R. ter Haar, 407–456. Chichester: John Wiley & Sons.

ter Haar, G. R. 1986b. Effects of increased temperature on cells, on membranes and on tissues. In: *Hyperthermia*, eds. D. J. Watmough and W. M. Ross, 14–41. Glasgow: Blackie & Son Ltd.

Wong, R. S. L. and Dewey, W. C. 1986. Effect of hyperthermia on DNA synthesis. In: *Hyperthermia in Cancer Treatment: Volume 1*, eds. L. J. Anghileri and J. Robert, 79–92. Boca Raton: CRC Press.

Wong, S. H., Watkins, R. D., Kupnik, M., Pauly, K. B., and Khuri-Yakub, B. T. 2008. Feasibility of MR-temperature mapping of ultrasonic heating from a CMUT. *IEEE Trans. Ultrason. Ferroelectr. Freq. Control* 55:811–818.

Wu, J. 2001. Shear stress in cells generated by ultrasound. *Progress in Biophysics and Molecular Biology* 93:363–373.

Wu, J. 2002. Theoretical study on shear stress generated by microstreaming surrounding contrast agents attached to living cells. *Ultrasound Med. Biol.* 28:125–129.

Zeqiri, B. and Barrie, J. 2008. Evaluation of a novel solid-state method for determining the acoustic power generated by physiotherapy ultrasound transducers. *Ultrasound Med. Biol.* 34:1513–1527.

Zipser, L. and Franke, H. 2004. Laser-scanning vibrometry for ultrasonic transducer development. *Sensors and Actuators A: Physical* 110:264–268.

Numerical Modeling for Simulation and Treatment Planning of Thermal Therapy: Ultrasound

Robert J. McGough
Michigan State University

6.1 Introduction

Pressure fields generated by single and multiple element transducers are routinely simulated during the initial design of an ultrasound applicator and also throughout the subsequent characterization and optimization of power depositions, temperature distributions, and thermal doses produced by these thermal therapy applicators. Numerical models describe the diffraction of ultrasound produced by single transducers, fixed-phase multiple transducer configurations, and ultrasound phased arrays. Nonlinear effects are also incorporated into some of these numerical models. The resulting power deposition then provides the input to the bioheat transfer equation. Temperatures are computed for most applications, and thermal doses are also calculated, especially for simulations of ablation therapy. In more advanced models, the effects of intervening tissue heating are considered, beamforming algorithms are evaluated, and more complicated issues such as patient anatomy and tissue inhomogeneities are also included. Patient treatment planning then combines several of these models in an effort to optimize the temperature distribution or thermal dose in the tumor while sparing sensitive normal tissues.

6.2 Models of Ultrasound Propagation

Simulated pressures generated by ultrasound applicators are often computed with a linear propagation model. These numerical models capture the effects of diffraction in the near-field region. Pressures are typically calculated with the Rayleigh-Sommerfeld integral or with expressions that are closely related to the Rayleigh-Sommerfeld integral, where each approach has an associated numerical accuracy, computation time, and algorithmic complexity. These numerical models are applicable to the three transducer geometries that are most often encountered in thermal therapy, namely the circular piston, the rectangular piston, and the spherical shell. Other pressure simulations include nonlinear propagation effects, especially when higher intensities are generated by a focused ultrasound transducer or a large aperture phased array. Some of these numerical models for linear and nonlinear ultrasound propagation are implemented in publicly available software. After the pressure is computed with one of these numerical models, power depositions are also calculated for subsequent bioheat transfer simulations.

6.2.1 Rayleigh-Sommerfeld Integral

The Rayleigh-Sommerfeld integral is the most common model for linear ultrasound propagation in simulations of thermal therapy. The numerical formulas derived from the Rayleigh-Sommerfeld integral superpose contributions from a point source surrounded by a rigid baffle radiating into an infinite half-space. The frequency-domain Green's function describing the contribution from each point source is $e^{-jkR}/(2\pi R)$, where the 2π in the denominator indicates a baffled point source radiating into an infinite half-space (as opposed to 4π in the denominator, which indicates an unbaffled source radiating in all directions in an infinite space). The main advantages of the Rayleigh-Sommerfeld integral are: (1) this intuitive formulation is based on the superposition of Green's functions; and (2) this integral formula is easily evaluated with a computer program. Disadvantages of the Rayleigh-Sommerfeld integral include relatively long computation times and large errors in the near-field region due to slow convergence.

The Rayleigh-Sommerfeld integral describing the pressure radiated from a finite transducer aperture driven by a time-harmonic input signal is

$$p(x,y,z,t) = \frac{j\omega\rho_0}{2\pi} U_0 e^{j\omega t} \int_S \frac{e^{-jk|\mathbf{r}-\mathbf{r}'|}}{|\mathbf{r}-\mathbf{r}'|} dS, \quad (6.1)$$

where p represents the pressure, (x,y,z) are the observation coordinates in Cartesian space, t is the time variable, j is $\sqrt{-1}$, ω is the radian frequency of the excitation, ρ_0 is the density of the medium, U_0 represents the uniform normal particle velocity on the face of the transducer, $e^{j\omega t}$ denotes a time-harmonic excitation, S indicates that the integration is performed over the surface of the radiating source, $\mathbf{r}$ represents the coordinates of the observation point, $\mathbf{r}'$ represents the coordinates of the source point, $|\mathbf{r}-\mathbf{r}'|$ is the distance from a source point on the radiating aperture to the observation coordinates, k is the wave number, and the integrand is the frequency-domain Green's function for a point source surrounded by a rigid baffle. The formula in Equation 6.1 admits direct discretization and evaluation for various piston shapes, including the circular piston, the rectangular piston, and the spherical shell.

6.2.1.1 Circular Piston

The Rayleigh-Sommerfeld integral for a circular transducer (Figure 6.1) excited with a time-harmonic and spatially uniform normal particle velocity U_0 is given by

$$p(x,y,z,t) = \frac{j\omega\rho_0}{2\pi} U_0 e^{j\omega t} \int_0^{2\pi} \int_0^a \frac{e^{-jk|\mathbf{r}-\mathbf{r}'|}}{|\mathbf{r}-\mathbf{r}'|} \sigma' d\sigma' d\theta', \quad (6.2)$$

where the source points are given in cylindrical coordinates and the observation points are given in Cartesian coordinates with $|\mathbf{r}-\mathbf{r}'| = \sqrt{(x-\sigma'\cos\theta')^2 + (y-\sigma'\sin\theta')^2 + z^2}$. In Equation 6.2, primed coordinates are source coordinates, unprimed coordinates are observation coordinates, σ' is the radial coordinate of the source point, θ' is the angle between the source point and the x axis, and $|\mathbf{r}-\mathbf{r}'|$ represents the distance from each sampled source point to an observation point. The normal of the circular piston is coincident with the positive z axis, and the center of the piston is located at the origin of the polar coordinate system.

Pressures in the near-field region are often numerically evaluated with the discretized version (Zemanek 1971) of the Rayleigh-Sommerfeld integral in Equation 6.2 via

$$p(x,y,z,t) = \frac{j\omega\rho_0}{2\pi} U_0 e^{j\omega t} \sum_{i_{\sigma'}=1}^{N_{\sigma'}} \sum_{i_{\theta'}=1}^{N_{\theta'}} \frac{e^{-jk|\mathbf{r}-\mathbf{r}'|}}{|\mathbf{r}-\mathbf{r}'|} \sigma' \Delta\sigma' \Delta\theta', \quad (6.3)$$

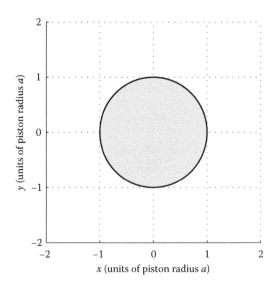

FIGURE 6.1 Circular piston with radius a centered in the $z = 0$ plane at $(x,y,z) = (0,0,0)$. The normal at the center of the circular piston is coincident with the z axis.

where $\Delta\sigma' = a/N_{\sigma'}$ and $\Delta\theta' = 2\pi/N_{\theta'}$ are defined for numerical integration with the midpoint rule, $N_{\sigma'}$ is the number of abscissas in the radial direction, and $N_{\theta'}$ is the number of abscissas in the θ' direction. The double sum in Equation 6.3 evaluates contributions from the Green's function for a point source in an infinite baffle, where the distance $|\mathbf{r}-\mathbf{r}'|$ is evaluated from the center $\mathbf{r}'$ of each sector on the piston face to an observation point $\mathbf{r}$. The contribution from the Green's function is then multiplied by the area $\sigma'\Delta\sigma'\Delta\theta'$ of each sector and superposed.

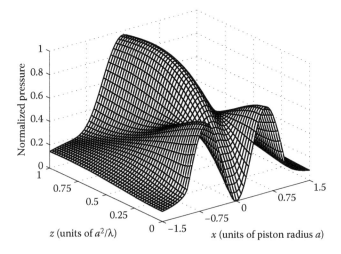

FIGURE 6.2 Simulated near-field pressure obtained with the Rayleigh-Sommerfeld integral for a circular piston with radius $a = \lambda$. The Rayleigh-Sommerfeld integral achieves a maximum error of 1% in 177.19 seconds when evaluated with $N = N_{\sigma'} = N_{\theta'} = 212$ on this 121-point by 101-point grid.

For on-axis near-field pressures generated by a circular piston, Equation 6.2 reduces to an analytical expression without any approximations. The on-axis time-harmonic pressure generated by a circular piston excited with uniform normal particle velocity U_0 is given by

$$p(z,t) = \rho_0 c U_0 e^{j\omega t} \left(e^{-jkz} - e^{-jk\sqrt{a^2+z^2}} \right). \quad (6.4)$$

In Equation 6.4, c represents the speed of sound and z indicates the distance along the positive z axis with $x = y = 0$.

In the far-field region, the approximate pressure generated by a circular piston is given by

$$p(\theta, R, t) \approx \frac{j\omega\rho_0}{2R} a^2 U_0 e^{j\omega t - jkR} \left[\frac{2J_1(ka\sin\theta)}{ka\sin\theta} \right], \quad (6.5)$$

where $R = \sqrt{x^2 + y^2 + z^2}$ is the distance from the center of the piston to the observation point (x,y,z), $\theta = \tan^{-1}(\sqrt{x^2 + y^2}/z)$ is the angle between the element normal (which is coincident with the positive z axis) and the vector that connects the center of the piston to the observation point (x,y,z), and $J_1(\cdot)$ is the Bessel function of the first kind of order one. The expression in square brackets is also known as the "jinc" function (analogous to the "sinc" function associated with rectangular apertures), where the factor of 2 in the numerator normalizes the result such that the peak value of the jinc function is equal to one when the argument $ka\sin\theta = 0$ (i.e., when $\theta = 0$). Furthermore, when $\text{jinc}(x) = 2J_1(x)/x$ as above, the zeros of $\text{jinc}(x)$ away from the peak value at $x = 0$ are the same as the zeros of $J_1(x)$.

Near-field and far-field pressures are often plotted with axial distances normalized with respect to the far-field transition distance, which is defined as

$$\frac{d^2}{4\lambda} \quad (6.6)$$

with $d = 2a$ for a circular transducer. The far-field transition distance is often misinterpreted as the distance at which the far-field approximation in Equation 6.5 is accurate. In fact, the far-field approximation in Equation 6.5 is only accurate at distances that satisfy $R \gg d^2/(4\lambda)$, and other methods for computing the pressure are required closer to the piston.

An example of the near-field pressure computed with the Rayleigh-Sommerfeld integral for a circular piston with radius $a = \lambda$ is shown in Figure 6.2. The circular piston is located in the $z = 0$ plane and centered at $(x,y,z) = (0,0,0)$ as illustrated in Figure 6.1. The pressure is calculated on a 121-point by 101-point grid with a simplified version of Equation 6.3 that exploits the symmetry of the circular piston by computing the integral from 0 to π and then doubling the result. Subject to the restriction $N = N_{\sigma'} = N_{\theta'}$, the smallest value of N that achieves a normalized error less than or equal to 1% throughout the entire grid defined for this pressure calculation is $N = 212$. The computational grid extends out to the far-field transition distance a^2/λ (defined in Equation 6.6) in the axial direction, and the grid extends out to $x = \pm 1.5a$ laterally. Numerical calculations of the

Rayleigh-Sommerfeld integral for a circular piston are evaluated on a desktop computer running the Windows XP operating system with a 2.4GHz Intel Core2 CPU 6600 and 4GB of RAM. The Rayleigh-Sommerfeld integral is implemented in optimized C++ code that is called from MATLAB® as an executable program through the MATLAB® MEX-file interface. The computation time for a 121-point by 101-point grid with $N = 212$ is 177.19 seconds (just under 3 minutes).

6.2.1.2 Rectangular and Square Pistons

For a sinusoidally excited rectangular aperture with uniform normal particle velocity U_0, the Rayleigh-Sommerfeld integral in Equation 6.1 becomes

$$p(x,y,z,t) = \frac{j\omega\rho_0}{2\pi} U_0 e^{j\omega t} \int_{-a}^{a} \int_{-b}^{b} \frac{e^{-jk|\mathbf{r}-\mathbf{r}'|}}{|\mathbf{r}-\mathbf{r}'|} dx'dy', \quad (6.7)$$

for calculations in Cartesian coordinates where $|\mathbf{r}-\mathbf{r}'| = \sqrt{(x-x')^2 + (y-y')^2 + z^2}$ is the distance from each point on the radiating source to the observation point (x,y,z), and primed coordinates (x',y') represent the coordinates of the radiating aperture. In Equation 6.7, the width of the rectangular piston is $2a$, and the height is $2b$. The normal at the center of the rectangular piston is coincident with the positive z axis, and the center of the piston is located at the origin of the Cartesian coordinate system. Numerical calculations of the pressure generated by a rectangular source in the near-field region replace the double integral with a double sum, and numerical evaluations of these integrals with the midpoint rule replace dx' and dy' with $\Delta x' = 2a/N_{x'}$ and $\Delta y' = 2b/N_{y'}$ (where $N_{x'}$ is the number of samples, or abscissas, in the x direction, and $N_{y'}$ is the number of abscissas in the y direction), yielding

$$p(x,y,z,t) = \frac{j\omega\rho_0}{2\pi} U_0 e^{j\omega t} \sum_{i_{x'}=1}^{N_{x'}} \sum_{i_{y'}=1}^{N_{y'}} \frac{e^{-jk|\mathbf{r}-\mathbf{r}'|}}{|\mathbf{r}-\mathbf{r}'|} \Delta x' \Delta y', \quad (6.8)$$

where $|\mathbf{r}-\mathbf{r}'|$ is the distance from each sampled source point to each observation point. In Equation 6.8, the $\Delta x'\Delta y'$ represents the area of a small rectangular patch on the piston face. The source coordinate $\mathbf{r}'$ indicates the location of the center of this patch.

In the far-field region, the approximate pressure generated by a rectangular piston is given by

$$p(x,y,z,t) \approx \frac{j\omega\rho_0}{\pi R} 2ab U_0 e^{j\omega t - jkR} \text{sinc}\left[\frac{kax}{R}\right] \text{sinc}\left[\frac{kby}{R}\right], \quad (6.9)$$

where $R = \sqrt{x^2 + y^2 + z^2}$ and the sinc function is defined as $\text{sinc}(x) = \sin(x)/x$. This definition of the sinc function is normalized such that the peak value is equal to one at $x = 0$ and the zeros of this function are coincident with the zeros of the sine function elsewhere. The product of sinc functions in Equation 6.9 is the far-field directional factor for a rectangular element, and the jinc function in Equation 6.5 is the far-field directional factor for a circular element. Alternate forms of the sinc and jinc functions

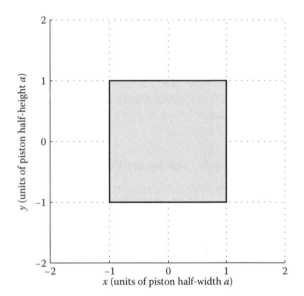

FIGURE 6.3 Square $2a$ by $2a$ piston centered in the $z = 0$ plane at $(x,y,z) = (0,0,0)$. The normal at the center of the square piston is coincident with the z axis.

given in Bracewell (2000) have convenient Fourier and Hankel transforms, respectively, for signal processing applications, but the sinc and jinc expressions used in Equations 6.9 and 6.5 are more commonly accepted for the analysis of diffracted pressure fields generated by finite apertures.

An example of the near-field pressure computed with the Rayleigh-Sommerfeld integral for the $2a$ by $2a$ square piston illustrated in Figure 6.3 is shown in Figure 6.4 for $a = \lambda$. This pressure calculation evaluates Equation 6.8 on a 121-point by 101-point grid. The smallest value of $N = N_{x'} = N_{y'}$ that achieves

a normalized error less than or equal to 1% throughout the entire grid is $N = 176$. The computational grid extends out to a^2/λ in the axial direction (which is equal to the far-field transition distance defined in Equation 6.6 with $d = 2a$), and the lateral grid boundary is located at $x = \pm 1.5a$. Numerical calculations of the Rayleigh-Sommerfeld integral for this square piston are evaluated on the same desktop computer described previously for near-field pressure simulations with a circular piston. The computation time for a 121-point by 101-point grid with $N = 176$ is 168.74 seconds (just under 3 minutes).

6.2.1.3 Spherical Shell

The spherically focused piston or spherical shell, which enhances the peak intensity generated by a single ultrasound transducer through geometrical focusing, is an ideal transducer shape for thermal therapy. An example of a spherical shell with radius a and radius of curvature $R = a\sqrt{2}$ is shown in Figure 6.5. The mathematical description of this curved transducer geometry is obtained when a plane slices through a sphere with radius R. The intersection of the sphere and the plane describes a circle with radius a, and the smaller of the two pieces is the spherical cap or spherical shell. For a spherically focused source with uniform normal particle velocity U_0, aperture radius a, and radius of curvature R, the Rayleigh-Sommerfeld integral for a time-harmonic excitation is given by

$$p(x,y,z,t) = \frac{j\omega\rho_0}{2\pi} U_0 e^{j\omega t} \int_0^{2\pi} \int_{\pi - \sin^{-1}(a/R)}^{\pi} \frac{e^{-jk|\mathbf{r}-\mathbf{r}'|}}{|\mathbf{r}-\mathbf{r}'|} R^2 \sin\phi' d\phi' d\theta',$$

(6.10)

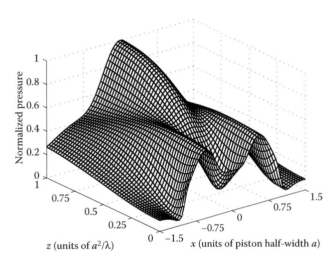

FIGURE 6.4 Near-field pressure computed with the Rayleigh-Sommerfeld integral for a square $2a$ by $2a$ piston with $a = \lambda$. The Rayleigh-Sommerfeld integral achieves a maximum error of 1% in 168.74 seconds when evaluated with $N_{x'} = N_{y'} = 176$ on this 121-point by 101-point grid.

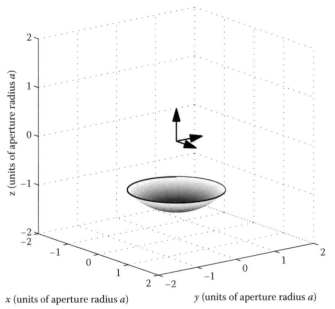

FIGURE 6.5 Spherical shell with aperture radius a and radius of curvature $R = a\sqrt{2}$. The spherical shell intersects the z axis at $z = -R$, the geometric center is located at $(x,y,z) = (0,0,0)$, and the normal evaluated at the center of the spherical shell is coincident with the z axis.

where the integral is evaluated in spherical coordinates and the pressure is computed in Cartesian coordinates with $|\mathbf{r} - \mathbf{r'}| = \sqrt{(x - R\cos\theta'\sin\phi')^2 + (y - R\sin\theta'\sin\phi')^2 + (z - R\cos\phi')^2}$. In Equation 6.10, primed coordinates are source coordinates, unprimed coordinates are observation coordinates, (R, θ', ϕ') is the spherical coordinate of a source point, and $|\mathbf{r} - \mathbf{r'}|$ represents the distance from each sampled source point to the observation point. The limits of integration for the ϕ' coordinate are selected such that the center of the spherical shell intersects the z axis at $z = -R$ (i.e., on the negative z axis), and the origins of both the source and observer coordinate systems are located at the geometric focus of the spherical shell. The spherical coordinate system defined for a spherically focused transducer is shifted relative to the coordinate systems defined in the center of the piston face for circular and rectangular pistons in Equation 6.2 and Equation 6.7, respectively.

The discretized version of the Rayleigh-Sommerfeld integral in Equation 6.10 is

$$p(x,y,z,t) = \frac{j\omega\rho_0}{2\pi} U_0 e^{j\omega t} \sum_{i_{\phi'}=1}^{N_{\phi'}} \sum_{i_{\theta'}=1}^{N_{\theta'}} \frac{e^{-jk|\mathbf{r}-\mathbf{r'}|}}{|\mathbf{r}-\mathbf{r'}|} R^2 \sin\phi' \Delta\phi' \Delta\theta', \quad (6.11)$$

where $\Delta\phi' = \sin^{-1}(a/R)/N_{\phi'}$, $\Delta\theta' = \pi/N_{\theta'}$, $N_{\phi'}$ is the number of abscissas in the ϕ' direction, and $N_{\theta'}$ is the number of abscissas in the θ' direction. In Equation 6.11, $R^2\sin\phi'\Delta\phi'\Delta\theta'$ represents the area of a small patch on the spherical shell, the source coordinate $\mathbf{r'}$ indicates the location of the center of this patch, and $\mathbf{r}$ indicates the location of the observation point.

The spherical shell also admits exact expressions for the on-axis pressure as well as approximate expressions for the pressure in the focal plane. The on-axis time-harmonic pressure generated by a spherical shell excited with uniform normal particle velocity U_0 is given by

$$p(r=0,z,t) = \rho_0 c U_0 \frac{R}{z} e^{j\omega t} \left(e^{-jk\sqrt{z^2 + 2z\sqrt{R^2-a^2}+R^2}} - e^{-jk|R+z|} \right) \quad (6.12)$$

for all points where $r = 0$ and $z \neq 0$ (i.e., at all points on the axis of symmetry other than the geometric focus) when pressures are evaluated in cylindrical coordinates (r,z) with the radial coordinate defined as $r = \sqrt{x^2 + y^2}$. The expression for the on-axis pressure in Equation 6.12 is exactly equivalent to the on-axis expression in (O'Neil 1949). At the geometric focus where $r = 0$ and $z = 0$, the time harmonic pressure generated by a spherical shell is

$$p(r=0,z=0,t) = j\omega\rho_0 U_0 e^{j\omega t - jkR}(R - \sqrt{R^2-a^2}). \quad (6.13)$$

An expression from (O'Neil 1949) for the approximate pressure in the focal plane ($z = 0$) is

$$p(r,z=0,t) \approx j\omega\rho_0 \frac{R(R - \sqrt{R^2-a^2})}{\sqrt{R^2+r^2}}$$

$$\times U_0 e^{j\omega t - jk\sqrt{R^2+r^2}} \left[\frac{2J_1(kar/\sqrt{R^2+r^2})}{kar/\sqrt{R^2+r^2}} \right], \quad (6.14)$$

where the expression in square brackets is the jinc function. For pressure evaluations on-axis where $r = 0$, the jinc function is equal to one, so Equation 6.14 is exactly equal to the expression in Equation 6.13 when evaluated at the geometric focus. Equation 6.14 provides an accurate numerical estimate of the main lobe width for all values of $a < R$. Equation 6.14 also accurately represents the sidelobe levels as well as the locations of the first few nulls and sidelobe peaks when the $f/\#$, defined for a spherical shell as $R/(2a)$, is greater than one, where much better agreement over a wider range of r values is achieved with a larger $f/\#$.

An example of the pressure computed with the Rayleigh-Sommerfeld integral for a spherically focused transducer with aperture radius $a = 20\lambda$ and radius of curvature $R = 20\lambda\sqrt{2}$ is shown in Figure 6.6. The geometric center of the spherically focused transducer is located at $(x,y,z) = (0,0,0)$, and the normal at the center of the spherical shell is coincident with the z axis. Furthermore, the spherical shell intersects the z axis at $z = -R$, and the circle that defines the aperture opening is located at $z = -\sqrt{R^2-a^2} = -a$ as illustrated in Figure 6.5. The pressure is evaluated on a 101-point by 201-point grid that extends from $x = -a$ to $x = a$ laterally and from $z = -a$ to $z = a$ axially, where the lateral extent of the grid is equal to the aperture diameter and the minimum axial value is determined by the coordinate of the aperture opening. The pressure calculations are performed with a simplified version of Equation 6.11 that exploits the cylindrical symmetry of the spherically focused transducer and the computational grid by computing the integral from 0 to π and then doubling the result. The number of abscissas in both directions is again restricted such that $N = N_{\phi'} = N_{\theta'}$. The smallest value of N that achieves a normalized error less than or

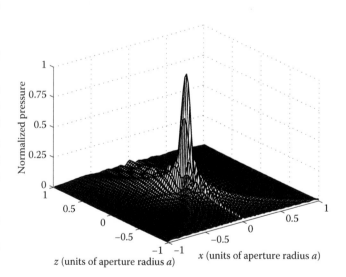

FIGURE 6.6 Simulated pressure obtained with the Rayleigh-Sommerfeld integral for a spherically focused transducer with aperture radius $a = 20\lambda$ and radius of curvature $R = 20\lambda\sqrt{2}$. The Rayleigh-Sommerfeld integral achieves a maximum error of 1% in 88.22 seconds when evaluated with $N = N_{\sigma'} = N_{\theta'} = 109$ on this 101-point by 201-point grid.

equal to 1% throughout the entire grid defined for this pressure calculation is $N = 109$. Although the aperture of the spherically focused transducer simulated here is much larger than that simulated for a circular piston in Figure 6.2, fewer abscissas are required for the spherically focused transducer because the computational grid is only adjacent to the spherically focused transducer at the very edge of the grid where $(x,z) = (\pm a, -a)$, whereas the pressure in Figure 6.2 is computed across the entire face of the circular piston. Using the same computer configuration described previously, the computation time for the Rayleigh-Sommerfeld integral evaluated on a 101-point by 201-point grid with $N = 109$ is 88.22 seconds (just under 1.5 minutes).

6.2.2 Rectangular Radiator Method

The rectangular radiator method described in (Ocheltree and Frizzell 1989) subdivides a rectangular transducer into small subelements, calculates the pressures generated by the small rectangular subelements, and superposes the results. The number of subdivisions is chosen such that the observation point is in the far field of each subelement, and then the pressure from each subelement is computed with the far-field approximation in Equation 6.9. The main advantage of the rectangular radiator method over the Rayleigh-Sommerfeld integral is the adaptive selection of the number of subelements, which reduces the number of subelements far from the piston face and likewise increases the number of subelements near the piston face as needed. The number of subelements is determined by the subdivision parameter F, which is defined such that the subelement width Δw (and height Δh) satisfies

$$w \le \sqrt{\frac{4\lambda z}{F}} \qquad (6.15)$$

at each axial distance z from the piston face. In (Ocheltree and Frizzell 1989), the choice of $F = 10$ is intended to guarantee that the observation point is at least 10 far-field distances from the subelement. However, this value of F is associated with discontinuities in the computed pressure field that are clearly evident when plotted on a linear scale. In an effort to avoid these numerical artifacts, $F = 20$ was used in Moros et al. (1993).

The calculation times associated with this approach represent another limitation of the rectangular radiator method. By subdividing the aperture in both directions, the rectangular radiator method numerically evaluates a two-dimensional integral. Thus, the Rayleigh-Sommerfeld integral and the rectangular radiator method are inherently $O(n^2)$, where n represents the number of subelements in each direction. Consequently, the computation times associated with these $O(n^2)$ methods increase quadratically as the number of subelements defined for these calculations increases. When pressure calculations are evaluated for several elements on a large grid in the near-field region, this quadratic increase translates into relatively long computation times.

6.2.3 Fast Near-Field Method

The fast near-field method (FNM), which evaluates one or more one-dimensional integrals that subtract a singularity from the integrand, addresses the main deficiencies of the Rayleigh-Sommerfeld integral and the rectangular radiator approach. By subtracting a singularity from the numerator of the integrand, the otherwise slow convergence is eliminated near the piston face and along the boundaries defined for impulse response calculations. As a result, the errors in the near-field region are much smaller with the fast near-field method than with these other methods. Furthermore, the fast near-field method evaluates one or more one-dimensional integrals (instead of a two-dimensional integral), so the fast near-field method is $O(n)$. Thus, the time required for calculations with the fast near-field method only increases linearly as the number of abscissas increases. The rapid convergence achieved by subtracting the singularity in one or more one-dimensional integrals enables the fast near-field method to achieve smaller numerical errors in less time compared to these other methods, especially for pressure calculations in the near-field region.

In contrast to the Rayleigh-Sommerfeld integral, where the integrand is consistently the same but the coordinates of the source points, the expressions describing the incremental areas, and the two-dimensional limits of integration are specific to each different piston shape, the fast near-field method defines unique one-dimensional integral expressions for each piston shape. By optimizing the integral expressions for flat circular, flat rectangular, and spherically focused transducers, near-field calculations with the fast near-field method achieve much smaller numerical errors in much less time while eliminating the numerical problems inherent to the Rayleigh-Sommerfeld integral near the piston face. The FNM expressions are evaluated with Gauss quadrature, which converges much faster than most other numerical integration schemes when applied to these calculations.

6.2.3.1 Circular Piston

The FNM expression for the time-harmonic pressure generated by a circular piston (McGough et al. 2004) in cylindrical coordinates is

$$p(r,z,t) = \rho_0 c U_0 e^{j\omega t} \frac{a}{\pi} \int_0^\pi \frac{r\cos\psi - a}{r^2 + a^2 - 2ar\cos\psi}$$

$$\times \left(e^{-jk\sqrt{r^2 + a^2 - 2ar\cos\psi + z^2}} - e^{-jkz} \right) d\psi, \qquad (6.16)$$

where a is the radius of the circular piston, (r,z) is the coordinate of the observation point, and ψ is the variable of integration. When the pressure is evaluated on the z axis (i.e., when $r = 0$), all ψ dependence is eliminated from the integrand, and the analytical expression for the on-axis pressure in Equation 6.4 is obtained. Numerical evaluations of Equation 6.16 discretize the one-dimensional integral with respect to the angle ψ, and the result is computed using Gauss quadrature.

Near-field pressures are computed with the FNM expression for a circular piston, and the results are evaluated on the same 121- by 101-point grid shown in Figure 6.2. For a circular piston with radius $a = \lambda$, the FNM expression in Equation 6.16 requires only $N = 8$ Gauss abscissas to achieve errors less than or equal to 0.71% throughout the grid. This result, which is calculated with optimized C++ code called from MATLAB® using the same computer described previously, is computed in 0.0342 seconds. This example shows that near-field pressure calculations with the fast near-field method expression for a circular piston achieves smaller errors in less time than the Rayleigh-Sommerfeld integral. For this combination of piston radius and computational grid, where the errors obtained with both methods are less than 1%, the FNM is more than 5,000 times faster than the Rayleigh-Sommerfeld integral.

6.2.3.2 Rectangular Piston

The FNM expression for the time-harmonic pressure generated by a rectangular piston (McGough 2004) in Cartesian coordinates is

$$p(x,y,z,t) = \frac{\rho_0 c U_0 e^{j\omega t}}{2\pi} \left[(x-a) \int_{y-b}^{y+b} \frac{e^{-jk\sqrt{z^2+(y')^2+(x-a)^2}} - e^{-jkz}}{(y')^2 + (x-a)^2} dy' \right.$$

$$- (x+a) \int_{y-b}^{y+b} \frac{e^{-jk\sqrt{z^2+(y')^2+(x+a)^2}} - e^{-jkz}}{(y')^2 + (x+a)^2} dy'$$

$$+ (y-b) \int_{x-a}^{x+a} \frac{e^{-jk\sqrt{z^2+(x')^2+(y-b)^2}} - e^{-jkz}}{(x')^2 + (y-b)^2} dx'$$

$$\left. - (y+b) \int_{x-a}^{x+a} \frac{e^{-jk\sqrt{z^2+(x')^2+(y+b)^2}} - e^{-jkz}}{(x')^2 + (y+b)^2} dx' \right], \quad (6.17)$$

where $2a$ is the piston width, $2b$ is the piston height, and (x,y,z) is the coordinate of the observation point. In Equation 6.17, the primed coordinates are the variables of integration, and the limits of integration represent the x and y components of the distances from the observation point to the four vertices of the rectangular piston. The center of the rectangular piston is located at $x = 0, y = 0, z = 0$, which is the center of the Cartesian coordinate system. Equation 6.17 evaluates the near-field pressure by superposing the contributions from four triangles, where the base of each triangle is one side of the rectangular piston. In Equation 6.17, each integral represents the contribution from a single triangle. When the observation point is located within the paraxial region where $|x| < a$ and $|y| < b$, the contributions from the four triangles are added, and the sum of the areas of the triangles is equal to the area of the rectangular piston. When the observation point is located outside of the paraxial region such that $|x| > a$ and $|y| > b$, the contributions from the two largest triangles are added and the contributions from the two smallest

triangles are subtracted, where the difference in these areas of these triangles is once again equal to the area of the rectangular piston. The superposition of these integrals is handled seamlessly by Equation 6.17, where the leading terms associated with each expression determine whether a contribution from a triangle is added or subtracted at every point in space. For numerical calculations, all four of the one-dimensional integrals in Equation 6.17 are discretized (with respect to the distance x' or y', as appropriate) and then the result is computed using Gauss quadrature. In the paraxial region where $|x| < a$ and $|y| < b$, the numerical performance is further improved when each triangle in Equation 6.17 is replaced with two right triangles and contributions from eight integrals are instead evaluated.

When near-field pressures are computed for a square 2λ by 2λ piston on the same 121- by 101-point grid shown in Figure 6.4, the FNM expression in Equation 6.17 requires only $N = 5$ Gauss abscissas to achieve errors less than or equal to 0.34% throughout the entire computational grid. The FNM result for a square piston, evaluated on the same computer running optimized C++ code called from MATLAB®, is obtained in 0.1291 seconds. Once again, the fast near-field method expression achieves smaller errors in less time than the Rayleigh-Sommerfeld integral. For this square piston and computational grid, where the errors obtained with both methods are less than 1% at all points on the grid, the FNM expression is more than 3,000 times faster than the Rayleigh-Sommerfeld integral.

6.2.3.3 Spherical Shell

The FNM expression for the time-harmonic pressure generated by a spherically focused transducer (McGough 2004) in cylindrical coordinates is

$$p(r,z,t) = \rho_0 c U_0 e^{j\omega t}$$

$$\times \frac{2aR}{\pi} \int_0^{\pi} \left\{ \frac{(r\cos\psi\sqrt{R^2-a^2}+za)}{(R^2 r^2 + 2ar\cos\psi\sqrt{R^2-a^2}z - a^2 r^2\cos^2\psi + z^2 a^2)} \right.$$

$$\left. \times [e^{-jk\sqrt{R^2+r^2+z^2-2ar\cos\psi+2z\sqrt{R^2-a^2}}} - e^{-jk(R+\text{sign}(z)\sqrt{r^2+z^2})}] \right\} d\psi,$$

$$(6.18)$$

where a is the aperture radius and R is the radius of curvature of the spherical shell. In cylindrical coordinates (r,z), the center of the sphere is located at $r = 0$ and $z = 0$, the normal at the center of the spherical shell is coincident with the z axis, and the spherical shell intersects the z axis at $z = -R$. In Equation 6.18, ψ is the variable of integration. As for a circular piston, when the pressure is evaluated on the z axis and therefore $r = 0$, all ψ dependence is eliminated from the integrand, and the analytical expressions for the on-axis pressure in Equations 6.12 and 6.13 are obtained. For numerical calculations, the one-dimensional integral in Equation 6.18 is discretized with respect to the angle ψ and evaluated with Gauss quadrature.

Pressures computed with the FNM expression for a spherical shell with an aperture radius $a = 20\lambda$ and a radius of curvature $R = 20\lambda\sqrt{2}$ are evaluated on the same 101- by 201-point grid shown in Figure 6.6. The FNM expression in Equation 6.18 converges to an error less than or equal to 1% throughout the grid with $N = 30$ Gauss abscissas. The result is calculated with optimized C++ code called from MATLAB® on the same computer in 0.3588 seconds. With additional abscissas, the peak FNM errors are even smaller at the expense of a small increase in computation time. Thus, relative to the Rayleigh-Sommerfeld integral, the fast near-field method achieves equal or smaller errors in less time for calculations of pressures generated by spherically focused transducers. For this combination of piston and grid parameters, where the peak errors obtained with both methods are equal to 1%, the FNM is more than 200 times faster than the Rayleigh-Sommerfeld integral.

6.2.3.4 Gauss Quadrature

FNM calculations are typically evaluated with Gauss quadrature (Davis and Rabinowitz 1975), which converges rapidly for the expressions in Equations 6.16, 6.17, and 6.18. The Gauss abscissas g_i and weights w_i are computed with the N point Gauss-Legendre quadrature rule for the interval $(-1,1)$, then the Gauss abscissas are linearly mapped (Abramowitz and Stegun 1972) to the limits of integration $(\gamma_{min}, \gamma_{max})$ according to

$$\gamma_i = \frac{\gamma_{max} - \gamma_{min}}{2} g_i + \frac{\gamma_{max} + \gamma_{min}}{2} \qquad (6.19)$$

and the Gauss weights w_i are scaled by $(\gamma_{max} - \gamma_{min})/2$. When applied to FNM calculations for pistons exhibiting cylindrical symmetry (i.e., circular pistons and spherical shells), $\gamma_{min} = 0$ and $\gamma_{max} = \pi$, so the linearly mapped Gauss abscissas are

$$\psi_i = \frac{\pi}{2} g_i + \frac{\pi}{2} \qquad (6.20)$$

and the scaled Gauss weights are given by $d\psi_i = (\pi/2)w_i$. FNM calculations for rectangular pistons define limits of integration that vary based on the location of the observation point, and separate mappings are defined for the variables of integration x_i' and y_i'. Integration with respect to the x_i' variable is performed with abscissas defined by

$$x_i' = ag_i + x \qquad (6.21)$$

and with weights $dx_i' = aw_i$, where a is the half-width of the rectangular piston and x represents the x coordinate of the observation point. Likewise, integration with respect to the y_i' variable is performed with abscissas defined by

$$y_i' = bg_i + y \qquad (6.22)$$

and with weights $dy_i' = bw_i$, where b is the half-height of the rectangular piston and y represents the y coordinate of the observation point.

6.2.4 Angular Spectrum Approach

The angular spectrum approach accelerates computations of diffracting pressure fields evaluated in parallel planes using 2D fast Fourier transforms (FFTs). The angular spectrum approach is especially valuable for simulating large 3D pressure fields generated by ultrasound phased arrays. When applied properly, the angular spectrum approach achieves a substantial reduction in the computation time without significantly increasing the numerical error.

The angular spectrum approach is analytically equivalent to the Rayleigh-Sommerfeld integral, which convolves the spatial distribution of the normal particle velocity in the plane containing the transducer with the Green's function for a point source surrounded by a rigid baffle. The Green's function for a point source surrounded by a rigid baffle radiating into a semi-infinite half-space is

$$h_u(x, y, z) = \frac{e^{-jkR}}{2\pi R}, \qquad (6.23)$$

where $R = \sqrt{x^2 + y^2 + z^2}$ is the distance from a point source located at $(0,0,0)$ to an observation point located at (x,y,z), the subscript u designates a normal particle velocity input, and lower case h indicates that the Green's function is evaluated in the spatial domain. Similarly, the normal particle velocity for an ultrasound transducer or an array of transducers strictly located in the $z = 0$ plane is represented by $u(x,y,0)$. The Green's function $h_u(x,y,z)$ convolved with the normal particle velocity $u(x,y,0)$ is proportional to the pressure

$$p(x, y, z, t) = j\omega\rho_0 e^{j\omega t} u(x, y, 0) **_{x,y} h_u(x, y, z), \qquad (6.24)$$

where $**_{x,y}$ indicates 2D convolution with respect to the spatial variables x and y. The Fourier transform of Equation 6.24, evaluated with respect to x and y in a plane orthogonal to the z axis at a distance z from the transducer face, is

$$P(k_x, k_y, z) = j\rho_0 \omega e^{j\omega t} U(k_x, k_y, 0) H_u(k_x, k_y, z). \qquad (6.25)$$

In Equation 6.25, the spatial frequencies are represented by k_x and k_y, the 2D Fourier transform of the normal particle velocity is $U(k_x, k_y, 0) = \mathcal{F}_{x,y}\{u(x,y,0)\}$, the 2D Fourier transform of the Green's function evaluated in a plane orthogonal to the z axis is $H_u(k_x, k_y, z) = \mathcal{F}_{x,y}\{h_u(x,y,z)\}$, and convolution in the spatial domain is replaced by multiplication in the spatial frequency domain (i.e., k-space). The analytical expression for $H_u(k_x, k_y, z)$ is

$$H_u(k_x, k_y, z) = \begin{cases} \dfrac{e^{-jz\sqrt{k^2 - k_x^2 - k_y^2}}}{j\sqrt{k^2 - k_x^2 - k_y^2}} & \text{for } k_x^2 + k_y^2 \leq k^2 \\[4mm] \dfrac{e^{-z\sqrt{k_x^2 + k_y^2 - k^2}}}{\sqrt{k_x^2 + k_y^2 - k^2}} & \text{for } k_x^2 + k_y^2 > k^2 \end{cases}, \qquad (6.26)$$

where $k = \omega/c$ is the wavenumber for a monochromatic excitation with radian frequency ω, the spatial frequency components corresponding to propagation in the x, y, and z directions are k_x, k_y, and k_z, respectively, and $k_x^2 + k_y^2 + k_z^2 = k^2$. In Equation 6.26, the term in the numerator is purely real and exponentially decaying for combinations of spatial frequencies outside of the circle with radius k, and the numerator is complex and oscillatory for combinations of spatial frequencies inside of the circle with radius k. Thus, spatial frequencies inside this circle correspond to propagating waves, and spatial frequencies outside this circle correspond to evanescent waves.

For certain transducer geometries, the Fourier transform $U(k_x,k_y,0)$ of the normal particle velocity $u(x,y,0)$ can be evaluated analytically, but in general, $U(k_x,k_y,0)$ is obtained from the 2D FFT of $u(x,y,0)$. The expression for $H_u(k_x,k_y,z)$ can also be obtained from the 2D FFT of the Green's function $h_u(x,y,z)$. After $U(k_x,k_y,0)$ and $H_u(k_x,k_y,z)$ are computed, the product is evaluated, and the pressure is then calculated with an inverse 2D FFT. The result describes the diffracted pressure within a single plane. If additional planes are required (for example, to model ultrasound propagation in a large 3D volume), then $H_u(k_x,k_y,z)$ is calculated for each additional plane, and then the inverse 2D FFT is evaluated for the product of $U(k_x,k_y,0)$ and $H_u(k_x,k_y,z)$ in as many planes as needed.

The angular spectrum approach also defines a transfer function in the spatial frequency domain that relates the Fourier transform of the pressure in one plane to the Fourier transform of the pressure in another parallel plane. This transfer function is designated by $H_p(k_x,k_y,\Delta z)$, where the subscript p indicates that the transfer function input is the Fourier transform of the pressure (as opposed to the Fourier transform of the normal particle velocity for the transfer function H_u), k_x and k_y are the spatial frequency components defined previously, and Δz is the orthogonal distance between the two parallel planes. The transfer function $H_p(k_x,k_y,\Delta z)$ describes forward propagation between an input pressure plane located at $z = z_0$ and an output pressure plane located at $z = z_0 + \Delta z$ according to

$$P(k_x,k_y,z) = P_0(k_x,k_y,z_0)H_p(k_x,k_y,\ z), \qquad (6.27)$$

where $P_0(k_x,k_y,z_0)$ is the 2D Fourier transform of the input pressure plane $p_0(x,y,z_0)$ located at $z = z_0$, and $P(k_x,k_y,z_0)$ is the 2D Fourier transform of the output pressure plane $p(x,y,z)$ located at $z = z_0 + \Delta z$. In Equation 6.27, the transfer function $H_p(k_x,k_y,\Delta z)$ is defined by the analytical expression

$$H_p(k_x,k_y,\ z) = \begin{cases} e^{-j\ z\sqrt{k^2 - k_x^2 - k_y^2}} & \text{for} \quad k_x^2 + k_y^2 \leq k^2 \\ e^{-\ z\sqrt{k_x^2 + k_y^2 - k^2}} & \text{for} \quad k_x^2 + k_y^2 > k^2 \end{cases}, \qquad (6.28)$$

and the input pressure plane $p_0(x,y,z_0)$ is typically calculated with the Rayleigh-Sommerfeld integral or the fast near-field method. Therefore, to compute the forward propagating ultrasound field, the 2D FFT of the input pressure plane $p_0(x,y,z_0)$ is

evaluated and multiplied by $H_p(k_x,k_y,\Delta z)$ from Equation 6.28, and then an inverse 2D FFT is applied to the result to obtain the output pressure plane $p(x,y,z)$ at $z = z_0 + \Delta z$.

In pressure calculations with the angular spectrum approach, each transfer function is associated with a specific error distribution. For example, angular spectrum calculations that evaluate the Green's function $h_u(x,y,z)$ in the spatial domain and then numerically calculate the 2D FFT of the result to obtain $H_u(k_x,k_y,z)$ are accurate in the central portion of the computational grid; however, the errors are large near the edges of the computed field due to wraparound errors that occur in the spatial domain (Zeng and McGough 2008). Computing $H_u(k_x,k_y,z)$ directly in the spatial frequency domain with Equation 6.26 avoids the problems with wraparound errors near the edge, but when the analytical expression $H_u(k_x,k_y,z)$ is sampled in the spatial frequency domain, the spectrum is inherently aliased due to the rapid oscillations that occur near $k^2 = k_x^2 + k_y^2$. This frequency-domain aliasing causes errors in the spatial domain that appear as unwanted high-frequency signals in the simulated pressure field. Thus, instead of errors that are confined to the edge for angular spectrum calculations with the Green's function $h_u(x,y,z)$, the errors are distributed across the entire computed pressure field for angular spectrum calculations with the analytical expression $H_u(k_x,k_y,z)$. These same errors are also caused by the rapid oscillations near $k^2 = k_x^2 + k_y^2$ that occur in the analytical transfer function $H_p(k_x,k_y,\Delta z)$ in Equation 6.28. In calculations with the analytical expressions $H_u(k_x,k_y,z)$ and $H_p(k_x,k_y,\Delta z)$, these errors are reduced with a low-pass filter in k-space (Christopher and Parker 1991a; Wu et al. 1997). The errors are also reduced when the medium is sufficiently attenuative or if the aperture is apodized (Zeng and McGough 2008).

Comparisons between angular spectrum simulations evaluated with $h_u(x,y,z)$ in Equation 6.23, $H_u(k_x,k_y,z)$ in Equation 6.26, and $H_p(k_x,k_y,\Delta z)$ in Equation 6.28 indicate that each approach has distinct advantages and disadvantages. Of these, angular spectrum calculations using the analytical expression for $H_u(k_x,k_y,z)$ in Equation 6.26 are the fastest, requiring minimal computational effort during initialization and the fewest 2D FFT calculations. The other two approaches are more time consuming, where simulations with $H_p(k_x,k_y,\Delta z)$ in Equation 6.28 require additional time to compute $p_0(x,y,z_0)$ in the input pressure plane, and angular spectrum calculations with $h_u(x,y,z)$ in Equation 6.23 require an additional 2D FFT in each plane where the pressure is computed. In contrast, the smallest errors are achieved by the analytical expression for $H_p(k_x,k_y,\Delta z)$ in Equation 6.28 when the errors due to frequency domain aliasing are eliminated with a low-pass filter applied in k-space (or through attenuation or apodization) and the input pressures are computed with the fast near-field method in a sufficiently large, well-sampled plane located near the transducer or array face (Zeng and McGough 2009). The errors produced by angular spectrum calculations with normal particle velocity inputs and the Green's function $h_u(x,y,z)$ or the transfer function $H_u(k_x,k_y,z)$ are reduced with increased spatial sampling rates; however, this significantly increases the amount of computer memory and the computation time. Furthermore, direct access to the normal particle velocity is restricted to planar transducer and array geometries. Thus, the utility of $h_u(x,y,z)$ in Equation 6.23 or $H_u(k_x,k_y,z)$ in Equation 6.26 for angular spectrum simulations of pressure fields evaluated for thermal therapy is limited, especially if the numerical errors are considered.

With the proper collection of input parameters, the angular spectrum approach rapidly and accurately computes pressures in large 3D volumes for simulations of thermal therapy. One important parameter for these simulations is the distance between adjacent spatial grid points (i.e., the spatial sampling rate). For time harmonic pressure simulations, the distance between two adjacent spatial samples must be no greater than $\lambda/2$, where λ is the wavelength in the medium. If the distance between adjacent spatial samples is any larger, then the pressure field will be aliased, and the temperatures computed with the bioheat transfer equation will be erroneous. This restriction on the spatial sampling, which is not specific to the angular spectrum approach, must be satisfied by any pressure field calculation that involves subsequent temperature calculations. Other parameters include the size of the computational grid and the location of the input pressure plane. For computations in large 3D volumes, the input pressure plane for angular spectrum calculations should be located about one wavelength (λ) from the phased array, and the source pressure plane should be larger than the transducer or phased array (Zeng and McGough 2009). This combination of parameters, when applied to the angular spectrum calculations, typically achieves errors less than or equal to 1% throughout the computational grid when the input pressure plane is computed using the fast near-field method. The reduction in the computation time achieved when the angular spectrum approach is combined with the fast near-field method can reach several orders of magnitude relative to the Rayleigh-Sommerfeld integral or the rectangular radiator method, especially for large phased arrays evaluated on large computational grids. This reduction in computation time is highly desirable for thermal therapy calculations, which can be very time consuming.

The angular spectrum approach is also amenable to certain enhancements. For example, if the input pressure plane calculations for each element in a phased array are stored in advance, then the total input pressure planes for a wide variety of focal patterns are quickly obtained via superposition (Vyas and Christensen 2008). This approach significantly reduces the computation time when evaluating a large number of focal patterns as in (Jennings and McGough 2010, Zeng et al. 2010). The angular spectrum approach also models wave propagation in flat, layered media (Clement and Hynynen 2003, Vecchio et al. 1994). Angular spectrum models in layered media typically only retain the transmitted component at planar boundaries, discarding the reflected component. This approach provides an estimate of the transmitted power under the assumption that the reflected power is relatively small and that multiple reflections are negligible. Angular spectrum models of propagation in inhomogeneous tissue are also under development. One such model is the hybrid angular spectrum approach (Vyas and Christensen 2008), which considers different values for the speed of sound, absorption, and density in each voxel. The hybrid angular spectrum approach propagates back and forth between the pressure evaluated in the spatial domain and the 2D Fourier transform of the pressure to account for the differences in the material properties for each voxel. This approach enables propagation through curved layers and inhomogeneous tissues.

6.2.5 Nonlinear Ultrasound Propagation

Nonlinear propagation effects occur in many therapeutic ultrasound applications, where these effects are described by several different numerical models. Some nonlinear models simulate diffraction with the angular spectrum approach (Christopher and Parker 1991b, Zemp et al. 2003), and then attenuation and nonlinear interactions are calculated separately. These nonlinear pressure simulations incorporate a relatively small number of harmonics (5–10) to describe nonlinear propagation without shock wave formation and a much larger number (at least 30 to 50) when modeling shock waves (Christopher and Parker 1991b). Nonlinear ultrasound propagation is also described by the KZK equation, which is a popular model for therapeutic ultrasound that simulates transient and time-harmonic nonlinear propagation for axisymmetric ultrasound sources (Lee and Hamilton 1995). The KZK equation is solved either in the frequency domain (Aanonsen et al. 1984, Khokhlova et al. 2001) or the time domain (Cleveland et al. 1996, Lee and Hamilton 1995), where the effects of diffraction, frequency-dependent attenuation and dispersion, and nonlinearity are computed separately at each propagation step. The KZK equation has also been extended to 3D Cartesian geometries (Yang and Cleveland 2005). Nonlinear ultrasound propagation for non-axisymmetric sources is also simulated with the Westervelt equation (Huijssen and Verweij 2010), which models full-wave nonlinear propagation. Nonlinear simulations are very time consuming, especially when compared to linear simulations, and for this reason, most simulations of nonlinear ultrasound are evaluated for 2D problems with cylindrical symmetry. Nonlinear ultrasound simulations are also evaluated in 3D despite the long simulation times.

6.2.6 Software Programs

Several software programs are available for simulations of therapeutic ultrasound. Some of these are commercial products, and others were developed by academic researchers. Most simulation programs were originally developed for other applications and then adapted to therapeutic ultrasound. One commercial product, PZFlex (Wojcik et al. 1993), is a finite element package that models the electromechanical behavior of ultrasound transducers, and SPFlex is a separate module that calculates the pressure fields generated by these transducers (Mould et al. 1999). Many other commercial finite element packages exist, but these typically have steep learning curves, and significant user effort is required when extending them for therapeutic ultrasound simulations.

A few software programs have been developed for ultrasound imaging simulations. The most popular of these is Field II (Jensen 1996), which is a freely downloadable program (http://server .oersted.dtu.dk/personal/jaj/field/) that runs within MATLAB®. Field II consists of a well-developed and well-documented set of routines and user interfaces that were primarily designed for diagnostic ultrasound simulations. Field II is also sometimes used for thermal therapy simulations that require calculations of time-harmonic pressure fields. Other freely available software for diagnostic imaging simulations includes the DREAM toolbox (http://www.signal.uu.se/Toolbox/dream/) and the Ultrasim package (http://folk.uio.no/ultrasim/). Although DREAM and Ultrasim generate useful results for other ultrasound applications, these programs are infrequently applied to simulations of thermal therapy.

FOCUS is a software package that is presently under development for both therapeutic and diagnostic ultrasound simulations. FOCUS (http://www.egr.msu.edu/~fultras-web) is a "fast object-oriented C++ ultrasound simulator" that quickly performs linear simulations of time-harmonic and transient pressure fields. Time-harmonic pressures are computed with the fast near-field method and the angular spectrum approach (Zeng and McGough 2009), and transient pressures are computed with the fast near-field method and time-space decomposition (Kelly and McGough 2006). In FOCUS, these time-harmonic and transient pressure calculations are implemented for circular, rectangular, and spherically focused transducers. The main routines in FOCUS compute pressure fields much faster than the Rayleigh-Sommerfeld integral, the rectangular radiator method, Field II, DREAM, and Ultrasim for both time-harmonic and transient excitations, where the greatest advantage is achieved in the near-field region. FOCUS also provides routines for the Rayleigh-Sommerfeld integral to facilitate comparisons and to provide references as needed. New features, including nonlinear propagation, are also under development.

HIFU Simulator is a software package that simulates nonlinear propagation of ultrasound (Soneson 2009). This software, which presently models axisymmetric 2D pressure distributions, simulates nonlinear ultrasound propagation for time-harmonic inputs with the KZK equation and the resulting temperature response with the Pennes bioheat transfer equation (BHTE). Support for layered tissue models is also included in HIFU Simulator. Routines that calculate nonlinear 3D pressure fields are presently under development.

6.2.7 Intensity and Power Calculations

Temperature simulations require intermediate calculations of the power deposition values, which in turn depend on the intensity. Thermal therapy simulations typically utilize the plane wave approximation for the intensity, which is given by

$$I(x,y,z) = \frac{|p(x,y,z)|^2}{2\rho_0 c}. \tag{6.29}$$

In Equation 6.29, the intensity at a point in space is represented by $I(x,y,z)$, the absolute value of the peak pressure is $|p(x,y,z)|$, the density is ρ_0, and the speed of sound is c. The units for the intensity $I(x,y,z)$ are watts per square meter (W/m²). The pressure in Equation 6.29 is typically evaluated in a lossy medium, which is modeled for a time-harmonic excitation by the complex wave number

$$k = \frac{\omega}{c} - ja. \tag{6.30}$$

Equation 6.30 describes the wave number k for an outward propagating (or left to right propagating) wave where the positive time convention indicated by $e^{j\omega t}$ represents the time-harmonic component of the pressure. In Equation 6.30, ω is the radian frequency of the time-harmonic excitation, and a is the attenuation in nepers per meter (Np/m). For pressure calculations in lossy media, the complex wavenumber k is inserted into any of the integral expressions given in previous sections, and the integral is evaluated using the same approaches described previously without any other changes. Pressure calculations with the angular spectrum approach also modify the propagation term to represent the loss (Zeng and McGough 2008).

The power deposition Q is twice the product of the intensity and the absorption coefficient (Nyborg 1981)

$$Q = 2\alpha I, \tag{6.31}$$

where α represents the absorption in Np/m and the units of the power deposition Q are W/m³. The absorption coefficient α describes the component of the attenuated ultrasound waveform that is converted into heat, whereas the remaining attenuation components are scattering, reflection, and refraction (Cobbold 2007). In some soft tissue models, the attenuation coefficient a and the absorption coefficient α are equal (Nyborg and Steele 1983), and in others, the ratio α/a is less than 1/2 (Goss et al. 1979).

6.3 Thermal Modeling and Treatment Planning

Patient treatment planning determines an optimal combination of treatment parameters that maximize the effectiveness of a treatment while minimizing undesirable side effects. For ultrasound-based thermal therapies, treatment planning attempts to optimize conformal delivery of hyperthermic temperatures or ablative thermal doses while minimizing thermal or mechanical damage to healthy tissues. For mechanically scanned ultrasound systems, treatment planning involves calculations of applied powers and applicator locations/orientations, and for ultrasound phased arrays, treatment planning optimizes the phase and amplitude of the driving signal for each array element. Treatment planning is often a computationally intensive process, and for therapeutic ultrasound, this generally includes linear or nonlinear pressure field calculations along with some combination of bioheat transfer modeling, thermal dose calculations, applicator path and orientation optimization, phased array beamforming, and visualization. Most treatment planning simulations predict the pressure and temperature fields generated during a treatment, and more detailed evaluations also optimize various treatment parameters.

6.3.1 Bioheat Transfer Model

The bioheat transfer equation (BHTE) (Pennes 1948) is a simplified model for heat transfer in biological systems (Strohbehn and Roemer 1984) that provides a comparative basis for thermal treatment evaluation (Roemer 1990). The transient bioheat transfer equation,

$$\nabla \cdot (\kappa \nabla T) - W C_b (T - T_a) + Q - \rho C \frac{\partial T}{\partial t} = 0 \qquad (6.32)$$

describes the time-dependent tissue temperature T in °C produced by the power distribution Q. In Equation 6.32, T_a is the arterial temperature, W_b represents the blood perfusion rate, C_b indicates the specific heat of blood, κ represents the thermal conductivity of tissue, ρ is the density of tissue, and the specific heat of tissue is indicated by C. The transient bioheat transfer equation in Equation 6.32 calculates changes in the tissue temperature in response to time-dependent variations in the blood perfusion or power deposition.

The steady-state bioheat transfer equation, evaluated for a constant thermal conductivity κ, is given by

$$Q - W_b C_b (T - T_a) + \kappa \nabla^2 T = 0. \qquad (6.33)$$

Equation 6.33 calculates the tissue temperature T for fixed values of the blood perfusion and the power deposition. Equation 6.33 also calculates temperatures for equivalent steady-state power depositions when the ultrasound applicator switches between field patterns much more rapidly than the thermal time constant of the medium (Moros et al. 1988).

6.3.2 Thermal Dose Calculations

The transient and steady-state temperatures computed with the bioheat transfer equations in Equation 6.32 and Equation 6.33, respectively, also provide the input for thermal dose calculations. The thermal dose is computed as (Sapareto and Dewey 1984)

$$t_{43°C} = \sum_{n=1}^{N} \Delta t \cdot R^{(T_n - 43°C)}, \qquad (6.34)$$

where T_n is the average temperature during the time interval n, N is the number of temperature samples collected, and Δt is the duration of each time interval. The parameter R is equal to 2 for temperatures less than 43°C, and R is equal to 4 for temperatures greater than or equal to 43°C. The thermal dose $t_{43°C}$ represents the equivalent time at which the tissue temperature is maintained

at 43°C. A commonly accepted objective for hyperthermia treatments is 60 minutes at 43°C, although much smaller values for the thermal dose are often achieved with clinical hyperthermia applicators. The thermal dose of 240 equivalent minutes is a standard threshold for tissue coagulation (McDannold et al. 1998), where tissues reaching or exceeding this value define the expected size of the ablated volume. For both hyperthermia and thermal ablation, Equation 6.34 calculates the thermal dose for prospective treatment evaluation and optimization.

6.3.3 Thermal Therapy Planning

Patient treatment planning attempts to determine treatment parameters that maximize treatment efficacy while minimizing normal tissue toxicity. To achieve these objectives, treatment planning for therapeutic ultrasound combines pressure and temperature simulations with optimization and visualization in an effort to conform therapeutic temperatures or thermal doses to the tumor volume. Simultaneously, treatment planning attempts to identify and avoid sources of patient pain and normal tissue damage. These problems are caused by bone heating (Fessenden et al. 1984), reflections at tissue/bone interfaces (Hynynen and DeYoung 1988), reflections at tissue/air interfaces (Hynynen 1990b), "hot spots" or excessive thermal doses caused by surface heating or intervening tissue heating (Damianou and Hynynen 1993, Moros et al. 1990, Wang et al. 1994), and cavitation in normal healthy tissues (Hynynen 1990a, Hynynen and Lulu 1990). These sources of normal tissue toxicity can be treatment limiting, so most treatment plans evaluate pressures and temperatures in both diseased and normal tissues.

For patient treatment planning, pressure and temperature distributions are ideally computed and evaluated in a detailed patient-specific model. However, the capabilities of most simulation models of therapeutic ultrasound are limited by the computation time, numerical error, and computer memory. Therapeutic ultrasound simulations are typically completed in seconds, minutes, or hours, where the numerical errors achieved in models of homogeneous tissue are preferably less than or equal to 1%. The available computer memory determines the size of the problem that can be simulated, and the number of points in a 3D computational grid defined for patient treatment planning is very large due to the relatively small wavelength of therapeutic ultrasound in tissue (about 1.5 mm at 1 MHz). These large grids, which routinely extend hundreds of wavelengths in each direction, are presently incompatible with 3D finite element and finite difference methods when evaluated on a fully equipped desktop computer. Although some finite element and finite difference models sample the pressure field 10 times per wavelength, much smaller numerical errors are obtained with these methods at 40 samples per wavelength, especially when complicated tumor and organ geometries, each with different tissue parameters, are incorporated into the simulation model. Unfortunately, maintaining such a large grid of pressure values is beyond the memory capacity of most modern desktop computers, and the computation times associated with finite element and finite difference methods, when applied to very large grids, are excessively long. In contrast, the approaches outlined in previous sections require only two samples per wavelength, and sufficient computer memory is only needed for at most one or two planes of pressure field values sampled at this lower rate. Thus, to satisfy these time, error, and memory requirements, certain compromises are required. Most patient treatment planning calculations use homogeneous or layered tissue models for pressure field simulations, although some newer approaches model more complicated geometries. These homogeneous or layered tissue models often provide a useful starting point for thermal therapy planning.

6.3.3.1 Ultrasound Hyperthermia with Fixed-Phase Applicators

Several numerical models have been developed for fixed-phase ultrasound hyperthermia applicators (Strohbehn and Roemer 1984). One such model computes the SAR and the temperature response for a stationary spherically focused transducer (Swindell et al. 1982). The pressure calculations for this model evaluate the Rayleigh-Sommerfeld integral by superposing the contributions from thin annular strips. This approach converts the 2D Rayleigh-Sommerfeld integral into an equivalent 1D integral. For a fixed transducer, SAR values and temperature maps are simulated for excitation frequencies of 500 kHz and 2 MHz. The simulation results demonstrate that the isothermal contours move closer to the transducer surface at 2 MHz due to increased attenuation. Simulated temperature maps also show that approximately spherical isothermal regions are produced with a single circular scan and with multiple concentric scans. In this model, the mechanically scanned ultrasound transducer rapidly traverses each concentric ring, uniformly depositing power along the scan trajectory. The time-averaged power is then computed, and the approximate temperature map is calculated with the steady-state bioheat transfer equation.

Another numerical model computes the impulse response for a scanned focused ultrasound transducer, approximates the intensity and the effect of attenuation with separate plane wave models, and then calculates the cylindrically symmetric intensity of the scanned pattern with a Hankel transform (Dickinson 1984). The intensities are then converted into a specific absorption rate (SAR), which equals the power deposition Q divided by the density ρ according to $SAR = Q/\rho$. This result provides the input to a cylindrically symmetric 2D bioheat transfer model consisting of planar layers of skin, fat, and muscle. The bioheat transfer model also includes a cylindrical tumor that consists of a well-perfused outer layer and a poorly perfused core. The temperature output of the bioheat transfer model is then converted into thermal dose specified in terms of equivalent minutes at $43°C$.

A related model considers the effect of normal tissue and tumor blood perfusion on the selection of mechanically scanned transducer parameters (Roemer et al. 1984). The pressures are once again computed by superposing the contributions from thin curved strips on the transducer surface, and the mechanically scanned intensity pattern is obtained from a circularly symmetric convolution. Temperatures are then computed for a cylindrically symmetric layered tissue model combined with a cylindrically symmetric tumor model containing three concentric regions.

Two different blood perfusion models are evaluated in Roemer et al. (1984), where one is uniformly perfused and the other is highly perfused in the periphery, moderately perfused in an intermediate region, and necrotic with no blood flow in the core. The parameters in these simulations include tumor location and size, scanning pattern radius, power deposition, transducer frequency, and the transducer f/#. The simulated temperatures are compared according to an algorithm that defines minimum and maximum tumor temperatures as well as maximum normal tissue temperatures. The results show that a mechanically scanned transducer operating at 500 kHz with an f/# of 0.8 satisfies the treatment objectives when the minimum tumor and maximum tumor temperatures are defined as $42°C$ and $60°C$, respectively.

In another simulation model, the temperature fluctuations (Moros et al. 1988) produced by mechanically scanned ultrasound hyperthermia applicators are evaluated. The focal plane temperature fluctuations are simulated with a 3D transient bioheat transfer model, where the input parameters are the f/#, the blood perfusion, the scan time, and the scan diameter for single circular scans and multiple concentric scans. The largest fluctuations occur in the focal plane, and the magnitude of these fluctuations increases linearly as a function of the scan time. Acceptably small temperature fluctuations are demonstrated with scan times of 10 seconds or less for a mechanically scanned transducer with an f/# of 2.6. Larger temperature fluctuations are also observed with smaller f/# transducers, and these require even smaller scan times to achieve comparable temperature fluctuations. Multiple concentric scan paths yield relatively uniform temperature distributions when the difference in the diameter of adjacent scans is equal to the focal spot diameter.

In a similar parametric study, pre-focal plane heating produced by mechanically scanned ultrasound hyperthermia applicators is characterized with a steady state bioheat transfer model (Moros et al. 1990). The frequency, the f/#, the focal depth, the blood perfusion, and the scan pattern diameter are the input parameters, and the resulting temperature distributions are evaluated for single and multiple transducer configurations. The results demonstrate that single and multiple circular scans produced excessive temperatures between the applicator and the tumor target. Although the peak power depositions are generated in the focal plane, the overlapping power depositions along the central axis cause the intervening tissue heating. Simulations of a mechanically scanned configuration consisting of four tilted transducers with overlapping foci also encounter the same problem unless the transducer closest to the central axis is turned off during the scan (Figure 6.7). This eliminates the excessive pre-focal plane heating by increasing the size of the effective acoustic aperture, which effectively spreads the incident energy across a wider area.

Related simulations are evaluated for the scanning ultrasound reflector linear array system, or SURLAS (Moros et al. 1997). This device, which facilitates concurrent delivery of ultrasound hyperthermia and radiation therapy, features two fixed-phase linear ultrasound arrays combined with a reflector that is scanned along a linear path to achieve variable depth control over the power deposition. The numerical model for the

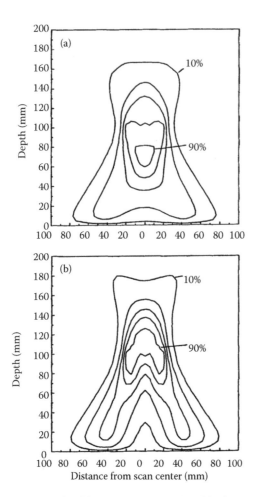

FIGURE 6.7 Simulated heating patterns produced by four mechanically scanned ultrasound transducers. (a) All four transducers are excited throughout the scan, which causes intervening tissue heating. (b) The transducer nearest the central axis is turned off, and the intervening tissue heating is significantly reduced. (After E. G. Moros, R. B. Roemer, and K. Hynynen, *Int. J. Hypertherm.*, 6, 2, 1990.)

the tumor, relevant organs, and the skin surface, are identified. The 3D model created by this step enables viewing of anatomical structures from different directions, which facilitates visualization of the target, sensitive normal tissues, and the acoustic window. After the 3D models of these anatomical structures are established, ultrasound parameter values, including the speed of sound, density, attenuation, and absorption, are assigned to each tissue type. The optimization step then determines ultrasonic, geometric, and electromechanical treatment parameters. Ultrasonic parameters, as defined for these ultrasound hyperthermia applicators, include the excitation frequency and lens focal length, which are specified by the tumor depth, locations of sensitive normal tissues, and the transducer clearance above the acoustic window. Next, the geometric parameters specify the orientation and the scan path of the ultrasound transducer. These parameters, which are selected in an effort to avoid excessive intervening tissue heating, are specifically determined by the geometric description of overlapping pressure beams. Similarly, the electromechanical parameters describe the limitations of the scanning machinery when interfaced to a human patient. The electromechanical parameters are restricted by the coupling bolus or coupling water bath, the acoustic window, the scan path, and the patient position. Thus, the geometric parameters, as defined here, primarily consider ultrasonic energy deposition within the tumor and in normal tissues, whereas the electromechanical parameters are essentially concerned with obstacles that might otherwise impede the motion of the mechanical scanning apparatus. The thermal modeling step estimates the temperature distribution within the patient, and then the results are combined with the patient anatomical model (Figure 6.8). The initial applicator settings are thus determined by these steps, and these settings are updated based on feedback collected during the treatment. Temperatures are also monitored and stored for retrospective comparison and evaluation.

6.3.3.2 Ultrasound Hyperthermia with Phased Array Applicators

Numerical modeling of ultrasound phased arrays involves many of the same calculations described for mechanically scanned hypthermia applicators. The main difference is that phased arrays steer and focus the pressure field electronically, whereas mechanically scanned fixed phase applicators are tilted and translated. Electronic steering and focusing enables phased array beamforming, which optimizes the phase and amplitude settings of the individual phased array elements. Beamforming calculations are often performed in conjunction with temperature simulations, and the outcome of these computations is also integrated with other aspects of treatment planning.

By adjusting the phases and amplitudes of the driving signals applied to each of the transducer elements, ultrasound phased arrays generate single focus patterns, annular patterns, and multiple focus patterns. For a single-focus beam pattern, the optimal driving signal phases are computed with field conjugation (Ibbini and Cain 1989). This approach excites each element with a sinusoidal signal and calculates the phase of the complex

pressures generated by the SURLAS superpose the contributions from the individual array elements multiplied by a plane wave approximation for the attenuation, and then temperatures are computed with a bioheat transfer model. Contributions from individual arrays are evaluated at 1 MHz, 3 MHz, and 5 MHz, and combined power depositions are computed for simultaneous excitations at 1 MHz and 5 MHz. The simulation results show that the penetration depth is controlled by adjusting the ratio of powers applied to the parallel opposed linear arrays.

When several of these numerical computations are combined with other related calculations, the resulting numerical models define a treatment planning approach for hyperthermia with a single mechanically scanned ultrasound transducer (Lele and Goddard 1987). This approach defines several steps, and the first three, namely visualization, optimization, and thermal modeling, are specifically for prospective patient treatment planning. In the visualization step, a 3D patient anatomical model is constructed from 2D CT scans, and various tissue types, including

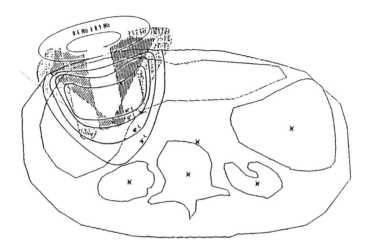

FIGURE 6.8 Early example of treatment planning for hyperthermia with a single mechanically scanned ultrasound transducer. The isotherms are represented by contours, the focused pressure beams are the shaded regions, and both of these are overlaid on contours that represent bone and tissue structures. (After P. P. Lele and J. Goddard, *Proc. 9th IEEE EMBS Conf.*, IEEE, New York, 1987.)

pressure at a specified point in space, where the phase of the calculated pressure is evaluated relative to the phase of the input signal. The relative phase of the pressure at this field point is calculated for each element, and when the phase of the sinusoidal driving signal applied to each element is defined as the negative of each relative phase value, a focused pressure field is generated. The phase of each excitation signal, which is also the phase of the conjugated pressure field produced by each array element at the focal point, maximizes the constructive interference in this location. This single focus beamforming approach is applicable to any phased array geometry.

A single focus generated by an ultrasound phased array is much smaller than the tumor target, so electronic scanning or some other method is required to improve tumor coverage. One such approach involves single spot scanning, which focuses ultrasonic energy at discrete points within the tumor volume. Single spot scanning is relatively straightforward in terms of driving signal calculations and other implementation details, but spot scanning is also potentially limited by intervening tissue heating (Sleefe and Lele 1985), which is increasingly problematic for larger tumors and/or smaller array apertures (Wang et al. 1994). With single spot scanning, intervening tissue heating is caused by overlapping power contributions from the individual focused beams, which is effectively the same problem that causes pre-focal heating with mechanically scanned ultrasound applicators. Whereas mechanical scanning reduces problems with intervening tissue heating by turning off the transducer that is closest to the central axis, ultrasound phased arrays achieve similar results by cancelling axial pressure fields. This approach, which is enabled through electronic control of the driving phases and amplitudes transmitted to each array element, prevents ultrasonic energy from preferentially accumulating along the central axis.

The annular patterns produced by the sector-vortex array (Cain and Umemura 1986) are examples of focal patterns that broaden the focal pattern while cancelling the on-axis pressures. The annular patterns generated by the sector-vortex array are obtained by adding a constant phase increment to the driving signal for each element in a ring, where the phase offset defined for one revolution around a ring of elements is an integer multiple M of 2π radians. The resulting pressure distribution in the focal plane produced by a single ring of elements is approximately represented by a Bessel function of order M. When $M = 0$, the sector-vortex array generates a single axial focus. For integer $M > 0$, annular patterns that cancel the on-axis pressure fields are produced, where increasing the value of M increases the radius of the annular pattern generated by a single ring of elements.

Relative to a single focal spot, multiple focus patterns also increase the size of the heated volume, and some of these patterns cancel the on-axis pressures simultaneously. One multiple focusing method evaluates the transfer matrix between the array elements and the focal points (and nulls) and computes the minimum norm pseudo-inverse of this matrix to determine the phased array driving signals (Ebbini and Cain 1989). The pseudo-inverse method, when combined with an iterative weighting algorithm, improves the excitation efficiency of the array so that the amplitudes of the excitations applied to each element are approximately equal. An extension of the pseudo-inverse method defines an objective function to maximize the intensity gain of the focal pattern generated by the phased array (Ebbini and Cain 1991). By maximizing the intensity gain, the heat localization is improved and problems with intervening tissue heating are reduced. Examples of gain-maximized multiple focus intensity patterns are shown in Figure 6.9. A further reduction in axial heating is achieved with mode scanning (McGough et al. 1994), which produces symmetric multiple focus patterns that simultaneously cancel the on-axis pressure field. Mode scanning exploits the planar symmetry in most phased array structures to produce rotating pressure fields similar to the annular

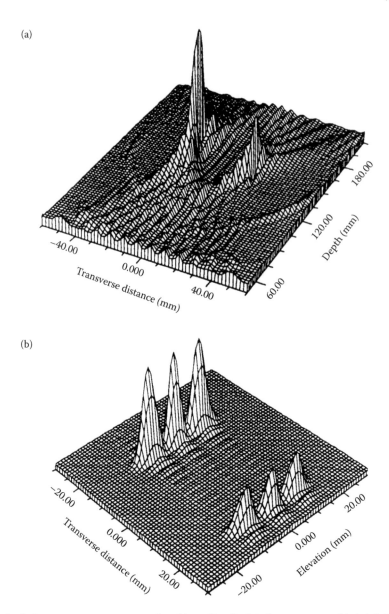

FIGURE 6.9 Simulated multiple focus intensity patterns produced by a 2D cylindrical section array: (a) Axial plane and (b) focal plane. (After E. S. Ebbini and C. A. Cain, *Int. J. Hypertherm.*, 7, 6, 1991.)

patterns generated by the sector vortex array. The multiple focus patterns produced by mode scanning, which typically consist of two or four discrete focal spots in a single focal plane, are also consistent with the planar symmetry of the phased array.

Although multiple focusing increases the size of the heated volume relative to a single focus, most multiple focus patterns cover only a fraction of the tumor volume. To heat the entire tumor, several single and/or multiple focus patterns are needed, but optimally distributing the foci among several beam patterns is a challenging problem. One approach solves the bioheat transfer equation for any number of single focus (or multiple focus) patterns and defines a transfer matrix that describes the temperature generated by each focal pattern at each control point (McGough et al. 1992). After the temperatures are specified

at each of these control points, a least-squares solution that describes the power weighting of each focal pattern is obtained. Another heating strategy defines a spiral focal point trajectory and generates approximately uniform temperatures in a circular region (Salomir et al. 2000). This is achieved in two steps, where therapeutic temperatures in the interior of the circular region are established with the spiral scan pattern, and then the temperature is maintained with a circular scan that focuses the ultrasound on the edge of the circular region.

The overlapping energy contributions generated by several beam patterns, each with multiple foci, are optimized in the tumor and in sensitive normal tissues with waveform diversity beamforming (Zeng et al. 2010). Waveform diversity calculations minimize a cost function with constraints using semi-definite programming, and the result describes the array excitation covariance matrix. The cost function defined for waveform diversity beamforming maximizes the difference between the power deposited at a representative tumor control point and the power deposited at the control points in sensitive normal tissues. Some of the constraints guarantee that the power deposited at the remaining tumor control points is comparable to that at the representative control point, and other constraints specify the total acoustic power generated by the ultrasound phased array while maintaining positive power values for each transducer element. The array excitation covariance matrix describes the optimal driving signals that satisfy the constraints while minimizing the cost function. The optimal number of multiple focus patterns is defined as the rank of the covariance matrix, and the driving signals are obtained from the singular value decomposition of the covariance matrix. When waveform diversity calculations are combined with mode scanning, all of the multiple focus patterns achieve on-axis cancellation while the computation time and the amount of computer RAM required are significantly reduced. This combination of waveform diversity and mode scanning is especially important for computationally intensive calculations with large ultrasound phased arrays with thousands of elements. Further improvements in conformal tumor heating are obtained when waveform diversity and mode scanning are combined with an adaptive algorithm that removes focal points from the interior of the tumor volume to improve conformal tumor heating (Jennings and McGough 2010). When focal points are distributed throughout the tumor, the interior temperatures are higher, and when focal points are removed from the interior, more uniform heating is achieved. With this approach, the focal points heat the tumor periphery, and the overlapping contributions from the multiple focus patterns heat the interior portion of the tumor.

Beamforming and bioheat transfer simulations, when combined with geometric aperture optimization and 3D visualization, also define a treatment planning framework for hyperthermia with ultrasound phased arrays (McGough et al. 1996). For patient treatment planning, all of these calculations are interfaced to a 3D patient model that is extracted from CT or MRI scans. Based on this model, the geometric aperture is optimized in an effort to reduce interactions between the ultrasonic power deposition and air or bone obstructions. If the results of geometric aperture optimization indicate that the acoustic window is smaller than the available array aperture, then the array is centered relative to the acoustic window and excess array elements are turned off. By deactivating these blocked elements, the phased array aperture is effectively conformed to the available acoustic window (Figure 6.10). The geometrically optimized array aperture then provides the input for beamforming and thermal optimization procedures. Once the optimal beam patterns are determined, simulated power depositions and temperature distributions are projected onto the tumor and other important anatomical structures and visualized in three dimensions. Results of different heating strategies are then evaluated and compared. In the context of prospective treatment planning,

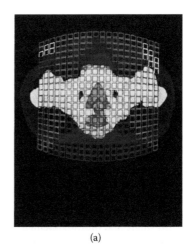

(a)

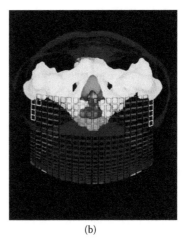

(b)

FIGURE 6.10   (See color insert.) (a) Unoptimized and (b) optimized ultrasound phase array apertures for hyperthermia in the prostate. (After R. J. McGough, M. L. Kessler, E. S. Ebbini, and C. A. Cain, *IEEE Trans. Ultrason. Ferroelect. Freq. Contr.*, 43, 6, 1996.)

3D visualization is useful for identifying the locations of potential bone heating, normal tissue hot spots, and tumor cold spots, and when realistic patient anatomical models are combined with 3D visualization of isothermal surfaces, the simulated size of the tumor volume that reaches a therapeutic temperature or thermal dose target value is also demonstrated. The 3D visualization provides valuable feedback during patient treatment planning, where the results are incorporated into the final treatment plan as needed.

A related treatment planning strategy has also been developed for hyperthermia with transurethral applicators (Chen et al. 2010). These hyperthermia applicators, which are placed within brachytherapy implants, consist of linear and sectored arrays with variable power control along the length of the device and in different directions. Treatment planning for these applicators combines a 3D bioheat transfer model with a temperature optimization algorithm and a geometric planning approach that includes 3D visualization of the anatomy, applicator, and temperature fields. The temperature optimization algorithm defined for these transurethral applicators determines the applicator power levels that maximize temperature coverage in the tumor subject to constraints on the maximum tumor temperature, the temperature in normal tissues, applicator positioning, and applicator orientation. The results of these treatment planning calculations determine the optimal combination of transurethral applicator parameters, including the sector angle and number of sections defined along the length of the device, for a given patient geometry.

6.3.3.3 Ablation Therapy with Fixed-Phase Applicators

Treatment planning for ablation therapy with fixed-phase ultrasound applicators is closely related to treatment planning for hyperthermia with single and multiple element applicators. For example, the applicators are similar, pressures and temperatures are often calculated with the same approaches, and power depositions are optimized with similar algorithms. However, some specific features influence the treatment planning strategy for ablation therapy with these devices. In particular, simulations of ultrasound ablation more frequently emphasize thermal dose calculations to determine the volume of the ablated region, some treatment planning calculations for ablation therapy consider changing ultrasound attenuation and blood perfusion values during the treatment, and other simulations evaluate the potential for irreversible thermal damage in sensitive normal tissues.

Simulations of the thermal doses produced by interstitial ultrasound applicators for ablation therapy combine transient bioheat transfer calculations with an approximate expression for the power deposition (Tyreus and Diederich 2002). The transient

temperatures generated by these applicators are evaluated with a cylindrically symmetric 2D finite element model that varies the acoustic attenuation and blood perfusion as a function of the thermal dose. For these calculations, the acoustic attenuation is constant for thermal doses less than or equal to 100 equivalent minutes at 43°C, then the attenuation increases linearly with the logarithm of the thermal dose until a threshold value for the attenuation is reached at a dose of 10,000,000 equivalent minutes at 43°C. The attenuation and absorption values are equal in these simulations. Similarly, the blood perfusion is modeled as constant up to a thermal dose of 300 equivalent minutes at 43°C, and then for larger thermal dose values, the blood perfusion is set to zero. The lethal thermal dose threshold for lesion formation in these calculations is defined as 600 equivalent minutes at 43°C. The results of this simulation model show that the dimensions and evolution of predicted lesions are an excellent agreement with experimentally measured temperatures and lesion dimensions (Tyreus and Diederich 2002). This agreement between the simulated temperatures and measured lesion shapes in the axial direction is demonstrated in Figure 6.11. Other numerical simulations with this model evaluate the size of lesions produced by applicators with different diameters (Tyreus et al. 2003) and establish strategies to avoid bone heating while ablating the prostate (Wootton et al. 2007).

Numerical simulations of pressures and temperatures produced by a mechanically rotated interstitial ultrasound applicator are also evaluated for thermal ablation in the prostate (Chopra et al. 2005). This interstitial applicator consists of a fixed-phase linear array of flat rectangular ultrasound transducers with

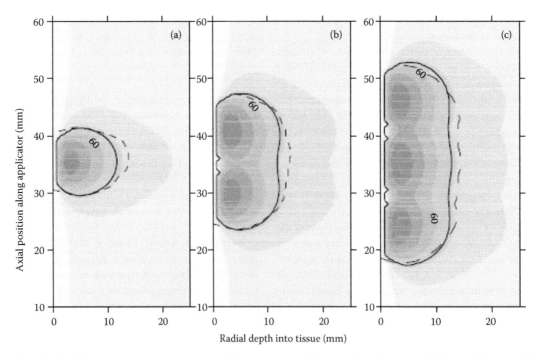

FIGURE 6.11 Simulated axial temperature contours produced by transurethral ultrasound applicators with (a) one, (b) two, and (c) three active segments. Representative measured lesions are indicated with dashed lines. (After P. D. Tyreus and C. J. Diederich, *Phys. Med. Biol.*, 47, 7, 2002.)

operating frequencies of 4.7 and 9.7 MHz. The applicator is continuously rotated within the urethra throughout the simulated treatment, where the power settings and excitation frequencies are determined by a feedback algorithm that controls the temperatures in a single 2D plane. In these simulations, the pressures are simulated with the rectangular radiator method, and temperatures are computed with a finite difference implementation of a 3D bioheat transfer model. The bioheat transfer simulations demonstrate that this applicator, when combined with the control algorithm, conforms thermal damage patterns to the outer boundary of the prostate (Figure 6.12). These simulation results also successfully predict the extent of thermal damage obtained in measurements with tissue-mimicking gel phantoms. Related simulation studies show that when the control algorithm is extended to 3D, conformal thermal coagulation is achieved in realistic 3D prostate models (Burtnyk et al. 2009). Other simulations evaluate the thermal dose in the rectum, the pelvic bone, the neurovascular bundle, and the urinary sphincters during transurethral ultrasound ablation of the prostate, and the results

of these simulations suggest treatment planning strategies that can reduce thermal injury to these sensitive structures (Burtnyk et al. 2010).

Treatment planning for transrectal HIFU defines 3D anatomical models of the prostate target and other tissue structures (Seip et al. 2004) and optimizes parameters for thermal ablation (Fedewa et al. 2005). The 3D anatomical models of the prostate, urethra, and rectal wall are obtained from diagnostic ultrasound images that are manually segmented and then parameterized. The resulting 3D anatomical structures, which can be combined with other models that describe the location of the neurovascular bundle and other sensitive normal tissues, provide the input for treatment optimization and 3D visualization. The treatment plan then optimizes the locations of the lesions subject to multiple constraints on the spacing between adjacent lesions, the extent of the treated margin beyond the prostate capsule, and the maximum treatment angle of the mechanically rotated ultrasound transducer. The optimization procedure, which calculates treatment parameters for two different

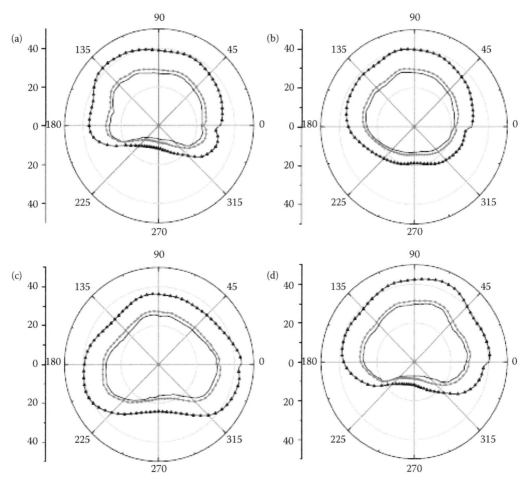

FIGURE 6.12 Simulated ultrasound ablation with a rotating transurethral applicator evaluated in four different prostate models. In each plot, the innermost contour represents the outer prostate boundary, the contour with circle markers located just outside of the prostate indicates the 100% cell kill boundary, and the outermost contour with triangle markers indicates the 0% cell kill boundary. (After R. Chopra, M. Burtnyk, M. A. Haider, and M. J. Bronskill, *Phys. Med. Biol.*, 50, 21, 2005.)

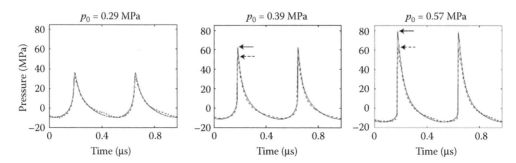

FIGURE 6.13 Simulated and measured nonlinear pressure waveforms generated at the focus of a spherical ultrasound transducer evaluated for different pressure normalization factors. (After M. S. Canney, M. R. Bailey, L. A. Crum, V. A. Khokhlova, and O. A. Sapozhnikov, *J. Acoust. Soc. Am.*, 124, 4, 2008.)

rectal wall distances and for two ultrasound transducers with different focal distances, selects the transducer and determines the power settings, the excitation frequency, and the exposure time. By combining parameterized 3D anatomical models with an algorithm that optimizes the lesion locations, patient treatment planning reduces the treatment time for these transrectal ultrasound applicators with an approach that is readily adapted to different probe parameters.

Other related numerical models simulate HIFU for ablation therapy and hemostasis. One simulation model, which is compared to measurements in water and in a gel phantom, computes the nonlinear pressure fields generated by a spherically focused HIFU transducer with the KZK equation (Canney et al. 2008). The simulation results in Figure 6.13 show that as the scale factor for the input pressure increases, the calculated shock fronts are increasingly steep with sharper pressure peaks. The simulated pressure waveforms demonstrate close agreement with the measured waveforms for lower focal intensity values, but some differences are observed at higher focal intensities. This nonlinear model is ultimately intended for pressure and intensity calculations in tissue. Another simulation model combines the KZK equation with a bioheat transfer model and evaluates the influence of nonlinear ultrasound propagation on transient and steady state temperatures (Curra et al. 2000). The simulation results for homogeneous tissue show that nonlinear propagation strongly influences the peak temperature and the rate of temperature increase in the focal zone and that smaller differences relative to linear ultrasound propagation models are observed elsewhere. When a large blood vessel is included in the bioheat transfer model, the temperature values decrease and the location of the peak temperature shifts in the linear and the nonlinear pressure simulations. The large blood vessel, which is located coaxially with respect to the spherically focused transducer, also eliminates the enhancement of the temperature deposition due to nonlinear ultrasound propagation (Figure 6.14).

6.3.3.4 Ablation Therapy with Ultrasound Phased Arrays

The thermal lesions generated by an ultrasound phased array system are often guided by thermometry obtained from magnetic resonance images, which provides temperature feedback during a treatment. Some examples of magnetic resonance-compatible phased arrays constructed for ablation therapy include a concentric-ring array (Fjield et al. 1996) and a spherical section array (Daum and Hynynen 1999). These arrays generate pressure fields that are numerically modeled with the Rayleigh-Sommerfeld integral, the resulting temperatures are modeled with the bioheat transfer equation, and then the thermal dose is computed.

As ultrasound phased array systems with larger apertures are developed for thermal ablation, the effect of aberrating layers grows increasingly important, and numerical models that account for aberrating bone and tissue structures are needed. One model of curved tissue layers subdivides curved surfaces into smaller planar radiators, and the contributions from these planar radiators are superposed (Fan and Hynynen 1994). This numerical approximation is applied to linear ultrasound propagation through a human skull model obtained from magnetic resonance images, and the results of phase aberration correction is evaluated for several spherically focused phased arrays (Sun and Hynynen 1998). For these simulations, the diameter and radius of curvature are constants, and the number of elements within the array aperture is varied. The results show that refocusing is more effective when more array elements are defined within the available aperture and that phase aberration correction is less effective with fewer array elements. Another simulation model extracts the skull geometry from CT scans of the head and then computes the pressure in a plane directly above the skull surface by backward projecting pressures calculated with the Rayleigh-Sommerfeld integral using the angular spectrum approach (Clement and Hynynen 2002). The resulting pressure is then forward propagated through the skull and into the focal plane for each array element, and the effects of phase aberration correction are reduced with a refocusing algorithm. These simulation models also describe the effect of phase aberration on focused ultrasound fields produced during ablation of uterine fibroids (Liu et al. 2005, White et al. 2008). These models of ultrasound propagation in the abdomen, which are obtained from segmented magnetic resonance images, demonstrate that focal patterns are distorted by the tissue layers and that the distortion is significantly reduced by phase aberration correction.

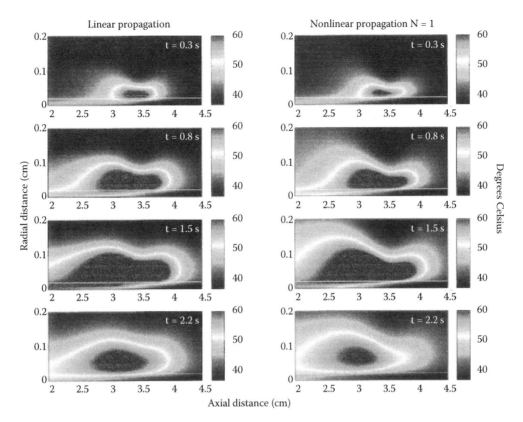

FIGURE 6.14 Simulated temperatures generated with a spherically focused transducer that is coaxially aligned with a 0.4 mm diameter blood vessel. Plots on the left are computed with a linear ultrasound propagation model, and plots on the right are computed with a nonlinear ultrasound propagation model. (After F. P. Curra, P. D. Mourad, V. A. Khokhlova, R. O. Cleveland, and L. A. Crum, *IEEE Trans. Ultrason. Ferroelect. Freq. Contr.*, 47, 4, 2000.)

Treatment planning for ultrasound ablation of uterine fibroids optimizes the path of the ultrasound beam and the distribution of lesions within the target volume (Tempany et al. 2003). These treatment plans, which are based on T1-weighted and T2-weighted magnetic resonance images collected prior to the treatment, define the size, volume, and location of the fibroids and other structures. The orientation of the phased array is established prior to the treatment with software that simultaneously displays magnetic resonance images and a simplified model of a focused ultrasound beam. This combination facilitates avoidance of bowel loops and scars in the beam path, where the applicator is translated and/or tilted as needed to avoid these structures. Figure 6.15 contains an example of 3D treatment planning for MR-guided ultrasound ablation of a uterine fibroid (Shen et al., 2009). The left panel shows an example of a beam that was tilted to change the beam angle on the sacrum, the center panel shows an orthogonal cross section and the lesion sizes in that cross section, and the right panel indicates the spacing between sparsely packed lesions.

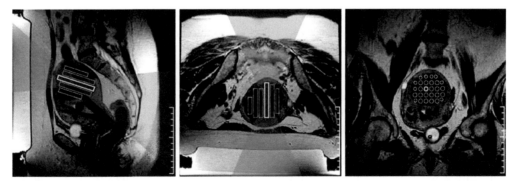

FIGURE 6.15 An example of geometric treatment planning, where the ultrasound phased array is tilted to avoid bowel loops and scars. (After S. H. Shen, F. Fennessy, N. McDannold, F. Jolesz, and C. Tempany, *Semin. Ultrasound CT MR*, 30, 2, 2009.)

Numerical simulations of ablation therapy with ultrasound phased arrays are encountered in several other applications. For example, a simulation model of a phased array focusing through the ribs is described in (Aubry et al. 2008), and pressures are also simulated for a related fixed-phase multiple element ultrasound applicator in (Civale et al. 2006). These models, when combined with experimental results, suggest strategies that reduce problems with phase aberration and bone heating. Another simulation model illustrates the effect of cavitation on the size of the lesion generated during HIFU (Chavrier et al. 2000). Other simulations calculate the temperature distribution from estimated power depositions while controlling the thermal dose for HIFU (Blankespoor et al. 2009).

6.4 Summary

Several different numerical models, including the Rayleigh-Sommerfeld integral and other faster approaches, are described for linear simulations of therapeutic ultrasound generated by single element and multiple element applicators. Methods for simulating nonlinear pressure propagation are summarized, a short survey of the available software for linear and nonlinear pressure calculations is provided, and the equations for bioheat transfer modeling and thermal dose calculations are quickly reviewed. These simulations, when combined with optimization and 3D visualization, provide the framework for patient treatment planning. Examples of treatment planning strategies are described for mechanically scanned ultrasound applicators and ultrasound phased array systems applied to hyperthermia and ablation therapy.

References

S. I. Aanonsen, T. Barkve, J. Naze Tjotta, and S. Tjotta. Distortion and harmonic generation in the nearfield of a finite amplitude sound beam. *J. Acoust. Soc. Am.*, 75(3):749–768, 1984.

M. Abramowitz and I. A. Stegun. *Handbook of mathematical functions, with formulas, graphs, and mathematical tables*, 887–889 and 916–919. Dover Publications, Inc., New York, 1972.

J. F. Aubry, M. Pernot, F. Marquet, M. Tanter, and M. Fink. Transcostal high-intensity-focused ultrasound: *Ex vivo* adaptive focusing feasibility study. *Phys. Med. Biol.*, 53(11):2937–2951, 2008.

A. Blankespoor, A. Payne, N. Todd, M. Skliar, S. Roell, J. Roland, D. Parker, and R. Roemer. Model predictive control of HIFU treatments in 3D for treatment time reduction. *AIP Conference Proceedings*, 1113(1):215–219, 2009.

R. N. Bracewell. *The Fourier transform and its applications*. McGraw-Hill, New York, 2000.

M. Burtnyk, R. Chopra, and M. J. Bronskill. Quantitative analysis of 3-D conformal MRI-guided transurethral ultrasound therapy of the prostate: Theoretical simulations. *Int. J. Hypertherm.*, 25(2):116–31, 2009.

M. Burtnyk, R. Chopra, and M. Bronskill. Simulation study on the heating of the surrounding anatomy during transurethral ultrasound prostate therapy: A 3D theoretical analysis of patient safety. *Med. Phys.*, 37(6):2862–75, 2010.

C. A. Cain and S.-I. Umemura. Concentric-ring and sector-vortex phased-array applicators for ultrasound hyperthermia. *IEEE Trans. MTT*, MTT-34(5):542–551, 1986.

M. S. Canney, M. R. Bailey, L. A. Crum, V. A. Khokhlova, and O. A. Sapozhnikov. Acoustic characterization of high intensity focused ultrasound fields: A combined measurement and modeling approach. *J. Acoust. Soc. Am.*, 124(4):2406–2420, 2008.

F. Chavrier, J. Y. Chapelon, A. Gelet, and D. Cathignol. Modeling of high-intensity focused ultrasound-induced lesions in the presence of cavitation bubbles. *J. Acoust. Soc. Am.*, 108(1):432–440, 2000.

X. Chen, C. J. Diederich, J. H. Wootton, J. Pouliot, and I-C. Hsu. Optimisation-based thermal treatment planning for catheter-based ultrasound hyperthermia. *Int. J. Hypertherm.*, 26(1):39–55, 2010.

R. Chopra, M. Burtnyk, M. A. Haider, and M. J. Bronskill. Method for MRI-guided conformal thermal therapy of prostate with planar transurethral ultrasound heating applicators. *Phys. Med. Biol.*, 50(21):4957–4975, 2005.

P. T. Christopher and K. J. Parker. New approaches to the linear propagation of acoustic fields. *J. Acoust. Soc. Am.*, 90(1):507–521, 1991a.

P. T. Christopher and K. J. Parker. New approaches to nonlinear diffractive field propagation. *J. Acoust. Soc. Am.*, 90(1):488–499, 1991b.

J. Civale, R. Clarke, I. Rivens, and G. ter Haar. The use of a segmented transducer for rib sparing in HIFU treatments. *Ultrasound Med. Biol.*, 32(11):1753–1761, 2006.

G. T. Clement and K. Hynynen. A non-invasive method for focusing ultrasound through the human skull. *Phys. Med. Biol.*, 47(8):1219–1236, 2002.

G. T. Clement and K. Hynynen. Forward planar projection through layered media. *IEEE Trans. Ultrason. Ferroelect. Freq. Contr.*, 50(12):1689–1698, 2003.

R. O. Cleveland, M. F. Hamilton, and D. T. Blackstock. Time-domain modeling of finite-amplitude sound in relaxing fluids. *J. Acoust. Soc. Am.*, 99(6):3312–3318, 1996.

R. S. C. Cobbold. Foundations of biomedical ultrasound. *Oxford University Press*, 2007.

F. P. Curra, P. D. Mourad, V. A. Khokhlova, R. O. Cleveland, and L. A. Crum. Numerical simulations of heating patterns and tissue temperature response due to high-intensity focused ultrasound. *IEEE Trans. Ultrason. Ferroelect. Freq. Contr.*, 47(4):1077–1089, 2000.

C. Damianou and K. Hynynen. Focal spacing and near-field heating during pulsed high-temperature ultrasound therapy. *Ultrasound Med. Biol.*, 19(9):777–787, 1993.

D. R. Daum and K. Hynynen. A 256-element ultrasonic phased array system for the treatment of large volumes of deep seated tissue. *IEEE Trans. Ultrason. Ferroelect. Freq. Contr.*, 46(5):1254–1268, 1999.

P. J. Davis and P. Rabinowitz. *Numerical integration*, 73–76, 87–90, 139–140, and 369. Academic Press, New York, 1975.

R. J. Dickinson. An ultrasound system for local hyperthermia using scanned focused transducers. *IEEE Trans. Biomed. Eng.*, BME-31(1):120–125, 1984.

E. S. Ebbini and C. A. Cain. Multiple-focus ultrasound phased-array pattern synthesis: Optimal driving signal distributions for hyperthermia. *IEEE Trans. Ultrason. Ferroelect. Freq. Contr.*, 36(5):540–548, 1989.

E. S. Ebbini and C. A. Cain. Optimization of the intensity gain of multiple-focus phased-array heating patterns. *Int. J. Hypertherm.*, 7(6):953–973, 1991.

X. Fan and K. Hynynen. The effects of curved tissue layers on the power deposition patterns of therapeutic ultrasound beams. *Med. Phys.*, 21(1):25–34, 1994.

R. J. Fedewa, R. Seip, R. F. Carlson, W. Chen, N. T. Sanghvi, M. A. Penna, K. A. Dines, and R. E. Pfile. Automated treatment planning for prostate cancer HIFU therapy. In *Ultrasonics Symposium, 2005 IEEE*, volume 2, 1135–1138, 2005.

P. Fessenden, E. R. Lee, T. L. Anderson, J. W. Strohbehn, J. L. Meyer, T. V. Samulski, and J. B. Marmor. Experience with a multitransducer ultrasound system for localized hyperthermia of deep tissues. *IEEE Trans. Biomed. Eng.*, BME-31(1):126–135, 1984.

T. Fjield, X. Fan, and K. Hynynen. A parametric study of the concentric-ring transducer design for MRI guided ultrasound surgery. *J. Acoust. Soc. Am.*, 100(2):1220–1230, 1996.

S. A. Goss, L. A. Frizzell, and F. Dunn. Ultrasonic absorption and attenuation in mammalian tissues. *Ultrasound Med. Biol.*, 5(2):181–186, 1979.

J. Huijssen and M. D. Verweij. An iterative method for the computation of nonlinear, wide-angle, pulsed acoustic fields of medical diagnostic transducers. *J. Acoust. Soc. Am.*, 127(1):33–44, 2010.

K. Hynynen. Biophysics and technology of ultrasound hyperthermia. In M. Gautherie, editor, *Methods of External Hyperthermic Heating*, 61–115. Springer-Verlag, New York, 1990a.

K. Hynynen. Hot spots created at skin-air interfaces during ultrasound hyperthermia. *Int. J. Hypertherm.*, 6(6):1005–1012, 1990b.

K. Hynynen and D. DeYoung. Temperature elevation at muscle-bone interface during scanned, focussed ultrasound hyperthermia. *Int. J. Hypertherm.*, 4(3):267–279, 1988.

K. Hynynen and B. A. Lulu. Hyperthermia in cancer treatment. *Investigative Radiology*, 25(7):824–834, 1990.

M. S. Ibbini and C. A. Cain. A field conjugation method for direct synthesis of hyperthermia phased-array heating patterns. *IEEE Trans. Ultrason. Ferroelect. Freq. Contr.*, 36(2):3–9, 1989.

M. R. Jennings and R. J. McGough. Improving conformal tumour heating by adaptively removing control points from waveform diversity beamforming calculations: A simulation study. *Int. J. Hypertherm.*, 26(7):710–724, 2010.

J. A. Jensen. Field: A program for simulating ultrasound systems. *Med. Biol. Eng. Comp., 10th Nordic-Baltic Conference on Biomedical Imaging*, 4, Suppl. 1, Part 1(1):351–353, 1996.

J. Kelly and R. J. McGough. A fast time-domain method for calculating the near field pressure generated by a pulsed circular piston. *IEEE Trans. Ultrason. Ferroelect. Freq. Contr.*, 53(6):1150–1159, 2006.

V. A. Khokhlova, R. Souchon, J. Tavakkoli, O. A. Sapozhnikov, and D. Cathignol. Numerical modeling of finite-amplitude sound beams: Shock formation in the near field of a CW plane piston source. *J. Acoust. Soc. Am.*, 110(1):95–108, 2001.

Y. S. Lee and M. F. Hamilton. Time-domain modeling of pulsed finite-amplitude sound beams. *J. Acoust. Soc. Am.*, 97(2):906–917, 1995.

P. P. Lele and J. Goddard. Optimizing insonation parameters in therapy planning for deep heating by SIMFU. In *Proc. 9th IEEE EMBS Conf.*, 1650–1651, New York, 1987. IEEE.

H. L. Liu, N. McDannold, and K. Hynynen. Focal beam distortion and treatment planning in abdominal focused ultrasound surgery. *Med. Phys.*, 32(32):1270–1280, 2005.

N. McDannold, K. Hynynen, D. Wolf, G. Wolf, and F. Jolesz. MRI evaluation of thermal ablation of tumors with focused ultrasound. *Journal of Magnetic Resonance Imaging*, 8(1):91–100, 1998.

R. J. McGough. Rapid calculations of time-harmonic nearfield pressures produced by rectangular pistons. *J. Acoust. Soc. Am.*, 115(5):1934–1941, 2004.

R. J. McGough, E. S. Ebbini, and C. A. Cain. Direct computation of ultrasound phased-array driving signals from a specified temperature distribution for hyperthermia. *IEEE Trans. Biomed. Eng.*, 39(8):825–835, 1992.

R. J. McGough, H. Wang, E. S. Ebbini, and C. A. Cain. Mode scanning: Heating pattern synthesis with ultrasound phased arrays. *Int. J. Hypertherm.*, 10(3):433–442, 1994.

R. J. McGough, M. L. Kessler, E. S. Ebbini, and C. A. Cain. Treatment planning for hyperthermia with ultrasound phased arrays. *IEEE Trans. Ultrason. Ferroelect. Freq. Contr.*, 43(6):1074–1084, 1996.

R. J. McGough, T. V. Samulski, and J. F. Kelly. An efficient grid sectoring method for calculations of the nearfield pressure generated by a circular piston. *J. Acoust. Soc. Am.*, 115(5):1942–1954, 2004.

E. Moros, R. Roemer, and K. Hynynen. Simulations of scanned focussed ultrasound hyperthermia: The effect of scanning speed and pattern on the temperature fluctuations at the focal depth. *IEEE Trans. Ultrason. Ferroelect. Freq. Contr.*, 35(5):552–560, 1988.

E. G. Moros, R. B. Roemer, and K. Hynynen. Pre-focal plane high-temperature regions induced by scanning focused ultrasound beams. *Int. J. Hypertherm.*, 6(2):351–366, 1990.

E. G. Moros, R. J. Myerson, and W. L. Straube. Aperture size to therapeutic volume relation for a multielement ultrasound system: Determination of applicator adequacy for superficial hyperthermia. *Med. Phys.*, 20(5):1399–1409, 1993.

E. G. Moros, X. Fan, and W. L. Straube. An investigation of penetration depth control using parallel opposed ultrasound arrays and a scanning reflector. *J. Acoust. Soc. Am.*, 101(3):1734–1741, 1997.

J. C. Mould, G. L. Wojcik, L. M. Carcione, M. Tabei, T. D. Mast, and R. C. Waag. Validation of FFT-based algorithms for large-scale modeling of wave propagation in tissue. *Proceedings of the IEEE Ultrasonics Symposium*, 1551–1556, 1999.

W. L. Nyborg. Heat generation by ultrasound in a relaxing medium. *J. Acoust. Soc. Am.*, 70(2):310–312, 1981.

W. L. Nyborg and R. B. Steele. Temperature elevation in a beam of ultrasound. *Ultrasound Med. Biol.*, 9(6):611–620, 1983.

K. B. Ocheltree and L. A. Frizzell. Sound field calculation for rectangular sources. *IEEE Trans. Ultrason. Ferroelect. Freq. Contr.*, 36(2):242–248, 1989.

H. T. O'Neil. Theory of focusing radiators. *J. Acoust. Soc. Am.*, 21(5):516–526, 1949.

H. H. Pennes. Analysis of tissue and arterial blood temperatures in the resting human forearm. *J. Appl. Physiol.*, 1(2):93–122, 1948.

R. B. Roemer. Thermal dosimetry. In M. Gautherie, ed., *Thermal Dosimetry and Treatment Planning*, 119–214. Springer-Verlag, New York, 1990.

R. B. Roemer, W. Swindell, S. T. Clegg, and R. L. Kress. Simulation of focused, scanned ultrasonic heating of deep-seated tumours: the effect of blood perfusion. *IEEE Trans. Sonics Ultrason.*, SU-31(5):457–466, 1984.

R. Salomir, J. Palussiere, F. C. Vimeux, J. A. de Zwart, B. Quesson, M. Gauchet et al. Local hyperthermia with MR-guided focused ultrasound: Spiral trajectory of the focal point optimized for temperature uniformity in the target region. *J. Magn. Reson. Imaging*, 12 (4):571–583, 2000.

S. A. Sapareto and W. C. Dewey. Thermal dose determination in cancer therapy. *Int. J. Radiat. Oncol. Biol. Phys.*, 10(6):787–800, 1984.

R. Seip, R. F. Carlson, W. Chen, N. T. Sanghvi, and K. A. Dines. Automated HIFU treatment planning and execution based on 3D modeling of the prostate, urethra, and rectal wall. In *Proceedings of the IEEE Ultrasonics Symposium, 2004 IEEE*, volume 3, 1781–1784, 2004.

S. H. Shen, F. Fennessy, N. McDannold, F. Jolesz, and C. Tempany. Image-guided thermal therapy of uterine fibroids. *Semin. Ultrasound CT MR*, 30(2):91–104, 2009.

G. E. Sleefe and P. P. Lele. The limitations of ultrasonic phased arrays for the induction of local hyperthermia. In *Proceedings of the IEEE Ultrasonics Symposium*, 949–952, New York, 1985. IEEE.

J. E. Soneson. A user-friendly software package for HIFU simulation. In E. S. Ebbini, editor, *American Institute of Physics Conference Series*, volume 1113, 165–169, 2009.

J. W. Strohbehn and R. B. Roemer. A survey of computer simulations of hyperthermia treatments. *IEEE Trans. Biomed. Eng.*, BME-31(1):136–149, 1984.

J. Sun and K. Hynynen. Focusing of therapeutic ultrasound through a human skull: A numerical study. *J. Acoust. Soc. Am.*, 104(3):1705–1715, 1998.

W. Swindell, R. B. Roemer, and S. T. Clegg. Temperature distributions caused by dynamic scanning of focused ultrasound transducers. In *1982 Ultrasonics Symposium*, 750–753, 1982.

C. M. Tempany, E. A. Stewart, N. McDannold, B. J. Quade, F. A. Jolesz, and K. Hynynen. MR imaging-guided focused ultrasound surgery of uterine leiomyomas: A feasibility study. *Radiology*, 226(3):897–905, 2003.

P. D. Tyreus and C. J. Diederich. Theoretical model of internally cooled interstitial ultrasound applicators for thermal therapy. *Phys. Med. Biol.*, 47(7):1073–1089, 2002.

P. D. Tyreus, W. H. Nau, and C. J. Diederich. Effect of applicator diameter on lesion size from high temperature interstitial ultrasound thermal therapy. *Med. Phys.*, 30(7):1855–1863, 2003.

C. J. Vecchio, M. E. Schafer, and P. A. Lewin. Prediction of ultrasonic field propagation through layered media using the extended angular spectrum method. *Ultrasound Med. Biol.*, 20(7):611–622, 1994.

U. Vyas and D. Christensen. Ultrasound beam propagation using the hybrid angular spectrum method. *Conf Proc IEEE Eng. Med. Biol. Soc.*, 2526–2529, 2008.

H. Wang, R. McGough, and C. Cain. Limits on ultrasound for deep hyperthermia. In *Proceedings of the IEEE Ultrasonics Symposium*, volume 3, 1869–1872, New York, 1994. IEEE.

P. J. White, B. Andre, N. McDannold, and G. T. Clement. A pre-treatment planning strategy for high-intensity focused ultrasound (HIFU) treatments. In *Ultrasonics Symposium, 2008. IUS 2008. IEEE*, 2056–2058, 2008.

G. L. Wojcik, D. K. Vaughan, N. Abboud, and J. Mould Jr. Electromechanical modeling using explicit time-domain finite elements. In *Ultrasonics Symposium, 1993. Proceedings., IEEE 1993*, 1107–1112 vol. 2, 1993.

J. H. Wootton, A. B. Ross, and C. J. Diederich. Prostate thermal therapy with high intensity transurethral ultrasound: The impact of pelvic bone heating on treatment delivery. *Int. J. Hypertherm.*, 23(8):609–622, 2007.

P. Wu, R. Kazys, and T. Stepinski. Optimal selection of parameters for the angular spectrum approach to numerically evaluate acoustic fields. *J. Acoust. Soc. Am.*, 101(1):125–134, 1997.

X. Yang and R. O. Cleveland. Time domain simulation of non-linear acoustic beams generated by rectangular pistons with application to harmonic imaging. *J. Acoust. Soc. Am.*, 117(1):113–123, 2005.

J. Zemanek. Beam behavior within the nearfield of a vibrating piston. *J. Acoust. Soc. Am.*, 49(1):181–191, 1971.

R. J. Zemp, J. Tavakkoli, and R. S. Cobbold. Modeling of nonlinear ultrasound propagation in tissue from array transducers. *J. Acoust. Soc. Am.*, 113(1):139–152, 2003.

X. Zeng and R. J. McGough. Evaluation of the angular spectrum approach for simulations of near-field pressures. *J. Acoust. Soc. Am.*, 123(1):68–76, 2008.

X. Zeng and R. J. McGough. Optimal simulations of ultrasonic fields produced by large thermal therapy arrays using the angular spectrum approach. *J. Acoust. Soc. Am.*, 125(5):2967–2977, 2009.

X. Zeng, J. Li, and R. J. McGough. A waveform diversity method for optimizing 3-d power depositions generated by ultrasound phased arrays. *IEEE Trans. Biomed. Eng.*, 57(1):41–47, 2010.

7

Numerical Modeling for Simulation and Treatment Planning of Thermal Therapy

Esra Neufeld
Foundation for Research on Information Technologies in Society (IT'IS) and Swiss Federal Institute of Technology (ETHZ)

Maarten M. Paulides
Erasmus MC Daniel den Hoed Cancer Center

Gerard C. van Rhoon
Erasmus MC Daniel den Hoed Cancer Center

Niels Kuster
Foundation for Research on Information Technologies in Society (IT'IS) and Swiss Federal Institute of Technology (ETHZ)

7.1 Need for Treatment Planning in Thermal Therapy

In the context of thermal therapy, numerical modeling and treatment planning are needed for various reasons:

- Determination of the correct dose for a patient is not straightforward.
- The commonly used antenna arrays are often hard to steer and give rise to complex energy deposition patterns that cannot easily be controlled [158]. Hot spots during treatment are hard to predict and can limit the amount of energy that can be deposited in a patient. They are painful and can ultimately damage healthy tissue. Treatment planning helps to reduce hot spots [186] and thus helps to reduce the risk of damage to thermally sensitive healthy tissues such as the spinal cord [138].
- Monitoring temperature during treatment is difficult. Invasive thermometry increases the risk of complications [172]. Noninvasive magnetic resonance imaging (MRI) thermometry is a promising alternative [176], but, even though currently being refined for this application, suffers from the following limitations: it is costly, not widely available, still not applicable for all tissue types, not yet sufficiently accurate and stable, and the MR adds to the claustrophobia issue already

associated with hyperthermia. It should also be noted that even with prefect temperature mapping the problem of optimal excitation would still not be solved, although approaches exist for using multiple MR temperature maps from different antennae settings as a basis for optimization [31].

- Treatment planning software can be used to train staff, to develop new applicators [138], and to study the impact of various physical [44, 171] and physiological parameters (e.g., tissue characteristics, etc).
- Treatment planning provides visual information about the treatment and might help further the understanding and increase the acceptance of hyperthermia as a treatment modality.

Figure 7.1 shows how hyperthermia treatment planning (HTP) can be embedded in the clinical flow. The following steps have to be performed during hyperthermia treatment planning (see Figure 7.2):

- Segmentation of medical images to generate a patient model. In contrast to radiotherapy, all tissues have to be segmented in accordance with their tissue type in order to select the correct properties (electrical and physiological).
- Simulation of the electromagnetic (EM) fields generated by the applicator antennas.
- Simulation of the induced temperature changes.

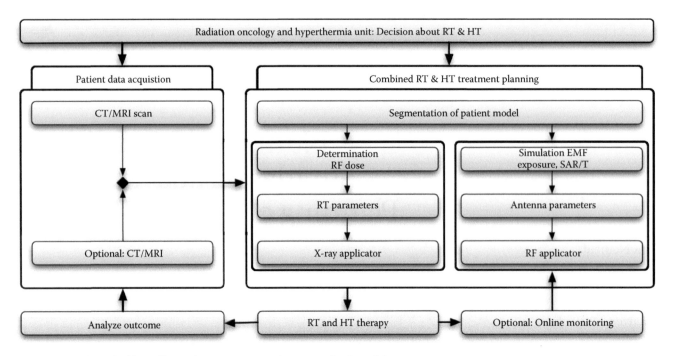

FIGURE 7.1 The embedding of hyperthermia treatment planning in the clinical flow.

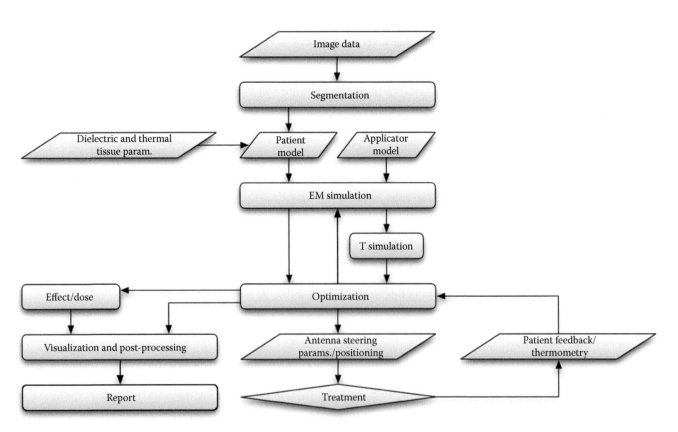

FIGURE 7.2 The steps and data involved in hyperthermia treatment planning.

- Optimization of the steering parameters.
- Determination of the resulting treatment effects.
- Visualization and post-processing.

It has been established that high-resolution treatment planning is required to correctly capture hot spots and reliably predict field distributions [166]. Only 3D simulations can correctly describe the temperature increase distribution generated by complex applicators [158]. While temperature increase is likely to be the relevant factor and the specific absorption rate (SAR) distribution does not correlate well with the temperature increase steady state [94] (due to thermal conduction, tissue specific perfusion, and boundary effects), temperature simulations carry additional uncertainties such as correct thermal model, large inter-patient variability of thermal tissue properties, etc. It is therefore unclear, and disputed, whether SAR-based or temperature-based planning is preferable [158].

7.2 Hyperthermia Treatment Planning (HTP)

HTP [102, 103], while recommended by European Society for Hyperthermic Oncology (ESHO) quality assurance guidelines [104,180], is currently rarely performed clinically. Most of the existing codes for HTP have been written and used for research purposes [82,160,165,192,198]. They are usually optimized for the variant of hyperthermia used at the specific institute and cannot be easily generalized.

The commercially available treatment planning tool HyperPlan is based on the AMIRA medical image analysis software and has been developed mostly by the group of Professor Peter Wust at the Charite Berlin and the Konrad Zuse Institute Berlin (ZIB). It is designed for and sold with the BSD deep hyperthermia systems. A variant of it has been adapted for nanofluid-based hyperthermia [69]. HyperPlan has also been adapted to offer an interface to MRI thermometry.

An alternative HTP tool is HYCAT. This tool was developed by the IT'IS Foundation [122] and has been introduced clinically by the HT group of Dr. Van Rhoon (Erasmus MC-Daniel den Hoed, Rotterdam). HYCAT is based on the EM simulation engine of SEMCAD x (SPEAG, Switzerland) and is complemented with the segmentation tool iSEG (Zurich Med Tech AG, Switzerland), thermo simulation software, multiple field optimizers, a Python scripting framework, and tools for effect quantification and post-processing, as well as wizards to reduce human error and simplify the setup of simulations. HYCAT has been applied in treatment planning [44, 135], improving existing and developing novel applicators [136], investigating RF and HIFU ablation, and helping to develop treatment guidelines (e.g., water bolus shape and temperature) [44, 171].

7.3 Segmentation

The accurate patient model required for EM and thermal simulations is obtained by segmentation of three-dimensional medical images obtained by MR, CT, etc. Segmentation is the task of identifying the image regions belonging to a tissue and is an important topic in the field of computer vision. A vast range of literature on this subject exists (e.g., for overviews see [80, 144] and for more detailed reviews on vessel segmentation see [47, 58, 89, 161, 162]). Compared to radiotherapy treatment planning, segmentation for hyperthermia treatment planning is far more demanding as more structures have to be distinguished with high accuracy (the dielectric and thermal parameters vary strongly between tissues). Also, the exact shape of tissue interfaces can be highly relevant for hot spot prediction, especially in regions with resonances such as the pelvis. The only alternative to detailed segmentation would be to determine corresponding tissue parameter maps for the patient using medical imaging, which has not been achieved to date.

Segmentation methods can be classified according to many criteria: (1) how automatic or interactive they are, (2) whether they work on slices or volumes, (3) whether they identify regions by using a homogeneity criterion (gray level/texture based) or if instead they try to find borders, (4) whether some prior or statistical knowledge is used, (5) whether they are based on local or global image information, (6) whether they identify single regions or use competitive approaches to identify multiple regions at the same time, (7) whether they find a static solution or a dynamic solution such as a curve evolution over time, etc. Which method is best suited depends on the image type (e.g., MRI, CT) and quality, the regions that have to be identified, the required accuracy, and the available time.

Requirements for segmentation tools in the context of HTP are the robustness of the segmentation approach, the necessity of generating detailed models (far more important than in radiotherapy as the differences in tissue parameters are considerable), the quality of the interface to the simulation tools, the ease of use and no requirement for highly specialized staff, and that they work with commonly available image data.

In the context of hyperthermia treatment planning, several segmentation techniques have been used and suggested:

AMIRA HyperPlan offers a choice between manual segmentation with a brush, a live-wire type delineation, manual thresholding, a simple thresholded region growing with the possibility of user-specified limits, and a type of evolving boundary method (perhaps based on fuzzy connectedness or a fast marching method). Interpolation can be used between slices, various filters for pre-processing the image data are available, and surfaces can be extracted and simplified based on the segmentation results. [191] compared the speed and accuracy of manual, live-wire, region/volume growing and watershed based segmentation. They concluded that while region/volume growing and watershed based segmentation are suitable, a well-interfaced manual segmentation can sometimes be faster as well as more accurate. [191] further concluded that 3D segmentation can be faster than 2D segmentation if good interactions and correction routines are provided. Contouring was also used in [86]. An automatic segmentation technique for the thigh to be used in HTP was presented in [141]. [165] used a thresholding technique.

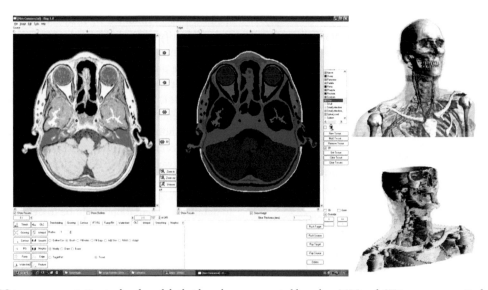

FIGURE 7.3   iSEG image segmentation tool and models that have been segmented based on MRI and CT images, respectively.

As the segmentation tool of HYCAT that has been developed specifically for the purpose of hyperthermia treatment planning, iSEG (see Figure 7.3) offers the possibility of flexibly combining various segmentation techniques ranging from highly automatic to highly interactive. The goal is that simpler tissues (e.g., bones in CT) should be automatically segmentable while more complex ones can be segmented with more user control. Segmentation techniques include interactive watershed transformation, level set methods, fuzzy connectedness, live-wire (competitive, hysteretic) region growing, image foresting transformation, automatic thresholding, brush drawing, and contouring [122, 127]. Topologically flexible, adaptive inter- and extrapolation is available to automatically suggest a segmentation based on previously segmented nearby slices. Support for multimodal image data is provided. Dedicated vessel segmentation is available for contrast enhanced CT. Noise reduction and boundary enhancement filters are available for preprocessing. Post-processing includes hole and gap closing routines, surface and contour extraction and simplification, skin adding, outline correction tools, connected component analysis, etc. The conclusions of [122] are that different methods should be used for different tissues: some tissues require interaction, but perhaps not on every slice when interpolation is applicable; segmentation, except for the simplest tissues, is mostly performed in 2D due to speed concerns and the higher likelihood of volumes to include regions where leakage can occur; hybrid methods combining contour and area-based information behave more robustly; robustness is also increased by using competitive region growing methods; and a predefined scheme should be developed that specifies the order of steps to be followed by a user to allow segmentation to be performed on a routine basis by technical staff.

The choice of the segmentation technique does not solely determine the quality and speed of the segmentation process. It is equally important to pay proper attention to the type of user interaction that is required and supported [133]. [67, 191, 193] discuss the impact of segmentation on treatment planning results (e.g., accuracy, continuous properties vs. tissue specific properties, and number of distinguished tissues.)

7.4 Electromagnetic Simulations

In the case of hyperthermia above a few 100 kHz, the spatial distribution of the power dissipation in different tissues due to ohmic losses respective to dielectric relaxation (molecular friction) can only be obtained by a full-wave solution of Maxwell's equations for the entire system (i.e., environment, antenna, water bolus, and patient). The dissipated power or the specific absorption rate (SAR; see Figure 7.4) is calculated as

$$SAR = \sigma E^2 / \rho$$

where σ is the electric conductivity, E the root-mean-square (RMS) value of the total electric field, and ρ the tissue density.

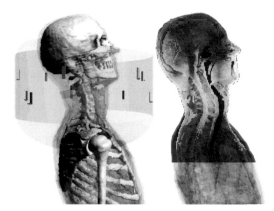

FIGURE 7.4   (Left) Model of a patient in a head-and-neck applicator and (right) the calculated SAR distribution for an unoptimized set of antenna steering parameters (see Figure 7.5 for an optimized distribution). Notice the complex SAR distribution and the undesired SAR hot spots in the vicinity of the spinal cord.

As the phases and amplitudes of the individual antennas are not known in advance (this is the task of the optimization step), the common approach is to individually compute the fields induced by each antenna and compute the total field for the entire applicator by linear superposition of the scaled and correctly phase-shifted fields of each antenna. The remaining challenge is to take into account the mutual coupling of the antennas and their interaction with the amplifiers.

Various techniques have been used to solve the Maxwell equations: the Finite-Difference Time-Domain (FDTD) method [29, 82, 141, 149, 160, 165] and the Finite-Element method (FEM) [140, 189] (based on the differential formulation of the Maxwell equations), the Weak Form of the Conjugate Gradient Fast Fourier Transformation (WF-CGFFT) method [184, 198], and the Volume Surface Integral Equation (VSIE) method [192] (based on the Integral formulation of the Maxwell equations). FDTD and WF-CGFFT mostly use rectangular meshes, while FEM and VSIE typically apply tetrahedral grids. FDTD on rectilinear meshes is ideally suited for highly inhomogeneous meshes and scales well with the mesh size. It offers simple parallelization and hardware acceleration. Applying transient excitations, it provides information for multiple frequencies in one run and potentially reduces convergence issues related to high-Q applicators. FEM on unstructured meshes is naturally conformal, thus avoiding issues related to staircasing. It furthermore allows for simple error estimation. FDTD and FEM results have been compared in [158].

In order to correctly capture interface effects and hot spots, it has been shown that high-resolution modeling is necessary, much finer than the coarse discretizations of only 10 mm often applied in HTP [167]. [128] uses graded rectilinear meshes that allow for sub-millimeter resolution where required (e.g., antenna feed points) without resulting in excessive numbers of voxels (volume pixels), resulting in a computational effort in the order of minutes at 435 MHz using moderately priced personal computers with GPU-based hardware acceleration and optimized grid generation [122, 128].

Other techniques speed up FDTD simulations by allowing larger time steps (e.g., using Alternating-Direction Implicit [ADI] time integration [28, 122, 121]) or by calculating on a coarser grid and using special post-processing techniques to approximate high-resolution information (e.g., using quasistatic zooming [166–169] or a special interpolation technique that handles interfaces in a physically correct manner [119]). Analytical local solutions can be used to improve thin antenna handling without resorting to high-resolution simulations [43, 54, 118, 122].

One of the major sources of error in predicting the SAR pattern is insufficient accounting for the scattered field on the antenna elements. The feeding networks with the matching circuits have to be considered correctly and methods have been developed to couple simulations with a model of the feeding network [116, 117, 122]. These methods improve the predictive value of simulations, but the feeding network still remains a major source of uncertainty. Another source of error is the staircasing errors of FDTD [1, 20]. Conformal subcell techniques increase the accuracy and help reduce staircasing errors [12, 122]. Various other

problems have been encountered that compromise the accuracy of energy deposition predictions. For example, the high-Q cavity formed by an annular phased array hyperthermia applicator can lead to an extremely long convergence time, mode flipping can occur before convergence [149], and a whispering gallery effect has been reported where the water bolus acts in a lens-like manner, refocusing the energy from one antenna on the opposite patient side. The last point is a physical effect and has since been demonstrated experimentally.

7.5 Thermal Simulations

A large number of thermal models have been developed (see reviews in [3, 37, 104, 188]) that essentially differ in their description of the impact of perfusion. Most of the approaches are based on early work by Pennes [143], which links temperature increase over time to external heat sources, metabolic heat generation, heat diffusion, and a homogeneous heat sink term that is proportional to tissue perfusion as well as the difference between local temperature and perfusing blood temperature. Pennes interprets this homogeneous heat sink term as being due to rapid equilibration occurring at the level of the microvasculature. This assumption has since been questioned as much of the heat exchange already occurs between larger counter-current vessels. This has led to the development of more complex heat equations (e.g., the Weinbaum-Jiji model, see [3]) that often couple a series of differential equations describing the heat distribution of the tissue as well as the arterial and the venous blood [3, 68]. Under certain conditions, the effect of counter-current vessels can be modeled by replacing the thermal conductivity with an effective thermal conductivity (e.g., the simplified Weinbaum-Jiji model, see [3] and a critical analysis of this approach in [187]). The predictions of the Pennes model can differ significantly from those obtained using an effective thermal conductivity approach [95, 188]. Various authors have proposed models interpolating between the Pennes and the effective thermal conductivity approaches with weights depending on the location or the local perfusion [18, 188]. The Pennes term has been reinterpreted to actually model heat exchange due to bleed-off, and is currently believed to be valid in the vicinity of large nonequilibrated vessels in addition to tissue where only microvasculature is present. A detailed multiscale analysis has provided additional justification for the Pennes term [48, 49]. Despite its shortcomings, the Pennes model continues to be the one used most frequently, which can be partly justified by the large amount of available experimentally determined tissue parameters for the Pennes model. Each model has its own range of validity [3]. [197] has studied vessels with diameters in the order of 0.1 mm and has shown that the detailed vessel branching pattern has an important influence on how the perfusion effect should be modeled. Others have criticized the Pennes equation for ignoring the directivity of blood flow. Variants including a flow-field term have therefore been introduced [3], sometimes based on a porous medium assumption [68, 99]. Another important fact ignored by the Pennes equation is the discreteness of blood vessels and the impact of large vessels with relatively long equilibration lengths. [40, 98] have analyzed various types of vessels, including counter-current networks, by modeling them discretely, and have established some relations concerning the equilibration and entrance lengths. They consequently developed the DIscrete VAsculature (DIVA) model [98] as they concluded that it can be important to model vessels individually. The DIVA model has been extended [129] to also cover highly thermo-conductive thin structures such as pacemaker wires (resulting in faster and more accurate simulations) and 1D boundary conditions (e.g., water-cooled ablation catheters).

Although it has been shown that detailed vessel networks can be relevant at more than just the local scale [173], the necessary information for segmentation is not readily available from standard medical imaging. It has been proposed that incomplete segmented vessel information from a specific patient can be extended by artificially grown (counter-current) vessel trees [38, 98] using fractals or a potential steered growth approach.

Large vessels have been modeled in [84, 109, 114, 177, 183], sometimes up to explicitly solving the Navier-Stokes equation to obtain a realistic flow distribution. [183] has even developed a special meshing technique for this purpose. [63] has shown that the impact of vessels can be interpreted as superdiffusion and suggests modeling it with a nonlocal term. [36] has studied how the convection coefficient relating the blood temperature in the vessel to the temperature distribution in the neighboring tissue depends on the size of the heated region and the flow rate.

Multiple experiments have been performed to compare various blood-flow models with reality. Some of them are reviewed in [97], which itself presents experiments testing the Pennes model and an effective conductivity model (extending the numerical analysis in [96]) and finds supporting evidence for both approaches. The importance of large vessels is analyzed, and the concept of thermally significant vessels is examined in detail. [27] compares the Pennes, the Weinbaum-Jiji, and the simplified Weinbaum-Jiji models to experimental data, concluding that the Pennes model should be used in the vicinity of vessels with diameters larger than 0.5 mm, while elsewhere the Weinbaum-Jiji models, which are found to deliver comparable results, are to be preferred. Similar results have been obtained by [2]. [78] deduces from experiments that vessels with diameters larger than 0.2 mm are thermally significant. A detailed numerical experiment [155] simulating a branching vasculature model with 10 generations of vessels concludes that (1) vessels have a strong impact, (2) the flow rate and influx temperature are important, while the diameter and Nusselt number play a lesser role, (3) arteries have a stronger impact than veins, and (4) which vessels are relevant has to be decided on a case-by-case basis. [39] compares the modeling of vessels using convective boundary conditions and detailed flow simulation with experiments, finding only minor differences for the specific case studied. [148] has compared the Pennes model with the model by Jain (perfusion dependent conductivity [85]) and a simple Dirichlet boundary condition approach at vessel surfaces, concluding that the Dirichlet boundary might be appropriate

for large vessels, while the Jain model is most suited for small vasculature.

The thermal solver implemented in HYCAT allows the flexible combination of convective flow distributions (from CFD calculations) or variable boundary conditions in major vessels with complex flow patterns (e.g., aorta), DIVA-like discrete vessel modeling for smaller vessels, and an effective thermal conductivity model for the microvasculature [122]. MRI perfusion maps can be used to account for inhomogeneous tissue perfusion within a tissue.

The temperature dependence of tissue parameters is discussed in [14, 23, 79, 106,122, 159]. [122] suggest a first-order approximation to the impact of temperature on the SAR distribution. The impact of skin temperature and the temperature of the hypothalamus on the perfusion rate is considered in [14, 107], including sweating. The influence of tissue damage on the tissue parameters can be found in [79, 159]. Refer to Section 7.9 in this chapter for more information on the modeling of temperature and tissue damage dependent tissue parameters, where the effects related to evaporation are also discussed. [195] describes how the tissue water content varies with temperature. [134] has developed a thermo-pharmacokinetic model describing the interaction between temperature and the concentration of vasodilating agents and their effects.

Whole-body models have been proposed [25, 26, 59, 88] to account for nonlocal thermoregulation, heart rate dynamics, heat radiation at the surface, moisture at the skin's surface and sweating, clothing, skin blood flow effects, and changes of the body core temperature. [26] has presented an interesting model of a limb heated by a hyperthermia applicator. While the limb is simulated in detail (including a separate simulation of arterial and venous blood temperature along the limb, sweating, and temperature-dependent tissue parameters, with muscle perfusion reacting with a time delay to temperature increases and the skin perfusion depending on the local as well as the averaged skin temperature), the rest of the body is considered by coupling the limb to a whole-body model. However, such whole-body models require a large number of additional parameters, introducing further uncertainty and making their routine application tedious. A simple model for body core heating due to the absorption of EM energy by blood has been proposed by [81] and is implemented in the thermal solver of HYCAT. Such a model could easily be coupled to more complex models of whole-body thermoregulation.

Various numerical methods can be used to obtain the temperature distribution. If only the steady state is required, the problem is essentially reduced to solving a Poisson equation. Various methods are routinely applied to solve the discretized equation [46]: Jacobi's method, successive over relaxation, conjugate gradient, or fast Fourier transformation based methods. [50] has developed a nonlinear, elliptic, multilevel FEM method to solve nonlinear heat equations. For the discretization, finite differences, finite element, and boundary element [164] methods are used. [17] uses a hybrid approach, modeling the vessels with FDTD and the tissues with FEM. To obtain the transient temperature field evolution, FDTD or FEMTD (Finite Element

Method in the Time-Domain) is used. Unconditionally stable ADI-FD (ADI Finite-Differences) schemes have been devised to allow arbitrarily large stable time steps [142, 181]. Discretization and staircasing errors in FDTD have been studied in [150, 123], and various methods have been presented to reduce staircasing errors [131, 150, 196]. Of particular interest is a conformal correction scheme that has been applied in the context of hyperthermia [123]. It was found that the impact of staircasing in the context of hyperthermia for a relevant example case accounted for 1.5°C [122].

Various techniques have been used to speed up thermal simulations. These include hardware acceleration (GPU and CELL BE chip [33]), parallelization [33], and numerical techniques such as adaptive mesh refinement [56], region decoupling [122], narrow-band update schemes for temperature-dependent tissue parameters and phase transitions [122], and steady-state estimation-based initialization [122].

7.6 Field Optimization

Usually it is only the antenna steering parameters that are optimized, whereas the patient position with respect to the antennas and the water bolus temperature are often fixed based on guidelines (e.g., [44, 171]) as repeated performance of the EM and thermal simulations would be required to optimize them. It is expected that, with the increase in computational power, the latter could also become standard. As optimization is a large area, only optimization techniques that have been applied in the field of HTP will be discussed.

Both SAR [11, 91, 139, 156, 185] and temperature (increase) [42, 91, 93, 94, 100, 101, 132, 164] based optimization approaches have been investigated (see Figure 7.5). The EM field for each antenna, or the temperature increase field for each pair of antennas (see [91]), on which the optimization is based, must usually be precomputed. However, [42, 100] base their optimization on temperature distributions in the patient for various antenna settings measured by MRI thermometry. The techniques that

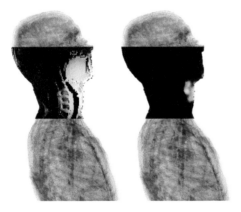

FIGURE 7.5 (Left) Optimized SAR distribution and (right) resulting temperature increase prediction (see Figure 7.4 for a better view of the model). Notice how the temperature increase is not perfectly correlated with the SAR distribution.

have been applied range from genetic algorithms [124, 156] to the generalized Eigenvalue method [91, 92, 124], particle swarm optimization, interior point optimization [182, 32], constrained sequential quadratic programming [93, 164], and a modified Newton Method [132]. The generalized Eigenvalue method is particularly fast but puts strong constraints on the possible form of the optimization functional and allows for virtually no additional constraints (e.g., load balancing between multiple amplifiers/antennas). These disadvantages can be overcome with genetic algorithm-based methods [124]. PDE constraint interior point optimization [182, 32] allows the coupled (nonlinear) thermal simulation/optimization problem to be solved while providing maximal flexibility with regard to the functional and allowing for nonlinear thermal models (e.g., thermoregulation). Different optimization functionals have been used [75, 94]. Some have tried to minimize the difference to a target distribution; others have tried to maximize the ratio of the averaged exposure in the target region to the averaged exposure of the other areas. A third approach tries to maximize the exposure of the target regions while restricting the tolerated exposure of healthy tissue. A detailed review of many optimization functionals and their advantages is given in [21].

Approaches that require repeated evaluation of the functional have attempted to speed up the calculation process by bringing it into a form that can be largely precomputed (e.g., [122, 125, 185]). Some functionals include the weighting of different tissues with specific sensitivities to heat. [91, 124] perform a first optimization to identify likely hot spot locations and then repeat the optimization process with increased weight for these problematic regions—a process called "hot spot suppression." When the contributions of different regions are precomputed, rapid reoptimization with reweighted regions (e.g., based on patient feedback such as pain complaints as suggested by [22]) can be performed, thus allowing treatment planning to make the step into the treatment room.

Various techniques have been used to speed up the optimization process by model reduction, e.g., (1) by grouping points that react in a similar manner to antenna setting changes [41], (2) by performing a Karhunen-Loeve transformation and only using the vectors contributing most to the variance, or (3) by using Eigenvalue-based optimization to find the best settings for individual targets and then combining these settings [11]. Another approach [93] has been proposed that performs high-resolution optimization based on low resolution field calculations that have been interpolated using quasistatic zooming and an associated temperature interpolation method. Most approaches optimize only the steady state. However, some consider the transient effects and even the cool down period [4, 5, 30], concluding that sometimes the ideal heating strategy should not aim at generating homogeneous heating in the tumor, but rather a higher exposure of the tumor border, and that the most intense treatment is not necessarily the best or the shortest one. Inverse methods have been used [61] to determine the ideal EM potential boundary condition at the patient circumference for optimal heating. Various publications [4–6, 90, 100] have studied the possibility of implementing

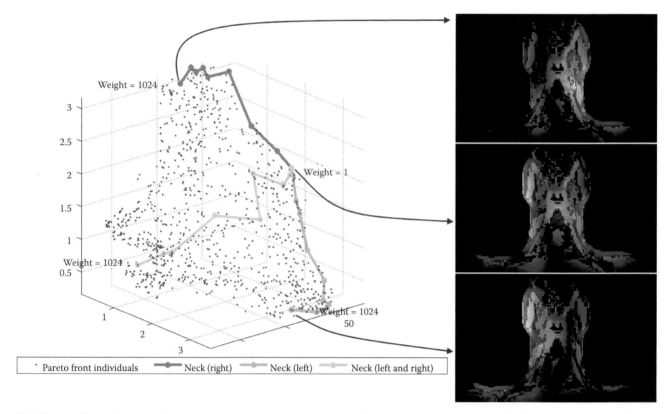

FIGURE 7.6 (See color insert.) Pareto front of optimized antenna settings for hyperthermia treatment of the head-and-neck area. Each point corresponds to an optimized antenna setting, and which one is selected is decided based on the weighting of the different goals (heat the tumor, avoid exposing the left side of the neck, avoid exposing the right side of the neck). Changing the weighting corresponds to gliding along the Pareto front and translates into shifting the energy deposition.

controllers that use feedback from probes or MRI thermometry, and sometimes information from an internal simulation model, to regulate the antenna steering over time, and optimize a thermal dose goal or a target distribution while respecting constraints or reducing treatment time. These approaches are, however, currently limited to 1D models. Time modulated hyperthermia can be used to achieve better coverage of large (or multiple) tumors and to obtain a reduced exposure time for individual hot spots [130]. [130] uses a multi-goal optimization approach to obtain the Pareto front of optimal solutions (see Figure 7.6), thus providing the physician with a selection of optimized treatment settings and an interactive way of selecting the most suitable one (taking into account patient and measurement probe feedback).

7.7 Biological Effect Determination

The administered heat results in biological effects, such as coagulation, necrosis, improved survival rates, etc. To quantify this, two approaches are commonly used: dose concepts and tissue damage calculations. They have been reviewed in [51, 76, 103], and some biological and chemical background is provided in [8]. Various dose concepts exist, ranging from T90 (the 90th percentile of the tumor temperature distribution, a measure for tumor

coverage) to CEM43 (cumulative equivalent minutes at 43°C, a dose specifying how many minutes of constant heating at 43°C would be needed to achieve the same effect). CEM43 is based on the observation that every additional degree above 43°C doubles the cell killing rate, while every degree below 43°C reduces it by a factor of four. This type of Arrhenius-like behavior lies at the heart of the tissue damage approach, which directly calculates the achieved biological effect using a kinetics-based Arrhenius model. A dose concept based on nonequilibrium theory and extending the CEM43 approach, while suggesting an explanation for the observed temperature dependency, has been proposed in [163].

Clinical and experimental data has shown dose-effect (cure, coagulation volumes, etc.) relationships [16, 64, 154]. Quantities specifying how much the tumor region with the poorest heating was exposed (CEM43T90, TDmin) seem to correlate best with treatment outcome. For high temperature treatments (refer to Section 7.10), the evolution of necrosis and its impact on perfusion has been studied extensively in [9, 24, 159]. It has been found that temperature isotherms are bad predictors for lesion size, especially for short- and long-duration treatments [24, 74], and that it is important to consider transient effects in the dose/ damage calculation [122].

7.8 Thermometry and Experimental Validation

Various approaches have been used to measure the EM and temperature distributions achieved during hyperthermia treatment. Surface temperatures can be measured with IR cameras and thermal sensors in contact with the skin [7]. In the presence of water bolus cooling or heating, surface temperature measurement can be problematic as the probe measurement will generally be affected by the proximity to the bolus when not sufficiently isolated. This problem can be reduced by correcting for the error using predetermined correction factors or by using directional sensors. Direct measurement in the patient has previously been restricted to interstitial and sometimes intraluminal catheters (e.g., thermocouples [suffer from self-heating problems], thermistor probes, or fiber optic probes for temperature measurements). This reduces the number of accessible data points, limiting the usefulness. Furthermore, the catheters can cause complications that lead to disagreement about whether interstitial thermometry should be used on a regular basis [157, 172]. The attractive alternative of MRI thermometry is currently emerging, offering noninvasive 3D thermometry. It requires MR-compatible applicators that are now being developed in several groups. While MR thermometry is hardly possible when patient movement is present (e.g., due to breathing), it has been shown that MR thermometry can even monitor temperatures near vessels [39]. Temperature resolution in the order of 0.5–1.0°C [66] can be achieved (at least in phantoms under controlled conditions), but only for tissues with high water content, whereas hot spots often occur in bone, cartilage, and fat tissues due to the dielectric contrast and low perfusion.

MR thermometry during hyperthermia has been validated extensively using a heterogeneous phantom [64, 66, 120, 141]. Various homogeneous and heterogeneous phantoms [66, 120, 122, 153] have been developed to validate simulations and for equipment quality assurance purposes. The heterogeneous phantoms include pieces of bone, chalk, or plastic to mimic complex bone shapes and hollow tubes to simulate the esophagus. The EM or temperature distribution in the phantoms is determined using embedded probes at selected, sensitive locations [66], IR cameras [137, 146], current- [174, 175] or LED-sheets [151, 191], or gels with temperature-dependent color transmission properties. Several treatment planning tools (most extensively HyperPlan) have been validated experimentally [65, 67, 94, 117, 158, 160, 184, 189] using patient data and phantoms. All of the previously mentioned thermometry methods have been applied. Generally, the validations concluded that while good qualitative agreement had been achieved, quantitative agreement for complex setups was limited. This is sometimes attributed to an inability to correctly simulate cross talk and feeding networks, the sensitivity with regard to patient positioning (which has been studied in [65]), and the inability to capture interface as well as antenna behavior with coarse meshes in FDTD (grid steps of 10 mm are often used and subsequently refined using post-processing methods). These problems can be overcome by the new developments in HTP tools described before and improved applicators.

For the comparison of field distributions, a new alternative to interpolation to a common mesh and subsequent subtraction is emerging [45]. It is based on the Gamma method, which has been developed to assess dose distributions during radiotherapy [111, 112]. The Gamma method accepts a value tolerance and a spatial tolerance and computes either the proportion of agreement or a deviation distribution.

7.9 Tissue Parameters

Various publications have presented tables with EM and thermal tissue properties. Tissue properties are assigned to a tissue that has been segmented, or are based directly on Hounsfield units from CT scans [67, 87] or normalized MRI gray values [57, 113]. It has been shown that CT-based property assignment results in serious misclassifications and should not be used as an automatic procedure [67, 193]. MRI data has also been used to estimate local perfusion [35]. Sensitivity studies have shown that the impact of dielectric parameter uncertainties on the EM and thermal distribution is relatively small for moderate hyperthermia [170]. The impact of perfusion variations has been studied in [34, 61] and can be significant when the problem is not conduction dominated. Far less information exists about tumor parameters (e.g., [72] for hepatic tumors), which can vary strongly depending on the tumor type and size as well as the patient. Various studies have claimed that tumors have properties resulting in preferential tumor heating [55, 75]. Measurements have been conducted to study the thermal conductivity anisotropy in muscle and the dependence of the effective thermal conductivity on perfusion [15]. The influence of water content changes with temperature due to diffusion, and evaporation has been addressed in [159, 194, 195]. Information about the temperature dependence of other tissue parameters can be found in [13, 23, 106, 159]. Reversible and irreversible effects from heat have been studied in [145] for dielectric properties and in [15] for thermal conductivity. Many parameters have been gained from excised tissue. It is relevant to know that postmortem changes occur that modify the properties [71]. Anesthesia is also known to have an impact on tissue properties.

7.10 Related Treatments

The simulation environment can be applied in a useful manner to other thermal treatments. Examples besides classical hyperthermia include radio-frequency (RF) or microwave (MW) ablation, RF surgery, and high intensity focused ultrasound (HIFU) ablation. All three of these are common in that they reach higher temperatures than common hyperthermia treatments (typically 60–80°C, sometimes over 100°C, such that cell water evaporation can be relevant) and produce more localized heating that destroys the tissue locally (mostly through coagulation).

To avoid charring close to the electrodes, which would reduce tissue conductivity, isolate the catheter, and thereby hamper the ability of achieving a larger necrotic region, internally cooled applicators or applicators that irrigate their surroundings with a saline solution are sometimes used.

Excellent reviews on the simulation of RF ablation have been published [13, 79]. Many simulations and studies have been performed (e.g., [10, 19, 24, 73, 75, 77, 83, 84, 105, 109, 114, 147, 148, 152, 159]). They consider effects such as tissue damage and resulting tissue property changes (sometimes assumed to be continuous, sometimes threshold triggered), temperature-dependent parameters, cell water evaporation and diffusion, convective cooling, the impact of blood vessels (by imposing a convective boundary condition, simulating laminar flow, or solving the Navier-Stokes equations for the full-flow picture, e.g., in branching vessels or the heart lumen), state changes (e.g., vapor generation and the involved latent heat requirements), internal cooling, and saline perfusion. Transient effects are only rarely considered, and many simulations are performed in 2D considering the rotational symmetry of the ablation catheter. The lesion size is either estimated using an isotherm (50–55°C) criterion or an Arrhenius-based tissue damage model. Finite element and finite differences methods (sometimes locally refined with special, conformal voxels) as well as hybrid models (finite differences for vessels, FEM for tissue) have been applied. The SAR distribution is usually obtained by solving the quasistatic equation [159]. Some simulations include a model of an impedance-based source voltage controller.

It was found that vessels can have a strong impact on the temperature distribution, that different models can result in very different predictions of the vessel influence (see [148]), and that the influence depends in a complex manner on the distance to the vessel, the flow rate, the vessel geometry and flow pattern, as well as the size of the heated region (e.g., [36, 109]).

Many of the previously mentioned publications include experimental validation. Especially noteworthy is [195], which experimentally analyzes the processes of cell water evaporation and diffusion, and presents an empirical relationship between temperature and water content during RF ablation.

Only a few comprehensive treatment planning software tools exist [46, 108, 147, 178]. [178] focuses on segmentation, automatic probe placement (which optimizes tumor coverage by assuming a given, ellipsoidal lesion size and finding the best position and orientation), and the identification of unacceptable access trajectories. No physical simulations are performed, no risk estimation is provided, and the impact of vessels is only considered by locally deforming the lesion ellipsoid near major vessels [179]. [108] and [60] present a very complete treatment planning tool, but the heating is triggered by a laser source and not by RF fields. Nevertheless, much of this work can be applied to RF ablation. In addition, [60] contains important work on HPC-based real time assessment, MR thermometry, and control.

7.11 Challenges

Major progress in HTP has been made during the last decade, resulting in tools such as HyperPlan and HYCAT. Nevertheless, the following challenges should be addressed to further increase the reliability of exposure prediction:

- EM tissue parameters are typically based on the values compiled by [62], which provides frequency-dependent models of the dielectric parameters for various tissues. Thermal properties originate from various literature sources and measurements. These parameters are complemented by models of thermoregulation (both local and whole body). However, the thermal properties in particular show a large inter- and intrasubject variability that is mostly not considered. A future goal should be to find ways of providing patient-specific parameters (e.g., perfusion based on MRI perfusion maps).

- Temperature monitoring is still needed during treatment at present. However, it only provides very sparse and often unreliable (e.g., intraluminal probes) information, increases the risk of complications in the case of invasive monitoring, and comes at high financial cost, as in the case of MRI. Due to the great progress in HTP, it appears achievable that—at least for fixed tissue geometries— reliable prediction is feasible supported by very sparse experimental information.

- Quality assurance is not sufficiently addressed in hyperthermia, in particular with respect to applicator characterization resulting in an unacceptably large interapplicator variability. The possibility of positioning the patient with sufficient precision is often not provided. Procedures and protocols vary widely between clinics. In the future, quality assurance guidelines should be extended (efforts are ongoing within ESHO) and technology provided to overcome the aforementioned shortcomings.

- Another challenge is clinical integration. Current tools have limited interfaces with radiotherapy planning (HYCAT supports RTSTRUCT data). They bring their own graphical interfaces, are badly interfaced to the applicator (related partly to regulatory issues), and have no interface to hospital information systems. All of this has to be improved in order to obtain acceptance in the standard clinical environment.

- Automatization needs to be improved to reduce human error and obtain acceptable treatment planning times. For this, an extended wizard approach together with more sophisticated segmentation functionality (e.g., automatization based on prior knowledge) might be the solution. Automatic segmentation is, however, very difficult to achieve for hyperthermia due to the many different tissues that have to be distinguished.

- Improved validation of the treatment planning tools, particularly in vivo, is of major importance (e.g., using the power offered by MRI thermometry). Ongoing clinical trials comparing treatment outcome with therapy planning to outcome without individualized planning might already offer first valuable data.

- At the moment, hyperthermia treatment is studied independently of the context. In the future, models should also look at the interaction with radiotherapy treatment or chemotherapy. Initial approaches exist [126], but a lot of effort is still needed in this direction.

7.12 Conclusions

Treatment planning has progressed greatly in recent times. Novel developments in segmentation, simulation, and optimization have helped to increase the reliability, realism, accuracy, and speed of treatment planning. Recent developments have even permitted treatment planning to make the jump into the treatment room as well as a step toward feedback-based treatments. Treatment planning is being increasingly used in the clinic to provide patient-specific treatments and is believed to contribute to the quality and outcome of hyperthermia treatments.

The two most commonly used commercial treatment planning tools, HyperPlan (based on AMIRA) and HYCAT (based on SEMCAD X), provide the following features and shortcomings:

HyperPlan is currently the treatment planning software most commonly used in the clinic. It is exclusively designed to work with BSD deep hyperthermia applicators and has detailed built-in models of their feeding networks. The use of unstructured mesh-based FEM improves interface handling but limits the achievable resolution and requires the complex task of meshing. The lower resolution, however, allows faster (particularly temperature-based) optimization. HyperPlan offers the possibility of working with MR thermometry images, and versions exist that can use MRT to correct the simulation predictions. It is integrated in AMIRA, which has powerful visualization functionality for medical image data.

HYCAT is geared toward the efficient simulation of high-resolution models. It offers advanced models for perfusion and thermoregulation, and flexible optimization functionality, including functionality to quickly reoptimize treatments based on patient feedback. Its segmentation software (iSEG) combines a wide range of segmentation methods, but patient-specific model generation still requires 2–6 hours, thus exceeding by far the interaction time common in radiotherapy. HYCAT offers flexible modeling (including special cases such as presence of implants), access to many parameters, and some user adaptable "wizards" to simplify the setup of simulations for the operator and to automatically generate reports. Its flexibility makes it particularly useful in research-oriented contexts. However, the software is currently not well integrated into the clinical flow

and is therefore not suitable for general clinical needs. In addition, the high-resolution approach leads to comparatively long simulation duration, despite hardware acceleration support. HYCAT is integrated in SEMCAD X, which offers powerful simulation, post-processing, and scripting functionality.

At this moment, hyperthermia treatment planning is starting to enter the clinic, especially as a complementary tool. When used for predictive pretreatment and online treatment planning, both commercial packages require extensive in vivo validation. Fortunately, the recent development of MRT has provided the technology required for robust 3D validations. The time needed for treatment planning must be reduced, and the development of modelable applicators (e.g., predictable loading effect, cross coupling, reflections, environment influence) requires attention to reduce the number of uncertainties during hyperthermia application and to produce well-controlled fields. Lastly, more research is needed into patient-specific thermal parameters (e.g., by specific 3D maps of thermal properties). It remains unclear what the exact number and extent of the uncertainties are, but HYCAT and HyperPlan both provide the opportunity for accurate risk assessments illustrating the worst-case patterns and the risk of underexposure, as is currently standard in radiation oncology.

References

1. A. Akyurtlu, D. H. Werner, V. Veremey, D. J. Streich, and K. Aydin. Staircasing errors in FDTD at an air-dielectric interface. *IEEE Microw Guid Wave Lett*, 9(11):444–446, 1999.

2. S. I. Alekseev, A. A. Radzievsky, I. Szabo, and M. C. Ziskin. Local heating of human skin by millimeter waves: Effect of blood flow. *Bioelectromagnetics*, 26(6):489–501, 2005.

3. H. Arkin, L. X. Xu, and K. R. Holmes. Recent developments in modeling heat transfer in blood perfused tissues. *IEEE Trans Biomed Eng*, 41(2):97–107, 1994.

4. D. Arora, D. Cooley, T. Perry, J. Guo, A. Richardson, J. Moellmer, R. Hadley, D. Parker, M. Skliar, and R. B. Roemer. MR thermometry-based feedback control of efficacy and safety in minimum-time thermal therapies: Phantom and in-vivo evaluations. *Int J Hyperthermia*, 22(1):29–42, 2006.

5. D. Arora, M. Skliar, and R. B. Roemer. Model-predictive control of hyperthermia treatments. IEEE *Trans Biomed Eng*, 49(7):629–39, 2002.

6. D. Arora, M. Skliar, and R. B. Roemer. Minimum-time thermal dose control of thermal therapies. *IEEE Trans Biomed Eng*, 52(2):191–200, 2005.

7. K. Arunachalam, P. Maccarini, T. Juang, C. Gaeta, and P. R. Stauffer. Performance evaluation of a conformal thermal monitoring sheet sensor array for measurement of surface temperature distributions during superficial hyperthermia treatments. *Int J Hyperthermia*, 24(4):313–25, 2008.

8. E. R. Atkinson. Hyperthermia dose definition. *Microwave Symposium Digest, MTT-S International*, 77(1):251, 1977.

9. P. Badini, P. De Cupis, G. Gerosa, and M. Giona. Necrosis evolution during high-temperature hyperthermia through implanted heat sources. *IEEE Trans Biomed Eng*, 50(3):305–15, 2003.

10. S. A. Baldwin, A. Pelman, and J. L. Bert. A heat transfer model of thermal balloon endometrial ablation. *Ann Biomed Eng*, 29(11):1009–18, 2001.

11. F. Bardati, A. Borrani, A. Gerardino, and G. A. Lovisolo. SAR optimization in a phased array radiofrequency hyperthermia system. *IEEE Trans Biomed Eng*, 42(12):1201–7, 1995.

12. S. Benkler, N. Chavannes, and N. Kuster. A new 3-D conformal PEC FDTD scheme with user-defined geometric precision and derived stability criterion, *IEEE Trans Ant Prop*, 54(6):1843–49, 2006.

13. E. J. Berjano. Theoretical modeling for radiofrequency ablation: State-of-the-art and challenges for the future. *Biomed Eng Online*, 5:24, 2006.

14. P. Bernardi, M. Cavagnaro, S. Pisa, and E. Piuzzi. Specific absorption rate and temperature elevation in a subject exposed in the far-field of radio-frequency sources operating in the 10–900 MHz range. *IEEE Trans Biomed Eng*, 50(3):295–304, 2003.

15. A. Bhattacharya and R. L. Mahajan. Temperature dependence of thermal conductivity of biological tissues. *Physiol Meas*, 24(3):769–83, 2003.

16. P. Bhowmick, J. E. Coad, S. Bhowmick, J. L. Pryor, T. Larson, J. De La Rosette, and J. C. Bischof. In vitro assessment of the efficacy of thermal therapy in human benign prostatic hyperplasia. *Int J Hyperthermia*, 20(4):421–39, 2004.

17. C. H. Blanchard, G. Gutierrez, J. A. White, and R. B. Roemer. Hybrid finite element-finite difference method for thermal analysis of blood vessels. *Int J Hyperthermia*, 16(4):341–53, 2000.

18. H. Brinck and J. Werner. Efficiency function: Improvement of classical bioheat approach. *J Appl Physiol*, 77(4):1617–22, 1994.

19. F. Burdio, E. J. Berjano, A. Navarro, J. M. Burdio, A. Guemes, L. Grande, R. Sousa, J. Subiro, A. Gonzalez, I. Cruz, T. Castiella, E. Tejero, R. Lozano, and M. A. de Gregorio. RF tumor ablation with internally cooled electrodes and saline infusion: What is the optimal location of the saline infusion? *Biomed Eng Online*, 6:30, 2007.

20. A. C. Cangellaris and D. B. Wright. Analysis of the numerical error caused by the stair-stepped approximation of a conducting boundary in FDTD simulations of electromagnetic phenomena. *IEEE Trans Antennas Propagation*, 39(10):1518–1525, 1991.

21. R. A. M. Canters, P. Wust, J. F. Bakker, and G. C. Van Rhoon. A literature survey on indicators for characterisation and optimisation of SAR distributions in deep hyperthermia, a plea for standardization, *Int J Hyperthermia*, 25(7):593–608, 2009.

22. R. A. M. Canters, M. Franckena, J. Zee, and G. C. Van Rhoon. Complaint-adaptive power density optimization as a tool for HTP-guided steering in deep hyperthermia treatment of pelvic tumors, *Phys Med Biol*, 53: 6799ff, 2008.

23. I. Chang. Finite element analysis of hepatic radiofrequency ablation probes using temperature-dependent electrical conductivity. *Biomed Eng Online*, 2:12, 2003.

24. I. A. Chang and U. D. Nguyen. Thermal modeling of lesion growth with radiofrequency ablation devices. *Biomed Eng Online*, 3(1):27, 2004.

25. C. K. Charny and R. L. Levin. Simulations of MAPA and APA heating using a whole body thermal model. *IEEE Trans Biomed Eng*, 35(5):362–71, 1988.

26. C. K. Charny and R. L. Levin. A three-dimensional thermal and electromagnetic model of whole limb heating with a MAPA. *IEEE Trans Biomed Eng*, 38(10):1030–9, 1991.

27. C. K. Charny, S. Weinbaum, and R. L. Levin. An evaluation of the Weinbaum-Jiji bioheat equation for normal and hyperthermic conditions. *J Biomech Eng*, 112(1):80–7, 1990.

28. C. C. P. Chen, T. W. Lee, N. Murugesan, and S. C. Hagness. *Geneneralized FDTD-ADI: An unconditionally stable fullwave Maxwell's equations solver for VLSI interconnect modeling.* ICCAD, pp. 156–163, 2000.

29. J. Y. Chen and O. P. Gandhi. Numerical simulation of annular phased arrays of dipoles for hyperthermia of deep-seated tumors. *IEEE Trans Biomed Eng*, 39(3):209–16, 1992.

30. K. S. Cheng and R. B. Roemer. Optimal power deposition patterns for ideal high temperature therapy/hyperthermia treatments. *Int J Hyperthermia*, 20(1):57–72, 2004.

31. K. S. Cheng, M. W. Dewhirst, P. R. Stauffer, and S. Das. Effective learning strategies for real-time image-guided adaptive control of multiple-source hyperthermia applicators. *Med Phys*, 37:1285ff, 2010.

32. M. Christen, O. Schenk, and H. Burkhart. Large-scale PDE-constrained optimization in hyperthermia cancer treatment planning, SIAM Conference on Parallel Processing for Scientific Computing, March, 2008, Atlanta, Georgia.

33. M. Christen, O. Schenk, P. Messmer, E. Neufeld, and H. Burkhart. Accelerating stencil-based computations by increased temporal locality on modern multi- and many-core architectures. Proceedings of the First International Workshop on New Frontiers in High-Performance and Hardware-Aware Computing (HipHaC'08). IEEE/ACM International Symposium on Microarchitecture (MICRO-41), 47ff, 2008, Lake Como, Italy.

34. S. T. Clegg, T. V. Samulski, K. A. Murphy, G. L. Rosner, and M. W. Dewhirst. Inverse techniques in hyperthermia: A sensitivity study. *IEEE Trans Biomed Eng*, 41(4):373–82, 1994.

35. D. J. Collins and A. R. Padhani. Dynamic magnetic resonance imaging of tumor perfusion. Approaches and biomedical challenges. *IEEE Eng Med Biol Mag*, 23(5):65–83, 2004.

36. L. Consiglieri, I. dos Santos, and D. Haemmerich. Theoretical analysis of the heat convection coefficient in large vessels and the significance for thermal ablative therapies. *Phys Med Biol*, 48(24):4125–34, 2003.

37. O. I. Craciunescu. Hyperthermia. www.duke.edu/~dr3/hyperthermia_general.html. 1997.

38. O. I. Craciunescu, S. K. Das, J. M. Poulson, and T. V. Samulski. Three-dimensional tumor perfusion reconstruction using fractal interpolation functions. *IEEE Trans Biomed Eng*, 48(4):462–73, 2001.

39. O. I. Craciunescu, T. V. Samulski, J. R. MacFall, and S. T. Clegg. Perturbations in hyperthermia temperature distributions associated with counter-current flow: Numerical simulations and empirical verification. *IEEE Trans Biomed Eng*, 47(4):435–43, 2000.

40. J. Crezee and J. J. Lagendijk. Temperature uniformity during hyperthermia: The impact of large vessels. *Phys Med Biol*, 37(6):1321–37, 1992.

41. S. K. Das, S. T. Clegg, and T. V. Samulski. Computational techniques for fast hyperthermia temperature optimization. *Med Phys*, 26(2):319–28, 1999.

42. S. K. Das, E. A. Jones, and T. V. Samulski. A method of MRI-based thermal modelling for a RF phased array. *Int J Hyperthermia*, 17(6):465–82, 2001.

43. J. de Bree, J. F. van der Koijk, and J. J. Lagendijk. A 3-D SAR model for current source interstitial hyperthermia. *IEEE Trans Biomed Eng*, 43(10):1038–45, 1996.

44. M. de Bruijne, T. Samaras, J. F. Bakker, and G. C. van Rhoon. Effects of waterbolus size, shape and configuration on the SAR distribution pattern of the lucite cone applicator. *Int J Hyperthermia*, 22(1):15–28, 2006.

45. M. de Bruijne, T. Samaras, N. Chavannes, and G. C. van Rhoon. Quantitative validation of the 3D SAR profile of hyperthermia applicators using the gamma method. *Phys Med Biol*, 52(11):3075–88, 2007.

46. J. Demmel. Berkley lecture CS267 (online): Solving the discrete Poisson equation using Jacobi, SOR, conjugate gradients and the FFT. 1996.

47. T. Deschamps. Extraction de courbes et surfaces par methodes de chemins minimaux et ensembles de niveaux. Applications en imagerie medicale 3D. Thesis Universite de Paris-Dauphine, 2001.

48. P. Deuflhard and R. Hochmuth. Multiscale analysis of thermoregulation in the human microvascular system. ZIB-Report 02-31, 2002.

49. P. Deuflhard and R. Hochmuth. Multiscale analysis for the bioheat transfer equation. ZIB-Report 03-08, 2003.

50. P. Deuflhard, M. Weiser, and M. Seebass. A new nonlinear elliptic multilevel FEM applied to regional hyperthermia. ZIBPreprint SC 98-35, 1999.

51. M. W. Dewhirst, B. L. Viglianti, M. Lora-Michiels, M. Hanson, and P. J. Hoopes. Basic principles of thermal dosimetry and thermal thresholds for tissue damage from hyperthermia. *Int J Hyperthermia*, 19(3):267–94, 2003.

52. M. W. Dewhirst, Z. Vujaskovic, E. Jones, and D. Thrall. Resetting the biologic rationale for thermal therapy. *Int J Hyperthermia*, 21(8):779–90, 2005.

53. C. J. Diederich. Thermal ablation and high-temperature thermal therapy: Overview of technology and clinical implementation. *Int J Hyperthermia*, 21(8):745–53, 2005.

54. F. Edelvik. A new technique for accurate and stable modeling of arbitrarily oriented thin wires in the FDTD method. *IEEE Trans Electrom Comp*, 45(2):416–423, 2003.

55. V. Ekstrand, H. Wiksell, I. Schultz, B. Sandstedt, S. Rotstein, and A. Eriksson. Influence of electrical and thermal properties on RF ablation of breast cancer: Is the tumour preferentially heated? *Biomed Eng Online*, 4:41, 2005.

56. B. Erdmann, J. Lang, and M. Seebass. Adaptive solutions of nonlinear parabolic equations with application to hyperthermia treatments, Proc. Int. Symp. on Advances in Computational Heat Transfer, Cesme, 1997.

57. P. Farace, R. Pontalti, L. Cristoforetti, R. Antolini, and M. Scarpa. An automated method for mapping human tissue permittivities by MRI in hyperthermia treatment planning. *Phys Med Biol*, 42(11):2159–74, 1997.

58. P. Felkel, R. Wegenkittl, and A. Kanitsar. Vessel tracking in peripheral CTA: An overview. Proc. 17th spring conf. Comp. graph., pp. 232–239, 2001.

59. D. Fiala, K. J. Lomas, and M. Stohrer. A computer model of human thermoregulation for a wide range of environmental conditions: The passive system. *J Appl Physiol*, 87(5):1957–72, 1999.

60. D. Fuentes, J. T. Oden, K. R. Diller, J. D. Hazle, A. Elliott, A. Shetty, and R. J. Stafford. Computational modeling and real-time control of patient-specific laser treatment of cancer. *Ann Biomed Eng*, 37(4):763ff, 2009.

61. S. Fujita, M. Tamazawa, and K. Kuroda. Effects of blood perfusion rate on the optimization of RF-capacitive hyperthermia. *IEEE Trans Biomed Eng*, 45(9):1182–6, 1998.

62. S. Gabriel, R. W. Lau, and C. Gabriel. The dielectric properties of biological tissues: III. Parametric models for the dielectric spectrum of tissues. *Phys Med Biol*, 41(11):2271–93, 1996.

63. V. V. Gafiychuk, I. A. Lubashevsky, and B. Y. Datsko. Fast heat propagation in living tissue caused by branching artery network. *Phys Rev E Stat Nonlin Soft Matter Phys*, 72(5 Pt 1):051920, 2005.

64. J. Gellermann, B. Hildebrandt, R. Issels, H. Ganter, W. Wlodarczyk, V. Budach, R. Felix, P. U. Tunn, P. Reichardt, and P. Wust. Noninvasive magnetic resonance thermography of soft tissue sarcomas during regional hyperthermia: Correlation with response and direct thermometry. *Cancer*, 107(6):1373–82, 2006.

65. J. Gellermann, M. Weihrauch, C. H. Cho, W. Wlodarczyk, H. Fahling, R. Felix, V. Budach, M. Weiser, J. Nadobny, and P. Wust. Comparison of MR-thermography and planning calculations in phantoms. *Med Phys*, 33(10):3912–20, 2006.

66. J. Gellermann, W. Wlodarczyk, H. Ganter, J. Nadobny, H. Fahling, M. Seebass, R. Felix, and P. Wust. A practical approach to thermography in a hyperthermia/magnetic resonance hybrid system: Validation in a heterogeneous phantom. *Int J Radiat Oncol Biol Phys*, 61(1):267–77, 2005.

67. J. Gellermann, P. Wust, D. Stalling, M. Seebass, J. Nadobny, R. Beck, H. Hege, P. Deuflhard, and R. Felix. Clinical evaluation and verification of the hyperthermia treatment planning system hyperplan. *Int J Radiat Oncol Biol Phys*, 47(4):1145–56, 2000.

68. S. Gerber. Perfusionsmodellierung in menschlichen Tumoren. Masterthesis Freie Universität Berlin, 2007.

69. U. Gneveckow, A. Jordan, R. Scholz, J. Gellermann, A. Feussner, C. H. Cho, M. Johannsen, K. Maier-Hauff, and P. Wust. 3-dimensional calculation of the temperature distribution during thermotherapy with magnetic nanoparticles. Book of Abstracts of the 23rd Annual Meeting of the ESHO, Berlin, Germany (ESHO-06), p. 51, 2006.

70. D. Haemmerich and P. F. Laeseke. Thermal tumour ablation: Devices, clinical applications and future directions. *Int J Hyperthermia*, 21(8):755–60, 2005.

71. D. Haemmerich, R. Ozkan, S. Tungjitkusolmun, J. Z. Tsai, D. M. Mahvi, S. T. Staelin, and J. G. Webster. Changes in electrical resistivity of swine liver after occlusion and postmortem. *Med Biol Eng Comput*, 40(1):29–33, 2002.

72. D. Haemmerich, S. T. Staelin, J. Z. Tsai, S. Tungjitkusolmun, D. M. Mahvi, and J. G. Webster. *In vivo* electrical conductivity of hepatic tumours. *Physiol Meas*, 24(2):251–60, 2003.

73. D. Haemmerich, S. Tungjitkusolmun, S. T. Staelin, F. T. Lee Jr., D. M. Mahvi, and J. G. Webster. Finite-element analysis of hepatic multiple probe radio-frequency ablation. *IEEE Trans Biomed Eng*, 49(8):836–42, 2002.

74. D. Haemmerich, J. G. Webster, and D. M. Mahvi. Thermal dose versus isotherm as lesion boundary estimator for cardiac and hepatic radio-frequency ablation. Proc. 25th Ann. Intern. Conf. IEEE EMBS, pp. 134–137, 2003.

75. D. Haemmerich and B. J. Wood. Hepatic radiofrequency ablation at low frequencies preferentially heats tumour tissue. *Int J Hyperthermia*, 22(7):563–74, 2006.

76. J. W. Hand. The current status of microwave induced hyperthermia and radiotherapy for the treatment of recurrent breast cancer. Application of Microwaves in Medicine, IEE Colloquium on, pp. 1–6, 1995.

77. L. J. Hayes, K. R. Diller, J. A. Pearce, M. R. Schick, and D. P. Colvin. Prediction of transient temperature fields and cumulative tissue destruction for radio frequency heating of a tumor. *Med Phys*, 12(6):684–92, 1985.

78. Q. He, L. Zhu, D. E. Lemons, and S. Weinbaum. Experimental measurements of the temperature variation along artery-vein pairs from 200 to 1000 microns diameter in rat hind limb. *J Biomech Eng*, 124(6):656–61, 2002.

79. X. He and J. C. Bischof. Quantification of temperature and injury response in thermal therapy and cryosurgery. *Crit Rev Biomed Eng*, 31(5–6):355–422, 2003.

80. T. Heinonen and P. Dastidar. Segmentation of voxel based medical images. *Int J BioEM*, 3(2), 2001.

81. A. Hirata, T. Asano, and O. Fujiwara. FDTD analysis of human body-core temperature elevation due to RF far-field energy prescribed in the ICNIRP guidelines. *Phys Med Biol* 52:5013–23, 2007.

82. S. N. Hornsleth. Radiofrequency regional hyperthermia. Thesis, Aalborg University, 1996.

83. S. Humphries, K. Johnson, K. Rick, N. Goldberg, and Z. J. Liu. Three-dimensional finite-element code for electrosurgery and thermal ablation simulations. *Progress in Biomedical Optics and Imaging*, 6:181, 2005.

84. M. K. Jain and P. D. Wolf. A three-dimensional finite element model of radiofrequency ablation with blood flow and its experimental validation. *Ann Biomed Eng*, 28(9):1075–84, 2000.

85. R. K. Jain, F. H. Grantham, and P. M. Gullino. Blood flow and heat transfer in Walker 256 mammary carcinoma. *J Natl Cancer Inst*, 62(4):927–33, 1979.

86. B. J. James and D. M. Sullivan. Creation of three-dimensional patient models for hyperthermia treatment planning. *IEEE Trans Biomed Eng*, 39(3):238–42, 1992.

87. B. J. James and D. M. Sullivan. Direct use of CT scans for hyperthermia treatment planning. *IEEE Trans Biomed Eng*, 39(8):845–51, 1992.

88. D. N. Kinsht. Modeling of processes of heat transfer in wholebody hyperthermia. *Biofizika*, 51(4):738–42, 2006.

89. C. Kirbas and F. K. H. Quek. A review of vessel extraction techniques and algorithms. *ACM Computing Surveys*, 36(2):81–121, 2004.

90. M. Knudsen and U. Hartmann. Optimal temperature control with phased array hyperthermia system. *IEEE Trans Microw Theory Techn*, 34(5):597–603, 1986.

91. T. Köhler. Effiziente Algorithmen für die Optimierung der Therapie-Planung zur regionalen Hyperthermie. Dissertation Universitaet Potsdam, 1998.

92. T. Köhler, P. Maass, P. Wust, and M. Seebass. A fast algorithm to find optimal controls of multiantenna applicators in regional hyperthermia. *Phys Med Biol*, 46(9):2503–14, 2001.

93. H. P. Kok, P. M. Van Haaren, J. B. Van de Kamer, J. Wiersma, J. D. Van Dijk, and J. Crezee. High-resolution temperature based optimization for hyperthermia treatment planning. *Phys Med Biol*, 50(13):3127–41, 2005.

94. H. P. Kok, P. M. A. van Haaren, J. B. van de Kamer, P. J. Zum Vrde Sive Vrding, J. Wiersma, M. C. C. M. Hulshof, E. D. Geijsen, J. J. B. van Lanschot, and J. Crezee. Prospective treatment planning to improve locoregional hyperthermia for oesophageal cancer. *Int J Hyperthermia*, 22(5):375–89, 2006.

95. M. C. Kolios, M. D. Sherar, and J. W. Hunt. Large blood vessel cooling in heated tissues: A numerical study. *Phys Med Biol*, 40(4):477–94, 1995.

96. M. C. Kolios, M. D. Sherar, and J. W. Hunt. Large blood vessel cooling in heated tissues: A numerical study. *Phys Med Biol*, 40(4):477–94, 1995.

97. M. C. Kolios, A. E.Worthington, M. D. Sherar, and J.W. Hunt. Experimental evaluation of two simple thermal models using transient temperature analysis. *Phys Med Biol*, 43(11):3325–40, 1998.

98. A. N. Kotte, G. M. van Leeuwen, and J. J. Lagendijk. Modelling the thermal impact of a discrete vessel tree. *Phys Med Biol*, 44(1):57–74, 1999.

99. H. S. Kou, T. C. Shih, and W. L. Lin. Effect of the directional blood flow on thermal dose distribution during thermal therapy: An application of a Green's function based on the porous model. *Phys Med Biol*, 48(11):1577–89, 2003.

100. M. E. Kowalski, B. Behnia, A. G. Webb, and J. M. Jin. Optimization of electromagnetic phased-arrays for hyperthermia via magnetic resonance temperature estimation. *IEEE Trans Biomed Eng*, 49(11):1229–41, 2002.

101. M. E. Kowalski and J. M. Jin. Model-order reduction of nonlinear models of electromagnetic phased-array hyperthermia. *IEEE Trans Biomed Eng*, 50(11):1243–54, 2003.

102. H. Kroeze, J. B. van de Kamer, A. A. de Leeuw, M. Kikuchi, and J. J. Lagendijk. Treatment planning for capacitive regional hyperthermia. *Int J Hyperthermia*, 19(1):58–73, 2003.

103. J. J. Lagendijk. Hyperthermia treatment planning. *Phys Med Biol*, 45(5):R61–76, 2000.

104. J. J. Lagendijk, G. C. Van Rhoon, S. N. Hornsleth, P. Wust, A. C. De Leeuw, C. J. Schneider, J. D. Van Dijk, J. Van Der Zee, R. Van Heek-Romanowski, S. A. Rahman, and C. Gromoll. ESHO quality assurance guidelines for regional hyperthermia. *Int J Hyperthermia*, 14(2):125–33, 1998.

105. Y. C. Lai, Y. B. Choy, D. Haemmerich, V. R. Vorperian, and J. G. Webster. Lesion size estimator of cardiac radiofrequency ablation at different common locations with different tip temperatures. *IEEE Trans Biomed Eng*, 51(10):1859–64, 2004.

106. J. Lang, B. Erdmann, and M. Seebass. Impact of nonlinear heat transfer on temperature control in regional hyperthermia. *IEEE Trans Biomed Eng*, 46(9):1129–38, 1999.

107. G. Lazzi. Thermal effects of bioimplants. *IEEE Eng Med Biol Mag*, 24(5):75–81, 2005.

108. A. Littmann, A. Schenk, B. Preim, A. Roggan, K. Lehmann, J. P. Ritz, C. T. Germer, and H. O. Peitgen. Kombination von Bildanalyse und physikalischer Simulation für die Planung von Behandlungen maligner Lebertumoren mittels laserinduzierter Thermotherapie. Bildverarbeitung für die Medizin, pp. 428–432, 2003.

109. Y. J. Liu, A. K. Qiao, Q. Nan, and X. Y. Yang. Thermal characteristics of microwave ablation in the vicinity of an arterial bifurcation. *Int J Hyperthermia*, 22(6):491–506, 2006.

110. W. E. Lorensen and H. E. Cline. Marching cubes: A high resolution 3D surface construction algorithm. *Computer Graphics* (Proceedings of SIGGRAPH '87), 21:163–169, 1987.

111. D. A. Low and J. F. Dempsey. Evaluation of the gamma dose distribution comparison method. *Med Phys*, 30(9):2455–64, 2003.

112. D. A. Low, W. B. Harms, S. Mutic, and J. A. Purdy. A technique for the quantitative evaluation of dose distributions. *Med Phys*, 25(5):656–61, 1998.

113. M. Mazzurana, L. Sandrini, A. Vaccari, C. Malacarne, L. Cristoforetti, and R. Pontalti. A semi-automatic method for developing an anthropomorphic numerical model of dielectric anatomy by MRI. *Phys Med Biol*, 48(19):3157–70, 2003.

114. Y. Mohammed and J. F. Verhey. A finite element method model to simulate laser interstitial thermo therapy in anatomical inhomogeneous regions. *Biomed Eng Online*, 4(1):2, 2005.

115. P. Monk and E. Suli. Error estimates for Yee's method on nonuniform grids. *IEEE Trans Magn*, 30(5):3200–3203, 1994.

116. T. P. Montoya and G. S. Smith. Modeling transmission line circuit elements in the FDTD method. *Microw Opt Technol Lett*, 21:105–114, 1999.

117. J. Nadobny, H. Fahling, M. J. Hagmann, P. F. Turner, W. Wlodarczyk, J. M. Gellermann, P. Deuflhard, and P. Wust. Experimental and numerical investigation of feed-point parameters in a 3-D hyperthermia applicator using different FDTD models of feed networks. *IEEE Trans Biomed Eng*, 49(11):1348–59, 2002.

118. J. Nadobny, R. Pontalti, D. Sullivan, W. Wlodarczyk, A. Vaccari, P. Deuflhard, and P. Wust. A thin-rod approximation for the improved modeling of bare and insulated cylindrical antennas using the FDTD method. *IEEE Trans Antennas and Propagation*, 51(8):1780–1796, 2003.

119. J. Nadobny, D. Sullivan, P. Wust, M. Seebass, P. Deuflhard, and R. Felix. A high-resolution interpolation at arbitrary interfaces for the FDTD method. *IEEE Trans Microw Theory Techn*, 46(11):1759–1766, 1998.

120. J. Nadobny, W. Wlodarczyk, L. Westhoff, J. Gellermann, R. Felix, and P. Wust. A clinical water-coated antenna applicator for MR-controlled deep-body hyperthermia: A comparison of calculated and measured 3-D temperature data sets. *IEEE Trans Biomed Eng*, 52(3):505–19, 2005.

121. T. A. Namiki. New FDTD algorithm based on alternating direction implicit method. *IEEE Trans Microw Theory Techn*, 47(10):2003–2007, 1999.

122. E. Neufeld. High resolution hyperthermia treatment planning. Thesis 17947 ETHZ, 2008.

123. E. Neufeld, N. Chavannes, T. Samaras, and N. Kuster. Novel conformal technique to reduce staircasing artifacts at material boundaries for FDTD modeling of the bioheat equation. *Phys Med Biol* 52:4371ff, 2007.

124. E. Neufeld, N. Chavannes, and N. Kuster. Fast and flexible high resolution field optimizer for SAR and temperature distributions, 24th Annual Meeting of the European Society for Hyperthermic Oncology, Book of Abstracts, Prague, Czech Republic, June 2007, p. 104.

125. E. Neufeld, N. Chavannes, M. M. Paulides, G. van Rhoon, and N. Kuster. Fast (re-) optimization for hyperthermia: Bringing treatment planning into the treatment room, 1st

ESHO Educational School on Clinical Hyperthermia and the 10th International Congress on Hyperthermic Oncology, Book of Abstracts, Munich, Germany, April 2008.

126. E. Neufeld, D. Szczerba, S. Hirsch, G. Szekely, and N. Kuster. In silico model of tumour growth and treatment with Doxorubicin and heat, 25th Annual Meeting of the European Society for Hyperthermic Oncology, Book of Abstracts, Verona, Italy, 2008.

127. E. Neufeld, T. Samaras, N. Chavannes, and N. Kuster. Robust medical image segmentation for hyperthermia treatment planning, 22nd Annual Meeting of the European Society for Hyperthermic Oncology, Book of Abstracts, Graz, Austria, June 2005, pp. 56–57.

128. E. Neufeld, T. Samaras, N. Chavannes, and N. Kuster. Important aspects of electromagnetic modeling in hyperthermia treatment planning using non-uniform FDTD, 22nd Annual Meeting of the European Society for Hyperthermic Oncology, Book of Abstracts, Graz, Austria, June 2005, pp. 47–48.

129. E. Neufeld, T. Samaras, N. Chavannes, G. Szekely, and N. Kuster. From vessels to wires and 1D boundaries: A technique for simulating the thermal impact of thin structures, 1st ESHO Educational School on Clinical Hyperthermia and the 10th International Congress on Hyperthermic Oncology, Book of Abstracts, Munich, Germany, April 2008.

130. E. Neufeld, A. Kyriacou, M. Paulides, G. van Rhoon, and N. Kuster. Enabling operator interaction in hyperthermia treatment planning using pareto optimization, Annual Meeting of the European Society for Hyperthermic Oncology, Book of Abstracts, Rotterdam, Netherlands, 2010.

131. E. Neufeld, N. Chavannes, T. Samaras, and N. Kuster. Novel conformal technique to reduce staircasing artifacts at material boundaries for FDTD modeling of the bioheat equation. *Phys Med Biol*, 52(15):4371–81, 2007.

132. K. S. Nikita, N. G. Maratos, and N. K. Uzunoglu. Optimal steady-state temperature distribution for a phased array hyperthermia system. *IEEE Trans Biomed Eng*, 40(12):1299–306, 1993.

133. S. D. Olabarriaga and A. W. M. Smeulders. Interaction in the segmentation of medical images: A survey. *Med Img Analysis*, 5:127–142, 2001.

134. C. Chen and R. B. Roemer. A thermo-pharmacokinetic model of tissue temperature oscillations during localized heating. *Int J Hyperthermia*, 21(2):107–24, 2005.

135. M. M. Paulides, J. F. Bakker, M. Linthorst, J. van der Zee, Z. Rijnen, E. Neufeld et al. The clinical feasibility of deep hyperthermia treatment in the head and neck: New challenges for positioning and temperature measurement. *Phys Med Biol*, 55:2465ff, 2010.

136. M. M. Paulides, J. F. Bakker, E. Neufeld, J. van der Zee, P. Jansen, P. C. Levendag, G. C. van Rhoon. The HYPERcollar: A novel applicator for hyperthermia in the head and neck. *Int J Hyperthermia*, 23(7):567–76, November 2007.

137. M. M. Paulides, J. F. Bakker, and G. C. van Rhoon. Electromagnetic head-and-neck hyperthermia applicator: Experimental phantom verification and FDTD model. *Int J Radiat Oncol Biol Phys*, 68(2):612–20, 2007.

138. M. M. Paulides, J. F. Bakker, A. P. Zwamborn, and G. C. Van Rhoon. A head and neck hyperthermia applicator: Theoretical antenna array design. *Int J Hyperthermia*, 23(1):59–67, 2007.

139. K. D. Paulsen, S. Geimer, J. Tang, and W. E. Boyse. Optimization of pelvic heating rate distributions with electromagnetic phased arrays. *Int J Hyperthermia*, 15(3):157–86, 1999.

140. K. D. Paulsen, X. Jia, and Jr. Sullivan, J. M. Finite element computations of specific absorption rates in anatomically conforming full-body models for hyperthermia treatment analysis. *IEEE Trans Biomed Eng*, 40(9):933–45, 1993.

141. M. J. Piket-May, A. Taflove, W. C. Lin, D. S. Katz, V. Sathiaseelan, and B. B. Mittal. Initial results for automated computational modeling of patient-specific electromagnetic hyperthermia. *IEEE Trans Biomed Eng*, 39(3):226–37, 1992.

142. S. Pisa, M. Cavagnaro, E. Piuzzi, P. Bernardi, and J. C. Lin. Power density and temperature distribution produced by interstitial arrays of sleeved-slot antennas for hyperthermia cancer therapy. *IEEE Trans Microw Theory Techn*, 51(12):2418–2425, 2003.

143. H. H. Pennes. Analysis of tissue and arterial blood temperatures in the resting human forearm. *J Appl Physiol*, 85(1):5–34, 1948.

144. R. Pohle. Computerunterstuetzte Bildanalyse zur Auswertung medizinischer Bilddaten. Habilitationsschrift Otto-von-Guericke-Universität Magdeburg, 2004.

145. M. Pop, A. Molckovsky, L. Chin, M. C. Kolios, M. A. Jewett, and M. D. Sherar. Changes in dielectric properties at 460 kHz of kidney and fat during heating: Importance for radio-frequency thermal therapy. *Phys Med Biol*, 48(15):2509–25, 2003.

146. A. W. Preece and J. L. Murfin. The use of an infrared camera for imaging the heating effect of RF applicators. *Int J Hyperthermia*, 3(2):119–22, 1987.

147. T. Preusser, H. O. Peitgen, F. Liehr, M. Rumpf, U. Weikard, and S. Sauter. Simulation of radio-frequency ablation using composite finite element methods. Perspective in image-guided surgery, Proc. Scientific Workshop on Medical Robotics Navigation and Visualization, pp. 303–310, 2004.

148. T. Preusser, A. Weihusen, and H. O. Peitgen. On the modeling of perfusion in the simulation of RF-ablation. *Simulation and Visualization*, SCS, pp. 259–269, 2005.

149. C. E. Reuter, A. Taflove, V. Sathiaseelan, M. Piket-May, and B. B. Mittal. Unexpected physical phenomena indicated by FDTD modeling of the Sigma-60 deep hyperthermia applicator. *IEEE Trans Microw Theory Techn*, 46(4):313–319, 1998.

150. T. Samaras, A. Christ, and N. Kuster. Effects of geometry discretization aspects on the numerical solution of the bio-heat transfer equation with the FDTD technique. *Phys Med Biol*, 51(11):N221–9, 2006.

151. C. Schneider and J. D. Van Dijk. Visualization by a matrix of light-emitting diodes of interference effects from a radiative four applicator hyperthermia system. *Int J Hyperthermia*, 7(2):355–66, 1991.

152. A. V. Shahidi and P. Savard. A finite element model for radiofrequency ablation of the myocardium. *IEEE Trans Biomed Eng*, 41(10):963–8, 1994.

153. H. Shantesh, T. Naresh, Mekala, and H. Nagraj. Thermometry studies of radio-frequency induced hyperthermia on hydrogel based neck phantom. *J Cancer Research and Therapeutics*, 1(3):162–167, 2005.

154. M. Sherar, F. F. Liu, M. Pintilie, W. Levin, J. Hunt, R. Hill, J. Hand, C. Vernon, G. van Rhoon, J. van der Zee, D. G. Gonzalez, J. van Dijk, J. Whaley, and D. Machin. Relationship between thermal dose and outcome in thermoradiotherapy treatments for superficial recurrences of breast cancer: Data from a phase III trial. *Int J Radiat Oncol Biol Phys*, 39(2):371–80, 1997.

155. D. Shrivastava and R. B. Roemer. Readdressing the issue of thermally significant blood vessels using a countercurrent vessel network. *J Biomech Eng*, 128(2):210–6, 2006.

156. N. Siauve, L. Nicolas, C. Vollaire, and C. Marchal. Optimization of the sources in local hyperthermia using a combined finite element-genetic algorithm method. *Int J Hyperthermia*, 20(8):815–33, 2004.

157. P. K. Sneed, M. W. Dewhirst, T. Samulski, J. Blivin, and L. R. Prosnitz. Should interstitial thermometry be used for deep hyperthermia? *Int J Radiat Oncol Biol Phys*, 40(5):1015–7, 1998.

158. G. Sreenivasa, J. Gellermann, B. Rau, J. Nadobny, P. Schlag, P. Deuflhard, R. Felix, and P. Wust. Clinical use of the hyperthermia treatment planning system HyperPlan to predict effectiveness and toxicity. *Int J Radiat Oncol Biol Phys*, 55(2):407–19, 2003.

159. T. Stein. Untersuchungen zur Dosimetrie der hochfrequenzstrominduzierten interstitiellen Thermotherapie in bipolarer Technik. Ecomed, Landsberg, 2000.

160. D. M. Sullivan, R. Ben-Yosef, and D. S. Kapp. Stanford 3D hyperthermia treatment planning system. Technical review and clinical summary. *Int J Hyperthermia*, 9(5):627–43, 1993.

161. J. S. Suri, K. Liu, L. Reden, and S. Laxminarayan. A review on MR vascular image processing algorithms: Acquisition and prefiltering: Part I. *IEEE Trans Inf Technol Biomed*, 6(4):324–37, 2002.

162. J. S. Suri, K. Liu, L. Reden, and S. Laxminarayan. A review on MR vascular image processing: Skeleton versus nonskeleton approaches: Part II. *IEEE Trans Inf Technol Biomed*, 6(4):338–50, 2002.

163. A. Szasz, G. Vincze, O. Szasz, and N. Szasz. Dose concept of oncological hyperthermia: Heat-equation considering the cell destruction, preprint.

164. C. Tiebaut and D. Lemonnier. Three-dimensional modelling and optimisation of thermal fields induced in a human body during hyperthermia. *Int J Therm Sci*, 41:500–508, 2002.

165. J. B. Van de Kamer, A. A. De Leeuw, S. N. Hornsleth, H. Kroeze, A. N. Kotte, and J. J. Lagendijk. Development of a regional hyperthermia treatment planning system. *Int J Hyperthermia*, 17(3):207–20, 2001.

166. J. B. Van de Kamer, A. A. De Leeuw, H. Kroeze, and J. J. Lagendijk. Quasistatic zooming for regional hyperthermia treatment planning. *Phys Med Biol*, 46(4):1017–30, 2001.

167. J. B. Van de Kamer and J. J. Lagendijk. Computation of high resolution SAR distributions in a head due to a radiating dipole antenna representing a hand-held mobile phone. *Phys Med Biol*, 47(10):1827–35, 2002.

168. J. B. Van de Kamer, J. J. Lagendijk, A. A. De Leeuw, and H. Kroeze. High-resolution SAR modelling for regional hyperthermia: Testing quasistatic zooming at 10 MHz. *Phys Med Biol*, 46(1):183–96, 2001.

169. J. B. van de Kamer, M. van Vulpen, A. A. de Leeuw, H. Kroeze, and J. J. Lagendijk. CT-resolution regional hyperthermia treatment planning. *Int J Hyperthermia*, 18(2):104–16, 2002.

170. J. B. Van de Kamer, N. Van Wieringen, A. A. De Leeuw, and J. J. Lagendijk. The significance of accurate dielectric tissue data for hyperthermia treatment planning. *Int J Hyperthermia*, 17(2):123–42, 2001.

171. M. L. Van der Gaag, M. De Bruijne, T. Samaras, J. Van der Zee, and G. C. Van Rhoon. Development of a guideline for the water bolus temperature in superficial hyperthermia. *Int J Hyperthermia*, 22(8):637–56, 2006.

172. J. van der Zee, J. N. Peer-Valstar, P. J. Rietveld, L. de Graaf-Strukowska, and G. C. van Rhoon. Practical limitations of interstitial thermometry during deep hyperthermia. *Int J Radiat Oncol Biol Phys*, 40(5):1205–12, 1998.

173. G. M. Van Leeuwen, J. J. Lagendijk, B. J. Van Leersum, A. P. Zwamborn, S. N. Hornsleth, and A. N. Kotte. Calculation of change in brain temperatures due to exposure to a mobile phone. *Phys Med Biol*, 44(10):2367–79, 1999.

174. G. C. van Rhoon, A. Ameziane, W. M. Lee, D. J. van der Heuvel, H. J. Klinkhamer, C. Barendrecht, K. Volenec, and P. J. Rietveld. Accuracy of electrical field measurement using the flexible Schottky diode sheet at 433 MHz. *Int J Hyperthermia*, 19(2):134–44, 2003.

175. G. C. Van Rhoon, D. J. Van Der Heuvel, A. Ameziane, P. J. Rietveld, K. Volenec, and J. Van Der Zee. Characterization of the SAR-distribution of the Sigma-60 applicator for regional hyperthermia using a Schottky diode sheet. *Int J Hyperthermia*, 19(6):642–54, 2003.

176. G. C. van Rhoon and P. Wust. Introduction: Non-invasive thermometry for thermotherapy. *Int J Hyperthermia*, 21(6):489–95, 2005.

177. J. F. Verhey, Y. Mohammed, A. Ludwig, and K. Giese. Implementation of a practical model for light and heat distribution using laser-induced thermotherapy near to a large vessel. *Phys Med Biol*, 48(21):3595–610, 2003.

178. C. Villard, L. Soler, N. Papier, and V. Agnus. RF-Sim: A treatment planning tool for radiofrequency ablation of hepatic tumors. Proceedings of the Seventh International Conference on Information Visualization, p. 561, 2003.

179. C. Villard, L. Soler, N. Papier, V. Agnus, S. Thery, A. Gangi, D. Mutter, and J. Marescaux. Virtual radiofrequency ablation of liver tumors. Lecture Notes in Computer Science, 2673:1003, 2003.

180. A. G. Visser and G. van Rhoon. Technical and clinical quality assurance, In: M. H. Seegenschmiedt, P. Fessenden, and C. C. Vernon, eds. *Principles and practice of thermoradiotherapy and thermo-chemotherapy.* Springer Verlag, Berlin, 1:453–472, 1995.

181. T. Y. Wang and C. C. P. Chen. 3-D thermal-ADI: A linear-time chip level transient thermal simulator. *IEEE Trans. Computer-Aided Design of Intergrated Circuits and Systems*, 21(12):1434–1445, 2002.

182. M. Weiser and A. Schiela. Function space interior point methods for PDE constrained optimization. ZIB Report 04-27, Zuse Institute Berlin, 2004.

183. J. A. White, A. W. Dutton, J. A. Schmidt, and R. B. Roemer. An accurate, convective energy equation based automated meshing technique for analysis of blood vessels and tissues. *Int J Hyperthermia*, 16(2):145–58, 2000.

184. J. Wiersma and J. D. Van Dijk. RF hyperthermia array modelling; Validation by means of measured EM-field distributions. *Int J Hyperthermia*, 17(1):63–81, 2001.

185. J. Wiersma, R. A. van Maarseveen, and J. D. van Dijk. A flexible optimization tool for hyperthermia treatments with RF phased array systems. *Int J Hyperthermia*, 18(2):73–85, 2002.

186. J. Wiersma, N. van Wieringen, H. Crezee, and J. D. van Dijk. Delineation of potential hot spots for hyperthermia treatment planning optimisation. *Int J Hyperthermia*, 23(3):287–301, 2007.

187. E. H. Wissler. Pennes' 1948 paper revisited. *J Appl Physiol*, 85(1):35–41, 1998.

188. J. Wren, M. Karlsson, and D. Loyd. A hybrid equation for simulation of perfused tissue during thermal treatment. *Int J Hyperthermia*, 17(6):483–98, 2001.

189. P. Wust, R. Beck, J. Berger, H. Fahling, M. Seebass, W. Wlodarczyk, W. Hoffmann, and J. Nadobny. Electric field distributions in a phased-array applicator with 12 channels: Measurements and numerical simulations. *Med Phys*, 27(11):2565–79, 2000.

190. P. Wust, H. Fahling, A. Jordan, J. Nadobny, M. Seebass, and R. Felix. Development and testing of SAR-visualizing phantoms for quality control in RF hyperthermia. *Int J Hyperthermia*, 10(1):127–42, 1994.

191. P. Wust, J. Gellermann, J. Beier, S. Wegner, W. Tilly, J. Troger, D. Stalling, H. Oswald, H. C. Hege, P. Deuflhard, and R. Felix. Evaluation of segmentation algorithms for generation of patient models in radiofrequency hyperthermia. *Phys Med Biol*, 43(11):3295–307, 1998.

192. P. Wust, J. Nadobny, M. Seebass, J. M. Dohlus, W. John, and R. Felix. 3-D computation of E fields by the volume-surface integral equation (VSIE) method in comparison with the finite integration theory (FIT) method. *IEEE Trans Biomed Eng*, 40(8):745–59, 1993.

193. P. Wust, J. Nadobny, M. Seebass, D. Stalling, J. Gellermann, H. C. Hege, P. Deuflhard, and R. Felix. Influence of patient models and numerical methods on predicted power deposition patterns. *Int J Hyperthermia*, 15(6):519–40, 1999.

194. D. Yang, M. C. Converse, D. M. Mahvi, and J. G. Webster. Expanding the bioheat equation to include tissue internal water evaporation during heating. *IEEE Trans Biomed Eng*, 54(8):1382–8, 2007.

195. D. Yang, M. C. Converse, D. M. Mahvi, and J. G.Webster. Measurement and analysis of tissue temperature during microwave liver ablation. *IEEE Trans Biomed Eng*, 54(1):150–5, 2007.

196. T. V. Youltsis, T. I. Kosmanis, E. P. Kosmidou, T. Zygiridis, N. V. Kantartszis, T. D. Xenos, and T. D. Tsiboukis. A comparative study of the biological effects of various mobile phone and wireless LAN antennas. *IEEE Trans Magn*, 38(2):777–780, 2002.

197. L. Zhu, L. X. Xu, Q. He, and S. Weinbaum. A new fundamental bioheat equation for muscle tissue–part II: Temperature of SAV vessels. *J Biomech Eng*, 124(1):121–32, 2002.

198. A. P. M. Zwamborn, P. M. van den Berg, J. Mooibroek, and F. T. C. Koenis. Computation of three dimensional electromagnetic-field distributions in a human body using the weak form of the CGFFT method. *Appl Comp Electromagnetic Soc*, 7:26–42, 1992.

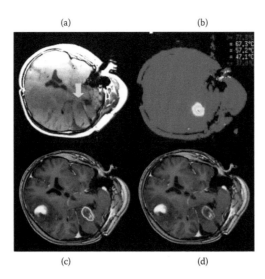

(a)　　　　　　　　　(b)

(c)　　　　　　　　　(d)

FIGURE 3.1   MRI guidance of laser ablation of brain lesion. In addition to planning the procedure, MRI is useful for targeting the volume of interest and verifying correct location (green arrow) of devices for therapy delivery (a). MR temperature imaging provides a spatiotemporal map of the temperature that can be used to help control delivery for safety and efficacy (b). The temperature history can be integrated with biological models that predict damage (orange contour) as shown in (c), which may be used as a surrogate for predicting the treatment endpoints during the course of therapy as opposed to more time-consuming posttreatment verification imaging, such as contrast-enhanced imaging, which demonstrates the perfusion deficit left by therapy and enhancing ring of edema (d).

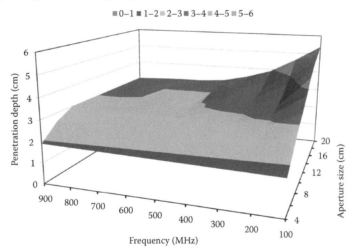

FIGURE 4.6   Dependence of effective penetration depth d_{eff} into muscle-like tissue associated with square aperture sources upon frequency f and aperture size a. The surface $d_{eff}(f, a)$ is colored according to 1 cm increments in d_{eff}.

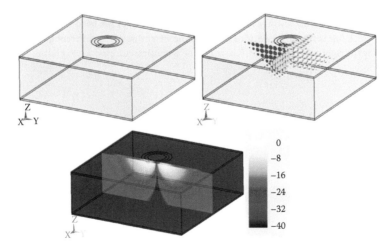

FIGURE 4.9   Top left: three-turn induction coil modeled as three concentric circular loops (radii 10, 8, and 6 cm) placed 3 cm above a plane-layered fat-muscle-fat phantom. Each loop is driven by a current source at 30 MHz. Top right: vector plot of current density J in planes $\times = 0$ and $y = 0$. Bottom: The resulting SAR distribution (in dB relative to the maximum SAR) in the $\times = 0$ plane.

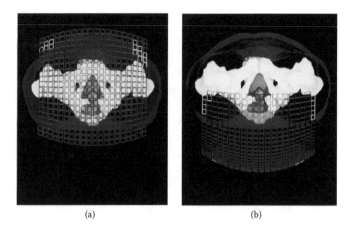

FIGURE 6.10 (a) Unoptimized and (b) optimized ultrasound phase array apertures for hyperthermia in the prostate. (After R. J. McGough, M. L. Kessler, E. S. Ebbini, and C. A. Cain, *IEEE Trans. Ultrason. Ferroelect. Freq. Contr.*, 43, 6, 1996.)

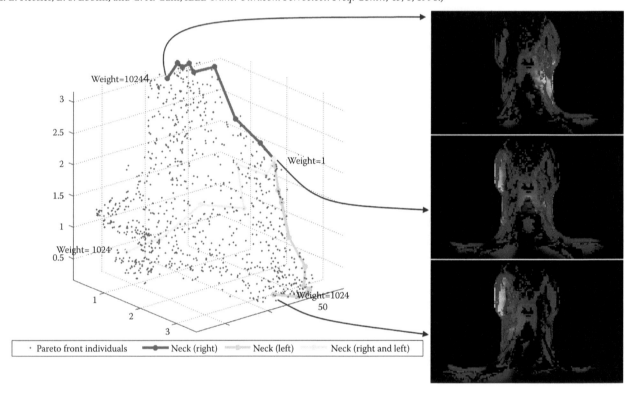

FIGURE 7.6 Pareto front of optimized antenna settings for hyperthermia treatment of the head-and-neck area. Each point corresponds to an optimized antenna setting, and which one is selected is decided based on the weighting of the different goals (heat the tumor, avoid exposing the left side of the neck, avoid exposing the right side of the neck). Changing the weighting corresponds to gliding along the Pareto front and translates into shifting the energy deposition.

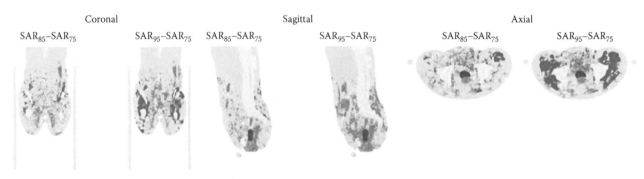

FIGURE 8.15 Comparison of the SAR distributions of 75 MHz to that of 85 and 95 MHz in the same patient. Changes in local SAR by >5 W/kg, increase in red and decrease in blue are presented for an average patient with locally advanced cervical cancer.

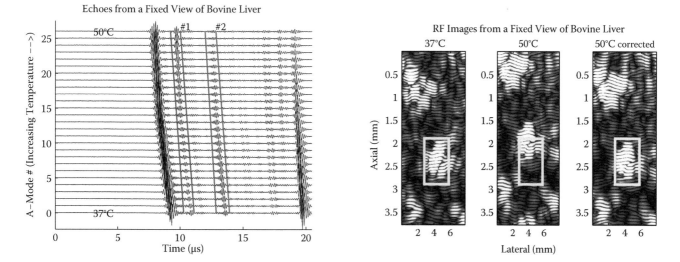

FIGURE 13.1 Changes in backscattered ultrasound with temperature. (Left) Echoes measured from a single site in a 1 cm thick sample of fresh bovine liver at temperatures from 37 to 50°C. The two delineated echoes (indicated by bands marked #1 and #2) shift with temperature and have energies that appear to change with temperature (similar to Figure 13.4 in Arthur et al.[42]). (Right) RF images of a fixed region in bovine liver showing apparent motion from 37 to 50°C. The right panel shows the image at 50°C after motion compensation (similar to Figure 13.3 in Arthur et al.[43]).

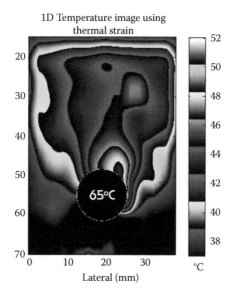

FIGURE 13.7 1D thermal strain temperature image from axial echo shifts in turkey breast muscle during nonuniform heating using the fixture in Figure 13.5. Tissue was surrounded by water at 37°C. At time = 0 seconds 65°C water was pumped through the tissue center (black disk). Compare to the 3D CBE image in Figure 13.14 from a different specimen of turkey. Thermocouple readings in planes adjacent to the image were used to scale the strain image to give temperature estimates.

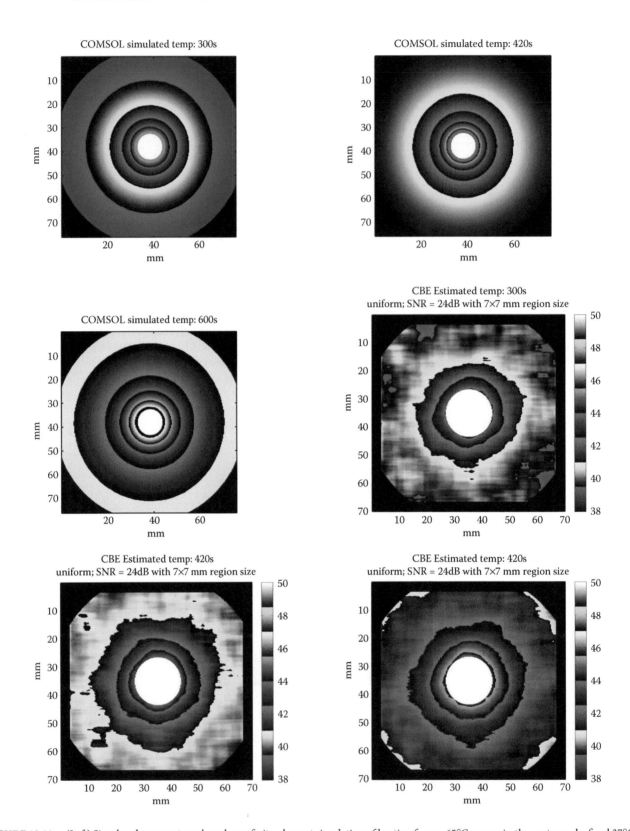

FIGURE 13.11 (Left) Simulated temperatures based on a finite-element simulation of heating from a 65°C source in the center and a fixed 37°C surrounding temperature. (Right) Temperature images estimated with CBE from simulated images for a homogeneous scatterer distribution over the heated region with an SNR of 24 dB. The mean estimation error for was 0.4 ± 0.2°C. (From Basu, D., *3D Temperature Imaging Using Ultrasonic Backscatter Energy during Non-Uniform Tissue Heating*, PhD thesis, Washington University, St. Louis, MO, 2010. With permission.)

II

Clinical Thermal Therapy Systems

8

External Electromagnetic Methods and Devices

Gerard C. van Rhoon
Erasmus MC Daniel den
Hoed Cancer Center

8.1 Introduction

Within the domain of thermal therapy, external electromagnetic devices are exclusively used to apply hyperthermia (i.e., inducing temperatures in the range of 40–45°C for durations of 60–90 minutes). Heating tissue involves complex technology whereby the attained tissue temperature is affected by the temperature-dependent physiological response of the tissue. Moreover, thermal goals for hyperthermia (HT) are complicated as hyperthermia causes a multitude of different biological effects on tissue, which are both time and temperature dependent.

Hyperthermia is always implemented as part of a multimodal, oncological strategy, (i.e., in combination with radiotherapy or chemotherapy). Despite the technological and biological challenges, hyperthermia has proven to possess a great potential to improve outcome of cancer treatment when combined with radiotherapy or chemotherapy. "The Kadota Fund International Forum 2004-Clinical Group Consensus" report published in March 2008 by the *International Journal of Hyperthermia* reflects a recent and authoritative opinion on clinical HT [1]. In this report, 68 clinicians and biologists from all over the world conclude that HT is an effective complementary treatment to, and a strong sensitizer of, radiotherapy (RT) and many cytotoxic drugs. It includes a table of 19 randomized trials, all showing significantly better results in the treatment arm with HT. Since then, new clinical results have been reported. The long-term follow-up of the Dutch Deep Hyperthermia study [2] showed that the improved 3-year survival rate for the RT + HT arm (i.e., 27–51%) was still present after 12 years follow-up [3]. Further, Franckena et al. [4] demonstrated in 378 patients

that the good results of RT + HT are sustainable in a routine clinical setting. In addition, two new positive phase III studies were published. Issels [5] reported in 340 patients with high-grade soft tissue sarcoma a doubling of the median disease free survival (16.2 to 31.7 months) when regional hyperthermia was added to the standard treatment for this disease. Hua et al. [6] reported a nearly 10% increase of the 5-year progression free survival (63.1% for RT vs. 72.7% for RT + HT) in their randomized study of 180 patients with nasopharyngeal cancer after adding intracavity hyperthermia to the conventional radiotherapy-chemotherapy treatment schedule. Remarkably, most of the positive results in recent trials are achieved without increased toxicity.

8.2 Impact of the Quality of the Hyperthermia Treatment on Clinical Outcome

Ample literature is becoming available revealing the close relation between HT treatment quality and treatment outcome. Over time a wide variety of dose parameters have been investigated. In retrospective studies, dose effect relationships were found for penetration depth [7], coverage by the 25% iso-SAR contour [8], and thermal dose expressed in various dose parameters [9, 10, 11]. Demonstrating a thermal-dose effect relationship in prospective trials is more difficult. Thrall et al. [12] showed that thermal dose is related to duration of local control in canine sarcomas treated by RT + HT. Maquire et al. [13] was less successful. Highly relevant for the future development of hyperthermia technology is the recent finding of a thermal-dose effect relationship in 420 patients with locally advanced cervical cancer (LACC) treated with RT + HT [14]. Even after adjustment for other correlating factors in the multivariate analysis (RT dose, tumor stage, size, performance status), the intraluminally measured thermal dose parameter remains significantly correlated with response and survival.

All these findings on the impact of quality of the hyperthermia treatment on clinical outcome clearly indicate that there is only one direction to go in hyperthermia! To obtain the highest probability of tumor control and enhance our ability to verify whether a specific biological mechanism of hyperthermia indeed is active, we must increase our ability to deliver a specified, highly controlled, and quantitatively plus objectively documented quality of the hyperthermia treatment.

8.3 Requirements of a Modern External Electromagnetic Heating Device

The clinical and biological need for enhanced control and uniform quantitative documentation of the quality of the applied hyperthermia treatment as defined in the previous paragraph provides a clear direction for engineers, physicists, and companies designing and selling hyperthermia equipment. If hyperthermia is to be accepted as an integral part of the oncological treatment pallet, care must be taken that the involved clinicians (radiation and medical oncologists) have control of the prescription of the thermal dose and are accurately informed of the thermal dose actually delivered. Furthermore, hyperthermia devices must be designed such that they provide the hyperthermia technologists with an easy-to-use instrument, making it a joy to deliver the prescribed thermal dose in a reliable manner. Consequently, innovation of hyperthermia technology should focus on improving the quality of treatment through enhanced control of targeting the RF energy to the tumor.

In principle the ideal hyperthermia system consists of an applicator holding multiple antenna elements to transfer the electromagnetic energy into the tissue, allowing excellent spatial and temporal control, an integrated water bolus to control skin temperature, and a temperature measuring system that provides real-time, 3D information on the temperature distribution. Of course the whole system has a computer-controlled feedback loop in combination with an intelligent graphical user interface, watchdog functions, and automatic data analysis. In reality a number of practical limitations is prohibiting the construction of such an ideal system, although the systems currently in use are slowly converging to these highly advanced levels. You can only shape the future if you know your history. Therefore, the next sections will briefly summarize the historical developments and experiences followed by discussions concerning the main parts identifiable in a hyperthermia system. When possible,

references are made to quantitative values of the desired requirements. Also, a number of associated practical limitations ("lessons learned"), which have to be solved in order to achieve the required level of control in hyperthermia treatment quality, will be addressed.

8.4 Historical Perspective of the Development of External Electromagnetic Devices

Since the early years of clinical hyperthermia, research has been directed at the development of either quasi-static or radiative techniques to apply and control hyperthermia using electromagnetic energy. Around 1975, electromagnetic fields were used only in physical therapy to increase local tissue perfusion in order to stimulate faster recovery of muscle or ligament injuries. For practical reasons, the first devices evaluated for hyperthermia came from the departments of physical therapy. Characteristics of the physical therapy devices were that they operated only at the ISM (Industry Science and Medicine) frequencies, had a single radiofrequency power source, one applicator, and a limited size of the treatment field. The applicators used both the magnetic and electric field component to transfer electromagnetic energy from the antenna/electrode to the patient. For physical therapy an essential advantage of their applicator design was that a direct contact between the applicator and the patient was not needed as energy transfer goes through air, the latter being highly convenient for the patient. Applying the devices for physical therapy for hyperthermia treatment quickly showed their limitations in the degree of freedom to adapt the energy distribution in tissue. Also skin cooling by air proved to be heterogeneous, thus ineffective for hyperthermia, and was abandoned in favor of cooling using a water bolus.

The experienced shortcomings in the existing equipment for physical therapy constituted the starting point for the development of more advanced heating systems for hyperthermia. It must be noted that, as advanced computer models were lacking, all research efforts were based on translating fundamental and experimental knowledge of electromagnetic fields into an empirical design of a hyperthermia applicator. For a good understanding of why the Hyperthermia Society has arrived at the currently applied technology (see Section 8.5), it helps to briefly review the early experiences with the different approaches.

8.4.1 Radiofrequency Inductive Heating

Transferring electromagnetic energy using the magnetic field component initially appeared attractive for the earlier mentioned possibility of avoiding direct contact between the applicator and skin. The most basic applicator is an inductive concentric coil that is placed around the patient. However, inductive concentric coil devices have zero power deposition at the center of the patient and thereby restrict the clinical use to eccentric tumors [15,16,17]. As a solution to this problem, a helical coil applicator and a coaxial pair magnetic system were constructed. A helical coil applicator operates at its resonance frequency and creates eddy currents by the inductive field component (with zero energy deposition at the center) as well as a longitudinal electric field (due to the resonance) with associated high energy deposition along this axis. Together this resulted in a fairly homogeneous energy deposition in cylindrical tissue configurations [18,19,20]. With the coaxial pair magnetic system, the setup is such that one coil was positioned above and one coil was positioned below the target volume. Although the zero power deposition in the center of the coil still exists, this problem was solved by moving the applicator around its longitudinal axis [21,22]. Due to the limited spatial control of the energy deposition, the ability to selectively heat the tumor was largely dependent on the differences in blood perfusion between tumor and normal tissue.

8.4.2 Radiofrequency Capacitive Heating

Capacitive systems also have a straightforward design and require relatively simple technology. In capacitive systems the electric field is used to transfer the electromagnetic energy to the tissue, and in all setups one applicator is placed below and one above the target volume (tumor). By using different sizes of electrodes it is possible to create a higher energy deposition near the smallest electrode. Due to the design of this device the electric field is always directed from one electrode to the other. Hence, in the majority of the clinical application the electric field is perpendicular to the body axis, and subsequently the energy distribution is characterized by high power deposition in the subcutaneous fatty tissue (see Chapter 4, this book). In Asian patients, preferential heating of the fat layer may be adequately counteracted [23,24] with efficient precooling, provided that the fat layer does not exceed a thickness of 1.5–2.0 cm. In European patients, who generally have thicker fat layers, the effect of precooling is limited [25]. A major disadvantage of capacitive systems is that the energy distribution cannot be adapted other than by replacing the electrodes or changing their size. Both methods require an interruption of the hyperthermia treatment. A study investigating the feasibility of SAR steering with a three-plate capacitor system was not successful [26]. Hence, also for capacitive systems, the ability to selectively heat the tumor is largely dependent on the differences in blood perfusion between tumor and normal tissue.

8.4.3 Radiofrequency Radiating Heating

The available equipment from physical therapy using radiating electromagnetic fields was extremely limited and consisted of a few applicators. In Europe, physical therapy used more or less one cylindrical waveguide (diameter around 6 cm) operating at 2450 MHz and one folded dipole antenna operating at 433 MHz. Although both applicators have been used for hyperthermia, their use was hampered by the small volume that could be heated with the 2450 MHz waveguide or the low directivity of the

433 MHz antenna. Hence, engineers and physicists immediately started to develop new antennas to radiate the electromagnetic energy into tissue. The penetration of electromagnetic energy is strongly dependent on the frequency of the electromagnetic field. Therefore, the application of hyperthermia using radiating electromagnetic heating is traditionally divided in equipment for superficial hyperthermia (i.e., heating tumors extending to a depth of 3–4 cm) and loco-regional or deep heating (i.e., tumors with a depth larger than 4 cm or situated centrally in the body). Clearly, the design strategy for superficial hyperthermia strongly deviates from that for loco-regional or deep hyperthermia. The division in superficial and loco-regional hyperthermia is used also in this chapter.

For superficial hyperthermia, antenna development focused initially strongly on rectangular or cylindrical waveguide-type applicators. Intrinsically, the aim of the early applicator designs was to limit the number of antenna elements. This early view was justified by the assumption that tumors had a lower blood perfusion than normal tissue, resulting in preferential heating of the tumor. Further, this view was supported by practical considerations as miniaturization of electromagnetic components, such as transistorized amplifiers, was hardly available or only at high costs. Clinical experience quickly demonstrated that all different types of tumor sizes and locations were referred for superficial hyperthermia, whereby the majority of the patients had recurrent breast cancer at the chest wall extending over a large area. The ability to heat such areas was extremely limited due to the poor characteristics of the available first-generation waveguides. Typically, a conventional waveguide with a homogeneous dielectric load (air, water, or other high permittivity material), has in muscle-equivalent tissue an effective field size of about 30–40% of the antenna aperture size (i.e., the area at 1 cm depth with a $SAR \geq 50\% \ SAR_{max}$). Hence, all treatments needed ample space around the target in order to facilitate the part of the waveguide extending outside the treatment field. For many patients this space was not available, while for others the size of the effective field was too small to treat the tumor area in one setting. In some designs moving the applicator was introduced to extend the size of the field of heating or to obtain a more homogeneous field distribution. The limited speed of movement caused this approach to vanish. Alternatively, multi-element applicators were applied. The latter more clearly demonstrated the disadvantage of the small effective field size in an array of conventional waveguides as it enhanced the heterogeneity of the SAR distribution. The need for improved applicators stimulated research into different types of antenna designs for superficial hyperthermia and resulted in various designs of single- and multi-element arrays, each with their own specific characteristics [27,42,43,51,52,56]. Differences between the antennas are:

- Circular or linear field polarization, and energy coupling: *E*- or *H*-field (spiral applicator: circular and *E*-field; current sheet applicator: linear and *H*-field; Lucite cone applicator: linear and *E*-field; dual concentric conductor microstrip antennas: linear and *E*-field)

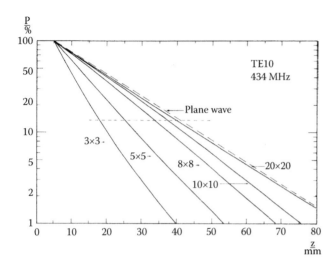

FIGURE 8.1 Absorbed power distribution as function of depth in a muscle equivalent medium for 434 MHz and different applicator sizes [28].

- Aperture size of the elements. Small elements will provide high spatial resolution of power control and are capable of conforming to body contour. As is well known and generally accepted [28,29], the penetration depth decreases with smaller aperture size. As can be seen in Figure 8.1 the penetration depth for a single applicator with an aperture of 10×10 or 5×5 cm^2 decreases from 22 to 14 mm for the 25% SAR value and from 13 to 8 mm for the 50% SAR value, respectively.

Eventually, these investigations brought us to the various types of applicators for superficial hyperthermia that are currently used by several university groups and/or are sold by the manufacturers of hyperthermia equipment.

For deep hyperthermia there was no equipment, other than the capacitive and inductive devices from physical therapy, available. From basic physics it was clear that the frequency of the electromagnetic field should be low in order to obtain sufficient penetration depth. Further, despite the increased penetration depth it was also clear that a single antenna element would not provide sufficient energy deposition at the center of the body. Consensus was rapidly reached that for "Western" patients (i.e., with often a thick fat layer), only those electromagnetic devices that generate a circumferential electrical field distribution around the patient, directed parallel to the body axis, are capable of producing loco-regional deep heating. Several different deep heating systems have been designed taking into consideration the previously mentioned existing views and limitations of equipment of the 1980s. One of the earliest radiative systems available for deep heating was the BSD-1000 system. It consisted of 16 waveguide antennas operating at a frequency of 70 MHz and with synchronous settings (i.e., amplitude and phase of all 16 antenna elements were similar, leaving only control of the total amount of energy delivered to the patient). In practice the first users rapidly adapted the cable length between the antennas such that they could dislocate the point of interference.

From the beginning clinical studies were performed to evaluate the clinical performance of the various deep heating systems. In a direct clinical comparison, it was shown that the annular array (AA) was superior to the concentric coil [30], and had an equal heating efficiency to the Thermotron R.F. capacitive system when used for patients with a fat thickness less than 1.5 cm [31]. In a small study of eight patients, Shimm et al. [32] found that with the air-coupled CDRH Helix (resonant helical coil) system similar temperatures were measured as with the BSD-1000 annular phased system. Despite all efforts, in approximately 90% of the clinical treatments with the BSD-1000 system, local pain, general discomfort, and rise of normal tissue temperature was power limiting with subsequently too low tumor temperatures [32,33,34,35,36]. These findings quickly initiated new applicator designs that all addressed in one or more ways the reported limitations of the BS-D1000 Annular Phased Array system.

The coaxial TEM applicator [37] was unique concerning its open water bolus; due to the lack of water pressure on the skin of the patient, a better treatment tolerance was realized. Due to its simple electromagnetic design the coaxial applicator could be used over a broad frequency range and needed only a single high power generator. This of course was beneficial for quality assurance and costs of equipment. Steering of the energy distribution was realized by moving the patient (i.e., the target into the center of the applicator where the maximum energy deposition was located). Salt water boli were used to reduce the amplitude of the electromagnetic field at specified locations. The ring applicator system as proposed by Franconi [38] and van Rhoon et al. [39,40] was designed along the same principle, but using lumped elements to couple the energy to the applicator. With this ring applicator a circumferential E-field distribution can be created from very low frequencies (27 MHz) to frequencies above 70 MHz. Like the TEM applicator [37], the ring applicator has the capability to adapt the axial extent [39,40] of the energy distribution. An important advantage of the ring applicator is its small size, which is comparable to that of the Sigma-60 applicator. Additionally, elliptical applicators have been studied to improve the SAR distribution [41]. The introduction around 1990 of the BSD-2000 deep hyperthermia system with the Sigma-60 applicator as the successor of the BSD-1000 system constituted a major improvement in equipment. At the same time, the Amsterdam group introduced the AMC four-waveguide system for deep hyperthermia [73].

8.5 Currently Available Systems[*]

8.5.1 Devices for Superficial Hyperthermia

Patients referred for superficial hyperthermia represent many different types of tumors (i.e., pathology, size, and location) (Figure 8.2). Superficial tumor locations for which hyperthermia

---

[*] This section on currently available systems is restricted to radiative electromagnetic applicator systems only!

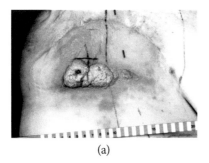

(a)

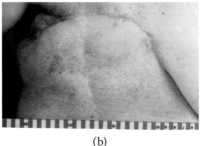

(b)

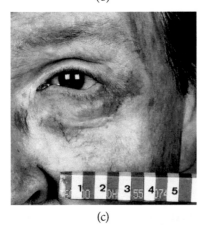

(c)

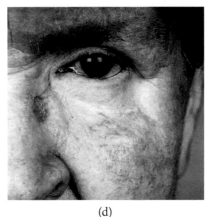

(d)

FIGURE 8.2 Two examples of patients referred for combined treatment with radiotherapy (re-irradiation) and superficial hyperthermia to demonstrate the large difference in location and thus requirements on the equipment: (a) and (c) before treatment, (b) and (d) after treatment.

treatment has been reported include all parts of the body (i.e., primary and recurrent tumors located at the skull, neck, legs, arms, vulva, groin, back, and chest wall). Especially for

recurrent breast cancer at the chest wall, the conditions can put high constraints on the required equipment. Chest wall recurrences can be very localized (e.g., a treatment area of less than 100 cm²), or very huge with tumor growth extending over a large part of or even over the whole chest wall (in case of cancer "en curasse" even large parts of the back can be involved). In the latter case it might be necessary to configure a treatment strategy consisting of several sequential sessions in order to heat the whole tumor volume. Frequently, the skin surface is highly irregular due to previous surgery or tumors protruding through the skin and with alternating tissue thicknesses varying between just skin and bone (ribs) to several centimeters of heterogeneous fat-muscle tissue. Depending on the frequency of the microwave fields, the rule of thumb is that the tumor should not extend deeper than 2 or 4 cm for systems operating at 915 or 434 MHz, respectively.

In addition, the strong spatial variations in tissue type, blood perfusion, and anatomy require a high degree of control on the spatial energy deposition. This can be achieved by using multi-element antenna systems with power control per element.

Fortunately, modern clinical hyperthermia devices using electromagnetic energy in the frequency range of 400–1000 MHz do provide the flexibility required to adequately and comfortably heat this large variation in target volumes. Research during the last three decades has delivered a number of antenna designs that, when used with proper judgment, provide adequate heating characteristics for the circumstances. In general the somewhat bulky and more ridged applicator types like waveguides and current sheet antenna arrays operating at 434 MHz are considered better suited to heat large nodular types of tumors (exophytic and ulcerating) as well as recurrent breast cancer with tumors growing to >2 cm depth, as they provide a higher energy deposition at depth. Conformal microwave arrays are on the other hand the most suited applicator type to heat large contoured areas of the torso, avoiding discomfort from underlying ribs.

Academic and commercial devices for superficial hyperthermia that are successfully used for clinical application are reported in Section 8.5.1.1.

8.5.1.1 Academy-Based Applicators

8.5.1.1.1 Lucite Cone Applicator [42,43]

The Lucite cone applicator (LCA) is a 434 MHz water-filled horn applicator designed for external heating of superficial malignancies (superficial hyperthermia). A conventional applicator (CA) consists of a water-filled rectangular waveguide that ends in a horn antenna. The LCA is a modification of the CA. The waveguide is made of brass and operates at the TE$_{10}$ mode at 434 MHz. The waveguide dimensions are 3 × 5 cm, and the aperture size of the radiating antenna is 10 × 10 cm². The two diverging metal walls of the horn antenna, which are parallel to the electric field, are replaced by Lucite walls. Additionally, a PVC cone with a height of 5.5 cm is inserted in the applicator at the center of the aperture. Compared to conventional

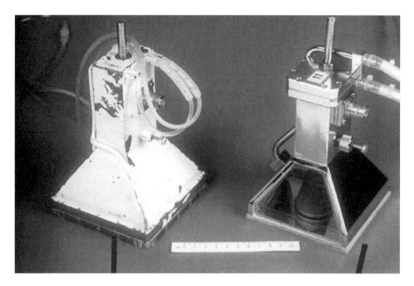

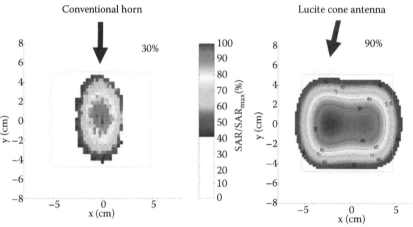

FIGURE 8.3 Conventional waveguide and Lucite cone waveguide antennas with the different relative SAR distribution as measured at 1 cm depth in a muscle-equivalent tissue indicated below the applicator [42].

waveguide applicators, the LCA distinguishes itself by the high ratio of effective field size to the aperture opening (area with a relative SAR > 50% divided by the aperture area of the waveguide; see Figure 8.3). For the LCA this ratio is around 80% with the 50% iso-SAR contour extending outside the aperture area at the side of the Lucite wall [44]. In comparison to the effective field size of a conventional waveguide applicator, the heating area of the LCA is ± 2.5 times larger [45], consequently the applicator target volume of a single LCA is $10 \times 10 \times 4$ cm^3. LCAs can be combined in array configurations to effectively cover the whole radiotherapy field (from 10×10 to 20×30 cm^2).

The LCA forms the backbone of the Rotterdam superficial hyperthermia system, which consists of six remotely controlled, solid-state amplifiers operating at 434 MHz, each capable of delivering an RF output of 200 W. Temperature is continuously measured with 32 fiber-optic temperature sensors. The graphical user interface available to the HT technician during treatment uses graphs to shows the temperature evolution over time and a projection of measured temperatures on the patient target area.

For recurrent breast tumors with depths up to 4 cm and extending over large areas of the chest wall, arrays of Lucite cone or current sheet applicators (see Section 8.5.1.1.2) are the only applicators able to heat such tumors with good spatial SAR control.

Advantages: robust applicator with reliable performance; effective field area approaches aperture size; spatial power control in array; penetration depth comparable to that of free space at 434 MHz; uniform performance between applicators; high RF-power output possible; versatile use over all body parts.

Disadvantages: rigid and large applicator, size may compromise setup; applicator mounting system required.

8.5.1.1.2 Current Sheet Applicators

The current sheet applicator (CSA) has a very compact design, and the radiofrequency (RF) electromagnetic energy is inductively coupled to the tissue. The basic principle of the CSA was

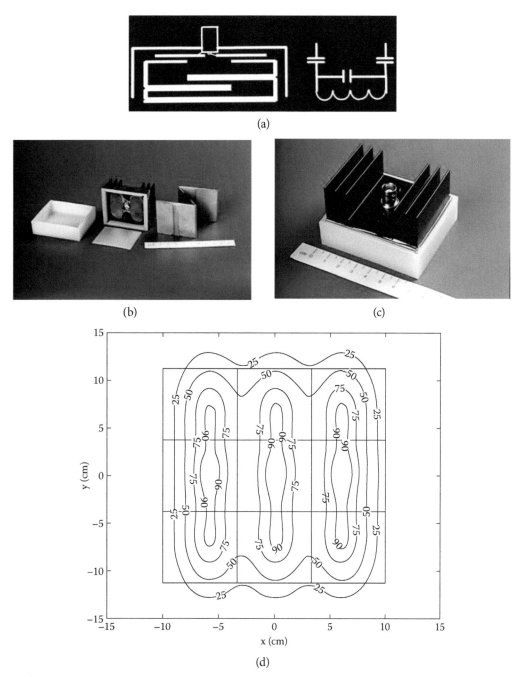

(a)

(b)   (c)

(d)

FIGURE 8.4 (a) Schematic drawing of the current sheet applicator (CSA) and its electronic representation. (b) Disassembled CSA (Rotterdam version) showing the capacitive electrodes, the current sheet is not shown (at the back). (c) Completely assembled CSA with cooling device to allow higher RF powers. (d) Gaussian beam predicted relative SAR distribution at 1 cm depth in muscle-equivalent tissue for a 3 × 3 CSA array (E-fields parallel) [52].

provided by Bach Andersen et al. [46] and Johnson et al. [47,48], while later Lumori et al. [49,50] and Gopal et al. [51] developed it into a clinical version. The latest clinical CSA design is by Rietveld et al. [52] (Figure 8.4) and is made of sticky copper foil (0.05 mm) and polyethene (PE). The RF power is capacitively coupled to the "current sheet" (55 × 49 mm²) creating a resonant circuit at 434 MHz. The total is embedded in a Teflon block; outer dimensions of the antenna are 75 × 65 mm² (l × w)

and 19 mm (h). The current flows along the longer dimension of the sheet and so the dominant E-field is linearly polarized parallel to the length of the CSA. Advantages offered by current sheet applicators for tissue heating include compact size, a linear polarization of the induced electric field, and relatively large heating area. It is shown that the effective field produced by a pair of these elements is continuous regardless of whether the

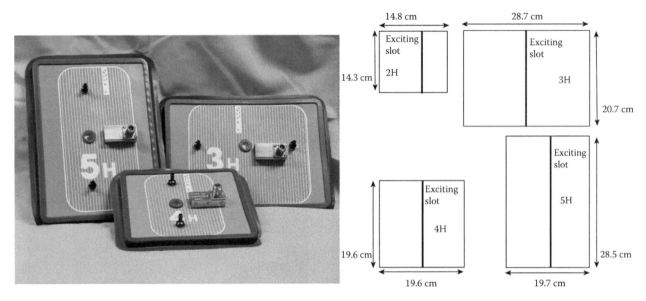

FIGURE 8.5 Contact flexible microstrip applicators, types 3H, 4H, and 5H. On the right the size of the applicator is indicated. (From Kok, H.P., Correia, D., De Greef, M., Van Stam, G., Bel, A., Crezee, J., *Int J Hyperthermia* 26, 2010.)

common edges of the elements are perpendicular or parallel to the direction of impressed current (see Figure 8.4).

Advantages: robust applicator with reliable performance; effective field area approaches aperture size; spatial power control in array; H-field energy coupling; uniform performance between applicators; versatile use over all body parts.

Disadvantages: power output limited by self-heating of the applicator, cooling required; rigid applicator; applicator mounting system required.

8.5.1.1.3 Contact Flexible Microstrip Applicators (CFMA)

Contact flexible microstrip applicators (see Figure 8.5) have been designed by Gelvich et al. [53]. They are available as single element applicators with different sizes: 2H: 14.8×14.3 cm²; 3H: 28.7×20.7 cm²; 4H: 19.6×19.6 cm²; and 5H: 28.5×19.7 cm². The applicators are available from SRPC "Istok" in Moscow, Russia. The CFMA consist of two active, coplanar electrodes and a shield electrode. The electrodes are mounted on a substrate of fluoroplastic with a thickness of 0.5–1.5 mm depending

on their operating frequency. In Europe the CFMA for superficial hyperthermia operates at 434 MHz. The two active electrodes are excited by the feeding pin of the coaxial cable as a plane dipole-like microstrip antenna (5 mm spacing between the electrodes). A short circuit is positioned at about ¼λ from the exciting slot [54]. The orientation of the principal E-field direction is across the slot and perpendicular to the lines indicated on the surface of each applicator. Bending is only possible around the axis perpendicular to the exciting slot. The whole applicator is incorporated in a rubber frame in which a thin rubber water bolus is integrated. The surface of the rubber frame is covered with 5 mm high lugs to prevent collapse of the water bolus. The thickness of the water bolus is 1 cm measured from the highest rim of the rubber frame. A special feature of the CFMA is that it can be bent around the axis perpendicular to the exciting slot in order to follow the curvature of the body.

Following the successful use of the single element applicators, SRPC "Istok" developed the CFMA-12, consisting of 12 capacity-type mini-antennas [55]. The 12-element microstrip applicator array adds spatial control of the energy distribution by amplitude modulation for each element (see Figure 8.6). The

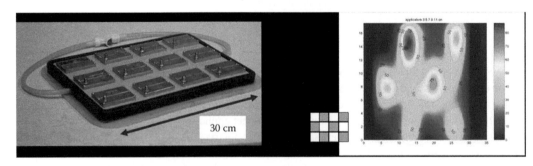

FIGURE 8.6 Twelve-element contact flexible microstrip applicator, type CFMA-12. On the right an example of the measured relative SAR distribution at 1 cm depth in a muscle-equivalent phantom with elements activated as indicated [55].

aperture size of the whole CFMA-12 is 20.8 × 30 cm². The aperture size of one single antenna element is 6.4 × 7.0 cm². The small dimensions are achieved due to the relatively high magnitude of the lumped inductance, which can be varied in wide limits. Each mini-antenna of the CFMA-12 array operates in the same way as the single element CFMA. The CFMA-12 antenna elements are close to the skin surface, hence a large Ez-component of the electric field may cause preferential fat heating.

Advantages: flexible substrate; quick and easy to apply to the treatment region; integrated water bolus; applicator size does not compromise its use; comfortable; CFMA-12 spatial power control; high RF-power output possible.

Disadvantages: single element CFMA: no power control; bending affects SAR pattern, effective heating depth increases and decreases for bent applicator with very local variations; SAR distribution is not similar for all applicators; resonance effect in water bolus; CFMA-12 substantial variation in power efficiency between single elements.

8.5.1.1.4 Dual Concentric Conductor

The dual concentric conductor (DCC) applicator was developed during the 1990s to specifically address the clinical demands for a flexible device that could uniformly heat large area superficial disease overlying complex contoured anatomy. In its most basic design the DCC applicator consists of square dual concentric conductor apertures etched from the front copper surface of double-sided flexible PCB material [56,57]. Each square element of the array has an aperture side of 4 cm and is separated from the 3 cm central patch by a 5 mm radiating gap. The DCC originally operated at a frequency of 915 MHz, the ISM frequency band for medical purposes in the United States, but later an additional DCC design was presented that operates at 433 MHz. Detailed investigation of the electrical field (x,y,z) components is highly relevant to fully analyze the near-field characteristics of a microstrip applicator. Due to the close proximity of the resonant patch to ground electrode, often a strong electrical field (z-component) perpendicular to the face of the applicator exists. A strong E_z component gives rise to a high SAR in the superficial tissues directly in front of the applicator. For the DCC, clinically desirable uniformity and penetration of SAR can be obtained for square apertures in the range of 4–6 cm per side at 433 MHz, using 9.5–12.5 mm thick water coupling boluses, as compared to the optimum 2–4.5 cm square apertures and 5–10 mm water bolus dimensions determined previously for 915 MHz arrays. With an accompanying reduction in the lateral adjustability of power control, the 433 MHz applicators can provide a significantly larger heating area than arrays with the same number of elements driven at a frequency of 915 MHz [58]. From theoretical comparison between DCCs operating at 915 and 433 MHz, it is concluded that the penetration depth is similar for optimally sized apertures at both frequencies of operation,

providing a maximum heating depth of 1–2 cm for the DCC array applicators.

In a very recent review on conformal microwave array applicators for hyperthermia of diffuse chest wall recurrences, Stauffer et al. reported the development of a "hyperthermia-vest" applicator consisting of up to 35 DCC elements [59]. At present this is the largest applicator available that is lightweight and sufficiently flexible to follow the contour of the large tumor spread along the chest wall. Moreover, by integrating a water bolus and thermal mapping catheters in the applicator design an important step has been made to improve comfort by rapid and accurate application to the patient's body (see Figure 8.7).

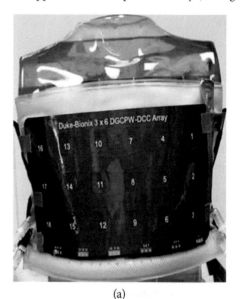

(a)

(b)

FIGURE 8.7 Conformal Microwave Array applicator. (a) Photo of 18-element dual concentric conductor array coupled with 9 mm-thick water bolus to a torso phantom. (b) 18-element DCC array carried by patient and kept at its position by an elastic outer support vest. (From Stauffer, P.R., Maccarini, P., Arunachalam, K., Craciunescu, P., Diederich, C., Juang et al., *Int J Hyperthermia* 26, 2010.)

The ability of the patient to walk around with the applicator mounted is unique. An interesting option that still needs further investigation is that the DCC applicator opens the possibility to simultaneous radiation and hyperthermia as the applicator is radiation transparent.

Advantages: conformal applicator; high comfort for the patient; rapid and accurate application to the treatment region; integrated water bolus; applicator size does not compromise its use; excellent spatial power control.

Disadvantages: penetration depth ≤2 cm; SAR pattern sensitive to air enclosures in water bolus.

8.5.1.2 Commercially Available Superficial Hyperthermia Systems

8.5.1.2.1 *Alba*

The ALBA system for superficial hyperthermia is a compact system with all required technology integrated in a single cabinet, is fully computerized, operates at 434 MHz, has a maximum RF-power output of 200 W, and includes a temperature and flow-controlled water bolus system [60]. There is a choice between three contact microstrip applicators of different size as specified in Table 8.1; all are slightly curved to easily conform to the body surface; the water bolus is integrated with the applicator (see Figure 8.8). Temperature is measured by miniature type T-thermocouples (4 to 16 channels, 50 μm diameter, Oxford Optronix Ltd.) using a power-pulse technique introduced by the Leeuw et al. (61) to minimize the effect of RF interference and probe self-heating on resulting temperature reading. Additional features of the system include an ultrasound scanner (HS 2000, Honda, Japan) for accurate verification of the position of the thermocouple probes with reference to tumor tissue, as well as

TABLE 8.1   Contact Curved Microstrip Applicators of Alba Hyperthermia System

Applicator Type	Aperture Size (cm)	EFS (cm²)	HC	Weight (kg)
α	7.3 × 19.8	3.6 × 15.8	0.26	0.7
β	14.8 × 14.4	8 × 11.5	0.32	0.9
γ	19.8 × 19.8	11.3 × 18	0.34	1.5

EFS = effective field size (area enclosed by 50% is oSAR contour line at 1 cm depth); HC = homogeneity coefficient (ratio between the area corresponding to 75% of SAR_{max} and the area corresponding to 25% of SAR_{max}).

an Oxford Optronix Laser Doppler Probe Oxylab (Oxford, UK) to collect information on the local blood perfusion in the microcirculation bed.

8.5.1.2.2 *BSD-500*

The BSD-500 system is a complete and versatile system that offers all instrumentation needed for interstitial hyperthermia and superficial hyperthermia with three different-sized single waveguide type applicators as well as two multi-element spiral applicators. The system has eight independently controlled RF-power output channels at a frequency of 915 MHz that can be operated asynchronously for superficial hyperthermia or in an electronically controlled synchronous phase mode for interstitial hyperthermia. For thermometry the system is equipped for maximal eight RF-immune Bowman temperature probes.

Originally, the BSD-500 came with the MA-100 and MA-120 side-loaded waveguides, and the MA-151 mini-dual-ridge waveguide (FDA approved, Figure 8.9). Details on the applicator specifications are presented in Table 8.2. Side loading of waveguides with high dielectric slabs is historically well

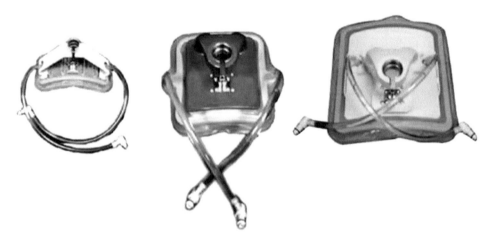

FIGURE 8.8   Contact curved microstrip applicators: α, β, and γ from the ALBA system for superficial hyperthermia.

FIGURE 8.9 BSD-500 applicators type MA-151, MA-120, and MA-100.

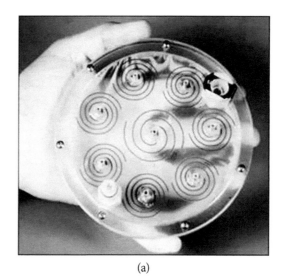

(a)

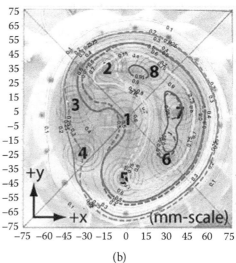

(b)

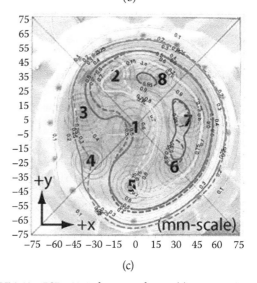

(c)

known as an effective method to improve the uniformity of the E-field distribution across the aperture of a rectangular waveguide applicator [62,63]. The drawback of such a waveguide design is the difference in wave velocity in the dielectric slab at the side walls with that at the center of the waveguide, making it difficult to combine these waveguides in an array setup. Similarly, the ridge waveguide design (single or dual ridge) is a known technique to concentrate the RF energy to the aperture area near the ridge. The MA-100, MA-120, and MA-151 are advised for heating nodular type of tumors fitting within the heating area. In accordance with the 915 MHz operating frequency, the recommended heating depths are 2.5, 2.5, and 2.0 cm, respectively.

Recently, BSD Medical Corporation has extended their applicator range for the BSD-500 with two multi-element applicators: an 8-element rigid and a 24-element flexible array applicator at 915 MHz. The SA-812 applicator is a closely spaced array of 8 dual-armed Archimedean spiral antennas (Figure 8.10). The array is configured as seven closely spaced 3.35 cm diameter spirals surrounding a central 4.05 cm diameter spiral. For each antenna the spiral traces are deposited on a 12 cm diameter plexiglass substrate with an integral water bolus. The overall diameter of the applicator is 14 cm and the thickness varies between 3.0 and 4.5 cm (maximum water bolus). The structure is such that it can be utilized with a mechanical support arm to maintain its position over the patient, but is considerably lighter in weight and more easily positioned near complex

FIGURE 8.10 BSD-500: 8-element applicator. (a) Prototype SA-812 with 8 dual-armed Archimedean spiral antennas; (b) Uniform SAR pattern; (c) "C-shape" SAR pattern. (From Johnson, J.E., Neuman, D.G., Maccarini, P.F., Juang, T., Stauffer, P.R., Turner, P., *Int J Hyperthermia* 22, 2006.)

TABLE 8.2 BSD-500 Waveguide Applicators[a]

Applicator Type	MA-100	MA-120	MA-151
Aperture size (cm²)	10 × 13	18 × 24	4 × 5
Heating area (cm²)	8 × 10	12.5 × 19.5	2.5 × 2.5
Heating depth	2.5	2.5	2.0

[a] All values as provided by BSD Medical Corporation.

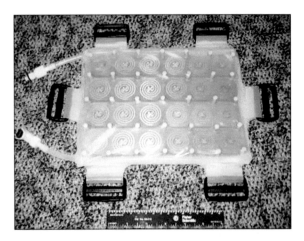

FIGURE 8.11   BSD-500. 24 dual-armed Archimedean spiral antenna flexible applicator.

patient anatomy than larger waveguide applicators. Important features of the applicator are its ability to generate uniform heating of 12 cm diameter regions or to fit the SAR distribution to irregularly shaped tumors by utilizing appropriate power combinations for the individual antennas [64].

The 24-element flexible array applicator of BSD Medical Corporation is the commercial version of the "microwave blanket" as earlier developed by Lee et al. [65]. The applicator consists of an array of 24 dual-armed Archimedean spiral antennas mounted on a rectangular, flexible silicone carrier that provides the required flexibility for the applicator to follow the tissue surface, and also a sufficient frame to keep a constant water bolus thickness (Figure 8.11). The applicator was specially developed for treatment of chest wall recurrence and can either be placed free at the target region or strapped to the body to enhance secure positioning. The spiral antennas are connected in pairs of three to one of the eight RF power amplifiers of the BSD-500 system. By controlling the output of each RF-power amplifier, the SAR pattern can be modified to adapt the SAR distribution to the contours of the tumor and to provide spatial SAR control to respond to patient complaints. Based on the operating frequency of 915 MHz, both multi-element array applicators are recommended for hyperthermia treatment of superficial tissue disease <2 cm deep.

8.5.2 Loco-Regional Hyperthermia Devices

Loco-regional hyperthermia is most commonly applied for advanced tumors located in the lower pelvis or in the abdomen. Children and young adolescents represent a special group of patients for loco-regional hyperthermia with special demands on the equipment. In recent years new equipment has been developed to extend loco-regional hyperthermia also to tumors in the head and neck regions. Considering the demonstrated thermal dose–effect relationships, the ability to control the 3-dimensional energy distribution is a mandatory requirement of a loco-regional hyperthermia system. For electromagnetic heating devices, the

only way to fulfill this criterion is by using a multi-antenna phased array system. Such a system generates a circumferential electrical field distribution around the patient with the ability to move the focus point through the body by selecting for all antennas the appropriate setting of amplitude and phase of the RF-signal. SAR steering is preferentially supported by extensive hyperthermia treatment planning [66,67,68]. To obtain a sufficient penetration depth, the current systems operate at frequencies ranging from 70 to 120 MHz. A physical consequence of the low frequencies selected is that the focal spot size in homogenous muscle or abdomen tissue will be large (10–15 cm in diameter), and the tissue is always located in the near field of the antennae.

Extensive electromagnetic modeling studies have indicated a need for higher frequencies and more degrees of freedom in order to better focus the energy to the target region and thus to increase tumor temperature [69,70]. The only exception might be cervical cancer [71,72]. Seebass et al. [71] showed for a patient model with cervical carcinoma a substantially better hyperthermia quality for an applicator of three antenna rings, with four paired dipoles per ring (= Sigma Eye) than for a single ring with four paired dipoles (= Sigma-60). Note, however, that they also showed that for three rings, each with each up to 12 independent antennas, increasing the frequency to 150 or 200 MHz did not further improve hyperthermia quality.

At present, three "radiative" devices are used for the clinical application of loco-regional hyperthermia—two academic and one commercial device.

8.5.2.1 The "AMC" Loco-Regional Hyperthermia System

The AMC-4 [73] and AMC-8 systems consist of four and eight 70 MHz waveguides organized in one and two rings, respectively (Figure 8.12). Each waveguide has an aperture size of 20.2 × 34.3 cm² and can be translated independently in the direction normal to its aperture [74]. In the clinic, the position of the waveguides is chosen such that a gap of 5 cm exists between the waveguides and the patient. Water boluses are placed between these gaps to ensure adequate coupling of the incident electromagnetic field. Furthermore, these water boluses circulated with distilled water provide superficial cooling that is essential to prevent overheating. For the AMC-8 system the distance between the two rings can be varied. The volume of the patient that is heated with both the AMC-4 and the AMC-8 system is relatively large compared to the target volume, however this is common for all loco-regional hyperthermia devices.

Loco-regional hyperthermia treatment with the 3D AMC-8 system can lead to a clinically relevant increase (plus 0.5°C in T90 and T50) of the target temperature compared to treatment with the 2D AMC-4 system. However, patient variability is high. The increase in temperature is associated with a substantial (36% to 71%) increase in applied RF-power for the AMC-4 system. Clearly, this can be explained by the fact that the heated volume is significantly larger for the AMC-8 system. This in turn leads to the advice to keep track of the increase in systemic temperature [75], which is common during loco-regional hyperthermia.

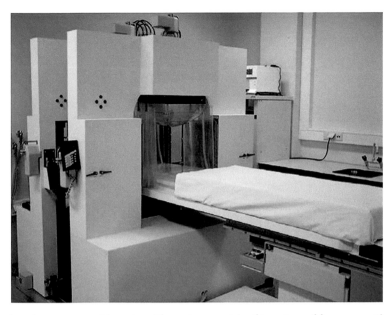

FIGURE 8.12 The AMC-8 phased-array waveguide system. The system consists of two rings of four waveguides operating at 70 MHz. Every waveguide has a separate water bolus that provides superficial cooling of the patient and coupling of the incident electromagnetic field into the patient. The two bottom waveguides share the same water bolus. The distance between the two rings is adjustable [73,74].

Ring-to-ring distance is of minor importance but should be known for proper optimization within 2 cm. For the AMC-8 system, the achievable thermal dose is stable over a range of −8 cm to +8 cm shifting the patient in the caudal − cranial direction. Axial position should be known within 1 − 2 cm both for the AMC-4 and the AMC-8 system.

A special feature of the design of the AMC-4 waveguide system is that the waveguide applicator has a relatively small width. Consequently, the overlap of the waveguide aperture in relation to the target volume is very limited in comparison with, for instance, the Sigma-60 applicator. Clinically, this feature is very important when heating tumors in the upper thorax; it allows heating of esophageal tumors with the AMC-4 loco-regional hyperthermia system [66,76].

8.5.2.2 The BSD-2000* Family of Loco-Regional Hyperthermia System

8.5.2.2.1 BSD-2000

The BSD-2000 system [77] is the most common deep hyperthermia system in clinical use and is available with the Tetra or Dodek amplifier set. The TETRA 4-channel solid-state amplifier operates in the frequency range of 75 to 140 MHz and delivers up to 1200 W to the applicator. The Dodek amplifier is a 12-channel solid-state amplifier with typically 150 W per single channel, total power 1800 W, over a frequency range of 50–250 MHz. When the Dodek amplifier is used in combination with a 4-channel applicator, the power output of three single channels is combined with an RF-power combiner and fed into the dual-dipole antenna. The

Sigma Treatment Base Unit includes both patient and applicator support systems. The patient is placed on the sling in preparation for the therapy, then the applicator is positioned over the tumor area and the water bolus filled. A large water reservoir mounted in the base unit maintains the bolus water at the desired temperature throughout the treatment.

For temperature measurement the BSD-2000 system is equipped with eight Bowman thermometers. Due to the high resistance carbon wires the Bowman probes are transparent for the electromagnetic field, and hence temperature measurement is continuous. In addition the system can be extended with eight probes to measure the external electrical field (E-field) at the surface of the patient body. This option is, however, used by only a few centers.

The BSD-2000 family of applicators includes the Sigma 60, Sigma 40, Sigma 30, and the Sigma 60-E (see Figures 8.13 and 8.14). The Sigma 30- to Sigma-60 applicators consist of a Lucite cylinder with integrated water bolus and four sets of dual dipole antennas mounted on the inner side of the cylinder (an antenna consists of two identical dipole arms built from tapered copper that are glued to the Lucite cylinder wall). The dipole is designed to operate in water and the dielectric contrast seen by the dipole (water on one side and air/Lucite on the backside) causes the dipole to emit the energy preferentially to the patient (and to keep stray radiation at an acceptable level). The water bolus serves also to cool the skin surface. The various applicators provide the opportunity to select an applicator size in accordance with treatment region, for example, for children and adults.

The BSD-2000 provides the possibility of SAR steering by phase and amplitude control, and by SAR steering the maximum energy deposition can be moved from the center of the body to the periphery (2D-SAR steering) [78]. Extensive characterization of the performance of the Sigma-60 applicator

---

* The BSD-2000, BSD-2000/3D, and the BSD-2000/3D/MR are not approved by the FDA for any kind of use within the United States.

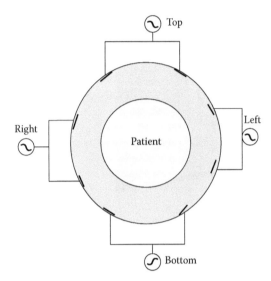

FIGURE 8.13 Schematic configuration of the Sigma applicator series. Each dual dipole antenna is connected to an RF power source.

demonstrates that the SAR distribution induced at the central cross-section of a homogeneous abdomen equivalent phantom possesses predominantly a Gaussian shape for the complete operational frequency range [79]. The longitudinal length of the 50% iso-SAR area is 21 cm at 70 MHz and roughly 19 cm at 120 MHz (i.e., nearly independent of the frequency). The reduction of the radial length of the 50% iso-SAR area from 15 to 9 cm for a frequency increasing from 70 to 120 MHz reflects the focusing effect at higher frequencies. This effect is visually demonstrated for a patient in Figure 8.15. It shows the changes by >5 W/kg (blue [darkest grey] decrease and red [medium grey] increase) in the predicted SAR distribution for an average patient with locally advanced cervical cancer after increasing the operating frequency from 75 MHz to 85 and 95 MHz, respectively. Increasing the frequency causes a more focused central heating (red [medium grey]) and less heating at the lateral parts of the patient (blue [darkest grey]). The figure also shows more energy deposition cranially of the target (i.e., tumor indicated in pink [dark grey in center of body]). Careful comparison is needed to

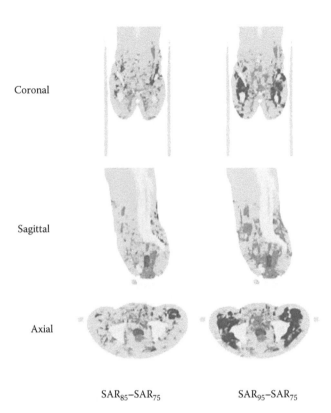

$SAR_{85}-SAR_{75}$ $\qquad$ $SAR_{95}-SAR_{75}$

FIGURE 8.15 **(See color insert.)** Comparison of the SAR distributions of 75 MHz to that of 85 and 95 MHz in the same patient. Changes in local SAR by >5 W/kg, increase in red and decrease in blue are presented for an average patient with locally advanced cervical cancer.

decide whether the increased SAR in the target is not limited by the unwanted increased SAR deposition cranially of the tumor.

A study by van Rhoon et al. [79] also confirmed the excellent SAR steering feasibility of the BSD-2000 system with the Sigma 60 under laboratory conditions. The location of the maximum SAR in the measured radial SAR profiles follows closely the selected target position. Small differences between the requested target position and the measured SAR maximum are to be expected due to the simplified theoretical algorithm used to calculate the required phase setting of each dipole pair

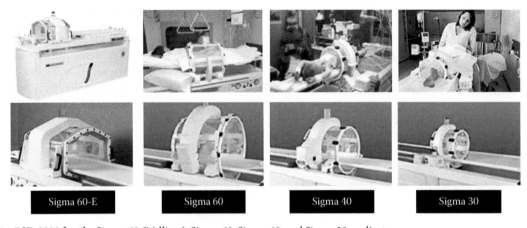

| Sigma 60-E | Sigma 60 | Sigma 40 | Sigma 30 |

FIGURE 8.14 BSD-2000 family: Sigma-60-E (ellipse), Sigma-60, Sigma-40, and Sigma-30 applicator.

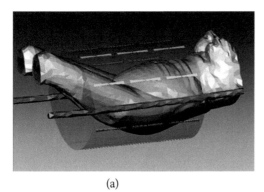

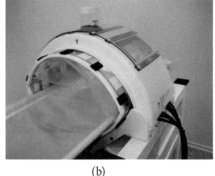

(a) (b)

FIGURE 8.16 Sigma Eye applicator (a) as implemented in the hyperplan treatment planning software, (b) the most recent version for clinical use.

to achieve maximum constructive interference at the site of the target position.

The Sigma 60-E (ellipse) differs only form the Sigma 60 in the fact that the Lucite cylinder is replaced by an octagonal-shaped Lucite container. The lower height of water above the abdomen should reduce water pressure and improve patient comfort. Fatehi et al. [80] demonstrate that at the lower frequencies of 75 to 90 MHz and at 100 MHz the SAR characteristics of the Sigma 60-E are almost identical to those of the Sigma-60 and Sigma Eye applicators, respectively.

8.5.2.2.2 BSD-2000/3D and BSD-2000/3D/MR

The addition of phase and amplitude control to the Sigma-60 applicator has proven its clinical utility to modify the SAR distribution in relation to patient complaints. However, when the aim is more specific, such as to increase a local target temperature, a quantum step toward improved SAR steering is required [81]. The Sigma Eye applicator was developed to provide this next level in three-dimensional power deposition at depth (see Figure 8.16). The Sigma Eye is eye-shaped and operates at 100 MHz, but includes three rings of eight dipoles each. With 12 independent phase and amplitude adjustments, it offers superior versatility to shape and steer power deposition peaks both axially and radially around the body interior. Furthermore, the Sigma Eye applicator provides an improved water bolus design (i.e., the length of the water bolus has been increased to provide contact with nearly the entire surface of the patient's body within the array). This prevents concentration of energy at the contacting edge of the water bolus. The Sigma Eye water bolus thickness, above the anterior patient's surface, is half to one-third that of the Sigma 60 bolus, which is expected to significantly improve patient comfort.

In several theoretical studies [69,70,71,72] the performance of the Sigma Eye applicator has been demonstrated to be superior to the Sigma-60 applicator in terms of ability to focus energy to the target and to obtain higher tumor temperatures. Canters et al. [82] demonstrated for 10 patients with locally advanced cervical cancer that use of the Sigma Eye with optimized phase and amplitude settings results in 8 of 10 patients with substantial higher temperature increase (Sigma Eye temperature is on average 1.5°C higher), in 1 of 10 patients no change, and in 1 of 10 patients with lower temperatures. In line with Seebass et al. [71], the data of Canters et al. [82] show that the increased number of degrees of freedom

of the Sigma Eye applicator translates in a higher gain potential of SAR targeting. However, in order to exploit this potential it is essential to find a solution for the accurate matching of the phase and amplitude used in the model to that applied in reality at the antenna feed points of the Sigma Eye applicator.

The BSD-2000/3D/MR is always delivered with the 12-channel Dodek amplifier with integrated phase-locked loop systems for accurate control of the phase and amplitude settings.

The improved 3D steering is particularly useful when implemented with a magnetic resonance system that is capable of noninvasive 3D imaging showing the temperature distribution in the heated regions [83]. The latter permits the 3D steering to more accurately target the energy to the tumor site. Using sophisticated microwave filtering and imaging software, the BSD-2000/3D/MR (Figure 8.17) allows an MRI system to be interfaced with and operate simultaneously with a BSD-2000/3D (BSD Medical Corporation). The MRI compatible version of the system facilitates pretreatment planning as well as real-time MRI monitoring of deep tissue temperature, perfusion, necrosis, and chemical changes during treatment [84,85,86]. Various temperature-sensitive MRI parameters exist and can be exploited for MR temperature mapping. The most popular parameters are proton resonance frequency shift, diffusion

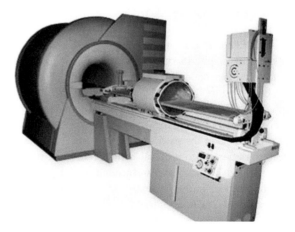

FIGURE 8.17 The BSD-2000/3D/MR system for loco-regional hyperthermia with simultaneous noninvasive temperature monitoring by MRI.

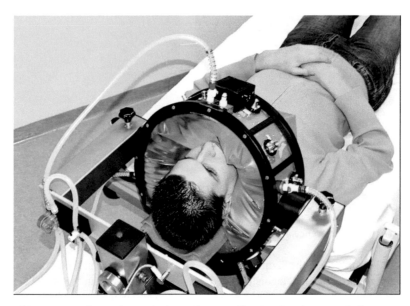

FIGURE 8.18   First version of the HYPERcollar applicator for hyperthermia of head and neck tumors [89].

coefficient D, longitudinal relaxation time T(1), and equilibrium magnetization M(0) [87]. The correlation of noninvasive MR thermography with direct tumor temperature measurements and clinical response has been demonstrated in clinical results [88]. Together, the BSD-2000/3D/MR offers optimized patient safety and easier hyperthermia treatment since there is no need for invasive thermometry catheters.

8.5.2.2.3 The HYPERcollar Loco-Regional Hyperthermia System

The HYPERcollar loco-regional hyperthermia system has been especially developed to provide adequate heating of tumors located centrally in the head and neck region. The design of the HYPERcollar differs from previous applicator design approaches in the fact that it was inversely developed. First, the target volume to be heated was defined, and then the applicator was designed by theoretically using advanced electromagnetic modeling. An important feature of such an approach is that a priori the translation of a predicted SAR distribution has a high probability to resemble the actual SAR distribution in vivo.

The HYPERcollar consists of a transparent Lucite cylinder (radius 40 cm, height 15 cm) covered with a fine conducting gauze forming the conducting backplanes required for the patch antennas (see Figure 8.18). Twelve probe-fed patch antennas are mounted in two rings [89] on the Lucite cylinder. The distance between the center planes of the two antenna rings is 6 cm. Details of the dimensions of the patch can be found in [90]. The entire applicator is attached to a movable trolley and can be rotated around the z-axis (patient-axis) and around the x-axis (left-right) for maximum positioning flexibility. An inflatable water bolus is attached to the Lucite cylinder for cooling of the skin and to enable an efficient transfer of the electromagnetic waves from the antennas into the patient. The system is equipped with 12 amplifiers with either incoherent or phase-controlled coherent 434 MHz signals. The maximum power at the antenna feeding points is 200 W per channel. The power setting is applied with a precision of 2 W and for the phase it is smaller than 0.1 degree [91].

The HYPERcollar is able to generate a central 50% iso-SAR focus of 35 ± 3 mm in diameter and about 100 ± 15 mm in length under clinical conditions. This SAR focus can be steered toward the desired location in the radial and axial directions with an accuracy of ~5 mm.

In a critical review Paulides et al. [92] reported on their first clinical experiences and demonstrate the pivotal role of hyperthermia treatment planning. Three representative patient cases (thyroid, oropharynx, and nasal cavity) were discussed. They reported that hyperthermia treatments for these tumor locations in the head and neck region, lasting as long as 1 h, were feasible and well tolerated and no acute treatment-related toxicity was observed. Maximum temperatures measured were in the range of those obtained during deep hyperthermia treatments in the pelvic region, but there is still ample room to improve mean temperatures. Further, they found that simulated power absorption correlated well with measured temperatures, illustrating the validity of a treatment approach of using energy profile optimizations to arrive at higher temperatures.

8.6 Summary and Future Directions

Considering the inexistence of specific hyperthermia equipment in the 1980s, it is fair to conclude that hyperthermia technology has come a long way. The currently available commercial and academic systems represent solid approaches for the clinical application of superficial and loco-regional hyperthermia with a good control of the heating quality.

The development of multi-element array applicators has brought a major improvement in the spatial control of energy deposition for both superficial and loco-regional hyperthermia. Impressive progress has also been made with regard to efficiency in applying the applicator setup to the patient as well as in the comfort of the applicator setup for the patient. Graphical user interfaces have matured to a "one-view" level (i.e., the hyperthermia technician is provided instantly with a good and clear overview of the temperature distribution in the target region). Thermometry systems have evolved from single points once per five minutes "emc" interfered measurement to continuous measuring 8-multi-sensor fiber-optic probes that are "emc" immune. The ultimate gain in thermometry is obtained in loco-regional hyperthermia where the development of noninvasive thermometry by magnetic resonance imaging provides 3-dimensional information on the temperature distribution. This technology is currently being tested for real-time feedback control to guide the delivery of a "prescribed" thermal dose.

The rapidly growing use of hyperthermia treatment planning systems will help strongly to advance high-quality application of hyperthermia treatments. Clearly, for a correct translation of predicted energy distribution from the monitor to the patient the patient needs to be precisely positioned, and the phase and amplitudes at the feed point for all antennas needs to be known accurately.

Heating tissue for hyperthermia in the coming years remains a procedure that involves complex technology, physiology, and biology. The great progress in equipment development has provided the hyperthermia community with important tools for a highly controlled quality of the clinical application of hyperthermia. This should stimulate physicians, biologists, and physicists to investigate new ways to exploit the potential benefits of hyperthermia for cancer treatment.

References

1. Van der Zee, J., Vujaskovic, Z., Kondo, M., Sugahara, T. 2008. The Kadota Fund International Forum 2004—clinical group consensus. *Int J Hyperthermia* 24: 111–122.
2. Van der Zee, J., González, D., Van Rhoon, G.C., Van Dijk, J.D.P., Van Putten, W.L.J., Hart, A.A.M. 2000. For the Dutch Deep Hyperthermia Group. Comparison of radiotherapy alone with radiotherapy plus hyperthermia in locally advanced pelvic tumours: A prospective, randomised, multicentre trial. *Lancet* 355: 119–1125.
3. Franckena, M., Stalpers, L.J., Koper, P.C., Wiggenraad, R.G., Hoogenraad, W.J., Van Dijk, J.D., Wárlám-Rodenhuis, C.C., Jobsen, J.J., Van Rhoon, G.C., Van der Zee, J. 2008. Long-term improvement in treatment outcome after radiotherapy and hyperthermia in locoregionally advanced cervix cancer: An update of the Dutch Deep Hyperthermia Trial. *Int J Radiat Oncol Biol Phys* 70:1176–82.
4. Franckena, M., Lutgens, L.C., Koper, P.C., Kleynen, C.E., Van der Steen-Banasik, E.M., Jobsen, J.J., Leer, J.W., Creutzberg, C.L., Dielwart, M.F., Van Norden, Y., Canters, R.A.M., Van Rhoon, G.C., Van der Zee, J. 2009. Radiotherapy and hyperthermia for treatment of primary locally advanced cervix cancer: Results in 378 patients. *Int J Radiation Oncology Biol Phys* 73: 242–250.
5. Issels, R.D., Lindner, L.H., Verweij, J., Wust, P., Richardt, P., Schem, B.C., Abdel-Rahman, S., Daugaard, S., Salat, C., Wendtner, C.M., Vujaskovic, Z., Wessalowski, R., Jauch, K.W., Dürr, H.R., Ploner, F., Baur-Melnyk, A., Mansmann, U., Hiddemann, W., Blay, J.Y., Hohenberger, P. 2010. Neo-adjuvant chemotherapy alone or with regional hyperthermia for localised high-risk soft-tissue sarcoma: A randomised phase 3 multicentre study. *Lancet Oncol* 11:561–570.
6. Hua, Y., Ma, S., Fu, Z., Hu, O., Wang, L., Pia, Y. 2010. Intracavity hyperthermia in nasopharyngeal cancer: A phase III clinical study. *Int J Hyperthermia* 1-7, early online.
7. Van der Zee, J., Van der Holt, B., Rietveld, P.J.M., Helle, P.A., Wijnmaalen, A.J., Van Putten, W.L.J., Van Rhoon, G.C. 1999. Reirradiation combined with hyperthermia in recurrent breast cancer results in a worthwhile local palliation. *Br J Cancer* 79: 483–490.
8. Lee, H.K., Antell, A.G., Perez, C.A., Straube, W.L., Ramachandran, G., Myerson, R.J., Emami, B., Molmenti, E.P., Buckner, A., Lockett, M.A. 1998. Superficial hyperthermia and irradiation for recurrent breast carcinoma of the chest wall: Prognostic factors in 196 tumors. *Int J Radiat Oncol Biol Phys* 40: 365–375.
9. Kapp, D.S. and Cox, R.S. 1995. Thermal treatment parameters are most predictive of outcome in patients with single tumor nodules per treatment field in recurrent adenocarcinoma of the breast [see comments]. *Int J Radiat Oncol Biol Phys* 33 887–899.
10. Oleson, J.R., Samulski, T.V., Leopold, K.A., Clegg, S.T., Dewhirst, M.W., Dodge, R.K., George, S.L. 1993. Sensitivity of hyperthermia trial outcomes to temperature and time: Implications for thermal goals of treatment. *Int J Radiat Oncol Biol Phys* 25: 289–297.
11. Sherar, M., Liu, F.F., Pintilie, M., Levin, W., Hunt, J., Hill, R., Hand, J., Vernon, C., Van Rhoon, G.C., Van der Zee, J., Gonzalez, D.G., Van Dijk, J., Whaley, J., Machin, D. 1997. Relationship between thermal dose and outcome in thermoradiotherapy treatments for superficial recurrences of breast cancer: Data from a phase III trial. *Int J Radiat Oncol Biol Phys* 39: 371–380.
12. Thrall, D.E., LaRue, S.M., Yu, D., Samulski, T., Sanders, L., Case, B., Rosner, G., Azuma, C., Poulson, J., Pruitt, A.F., Stanley, W., Hauck, M.L., Williams, L., Hess, P., Dewhirst, M.W. 2005. Thermal dose is related to duration of local control in canine sarcomas treated with thermoradiation-therapy. *Clin Cancer Res* 11: 5206–5214.
13. Maguire, P.D., Samulski, T.V., Prosnitz, L.R., Jones, E.L., Rosner, G.L., Powers, B., Layfiels, L.W., Brizel, D.M., Scully, S.P., Harrelson, J.M., Dewhirst, M.W. 2001. A phase II trial testing the thermal dose parameter CEM43°T90 as a predictor of response in soft tissue sarcomas treated with pre-operative thermoradiotherapy. *Int J Hyperthermia* 17: 283–290.
14. Franckena, M., Fatehi, D., De Bruijne, M., Canters, R.A.M., Van Norden, Y., Mens, J.W., Van Rhoon, G.C., Van der Zee, J. 2009 Hyperthermia dose-effect relationship in 420 patients with cervical cancer treated with combined radiotherapy and hyperthermia. Accepted for publication *Eur J Cancer* 45: 1969–1978.

15. Oleson, J.R. 1984. A review of magnetic induction methods for hyperthermia treatment of cancer. *IEEE Trans Biomed Eng* 31: 91–97.

16. Samulski, T.V. 1989. Heating deeply seated tumors using magnetic induction and annular phased array techniques. In *Hyperthermic Oncology*, eds. T. Suguhara and M. Saito, vol. 2, 648–651, Taylor & Francis, London.

17. Sapozink, M.D., Gibbs, F.A., Thomson, J.W., Eltringham, J.R., Stewart, J.R. 1985. A comparison of deep regional hyperthermia from an annular array and a concentric coil in the same patients. *Int J Radiat Oncol Biol Phys* 11: 179–190.

18. Ellinger, D.C., Chute, F.S., Vermeulen, F.E. 1989. Evaluation of a semi-cylindrical solenoid as an applicator for radiofrequency hyperthermia. *IEEE Trans Biomed Eng* 36: 987–993.

19. Hagmann, M.J. 1985. Coupling efficiency of helical coil hyperthermic applications. *IEEE Trans Biomed Eng* 32: 539–540.

20. Ruggera, P.S. and Kantor, G. 1984. Development of a family of RF helicoil applicators which produce transversely uniform axially distribution on cylindrical fat-muscle phantoms. *IEEE Trans Biomed Eng* 31: 98–106.

21. Corry, P.M., Jabboury, K., Kong, J.S., Armour, E.P., McGraw, F.J., LeDuc, T. 1988. Evaluation of equipment for hyperthermia treatment of cancer. *Int J Hyperthermia* 4: 53–74.

22. Corry, P.M. and Jabboury, K. 1989. Magnetic induction hyperthermia with transaxial coils. In *Hyperthermic Oncology*, eds. T. Suguhara and M. Saito, vol. 2, 595–597, Taylor & Francis, London.

23. Hiraoka, M., Jo, S., Akuta, K., Nishimura, Y., Takahashi, M., Abe, M. 1987. Radiofrequency capacitive hyperthermia for deep-seated tumors. I. Studies on thermometry. *Cancer* 60: 121–127.

24. Hiraoka, M., Jo, S., Akuta, K., Nishimura, Y., Takahashi, M., Abe, M. 1987. Radiofrequency capacitive hyperthermia for deep-seated tumors. II. Effects of thermoradiotherapy. *Cancer* 60: 128–135.

25. Van Rhoon, G.C., Van der Zee, J., Broekmeyer-Reurink, M.P., Visser, A.G., Reinhold, H.S. 1992. Radiofrequency capacitive heating of deep-seated tumours using pre-cooling of the subcutaneous tissues: Results on thermometry in Dutch patients. *Int J Hyperthermia* 8: 843–54.

26. Nussbaum, G.H., Sidi, J., Rouhanizadeh, N., Morel, P., Jasmin, C., Convert, G., Mabire, J.P., Azam, G. 1986. Manipulation of the central axis heating patterns with a prototype, three-electrode capacitive device for deep-tumor hyperthermia. *IEEE Trans Biomed Eng* 34: 620–625.

27. Lee, E.R. 1995. *Electromagnetic Superficial Heating Technology, Principles and Practice of Thermoradiotherapy and Thermochemotherapy, Biology, Physiology, Physics*, eds. M.H. Seegenschmiedt, P. Fessenden, and C.C. Vernon, Springer Verlag GmbH, Berlin, Heidelberg, New York, vol. 1, 193–217.

28. Nilsson, P. 1984. Physics and technique of microwave induced hyperthermia in the treatment of malignant tumours. PhD thesis, University of Lund.

29. Hand, J.W. 1990. Biophysics and technology of electromagnetic hyperthermia. In, *Methods of External Hyperthermic Heating*, ed. M. Gautherie, Springer Verlag, Berlin, Heidelberg, New York 1–59.

30. Sapozink, M.D., Gibbs, F.A., Thomson, J.W., Eltringham, J.R., Stewart, J.R. 1985. A comparison of deep regional hyperthermia from an annular phased array and a concentric coil in the same patients. *Int J Radiat Oncol Biol Phys* 11: 179–190.

31. Egawa, S., Tsukiyama, I., Akine, Y., Kajiura, Y., Ogino, T., Yamashita, K. 1988. Hyperthermic therapy of deep seated tumors: Comparison of the heating efficiencies of an APA and a capacitively coupled RF system. *Int J Radiat Oncol Biol Phys* 14: 521–528.

32. Shimm, D.S., Cetas, T.C., Hynynen, K.H., Beuchler, D.N., Anhalt, D.P., Sykes, H.F., Cassady, J.R. 1989. The CDRH helix, a phase I clinical trial. *Am J Clin Oncol Cancer Clin Trials* 12: 110–113.

33. Howard, G.C.W., Sathiaseelan, V., King, G.A., Dixon, A.K., Anderson, A., Bleehen, N.M. 1987. Hyperthermia and radiation in the treatment of superficial malignancy: An analysis of treatment parameters, response and toxicity. *Int J Hyperhtermia* 3: 1–8.

34. Sapozink, M.D., Gibbs, F.A., Gibbs, P., and Stewart, J.R. 1988. Phase I evaluation of hyperthermia equipment—University of Utah Institutional Report. *Int J Hyperthermia* 4: 117–132.

35. Kapp, D.S., Fessenden, P., Samulski, T.V., Bagshaw, M.A., Cox, R.S., Lee, E.R., Lohrbach, A.W., Meyer, J.L., and Prionas, S.D. 1988. Stanford University institutional report. Phase I evaluation of equipment for hyperthermic treatment of cancer. *Int J Hyperthermia* 4: 75–115.

36. Shimm, D.S., Cetas, T.C., Oleson, J.R., Cassady, J.R., and Sim, D.A. 1988. Clinical evaluation of hyperthermia equipment: The University of Arizona institutional report for the NCI hyperthermia equipment evaluation contract. *Int J Hyperthermia* 4: 39–51.

37. De Leeuw, A.A.C. 1993. The coaxial TEM regional hyperthermia system, design and clinical introduction. PhD thesis, University of Utrecht, The Netherlands.

38. Franconi, C. 1987. *Hyperthermia Heating Technology and Devices, in Physics and Technology of Hyperhtermia*, eds. S.B. Field and C. Franconi, Martinus Nijhof Publishers, Dordrecht, The Netherlands 80–122.

39. Van Rhoon, G.C., Visser, A.G., Van den Berg, P.M., Reinhold, H.S. 1988. Evaluation of ring capacitor plates for regional deep heating. *Int J Hyperthermia* 4: 133–142.

40. Van Rhoon, G.C., Sowinski, M.J., Van den Berg, P.M., Visser, A.G., Reinhold, H.S. 1990. A ring capacitor applicator in hyperthermia: Energy distributions in a fat-muscle layered model for different ring electrode configurations. *Int J Radiat Oncol Biol Phys* 18: 77–85.

41. Raskmark, P. and Andersen, J.B. 1984. Electronically steered heating of a cylinder, in Proceedings 4th International Symposium on Hyperthermic Oncology, ed. J. Overgaard, Taylor and Francis, London, UK 617–620.

42. Van Rhoon, G.C., Rietveld, P.J., Van der Zee, J. 1998. A 433 MHz Lucite cone waveguide applicator for superficial hyperthermia. *Int J Hyperthermia* 14: 13–27.

43. Rietveld, P.J., Van Putten, W.L., Van der Zee, J., Van Rhoon, G.C. 1999. Comparison of the clinical effectiveness of the 433 MHz Lucite cone applicator with that of a conventional waveguide applicator in applications of superficial hyperthermia. *Int J Radiat Oncol Biol Phys* 43: 681–687.

44. De Bruijne, M., Samaras, T., Bakker, J.F., Van Rhoon, G.C. 2006. Effects of waterbolus size, shape and configuration on the SAR distribution pattern of the Lucite cone applicator. *Int J Hyperthermia* 22: 15–28.

45. Samaras, T., Rietveld, P.J.M., Van Rhoon, G.C. 2000. Effectiveness of FDTD in predicting SAR distributions from the Lucite cone applicator. *IEEE Trans MTT* 48: 2059–2063.

46. Bach Andersen, J., Baun, A., Harmark, K., Heinzl, L., Raskmark, P., Overgaard, J. 1984. A hyperthermia system using a new type of inductive applicator. *IEEE Trans Biomed Eng* BME 31: 21–27.

47. Johnson, R.H., James, J.R., Hand, J.W., Hopewell, J.W., Dunlop, P.R.C., Dickinson, R.J. 1984. New low-profile applicators for local heating of tissues. *IEEE Trans Biomed Eng* BME 31:28–37.

48. Johnson, R.H., Preece, A.W., Green, J.L. 1990. Theoretical and experimental comparison of three types of electromagnetic hyperthermia applicator. *Phys Med Biol* 35: 761–779.

49. Lumori, M.L.D., Hand, J.W., Gopal, M.K., Cetas, T.C. 1990. Use of Gaussian beam model in predicting SAR distributions from current sheet applicators. *Phys Med Biol* 35: 387–397.

50. Lumori, M.L.D., Andersen, J.B., Gopal, M.K., Cetas, T.C. 1990. Gaussian beam simulations of fields emanating from aperture sources and propagating in bounded, homogeneous and layered lossy media. *IEEE Trans Microwave Theory and Techniques* 38: 1623–1630.

51. Gopal, M.K., Hand, J.W., Lumori, M.L.D., Alkhair, S., Paulsen, K.D., Cetas, T.C. 1992. Current sheet applicator arrays for superficial hyperthermia of chestwall lesions. *Int J Hyperthermia* 8: 227–240.

52. Rietveld, P.J., Stakenborg, J., Cetas, T.C., Lumori, M.L., Van Rhoon, G.C. 2001. Theoretical comparison of the SAR distributions from arrays of modified current sheet applicators with that of Lucite cone applicators using Gaussian beam modelling. *Int J Hyperthermia* 17: 82–96.

53. Gelvich, E.A. and Mazokhin, V.N. 2002. Contact flexible microstrip applicators (CFMA) in a range from microwaves up to short waves. *IEEE Trans Biomed Eng* 49: 1015–23.

54. Kok, H.P., Correia, D., De Greef, M., Van Stam, G., Bel, A., Crezee, J. 2010. SAR deposition by curved CFMA-434 applicators for superficial hyperthermia: Measurements and simulations. *Int J Hyperthermia* 26: 171–184.

55. Lee, W.M., Gelvich, E.A., Van der Baan, P., Mazokhin, V.N., Van Rhoon, G.C. 2004. Assessment of the performance characteristics of a prototype 12-element capacitive contact flexible microstrip applicator (CFMA-12) for superficial hyperthermia. *Int J Hyperthermia* 20: 607–624.

56. Stauffer, P.R., Rossetto, F., Leoncini, M., Gentili, G.B. 1998. Radiation patterns of dual concentric conductor microstrip antennas for superficial hyperthermia. *IEEE Trans Biomedical Eng* 45: 605–613.

57. Rossetto, F. and Stauffer, P.R. 1999. Effect of complex bolus-tissue load configurations on SAR distributions from dual concentric conductor applicators. *IEEE Trans Biomedical Eng* 46: 1310–1319.

58. Rossetto, F. and Stauffer, P.R. 2001. Theoretical characterization of dual concentric microwave applicators for hyperthermia at 433 MHz. *Int J Hyperthermia* 17: 258–270.

59. Stauffer, P.R., Maccarini, P., Arunachalam, K., Craciunescu, P., Diederich, C., Juang, T., Rosseto, F., Schlorff, J., Milligan, A., Hsu, J., Sneed, P., Vujaskovic, Z. 2010. Conformal microwave array (CMA) applicators for hyperthermia of diffuse chest wall recurrence. *Int J Hyperthermia* 26: 686–698.

60. Gabriele, P., Ferrara, T., Baiotto, B., Garibaldi, E., Marini, P.G., Penduzzu, G., Giovannini, V., Bardati, F., Guiot, C. 2009. Radio hyperthermia for re-treatment of superficial tumours. *Int J Hyperthermia* 25: 189–198.

61. De Leeuw, A.A.C., Crezee, J., Lagendijk, J.J.W. 1993. Temperature and SAR measurements in deep-body hyperthermia with thermocouple thermometry. *Int J Hyperthermia* 9: 685–697.

62. Tsandoulas, G.N. and Fitzgerald, W.D. 1972. Aperture efficiency enhancements in dielectrically loaded horns. *IEEE Transactions on Antennas and Propagation* 21: 69–74.

63. Kantor, G. 1981. Evaluation and survey of microwave and radiofrequency applicators. *Journal of Microwave Power* 16: 135–150.

64. Johnson, J.E., Neuman, D.G., Maccarini, P.F., Juang, T., Stauffer, P.R., Turner, P. 2006. Evaluation of a dual-arm Archimedean spiral array for microwave hyperthermia. *Int J Hyperthermia* 22: 475–490.

65. Lee, E.R., Wilsey, T.R., Tarczy-Hornoch, P., Kapp, D.S., Fessenden, P., Lohrbach, A.W., Prinoas, S.D. 1992. Body conformable 915 MHz microstrip array applicators for large surface area hyperthermia. *IEEE Trans Biomed Eng* 39: 470–483.

66. Kok, H.P., Van Haaren, P.M.A., Van de Kamer, J.B., Zum Vörde Sive Vörding, P.J., Wiersma, J., Hulshoff, M.C.C.M., Geijsen, E.D., Van Lanschot, J.J.B., Crezee, J. 2006. Prospective treatment planning to improve locoregional hyperthermia for oesophageal cancer. *Int J Hyperthermia* 22: 375–389.

67. Canters, R.A., Franckena, M., Paulides, M.M., Van Rhoon, G.C. 2009. Patient positioning in deep hyperthermia: Influences of inaccuracies, signal correction possibilities and optimization potential. *Phys Med Biol* 21: 3923–36.

68. Canters, R.A., Wust, P., Bakker, J.F., Van Rhoon, G.C. 2009. A literature survey on indicators for characterisation and optimisation of SAR distributions in deep hyperthermia, a plea for standardisation. *Int J Hyperthermia* 25: 593–608.

69. Wust, P. 1995. Electromagnetic deep heating technology. In *Principles and Practice of Thermoradiotherapy and Thermochemotherapy*, eds. M.H. Seegenschmiedt, P. Fessenden, and C.C. Vernon., Springer Verlag, Berlin 219–251.

70. Paulsen, K.D. 1995. Principles of power deposition models. In *Principles and Practice of Thermoradiotherapy and Thermochemotherapy*, eds. M.H. Seegenschmiedt, P. Fessenden, and C.C. Vernon,. Springer Verlag, Berlin, 399–423.

71. Seebass, M., Beck, R., Gellermann, J., Nadobny, J., Wust, P. Electromagnetic phased arrays for regional hyperthermia: Optimal frequency and antenna arrangement. *Int J Hyperthermia* 17: 4, 321–326.

72. Paulsen, K.D., Geimer, S., Tang, J., Boyse, W.E. 1999. Optimization of pelvic heating rate distributions with electromagnetic phased arrays. *Int J Hyperthermia* 15:3, 157–186.

73. Van Dijk, J.D.P., Schneider, C.J., Van Os, R.M., Blank, L.E., Gonzalez, D.G. 1990. Results of deep body hyperthermia with large waveguide radiators. *Adv Exp Med Biol* 267: 315–319.

74. De Greef, M., Kok, H.P., Bel, A., Crezee, J. 2011. 3D versus 2D steering in patient anatomies: A comparison using hyperthermia treatment planning. *Int J Hyperthermia* 27: 74–85.

75. Van Haaren, P.M., Hulshof, M.C., Kok, H.P., Oldenborg, S., Geijsen, E.D., Van Lanschot, J.J., Crezee, J. 2008. Relation between body size and temperatures during locoregional hyperthermia of oesophageal cancer patients. *Int J Hyperthermia* 24: 663–674.

76. Van Haaren, P.M., Hulshof, M.C., Kok, H.P., Oldenborg, S., Geijsen, E.D., Van Lanschot, J.J., Crezee, J. 2008. Relation between body size and temperatures during locoregional hyperthermia of oesophageal cancer patients. *Int J Hyperthermia* 24: 8, 663–74.

77. Turner, P.F. and Schaefermeyer, T. 1989. BSD-2000 approach for deep local and regional hyperthermia: Physics and technology. *Strahlenther Onkol* 165: 738–41.

78. Wust, P., Fähling, H., Felix, R., Rahman, S., Issels, R.D., Feldmann, H., Van Rhoon, G.C., Van der Zee, J. 1995. Quality control of the SIGMA applicator using a lamp phantom: A four-centre comparison. *Int J Hyperthermia* 11: 755–68.

79. Van Rhoon, G.C., Van der Heuvel, D.J., Ameziane, A., Rietveld, P.J.M., Volenec, K., Van der Zee, J. 2003. Characterization of the SAR-distribution of the Sigma-60 applicator for regional hyperthermia using a Schottky diode sheet. *Int J Hyperthermia* 19: 6, 642–654.

80. Fatehi, D. Van Rhoon, G.C. 2008. SAR characteristics of the Sigma-60-Ellipse applicator. *Int J Hyperthermia* 24: 4, 347–56.

81. Wust, P., Seebass, M., Nadobny, J., Deuflhard, P., Monich, G., Felix, R. 1996. Simulation studies promote technological development of radiofrequency phased array hyperthermia. *Int J Hyperthermia* 12: 477–94.

82. Canters, R.A.M., Franckena, M., Paulides, M.M., Van der Zee, J., Van Rhoon, G.C. 2011. Comparison of the potential increased SAR targeted to the tumor between the Sigma-60 and Sigma Eye applicator, paper submitted.

83. Wust, P., Hildebrandt, B., Sreenivasa, G., Rau, B., Gellermann, J., Riess, H., Felix, R., Schlag, P.M. 2002. Hyperthermia in combined treatment of cancer. *Lancet Oncol* 3: 487–97.

84. Gellermann, J., Wust, P., Stalling, D., Seebass, M., Nadobny, J., Beck, R., Hege, H., Deuflhard, P., Felix, R. 2000. Clinical evaluation and verification of the hyperthermia treatment planning system hyperplan. *International Journal of Radiation Oncology, Biology & Physics* 47: 1145–1156.

85. Wust, P., Gellermann, J., Seebass, M., Fahling, H., Turner, P., Wlodarczyk, W., Nadobny, J., Rau, B., Hildebrandt, B., Oppelt, A., Schlag, P.M., Felix, R. 2004. [Part-body hyperthermia with a radiofrequency multiantenna applicator under online control in a 1.5 T MR-tomograph]. *Rofo* 176: 363–374.

86. Gellermann, J., Wlodarczyk, W., Ganter, H., Nadobny, J., Fahling, H., Seebass, M., Felix, R., Wust, P. 2005. A practical approach to thermography in a hyperthermia/magnetic resonance hybrid system: Validation in a heterogeneous phantom. *International Journal of Radiation Oncology, Biology & Physics* 61: 267–277.

87. Lüdemann, L., Wlodarczyk, W., Nadobny, J., Weihrauch, M., Gellermann, J., Wust, P. 2010. Non-invasive magnetic resonance thermography during regional hyperthermia. *Int J Hyperthermia* 26: 273–82.

88. Gellermann, J., Wlodarczyk, W., Hildebrandt, B., Ganter, H., Nicolau, A., Rau, B., Tilly, W., Fahling, H., Nadobny, J., Felix, R., Wust, P. 2005. Noninvasive magnetic resonance thermography of recurrent rectal carcinoma in a 1.5 Tesla hybrid system. *Cancer Research* 65: 5872–5880.

89. Paulides, M.M., Bakker, J.F., Zwamborn, A.P.M., Van Rhoon, G.C. 2007. A head and neck hyperthermia applicator: Theoretical antenna array design. *Int J Hyperthermia* 23: 59–67.

90. Paulides, M.M., Bakker, J.F., Chavannes, N., Van Rhoon, G.C. 2007. A patch antenna design for application in a phased-array head and neck hyperthermia applicator. *IEEE Transactions on Biomedical Engineering* 54: 11, 2057–63.

91. Bakker, J.F., Paulides, M.M., Westra, A.H., Schippers, H., Van Rhoon, G.C. 2010. Design and test of a 434 MHz multichannel amplifier system for targeted hyperthermia applicators. *Int J Hyperthermia* 26: 158–70.

92. Paulides, M.M., Bakker, J.F., Linthorst, M., Van der Zee, J., Rijnen, Z., Neufeld, E., Pattynama, P.M., Jansen, P.P., Levendag, P.C., Van Rhoon, G.C. 2010. The clinical feasibility of deep hyperthermia treatment in the head and neck: New challenges for positioning and temperature measurement. *Phys Med Biol* 7(55): 2465–80.

9

Interstitial Electromagnetic Devices for Thermal Ablation

Dieter Haemmerich
*Medical University of
South Carolina*

Chris Brace
University of Wisconsin, Madison

9.1 Introduction of Thermal Tissue Ablation

The term *thermal ablation* refers to localized destruction of tissue via heating above ~50°C for typically a few minutes. Heating via radiofrequency (RF) electric current (RF ablation) with the goal of locally destroying tissue has been in use clinically for treatment of cardiac arrhythmia since the 1980s by destroying small tissue regions that are responsible for the arrhythmia. Currently, thermal ablation is clinically used to treat various other diseases including varicose veins and uterine bleeding (Haemmerich 2006a, Cooper 2004, Markovic 2009) as well as various forms of cancer. In this chapter we will discuss engineering and biophysics of interstitial electromagnetic devices for thermal ablation that employ heating either by radiofrequency electric current or by microwaves.

9.2 Biophysics of Radiofrequency Ablation

When electric current is applied to tissue, heating results due to resistive losses. This heating occurs independent of frequency of the current; at frequencies below ~10 kHz additional effects such

as stimulation of excitable tissue and electrochemical reactions (at DC) may occur. To avoid these effects, RF ablation-based heating employs higher frequencies in the range of 450–500 kHz (this particular range is also used because electrosurgical devices, which served as predicate devices, operate in the same frequency band). Figure 9.1 illustrates the setup of an RF ablation procedure, where an electrode is inserted into a target location in tissue to be destroyed. Materials used for RF electrodes include steel, platinum, and Ni-Ti alloys; parts of the RF electrode are typically electrically insulated (e.g., to avoid heating of the shaft region). Ground pads (dispersive electrodes) are placed on the patient's skin (often thigh or back) to serve as a return path for the RF current.

Inside the RF generator, cables, and RF electrode, electric current is carried by electrons. Inside tissue, however, ions (Na$^+$, K$^+$, Cl$^-$) serve as carriers of the electric current. The resulting oscillations of ions produce heat due to friction (i.e., electrically resistive heating).

The specific absorption rate (SAR [W/kg]) describes the amount of power deposited locally within tissue. SAR is typically high close to the interstitial applicator and drops rapidly with distance from the applicator—i.e., direct heating by RF current is limited to close proximity of the applicator, and thermal conduction allows more distant tissue regions to obtain sufficient temperatures (Schramm 2007). For interstitial RF

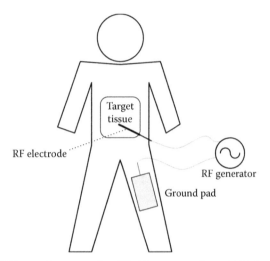

FIGURE 9.1 Overview of an RF ablation procedure. A RF electrode is inserted into the target tissue under imaging guidance. RF energy provided by a generator is applied to the electrode and results in tissue heating around the electrode. A ground pad placed on the patient's thighs or back serves as return path for the RF current. (Reproduced with permission from S. Vaezy and V. Zderic, eds. *Image-Guided Therapy Systems*, Artech House, Inc., Norwood, MA, 2009. © 2009 by Artech House, Inc.)

applicators, the SAR depends on electrical tissue conductivity and magnitude of electric current density generated around the electrode (Equation 9.1).

$$SAR = \frac{\sigma}{\rho}|E|^2 = \frac{1}{\sigma \times \rho}|J|^2 \qquad (9.1)$$

(σ = tissue electrical conductivity; ρ = tissue mass density; E = electric field strength; J = electric current density).

The process of tissue heating during RF ablation can be mathematically described by the following heat transfer equation where T represents spatially and temporally varying tissue temperature (Diller 2000):

$$\rho c \frac{\partial T}{\partial t} = \nabla \cdot k \nabla T + \frac{E^2}{\sigma} - Q_{perf} \qquad (9.2)$$

(k = tissue thermal conductivity; c = tissue specific heat; T = tissue temperature).

The term on the left-hand side represents the change in tissue temperature due to RF heating. The first term on the right-hand side describes thermal conduction, and the second term represents heat generated by RF current (equivalent to SAR). The term Q_{perf} represents heat losses due to cooling by blood perfusion. A number of ways have been suggested to mathematically model perfusion, and the most widely used model employs a distributed heat sink and was first proposed by Pennes more than 50 years ago (Pennes 1948), but is accurate only for tissue regions where small vessels (<1 mm) are present. Larger vessels have to

be included via other ways (e.g., by explicitly including these vessels in the model geometry). Whether tissue is destroyed due to heating depends on both temperature and time (see Chapter 2, this book). As an approximation, a few minutes are necessary to kill cells at 50°C, but only seconds at temperatures above 60°C (Dewhirst 2003). While there are some differences in thermal tolerance between cell types, these differences are not relevant for thermal ablation procedures due to the large temperature gradients. Since thermal ablation procedures happen in a time frame of several minutes, the 50°C isotherm is frequently used to approximate the ablation zone boundary (boundary of cell death; see Figure 9.2) (Berjano 2006).

During RF ablation, maximum tissue temperatures up to ~110°C can be achieved. Above this temperature, tissue vaporizes and prevents any further heating by RF current due to the electrically insulating properties of the vapor. In addition, there is the possibility of tissue carbonization (often referred to as "tissue charring") at locations of very high RF current densities—typically very close to the RF electrode. Tissue charring is an irreversible process and limits further RF energy deposition. The applied RF power therefore has to be controlled to keep tissue temperature in the desired range and limit both vaporization and charring.

Most commercial RF ablation devices employ one of the following three methods of controlling magnitude of applied RF power:

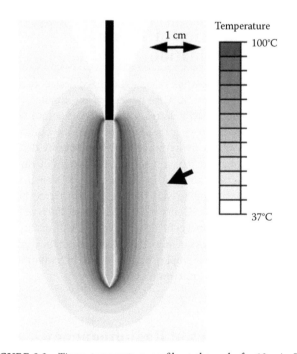

FIGURE 9.2 Tissue temperature profile at the end of a 12 min RF tumor ablation with a cooled needle electrode (same as shown in Figure 9.6B) from a computer simulation. The black part of the electrode is electrically insulated, and heating due to RF current results around the exposed metal electrode (electrode tip, shown in gray). Black arrowhead marks boundary of ablation zone (~50°C).

Power control: This is the simplest form of control, where constant power is applied during the ablation procedure.

Temperature control: Thermal sensors are integrated into the RF electrode, and applied power is controlled such that tissue temperature measured by these sensors is kept at a certain target value (typically near 100°C) during the ablation period.

Impedance control: Tissue impedance, which is measured between the RF electrode and the reference electrode (i.e., ground pad), represents a weighted average of the electrical resistivity (= 1/conductivity) of all tissue that serves as electrical pathway between the two electrodes. The weighting is such that tissue close to the RF electrode has highest influence (i.e., regions with high electric current densities have highest weighting; see also Figure 9.5). Therefore, changes in electrical conductivity around the electrode will have a considerable effect on measured impedance. As tissue temperature increases, the electrical conductivity rises, which results in a drop in impedance (note again that impedance is inversely related to electrical conductivity). As tissue vaporizes above 100°C, an increase in impedance results (details described in Section 9.2.2). After impedance exceeds a defined threshold level, RF power is discontinued for a certain time period during which tissue cools down, vapor settles, and impedance returns to baseline (see Figure 9.3). Subsequently, RF power is reapplied at a lower level.

9.2.1 Electrode Cooling

Cooling of the RF electrode is a commonly used method that allows for increased ablation zone dimensions. In the absence of active cooling of the electrode, highest temperature is observed next to the electrode (Figure 9.4). Through circulation of a coolant within the electrode (typically water), both the electrode and tissue in close proximity are cooled, and the location of highest

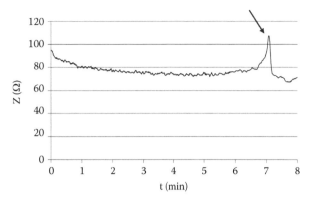

FIGURE 9.3 Impedance control of applied RF power. Once tissue impedance (*Z*) exceeds a defined threshold (arrow) as a result of vaporization, power is discontinued and reapplied after a brief period.

temperature is moved deeper into the tissue from the electrode surface (Figure 9.4) (Haemmerich 2003a). This results in an overall increase of temperature throughout the tissue, producing a larger ablation zone. During cardiac ablation, a similar convective cooling effect is achieved by the blood flowing by the electrode (for more details see Section 9.3.2 on cardiac ablation).

9.2.2 Electrical Tissue Conductivity at Radiofrequencies

Electrical conductivity quantifies how well a certain material conducts electric current. Electrical tissue conductivity is one of the properties affecting heating during RF ablation (see Equation 9.2), and generally depends on frequency (Gabriel 1996b). In this chapter we are concerned with the radiofrequency range of ~450–500 kHz at which RF ablation is performed, and in the following

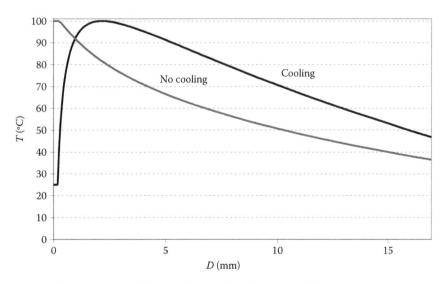

FIGURE 9.4 Tissue temperature (*T*) as a function of distance (*D*) from the RF electrode. Without cooling, location of maximum tissue temperature is adjacent to the electrode. With cooling, this location is moved further from the electrode, resulting in a larger ablation zone. (Reproduced from Haemmerich, D., Chachati, L., Wright, A.S., Mahvi, D.M., Lee, F.T. Jr., and Webster, J.G., *IEEE Transactions on Biomedical Engineering* 50, 2003a. © IEEE.)

TABLE 9.1 Electrical Tissue Conductivity at a Frequency of 500 kHz (= RF ablation frequency)

Tissue Type	Electrical Conductivity σ (S/m)	Reference
Normal Liver (rat, in vivo)	0.36	(Haemmerich 2003b)
Liver Tumor (rat, in vivo)	0.45	(Haemmerich 2003b)
Myocardium (porcine, in vivo)	0.54	(Tsai 2002)
Lung inflated (porcine, ex vivo, 37°C)	0.1	(Gabriel 1996b)
Fat (porcine, ex vivo, 22°C)	0.02	(Pop 2003)
Bone (porcine, ex vivo, 20°C)	0.03	(Gabriel 1996b)
Blood (rabbit, ex vivo, 20°C)	0.7	(Gabriel 1996a)
Vaporized Tissue	~1e-15	Assumed same as air

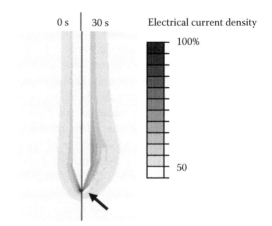

FIGURE 9.5 Electrical RF current density for cooled needle electrode at beginning of ablation (left) and after 30 s (right). Due to tissue vaporization around 100°C and associated drop in local electrical conductivity, current density drops after 30 s around the electrode tip (arrow), since at this location temperatures above 100°C are obtained first. Current density is shown as percentage of maximum. (Reproduced with permission from Haemmerich, D., and Wood, B.J., *Int J Hyperthermia* 22, 2006b. Note also relationship between current density and SAR [Equation 9.1].)

discussion we will always assume this frequency range when referring to electrical tissue conductivity. There is a large amount of literature where electrical conductivity has been measured experimentally in various tissues, typically from animals (Duck 1990, Foster 1989, Gabriel 1996a, Gabriel 1996b). It is important to note that most of these studies were performed in extracted tissue, since electrical conductivity changes rapidly after tissue removal from the body (Haemmerich 2002, Tsai 2002). In Table 9.1 we list the electrical conductivity of various tissues from measurements in live animals whenever such data is available.

Since RF ablation is performed in soft tissue organs such as liver, as well as in bone and lung, the differences in electrical conductivity between these tissues are of relevance. In particular, the low value reported for lung tissue (probably due to the presence of electrically insulating air inside lung alveoli) likely is one of the reasons why RF ablation creates smaller ablation zones in lung compared to other tissues (Brace 2009a). Also notable is that multiple studies have reported differences between conductivity of normal and cancer tissue (Esrick 1994, Lu 1992, Surowiec 1988, Swarup 1991, Haemmerich 2003b), where tumor tissue has about 1.3 times higher electrical conductivity, likely due to loss of cell membrane integrity associated with cell necrosis (i.e., cell death) often present in tumors (Haemmerich 2003b).

Electrical tissue conductivity varies considerably with temperature as already briefly discussed. An increased ion mobility with rising temperature results in a reversible increase in electrical conductivity with a temperature coefficient of approximately 1.5%/°C (Duck 1990). Once tissue temperatures exceed approximately 50°C, irreversible changes are observed. In one study electrical conductivity of kidney tissue was measured at temperatures up to 80°C, where permanent increase in conductivity was observed above 50°C (Pop 2003). Similar changes were observed in a recent study performed on surgically removed liver and tumor tissue from liver cancer patients,

where electrical conductivity was measured before and after tissue ablation (Haemmerich 2009). The causes for these irreversible changes are not completely clear, but tissue dehydration and changes in cell membrane properties may be two contributing factors. At temperatures above 100°C (boiling temperature of water), tissue begins to vaporize. Vapor has a very low electrical conductivity (i.e., it is insulating, see Table 9.1), and therefore has a large impact on RF current pathways. As shown in Figure 9.5, current density drops at locations where vapor starts to form, and heating at these locations is effectively eliminated.

9.3 Clinical Applications of Radiofrequency Ablation

9.3.1 Cancer Treatment (aka Radiofrequency Tumor Ablation)

When thermal ablation using RF current was first clinically used for cancer treatment, it was initially used for liver cancer (both primary and metastatic) since for these patients often surgery is not possible (~80% of patients) (Rhim 1999), and the other two standard therapies—chemo- and radiation therapy—are not effective due to biological reasons. Recent clinical studies on patient survival suggest that with proper patient selection similar patient survival rates as with surgical tumor removal (the current gold standard) are attainable (Gillams 2009, Livraghi 2008). In the last few years, tumor ablation has expanded to other cancer types such as lung, kidney, bone, and adrenal gland (Gillams 2008). Especially for lung cancer, tumor ablation has received considerable attention, as there are more than 200,000 new cases of lung cancer reported annually in the United States

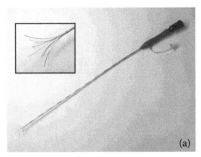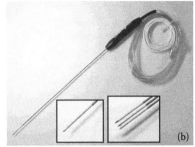

FIGURE 9.6 Commercial RF electrodes for tumor ablation. (a) Multi-prong electrode, with magnified electrode tip shown in insert. The prongs are extended once the catheter is placed in the tumor. (b) Cooled needle electrode, available as single, or three-needle cluster (see magnified inserts). Active electrode at the tip is 3 cm (single) or 2.5 cm long (cluster), and is internally cooled by circulating water. (Reproduced with permission from S. Vaezy and V. Zderic, eds. *Image-Guided Therapy Systems*, Artech House, Inc., Norwood, MA, 2009. © 2009 by Artech House, Inc.)

alone (Jemal 2007), and most patients are not surgical candidates; recent studies suggest that patients with inoperable lung cancer may benefit from RF ablation (Simon 2007, Gillams 2008). A tumor ablation procedure can be performed minimally invasively through a small incision in the skin (by an interventional radiologist), or during laparoscopy or open surgery (by a surgeon). During the procedure, the patient is typically under light general anesthesia or conscious sedation and can leave the hospital the same, or the next, day.

The treatment goal is to thermally ablate tissue in a zone that encompasses the under imaging visible tumor as well as a ~1 cm margin of normal tissue to ensure destruction of any cancer microsatellites that may surround the tumor (Sasaki 2005). For tumors larger than 3 cm, multiple overlapping ablations created either sequentially (Chen 2004) or simultaneously with multiple electrodes (Laeseke 2007) are typically required.

Limitations of current RF tumor ablation procedures include:

1. limited performance close to large vasculature, that may result in tumor recurrence due to inadequate temperatures;
2. inadequate intra-procedural imaging feedback on ablation zone growth;
3. the size of the ablation zone of a single ablation is often not adequate to treat large tumors (>3 cm diameter), resulting in prolonged procedural times and higher recurrence rates.

9.3.1.1 Radiofrequency Electrodes for Tumor Ablation

There are a number of different electrodes commercially available. Some electrodes have multiple tines that are extended after the electrode is inserted into the tissue (Figure 9.6a), while another kind uses needle electrodes that are internally cooled (Figure 9.6b). Most of the electrode shaft is insulated such that tissue heating is only produced at the most distal electrode portion (see inlets in Figure 9.6).

9.3.2 Cardiac Arrhythmia Treatment

RF ablation (cardiac RF catheter ablation) is frequently used for treatment of cardiac arrhythmia (i.e., irregular heart beats),

e.g., for different types of tachycardia (fast heart rhythm above 150 bpm) and atrial fibrillation (i.e., quivering of the atria) (Huang 2006, Wilber 2007). The spatiotemporal activation pattern of the heart is determined by a specialized conduction system, and abnormalities in electrical conduction in the heart can result in arrhythmia. The goal of cardiac RF ablation is to destroy a small region of heart tissue to normalize electrical activation in the heart (e.g., by destroying additional conduction pathways not present in a normal heart).

Cardiac ablation is performed in a specifically equipped interventional laboratory by an electrophysiologist. A catheter is inserted into a vessel (typically in the groin or neck) and steered into the heart (Figure 9.7). The procedure is guided by x-ray imaging. In addition, electrical measurements of local cardiac activity (similar to ECG) are performed via recording electrodes located on the RF catheter (Figure 9.8), as well as by additional specialized recording catheters placed at various locations in the heart. Through these local electrical activity measurements, temporal activation patterns of the heart can be determined to diagnose the cause of the arrhythmia and site that needs to be ablated. Subsequently, the RF catheter is steered to the target site and RF current is applied to the heart tissue for ~45–120 s, creating an ablation zone of ~5–10 mm in diameter (Figure 9.9).

For most commercially available catheters, the applied RF power is adjusted depending on temperature measured by a sensor located within the electrode tip (temperature control). Note, however, that electrode tip temperature is lower than the maximum temperature within the tissue (see Figure 9.9). This is of importance because one of the undesired effects that can occur is tissue perforation (often called "popping" due to the sound associated with this event), which is due to tissue vaporization above 100°C. Intracardiac blood flow (i.e., blood velocity at the electrode) considerably affects tissue heating and resulting size of the ablation zone (Figure 9.10). Usually, larger ablation zones are possible at locations with high blood flow (Cao 2001, Tungjitkusolmun 2001, Petersen 1999). To create ablation zones of varying dimensions, catheters of different lengths and diameters are commercially available. Some newer catheter designs

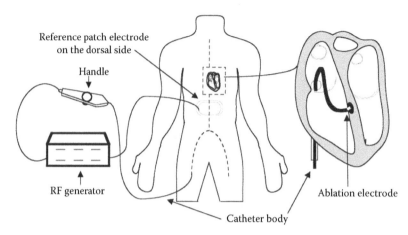

FIGURE 9.7 Schematics of cardiac RF ablation system. A cardiac RF catheter is inserted through a leg vein and steered to the target site inside the heart (right figure). A reference patch electrode (i.e., ground pad) is placed on the patient's back. The small black region around the RF electrode at the catheter tip depicts the ablation zone. (Reproduced with permission from Panescu, D., Whayne, J.G., Fleischman, S.D., Mirotznik, M.S., Swanson, D.K., and Webster, J.G., *IEEE Transactions on Biomedical Engineering* 42, 1995. © IEEE.)

use internal or external cooling to increase ablation zone size (see also Section 9.2).

One very common type of cardiac arrhythmia that affects ~2.2 million people in the United States (most prevalent among people older than 60 years) is atrial fibrillation, where the atria quiver instead of beat, thereby reducing the pumping performance of the heart, with a risk of blood clot formation (Wilber 2007). Cardiac ablation is increasingly used for treatment of atrial fibrillation in patients not responsive to medications. For treatment of atrial fibrillation, multiple linear, contiguous ablation zones have to be created. This is difficult and time consuming with current devices, resulting in long procedural times and high recurrence rates (Wilber 2007). New devices that allow more rapid and reliable creation of linear ablation zones are in development (Burkhardt 2009, Siklody 2010).

9.3.3 Other Applications of RF Ablation

RF ablation is clinically used for treatment of uterine bleeding in women who don't respond to standard therapies such as medication and scraping of the endometrium. During treatment, the whole endometrium (i.e., lining of the uterus) is ablated within typically 3–10 minutes. Different methods are employed to thermally ablate the endometrium, including RF ablation devices that employ mesh electrodes with power applied between two electrically insulated meshes (Cooper 2004).

RF and laser-based ablation devices are available as treatment modality for varicose veins (Markovic 2009) (endovascular ablation), which are visible, dilated and twisted veins near the skin surface. Varicose veins most often affect legs and thighs, due to blood pooling and vein enlargement resulting from insufficiencies in the venous valves. The treatment goal is typically closure of the affected veins. Methods to obtain vein closure include surgical stripping, injection of a drug that results in vein swelling and closure, and ablation. During ablation, a catheter is introduced into the vein, and the vessel wall is heated, resulting in collagen shrinkage and closure of the vein.

9.4 Biophysics of Microwave Ablation

Microwaves represent the portion of the electromagnetic spectrum between 300 MHz and 300 GHz. The Federal Communications Commission (FCC) or International Telecommunications Union (ITU) permit several unrestricted frequency bands for industrial, scientific, and medical use in several regions of the world, including those most commonly used for microwave ablation procedures today: 915 MHz and 2.45 GHz. In addition to these two frequencies, 433 MHz has been explored as an option for microwave hyperthermia in Europe, but not commonly for high-temperature focal microwave ablation. Frequencies higher than 2.45 GHz can also be used for some applications but have not been widely explored to date.

As with other thermal ablation modalities, microwave ablation describes the rapid destruction of tissue by microwave

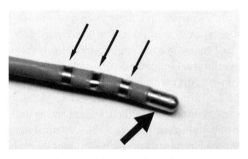

FIGURE 9.8 Cardiac RF ablation catheter (7F = 2.3 mm diameter). Several electrodes are located along the catheter: RF electrode of 4mm length (large arrow), as well as three smaller electrodes used for recording of local electrical activity in the heart (small arrows).

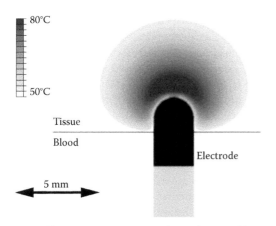

FIGURE 9.9 Tissue temperature around a cardiac RF ablation catheter (2.3 mm diameter) at the end of a 45 s ablation. Outermost 50°C boundary estimates boundary of ablation zone.

heating (Figure 9.11). In this respect, microwave ablation is the more acute and higher-temperature extension of microwave hyperthermia, which has a longer history in the literature and in clinical practice. Hyperthermia typically refers to the temperature range of 41°C to 46°C, where most temperature-induced physiological changes are reversible, while thermal ablation at temperatures over 50°C is associated with irreversible changes such as denaturation of cellular proteins and microvascular coagulation leading to rapid cell death (Dewhirst 2003). Despite the difference in target temperature range, the physics of heating tissues is relatively similar for both therapies. A notable exception is when tissue temperatures approach or exceed 100°C, when a series of temperature-induced changes in tissue properties must be considered (Brace 2010b, Brace 2008).

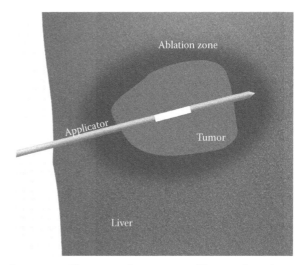

FIGURE 9.11 Illustration of a microwave ablation in liver. A zone of complete cellular necrosis is created by heat generated from the applicator antenna, which completely encompasses the tumor with a safety margin.

Electromagnetic energy propagation can be described by solving Maxwell's equations in a source-free lossless medium:

$$\cdot E = 0 \tag{9.3}$$

$$\times E = -\frac{\partial B}{\partial t} \tag{9.4}$$

$$\cdot B = 0 \tag{9.5}$$

$$\times B = \mu\varepsilon\frac{\partial E}{\partial t} \tag{9.6}$$

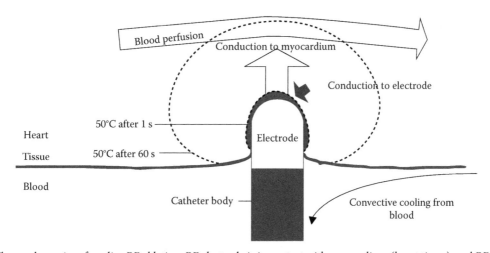

FIGURE 9.10 Thermodynamics of cardiac RF ablation: RF electrode is in contact with myocardium (heart tissue), and RF current results in tissue heating. Thermal conduction of heat into the tissue results in growth of the ablation zone (approximated by 50°C isotherm). Heat loss of tissue in proximity of the electrode is due to thermal conduction through the electrode (black arrow); electrode and tissue surface experience convective cooling from blood inside the chamber. Surface cooling produces the typical tear drop–shaped ablation zone. (Reproduced from Tungjitkusolmun, S., Finite element modeling of radiofrequency cardiac and hepatic ablation. PhD thesis University of Wisconsin, Madison, 2000.)

where E is the electric field vector (V/m), B is the magnetic flux density vector (Tesla), and μ and ε are the magnetic permeability (H/m) and dielectric permittivity (F/m) of the ambient medium, respectively. Equations 9.4 and 9.6 may then be combined using a vector identity to form the wave equations of electromagnetic propagation:

$$\frac{\partial E^2}{\partial t^2} = v_p^2 \,{}^2 E \tag{9.7}$$

$$\frac{\partial B^2}{\partial t^2} = v_p^2 \,{}^2 B \tag{9.8}$$

where v_p is the wave propagation velocity (m/s), equal to $(\mu\varepsilon)^{-1/2}$. Equations 9.3 through 9.8 imply two important physical processes: (1) time-varying electric or magnetic fields give rise to spatial variations in magnetic and electric fields, respectively, producing a self-propagating electromagnetic field; and (2) wave propagation characteristics are determined by the magnetic permeability and relative permittivity of the wave medium. The first process forms the basis for understanding how microwave propagation is created from an antenna, and the second process illustrates how biological tissues respond to applied microwave energy, including both the wavelength of propagation inside tissues, and the absorption of microwave energy.

9.4.1 Dielectric Properties of Biological Tissues

If the propagation medium is not lossless and has a finite conductivity, σ (S/m), then Equation 9.6 takes the form:

$$\nabla \times B = \mu\varepsilon\frac{\partial E}{\partial t} + J \tag{9.9}$$

where J is the electrical current density (A/m^2) equal to σE. Casting Equation 9.9 into time harmonic form ($\partial/\partial t \rightarrow j\omega$, where $\omega = 2\pi f$) with the current density replaced by σE gives:

$$\times B = j\omega\mu\varepsilon E + \sigma E = j\omega\mu\left(\varepsilon - j\frac{\sigma}{\omega}\right)E. \tag{9.10}$$

Equating Equations 9.6 and 9.10 reveals that the dielectric permittivity in a medium with finite conductivity is complex-valued:

$$\dot{\varepsilon} = \varepsilon' - j\frac{\sigma}{\omega} = \varepsilon' - j\varepsilon''. \tag{9.11}$$

Tissues are characterized by substantial complex-valued dielectric permittivities but magnetic permeabilities approximately equal to that of vacuum. Relative permittivity, ε_r, is the real part of the complex permittivity and quantifies the ability of a material to store electrical energy relative to a vacuum. It is frequently referred to as dielectric constant; however, since it is variable depending on frequency, temperature, and other

factors in biological tissues, the term *relative permittivity* will be used here. The effective conductivity of a material, σ, is defined from the imaginary part of the complex permittivity and is used to describe how well a material absorbs microwave energy. Materials with high effective conductivities such as tissue are said to be lossy. It is important to note that effective conductivity describes contributions from both moving charges (electrical currents) and time-varying electric fields (displacement currents; Balanis 1989). Displacement currents dominate for most biological tissues in the microwave spectrum and are produced by the rotation of polar molecules such as water in tissue.

The dielectric properties of tissues are most closely related to their water content (Schwan 1955, Schepps 1980). Tissues with high volumetric water content are characterized by higher relative permittivities and conductivities. Most organs and soft tissues are considered to be high water content (Gabriel, C. 1996, Gabriel, S. 1996a). As a result, the wavelength of an electromagnetic field is shorter and power loss (i.e., heating) is more rapid in these tissues. Conversely, tissues with lower water content such as adipose tissue are slower to generate heat. Note that there is a distinction between water content by mass, which is more commonly reported in the tissue property literature, and water content by volume. For example, consider aerated lung tissue. Lung is generally considered to have a relatively high water content by mass but low water content by volume due to its very low density (Duck 1990). Therefore, heat generated in lung tissue is primarily produced in the blood-filled spaces between alveoli, but not in the air-filled alveoli themselves. There are no universal models to describe the dependence of tissue permittivity on water content, but mixture models have been proposed for some tissues (Schwan 1977, Ng 2008, Steel 1986).

Tissue	100 MHz		1 GHz		10 GHz		100 GHz	
	ε_r	σ	εr	σ	εr	σ	εr	σ
Adipose	6.07	0.04	5.45	0.05	4.6	0.59	3.56	3.56
Bone (cortical)	15.3	0.64	12.4	0.16	8.12	2.14	3.3	8.66
Breast	5.69	0.3	5.41	0.05	3.88	0.74	2.59	1.84
Kidney	98.1	0.81	57.9	1.45	40.3	11.6	8.04	57.1
Liver	69	0.49	46.4	0.90	32.5	9.39	6.87	42.9
Lung (inflated)	31.6	0.31	21.8	0.47	16.2	4.21	4	21.4
Muscle	65.9	0.71	54.8	0.98	42.8	10.6	8.63	62.5

The dielectric properties of tissue are also dependent on frequency, temperature, and structure. The frequency dependence of each tissue can be described analytically using a multipole Cole-Cole model. Each pole, n, of the model describes the complex permittivity dispersion within a frequency range, with the summation of multiple dispersions describing a larger frequency spectrum:

$$\hat{\varepsilon}(\omega) = \varepsilon_\infty + \sum_n \frac{\varepsilon_{s,n} - \varepsilon_{\infty n}}{1 + (j\omega\tau_n)^{(1-\alpha_n)}} + \frac{\sigma_i}{j\omega\varepsilon_0} \; (\text{F/m}), \tag{9.12}$$

where ε_∞ is the relative permittivity at a frequency where $\omega\tau \gg 1$, ε_s is the relative permittivity at a frequency where $\omega\tau \ll 1$, τ is the relaxation time constant (s), α is an empirical distribution parameter that broadens the dispersion in each frequency range, σ_i is the static ionic conductivity, and ε_0 is the permittivity of free space. Relaxation, permittivity, and distribution parameters have been described in the literature for most common biological tissues (Gabriel 1996b). The dielectric properties of tissue are generally thought to be relatively isotropic, but this assumption may not be valid in some situations in tissue with anisotropic structure such as muscle (Epstein 1983).

The temperature dependencies of relative permittivity and conductivity are not yet as fully characterized as frequency dependence. Most of the reported data are clustered around the ambient room temperature (~20°C) or normal body temperature (~37°C). In the range of 10–60°C, temperature dependence for both relative permittivity and conductivity have been assumed to take a linear form (Duck 1990, Lazebnick 2006, Pop 2003, Bircan 2002, Chin 2001, Stauffer 2003). However, near temperatures of water phase change (0°C or 100°C), substantial deviations from the linear model have been reported (Brace 2008). At temperatures in excess of approximately 60°C, protein and cellular structure changes can also affect tissue permittivity. Such effects have not been well characterized in the available literature on bulk tissue, but some data suggest that protein denaturation may be detectable by a sudden slight change in the temperature curve of relative permittivity or conductivity in some tissues (Bircan 2002, Wall 1999).

Penetration of a microwave field into a tissue medium is also dependent on the dielectric properties of the tissue. For a plane wave in a homogenous isotropic medium, the penetration depth, δ (m), of an electromagnetic field is defined as the distance required for the electric field to attenuate to $1/e$ (~37%) of its initial value:

$$\delta = \frac{1}{\omega\sqrt{\mu\varepsilon}\left\{\frac{1}{2}\left[\sqrt{1+\left(\frac{\sigma}{\omega\varepsilon}\right)^2}-1\right]\right\}^{1/2}}(m). \qquad (9.13)$$

In the case of most tissues and microwave ablation frequencies, the penetration depth from Equation 9.13 can be approximated by assuming the tissue is a good dielectric ([σ/ωε]2 << 1):

$$\delta = \frac{2}{\sigma}\sqrt{\frac{\varepsilon}{\mu}}(m). \qquad (9.14)$$

It is important to note from Equation 9.14 that penetration depth is inversely related to conductivity. Deeper plane-wave penetration occurs with lower frequencies, which may be more desirable for non-focal heating applications such as regional hyperthermia (Fotopoulou 2010, van Rhoon 1998, Turner 1989). Higher frequencies with less energy penetration may be desirable in more superficial applications such as ablation of the endometrium (Feldberg 1998, Hodgson 1999).

One final point of consideration: much of the energy radiated by microwave antennas in lossy media such as biological tissues is absorbed in the near or Fresnel zones of the antenna and may not propagate as a plane wave. In this case, the aforementioned calculations of attenuation and penetration depth should be viewed as approximations. Electromagnetic simulation and, in particular, simulation of electromagnetic-thermal interactions can provide more accurate estimates of temporal heating for comparing devices and frequencies for microwave ablation.

9.4.2 Electromagnetic Interactions with Tissue

When electromagnetic waves propagate through a lossy medium, some of the energy is converted into heat. In biological tissues, dielectric hysteresis losses dominate, so heat is primarily generated by the rotation of polar water molecules. The rapid oscillation of these molecules effectively increases their kinetic energy and, hence, temperature. Heat generation is proportional to the energy applied (i.e., microwave power) and effective conductivity of the tissue medium:

$$Q_{EM} = \frac{\sigma}{2}|E|^2 \quad (\text{W m}^{-3}), \qquad (9.15)$$

where the square of the electric field intensity, $|E|^2$, is linearly proportional to the applied power. Comparing Equation 9.15 to Equation 9.14 reveals that while penetration depth is inversely proportional to the effective conductivity, the heat generation rate is directly proportional to effective conductivity. This makes intuitive sense from a conservation of energy perspective: the primary cause of field attenuation is the conversion of microwave energy to heat. Note also that since the effective conductivity of the tissue is related to frequency, temperature, water content, and other factors, the heat generation rate is also dependent on these same factors. Once heat is generated inside the tissue, heat transfer can be modeled using the so-called bioheat equation (Pennes 1948):

$$\rho C_p \frac{\partial T}{\partial t} = \nabla \cdot (k_T \nabla T) + Q_{EM} - Q_{perfusion} (\text{W m}^{-3}) \qquad (9.16)$$

where ρ is density (kg m⁻³), C_p is the specific heat capacity at constant pressure (J kg⁻¹ m⁻³), T is temperature (K), t is time (s), k_T is thermal conductivity (W/m K), Q_{EM} is the heat generation rate illustrated in Equation 9.15, and Q_p is the rate of heat lost to blood perfusion (W m⁻³). The term on the left-hand side of Equation 9.16 is the heating rate for a given volume of tissue, while the first term on the right-hand side describes thermal conduction inside the tissue. Heat lost to blood perfusion can be represented as a convective process using the following equation:

$$Q_{perf} = w_{bl}\rho_{bl}C_{p,bl}(T-T_0) \quad (\text{W m}^{-3}). \qquad (9.17)$$

This heat loss term encompasses microvascular blood perfusion and blood flow through large vessels, which can both be approximated as a convective heat transfer process (Pennes 1948).

The perfusion coefficient, w_{bl}, is often provided as a function of temperature. Inflammation due to mild hyperthermia increases local blood flow, which is captured by an increase in the perfusion coefficient from 40–46°C. However, at temperatures above 50°C, blood coagulation and tissue contraction lead to a rapid decrease in blood perfusion and eventual thrombosis (Craciunescu 2001, Tompkins 1994, Rawnsley 1994). This effect is frequently modeled using a Heaviside step or other sigmoidal function for the perfusion coefficient with respect to temperature.

9.4.3 Microwaves Compared to Other Sources of Thermal Therapy

Microwave energy has unique properties that make it an attractive choice for thermal ablation. The most-cited advantages for microwave ablation are a faster heating rate (Figure 9.12), greater volume of active or direct heating, higher temperatures inside the ablation zone, less susceptibility to ablation-induced changes such as charring or desiccation, larger zones of ablation, and improved performance with multiple applicators. These comparisons are often drawn with RF ablation, but may also apply to varying degrees in comparisons to laser sources, ultrasound sources, or cryoablation. The common thread to these potential advantages is the fact that microwave heating occurs as a result of electromagnetic propagation through biological tissues of all types. Radiofrequency, laser, and ultrasound energy sources can be significantly hampered by specific types of tissue. For example, lung and bone tissues are poor electrical conductors at RF frequencies, which limits the amount of current and, hence, power that can be deposited during RF ablation (Duck 1990). Charred and desiccated tissues are also poor conductors of RF electrical current, and present a barrier to laser light penetration and ultrasound wave propagation. Consequently, RF, laser, and ultrasound sources are typically modulated or tempered during ablation treatments to avoid generating temperatures over 100°C. On the other hand, microwaves propagate through all types of tissues, including water vapor and dehydrated, charred, or desiccated tissues created during the ablative process. The result is that microwaves can be applied continuously throughout the ablation procedure.

The wavelength and penetration depth of microwaves in tissue is also suitable for a variety of medical applications. At 915 MHz and 2.45 GHz, wave penetration is 2–4 cm in most tissues, which is commensurate with the majority of tumors targeted by thermal ablation. By contrast, power deposition attenuates rapidly away from RF electrodes and optical fibers used for laser ablation. As a result, microwaves may produce more active heating than these other energy sources. Active heating is thought to be more potent against heat loss to blood perfusion or large vascular heat sinks. Indeed, while RF ablation has been shown to be relatively ineffective near vessels 3 mm or larger in diameter, microwaves are effective against vessels at least 3 mm in diameter and have been shown to induce complete thrombosis of vessels up to 10 mm in diameter (Yu 2008, Lu 2002, Brace 2007a, Wright 2003). Other studies have shown that microwaves produce larger zones of ablation in liver, kidney, and lung tissues when compared to RF ablation, even when the applied power was held constant (Andreano 2010, Brace 2009a, Laeseke 2009). Other studies have also shown that by using higher powers and shorter treatment times, microwaves may actually be more effective in vivo than ex vivo, an effect not noted with other ablation energy types to date (Hines-Peralta 2006).

Microwave energy is not without drawbacks, however. As mentioned earlier, the 2–4 cm wavelength penetration of microwave energy in tissue means that heating precision may be sacrificed with microwaves. Rapid heating and high temperatures also need more critical safety evaluation, especially when applied for long times (several minutes) to a large volume of tissue. This is especially true from a monitoring standpoint, when rapid heating might be difficult to capture without real-time imaging. Microwave energy can also overheat the cabling used to transfer power from the generator to the applicator. This internal cable heating must be offset either by limiting energy transmission or by active cable cooling to prevent unwanted thermal damage to tissues along the cable length (He 2010). However, even when cooled, many antenna designs produce elongated ablations, which may not be desirable for many clinical applications (Tse 2009, Sun 2009, Brace 2007b). Many of these deficiencies may be overcome by improved system component design and optimized power application protocols.

9.5 Microwave Ablation Systems

The microwave ablation system consists of three main components: a microwave power generator, a system for delivering power to the applicator antenna, and the antenna itself. Of these, antenna design has been given the most discussion in the existing literature; however, techniques for distributing power to multiple antennas and the role of frequency have also been investigated. A more detailed discussion of each component follows.

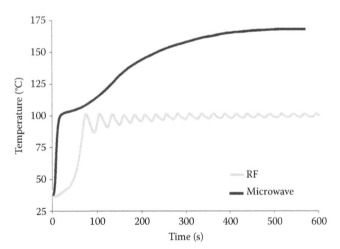

FIGURE 9.12 Temperatures measured 5 mm from a microwave (MW) and RF applicator during ablation of the renal cortex demonstrate more rapid and continuous heat generation.

9.5.1 Power Generation and Distribution

9.5.1.1 Generator Design

Microwave power in clinical ablation systems is generated using either solid-state semiconductor sources or vacuum-tube devices such as the magnetron. Magnetron sources have the advantage of relatively high-power conversion efficiency (typically greater than 70%), the ability to produce substantial powers from a single device, and a high tolerance to imperfect load matching without device failure. However, microwave generators that employ magnetron sources distribute power to multiple antennas by using passive splitters. Phase control of each channel can be achieved using passive phase adjusters, but these must be able to distribute the full power of each channel. Phase adjusters capable of distributing typical microwave ablation powers of 50–150 W tend to be relatively bulky and may not be ideal for clinical implementations.

More recently, solid-state devices have become available that may satisfy the requirements for clinical microwave ablation systems. A solid-state generator consists of a low-power frequency source, or oscillator, followed by control and amplification stages that boost power output by 4–5 orders of magnitude. In general, solid-state systems suffer from lower-power efficiency (typically less than 40%), less tolerance to high-reflected powers produced by imperfect load matching, and lower-output power per channel than magnetron sources. However, solid-state systems utilize smaller components, can be phase adjusted and power controlled in the preamplification stages, and can have a cleaner output spectrum than magnetron sources with comparable control features.

Both solid-state and magnetron generators are used in microwave ablation systems currently in clinical use. Some of these systems allow for a single antenna to be operated by a single-generator system, requiring additional generators to power each element of a multiple-antenna array. Other systems provide external power splitters to divide power between two or more antennas, but without any control of the relative phase between antennas. Still other systems provide multiple power channels from a single-generator system, with or without control of the relative phase between antennas. Even without the ability to control antenna phase, most single-generator, multiple-antenna systems produce a coherent output in each output channel.

9.5.1.2 Frequency Considerations

Energy penetration and heating rates in biological tissues depend on the frequency of the applied electromagnetic field. The simplest way to compare frequencies is to assume plane-wave propagation into the tissue. The plane-wave condition implies that the radiating wavefronts lie along parallel planes, which occurs in the far-field of the antenna. The far-field is typically defined as radial distances greater than $2D^2/\lambda$, where D is the largest dimension of the antenna and λ is the wavelength in tissue. This distance is about 4 cm from the antenna in tissue. The fact that 4 cm is greater than the radius of a typical ablation zone implies that the far-field condition is rarely achieved in practice; energy is absorbed instead in the near-field or Fresnel zones of the antenna, where waves must be treated as spherical. For this reason, comparing frequencies based upon plane-wave assumptions alone may not be appropriate. More investigation is needed to address this concern. As a result, reports comparing 915 MHz and 2.45 GHz systems have provided mixed results; some conclude that 915 MHz produces larger ablation volumes, while others suggest that equally large ablations can be achieved with 2.45 GHz (Hines-Peralta 2006, Sun 2009, Brace 2007b, Strickland 2002, Hope 2008). None of the existing literature has employed a well-controlled design to test equal powers delivered to the tissue. In one notable exception, frequencies as high as 9.2 GHz have been used to ablate the endometrial lining of the uterus. Development of that system demonstrated that 9.2 GHz energy provided an appropriate balance of energy penetration and rapid heating required for that particular application (Feldberg 1998).

9.5.1.3 Coaxial Cables

Microwave power is carried from the generator to the antenna through coaxial cables, due to their relative flexibility, excellent propagation characteristics, and ease of connectivity. Cable flexibility is a function of cable diameter and construction materials, with improved flexibility in braided conductors and low-density dielectrics. Power handling—the ability of a cable to safely transfer power without overheating or failure—is related to these same factors, as well as the frequency of the applied microwave power. Therefore, coaxial cables used to distribute microwave power may have a greater diameter and increased stiffness when compared to their counterparts in RF or laser ablation systems.

Coaxial cables also comprise the underlying structure of most interstitial microwave ablation antennas. Those intended for percutaneous use are 1.5 mm to 2.5 mm in diameter, while antennas greater than 5 mm in diameter have been reported for surgical applications or endometrial ablation (Feldberg 1998, Hines-Peralta 2006, Strickland 2002). Data from the biopsy literature suggest that larger needle diameters are associated with an increased risk of complications such as bleeding and pneumothorax, providing motivation to decrease percutaneous antenna diameter (Geraghty 2003). However, small-diameter coaxial cables absorb more microwave energy, which reduces power throughput and, in turn, increases heat generation within the cable. Such heating can lead to potentially dangerous thermal damage along the antenna shaft. Yet, increased power delivery has been associated with faster and potentially more effective microwave ablation treatments, particularly for large tumors. A balance between small cable diameter, high power throughput, and low internal heating must be achieved in the power delivery and antenna cables.

9.5.2 Microwave Ablation Antenna Design

The microwave ablation antenna can be defined in a number of different ways. In surgical and percutaneous applications, the antenna is typically defined as the entire applicator beyond the flexible coaxial power delivery cable. With this definition, the

FIGURE 9.13 Cartoon schematic of a typical microwave ablation antenna. Coaxial cable runs the length of the shaft, with the radiating element at the distal end of the antenna. Energy is produced around the radiating element.

antenna consists of a rigid shaft and a radiating section at the distal aspect (Figure 9.13). In catheter-based designs, the term *antenna* usually refers only to the distal radiating element, rather than the entire catheter.

Antenna properties relevant to thermal ablation include both the pattern of radiation and reflection coefficient, or return loss. In general, the lowest return loss is desirable to maximize energy transfer from the antenna into the tissue. Energy reflected from the antenna reduces tissue heating, while increasing unwanted heating of the antenna shaft. In extreme cases, high return loss may necessitate short ablation times to prevent thermal damage along the antenna shaft (Sato 1996). The desired radiation pattern is largely dependent on the clinical application. Most antennas in use currently radiate in the normal (broadside) mode, with propagation directed radially outward from the antenna. This is especially true of antennas designed for tumor ablation applications, where the ideal radiation pattern is focused and omnidirectional to match the approximately spherical shape of many tumors. Antennas designed in the axial (end-fire) mode have been developed for cardiac applications to produce localized heating of a spot at the distal tip of a catheter (Gu 1999).

To achieve the goals of low return loss and focused energy radiation, several designs have been proposed (Figure 9.14). Broadly, these can be classified as designs that utilize a linear element, coaxial slot, loop, or helix as their primary mode of radiation. Designs primarily comprised of a linear element include monopoles, dipoles, and triaxial antennas (Brace 2005, Labonte

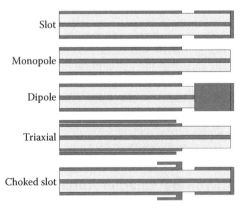

FIGURE 9.14 Schematic cross sections of five antenna designs for microwave ablation. From top to bottom: slot, monopole, dipole, triaxial, and choked slot. Metallic components are represented in dark gray, dielectric insulators in light gray.

1996, Hurter 1991, Ahn 2005). These antennas are highly efficient, coupling over 95% of input power into the tissue, with respectable broadside radiation patterns. However, they are relatively narrowband, and backwards radiation along the antenna shaft can lead to undesirably long ablation zones. Tip loading can be used to increase the electrical length of an antenna or alter its heating pattern.

Coaxial slot antennas radiate from one or more annular slots (Lin 1987, Ito 1990, Saito 2004). Since energy radiates from a smaller aperture than linear element designs, they may be desirable for ablation of small volumes. However, single-slot antennas also suffer from excessive backward heating along the antenna shaft without design modification. Multiple-slot antennas show promise for reducing this backwards heating (Saito 2004). Coaxial chokes have also been proposed to reduce unwanted backwards radiation by enforcing an open boundary at the antenna feed point, thereby eliminating current on the outer surface of the coaxial cable (Bertram 2006, Longo 2003, Lin 1996, Wong 1993). However, chokes add to the total antenna diameter, making them less attractive for use in percutaneous applications.

Looped or helical designs are less common in practice, but have been described in the literature (Gu 1999, Shock 2004, Liu 1996). Notably, looped designs operating in the axial mode have been described for cardiac ablation, while catheter-based helices have been proposed for microwave-assisted angioplasty. Single and multi-loop designs have also been tested for microwave ablation in the liver. Deployable loop designs are more problematic to use in practice since they are more difficult to visualize completely on ultrasound during guidance to the target and can be difficult to retract at the end of the procedure.

9.5.2.1 Antenna Cooling

Antenna cooling is an effective solution to the problem of excessive internal heating in the antenna shaft (Figure 9.15). Water and gas-cooled antenna designs have all been described in the literature, and most current clinical systems use some type of cooling along the antenna shaft (Kuang 2007, Yeh 1994, Knavel 2010). Sufficiently cooling the antenna shaft prevents unwanted thermal damage that may be produced by heat conduction from inside the antenna shaft, heat conduction from the hot ablation zone along the antenna shaft, or backward heating from the radiating segment. By eliminating excessive heat produced by these three sources, higher powers can be passed through even very small-diameter antennas. In some implementations, cryogenic gas cooling can also be used to create a small ice ball at the distal tip of the applicator to prevent applicator migration during placement or pre-procedural imaging (Knavel 2010).

9.5.3 Multiple-Antenna Arrays

9.5.3.1 Limitations to Single-Applicator Ablation

Despite technical advances and optimization, power delivery from a single antenna is inherently limited. First, consider

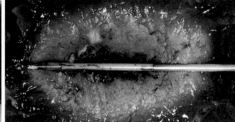

FIGURE 9.15 Microwave ablations created with uncooled (left) and water-cooled (right) applicators.

energy radiated from a single-point source. Ignoring any temperature or perfusion effects, a constant energy inflow would be expected to produce a constant spherical diffusion of heat—that is, a constant increase in ablation zone volume (much like the inflation of a balloon). However, since the radius of the ablation zone is proportional to the cube-root of the volume, diametric ablation zone growth slows as the ablation volume increases (Figure 9.16). There are diminishing returns to using a single radiation source for large-volume ablation for extended time periods.

Second, the temperature profile across any thermal ablation is not consistent, typically containing a peak near the applicator and exponential decay toward the periphery of the ablation zone. Most of the applied energy is also absorbed near the applicator. The result is that once an initial ablation zone has formed, additional energy primarily heats tissue that is already necrotic, contributing only to maintain or increase the temperature gradient across the ablation zone. This effect may be more or less pronounced because of inherent changes in the tissue properties of ablated tissue. Therefore, the concept of spatially distributing the available power by using deployable or multiple-applicator arrays has seen increased interest in recent years.

9.5.3.2 Antenna Arrays for Microwave Ablation

Studies have demonstrated that power can be more efficiently distributed by using an array of antennas, even when compared to a single antenna delivering the same total power (e.g., 90 W in a single antenna versus 30 W in three antennas spaced 1–2 cm apart in a triangular configuration; Laeseke 2010). In addition, heating produced simultaneously by multiple sources in proximity is known to produce ablation zones larger than might be expected from a sum of each source. Several recent studies have confirmed this *thermal synergy* when using arrays for RF and microwave ablation, producing ablations up to 7 cm in diameter (Brace 2009b, Oshima 2008, Laeseke 2006, Lee 2007, Yu 2006, Simon 2006, Wright 2003).

Numerous studies of phased-controlled antenna arrays for microwave tissue heating have been described in the hyperthermia literature (Jones 1989, Furse 1989, Lyons 1984, Turner 1984, Trembly 1985, Trembly 1986, Turner 1986). The objective of many of these studies was to produce a precise zone of low-temperature hyperthermia (41–45°C) in a target area without heating surrounding structures. Similar investigations for high-temperature microwave ablation are ongoing. One potentially detrimental effect of antenna interference can occur if the relative phase is not known or controllable, as with multiple-generator multiple-antenna systems. Under this condition, the relative phase between antennas is unknown, and the resulting interference may be constructive or destructive, leading to unpredictable results (Lubner 2010). It is currently unknown how much clinical impact random phase has on the final ablation zone. As an alternative, power may be switched between antennas in the array to eliminate interference, producing more predictable results (Brace 2007c). More study is needed to optimize and control power produced by antenna arrays for microwave ablation.

9.6 Microwave Ablation Conclusions

In conclusion, microwave ablation is playing an increasingly important role in the treatment of many medical conditions. In most cases, devices and equipment used clinically are in their first or second generation. Recent developments including cooled systems to increase power delivery, antenna arrays to distribute power spatially, and optimization of technology and technique based on specific tissue targets have already

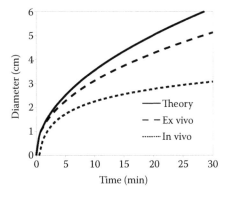

FIGURE 9.16 Ablation zone diameters predicted using a constant volumetric growth (theory), or simulations in ex vivo or in vivo liver. Note the trend of decreasing diametric growth with time.

impacted the clinical utilization of microwave ablation. Additional developments are likely to continue this trend and, as technologies improve, increased clinical utilization will likely follow.

References

Ahn, H., and Lee, K. 2005. Interstitial antennas tipped with reactive load. *IEEE Microwave Wireless Comp Lett* 15:215–20.

Andreano, A., Huang, Y., Meloni, M.F., Lee, F.T. Jr., and Brace, C.L. 2010. Microwaves create larger ablations than radiofrequency when controlled for power in ex vivo tissue. *Med Phys* 2010 37:2967–73.

Balanis, C.A. 1989. *Advanced Engineering Electromagnetics*. John Wiley & Sons, New York.

Berjano, E.J. 2006. Theoretical modeling for radiofrequency ablation: State-of-the-art and challenges for the future. *Biomed Eng Online* 5:24.

Bertram, J.M., Yang, D., Converse, M.C., Webster, J.G., and Mahvi, D.M. 2006. Antenna design for microwave hepatic ablation using an axisymmetric electromagnetic model. *Biomed Eng Online* February 27:5–15.

Bircan, C., and Barringer, S. 2002. Determination of protein denaturation of muscle foods using the dielectric properties. *J Food Sci* 67:202–05.

Brace, C.L., Laeseke, P.F., van der Weide, D.W. and Lee, F.T. Jr. 2005. Microwave ablation with a triaxial antenna: Results in ex vivo bovine liver. *IEEE Trans Microw Theory Tech* 53:215–20.

Brace, C.L., Laeseke, P.F., Sampson, L.A. et al. 2007a. Microwave ablation with multiple simultaneously powered small-gauge triaxial antennas: Results from an in vivo swine liver model. *Radiology* 244:151–56.

Brace, C.L., Laeseke, P.F., Sampson, L.A. et al. 2007b. Microwave ablation with a single small-gauge triaxial antenna: In vivo porcine liver model. *Radiology* 242:435–40.

Brace, C.L., Laeseke, P.F., Sampson, L.A., van der Weide, D.W., and Lee, F.T. Jr. 2007c. Switched-mode microwave ablation: Less dependence on tissue properties leads to more consistent ablations than phased arrays. *Radiological Society of North America Annual Meeting*, Chicago, IL.

Brace, C.L. 2008. Temperature-dependent dielectric properties of liver tissue measured during thermal ablation: Toward an improved numerical model. *Conf Proc IEEE Eng Med Biol Soc* 1:230–33.

Brace, C.L., Hinsahw, J.L., Laeseke, P.F. et al. 2009a. Pulmonary thermal ablation: Comparison of radiofrequency and microwave devices by using gross pathologic and CT findings in a swine model. *Radiology* 251:705–11.

Brace, C.L., Sampson, L.A., Hinsahw, J.L., Sandhu, N., and Lee, F.T. Jr. 2009b. Radiofrequency ablation: Simultaneous application of multiple electrodes via switching creates larger, more confluent ablations than sequential application in a large animal model. *J Vasc Interv Radiol* 20:118–24.

Brace, C.L. 2010a. Microwave tissue ablation: Biophysics, technology, and applications. *Crit Rev Biomed Eng* 38:65–78.

Brace, C.L., Diaz, T.A., Hinsahw, J.L. et al. 2010b. Tissue contraction caused by radiofrequency and microwave ablation: A laboratory study in liver and lung. *J Vasc Interv Radiol* 21:1280–86.

Burkhardt, J.D., and Natale, A. 2009. New technologies in atrial fibrillation ablation. *Circulation* 120:1533–41.

Cao, H., Vorperian, V.R., Tungjitkusolmun, S., Tsai, J.Z., Haemmerich, D., Choy, Y.B., and Webster, J.G. 2001. Flow effect on lesion formation in RF cardiac catheter ablation. *IEEE T Bio-Med Eng* 48:425–33.

Chen, M.H., Yang, W., Yan, K., Zou, M.W., Solbiati, L., Liu, J.B., and Dai, Y. 2004. Large liver tumors: Protocol for radiofrequency ablation and its clinical application in 110 patients—mathematic model, overlapping mode, and electrode placement process. *Radiology* 232:260–71.

Chin, L., and Sherar, M. 2001. Changes in dielectric properties of ex vivo bovine liver at 915 MHz during heating. *Phys Med Biol* 46:197–211.

Cooper, J., and Gimpelson, R.J. 2004. Summary of safety and effectiveness data from FDA: A valuable source of information on the performance of global endometrial ablation devices. *J Reprod Med* 49:267–73.

Craciunescu, O.I., Das, S.K., McCauley, R.L. et al. 2001. 3D numerical reconstruction of the hyperthermia induced temperature distribution in human sarcomas using DE-MRI measured tissue perfusion: Validation against non-invasive MR temperature measurements. *Int J Hyperthermia* 17:221–39.

Dewhirst, M.W., Viglianti, B.L., Lora-Michiels, M. et al. 2003. Basic principles of thermal dosimetry and thermal thresholds for tissue damage from hyperthermia. *Int J Hyperthermia* 19:267–94.

Diller, K.R., Valvano, J.W., and Pearce, J.A. 2000. *CRC Handbook of Thermal Engineering*, ed. F. Kreith (Boca Raton: CRC Press, 114–215).

Duck, F.A. 1990. *Physical Properties of Tissue: A Comprehensive Reference Book*. Academic Press, London.

Epstein, B.R., and Foster, K.R. 1983. Anisotropy in the dielectric properties of skeletal muscle. *Med Biol Eng Comput* 21:51–55.

Esrick, M.A., and McRae, D.A. 1994. The effect of hyperthermia-induced tissue conductivity changes on electrical impedance temperature mapping. *Phys Med Biol* 39:133–44.

Feldberg, I., and Cronin, N. 1998. A 9.2 GHz microwave applicator for the treatment of menorrhagia. *IEEE MTT-S* 2:755–58.

Foster, K.R., and Schwan, H.P. 1989. Dielectric properties of tissues and biological materials: A critical review. *Crit Rev Biomed Eng* 17:25–104.

Fotopoulou, C., Hee Cho, C., Kraetschell, R. et al. 2010. Regional abdominal hyperthermia combined with systemic chemotherapy for the treatment of patients with ovarian cancer relapse: Results of a pilot study. *Int J Hyperthermia* 26:118–26.

Furse, C.M., and Iskander, M.F. 1989. Three-dimensional electromagnetic power deposition in tumors using interstitial antenna arrays. *IEEE Trans Biomed Eng* 36:977–86.

Gabriel, C., Gabriel, S., and Corthout, E. 1996. The dielectric properties of biological tissues: I. Literature survey. *Phys Med Biol* 41:2231–49.

Gabriel, S., Lau, R.W., and Gabriel, C. 1996a. The dielectric properties of biological tissues: II. Measurements in the frequency range 10 Hz to 20 GHz. *Phys Med Biol* 41:2251–69.

Gabriel, S., Lau, R.W., and Gabriel, C. 1996b. The dielectric properties of biological tissues: III. Parametric models for the dielectric spectrum of tissues. *Phys Med Biol* 41:2271–93.

Geraghty, P.R., Kee, S.T., McFarlane, G., Razavi, M.K., Sze, D.Y., and Dake, M.D. 2003. CT-guided transthoracic needle aspiration biopsy of pulmonary nodules: Needle size and pneumothorax rate. *Radiology* 229:475–81.

Gillams, A. 2008. Tumour ablation: Current role in the liver, kidney, lung and bone. *Cancer Imaging* 8 Suppl A:S1–5.

Gillams, A.R., and Lees, W.R. 2009. Five-year survival in 309 patients with colorectal liver metastases treated with radiofrequency ablation. *Eur Radiol* 19:1206–13.

Gu, Z., Rappaport, C.M., Wang, P.J., and VanderBrink, B.A. 1999. A 2 1/4-turn spiral antenna for catheter cardiac ablation. *IEEE Trans Biomed Eng* 46:1480–82.

Haemmerich, D. 2006a. *Wiley Encyclopedia of Medical Devices and Instrumentation*, ed. J G Webster (New Jersey: John Wiley & Sons, 362–79).

Haemmerich, D., Chachati, L., Wright, A.S., Mahvi, D.M., Lee, F.T. Jr., and Webster, J.G. 2003a. Hepatic radiofrequency ablation with internally cooled probes: Effect of coolant temperature on lesion size. *IEEE Transactions on Biomedical Engineering* 50:493–500.

Haemmerich, D., Staelin, S.T., Tungjitkusolmun, S., Mahvi, D.M., and Webster, J.G. 2003b. In-vivo conductivity of hepatic tumors. *Physiological Measurement* 24:251–60.

Haemmerich, D., Ozkan, O.R., Tsai, J.Z., Staelin, S.T., Tungjitkusolmun, S., Mahvi, D.M., and Webster, J.G. 2002. Changes in electrical resistivity of swine liver after occlusion and postmortem. *Med Biol Eng Comput* 40:29–33.

Haemmerich, D., Schutt, D.J., Wright, A.W., Webster, J.G., and Mahvi, D.M. 2009. Electrical conductivity measurement of excised human metastatic liver tumours before and after thermal ablation. *Physiol Meas* 30:459–66.

Haemmerich, D., and Wood, B.J. 2006b. Hepatic radiofrequency ablation at low frequencies preferentially heats tumour tissue. *Int J Hyperthermia* 22:563–74.

He, N., Wang, W., Ji, Z., Li, C., and Huang, B. 2010. Microwave ablation: An experimental comparative study on internally cooled antenna versus non-internally cooled antenna in liver models. *Acad Radiol.* 17:894–99.

Hines-Peralta, A.U., Pirani, N., Clegg, P., Cronin, N., Ryan, T.P., Liu, Z., and Goldber, S.N. 2006. Microwave ablation: Results with a 2.45-GHz applicator in ex vivo bovine and in vivo porcine liver. *Radiology* 239:94–102.

Hodgson, D.A., Feldberg, I.B., Sharp, N., Cronin, N., Evans, M., and Hirschowitz, L. 1999. Microwave endometrial ablation: Development, clinical trials and outcomes at three years. *Br J Obstet Gynaecol* 106:684–94.

Hope, W.W., Schmelzer, T.M., Newcomb, W.L., Heath, J.J., Lincourt, A.E., Norton, H.J. et al. 2008. Guidelines for power and time variables for microwave ablation in an in vivo porcine kidney. *J Surg Res* 153:263–67.

Huang, S., and Wood, M. 2006. *Catheter Ablation of Cardiac Arrhythmias.* (Philadelphia: Saunders).

Hurter, W., Reinbold, F., and Lorenz, W. 1991. A dipole antenna for interstitial microwave hyperthermia. *IEEE Trans Microwave Theory Tech* 39:1048–54.

Ito, K., Hyodo, M., Shimura, M., and Kasai, H. 1990. Thin applicator having coaxial ring slots for interstitial microwave hyperthermia. *Ant Prop Soc Int Sym* 3:1233–36.

Jemal, A., Siegel, R., Ward, E., Murray, T., Xu, J., and Thun, M.J. 2007. Cancer statistics, 2007. *CA: A cancer journal for clinicians* 57:43–66.

Jones, K.M., Mechling, J.A., Strohbehn, J.W., and Trembly, B.S. 1989. Theoretical and experimental SAR distributions for interstitial dipole antenna arrays used in hyperthermia. *IEEE Trans Microw Theory Tech* 37:1200–09.

Knavel, E., Lubner, M., Hinshaw, J.L., Lee, F.T. Jr., and Brace, C.L. 2010. Thermal tumor ablation: Measurement of force required to remove applicators with and without securing functions. *World Conference on Interventional Oncology*, Philadelphia, PA.

Kuang, M., Lu, M.D., Xie, X.Y., Xu, H.X., Mo, L.Q., Liu, G.J. et al. 2007. Liver cancer: Increased microwave delivery to ablation zone with cooled-shaft antenna—experimental and clinical studies. *Radiology* 242:914–24.

Labonte, S., Blais, A., Legault, S., Ali, H., and Roy, L. 1996. Monopole antennas for microwave catheter ablation. *IEEE Trans Microwave Theory Tech* 44:1832–40.

Laeseke, P.F., Lee, F.T. Jr., Sampson, L.A., van der Weide, D.W., and Brace, C.L. 2006. Multiple-electrode radiofrequency ablation creates confluent areas of necrosis: In vivo porcine liver results. *Radiology* 241:116–24.

Laeseke, P.F., Frey, T.M., Brace, C.L., Sampson, L.A., Winter, T.C. 3rd, Ketzler, J.R., and Lee, F.T. Jr. 2007. Multiple-electrode radiofrequency ablation of hepatic malignancies: Initial clinical experience. *AJR Am J Roentgenol* 188:1485–94.

Laeseke, P.F., Lee, F.T. Jr., Sampson, L.A. et al. 2009. Microwave ablation versus radiofrequency ablation in the kidney: High-power triaxial antennas create larger ablation zones than similarly sized internally cooled electrodes. *J Vasc Interv Radiol* 20:1224–29.

Laeseke, P.F., Sampson, L.A., Lee, F.T. Jr., and Brace, C.L. 2010. Multiple-antenna microwave ablation: Spatially distributing power improves thermal profiles and reduces invasiveness. *Journal of Interventional Oncology* 2:65–72.

Lazebnik, M., Converse, M.C., Booske, J.H., and Hagness, S.C. 2006. Ultrawideband temperature-dependent dielectric properties of animal liver tissue in the microwave frequency range. *Phys Med Biol* 51:1941–55.

Lee, J.M., Han, J.K., Kim, H.C., Kim, S.H., Kim, K.W., Joo, S.M., and Choi, B.I. 2007. Multiple-electrode radiofrequency ablation of in vivo porcine liver: Comparative studies of consecutive monopolar, switching monopolar versus multipolar modes. *Invest Radiol* 42:676–83.

Lin, J.C., and Wang, Y.J. 1987. Interstitial microwave antennas for thermal therapy. *Int J Hyperthermia* 3:37–47.

Lin, J.C., and Wang, Y.J. 1996. The cap-choke catheter antenna for microwave ablation treatment. *IEEE Trans Biomed Eng* 43:657–60.

Liu, P., and Rappaport, C.M. 1996. A helical microwave antenna for welding plaque during balloon angioplasty. *IEEE Trans Microw Theory Tech* 44:1819–31.

Livraghi, T., Meloni, F., Di Stasi, M., Rolle, E., Solbiati, L., Tinelli, C., and Rossi, S. 2008. Sustained complete response and complications rates after radiofrequency ablation of very early hepatocellular carcinoma in cirrhosis: Is resection still the treatment of choice? *Hepatology* 47:82–9.

Longo, I., Gentili, G., Cerretelli, M., and Tosoratti, N. 2003. A coaxial antenna with miniaturized choke for minimally invasive interstitial heating. *IEEE Trans Microwave Theory Tech* 50:82–88.

Lu, Y., Li, B., Xu, J., and Yu, J. 1992. Dielectric properties of human glioma and surrounding tissue. *Int J Hyperthermia* 8:755–60.

Lu, D.S., Raman, S.S., Vodopich, D.J., Wang, M., Sayre, J., and Lassman, C. 2002. Effect of vessel size on creation of hepatic radiofrequency lesions in pigs: Assessment of the "heat sink" effect. *AJR Am J Roentgenol* 178:47–51.

Lubner, M.G., Brace, C.L., Hinsahw, J.L., and Lee, F.T. Jr. 2010. Microwave tumor ablation: Mechanism of action, clinical results, and devices. *J Vasc Interv Radiol* 21(8 Suppl):S192–203.

Lyons, B.E., Britt, R.H., and Strohbehn, J.W. 1984. Localized hyperthermia in the treatment of malignant brain tumors using an interstitial microwave antenna array. *IEEE Trans Biomed Eng* 31:53–62.

Markovic, J.N., and Shortell, C.K. 2009. Update on radiofrequency ablation. *Perspect Vasc Surg Endovasc Ther* 21: 82–90.

Ng, S.K., Ainsworth, P., Plunkett, A., Haigh, A.D., Gibson, A.A., Parkinson, G., and Jacobs, G. 2008. Determination of added fat in meat paste using microwave and millimetre wave techniques. *Meat Science* 79:748–56.

Oshima, F., Yamakado, K., Nakatsuka, A., Takaki, H., Makita, M., and Takeda, K. 2008. Simultaneous microwave ablation using multiple antennas in explanted bovine livers: Relationship between ablative zone and antenna. *Radiat Med* 26:408–14.

Panescu, D., Whayne, J.G., Fleischman, S.D., Mirotznik, M.S., Swanson, D.K., and Webster, J.G. 1995. Three-dimensional finite element analysis of current density and temperature distributions during radio-frequency ablation. *IEEE Transactions on Biomedical Engineering* 42: 879–90.

Pennes, H.H. 1948. Analysis of tissue and arterial blood temperatures in the resting human forearm. *J Appl Physiol* 1:93–122.

Petersen, H.H., Chen, X., Pietersen, A., Svendsen, J.H., and Haunso, S. 1999. Lesion dimensions during temperature-controlled radiofrequency catheter ablation of left ventricular porcine myocardium: Impact of ablation site, electrode size, and convective cooling. *Circulation* 99: 319–25.

Pop, M., Molckovsky, A., Chin, L., Kolios, M.C., Jewett, M.A., and Sherar M.D. 2003. Changes in dielectric properties at 460 kHz of kidney and fat during heating: Importance for radiofrequency thermal therapy. *Phys Med Biol* 48:2509–25.

Rawnsley, R.J., Roemer, R.B., and Dutton, A.W. 1994. The simulation of discrete vessel effects in experimental hyperthermia. *J Biomech Eng* 116:256–62.

Rhim, H., and Dodd, G.D. 3rd. 1999. Radiofrequency thermal ablation of liver tumors. *J Clin Ultrasound* 27:221–9.

Saito, K., Yoshimura, H., Ito, K., Aoyagi, Y., and Horita, H. 2004. Clinical trials of interstitial microwave hyperthermia by use of coaxial-slot antenna with two slots. *IEEE Trans Microw Theory Tech* 52:1987–91.

Sasaki, A., Kai, S., Iwashita, Y., Hirano, S., Ohta, M., and Kitano, S. 2005. Microsatellite distribution and indication for locoregional therapy in small hepatocellular carcinoma. *Cancer* 103:299–306.

Sato, M., Watanabe, Y., Ueda, S., Iseki, S., Abe, Y., Sato, N. et al. 1996. Microwave coagulation therapy for hepatocellular carcinoma. *Gastroenterology* 110:1507–14.

Schepps, J.L., and Foster, K.R. 1980. The UHF and microwave dielectric properties of normal and tumour tissues: Variation in dielectric properties with tissue water content. *Phys Med Biol* 25:1149–59.

Schramm, W., Yang, D., Wood, B.J., Rattay, F., and Haemmerich, D. 2007. Contribution of direct heating, thermal conduction and perfusion during radiofrequency and microwave ablation. *Open Biomedical Engineering Journal* 1:47–52.

Schwan, H.P., and Li, K. 1955. Measurements of materials with high dielectric constant and conductivity at ultrahigh frequencies. *AIEE Trans Comm Electronics* 74:603–07.

Schwan, H.P., and Foster, K.R. 1977. Microwave dielectric properties of tissue. Some comments on the rotational mobility of tissue water. *Biophys J* 17:193–97.

Shock, S.A., Meredith, K., Warner, T.F., Sampson, L.A., Wright, A.S., Winter, T.C. et al. 2004. Microwave ablation with loop antenna: In vivo porcine liver model. *Radiology* 231:143–49.

Siklody, C.H., Minners, J., Allgeier, M., Allgeier, H.J., Jander, N., Keyl, C., Weber, R., Schiebeling-Romer, J., Kalusche, D., and Arentz, T. 2010. Pressure-guided cryoballoon isolation of the pulmonary veins for the treatment of paroxysmal atrial fibrillation. *J Cardiovasc Electrophysiol* 21:120–5.

Simon, C.J., Dupuy, D.E., Iannitti, D.A., Lu, D.S., Yu, N.C., and Aswad, B.I. 2006. Intraoperative triple antenna hepatic microwave ablation. *AJR Am J Roentgenol* 187:W333–40.

Simon, C.J., Dupuy, D.E., DiPetrillo, T.A., Safran, H.P., Grieco, C.A., Ng, T., and Mayo-Smith, W.W. 2007. Pulmonary radiofrequency ablation: Long-term safety and efficacy in 153 patients. *Radiology* 243:268–75.

Stauffer, P.R., Rossetto, F., Prakash, M., Neuman, D.G., and Lee, T. 2003. Phantom and animal tissues for modeling the electrical properties of human liver. *Int J Hyperthermia* 19:89–101.

Steel, M.C., and Sheppard, R.J. 1986. Dielectric properties of lens tissue at microwave frequencies. *Bioelectromagnetics* 7:73–81.

Strickland, A.D., Clegg, P.J., Cronin, N.J., Swift, B., Festing, M., and West, K.P. 2002. Experimental study of large-volume microwave ablation in the liver. *Br J Surg* 89:1003–07.

Sun, Y., Wang, Y., Ni, X., Gao, Y., Shao, Q., Liu, L., and Liang, P. 2009. Comparison of ablation zone between 915- and 2,450-MHz cooled-shaft microwave antenna: Results in in vivo porcine livers. *AJR Am J Roentgenol* 192:511–14.

Surowiec, A.J., Stuchly, S.S., Barr, J.B., and Swarup, A. 1988. Dielectric properties of breast carcinoma and the surrounding tissues. *Ieee T Bio-Med Eng* 35: 257–63.

Swarup, A., Stuchly, S.S., and Surowiec, A. 1991. Dielectric properties of mouse MCA1 fibrosarcoma at different stages of development. *Bioelectromagnetics* 12:1–8.

Tompkins, D.T., Vanderby, R., Klein, S.A., Beckman, W.A., Steeves, R.A., Frye, D.M., and Paliwal, B.R. 1994. Temperature-dependent versus constant-rate blood perfusion modelling in ferromagnetic thermoseed hyperthermia: Results with a model of the human prostate. *Int J Hyperthermia* 10:517–36.

Trembly, B.S. 1985. The effects of driving frequency and antenna length on power deposition within a microwave antenna array used for hyperthermia. *IEEE Trans Biomed Eng* 32:152–57.

Trembly, B.S., Wilson, A.H., Sullivan, M.J., Stein, A.D., Wong, T.Z., and Strohbehn, J.W. 1986. Control of the SAR pattern within an interstitial microwave array through variation of antenna driving phase. *IEEE Trans Microw Theory Tech* 34:568–71.

Tsai, J.Z., Will, J.A., Hubbard-Van Stelle, S., Cao, H., Tungjitkusolmun, S., Choy, Y.B., Haemmerich, D., Vorperian, V.R., and Webster, J.G. 2002. In-vivo measurement of swine myocardial resistivity. *Ieee T Bio-Med Eng* 49:472–83.

Tse, H.F., Liao, S., Siu, C.W., Yuan, L., Nicholls, J., Leung, G. et al. 2009. Determinants of lesion dimensions during transcatheter microwave ablation. *Pacing Clin Electrophysiol* 32:201–08.

Tungjitkusolmun, S. 2000. Finite element modeling of radiofrequency cardiac and hepatic ablation. Phd thesis (Madison: University of Wisconsin).

Tungjitkusolmun, S., Vorperian, V.R., Bhavaraju, N., Cao, H., Tsai, J.Z., and Webster, J.G. 2001. Guidelines for predicting lesion size at common endocardial locations during radio-frequency ablation. *Ieee T Bio-Med Eng* 48:194–201.

Turner, P.F. 1984. Regional hyperthermia with an annular phased array. *IEEE Trans Biomed Eng* 31:106–14.

Turner, P.F. 1986. Interstitial equal-phased arrays for EM hyperthermia. *IEEE Trans Microw Theory Tech* 34:572–78.

Turner, P.F., Tumeh, A., and Schaefermeyer, T. 1989. BSD-2000 approach for deep local and regional hyperthermia: Physics and technology. *Strahlenther Onkol* 165:738–41.

van Rhoon, G.C., Rietveld, P.J., and van der Zee, J. 1998. A 433 MHz Lucite cone waveguide applicator for superficial hyperthermia. *Int J Hyperthermia* 14:13–27.

Wall, M.S., Deng, X.H., Torzilli, P.A., Doty, S.B., O'Brien, S.J., and Warren RF. 1999. Thermal modification of collagen. *J Shoulder Elbow Surg* 8:339–44.

Wilber, D.J., Packer, D.L., and Stevenson, W.G. 2007. *Catheter Ablation of Cardiac Arrhythmias: Basic Concepts and Clinical Applications* (Malden, MA: Blackwell Publishing).

Wong, T.Z., Jonsson, E., Hoopes, P.J., Trembly, B.S., Heaney, J.A., Douple, E.B., and Coughlin, C.T. 1993. A coaxial microwave applicator for transurethral hyperthermia of the prostate. *Prostate* 22:125–38.

Wright, A.S., Lee, F.T. Jr., and Mahvi, D.M. 2003. Hepatic microwave ablation with multiple antennae results in synergistically larger zones of coagulation necrosis. *Ann Surg Oncol* 10:275–83.

Yeh, M.M., Trembly, B.S., Douple, E.B., Ryan, T.P., Hoopes, P.J., Jonsson, E., and Heaney, J.A. 1994. Theoretical and experimental analysis of air cooling for intracavitary microwave hyperthermia applicators. *IEEE Trans Biomed Eng* 41:874–82.

Yu, N.C, Lu, D.S, Raman, S.S Dupuy, D.E, Simon, C.J, Lassman, C. et al. 2006. Hepatocellular carcinoma: Microwave ablation with multiple straight and loop antenna clusters—pilot comparison with pathologic findings. *Radiology* 239:269–75.

Yu, N.C., Raman, S.S., Kim, Y.J., Lassman, C., Chang, X., and Lu, D.S. 2008. Microwave liver ablation: Influence of hepatic vein size on heat-sink effect in a porcine model. *J Vasc Interv Radiol* 19:1087–92.

10

Clinical External Ultrasonic Treatment Devices

Lili Chen
Fox Chase Cancer Center

Faqi Li
Chongqing Medical University

Feng Wu
*Chongqing Medical University
and University of Oxford*

Eduardo G. Moros
*H. Lee Moffitt Cancer Center
and Research Institute*

10.1 Introduction

High intensity focused ultrasound (HIFU) or focused ultrasound surgery (FUS) is a completely noninvasive treatment modality. It has long been known to offer "trackless lesioning," and it has been identified as an "ideal surgical tool" for several decades. However, only after it was integrated with modern imaging methods has HIFU become a clinical reality. A high quality imaging technique was needed to provide visualization and localization of target tissue, monitoring of tissue changes during treatment, and outcome assessment after treatment.

FUS integrated with a modern imaging system has been used for treating both benign and malignant diseases, such as uterine fibroids, benign prostatic hyperplasia, prostate cancer, breast cancer, brain tumors, and palliative treatment of bone metastases to relieve pain (Gianfelice et al. 2008, Funaki et al. 2009, Furusawa et al. 2006, 2007, Liberman et al. 2009, McDannold et al. 2010, Zhang and Wang 2010). Its future potential is vast both in surgical and other interventions. For example, pulsed mode FUS has recently been used to enhance local drug delivery in animal models (Yuh et al. 2005, Rapoport et al. 2009, Nelson et al. 2002, Khaibullina et al. 2008, Chen et al. 2010). The imaging modalities that have been integrated for treatment guidance include both ultrasound (US) imaging and magnetic resonance (MR) imaging. The advantages and disadvantages of each imaging modality are discussed elsewhere in this book. MR guided focused ultrasound surgery (MRgFUS) has been used clinically for the treatment of uterine fibroids and is being studied in clinical trials for the treatment of patients with bone metastases, breast tumors, and brain tumors, as well as prostate cancer. MRgFUS is also being investigated for the treatment of liver tumors, stroke, epilepsy, movement disorders, and pancreatic and kidney tumors. Ultrasound guided FUS is currently being used clinically in the treatment of kidney, liver, and pancreatic cancers in several countries (Wu et al. 2004, Ritchie et al. 2010, Zhang et al. 2010). Several manufacturers are currently in the process of developing commercial clinical FUS systems. This chapter focuses on the introduction of clinical externally applied ultrasound devices that have been approved by the Food and Drug Administration (FDA) in the United States and/or have the *Conformité Européenne* (CE) mark of approval for clinical application, which consist of the ExAblate system (InSightec-TxSonics, Haifa, Israel, and Dallas, Texas) with MR guidance and the Haifu system (Chongqing Haifu [HIFU] Technology Co., Ltd, China) with ultrasound guidance. The Philips Sonavelle MR-HIFU system that is expected to be the next clinical system to be cleared by the FDA/CE is also presented briefly.

10.2 InSightec Systems

10.2.1 ExAblate 2000/ExAblate®One

The ExAblate 2000 provides noninvasive surgery using MR guidance. It is mainly designed for treating uterine fibroids and bone metastases, but can also be used for treating other sites such as breast cancer, liver cancer, and pancreatic cancer. Moreover, recent reports demonstrate that the system can also be used for small animal research studies on the enhancement of local drug delivery using pulsed focused ultrasound (Chen et al. 2010, Mu et al. 2012).

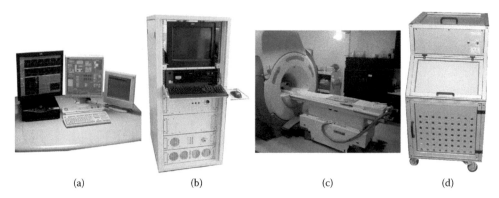

(a)   (b)   (c)   (d)

FIGURE 10.1 The main components of the ExAbalte 2000 system: (a) the operator console, (b) the equipment cabinet, (c) the patient treatment table, and (d) the chiller.

ExAblate 2000 can be integrated as an add-on to a 1.5 or 3.0 Tesla MR Scanner from GE Medical Systems. The ExAblate 2000 received the CE mark in 2002 and FDA clearance in 2004 for the treatment of symptomatic uterine fibroids. In 2007, it received the CE mark for pain palliation of bone metastases, and in 2010 for adenomyosis. This device has already treated thousands of patients around the world and is currently being investigated in clinical trials for painful bone metastases (phase III), breast cancer (phase I), liver cancer (phase I), and prostate cancer (phase I).

10.2.1.1 Main Components

The main components of the system include the patient treatment table containing the ultrasonic array and integrated with the MR scanner (GE 1.5 T or 3.0 T), the operator console, the equipment cabinet, and the chiller (Figure 10.1a–d). The safety devices are also installed in the clinical focused ultrasound system.

10.2.1.1.1 Patient Treatment Table

The patient treatment table is a modified GE SIGNA MRI table. It is detachable and can be docked to a GE 1.5 T or 3.0 T MR scanner in the same way that the standard MR table docks; it is connected with a single quick-connect socket. A phased array spherically curved transducer with 208 elements (Figure 10.2) is housed in a sealed degassed water tank in the patient treatment table and is connected to an electronic motioning and positioning system controlled by a computer. The transducer can be moved in the X and Y directions (along and horizontally across the scanner's longitudinal axis, respectively) but not in the Z direction (vertically up and down); the transducer can also be tilted. The geometric focal length of the array is 16 cm (without

electronic steering). However, since the transducer is a phased array, the phases of the radiofrequency signals to each array element can be adjusted to steer the acoustic focal length from 6 cm to 22 cm from the transducer's center. The size of the focal zone (lesions) may vary depending on the size and depth of the volume being treated, ranging from 1 mm to 10 mm in diameter and from 8 mm to 45 mm in length. The operating frequency is around 1 MHz, ranging from 0.95 to 1.35 MHz. For bone treatments the frequency used is 1 MHz.

10.2.1.1.2 Equipment Cabinet

The equipment cabinet is located in an adjacent control room. It consists of a main power switch and the electrical components that drive the transducer's positioning system and the ultrasound phased array.

10.2.1.1.3 Operator Console

The console is located in the control room next to the GE SIGNA workstation. It includes a flat panel display, keyboard, mouse, and stop-sonication button. It controls the entire treatment process including treatment planning, the treatment table position, the transducer's position and tilt angle, sonication power and time, and MRI monitoring and assessment of the treatment. The latter includes both conventional MR imaging sequences and MR temperature mapping estimation sequences.

10.2.1.1.4 Chiller

A cooling system rack, which includes the chiller and associated electronics, is also installed in the equipment room. It is used for cooling the water in the tank and also for breast treatments to prevent skin burns produced by ultrasound absorption.

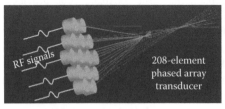

FIGURE 10.2 The phased array transducer. (Courtesy of InsighTec, Inc.)

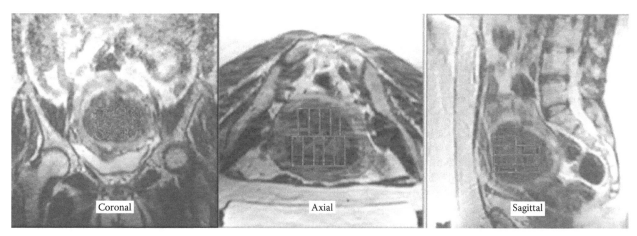

FIGURE 10.3 An example of the treatment planning screen. (Courtesy of InsighTec, Inc.)

10.2.1.1.5 Safety Devices

Safety devices include a sonication lamp, which is installed in a prominent position on the MRI scanner and lights up during treatment sonications, and three stop-sonication switches, which instantaneously stop the delivery of ultrasound energy to the patient. One is held by the patient who is instructed to squeeze it in case of sudden discomfort or emergency; another is mounted on the scanner for use by staff in the treatment room; and the third switch is located in the control room, next to the operator.

10.2.1.2 MR Imaging Guidance

Magnetic resonance imaging (MRI) is used for guidance before, during, and posttreatment with either a 1.5 T or 3.0 T MR scanner (GE Medical Systems, Milwaukee, Wisconsin). In FUS treatment, MRI provides a tool for precise visualization of the tumor (or treatment target), other organs, critical structures, and the ultrasound beam path. MR images enable the physician and physicist to contour the treatment target and plan the treatment in three dimensions. The beam path for each focus position and sonication can be visualized slice by slice to verify that it is not passing through any undesirable region such as nerve bundles, major blood vessels, bone, air/gas pockets, and others.

Successful thermal ablation of each focal point is the end point of a FUS treatment. FDA/CE approved MR thermometry using a proton resonant frequency shift method (Peters et al. 1998, Graham et al. 1999) provides the ability to get immediate feedback. Real time (with approximately a 3-second delay) thermal maps generated by the ExAblate and calculated thermal doses (Sapareto and Dewey 1984) superimposed on the MR anatomical images enable the treatment team to monitor heating of the target spot and determine immediately if the sonication was effective. It also allows the operator to adjust the FUS treatment parameters for each treatment spot to reach the designated temperature. Thus MR provides the physician with a wealth of anatomical, geometrical, and thermal information to ensure safety and efficacy.

10.2.1.3 Treatment Planning

To identify the treatment volume, the ExAblate uses conventional diagnostic MR images taken at the beginning of treatment. The treatment planning system allows the physician and physicist to define the target, plan energy parameters for each sonication, verify safe beam path, and edit location and type of individual treatment spots. The physician delineates the tumor and defines safe treatment pass-zones that will avoid energy passage through sensitive tissue (e.g., scar tissue) or critical structures as mentioned before. The physician selects an application-specific treatment protocol that determines the main attributes of the planned treatment. The system then computes a treatment plan, composed of 20–100 sonication (focus position) points that cover the specified target. Figure 10.3 shows an example of the treatment planning for a patient with a uterine fibroid.

10.2.1.4 Treatment Delivery

Prior to focused ultrasound therapy, the effective focal spot should be verified with lower energy (sublethal or sub-ablative sonication) and adjusted using MR thermometry. For each focal spot, the physician is able to select treatment parameters including the acoustic power and delivery duration. The spots are treated in sequence automatically determined by the ExAblate system for total treatment time and safety optimization. In addition, any spot can be picked up and treated manually by the physician. During a treatment sequence, the robotic system positions the transducer below the target point and delivers the planned energy. The ExAblate directs the MR to continuously acquire thermal images that include the point being treated and the surrounding anatomy. The system provides real-time monitoring of energy deposition and the essential feedback of where the energy was delivered and the temperature reached. This quantitative feedback allows the operator to control and adjust the treatment parameters in real time to ensure that the targeted tumor is fully treated and the surrounding tissue is spared. The workstation displays thermal images and computes and overlays thermal doses on the treated regions.

10.2.1.5 Quality Assurance

Quality Assurance (QA) of the various treatment systems (hardware, software, and patient safety features) must be done before the system is first used and periodically thereafter (Gorny et al. 2006, 2009). Specialized training on the use of both the MR scanner and the focused ultrasound system is critical for ensuring proper performance and safe operation. Briefly, the QA procedures include: (1) checking the water level in the cooling system reservoir; (2) checking the treatment table docking to the MR scanner; (3) connecting power to the treatment table (the water hose also needs to be connected to the treatment table for breast treatments); (4) testing of the MR coil and MR images; (5) calibrating the transducer position; (6) ensuring that the images shown on the treatment workstation are the correct images; (7) calibrating the effective focal spot in two directions (coronal and axial or coronal and sagittal) using MR thermometry; and (8) testing all the safety devices alluded to in Section 10.2.1.1.5.

10.2.2 ExAblate 2100/ExAblate®OR

The first MRgFUS treatment of a uterine fibroid was performed in 2001 using the ExAblate 2000 system. Since then, thousands of patients have been treated in more than 60 different hospitals globally. The knowledge accumulated from the clinical use of this new treatment technology has been implemented by vendors into software and hardware updates. It has also led to clinical guidelines for improved clinical outcomes and maintaining of a high level of safety. The updated ExAblate system was named the ExAblate®OR (Figure 10.4a). The major advances of the ExAblate 2100 system over the ExAblate 2000 system are: (1) the table contains additional devices for the integration of both the conformal bone transducer (Figure 10.4b) and the endorectal FUS transducer; (2) the transducer in the water tank can be moved in the vertical direction; and (3) various improvements for uterine

fibroid treatment, including optimized treatment planning, new types of spots, and electronically customized transducer apertures to allow precise energy delivery to tissues adjacent to bowels or to other acoustic obstacles.

10.2.2.1 Conformal Bone Transducer

The conformal bone transducer (with straps and patient support accessories) is an option integrable into the ExAblate 2100 system. The transducer is a phased array transducer with 1000 elements operating at a frequency of 550 kHz. The phase (and amplitude) to each transmitting element is digitally controlled and monitored during sonications. Therapeutic bursts of 1000 J to 2000 J of energy are typically delivered to coagulate the targeted bone tissue interface area using multiple sonications. The conformal transducer positioning and strapping features enable flexible patient positioning and access to any desired treatment site in the human body.

The built-in water circulation system includes active cooling to maintain the skin temperature at about 20°C. The water-impermeable membrane provides acoustic coupling from the phased array to the skin. The system monitors and controls the water temperature, protects against leaks, and prevents bubble formation. The conformal bone transducer comes with an integrated tracking coil, which allows it to locate the position and angulations of the transducer while inside the MR bore.

10.2.2.2 Endorectal FUS Transducer

An endorectal FUS transducer is also an available option. It can be integrated into the modified ExAblate 2100 system for prostate cancer clinical trials. Patients with biopsy proven low risk prostate cancers were treated using the MRgFUS ExAblate system in Italy, Russia, Singapore, and India. Figure 10.5 shows an

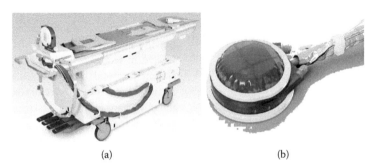

(a)   (b)

FIGURE 10.4   (a) The ExAblate 2100 UF V2 system, and (b) the conformal bone transducer. (Courtesy of InsighTec, Inc.)

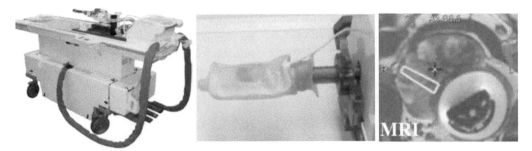

FIGURE 10.5   The Endorectal FUS transducer. (Courtesy of InsighTec, Inc.)

endorectal FUS transducer and a MR image of the transducer for prostate cancer treatment.

The other major components, such as the system control electronics, the operator's workstation, and the MR scanner, are standard, similar to the ExAblate 2000 described previously.

10.2.3 ExAblate 4000 (ExAblate Neuro Transcranial [tc]) MRgFUS System

The ExAblate 4000, later named the ExAblate Neuro, is a transcranial (tc) MRgFUS system (InSightec Ltd., Tirat Carmel, Israel). The system is currently under evaluation for clinical safety and efficacy in functional neurosurgery and tumor ablation, stroke, and targeted drug delivery. This system is designed for treating diseases in the brain only.

A hemispherical, helmet-like, multi-element (4000 elements) phased array transducer (Figure 10.6) enables focal targeting of brain tissue through the intact cranium. The system is integrated with a standard GE 1.5 T or 3 T MRI system using a detachable treatment table. In the treatment room, the patient lies on the table with his or her head immobilized in a stereotactic frame, with the helmet-like transducer positioned around his or her head. A sealed water system with an active cooling and degassing capacity maintains the skull and skin surface at a comfortably low temperature to protect intervening skin and skull from overheating.

A series of conventional MRI scans are displayed on the ExAblate workstation and analyzed by the physician/physicist to determine the targeted regions. Preoperative CT and interoperative MR scans can be fused for reconstructing a model of the skull and brain anatomy used in treatment planning. The treatment is based on multiple sonications that cover the targeted volume. Sublethal focal spots can be used to confirm target localization accuracy and patient comfort prior to ultrasound treatment via the measurement of temperature using MR. During energy delivery to each spot, thermal images were used for real-time feedback control of the treatment, allowing the physician/physicist to adjust parameters accordingly to reach the designated treatment effect. Posttreatment contrast imaging can be used to confirm the treatment effect.

The system has been used for treating patients with brain tumors through the intact skull (without a prior craniotomy procedure), and it has been demonstrated that the treatment is safe without skull heating (which might cause damage to the brain surface). Based on these results, the FDA approved a continuation of the study with emphasis on efficacy (McDonald et al. 2010).

10.3 Haifu Systems

Research and development of the Haifu systems began in Chongqing, China, in 1988. After extensive *ex vivo* and *in vivo* animal studies with the improvement of key techniques, a prototype of the ultrasound-guided extracorporeal HIFU system was successfully developed for clinical trials in Chongqing Medical University, China, in 1997. From December 1997 to October 2001, a total of 1038 patients with solid tumors were treated with HIFU in 10 Chinese hospitals through clinical trials. The tumors included primary and metastatic liver cancers, malignant bone tumors, breast cancers, soft tissue sarcomas, kidney cancers, pancreatic cancers, abdominal and pelvic malignant tumors, uterine fibroids, and benign breast tumors. In addition, the US-guided HIFU system (Model JC, Focused Ultrasound Tumor Therapeutic System, Chongqing Haifu [HIFU] Technology Company, Ltd., Chongqing, China) was approved by China's SFDA (the national agency similar to the FDA in the United States) in 1999, followed by CE mark in Europe in 2005. A second HIFU system, as shown in Figure 10.7, the JC200 Focused Ultrasound Tumor Therapeutic System, was approved by both China's SFDA and CE in 2007. To date, both US-guided extracorporeal HIFU systems have treated large numbers of patients in 15 countries, including China, Japan, Korea, the United Kingdom, Italy, Spain, Russia, Romania, Ukraine, and Saudi Arabia. The clinical results are promising and support the safety, feasibility, and efficacy of HIFU therapy for solid tumors.

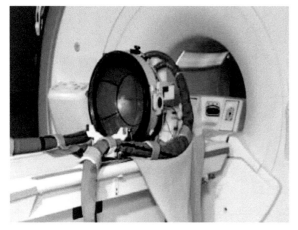

FIGURE 10.6 The ExAblate Neuro transcranial (tc) MRgFUS system. (Courtesy of InsighTec, Inc.)

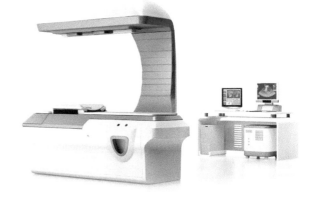

FIGURE 10.7 Model JC200 Focused Ultrasound Tumor Therapy System (Chongqing Haifu).

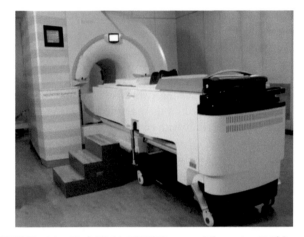

FIGURE 10.8   Model JM 2.5C HIFU system (Chongqing Haifu).

In 2005, Chongqing Haifu started an R&D project in collaboration with Siemens Medical Systems to develop an MRI-guided extracorporeal HIFU System (JM 2.5C HIFU System) shown in Figure 10.8. The JM 2.5C system was fully integrated with a 1.5 T MRI system (Symphony, Siemens, Germany). With the successful development of the device, clinical trials in China have recently been completed for the treatment of uterine fibroids. The results show that the JM 2.5C system is safe, reliable, and effective for the treatment of patients with large uterine fibroids. Using the JM 2.5C system, a higher ratio of complete ablation and a relief of symptoms related to fibroids have been observed in patients during a 2-year follow-up, with a significant shrinkage of the treated fibroids. These findings were reported at the First International Clinical Symposium of Therapeutic Ultrasound in October 2009. More than 120 clinicians and scientists from 19 countries attended this symposium, shared their clinical experiences in HIFU therapy, and discussed the potential of HIFU as a

minimally invasive therapeutic technique for future cancer therapy. *Ultrasonics Sonochemistry*, one of the international journals in ultrasound, reported on the findings at this symposium that patients with malignant tumors who were not suitable for surgery and were treated by ultrasonic noninvasive methods have survived for ten years. China is the only country in the world that has provided such data owing to its long-term research and clinical applications in this field

10.3.1 Main Components

The Ultrasound-Guided Extracorporeal HIFU system consists of three main parts: the treatment platform, the central operator console, and the auxiliary system (Figure 10.9). Each functional module is described next.

The *treatment platform* includes a power source, an integrated transducer, and a six-dimensional motion device for transducer movement. The power source provides high-frequency electrical signals that drive the HIFU transducer. The integrated transducer includes a 1.0 MHz HIFU transducer and a 3.5–5.0 MHz B-mode ultrasound imaging probe, which is embedded in the center of the HIFU transducer for real-time monitoring of the HIFU procedure, including localization of the target, treatment planning, observation of targeted tissue response in terms of tissue grayscale changes, and control of ultrasound energy delivery. The integrated transducer is inside a rubber water tank filled with circulating degassed and temperature-controlled water. The transducer can move in six directions, including translational motion in the x, y, and z axes, rotation of the imaging probe for 3D ultrasound imaging of the targeted area, and rotation of the integrated treatment transducer along the x and y axes.

The *central operator console* includes an ultrasonic (imaging) monitoring system and a treatment control system.

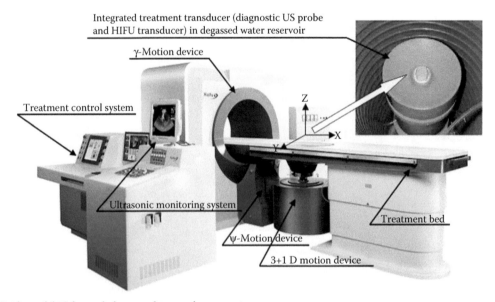

FIGURE 10.9   Haifu model JC focused ultrasound tumor therapy system.

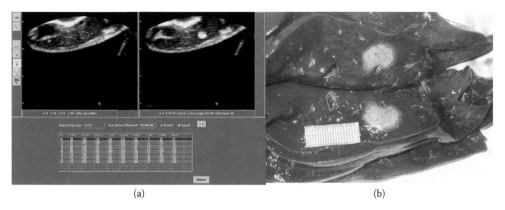

(a)   (b)

FIGURE 10.10 (a) Comparison of grayscale value of targeted liver tissue before (left) and after (right) HIFU exposure; (b) the range of hyperechoic area is good correspondence to the maximal area of coagulation necrosis.

The ultrasonic monitoring system is mainly used for locating a targeted tumor, monitoring treated-tissue response, and controlling the deposition of ultrasound energy during the HIFU procedure. The treatment control system is composed of a treatment planning system (TPS), digital evaluation of the grayscale changes within the targeted tissues before and after HIFU exposures, treatment data storage, and data analysis.

The *auxiliary system* includes a water management device for degassing and temperature controlling and a safe protection device. The water management device provides degassed water as a coupling medium allowing ultrasonic propagation from the transducer to the skin. Degassing is used to avoid skin burn that is caused by air bubbles from the water gathered on the skin surface (bubbles stop/scatter sound waves). The oxygen content is kept at ≤3 ppm. As sound attenuation is very low in water, the loss of ultrasound energy is not significant while the sound waves travel through the water tank. The safe protection devices ensure equipment stability and normal operation.

In addition, a telemedicine unit is embedded in the Haifu system in order to let users share their treatment database with other users or remote experts. It can also provide real-time consultation and diagnostic services for cancer patients, and timely repair and maintenance remotely.

10.3.2 US Imaging Guidance

As a noninvasive thermal ablation tool for tumor therapy, HIFU is performed under image guidance. Either diagnostic ultrasound or MRI is clinically used to guide the HIFU treatment procedure (depending on model). Both imaging modalities have similar functions for real-time guidance, such as positioning a target area, monitoring either temperature rise or grayscale changes in the target, evaluating therapeutic response, and controlling energy delivery. Benefits of a US-guided HIFU system include low cost, real-time imaging, and sound channel visibility. In addition, there are no image artifact problems when imaging moving internal organs (e.g., liver, kidney, and pancreas) due to respiration. Advantages of an MRI-guided HIFU

system include superior image quality and anatomic detail and real-time temperature measurement.

For US-guided HIFU systems, the evaluation of therapeutic response is based on echo changes in the targeted tissues before and immediately after HIFU exposures. Compared to US imaging before HIFU, there is a strong echo from the targeted tissue immediately after a HIFU exposure. With the extension of observation time, the echo gradually weakens, and the range of the strong echo significantly decreases. Using proprietary software, the strength of the echo is automatically analyzed in the targeted area before and after HIFU exposures to derive the changes in grayscale values. Coagulation necrosis is predicted based on these changes. There is good agreement between the range of the hyperechoic area and the maximal area of coagulation necrosis (Figure 10.10). During the HIFU procedure, the grayscale changes are helpful in determining whether coagulation necrosis occurs in the treated tissue. If the change cannot reach a certain value after one HIFU exposure in terms of grayscale differences, HIFU exposures are repeated until the grayscale values reach the level of coagulation necrosis. Although it is still unclear which mechanisms are involved in the development of the hyperechoic area induced by HIFU exposures, acoustic cavitation and tissue dehydration/boiling (evaporation) may be major reasons responsible for hyperechogenicity.

10.3.3 Quality Assurance

The quality assurance of HIFU systems is very important to ensure the safety of patients and the effectiveness of the treatments. China has been a leader in developing and establishing quality standards for clinical HIFU products. In September 2005, the national standard GB/T 19890-2005 "The measurement of acoustic power and field characteristics in high intensity focused ultrasound (HIFU)" was published by the Chinese government, making it the first of its kind in the world. In fact, it was published as a technical report by the International Electrotechnical Commission (IEC/TR 62649) in April 2010. China's SFDA issued the industry standards for HIFU therapy

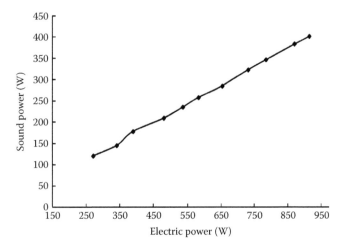

FIGURE 10.11 The electric-sound power curve diagram of an HIFU transducer.

systems (Standard YY0592-2005) in December 2005. It is the first national standard for HIFU products for clinical applications, which has been useful in regulating the manufacture and clinical use of HIFU systems.

HIFU QA includes the following aspects. The acoustic power from the HIFU transducer is measured using the radiation force method. The relationship between the measured ultrasonic power and the corresponding electric power must be known and stable (Figure 10.11). All data are stored in the central operator console platform, which are used as a reference during the HIFU procedure for real-time monitoring of the electric power delivered to the transducer, and for feedback control of the acoustic power and calibration procedures.

A calibrated PVDF hydrophone is used to obtain technical parameters of the HIFU acoustic field, such as sound frequency, the full width half maximum (FWHM) of the pressure field, maximum side-lobe level, and focal length. These data assist operators in the selection of an appropriate HIFU transducer for treatment. In a US-guided HIFU system, the focal length measured by the PVDF hydrophone in a water tank has to be verified *ex vivo* in both a tissue-mimic phantom and in a biological tissue sample such as ox liver. Immediately after a HIFU exposure, ultrasound imaging shows a hyperechoic change *ex vivo* in the focus, indicative of the place of coagulation necrosis in the tissue-mimic phantom and biological tissues. This is very useful in determining the actual focal length of the HIFU transducer in living tissues. The movement accuracy of the integrated transducer measured along the x, y, and z axes should be ±1 mm.

10.3.4 Treatment Planning

Comprehensive considerations were given to the design of the treatment planning system (TPS), especially with regard to biological factors for each patient that may affect the safety, effectiveness, and efficiency of HIFU therapy. The safety is assessed by whether there could be excessive energy deposition before or beyond the targeted area, the sensitivity of normal tissue around the target, and whether a patient could tolerate the entire HIFU procedure in terms of general conditions. The effectiveness depends on the range of the targeted area and whether there should be sufficient energy deposition for thermal ablation. The therapeutic efficiency is based on both the effectiveness and safety of HIFU therapy, which is determined by the rate of energy deposition in the targeted area.

Medical history, physical examination, histopathological diagnosis, CT/MRI imaging, B-mode, and color Doppler ultrasound and tumor/site characteristics are collected for treatment planning. Tumor characteristics include histological type, location, size, depth, blood supply, functional status, motion caused by respiration, the structures in the beam path, and surrounding organs. The patient's general condition, treatment status before HIFU, and the purpose of the HIFU treatment are also considered. After comprehensive considerations, a HIFU treatment plan is established for each patient. It is composed of the choice of the transducer, the 3D conformal scanning strategy, the selection of the therapeutic dose to be delivered to the tumor, and whether the patient will receive adjuvant cancer therapy such as chemotherapy after HIFU. Based on the treatment plan, the HIFU procedure is subsequently performed. However, the treatment plan can be revised during the HIFU procedure based on real-time imaging provided by the diagnostic ultrasound capability. After the HIFU treatment, radiological examinations are performed to assess the therapeutic effects. This is significant in identifying whether coagulation necrosis has occurred in the treated tumor, and whether a second HIFU session needs to be applied as shown in the HIFU treatment planning flow chart (Figure 10.12).

It is fundamental to use a 3D conformal treatment strategy. As the HIFU cigar-shaped focal volume is small, approximately 20 mm in length and 2 mm in width, the size of the coagulation necrosis caused by a single sonication is accordingly small in biological tissues. Therefore, it is essential to move the HIFU focus within the targeted tumor until the entire tumor is ablated. There are two methods to produce a line-shaped ablation. One is to scan the focus continuously while sonicating. The other is to use multiple separate sonications in line, and the location of each sonication is overlapped in order to avoid living tumor remaining between consecutive exposures. By scanning the focus in successive sweeps from the deep/distal to shallow/proximal regions of a tumor, the targeted regions on each slice can be completely ablated. This process is repeated slice by slice to achieve complete tumor ablation, as shown in Figure 10.13a. During focused ultrasound ablation of each slice, real-time US images obtained before and after each exposure are compared to check for echogenic changes, which indicate the extent of coagulation necrosis. Figure 10.13b shows that using HIFU 3D ablation scanning, a volume of coagulation necrosis is achieved from spot-, line- and slice-shaped tissue destruction in *ex vivo* ox liver. The margin between the treated and untreated liver is clear, and there is no living tissue within the treated area.

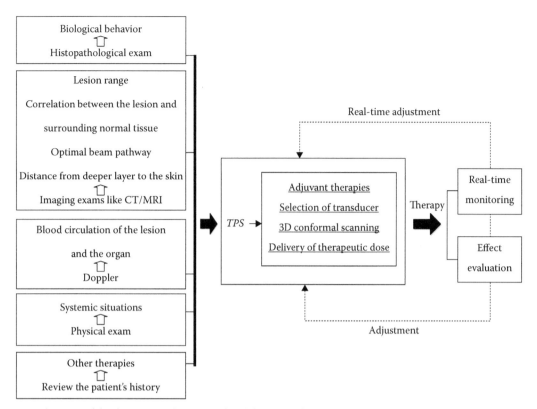

FIGURE 10.12   Development of the therapeutic plan system (TPS) for HIFU clinical application.

According to the surgical standards for tumor resection, it is important to obtain safe margins. The extent of the treated volume may vary because of different therapeutic purposes. Additionally, the margin size is also dependent on the histological type of tumor. Therefore, a 3D conformal HIFU treatment plan includes not only ablation of the tumor but also surrounding normal tissues in order to achieve the proper margin.

10.3.5 Treatment Dose Delivery

The measurement of the dose required for successful HIFU radiation is a controversial and evolving topic. It is difficult to quantify the biological effects of HIFU on living tissues by using ultrasonic energy deposited in the targeted tissue and the exposure time, or by theoretical calculations. Based on a large amount

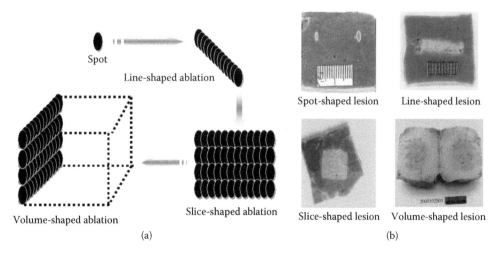

FIGURE 10.13   (a) Schematic diagram shows the conformal therapeutic plan for HIFU therapy, whose 3D treatment strategy is composed of spot, line, slice, and volume ablations; (b) after HIFU exposure, the spot-shaped, line-shaped, slice-shaped, and volume-shaped coagulation necrosis in *ex vivo* bovine liver.

of preclinical and clinical results, the guideline for HIFU dose delivery was established for the Haifu system. There are expert databases of the dosimetry and dose delivery guidelines in the central operator console, which were collected from a large volume of therapeutic data from patients treated. Moreover, ultrasound dose must also be controlled using ultrasound imaging as explained previously. The guidelines of dose delivery with the Haifu systems are outlined here:

10.3.5.1 HIFU Biological Focal Region (BFR)

One HIFU exposure or sonication of sufficient power produces a single spot of coagulation necrosis in soft tissue. In addition to heat, nonlinear effects such as cavitation are involved in the necrosis formation. Fry (1993) considered that the practicality of focused ultrasound depends on the tissue properties between the ultrasonic source and the focal region, and the energy accumulation in the focal region. Therefore, the treatment of a given target requires a special correction method depending on the various organs and tissues involved. The variation of the coagulation necrosis zone in targeted tissues must be determined as a requirement of clinical treatment procedures. In this context, Wang et al. (2003) termed the single coagulation necrosis induced by a single HIFU exposure as a HIFU Biological Focal Region (BFR), which is related to the Acoustic Focal Region (AFR) produced in a water tank. Detailed studies show that BFR is based on AFR and is a function of the acoustic intensity, exposure time, treatment depth (distance from the focus to the skin surface), tissue structure, and functional status. Mathematically, BRF can therefore be expressed as follows:

$$BFR = f(AFR, I, t, D, T_s, T_f)$$

where, I is the acoustic intensity, t is the exposure time, D is the depth of treatment, T_s is the structure of the biological tissue, and T_f is the functional state of the tissue.

10.3.5.2 The Energy Efficiency Factor (Energy Effect Factor, EEF)

The BFR reveals a general principle that there are multiple factors influencing the formation of coagulation necrosis. The aim of three-dimensional conformal treatment planning is to cover the volume of an entire tumor with coagulation necrosis. The concept of the energy efficiency factor, or EEF, is put forward for the determination of the ultrasound energy needed to be delivered to a target volume to cause coagulation necrosis. In other words, the dose-effect relationship is quantified by the EEF as follows,

$$EEF = {\eta Pt}/{V} \text{ (J/mm3)}$$

where η is the energy conversion coefficient of the HIFU transducer ($\eta \sim 0.7$), P is the acoustic power, t is the total time of treatment, and V is the volume of the target. Results show that using the same transducer, EEF is affected by several factors including

the depth of the targeted tumor (from the skin to the tumor), the structure and functional state of the biological tissues, and the focus scanning mode.

10.4 Philips Sonavelle MR-HIFU System

The systems presented previously are clinical in the sense that they have received FDA clearance in the United States or the CE mark in Europe for specific clinical indications in humans. In this section we briefly introduce a new system developed by Philips, which is expected to enter the clinical market in the near future.

The Sonalleve MR-HIFU system (Philips Healthcare, Vantaa, Finland) was designed with the main objective of improving treatment efficiency (Figure 10.14). This objective is achieved by a volumetric ablation technique reminiscent of focus scanning techniques used earlier in non-ablative scanned focused ultrasound hyperthermia treatments (Moros et al. 1988). Volumetric ablation is accomplished by scanning the focus in concentric circles rapidly and outwardly, and repeating the pattern until the target is ablated. This approach produces homogeneously ablated volumes in relative short times (Köhler et al. 2009). The same scanning technique is also being investigated for enhanced drug delivery via hyperthermia. Nominal target volumes (or cells) available are 4, 8, 12, 14, and 16 mm in diameter. The nominal length of the cylindrical volume varies with

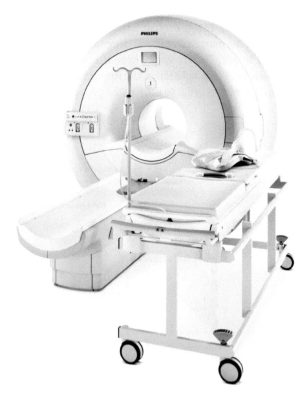

FIGURE 10.14 The Philips Sonavelle MR-HIFU system. (Courtesy of Philips Healthcare.)

cell diameter from 10 to 40 mm. The transducer (phased array) has a focal length of 140 mm and employs 256 elements operating at frequencies between 1.2 and 1.5 MHz. Another important feature of the Sonalleve system is its automatic temperature feedback control system, which modulates power to the acoustic sources while the focus is being scanned based on MR-measured temperature feedback (Enholm et al. 2010). Experiments have shown that under- and overtreatment can be prevented with this control system. Moreover, a recently completed multicenter prospective clinical trial showed that volumetric MR-HIFU is safe and feasible in the treatment of uterine fibroids (Voogt et al. 2012). Finally, this system is also being used by investigators to perform animal studies (Hijnen et al. 2012).

References

Chen L, Mu Z, Hachem P, Ma C-M, and Pollack A (2010). MR-guided focused ultrasound: Enhancement of intratumoral uptake of [3H]-docetaxel *in vivo*. *Phys. Med. Biol.* 55(24):7399–7410.

Enholm JK, Köhler MO, Quesson B, Mougenot C, Moonen CTW, and Sokka SD (2010). Improved volumetric MR-HIFU ablation by robust binary feedback control. *IEEE Trans Biomed Eng.* 57(1):103–13.

Fry FJ (1993). Intense focused ultrasound in medicine. Some practical guiding physical principles from sound source to focal site in tissue. *Eur Urol.* 23(Suppl 1):2–7.

Funaki K, Fukunishi H, and Sawada K (2009). Clinical outcomes of magnetic resonance-guided focused ultrasound surgery for uterine myomas: 24-month follow-up. *Ultrasound Obstet Gynecol.* 34:584–9.

Furusawa H, Namba K, Thomasen S, Akiyama F, Bendet A, Tanaka C, Yasuda Y, and Nakahara H (2006). Magnetic resonance-guided focused ultrasound surgery of breast cancer: Reliability and effectiveness. *J Am Coll Surg.* 203:54–63.

Furusawa H, Namba K, Nakahara H, Tanaka C, Yasuda Y, Hirabara E, Imahariyama M, and Komaki K (2007). The evolving non-surgical ablation of breast cancer: MR guided focused ultrasound (MRgFUS). *Breast Cancer* 14:55–8.

Gianfelice D, Gupta C, Kucharczyk W, Bret P, Havill D, and Clemons M (2008). Palliative treatment of painful bone metastases with MR imaging—guided focused ultrasound. *Radiology* 249: 355–63.

Gorny KR, Hangiandreou NJ, Hesley GK, Gostout BS, McGee KP, and Felmlee JP (2006). MR guided focused ultrasound: Technical acceptance measures for a clinical system. *Phys Med Biol.* 51(12):3155–73.

Gorny KR, Hangiandreou NJ, Ward HA, Hesley GK, Brown DL, and Felmlee JP (2009). The utility of pelvic coil SNR testing in the quality assurance of a clinical MRgFUS system. *Phys Med Biol.* 54(7):N83–91.

Graham SJ, Chen L, Leitch M, Peter RD, Bronskill MJ, Foster FS, Henkelman RM, and Plewes DB (1999). Quantifying tissue damage due to focused ultrasound heating observed by MRI. *Magn Res Med.* 41:321–8.

Hijnen NM, Heijman E, Köhler MO, Ylihautala M, Ehnholm GJ, Simonetti AW, and Grüll H (2012). Tumour hyperthermia and ablation in rats using a clinical MR-HIFU system equipped with a dedicated small animal set-up. *Int J Hyperthermia* 28(2):141–55.

Khaibullina A, Jang BS, Sun H, Le N, Yu S, Frenkel V, Carrasquillo JA, Pastan I, Li KC, and Paik CH (2008). Pulsed high-intensity focused ultrasound enhances uptake of radiolabeled monoclonal antibody to human epidermoid tumor in nude mice. *J Nucl Med.* 49:295–302.

Köhler MO, Mougenot C, Quesson B, Enholm J, Le Bail B, Laurent C, Moonen CTW, and Ehnholm GJ (2009). Volumetric HIFU ablation under 3D guidance of rapid MRI thermometry. *Medical Physics* 36(8):3521–35.

Liberman B, Gianfelice D, Inbar Y, Beck A, Rabin T, Shabshin N, Chander G, Hengst S, Pfeffer R, Chechick A, Hanannel A, Dogadkin O, and Catane R (2009). Pain palliation in patients with bone metastases using MR-guided focused ultrasound surgery: A multicenter study. *Ann Surg Oncol.* 16:140–6.

McDannold N, Clement GT, Black P, Jolesz F, and Hynynen K (2010). Transcranial magnetic resonance imaging-guided focused ultrasound surgery of brain tumors: Initial findings in 3 patients. *Neurosurgery* 66:323–32; discussion 332.

Moros EG, Roemer R, and Hynynen K (1988). Simulations of scanned focused ultrasound hyperthermia: The effect of scanning speed and pattern on the temperature fluctuations at the focal depth. *IEEE Transactions on Ultrasonics, Ferroelectrics and Frequency Control* 35:552–60.

Mu Z, Ma C-M, Chen X, Cvetkovic D, Pollack A, and Chen L (2012). MR guided pulsed high intensity focused ultrasound enhancement of docetaxel combined with radiotherapy for prostate cancer treatment. *Phys Med Biol* 57(2):535–45.

Nelson JL, Roeder BL, Carmen JC et al. (2002). Ultrasonically activated chemotherapeutic drug delivery in a rat model. *Cancer Research* 62:7280–3.

Peters RD, Hinks RS, and Henkelman RM (1998). *Ex vivo* tissue-type invariability in proton-resonance frequency shift MR thermometry. *Magn Res Med* 40:454–9.

Rapoport NY, Nam KH, Gao Z, and Kennedy A (2009). Application of ultrasound for targeted nanotherapy of malignant tumors. *Acoustical Physics* 55:594–601.

Ritchie RW, Leslie T, Phillips R, Wu F, Illing R, ter Haar G, Protheroe A, and Cranston D (2010). Extracorporeal high intensity focused ultrasound for renal tumours: A 3-year follow-up. *BJU Int.* 106:1004–9.

Sapareto SA and Dewey WC (1984). Thermal dose determination in cancer therapy. *Int.J.Radiat.Oncol Biol Phys.* 10:787–800.

Voogt MJ, Trillaud H, Kim YS, Mali WP, Barkhausen J, Bartels LW, Deckers R, Frulio N, Rhim H, Lim HK, Eckey T, Nieminen HJ, Mougenot C, Keserci B, Soini J, Vaara T, Köhler MO, Sokka S, and van den Bosch MA (2012). Volumetric feedback ablation of uterine fibroids using magnetic resonance-guided high intensity focused ultrasound therapy. *Eur Radiol.* 22(2):411–7.

Wang ZB, Bai J, Li FQ, Du YH, Wen S, Hu K, Xu GH, Ma P, Yin NG, Chen WZ, Wu F, and Feng R (2003). Study of a "biological focal region" of high-intensity focused ultrasound. *Ultrasound Med Biol.* 29(5):749–54.

Wu F, Wang ZB, Chen WZ, Zhu H, Bai J, Zou JZ, Li KQ, Jin CB, Xie FL, and Su HB (2004). Extracorporeal high intensity focused ultrasound ablation in the treatment of patients with large hepatocellular carcinoma. *Annals of Surgical Oncology* 11:1061–9.

Yuh EL, Shulman SG, Mehta SA, Xie J, Chen L, Frenkel V, Bednarski MD, and Li KC (2005). Delivery of systemic chemotherapeutic agent to tumors by using focused ultrasound: Study in a murine model. *Radiology* 234:431–7.

Zhang L and Wang ZB (2010). High-intensity focused ultrasound tumor ablation: Review of ten years of clinical experience. *Front Med China* 4:294–302.

Endocavity and Catheter-Based Ultrasound Devices

Chris J. Diederich
*University of California,
San Francisco*

11.1 Introduction

In contrast to extracorporeal systems, endocavity and catheter-based ultrasound devices have been developed for delivering hyperthermia and thermal ablation from placement within the body. These minimally invasive ultrasound techniques can be used to apply hyperthermia as an adjunct to radiation therapy and/or chemotherapy, or as a surgical alternative for tumor or tissue ablation. For these applications, the heating source is positioned directly within or adjacent to a deep target volume via placement within a body cavity, lumen, or by direct insertion. Even though these technologies are more invasive than extracorporeal systems, many of these approaches may be preferable for sites where energy localization from external devices is difficult or where localization of all power and energy propagation within the target tissue is critical.

There are many physical properties of ultrasound that make it a favorable energy modality for applications in this setting [1,2]: small wavelengths in the 1–15 MHz range combined with an ability to shape transducers or phased arrays yield small energy radiating platforms capable of precise and predictable spatial control of power deposition; favorable energy penetration and focusing capabilities allow therapeutic heating at distances away from the applicator, and opportunity to heat larger volumes quickly while protecting intervening tissues. Thus, in clinical practice, many of these ultrasound devices can selectively direct or conform the heating volume to a specified target area while protecting or avoiding other tissues. This enhanced spatial localization and energy penetration provides ultrasound with a significant advantage over the radio-frequency (RF) currents, microwave (MW),

laser, and cryotherapy technology as currently applied for interstitial and intracavitary hyperthermia and tumor ablation therapies [3–8]. Furthermore, real-time ultrasound [9–13] and MR imaging techniques [14–17] can be employed for many of these ultrasound applicator configurations to monitor thermal therapy and verify treatment. These advantages and significant potential of ultrasound technology for endocavity and catheter-based applications of thermal therapy are further highlighted in the following sections, which provide a contemporary and brief review of some technologies that are commercially available for clinical use or under development and recently implemented in clinical studies.

11.2 Transrectal Devices for Prostate Thermal Therapy

11.2.1 Transrectal Ultrasound Hyperthermia

Intracavitary ultrasound applicators specifically for delivering prostate hyperthermia from within the rectum have been developed and used successfully in clinical treatments [18,19]. Original applicators consist of a linear segmented array (4–8) of sectioned PZT tubes (180° sections, 10 mm long, 1.5 cm OD), each under separate power control and operating between 1 MHz and 2 MHz [20]. The transducers are mounted on a plastic structure that facilitates support and placement in the rectum, as well as temperature-regulated water flow within an expandable bolus. The cylindrical ultrasound transducers are sectored to shape and direct the heating field in an ~120° arc to the target volume. The heating energy is emitted radially from the length of each transducer segment and the power applied along the length of the applicator is

FIGURE 11.1 Transrectal ultrasound applicator for prostate hyperthermia demonstrating multi-sectored transducer segments that give angular and longitudinal control of heating directed to the prostate, and the water-cooled coupling balloon. (Photos courtesy of Dr. Kullervo Hynynen, Sunnybrook Health Sciences Centre, and Dr. Mark Hurwitz, Dana-Farber/Brigham & Women's Hospital, Harvard Medical School.)

adjustable for tailoring the heating distribution to fit the intended target region extending from apex to base. Studies have indicated that these applicators could therapeutically heat tissues 3–4 cm deep from the rectal cavity wall, which is sufficient to treat most prostate glands, while proper cooling of the bolus will maintain the rectal mucosa at subtherapeutic temperatures. Devices of this design scheme have been implemented in a phase I feasibility and toxicity trial [19], which evaluated transrectal ultrasound hyperthermia given with concurrent standard external beam irradiation in the treatment of locally advanced adenocarcinoma of the prostate. Therapeutic temperatures ranging between 40.6°C and 43.2°C were reported for a total of 14 patients. Advanced versions of this applicator have added four sectors on each tubular section (Figure 11.1), for 16 channels total, thus adding additional control of the heating in the angular expanse as well as longitudinal control [18]. MR compatible versions of this device and control algorithms for feedback control have been investigated and demonstrated that MR directed hyperthermia with this approach is feasible [21,22], including demonstration of heat-directed gene delivery [23] to selected regions of the prostate. Transrectal ultrasound hyperthermia applicators demonstrate improved heating penetration and spatial control of power deposition compared to capabilities of transrectal microwave techniques, providing an improved technique to heat more of the gland without complications to rectal tissue.

Delivering effective and therapeutic hyperthermia to the whole prostate gland is achievable with this investigational device, with no increase of rectal toxicity [24,25]. Recent analysis of clinical data by Hurwitz et al. has shown that endorectal hyperthermia delivered concurrently with external beam radiation yields improvement in survival over other therapies [26]. These applicators are ideally suited for applying conventional hyperthermia to the whole prostate gland, and may be useful for radiation or chemotherapy plus heat, and thermal targeted drug delivery directly to the prostate.

11.2.2 Transrectal HIFU for Prostate Therapies

11.2.2.1 Ultrasound Guided Systems

There are two commercial systems available for transrectal high intensity focused ultrasound (HIFU) under ultrasound (US) guidance for thermal ablation of prostate tissue: the Sonoblate 500 (Misonix, U.S.) and the Ablatherm (EDAP TMS, Lyon,

France) [27–29]. These transrectal HIFU systems utilize sharply focused ultrasound transducers that produce small intense focal patterns capable of producing selective or well-localized high-temperature thermal damage within the prostate while avoiding nontargeted surrounding tissues such as rectum and the neuro-vascular bundles (NVB).

The Sonoblate system consists of a mechanically scanned fixed focus HIFU applicator integrated with B-mode imaging capabilities (Figure 11.2a–c). The imaging component, referred to as split beam technology [30], yields transverse and longitudinal images for treatment setup and monitoring during treatment. The HIFU applicator consists of a 30 mm long × 22 mm wide curved transducer operating at 4 MHz with a 25–45 mm variable focal depth selected *a priori* based upon size and shape of the prostate. Typically 3–4 s high intensity sonications or periods of applied power produce ~ $3 \times 3 \times 10$–12 mm^3 coagulated tissue zones per shot, with a 6–12 s delay between shots. The Ablatherm endorectal applicator (Figure 11.2d–f) is similar, with a focused therapy transducer operating at 3 MHz with fixed focus at 40 mm depth that is robotically positioned and coupled with a 7.5 MHz imaging array. The Ablatherm produces coagulation zones of ~1.7 mm × 19–26 mm adjustable length for each shot. These short exposures to high temperatures generate lethal thermal doses that generate a well-defined zone of thermal coagulation and necrosis. The endorectal applicator on each of these devices is water-cooled within an expandable balloon to ~20°C to protect the rectum from excess thermal exposure as well as couple the ultrasound energy and adjust the depth of focus.

The ultrasound imaging is used for accurate anatomical positioning, tissue targeting, and real-time monitoring during treatment (Figure 11.2c). Based upon an image-based treatment plan, the power levels and positions of required sonication points and subsequent thermal lesions are mechanically stepped in time to thermally coagulate a larger contiguous treatment volume. Each of these systems utilizes different treatment strategies [29]: the Sonoblate 500 treats three coronal layers, from anterior to posterior; the Ablatherm targets four to six volume boundaries in axial slices extending from apex to base. During the treatment session, US imaging can be used to qualitatively assess the treatment progression and provide feedback as to generation of a thermal lesion based upon changes in backscatter or attenuation [13].

The Sonoblate was initially applied to treatment of benign prostatic hypertrophy (BPH) and later utilized for treating

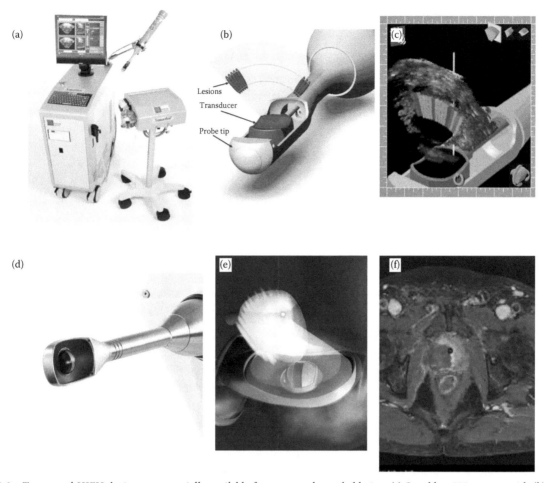

FIGURE 11.2 Transrectal HIFU devices commercially available for prostate thermal ablation: (a) Sonoblate 500 system, with (b) transducer motion and sequential thermal lesion formation through target regions (Koch et al. 2007 [30]) and (c) real-time US monitoring during procedure (Courtesy of Dr. Naren Sangvhi); (d) Ablatherm system with integrated imaging and therapy transducer, (e) transducer rotation and sequential lesion formation within axial slice, and (f) typical MR contrast image after treatment demonstrating complete and contiguous ablation of target zones through summation of smaller focal lesions. (Courtesy of EDAP TMS France, and reprinted with permission from Crouzet, S. et al., *Int J Hyperthermia*, 26, 8, 2010.)

cancer [31]. The Sonoblate has been used for treating BPH [32–35], where the target zone is defined as the anterior-lateral periurethral tissue between the bladder neck and verumontanum, and on some later studies has included the proximal 5 mm of the bladder neck. Both systems have shown strong clinical efficacy for treating localized prostate cancer, either primary, recurrent, or salvage therapy [29,36–40], and can target whole gland or regions of localized disease. Overall treatment durations range to ~2 hours for large prostate glands or target regions, while precisely avoiding damage to rectum, external sphincter, and NVB. Typical follow-up image post-treatment of T1-contrast enhanced MRI demonstrating complete and contiguous ablation zone defined by summation of the individual focal ablations is depicted in Figure 11.2f adapted from Crouzet [27].

11.2.2.2 MR-Guided Endorectal Phased Array

Phased array applicators have been investigated for transrectal delivery of HIFU [41–44] as a means to provide for faster dynamic electrical scanning (no applicator movement) and increased flexibility in focal shape, beam patterns, and positioning. These include MRI compatible intracavitary applicators that facilitate simultaneous MRI temperature imaging for treatment monitoring and control [45, 46].

Currently, the ExAblate Prostate System (InSightec LTD, Tirat Carmel, Israel) is the only commercially sponsored MRg phased array applicator available for endorectal treatment of prostate cancer. The transducer consists of 23 mm wide × 40 mm long piezocomposite array with 990 individual elements operating at 2.3 MHz (Figure 11.3). The transducer assembly can be translated and rotated within an endorectal housing, which is encased in an expandable balloon with water cooling to 10–15°C. The system can be integrated to the patient table and control system of a GE 1.5T or 3.0T MR imaging scanner. Standard MR imaging provides anatomic detail and geometric orientation for accurate treatment planning and guidance in the treatment position. MR temperature imaging [14,16,47] provides an accurate means

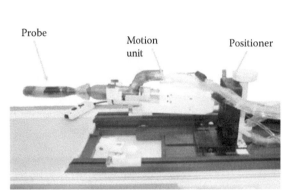
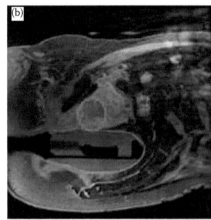
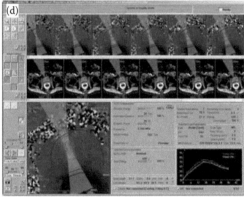
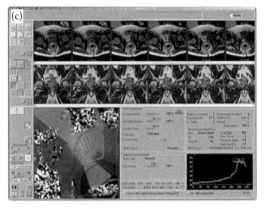

FIGURE 11.3 ExAblate Prostate System by InSightec for MR guided ablation of prostate cancer: (a) endorectal phased array applicator integrated with GE MRI system; (b) sagittal image of endorectal applicator positioned in rectum adjacent to prostate and treated region highlighted in the circle; (c, d) accurate spatial and temporal temperature distributions and thermal dose mapped in real time and correlated to treatment plan and target region. (Courtesy of Christopher Wai Sam Cheng, MD, Urology Department, Singapore General Hospital, Singapore, and Jin Wei Kwek, MD, Radiology Department, National Cancer Center, Singapore.)

of real-time temperature monitoring and feedback control, and running time thermal dose calculation can be used to verify lethal thermal exposures within treatment regions. Additional treatment verification immediately post-procedure can be obtained by T1-contrast enhanced image studies to identify perfusion deficits associated with treated areas. Based upon a patient specific treatment plan, regions in each axial slice of the prostate as indicated region of target are sonicated and monitored using temperature and thermal dose as feedback, sweeping out and conforming to target region (Figure 11.3c, d). Since accurate temperature monitoring and feedback control is provided, complex beam forming and electronic scanning for a series of "macro shot" sequences can be applied to sonicate over larger volumes and without cooling intervals to produce larger contiguous lesions more quickly, compared to current configuration of US guided systems.

11.3 Transurethral Devices for Prostate Therapy

Transurethral ultrasound applicators in various configurations (tubular, planar, and curvilinear transducer arrays; stationary or rotating) have been developed and evaluated *in vivo* and demonstrated significant potential for delivering controlled thermal ablation to the prostate gland for possible treatment of BPH and cancer [48–54], and can be integrated with MR temperature monitoring for treatment control and assessment [55, 56]. The general configuration of these devices currently under investigation consists of catheters or rigid devices placed within the urethra with 4–6.4 mm diameter delivery catheters and 2.5 mm to 4 mm wide by 2–3 cm long transducer arrays that would be positioned directly within the prostatic urethra. Techniques and devices have also been investigated for dual frequency operation to control penetration depth and treatment duration. The most extensive development of a transurethral applicator, including recent clinical evaluation in a pilot study, has been performed specifically for a planar array configuration; this ultrasound applicator couples rotation of the transducer array with MR temperature feedback to precisely sweep out a conformal thermal ablation zone along predefined boundaries [56,57]. The applicator configuration consisted of 4–5 transducer sections (3.5 mm × 5 mm long segments, operating at 8–9 MHz), mounted within a rigid plastic and brass 6.4 mm delivery applicator (Figure 11.4). Temperature regulated cooling water is circulated

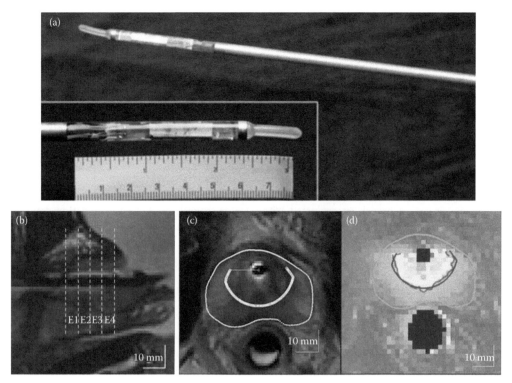

FIGURE 11.4 Transurethral ultrasound applicator with rotating multi-element planar transducer for MR guided prostate ablation: (a) distal end of applicator with transducer array; (b) sagittal image and alignment of MRI slices for monitoring temperature delivery and providing closed loop treatment control during rotation to conform to (c) preset outer boundary; (d) cumulative temperature and thermal dose map correlating to completed coagulation within target zone obtained in clinical pilot study. (Reprinted with permission from Chopra R. et al. *Int J Hyperthermia,* 26, 8, 2010.)

with the applicator and within an ancillary endorectal cooling device to thermally protect the rectum and urethral mucosa. The planar applicator without rotation can generate thermal lesions ~3 mm wide and extending 5–20 mm deep from the urethra. During treatment rotation, speed and applied power are both automatically controlled with an adaptive feedback control loop based upon the MR thermometry and target boundary. Once positioned within the prostate, the transducer assembly can be rotated (on average 15°/min) to sweep the heating zone (54–55°C outer contour line) and conform the coagulation to a larger target volume, emanating from the urethra outward toward the capsule boundary. The rotation rate is driven by an MRI compatible piezoceramic motor under computer control. Treatment modeling and canine experiments indicate that monitoring temperature with multiple slices through the target and automatic feedback control can maintain better than 1–3 mm accuracy in ablation zones [58,59]. A clinical feasibility study has recently been undertaken using this device, limited to two transducer elements active and one 10 mm MR monitoring slice, to target 30% of the prostate gland prior to prostatectomy. The results have indicated that transurethral ultrasound under MR treatment control is feasible and has potential to be a precise method to ablate prostate tissue [56,57].

11.4 Interstitial and Intraluminal Devices

11.4.1 Interstitial with Tubular Sources

Interstitial ultrasound applicators using tubular transducer arrays have been developed for applying hyperthermia in conjunction with brachytherapy and investigated for high-temperature thermal therapy alone [60–65]. As shown in Figure 11.5a, devices consist of a linear array of small tubular transducers, each wired for independent power control, which are inserted within a plastic implant catheter. These closed-ended plastic catheters are inserted directly into the target region during a surgical procedure, and can be either single or multiple catheter implants depending upon the size of the target region. Water flow within the catheter couples the ultrasound energy and extends penetration of the maximum temperature away from the catheter surface. The ultrasound energy emanating from each transducer section, typically 6.5–9 MHz, is collimated within the borders of each segment so that the axial length of the therapeutic temperature zone remains well defined by the number of active elements over a large range of treatment duration and applied power levels [63, 66]. The relative power levels to each transducer section can be adjusted during the treatment to tailor the heating zone along the applicator length and accommodate for dynamic changes in blood flow and real-time heating pattern control. The angular or rotational heating pattern of these arrays can be controlled during fabrication by scoring the transducer surface to isolate activate sectors (e.g., 90°, 180°, 270°, or 360°) to produce directional or angularly shaped heating patterns. The orientation of these directional applicators within an implant catheter can be used to target a treatment zone while protecting critical normal tissue. Multi-applicator implants of interstitial ultrasound applicators can be used to produce contiguous zones of therapeutic temperatures [62,67] between applicators with separation distances of 2–3 cm, while maintaining protection in nontargeted areas. These catheter-cooled devices suitable for percutaneous insertion have been evaluated with transducer diameters between 1.2 mm and 1.5 mm and outer catheter diameters between 2.1 mm (14 gauge) and 2.4 mm (13 gauge), respectively, with the latter being the most common configuration [63,68].

Interstitial ultrasound hyperthermia integrated with HDR brachytherapy for the treatment of locally advanced cervical and prostate cancer has been performed using the 13 gauge multi-transducer applicators [69] in a clinical pilot study. In this setting, the hyperthermia is delivered either immediately preceding or after the HDR radiation treatment using the same catheters. Additional catheters are often used for temperature measurement with multi-junction thermocouple probes to monitor and provide treatment control. Applicator configurations typically used for cervix and prostate treatment include transducer diameters of 1.5 mm within 13 gauge catheter, 10–15 mm transducer lengths, 1–4 active transducer segments, with 180° or 360° directional patterns. Given an implant pattern typical of HDR brachytherapy for these sites, the length of heating and directivity within each catheter is tailored *a priori* to best fit the clinical target volume. A 3D optimization-based treatment planning platform can determine best treatment configuration plus the initial starting power levels to each transducer segment [70]. The insertion depth and rotation angle of the applicator within the plastic catheter can be adjusted by sliding and rotating the device within the sealing hemostasis valve (Figure 11.5a). As mentioned before, the enhanced radial penetration of ultrasound allows larger applicator separation (2–3 cm) of heating devices, and directional applicators can be selected to either protect nontargeted tissues (e.g., bladder, rectum) or preferentially target eccentric tumor volumes. Examples of a typical patient setup and temperature distributions achieved are shown in Figure 11.5b for the treatment of prostate cancer using three applicators placed in the posterior lower portion of the gland; 180° directional applicators are placed and aimed anterior toward the hyperthermia target volume and away from the rectum, ensuring sparing of the thermally sensitive rectal wall. A second example is demonstrated in Figure 11.5c, for preferentially targeting a necrotic area of a cervical tumor. Spatial control and penetration characteristics allow tissue protection and thorough therapeutic temperatures within the target and protection. This spatial control and penetration is not possible with other interstitial hyperthermia modalities such as MW and RF [8].

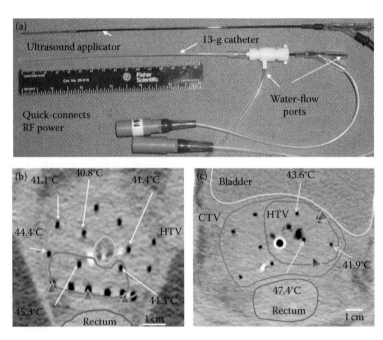

FIGURE 11.5 Catheter-based or interstitial ultrasound applicator based upon arrays of tubular transducers and clinical implementation for hyperthermia integrated with HDR: (a) multi-transducer applicator with 180° heating pattern and control of heating profile along length, within plastic implant catheter; (b) example prostate treatment with three directional heating applicators directing energy away from rectum toward target zone, with measured therapeutic temperatures; (c) example treatment of cervical cancer, applicators are within periphery and aligned to aim into the target zone, also shown with measured temperatures.

The general advantages of these devices noted for hyperthermia are also applicable for percutaneous thermal ablation whereby high temperature alone is used to destroy the tumor or target region. Demonstration of capabilities include substantial size thermal lesions *in vivo* [63,66,68,71] up to 21–25 mm *radial* distance from the applicator, within 5 min treatment times, while maintaining axial and angular control of lesion shape [72]. Furthermore, these devices are MRI compatible, and ability to monitor temperature elevation in real time using MRTI has been demonstrated *in vivo* [71,73]. Although not applied in clinical study as yet for thermal ablation, there are significant advantages such as dynamic spatial control in angle and length, penetration, and compatibility with MRTI that are currently being explored.

11.4.2 Endocervical Devices with Tubular Sources

An endocervical ultrasound applicator has been designed specifically for delivering hyperthermia to the uterine cervix and surrounding target regions [74]. The desired treatment region is typically 2–3 cm in length along the cervix and 2–4 cm diameter with lateral extension typical. These endocavity or intrauterine applicators (Figure 11.6a) are designed to be placed within the HDR brachytherapy tandem catheter (6 mm diameter) that is typically placed within the uterine cervix either through a ring applicator or a vaginal cylinder. Multiple tubular transducer segments (2–3 transducers, 3.5 mm diameter × 10 mm length each) can be activated to tailor heating to target regions at the cervix. Water cooling is applied for coupling and improved penetration. In contrast to the smaller interstitial devices, each of these larger tubes can have multiple sectors that can be independently activated (dual 180° activation typical). For devices using tubular transducers with multiple sectors, there are ~20–30° dead zones without acoustic energy output between each of the active sectors. For treatment of the cervical cancer, the bladder and rectum often confine the anterior and posterior treatment borders close to the cervix. The active portions can be directed laterally into the parametrium and power adjusted to extend penetration, while the acoustic dead zones between active sectors can be used to afford less heating to the thermally sensitive bladder and rectum. Studies have demonstrated the capabilities of these devices to generate shaped and penetrating heating patterns around the cervix extending >2 cm radial [74], and in practice they can be combined with interstitial to further boost regions of heating outside of the cervix [75] (as depicted in Figure 11.6 c).

11.4.3 Intraluminal Devices with Rotating Planar Sources

Intraluminal ultrasound applicators (3.8–4.0 mm OD) that utilize planar ultrasound transducer segments with rotational transducer control have been devised in efforts by Lafon et al. [76–79] and Chopra et al. [53], including a design that enables

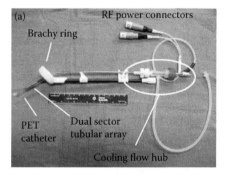

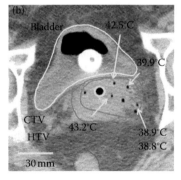
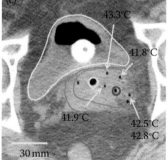

FIGURE 11.6 (a) Endocervical ultrasound applicator within tandem catheter and ring applicator as integrated system for HDR brachytherapy combined with hyperthermia; (b) dual-directional applicator within cervix is positioned to preferentially direct heating laterally extending 2 cm radially; (c) addition of 360° interstitial applicator in periphery further extends heating penetration. Inner contour is the hyperthermia target zone (HTV). Steady-state temperatures measured mid-target are labeled.

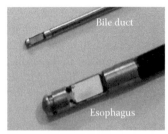
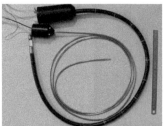

FIGURE 11.7   Intraluminal ultrasound devices with rotating planar transducers for endoscopic insertion and treatment within the biliary ducts and direct insertion within the esophagus for thermal ablation. (Courtesy of Drs. David Melodelima, Frederic Prat, and Cyril Lafon of Inserm, Lyon, France.)

the frequency of an interstitial planar transducer applicator to be varied dynamically to adjust the depth of thermal coagulation [80]. In contrast to the sectored tubular interstitial devices, these applicators produce narrow heating fields and rely on mechanical rotation of the ultrasound source to "sweep" out a larger treatment field that can conform to a predetermined outer boundary. *Ex vivo* and *in vivo* testing of these devices have yielded single-shot thermal lesions extending up to a radial depth of 10–20 mm, and can sweep out 360° coagulation zones in 5 minutes [79]. Intraductal or intraluminal high-intensity ultrasound devices have been configured with a rotating planar transducer segment (2.8 mm wide × 8 mm long, 10 MHz, 14 W/cm² intensity) at the distal end of a 4 mm flexible catheter [76], as shown in Figure 11.7. The transducer portion is covered with a coupling membrane that is inflated and water cooled when deployed. The device is specific for deployment from the working channel of an endoscope, which can be positioned under fluoroscopic guidance within tumor obstruction of the bile duct. This endoscopic device has been evaluated in a human pilot study with 10 patients, where varying applied power levels and rotation position were used to shape the thermal lesion greater than 10 mm radius over 360° at the site of treatment [81] and provide relief of obstructions with no adverse events reported. Similar applicator structures have also been demonstrated for interstitial ablation of deep-seated targets [4], with the possibility of using dual-mode arrays to both deliver and monitor the conformal thermal ablation [82].

Larger diameter (10 mm) applicators (Figure 11.7), based upon rotating planar (8 mm wide × 15 mm long) transducers, have been devised for direct insertion into the esophagus for treatment of esophageal cancer and palliation of associated strictures. The standard treatment protocol [83] was for device placement and positioning using fluoroscopy (Figure 11.7c), then 20 short sequential ablations with 18° rotation between them were applied for circumferential ablation; if sections of esophagus greater than ~15 mm were required to be treated, then the applicator was repositioned in a linear fashion and the process repeated for further treatment. This clinical pilot study demonstrated feasibility of the approach; the endogastric ultrasound applicator could induce localized tumor necrosis within the esophageal stricture, and can provide symptomatic improvement. More recent advanced phased arrays and MR compatible

devices suitable for MR guided procedures with fast MR temperature monitoring are under evaluation [84,85] and indicate potential for precision MR directed ablations within esophagus and biliary ducts, or other similarly accessible sites.

11.5 Summary

Ultrasound technology has many advantages over other modalities and devices as currently applied for interstitial and intracavitary thermal therapy for the treatment of cancer. Given the ability to integrate small or focused transducers within these applicators, and the enhanced penetration and spatial control of energy, these devices can afford precise and accurate delivery of thermal ablation and hyperthermia. Furthermore, these devices can be combined with ultrasound or MR imaging to dramatically improve targeting, treatment control, and treatment verification. Although not covered herein, there are many investigations currently underway to continue miniaturization and customization of ultrasound devices for site-specific treatment in other areas of interest. Many of the devices discussed herein provide a powerful tool that can be used for conformal thermal ablation therapy or hyperthermia, either alone or concurrent with chemotherapy or radiation therapy, or with thermally mediated drug delivery.

References

1. Diederich, C.J. and K. Hynynen, Ultrasound technology for hyperthermia. *Ultrasound in Medicine and Biology*, 1999. **25**(6): p. 871–887.
2. ter Haar, G., Ultrasound focal beam surgery. *Ultrasound in Medicine and Biology*, 1995. **21**(9): p. 1089–100.
3. Ahmed, M., C.L. Brace, F.T. Lee, Jr., and S.N. Goldberg, Principles of and advances in percutaneous ablation. *Radiology*, 2010. **258**(2): p. 351–69.
4. Lafon, C., D. Melodelima, R. Salomir, and J.Y. Chapelon, Interstitial devices for minimally invasive thermal ablation by high-intensity ultrasound. *Int J Hyperthermia*, 2007. **23**(2): p. 153–63.
5. Ryan, T.P., P.F. Turner, and B. Hamilton, Interstitial microwave transition from hyperthermia to ablation: Historical perspectives and current trends in thermal therapy. *Int J Hyperthermia*, 2010. **26**(5): p. 415–33.

6. Stafford, R.J., D. Fuentes, A.A. Elliott, J.S. Weinberg, and K. Ahrar, Laser-induced thermal therapy for tumor ablation. *Crit Rev Biomed Eng*, 2010. **38**(1): p. 79–100.

7. Diederich, C.J., Thermal ablation and high-temperature thermal therapy: Overview of technology and clinical implementation. *Int J Hyperthermia*, 2005. **21**(8): p. 745–53.

8. Stauffer, P.R., Diederich, C.J., Seegenschmiedt, M.H., Interstitial heating technologies, in *Principles and practices of thermoradiotherapy and thermochemotherapy*, M.H. Seegenschmiedt, P. Fessenden, and C.C. Vernon, Editors. 1995, Springer-Verlag: Berlin.

9. Lai, C.Y., D.E. Kruse, C.F. Caskey, D.N. Stephens, P.L. Sutcliffe, and K.W. Ferrara, Noninvasive thermometry assisted by a dual-function ultrasound transducer for mild hyperthermia. *IEEE Trans Ultrason Ferroelectr Freq Control*, 2010. **57**(12): p. 2671–84.

10. Liu, D. and E.S. Ebbini, Real-time 2-D temperature imaging using ultrasound. *IEEE Trans Biomed Eng*, 2010. **57**(1): p. 12–6.

11. Maleke, C. and E.E. Konofagou, *In vivo* feasibility of real-time monitoring of focused ultrasound surgery (FUS) using harmonic motion imaging (HMI). *IEEE Trans Biomed Eng*, 2010. **57**(1): p. 7–11.

12. Arnal, B., M. Pernot, and M. Tanter, Monitoring of thermal therapy based on shear modulus changes: I. shear wave thermometry. *IEEE Trans Ultrason Ferroelectr Freq Control*, 2011. **58**(2): p. 369–78.

13. Ribault, M., J.Y. Chapelon, D. Cathignol, and A. Gelet, Differential attenuation imaging for the characterization of high intensity focused ultrasound lesions. *Ultrason Imaging*, 1998. **20**(3): p. 160–77.

14. de Senneville, B.D., C. Mougenot, B. Quesson, I. Dragonu, N. Grenier, and C.T. Moonen, MR thermometry for monitoring tumor ablation. *Eur Radiol*, 2007. **17**(9): p. 2401–10.

15. Quesson, B., J.A. de Zwart, and C.T. Moonen, Magnetic resonance temperature imaging for guidance of thermotherapy. *J Magn Reson Imaging*, 2000. **12**(4): p. 525–33.

16. Rieke, V. and K. Butts Pauly, MR thermometry. *J Magn Reson Imaging*, 2008. **27**(2): p. 376–90.

17. Rivens, I., A. Shaw, J. Civale, and H. Morris, Treatment monitoring and thermometry for therapeutic focused ultrasound. *Int J Hyperthermia*, 2007. **23**(2): p. 121–39.

18. Hurwitz, M.D., I.D. Kaplan, G.K. Svensson, K. Hynynen, and M.S. Hansen, Feasibility and patient tolerance of a novel transrectal ultrasound hyperthermia system for treatment of prostate cancer. *Int J Hyperthermia*, 2001. **17**(1): p. 31–7.

19. Fosmire, H., K. Hynynen, G.W. Drach, B. Stea, P. Swift, and J.R. Cassady, Feasibility and toxicity of transrectal ultrasound hyperthermia in the treatment of locally advanced adenocarcinoma of the prostate. *Int J Radiat Oncol Biol Phys*, 1993. **26**(2): p. 253–9.

20. Diederich, C.J. and K. Hynynen, The development of intracavitary ultrasonic applicators for hyperthermia: A design and experimental study. *Med Phys*, 1990. **17**(4): p. 626–34.

21. Smith, N.B., M.T. Buchanan, and K. Hynynen, Transrectal ultrasound applicator for prostate heating monitored using MRI thermometry. *Int J of Radiat Oncol, Biol, Phys*, 1999. **43**(1): p. 217–25.

22. Smith, N.B., N.K. Merrilees, M. Dahleh, and K. Hynynen, Control system for an MRI compatible intracavitary ultrasound array for thermal treatment of prostate disease. *Int J Hyperthermia*, 2001. **17**(3): p. 271–82.

23. Silcox, C.E., R.C. Smith, R. King, N. McDannold, P. Bromley, K. Walsh et al., MRI-guided ultrasonic heating allows spatial control of exogenous luciferase in canine prostate. *Ultrasound Med Biol*, 2005. **31**(7): p. 965–70.

24. Hurwitz, M.D., I.D. Kaplan, J.L. Hansen, S. Prokopios-Davos, G.P. Topulos, K. Wishnow et al., Hyperthermia combined with radiation in treatment of locally advanced prostate cancer is associated with a favourable toxicity profile. *Int J Hyperthermia*, 2005. **21**(7): p. 649–56.

25. Hurwitz, M.D., I.D. Kaplan, J.L. Hansen, S. Prokopios-Davos, G.P. Topulos, K. Wishnow et al., Association of rectal toxicity with thermal dose parameters in treatment of locally advanced prostate cancer with radiation and hyperthermia. *Int J Radiat Oncol Biol Phys*, 2002. **53**(4): p. 913–8.

26. Hurwitz, M.D., J.L. Hansen, S. Prokopios-Davos, J. Manola, Q. Wang, B.A. Bornstein et al., Hyperthermia combined with radiation for the treatment of locally advanced prostate cancer: Long-term results from Dana-Farber Cancer Institute study 94-153. *Cancer*, 2011. **117**(3): p. 510–6.

27. Crouzet, S., F.J. Murat, G. Pasticier, P. Cassier, J.Y. Chapelon, and A. Gelet, High intensity focused ultrasound (HIFU) for prostate cancer: Current clinical status, outcomes and future perspectives. *Int J Hyperthermia*, 2010. **26**(8): p. 796–803.

28. Pichardo, S., A. Gelet, L. Curiel, S. Chesnais, and J.Y. Chapelon, New integrated imaging high intensity focused ultrasound probe for transrectal prostate cancer treatment. *Ultrasound Med Biol*, 2008. **34**(7): p. 1105–16.

29. Warmuth, M., T. Johansson, and P. Mad, Systematic review of the efficacy and safety of high-intensity focussed ultrasound for the primary and salvage treatment of prostate cancer. *Eur Urol*, 2010. **58**(6): p. 803–15.

30. Koch, M.O., T. Gardner, L. Cheng, R.J. Fedewa, R. Seip, and N.T. Sanghvi, Phase I/II trial of high intensity focused ultrasound for the treatment of previously untreated localized prostate cancer. *J Urol*, 2007. **178**(6): p. 2366–70; discussion 2370–1.

31. Madersbacher, S., M. Pedevilla, L. Vingers, M. Susani, and M. Marberger, Effect of high-intensity focused ultrasound on human prostate cancer in vivo. *Cancer Research*, 1995. **55**(15): p. 3346–51.

32. Madersbacher, S., C. Kratzik, M. Susani, and M. Marberger, Tissue ablation in benign prostatic hyperplasia with high intensity focused ultrasound. *J Urol*, 1994. **152**(6 Pt 1): p. 1956–60.

33. Sanghvi, N.T., F.J. Fry, R. Birhle, R.S. Foster, M.H. Phillips, J. Syrus et al., Noninvasive surgery of prostate tissue by high-intensity focused ultrasound. *IEEE Trans Ultrason Ferroelectr Freq Control* 1996. **43**(6): p. 1099–110.

34. Bihrle, R., R.S. Foster, N.T. Sanghvi, J.P. Donohue, and P.J. Hood, High intensity focused ultrasound for the treatment of benign prostatic hyperplasia: Early United States clinical experience. *J Urology*, 1994. **151**(5): p. 1271–5.

35. Sanghvi, N.T., R.S. Foster, R. Bihrle, R. Casey, T. Uchida, M.H. Phillips et al., Noninvasive surgery of prostate tissue by high intensity focused ultrasound: An updated report. *Euro J of Ultrasound*, 1999. **9**(1): p. 19–29.

36. Uchida, T., S. Shoji, M. Nakano, S. Hongo, M. Nitta, A. Murota et al., Transrectal high-intensity focused ultrasound for the treatment of localized prostate cancer: Eight-year experience. *Int J Urol*, 2009. **16**(11): p. 881–6.

37. Uchida, T., S. Shoji, M. Nakano, S. Hongo, M. Nitta, Y. Usui et al., High-intensity focused ultrasound as salvage therapy for patients with recurrent prostate cancer after external beam radiation, brachytherapy or proton therapy. *BJU Int*, 2011. **107**(3): p. 378–82.

38. Shoji, S., M. Nakano, Y. Nagata, Y. Usui, T. Terachi, and T. Uchida, Quality of life following high-intensity focused ultrasound for the treatment of localized prostate cancer: A prospective study. *Int J Urol*, 2010. **17**(8): p. 715–9.

39. Chaussy, C. and S. Thuroff, High-intensity focused ultrasound in the management of prostate cancer. *Expert Rev Med Devices*, 2010. **7**(2): p. 209–17.

40. Chaussy, C.G. and S.F. Thuroff, Robotic high-intensity focused ultrasound for prostate cancer: What have we learned in 15 years of clinical use? *Curr Urol Rep*, 2011. **12**(3): p. 180–7.

41. Gavrilov, L.R., J.W. Hand, P. Abel, and C.A. Cain, A method of reducing grating lobes associated with an ultrasound linear phased array intended for transrectal thermotherapy. *IEEE Trans Ultrason Ferroelectr Freq Control*, 1997. **44**(5): p. 1010–17.

42. Chapelon, J.Y., P. Faure, M. Plantier, D. Cathignol, R. Souchon, F. Gorry et al., The feasibility of tissue ablation using high intensity electronically focused ultrasound. *Proc of IEEE Ultrasonics Symp*, 1993. **2**: p. 1211–14.

43. Chapelon, J.Y., M. Ribault, F. Vernier, R. Souchon, and A. Gelet, Treatment of localised prostate cancer with transrectal high intensity focused ultrasound. *Euro J of Ultrasound*, 1999. **9**(1): p. 31–8.

44. Hutchinson, E.B., M.T. Buchanan, and K. Hynynen, Design and optimization of an aperiodic ultrasound phased array for intracavitary prostate thermal therapies. *Med Phys*, 1996. **23**(5): p. 767–76.

45. Hutchinson, E., M. Dahleh, and K. Hynynen, The feasibility of MRI feedback control for intracavitary phased array hyperthermia treatments. *Int J Hyperthermia*, 1998. **14**(1): p. 39–56.

46. Sokka, S.D. and K.H. Hynynen, The feasibility of MRI-guided whole prostate ablation with a linear aperiodic intracavitary ultrasound phased array. *Phys in Med and Biol*, 2000. **45**(11): p. 3373–83.

47. Pilatou, M.C., E.A. Stewart, S.E. Maier, F.M. Fennessy, K. Hynynen, C.M. Tempany et al., MRI-based thermal dosimetry and diffusion-weighted imaging of MRI-guided focused ultrasound thermal ablation of uterine fibroids. *J Magn Reson Imaging*, 2009. **29**(2): p. 404–11.

48. Diederich, C.J., R.J. Stafford, W.H. Nau, E.C. Burdette, R.E. Price, and J.D. Hazle, Transurethral ultrasound applicators with directional heating patterns for prostate thermal therapy: *In vivo* evaluation using magnetic resonance thermometry. *Med Phys*, 2004. **31**(2): p. 405–13.

49. Diederich, C.J. and E.C. Burdette, Transurethral ultrasound array for prostate thermal therapy: Initial studies. *IEEE Trans Ultrason Ferroelectr Freq Control*, 1996. **43**(6): p. 1011–22.

50. Ross, A.B., C.J. Diederich, W.H. Nau, H. Gill, D.M. Bouley, B. Daniel et al., Highly directional transurethral ultrasound applicators with rotational control for MRI guided prostatic thermal therapy. *Phys Med Biol*, 2004. **49**(1): p. 189–204.

51. Ross, A.B., C.J. Diederich, W.H. Nau, V. Rieke, R.K. Butts, G. Sommer et al., Curvilinear transurethral ultrasound applicator for selective prostate thermal therapy. *Med Phys*, 2005. **32**(6): p. 1555–65.

52. Kinsey, A.M., C.J. Diederich, V. Rieke, W.H. Nau, K.B. Pauly, D. Bouley et al., Transurethral ultrasound applicators with dynamic multi-sector control for prostate thermal therapy: *In vivo* evaluation under MR guidance. *Med Phys*, 2008. **35**(5): p. 2081–93.

53. Chopra, R., C. Luginbuhl, A.J. Weymouth, F.S. Foster, and M.J. Bronskill, Interstitial ultrasound heating applicator for MR-guided thermal therapy. *Phys Med Biol*, 2001. **46**(12): p. 3133–45.

54. Chopra, R., M. Burtnyk, M.A. Haider, and M.J. Bronskill, Method for MRI-guided conformal thermal therapy of prostate with planar transurethral ultrasound heating applicators. *Phys Med Biol*, 2005. **50**(21): p. 4957–75.

55. Pauly, K.B., C.J. Diederich, V. Rieke, D. Bouley, J. Chen, W.H. Nau et al., Magnetic resonance-guided high-intensity ultrasound ablation of the prostate. *Top Magn Reson Imaging*, 2006. **17**(3): p. 195–207.

56. Chopra, R., M. Burtnyk, A. N'Djin W, and M. Bronskill, MRI-controlled transurethral ultrasound therapy for localised prostate cancer. *Int J Hyperthermia*, 2010. **26**(8): p. 804–21.

57. Siddiqui, K., R. Chopra, S. Vedula, L. Sugar, M. Haider, A. Boyes et al., MRI-guided transurethral ultrasound therapy of the prostate gland using real-time thermal mapping: Initial studies. *Urology*, 2010. **76**(6): p. 1506–11.

58. Chopra, R., K. Tang, M. Burtnyk, A. Boyes, L. Sugar, S. Appu et al., Analysis of the spatial and temporal accuracy of heating in the prostate gland using transurethral ultrasound therapy and active MR temperature feedback. *Phys Med Biol*, 2009. **54**(9): p. 2615–33.

59. Burtnyk, M., R. Chopra, and M.J. Bronskill, Quantitative analysis of 3-D conformal MRI-guided transurethral ultrasound therapy of the prostate: Theoretical simulations. *Int J Hyperthermia*, 2009. **25**(2): p. 116–31.

60. Lee, R.J., M. Buchanan, L.J. Kleine, and K. Hynynen, Arrays of multielement ultrasound applicators for interstitial hyperthermia. *IEEE Trans Biomed Eng*, 1999. **46**(7): p. 880–90.

61. Hynynen, K., The feasibility of interstitial ultrasound hyperthermia. *Medical Physics*, 1992. **19**(4): p. 979–87.

62. Diederich, C.J., Ultrasound applicators with integrated catheter-cooling for interstitial hyperthermia: Theory and preliminary experiments. *Int J Hyperthermia*, 1996. **12**(2): p. 279–97.

63. Nau, W.H., C.J. Diederich, and E.C. Burdette, Evaluation of multielement catheter-cooled interstitial ultrasound applicators for high-temperature thermal therapy. *Med Phys*, 2001. **28**(7): p. 1525–34.

64. Diederich, C.J., W.H. Nau, and P.R. Stauffer, Ultrasound applicators for interstitial thermal coagulation. *IEEE Trans Ultrason Ferroelectr Freq Control*, 1999. **46**(5): p. 1218–28.

65. Deardorff, D.L. and C.J. Diederich, Ultrasound applicators with internal water-cooling for high-powered interstitial thermal therapy. *IEEE Tran on Biomed Engin*, 2000. **47**(10): p. 1356–65.

66. Deardorff, D.L., C.J. Diederich, and W.H. Nau, Control of interstitial thermal coagulation: Comparative evaluation of microwave and ultrasound applicators. *Med Phys*, 2001. **28**(1): p. 104–17.

67. Diederich, C.J. and K. Hynynen, Ultrasound technology for interstitial hyperthermia. In *Interstitial and intracavitary thermoradiotherapy*, M.H. Seegenschmiedt and R. Sauer, Editors. 1993, Springer-Verlag: Berlin, p. 55–61.

68. Tyreus, P.D., W.H. Nau, and C.J. Diederich, Effect of applicator diameter on lesion size from high temperature interstitial ultrasound thermal therapy. *Med Phys*, 2003. **30**(7): p. 1855–63.

69. Diederich, C., J. Wootton, P. Prakash, V. Salgaonkar, T. Juang, S. Scott et al., Catheter-based ultrasound hyperthermia with HDR brachytherapy for treatment of locally advanced cancer of the prostate and cervix. In *Energy Based Treatment of Tissue and Assessment VI, Proceedings of SPIE*. 2011. San Francisco, CA.

70. Chen, X., C.J. Diederich, J.H. Wootton, J. Pouliot, and I.C. Hsu, Optimisation-based thermal treatment planning for catheter-based ultrasound hyperthermia. *Int J Hyperthermia*, 2010. **26**(1): p. 39–55.

71. Kangasniemi, M., C.J. Diederich, R.E. Price, R.J. Stafford, D.F. Schomer, L.E. Olsson et al., Multiplanar MR temperature-sensitive imaging of cerebral thermal treatment using interstitial ultrasound applicators in a canine model. *J Magn Reson Imaging*, 2002. **16**(5): p. 522–31.

72. Deardorff, D.L. and C.J. Diederich, Axial control of thermal coagulation using a multi-element interstitial ultrasound applicator with internal cooling. *IEEE Trans Ultrason Ferroelectr Freq Control*, 2000. **47**(1): p. 170–8.

73. Nau, W.H., C.J. Diederich, A.B. Ross, K. Butts, V. Rieke, D.M. Bouley et al., MRI-guided interstitial ultrasound thermal therapy of the prostate: A feasibility study in the canine model. *Med Phys*, 2005. **32**(3): p. 733–43.

74. Wootton, J.H., I.C. Hsu, and C.J. Diederich, Endocervical ultrasound applicator for integrated hyperthermia and HDR brachytherapy in the treatment of locally advanced cervical carcinoma. *Med Phys*, 2011. **38**(2): p. 598–611.

75. Wootton, J.H., P. Prakash, I.C. Hsu, and C.J. Diederich, Implant strategies for endocervical and interstitial ultrasound hyperthermia adjunct to HDR brachytherapy for the treatment of cervical cancer. *Phys Med Biol*, 2011. **56**(13): p. 3967–84.

76. Lafon, C., J.Y. Chapelon, F. Prat, F. Gorry, J. Margonari, Y. Theillere et al., Design and preliminary results of an ultrasound applicator for interstitial thermal coagulation. *Ultrasound Med Biol*, 1998. **24**(1): p. 113–22.

77. Lafon, C., S. Chosson, F. Prat, Y. Theillère, J.Y. Chapelon, A. Birer et al., The feasibility of constructing a cylindrical array with a plane rotating beam for interstitial thermal surgery. *Ultrasonics*, 2000. **37**(9): p. 615–21.

78. Lafon, C., Y. Theillère, F. Prat, A. Arefiev, J.Y. Chapelon, and D. Cathignol, Development of an interstitial ultrasound applicator for endoscopic procedures: Animal experimentation. *Ultrasound in Med Biol*, 2000. **26**(4): p. 669–75.

79. Lafon, C., F. Prat, J.Y. Chapelon, F. Gorry, J. Margonari, Y. Theillere et al., Cylindrical thermal coagulation necrosis using an interstitial applicator with a plane ultrasonic transducer: *In vitro* and *in vivo* experiments versus computer simulations. *Int J Hyperthermia*, 2000. **16**(6): p. 508–22.

80. Chopra, R., C. Luginbuhl, F.S. Foster, and M.J. Bronskill, Multifrequency ultrasound transducers for conformal interstitial thermal therapy. *IEEE Trans Ultrason Ferroelectr Freq Control*, 2003. **50**(7): p. 881–9.

81. Lafon, C., L. de, Y. Theillere, F. Prat, J.Y. Chapelon, and D. Cathignol, Optimizing the shape of ultrasound transducers for interstitial thermal ablation. *Med Phys*, 2002. **29**(3): p. 290–7.

82. Owen, N.R., J.Y. Chapelon, G. Bouchoux, R. Berriet, G. Fleury, and C. Lafon, Dual-mode transducers for ultrasound imaging and thermal therapy. *Ultrasonics*, 2010. **50**(2): p. 216–20.

83. Melodelima, D., F. Prat, J. Fritsch, Y. Theillere, and D. Cathignol, Treatment of esophageal tumors using high intensity intraluminal ultrasound: First clinical results. *J Transl Med*, 2008. **6**: p. 28.

84. Melodelima, D., R. Salomir, J.Y. Chapelon, Y. Theillere, C. Moonen, and D. Cathignol, Intraluminal high intensity ultrasound treatment in the esophagus under fast MR temperature mapping: *In vivo* studies. *Magn Reson Med*, 2005. **54**(4): p. 975–82.

85. Melodelima, D., R. Salomir, C. Mougenot, C. Moonen, and D. Cathignol, 64-element intraluminal ultrasound cylindrical phased array for transesophageal thermal ablation under fast MR temperature mapping: An *ex vivo* study. *Med Phys*, 2006. **33**(8): p. 2926–34.

III

Physical Aspects of Emerging Technology for Thermal Therapy

Evolving Tools for Navigated Image-Guided Thermal Cancer Therapy

Kevin Cleary
Children's National Medical Center

Emmanuel Wilson
Children's National Medical Center

Filip Banovac
*Georgetown University
Medical Center*

12.1 Introduction

Image-guided interventions have become a mainstay of cancer diagnosis and treatment. By image-guided we mean the use of medical imaging such as fluoroscopy or computed tomography (CT) to carry out minimally invasive procedures. These procedures can be done through percutaneous instrument placement (through the skin) or through the vasculature using catheter-based approaches. Navigation systems to assist the physician in more precisely completing image-guided interventions have also been introduced in recent years by commercial companies and are an active area of research. Since we focus on navigation systems for cancer interventions in this chapter, we will call this Navigated Image-Guided Thermal Cancer Therapy (NIGTCT).

We will describe the technology of navigated image-guided interventions, review the literature where image-guided navigation has been applied to thermal therapy in the interventional oncology domain, and describe emerging trends relevant to the future of image-guided thermal cancer therapy.

12.2 Technology of Navigation Systems

12.2.1 Definition of Image-Guided Surgery and Historical Review

One definition of image-guided surgery is a surgical procedure facilitated through a correlation between preoperative and intraoperative imaging. In this regard, as intraoperative imaging is becoming more common in the clinical arena, more sophisticated image processing and guidance is increasingly employed for surgical procedures. Image-guided surgery can also refer to the use of mechanical devices or computers to enable a mapping from preoperative images to the intraoperative workspace. These developments began with the introduction of the stereotactic frame for neurosurgery. The successful adoption of mechanical digitizers and stereotactic frames for neurosurgical application was enabled through the availability of easily identifiable exterior landmarks such as the auditory canals and inferior orbital rims (Galloway et al. 2008).

The rapid improvement in computing capabilities, the development of systems capable of tracking objects in space, and improvements in tomographic imaging modalities, including CT and magnetic resonance imaging (MRI), led to a paradigm shift in the practice of image-guided surgery. As optical tracking systems became smaller and less expensive, their adoption in clinical practice led to frameless stereotaxy as a means of localizing surgical tools without a bulky mechanical reference frame. This progression enabled increasingly less invasive surgical interventions while simultaneously achieving improved clinical accuracy and outcome.

The shift toward minimally invasive surgery necessitates a greater reliance on image guidance and preoperative planning. Minimally invasive surgery can reduce patient trauma and lead to faster recovery times compared to conventional open surgery. The primary goal of image guidance and navigation is to assist the physician in precisely localizing the anatomy of interest. This goal is accomplished by overlaying the surgical tools on an augmented reality view of the patient anatomy.

A typical navigated image-guided surgery system consists of five key components as shown in Figure 12.1: (1) imaging, (2) tracking, (3) registration, (4) planning, and (5) navigation. A medical image dataset serves as the anatomical map upon which tracked surgical tools are displayed. Surgical tools are tracked in three-dimensional space by a tracking system. The image dataset and tracker information are referenced in different coordinate reference frames. The process of registration aligns these two coordinate frames to a single common frame. Once registered, the physician can manipulate the visualization and representation of the patient data. Navigation and guidance is achieved by constantly updating the position of the tracked surgical tools with respect to the patient image. A brief overview of each of these components is provided next.

12.2.2 Imaging

Within the context of an image-guided navigation system, the role of the imaging modality is primarily to provide a three-dimensional (3D) map of the patient anatomy. CT is the preferred modality for visualizing bony structures as well as lungs and organs in the chest cavity. MRI, however, provides better soft tissue contrast and is preferred for brain and spinal cord imaging or visualizing tendons and ligaments. While CT has the disadvantage of using ionizing radiation, the risk versus reward to the patient must always be considered. Image-guided navigation often reduces the number of CT images required to precisely target the anatomy and, therefore, may lead to a net reduction in total radiation dose when compared against conventional imaging without the use of navigated image guidance. Limitations with MRI include high cost, availability, and the need for compatible ablation tools and devices. Ultrasound has the advantages of being real time, low cost, and readily available, but the image discrimination of ultrasound for navigation purposes can be poor and ultrasound is usually deployed as a two-dimensional imaging modality.

12.2.3 Tracking

Tracking devices are an essential component of a navigated image-guided surgery system. These devices are used to track the position of instruments relative to the patient anatomy. Improvements in sophistication and accuracy of tracking systems in the last 10 years have been critical to the development of image-guided systems. Early tracking systems were essentially mechanical digitizers and proved to be too large and bulky for use in a clinical operating environment. Optical tracking systems were soon adopted due to their high accuracy and relatively large workspace. However, optical tracking systems require that a line-of-sight be maintained between the tracking device and

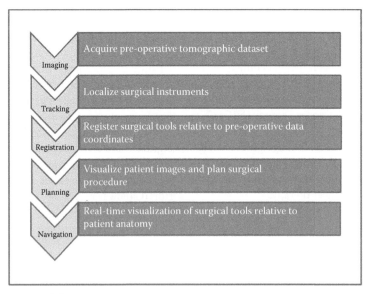

FIGURE 12.1 Image-guided system workflow.

instrument to be tracked. This is not always convenient and, furthermore, precludes tracking of flexible instruments inside the body. Electromagnetic tracking systems, refined for clinical use over the past five years, have no line-of-sight limitations and can track instruments such as catheters and the tips of needles inside the body. The choice of tracking system is highly application dependent and requires an understanding of the desired working volume and accuracy constraints.

Optical tracking systems can be characterized as passive or active depending on the type of markers used. While both varieties use infrared (IR) light to localize markers, the difference lies in the method used. Passive varieties use retroreflective spheres as markers with the tracking camera acting as the source and detector of IR light. Light emitted from the tracker camera is reflected off the passive markers and detected at the camera to triangulate marker location. Active varieties use

IR light-emitting diodes as the marker with the tracker camera acting only as a detector to localize the markers. Northern Digital Inc. (NDI) is a manufacturer of both passive and active marker tracking devices that have been used in medical applications. Other forms of passive tracking include videometric camera units that track the position of pattern-marked tools in video image sequences, as exemplified by the MicronTracker system from Claron Technology Inc. These active IR, passive IR, and videometric trackers are shown in Figure 12.2. While optical tracking systems offer high measurement accuracy in a relatively large workspace, they do have line-of-sight limitations, whereby valid position tracking is only possible as long as no occlusions occur between the tracker camera and tracked tools. In a congested operating environment and in interventional radiology applications, the line-of-sight limitation can be cumbersome.

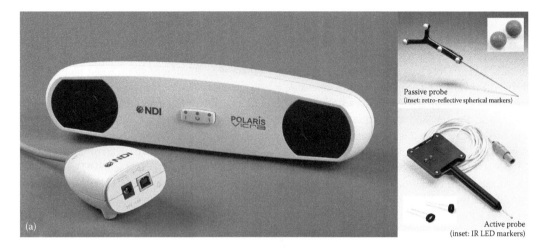

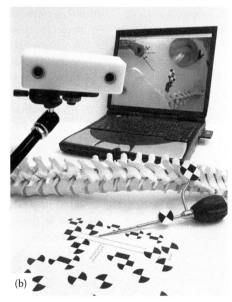

FIGURE 12.2 Commonly used optical tracking technologies: (a) NDI Vicra IR-based tracker camera, also shown: (top-right inset) passive probe with retroreflective markers and (bottom-right inset) active probe with active IR LED markers (With permission, © 2011 Northern Digital Inc.); (b) Claron MicronTracker videometric tracker with pattern-marked markers. (With permission, © 2011 Claron Technology Inc.)

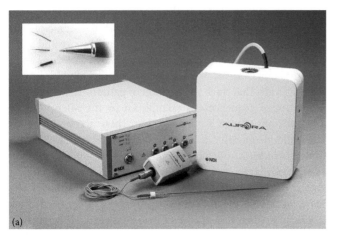
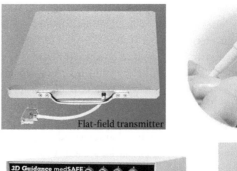
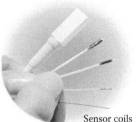

Flat-field transmitter

Sensor coils

3D Guidance medSAFE

medSAFE controller   Mid-range transmitter

(a)   (b)

FIGURE 12.3 Images of two leading electromagnetic tracking devices: (a) NDI Aurora (© 2011 Northern Digital Inc. With permission.); (b) Ascension 3D Guidance medSAFE™ device, shown with a selection of transmitters and sensor coils. (© 2011 Ascension Technology Corp. With permission.)

Electromagnetic tracking devices have found increasing use in medical applications in the last decade as tracker technologies have improved. Electromagnetic tracking sensors that can be embedded in medical instruments typically consist of small coils wound around a ferrite core. A field generator creates an electromagnetic field that induces a small current in the sensor coil as it is moved within the field. Position and orientation measurements of the sensor can then be determined within some working volume. The primary benefit of electromagnetic tracking is that there is no line-of-sight requirement as the electromagnetic fields are not significantly attenuated by the human body. The sensor coils can also be embedded at the tip of medical instruments such as catheters and needles to directly track tip placement in the clinical environment. However, electromagnetic tracking can be sensitive to metal objects in the working space and this fact must be taken into account when using electromagnetic tracking devices. Two leading vendors of electromagnetic tracking systems for the medical market are Northern Digital Inc. and Ascension Technology Corp., shown in Figure 12.3.

Irrespective of the tracking system used, its purpose is to provide a 3D Cartesian representation of the position and orientation of markers attached to surgical tools and the patient anatomy. Different tracker manufacturers provide device-specific application programmer interfaces (API) that can be used to control the tracker from a host computer and acquire the marker positions. This position information is typically output as position in millimeters and orientation in quaternions or Euler angles. The nature of the clinical application will determine the number of tools to be tracked simultaneously and whether optical or electromagnetic tracking is the best choice.

12.2.4 Registration

Registration is defined as the aligning of two disparate coordinate frames. In the context of image-guided interventions this implies alignment of the coordinate frame of the tracked surgical tools with that of the preoperative imaging modality. In the case of multimodality datasets this can mean the spatial alignment of two or more imaging modalities to a common reference coordinate. Registration of the tracked tools with the preoperative image dataset is typically a manual operation requiring landmarks defined in the preoperative dataset to be identified using a tracked tool or tracked stylus. The most common registration method is a paired point registration that aims to minimize the sum of least square error between the two coordinate frames (Arun et al. 1987; Horn 1987). Paired point registration is achieved through the use of multimodality markers affixed to the patient, shown in Figure 12.4, that serve as the common reference point between tomographic dataset and tracker space.

Rigid registration schemes have been employed in percutaneous procedures, where the procedure is conducted on the same bed and operating environment as used to acquire the preoperative image. In laparoscopic and open surgical procedures, organ motion due to respiration and displacement of the abdominal cavity can be significant. In such instances, nonrigid, deformable registration schemes may be applied, although at present this is primarily a research interest (Hawkes et al. 2005; Dandekar et al. 2007). In percutaneous applications, where accurate accounting of respiratory motion using deformable registration could be beneficial in improving needle placement accuracy, it is not clear how best to present that respiratory motion information to the physician during instrument placement.

Recent studies have shown that the fusion of abdominal images from different modalities can improve diagnosis and monitoring of disease progression (Kuehl et al. 2008; Giesel et al. 2009). As positron emission tomography (PET) gains in prominence as a cancer staging modality, the use of image fusion becomes more prevalent. Image fusion entails the overlaying of multiple imaging modalities in a spatially relevant form, for example, a functional image such as 18-FDG-PET or dynamic contrast enhanced (DCE) MRI with a preoperative CT dataset. In the case of RFA of hepatocellular carcinomas, where postoperative outcome shortly

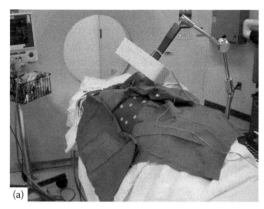
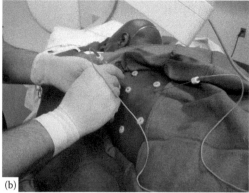

FIGURE 12.4 (a) Electromagnetic tracking used to guide percutaneous biopsy of a lung tumor in the CT suite. Doughnut markers on back of patient are used for paired point registration. (b) The physician is using a tracked biopsy needle to touch each of the doughnut markers for paired-point registration. These needles have an electromagnetic sensor at their tips and the wire coming out of each needle connects to each control station for the electromagnetic tracking system.

following the procedure is hard to gauge using CT or US, PET/CT has been shown to provide an effective means of determining procedural efficacy (Israel et al. 2001). Therefore, image fusion will most likely play an increasingly important role in conjunction with registration of multiple imaging modalities to complete a staging, planning, and postprocedural assessment function.

12.2.5 Planning

The success of a percutaneous tumor treatment depends on the choice of probe trajectory, maximal destruction of the tumor, and minimal harm of healthy tissue. Planning, therefore, needs to optimize these parameters. Typically, planning enables the physician to preview the patient dataset and plan an optimal path to the tumor while avoiding critical structures.

The tasks defined within the planning component have a significant dependence on the clinical procedure. Each therapy choice has technological limitations based on the physics of its operation and unique criteria that need to be optimized. With procedures such as RFA and cryoablation, ablation of a margin of tissue surrounding the tumor is critical to a successful clinical outcome. More sophisticated planning algorithms incorporate models that characterize tissue properties, ablation dynamics, and heat-sink systems. For RFA applications, the Pennes bioheat equation (Pennes 1948) determines the transfer of heat energy around an RFA probe in tissue. The planning component of an image-guided therapy application taking this into consideration can therefore provide a representation of the heat-induced necrosis region around the RFA probe (accounting for vasculature in the vicinity and the subsequent result of heat-sink), overlay this information on the preoperative dataset, and plan a trajectory to the target site such that the full tumor is covered and critical structures are avoided along the path of probe insertion.

12.2.6 Navigation

Navigation refers to the process of using the virtual display from an image-guided system to precisely place the therapeutic

instrument. Once the images have been acquired and registered and a path has been planned, the image-guided system can provide a navigation view, where the tracked position of the instrument is overlaid onto the patient anatomy. The navigation view of an open source image-guided system developed in our research group is shown in Figure 12.5. As described in the figure caption, the navigation system provides a four-quadrant display that includes several views of the anatomy and the planned path. The physician can then use this display to guide the instrument to the target. The system was constructed using the open source software package the Image-Guided Surgical Toolkit (Enquobahrie et al. 2007), which is freely available at igstk.org and can be used to prototype image-guided surgery applications.

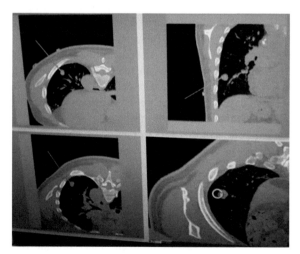

FIGURE 12.5 Navigation view for image-guided lung biopsy. The top-left window shows a true axial view, and it can be seen that a true axial path would hit the rib. Therefore, the off-axial path in the bottom-left window was chosen. The top-right window shows a sagittal view. A birds-eye view, looking down the needle shaft, is shown in the bottom-right window.

12.3 Clinical Applications

12.3.1 Introduction and Overview

In this section we will describe several clinical applications where navigation and image guidance have played a major role. In particular, navigation and image guidance have enabled minimally invasive approaches to thermal therapy. Although for many diseases, such as liver tumors, surgical resection and organ transplantation are seen as the gold standard for curative therapy, many patients are not candidates for surgery. For liver metastases alone, only 10–25% of patients are candidates for resection (Scheele et al. 1995). For the majority of nonsurgical candidates, image-guided percutaneous ablation is accepted as the best therapeutic choice. In these patients, thermal therapies have been proven to be an effective method of palliative care and improvement of quality of life.

As procedures become more minimally invasive, several difficulties in visualizing the tumor, planning instrument placement, and prediction of induced necrosis present themselves. Factors that determine successful clinical outcome are twofold. Appropriate detection and visualization of the neoplasm and staging is one concern. Accurate targeting and ablation of the complete tumor plus a margin is a second concern. The most critical function of any image-guided approach is to address these two concerns.

The imaging modality chosen is operator dependent and based on local availability of dedicated equipment such as CT and MR systems. Lesions, such as for many primary HCC that are typically best visualized in arterial-phase CT, would be hard to identify intra-procedurally. Currently, targeting of such lesions require mental 3D reconstruction of 2D images by an interventional radiologist, thus complicating the procedural outcome (Wood et al. 2007). The only imaging modality that at present has proven real-time intraoperative temperature monitoring of the efficacy of an ablative treatment is MRI using special protocols for thermal data acquisition, which can be cost prohibitive or unavailable in many institutions.

In patients with multiple tumors or where healthy organ volume is limited, the importance of defining an accurate margin that encompasses the tumor while sparing as much healthy tissue as possible is critical. Planning a best path to the tumor while avoiding critical structures is implicitly more difficult and assumes greater importance under a minimally invasive approach. In such cases, the need for preoperative planning is amplified and some researchers have developed prototype systems to address this need.

Image-guided cancer intervention is a rapidly developing field. These procedures are typically done in a multidisciplinary setting with collaborations between interventional radiologists, surgeons, hepatologists, oncologists, and other medical professionals. The increasing interest in this field is evident through development in the past five years of two related conferences: (1) the World Conference on Interventional Oncology (WCIO) and (2) the European Conference on Interventional Oncology (ECIO). In addition, specialized centers such as the Center for Image-Guided Interventions (CIGI) at Memorial Sloan-Kettering Cancer Center in New York City and the Center for Interventional Oncology at the NIH Clinical Center are beginning to appear.

Broadly speaking, image-guided minimally invasive cancer therapy can be grouped within four approaches: ablation, chemotherapy, radiation therapy, and gene therapy. The focus of this chapter will be on ablation and chemotherapy applications. In the remainder of this section we will review the literature in navigated image-guided cancer therapy for radiofrequency ablation (RFA), cryoablation, and transarterial chemoembolization (TACE) as follows:

- Section 12.3.2: Radiofrequency Ablation
- Section 12.3.3: Cryoablation
- Section 12.3.4: Transarterial Chemoembolization

12.3.2 Radiofrequency Ablation

12.3.2.1 Introduction

Radiofrequency ablation (RFA) is the most widely applied thermal ablation technology with over 100,000 estimated liver ablation procedures performed worldwide (Ahmed et al. 2011). While many other thermal modalities such as microwave, cyroablation, high-intensity ultrasound, irreversible electroporation, and interstitial laser have also evolved, radiofrequency ablation is the mainstay of most clinical practices and is considered the gold standard of local ablative therapy for nonresectable liver tumor patients. RFA has been used successfully in the management of hepatocellular carcinoma (HCC) (Rossi et al. 1998; Livraghi et al. 1999), hepatic metastases of colorectal cancer, renal, breast, osteoid osteomas, and more recently lung neoplasms (Pennathur et al. 2009). RFA is accepted as the best therapeutic solution for patients with early stage HCC who would not otherwise qualify for liver transplantation or surgical resection, or for patients with metastatic lesions (Crocetti et al. 2010).

An RFA system consists of an RF generator, introducer trocar needle, and a large dissipative electrode. The RF generator produces an alternating electric field (< 30MHz; typically 460–480 kHz) across the patient between the electrode pad(s) placed on the patient skin and the tip of the electrode probe. The electrode probe is placed within the tumor using ultrasound, CT, or MR image guidance. The RF signal between the electrode pad and RF probe causes tissue agitation as a result of oscillatory movement of ions. Tissue heating induces cellular death via thermal coagulation necrosis.

While earlier probes were monopolar devices, the development of multiple hooked (multi-tined) probes (LeVeen® needle electrode) has enabled the ablation of larger lesions (LeVeen 1997; Curley et al. 1999). Probes of this type are the most commonly employed for soft-tissue applications today. The probe consists of multiple curved tines that are deployed from a central cannula. Multiple tines help spread the area of coagulation

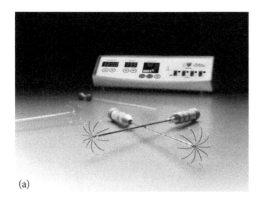
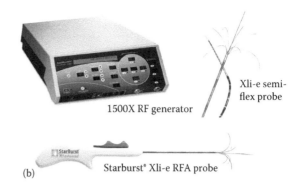
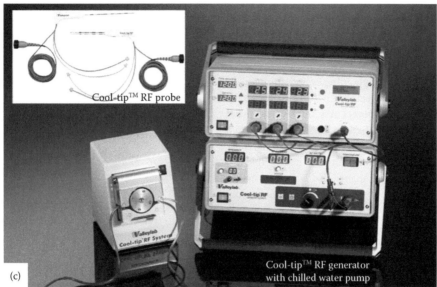

FIGURE 12.6 Images of three RFA devices for soft-tissue application: (a) Boston Scientific RF3000 (With permission, © 2011 Boston Scientific Corp.); (b) AngioDynamics StarBurst® devices shown with multi-tined LeVeen® probes (With permission, © 2011 AngioDynamics®); (c) Valleylab/Covidien Cool-tip™ device. (With permission, © 2010 Covidien.)

necrosis in a more complete, predictable, and spherical manner. The tines can also be partially deployed to adjust the size of the necrotic volume.

The major U.S. Food and Drug Administration (FDA)–cleared RFA devices in use for soft-tissue ablation include the RF3000 device (Boston Scientific, Natick, MA), the StarBurst® Talon device (AngioDynamics, Latham, NY, previously RITA Medical Inc.), and the Cool-tip™ RF device (Valleylab, Boulder, CO, now a subsidiary of Covidien). Multiple-tine LeVeen® probes are available for Boston Scientific and AngioDynamics systems. These devices, along with their respective probes, are shown in Figure 12.6.

12.3.2.2 Navigation Systems and RFA

Several researchers have investigated the use of image-guided navigation to assist in accurate placement of the radiofrequency ablation probe. The major studies investigating the use of navigation systems and RFA are shown in Table 12.1 below. From the table, it can be seen that the focus is on liver

applications, mirroring clinical practice. Both optical and electromagnetic tracking systems have been investigated, as well as CT and MRI guidance. While several studies were simulations based on clinical data sets, a few studies have incorporated navigation systems in clinical practice. Each of the studies listed in the table will now be described. One of the earliest groups to investigate navigation for RFA developed a 3D simulator and treatment planning system for optimal needle placement (Villard et al. 2005). The system included segmentation of the liver and the tumor, and allowed the user to virtually place the radiofrequency probe and observe a meshed spheroid representing the 60°C isosurface. The system also included the ability to simulate the placement of multiple needles to treat larger tumors and the optimal positioning of each needle to fully treat the lesion while minimizing damage to vital structures. Simulations were completed using clinical data from 12 patients having a single, small liver tumor.

The issues with current radiofrequency ablation techniques and the need for accurate image guidance were described by

TABLE 12.1   Studies Describing the Use of Image-Guided Navigation with Radiofrequency Ablation

Organ	Image Modality	Navigation	Study	Reference
Liver	CT	Planning only	Clinical data	Villard et al. (2005)
Liver	CT	Planning only	Clinical data	Bale et al. (2010)
Liver	CT	Robot assisted	Human trial	Solomon et al. (2006)
Liver	CT	Optical	Human trial	Mundeleer et al. (2008)
Liver, Kidney	CT	Electromagnetic	Human trial	Stone et al. (2007); Giesel et al. (2009)
Liver	MRI	Optical	Human trial	Maeda et al. (2009)

Bale et al. (2010). These authors provide an excellent overview of the difficulties in precisely placing the radiofrequency ablation probe in clinical cases where the lesions are close to major vessels or the size of the lesion requires overlapping probes. They discuss the potential of navigation and image guidance for planning and executing RF ablations and give an example case of six different trajectories for a large liver metastasis.

The development of a treatment planning system for overlapping ablations along with robotic assistance for accurate needle placement was described by Solomon et al. (2006). Customized software was developed to allow the user to overlay predicted zones of ablation onto the CT images from two commercially available RFA systems. The planning software was coupled to a surgical robot called Acubot that included a remote center of motion capability for needle positioning at any orientation. A human clinical study was completed on two patients with hepatic lesions larger than 3.0 cm that required two overlapping ablations. In both cases the overlapping ablations were performed through the same skin entry site, and no complications were observed.

A navigation system for radiofrequency ablation was developed by a group in Belgium (Mundeleer et al. 2008). The navigation system was based on optical tracking (Polaris, NDI), and passive retroreflective markers were attached to the handle of the RFA probe. Image registration was accomplished by a combination of landmark transforms (paired point matching) and an iterative closest point implementation. *Ex vivo* experiments were completed using a graduated box phantom and a veal liver, followed by two *in vivo* tests, one laparoscopic and one percutaneous. The results showed the feasibility of the approach and the potential of such a system in the clinical environment.

One pioneering group in the use of navigation for RFA has been the Center for Interventional Oncology at the NIH Clinical Center. This group has demonstrated the feasibility of fusing CT, MRI, and PET for treatment planning, navigation, and follow-up in RFA for kidney and liver tumors (Giesel et al. 2009). For RFA of kidney tumors, electromagnetic tracking (Aurora, Northern Digital) was used with multiplanar reconstructed CT images (Percunav, Traxtal Inc., Toronto, and Philips Medical Systems, Andover, MA) to guide probe placement (Figure 12.7).

While most of the work in navigation for RFA has been based on CT imaging, several researchers have investigated other imaging modalities, including MRI. A Japanese group performed ablation on 34 liver cancer patients using a navigation system based on open MRI (Maeda et al. 2009). An ultrasound probe was tracked with a Polaris optical tracking system, and during ablation, the ultrasound probe, needle, tumor, and MR image were displayed on the monitor (Figure 12.8). Registration between the real-time ultrasound images and the preoperative MR images was done using the open source 3D Slicer navigation and display software (Brigham and Women's Hospital, Boston, Massachusetts) (Gering et al. 2001).

12.3.3 Cryoablation

12.3.3.1 Introduction

While cyroablation has been part of medical practice for several decades, it is only in the last several years that technological advances have resulted in cyroprobes thin enough (down to 17 gauge) for percutaneous applications. The objective of cyroablation is to reduce the temperature of the tumor and a surrounding margin to below the lethal level of minus 20°C without damaging surrounding healthy tissue (Georgiades et al. 2011). The major applications of cyroablation are in the kidney and prostate, although the technology has also been employed in other organs.

There are two FDA-approved cyroablation systems available in the United States from the companies Endocare and Galil Medical. Endocare was recently acquired by HealthTronics and offers four cyroprobes, with the largest probe generating a −20°C teardrop-shaped isotherm of 44 mm in length and 24 mm in diameter. The Galil Medical system also offers four cyroprobes that also differ in the size and shape of ice ball created. These devices, along with a sampling of cryoprobes, are shown in Figure 12.9.

12.3.3.2 Navigation Systems and Cyroablation

While there have not been as many studies published on the use of navigation systems in cyroablation as in radiofrequency ablation, there have been several investigations worth noting as shown in Table 12.2. Most of the cryoablation studies have focused on the kidney, but there have been also been several studies investigating the liver and other organs. Each study listed in the table will now be described. The use of navigation in percutaneous renal tumor ablation was investigated in a clinical trial of 10 patients (Haber et al. 2010b). Patients with enhancing renal masses underwent a preoperative CT scan with a preplaced tracking sensor. The CT images were sent to the navigation platform for three-dimensional volume rendering. A handle that

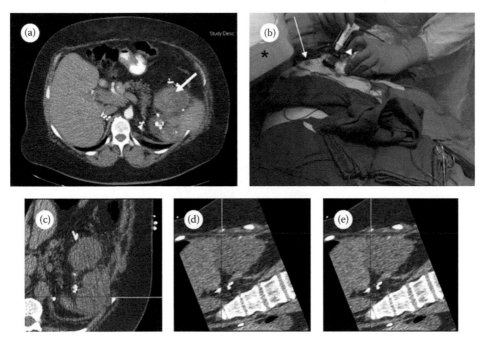

FIGURE 12.7 Electromagnetic tracking to guide renal tumor RFA. (a) Preprocedure contrast-enhanced axial CT demonstrates a left upper renal neoplasm. (b) Fiducial markers (long white arrow) on the patient's skin surface enable registration of electromagnetic tracking with those on corresponding CT images. (c-e) Real-time 3-dimensional visualization and navigation during RFA is made possible by tracking the tip of the guiding instrument using tiny sensors embedded in the electrode guide. Images from the graphic user interface demonstrating coregistered data, enabling real-time display of instrument position and orientation. The crosshairs generated by electromagnetic needle tracking are fused with multiplanar reconstructed CT images. (Reprinted from Stone, M. J., Venkatesan, A. M., Locklin, J. et al., *Tech Vasc Interv Radiol* 10, 2, 2007. © 2011, with permission from Elsevier.)

was tracked by an infrared camera was used to navigate and guide the needle percutaneously to the target. A mean decrease in fluoroscopy duration of 18 seconds was noted for each probe placed. The mean target registration error was 4.2 mm, and no complications were observed.

In a porcine model, this same group investigated the addition of real-time ultrasonography to CT imaging for perctuaneous renal cyroablation (Haber et al. 2010a). Thirty-five artificial renal tumors were created in 11 pigs. The cyroprobe was inserted percutaneously and ice ball formation was monitored continuously using real-time sonography. Synchronization between the ultrasound and CT images was successful in all cases.

MRI can also be used to monitor cyroablation and has the advantage of better soft tissue contrast. MRI-guided percutaneous cyroablation of liver cancer was completed in 32 patients with hepatocellular carcinoma at the diaphragm dome, the first hepatic hilum, and near the gallbladder (Wu et al. 2010). Cryoprobe insertion was guided by an optical navigation system (Panasee® MRI-compatible optical navigation system, XinAoMDT, China) and with continuous MRI scans (Signa® 0.35T open scanner, GEHealthcare, Milwaukee, U.S.). The number of cryoprobes used varied from 2 to 7. The conclusions were that 3D imaging, an arbitrary oblique scan plane, direct multi-plane reconstruction, and fusion with optical navigation information allowed for faster, safer, and more precise probe position.

Intraoperative visualization of 3D temperature maps along with 3D navigation for hepatic cyroablation was studied in a porcine model (Samset et al. 2005). An optical tracking system was integrated with a navigation system and graphical user interface based on the Qt application programming interface (Nokia, Finland). The software included the ability to segment the liver and display temperature maps (Figure 12.10). The concept was demonstrated in a single swine study, but no validation was completed.

An early paper described image navigation for MR-guided cryosurgery in 11 cases including five renal tumors, three uterine

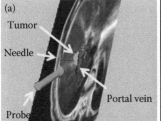
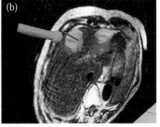

FIGURE 12.8 Navigation display based on the 3D Slicer user interface show the ultrasound probe, needle, tumor, and portal vein (a) and segmentation of the gall bladder displayed with the tumor (b). (Reprinted from Maeda, T., Hong, J., Konishi, K. et al., *Surg Endosc* 23,5, 2009. © 2011, with permission from Springer Science + Business Media.)

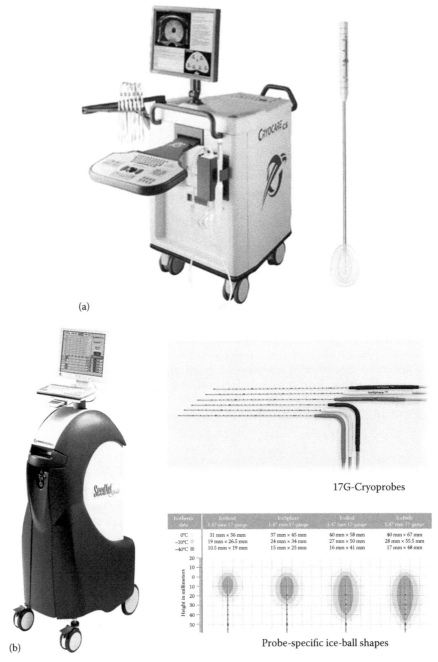

(a)

(b)

17G-Cryoprobes

Isotherm data	IceSeed 1.47 mm 17-gauge	IceSphere 1.47 mm 17-gauge	IceRod 1.47 mm 17-gauge	IceBulb 1.47 mm 17-gauge
0°C	31 mm × 36 mm	37 mm × 45 mm	40 mm × 58 mm	40 mm × 67 mm
–20°C	19 mm × 26.5 mm	24 mm × 34 mm	27 mm × 50 mm	28 mm × 55.5 mm
–40°C	10.5 mm × 19 mm	15 mm × 25 mm	16 mm × 41 mm	17 mm × 48 mm

Probe-specific ice-ball shapes

FIGURE 12.9 Images of two cryoablation devices for soft-tissue application: (a) Endocare Cryocare CS shown with illustration of ice ball and isotherms at tip of cryoprobe (© 2011 Endocare Inc. With permission.); and (b) Galil Medical SeedNet systems. (© 2008 Galil Medical. With permission.)

TABLE 12.2 Studies Describing the Use of Image-Guided Navigation with Cyroablation

Organ	Imaging Modality	Navigation	Study	Reference
Kidney	CT	Optical	Human study	Haber et al. (2010b)
Kidney	CT/ultrasound	Magnetic	Porcine model	Haber et al. (2010a)
Liver	MRI	Optical	Human study	Wu et al. (2010)
Liver	MRI	Optical	Porcine model	Samset et al. (2005)
Kidney, Liver, and Uterus	MRI	Optical	Human study	Mogami et al. (2002)

FIGURE 12.10 MR image in plane with the cryoprobe during a porcine study, showing the artificial tumor colored according to the temperature, and the surface of the liver. (Reprinted from Samset, E., Mala, T., Aurdal, L., and Balasingham, I., *Comput Med Imaging Graph* 29, 2005. © 2011, with permission from Elsevier.)

fibroids, and three metastatic liver tumors (Mogami et al. 2002). The MRI system was a 0.3T AIRIS II (Hitachi, Tokyo, Japan). The optical navigation system was based on a Polaris tracking system and a passive reference frame that was attached to the distal end of the cryoprobe. The size of the lesions ranged from 1.2 cm (metastatic liver tumor) to 9.0 cm (uterine fibroid). The cryoprobes used ranged 2 to 3 mm in diameter, and the number of probes used range from one to three. The navigation system was considered useful in the precise placement of the probes, although some disadvantages were noted, including the need to maintain a line of sight from the tracking system to the reference frame and the fact that needle bending cannot be accounted for since the tip of the cryoprobe is not tracked.

A team from Brigham and Women's Hospital in Boston looked at image registration of preprocedural MRI and intraprocedural CT to visualize renal tumors during CT-guided cyroablation procedures (Oguro et al. 2011). Both rigid and nonrigid registration methods were used. The registration methods were applied retrospectively to 11 CT-guided procedures, and the team concluded that the nonrigid method was more accurate and could be used to improve visualization.

12.3.4 Transarterial Chemoembolization

Transcatheter arterial chemoembolization (TACE) is a minimally invasive approach for treatment of unresectable hepatocellular carcinoma (HCC) that may provide symptom relief and prolong survival. TACE relies on inherent differences in blood supply to the hepatic tumor versus the normal hepatic parenchyma. Because hepatic tumors are supplied almost exclusively by the hepatic artery, high intratumoral chemotherapeutic concentration with low systemic exposure can be achieved through transcatheter arterial infusion into the vessels directly supplying the tumor. Recent developments in TACE include the use of drug eluting microspheres that provide simultaneous delivery of chemotherapy and embolization with controlled drug release over time (Liapi et al. 2010).

It should be noted that TACE is not a thermal therapy, however we have included it in this chapter as it is an important technique in the interventional oncology armamentarium, and the planning systems that have been developed for TACE as described following might be applied to guidance of thermal therapies in the future. In addition, we will describe the role of TACE in combination therapy in Section 12.4.

While navigation techniques have not been directly applied to TACE, there have been some reports of planning systems that have been developed to identify the vessels feeding a tumor or to evaluate TACE treatments (Deschamps et al. 2010). In a study of 18 consecutive patients undergoing TACE, the tumor feeding vessels were identified using three-dimensional analysis on a General Electric Advantage Workstation (Figure 12.11). The images were acquired on a General Electric Innova rotational angiography system that provides cone-beam CT imaging using a flat panel detector. The determination of feeding vessels was done by postprocessing of the images in three steps: (1) structure extraction; (2) target definition; and (3) feeding vessel selection. Structure extraction is accomplished by manual placement of a seed point at the entrance of the hepatic artery and applying thresholding and intensity-based segmentation to "grow" the vasculature. For target definition, the user adjusts a region of interest around the tumor. The feeding vessels are then highlighted, and background structures such as the spine can be hidden for better visualization. The study conclusion was that vessel segmentation and three-dimensional analysis has a higher sensitivity in determining sub-segmental feeding vessels than using two-dimensional imaging alone.

In a related study, contrast-enhanced ultrasound was used for postinterventional follow-up after TACE (Ross et al. 2010). The ultrasound images were fused with CT or MRI of the liver after treatment to evaluate the tumor vascularity and perfusion of HCC tumor lesions. The study conclusion was that image fusion provided a better visualization of the microcirculation and the residual tumor perfusion at an earlier time point than the usual modalities such as non-contrast enhanced CT. While there was no real-time navigation used during the procedure, both this study and the preceding study showed the value of providing more imaging information during the procedure, and one might speculate that navigation during the procedure could help as well.

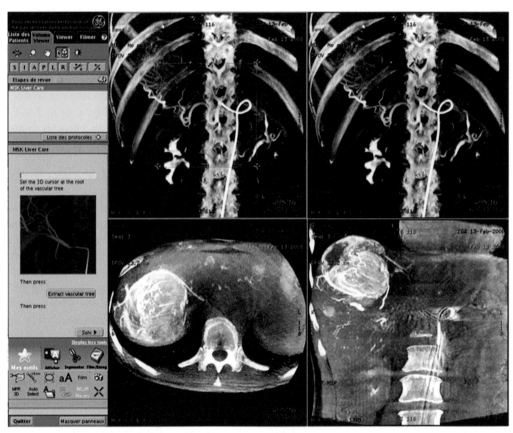

FIGURE 12.11 Image display on the Advantage Workstation. The upper views show a volume rendered reconstruction of the 3D image. Axial and coronal multiplanar reconstructions are shown in the lower views. (Reprinted from Deschamps, F. et al., *CardioVascular and Interventional Radiology* 33, 2010. © 2011, with permission from Springer Science + Business Media.)

12.4 Potential Future Developments

Just as medical imaging has continued to improve and enable more minimally invasive procedures, the authors believe that improvements in navigation systems and related technologies will lead to their increased use in precision thermal interventions. In this section, we present three areas where technology development may contribute to improved thermal therapies.

12.4.1 Joystick Navigation for Catheter Placement

Catheter ablation is a commonly performed procedure in cardiac interventions. Two remote catheter navigation systems have been introduced into clinical practice, both enabling remote joystick control of the ablation catheter (Ernst 2008). The Sensei X Robotic Catheter system from Hansen Medical is based on a steerable sheath that drives a remote catheter under joystick control. The other system, Niobe from Stereotaxis, is based on a steerable permanent magnet field that enables remote manipulation of the catheter. While these systems are currently focused on cardiac applications, the technology could be applied to TACE for steering of the catheter or perhaps to future thermal therapies with flexible applicators.

12.4.2 Robotically Assisted Needle Placement

Several different robotic systems have been developed to assist in needle placement under imaging modalities such as CT, MRI, or US. In collaboration with the URobotics Laboratory at Johns Hopkins, our group employed a needle driver robot for spinal nerve blocks at Georgetown University Hospital (Figure 12.12). The robot used was from the same laboratory as the system described by Solomon (Solomon et al. 2006) and includes a remote center of motion capability for positioning the needle at any angle. The needle driver is controlled by a joystick and touch screen interface. Such systems might be incorporated with planning systems in the future to more precisely carry out thermal ablations or be integrated with imaging devices for real-time feedback and the ability to modify thermal treatment plans on the fly for improved outcomes.

12.4.3 Combination Therapy and Navigation Assistance

An exciting area where navigation assistance and computer-based planning might play a significant role is combination therapy. One combination therapy approach is based on transcatheter arterial chemoembolization (TACE) and radiofrequency

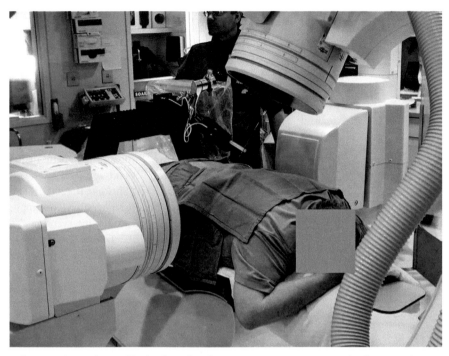

FIGURE 12.12 Robotically assisted spinal nerve blocks clinical trial at Georgetown University Medical Center. The interventional neuroradiologist used a joystick controlled robot to place the needle during the intervention.

ablation (RFA) for primary liver cancer, specifically hepatocellular carcinoma (HCC).

A common clinical scenario is that mostly hypervascular HCC lesions occasionally do not have an abundant arterial vascular supply, making them difficult to detect on angiography and making TACE difficult to perform clinically. Conversely, HCC lesions can have a sufficient vascular supply but be essentially invisible without the use of iodinated contrast agents. This makes the RFA procedure challenging as the lesions are hard or impossible to see during the CT-guided ablation procedure. Clinically, combination therapy is sometimes used where the TACE procedure is done first and ethiodol is injected transarterially into the liver segment containing the tumor. This traps the ethiodol in the tumor and allows subsequent visualization during the RFA.

Navigation and image guidance could enable a more quantitative and scientifically reproducible approach to this therapy by providing improved visualization of the tumor's vascular supply to guide catheter position during the TACE procedure, navigation assistance for precise placement of the RFA probe, and intra-procedural feedback regarding quantitative treatment assessment. The proposed workflow for such a system is shown in Figure 12.13.

12.5 Summary

Navigation and image guidance for thermal cancer therapy is an emerging area of research and clinical practice. Navigation through the use of tracking systems has been shown to assist with precision placement of instrumentation such as radiofrequency

ablation probes or cryoprobes. This method of augmenting the current clinical workflow is one potential path for improving clinical outcomes. In tandem, improvements in imaging can provide better visualization of tumor margins to enable more complete tumor resection while preserving healthy tissue. The concurrent use of reliable real-time feedback of the ablation procedure further maximizes the ability to obtain precise surgical margins. Temperature sensors and imaging thermometry in current use have their limitations. The evolution of these technologies to provide a more refined 3D view of the ablation process will further improve clinical outcome. Used in conjunction with this real-time feedback, simulation and modeling of the ablation process has been shown to accentuate clinical practice.

Precise tissue margins and more focused tumor resection can be made possible through the confluence of navigation, simulation, modeling, and real-time feedback. Just as tracking of tools has been shown to improve targeting accuracy, the use of robotics such as remote manipulation of catheter systems and robotic needle placement will further enhance this trend. The use of medical robotics has increased in clinical practice across the country in the last decade. Past the initial learning phase involved with the use of a new device, reduction in procedure times have been consistently noted. Planning and preoperative simulation will also enable more complex procedures.

The complexity of biochemical processes related to tumor growth and proliferation and the unique physiology of different types of cancers have resulted in an appreciation of combination therapy as an effective tool in the clinical armamentarium. While a single treatment modality applicable for a broad range of cancer types has remained elusive, and the notion now all but abandoned,

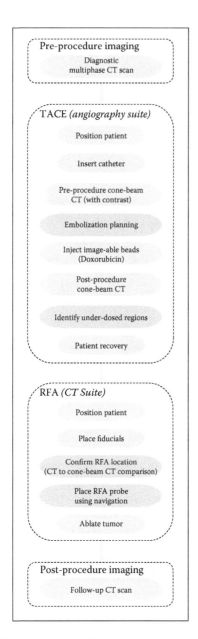

FIGURE 12.13 Proposed workflow for navigated image-guided combination therapy based on TACE and RFA.

technological improvements in application and sophistication of the various components of a navigated image guidance system can be expected to influence future clinical practice.

References

Ahmed, M. and Goldberg, S. N. (2011). Chapter 5: Liver ablation. In *Percutaneous Tumor Ablation: Strategies and Techniques*, eds. K. Hong and C. S. Georgiades, 52–72. New York: Thieme Medical Publishers, Inc.

Arun, K. S., Huang, T. S., and Blostein, S. D. (1987). "Least-squares fitting of two 3-D point sets." *IEEE Trans. Pattern Anal. Mach. Intell.* 9(5): 698–700.

Bale, R., Widmann, G., and Stoffner, D. I. R. (2010). "Stereotaxy: Breaking the limits of current radiofrequency ablation techniques." *Euro J Radiology* 75(1): 32–36.

Crocetti, L., de Baere, T., and Lencioni, R. (2010). "Quality improvement guidelines for radiofrequency ablation of liver tumours." *Cardiovascular and Interventional Radiology* 33(1): 11–17.

Curley, S. A., Izzo, F., Delrio, P., Ellis, L. M., Granchi, J. Vallone, P. et al. (1999). "Radiofrequency ablation of unresectable primary and metastatic hepatic malignancies: Results in 123 patients." *Annals of Surgery* 230(1): 1–8.

Dandekar, O. and Shekhar, R. (2007). "FPGA-accelerated deformable image registration for improved target-delineation during CT-guided interventions." *Biomedical Circuits and Systems, IEEE Transactions* 1(2): 116–127.

Deschamps, F., Solomon, S. B., Thornton, R. H., Rao, P., Hakime, A., Kuoch, V. et al. (2010). "Computed analysis of three-dimensional cone-beam computed tomography angiography for determination of tumor-feeding vessels during chemoembolization of liver tumor: A pilot study." *Cardiovascular and Interventional Radiology* (33): 1235–1242.

Enquobahrie, A., Cheng, P., Gary, K., Ibanez, L., Gobbi, D., Lindseth, F. et al. (2007). "The image-guided surgery toolkit IGSTK: An open source C++ software toolkit." *J Digit Imaging* 20 Suppl 1: 21–33.

Ernst, S. (2008). "Magnetic and robotic navigation for catheter ablation: "Joystick ablation"." *J Interv Card Electrophysiol* 23(1): 41–44.

Galloway, R. and Peters, T. M. (2008). Chapter 1: Overview and history of image-guided interventions. In *Image-Guided Interventions: Technology and Applications*, eds. T. M. Peters and K. R. Cleary, 1–21. Heidelberg, Germany: Springer.

Georgiades, C. S. and Marx, J. K. (2011). Chapter 2: Cryoablation: Mechanism of action and devices. In *Percutaneous Tumor Ablation: Strategies and Techniques*, eds. K. Hong and C. S. Georgiades, 15–26 Thieme Medical Publishers, Inc.

Gering, D. T., Nabavi, A., Kikinis, R., Hata, N., O'Donnell, L. J., Grimson, W. E. L. et al. (2001). "An integrated visualization system for surgical planning and guidance using image fusion and an open MR." *J Magn Reson Imaging* 13(6): 967–975.

Giesel, F. L., Mehndiratta, A., Locklin, J., McAuliffe, M. J., White, S., Choyke, P. L. et al. (2009). "Image fusion using CT, MRI and PET for treatment planning, navigation and follow up in percutaneous RFA." *Experimental Oncology* 31(2): 106–114.

Haber, G. P., Colombo, J. R., Remer, E., O'Malley, C., Ukimura, O., Magi-Galluzzi, C. et al. (2010a). "Synchronized real-time ultrasonography and three-dimensional computed tomography scan navigation during percutaneous renal cryoablation in a porcine model." *J Endourology* 24(3): 333–337.

Haber, G. P., Crouzet, S., Remer, E. M., O'Malley, C., Kamoi, K., Goel, R. et al. (2010b). "Stereotactic percutaneous cryoablation for renal tumors: Initial clinical experience." *J Urology* 183(3): 884–888.

Hawkes, D. J., Barratt, D., Blackall, J. M., Chan, C., Edwards, P. J., Rhode, K. et al. (2005). "Tissue deformation and shape models in image-guided interventions: A discussion paper." *Medical Image Analysis* 9(2): 163–175.

Horn, B. K. P. (1987). "Closed-form solution of absolute orientation using unit quaternions." *J Optical Society Amer* 4(4): 629–642.

Israel, O., Keidar, Z., Iosilevsky, G., Bettman, L., Sachs, J., and Frenkel, A. (2001). "The fusion of anatomic and physiologic imaging in the management of patients with cancer." *Seminars in Nuclear Medicine* 31: 191–205.

Kuehl, H., Antoch, G., Stergar, H., Veit-Haibach, P., Rosenbaum-Krumme, S., Vogt, F. et al. (2008). "Comparison of FDG-PET, PET/CT and MRI for follow-up of colorectal liver metastases treated with radiofrequency ablation: Initial results." *Euro J Radiology* 67(2): 362–371.

LeVeen, R. F. (1997). "Laser hyperthermia and radiofrequency ablation of hepatic lesions." *Seminars in interventional radiology,* Thieme. 14: 313–324.

Liapi, E. and Geschwind, J. F. H. (2010). "Chemoembolization for primary and metastatic liver cancer." *Cancer J* 16(2): 156–162.

Livraghi, T., Goldberg, S. N., Meloni, F., Solbiati, L., and Gazelle, G. S. (1999). "Small hepatocellular carcinoma: Treatment with radio-frequency ablation versus ethanol injection." *Radiology* 210(3): 655–661.

Maeda, T., Hong, J., Konishi, K., Nakatsuji, T., Yasunaga, T., Yamashita, Y. et al. (2009). "Tumor ablation therapy of liver cancers with an open magnetic resonance imaging-based navigation system." *Surg Endosc* 23(5): 1048–1053.

Mogami, T., Dohi, M., and Harada, J. (2002). "A new image navigation system for MR-guided cryosurgery." *Magn Reson Med Sci* 1(4): 191–197.

Mundeleer, L., Wikler, D., Leloup, T. and Warzee, N. (2008). "Development of a computer assisted system aimed at RFA liver surgery." *Comput Med Imaging Graph* 32(7): 611–621.

Oguro, S., Tuncali, K., Elhawary, H., Morrison, P. R., Hata, N., and Silverman, S. G. (2011). "Image registration of preprocedural MRI and intra-procedural CT images to aid CT-guided percutaneous cryoablation of renal tumors." *Int J Comput Assist Radiol Surg* 6(1): 111–117.

Pennathur, A., Abbas, G., Gooding, W. E., Schuchert, M. J., Gilbert, S., Christie, N. A. et al. (2009). "Image-guided radiofrequency ablation of lung neoplasm in 100 consecutive patients by a thoracic surgical service." *Annals of Thoracic Surgery* 88(5): 1601–1608.

Pennes, H. H. (1948). "Analysis of tissue and arterial blood temperatures in the resting human forearm." *J Applied Physiol* 1(2): 93–122.

Ross, C. J., Rennert, J., Schacherer, D., Girlich, C., Hoffstetter, P., Heiss, P. et al. (2010). "Image fusion with volume navigation of contrast enhanced ultrasound (CEUS) with computed tomography (CT) or magnetic resonance imaging (MRI) for post-interventional follow-up after transcatheter arterial chemoembolization (TACE) of hepatocellular carcinomas (HCC): Preliminary results." *Clinical Hemorheology and Microcirculation* 46(2): 101–115.

Rossi, S., Buscarini, E., Garbagnati, F., Di Stasi, M., Quaretti, P., Rago, M. et al. (1998). "Percutaneous treatment of small hepatic tumors by an expandable RF needle electrode." *Amer J Roentgenology* 170(4): 1015–1022.

Samset, E., Mala, T., Aurdal, L., and Balasingham, I. (2005). "Intra-operative visualisation of 3D temperature maps and 3D navigation during tissue cryoablation." *Comput Med Imaging Graph* 29(6): 499–505.

Scheele, J., Stang, R., Altendorf-Hofmann, A. and Paul, M. (1995). "Resection of colorectal liver metastases." *World J Surgery* 19(1): 59–71.

Solomon, S. B., Patriciu, A., and Stoianovici, D. S. (2006). "Tumor ablation treatment planning coupled to robotic implementation: A feasibility study." *J Vascul Interventional Radiol* 17(5): 903–907.

Stone, M. J., Venkatesan, A. M., Locklin, J., Pinto, P., Linehan, M., and Wood, B. J. (2007). "Radiofrequency ablation of renal tumors." *Tech Vasc Interv Radiol* 10(2): 132–139.

Villard, C., Soler, L., and Gangi, A. (2005). "Radiofrequency ablation of hepatic tumors: Simulation, planning, and contribution of virtual reality and haptics." *Computer Methods on Biomechanical and Biomedical Engineering* 8(4): 215–227.

Wood, B. J., Locklin, J. K., Viswanathan, A., Kruecker, J., Cebral, J., Sofer, A. et al. (2007). "Technologies for guidance of radiofrequency ablation in the multimodality interventional suite of the future." *J Vascul Interventional Radiol* 18(1 Pt 1): 9–24.

Wu, B., Xiao, Y. Y., Zhang, X., Zhang, A. L., Li, H. J., and Gao, D. F. (2010). "Magnetic resonance imaging-guided percutaneous cryoablation of hepatocellular carcinoma in special regions." *Hepatobiliary Pancreat Dis Int* 9(4): 384–392.

Temperature Imaging Using Ultrasound

R. Martin Arthur
Washington University in St. Louis

13.1 Introduction

Thermal therapies for the treatment of pathological tissues range from freezing with cryosurgery to high-temperature ablation.[1–11] The application and impact of these methods either alone as an alternative or adjuvant to radiotherapy or chemotherapy is the subject of ongoing investigation.[12–19] In particular, the significance and impact of high-temperature ablation therapies is growing rapidly.[4, 6–9,20–24]

A major limitation of thermal therapies, however, is the lack of detailed thermal information available to monitor and guide the therapy.[13, 14,16,25–29] Temperatures are routinely measured with sparse invasive measurements. The limited number of measurements may produce temperature distributions with less detail than is necessary to assess thermal dosimetry properly.[16, 26] With the advent of multi-element heating devices, there is increased need for temperature measurements that could provide detailed feedback about temperature distributions. This information in near real time would considerably improve the ability to deliver consistently effective temperature distributions.[30–34]

To meet the capability of present and forthcoming heating technologies for hyperthermia, a clinically useful method is needed to measure 3D temperature distributions to within 1°C with spatial resolution within 1 cm³ or better. A noninvasive method for volumetrically determining temperature distribution during treatment would greatly enhance the ability to

uniformly heat tumors at therapeutic levels in patients receiving thermal therapy.[35]

Many investigators have looked at ways of measuring temperature noninvasively. Possible methods include impedance tomography,[36] microwave radiometry,[37] and magnetic resonance imaging (MRI).[4,38] MRI temperature imaging appears to have the required accuracy and spatial resolution for many thermal therapy scenarios, but it is expensive, requires a fixed installation, and may be difficult to use along with some heating therapies.[38] Nevertheless, at present MRI is the most advanced clinical technology for noninvasive monitoring of thermal therapies.[39, 40]

On the other hand, ultrasound is a nonionizing, convenient, and inexpensive modality with relatively simple signal processing requirements. These attributes make it an attractive method to use for temperature estimation if a temperature-dependent ultrasonic parameter can be identified, measured, and calibrated. Methods for using ultrasound as a noninvasive thermometer fall into three categories: (1) those exploiting thermal strain, which are based on echo shifts due to changes in tissue thermal expansion and speed of sound (SOS), (2) those that use the measurement of acoustic attenuation coefficient, and (3) those that exploit the change in backscattered energy (CBE) from tissue inhomogeneities.

In this work, we explore ongoing efforts in temperature estimation with ultrasound using shifts in echo position, thermal strain, variations in ultrasonic attenuation, and changes in backscattered ultrasonic energy that occur due to thermal effects. This chapter is an extension of a review article published in 2005.[41] That article focused on the hyperthermia temperature range. We examine approaches that appear to have the greatest potential for monitoring temperatures in both the hyperthermia (41–45°C) range and in ablation border zones (>60°C).

13.2 Thermal Effects on Backscattered Ultrasound

As tissue is insonified during heating, at least two effects are easily seen in ultrasonic backscattered signals and images. They are a shift in apparent position of scattering regions and changes in signal strength from those regions. These changes are associated with thermal effects on SOS in tissue and on tissue attenuation and backscatter-coefficient properties.

13.2.1 Echo Shifts and Changes in Signal Strength

Amplitude-mode backscattered signals and radio-frequency (RF) images from samples of bovine liver during uniform heating in a water bath are shown in Figure 13.1. With temperature increase, the time it took for echoes from the liver specimen interfaces and scattering regions to reach the insonifying transducer changed in the A-mode echo signals. Apparent motion in the RF images occurred in part because SOS changes with temperature. SOS, however, was assumed fixed at 1540 m/s by the imaging system.

The rectangular region in the RF images of Figure 13.1 highlights an image feature that appears to move as the specimen was heated. Movement in the axial direction in images is similar to the shifts in the A-mode echo signals. Apparent motion toward the transducer is consistent with the change in SOS in the water bath. The images show, however, that there is also an apparent lateral movement, presumably due to changes in the tissue alone.

There are also clear changes in backscattered signal strength with temperature. This effect is seen in the delineated bands of the A-mode signals in Figure 13.1, which isolate scattering regions whose energy increased (band #1) or decreased (band #2) with temperature.

13.2.2 Nonthermal and Unwanted Thermal Effects

To rule out nonthermal sources in backscattered signals and images, changes in ultrasonic signals and images must be determined over the duration of a measurement paradigm when no heating occurs. Unwanted thermal effects include those on the measurement system itself. Spurious thermal effects may be seen in tissue as well. For example, in echo shift measurements, the heated region has a tendency to cause a "thermal lens" effect that distorts the image of the tissue beyond the heated region, where this effect can cause artifacts in temperature estimation.[44]

Thermal effects on a typical phased array transducer in a water bath have been seen in B-mode images of the stainless steel wires in the AIUM 100 mm test object.[45] Apparent motion toward the transducer occurred with an increase in temperature during heating from 37 to 50°C.[43] That motion, however, can be accounted for by the change in SOS in the water bath. The changes in backscattered signal level due to thermal effects

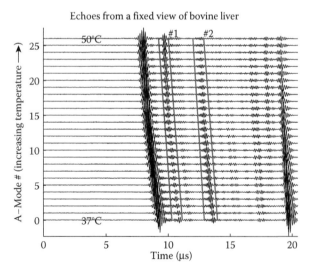

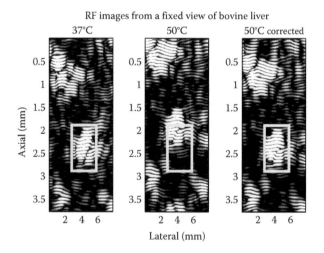

FIGURE 13.1 (**See color insert.**) Changes in backscattered ultrasound with temperature. (Left) Echoes measured from a single site in a 1 cm thick sample of fresh bovine liver at temperatures from 37 to 50°C. The two delineated echoes (indicated by bands marked #1 and #2) shift with temperature and have energies that appear to change with temperature (similar to Figure 13.4 in Arthur et al.[42]). (Right) RF images of a fixed region in bovine liver showing apparent motion from 37 to 50°C. The right panel shows the image at 50°C after motion compensation (similar to Figure 13.3 in Arthur et al.[43]).

on the AIUM wires and transducer, however, were small. The change in signal level from a wire in the AIUM test object over one hour was $< \pm 0.1$ dB at a constant temperature. Images at 37 and 50°C showed a change in signal level of about 0.3 dB. The small changes over time and with temperature due to effects on the test object and transducer are usually negligible compared to thermal effects in tissue.

We also tested our system under conditions for which we expected to see no change in backscattered energy. We measured the energy in backscattered signal levels in bovine liver over the time required for an experiment, but without heating, to determine the magnitude of nonthermal effects. Signal-level changes of $< \pm 0.2$ dB were typically seen over 80 minutes, twice the approximate duration of an experiment.[42]

13.3 Image Motion

Tracking the echoes from tissue interfaces and scattering regions is important for any temperature estimation method because a key challenge for *in vivo* studies is measurement of thermal effects in the presence of real motion in a perfused living system, in addition to apparent motion from thermal effects.

Quantifying the time dependence of echo positions is important because echo shift is the basis for much of the recent work on temperature estimation. In addition, in order to compare signal strengths under consistent conditions, motion of scattering regions must be tracked so that compensation for that motion can be applied as shown in Figure 13.2.

Techniques for motion estimation in ultrasonic images, that is, for estimation of the displacement between two image regions, comprise an active field of research[47–51] and have been used successfully for elasticity imaging, phase aberration correction, blood velocity estimation, and other applications. In these areas, motion estimation is typically called speckle tracking or time-delay estimation. Work based on exploiting thermal effects that induce tissue strain has been aimed at guiding focused ultrasound therapies, as well as estimation of temperature.[52–55]

Measurement of echo shifts and the tracking of real and apparent motion has been based on correlation techniques. Employing RF signals permits the use of cross-correlation as a similarity measure for automatic tracking of regions as a function of temperature. The result of maximizing 2D cross-correlation to estimate displacement and then shifting and resampling images to apply that displacement is shown in the RF images of

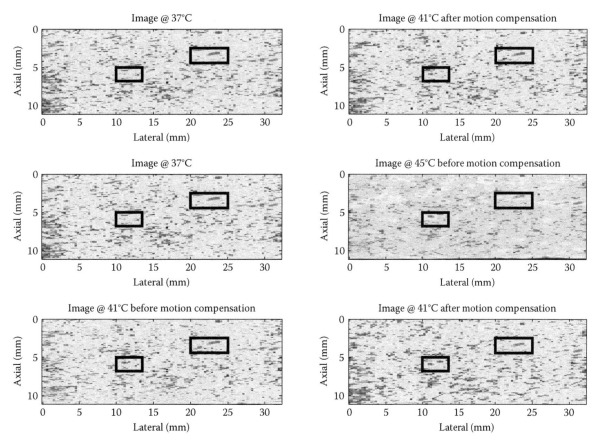

FIGURE 13.2 (Left) Conventional B-mode images at various temperatures from a specimen of turkey breast muscle before motion compensation. Apparent motion can be seen clearly in the boxes. (Right) Conventional B-mode images at the same temperatures after motion compensation. The patterns isolated in the boxes show both motion compensation and a change in the signal strength with temperature. (From Guo, Y., *A Framework for Temperature Imaging using the Change in Backscattered Ultrasonic Signals*, PhD thesis, Washington University, St. Louis, MO, 2009. With permission.)

Figure 13.1. Motion-compensated RF images can then be envelope detected to yield the motion-compensated conventional B-mode images of Figure 13.2.

Potential pitfalls with motion tracking and compensation techniques involve (1) propagation of errors in displacement estimation over the temperature range and (2) nonrigid image motion (i.e., apparent motion over temperatures that cannot be compensated by a simple shift of the images). Errors in displacement estimation depend on multiple factors, including: quantization errors that depend on the image sampling rate (or pixel size); decorrelation of the RF signals due to small changes in the underlying scattering structure;[56] signal-to-noise ratio and size of the region;[46, 57] and artifacts or features that appear in only one of the images.

There are two primary objectives for motion tracking in ultrasonic temperature imaging. For methods based on echo shifts, tracking is local and most effective in 1D. For methods that use signal strength changes, tracking is global, that is over a tissue volume and in 2D or 3D, where motion is likely to be nonrigid.

13.3.1 Tracking Echo Shifts

Of the ultrasonic thermometry methods explored to date, the use of echo shifts has received the most attention in the last decade. Most of these efforts have been geared toward RF ablation and high intensity focused ultrasound (HIFU) therapy, which typically heats small volumes of tissue to above 60°C.

Apparent and actual displacements of scattering regions are produced by changes in SOS and thermal expansion, respectively. Temperature estimation using these effects is based on measuring displacements in the direction of propagation z, which can be related to changes in SOS and to thermal expansion.[44, 58-61] The echo shift $t_\delta(z)$:

$$t_\delta(z) = t(z) - t_0(z) = 2 \int_0^z \left[\frac{[1 - \alpha(\zeta)]\,\partial\theta(\zeta)d\zeta}{c[\zeta, \theta(\zeta)]} - \frac{1}{c[\zeta, \theta(\zeta)]} \right] d\zeta, \quad (13.1)$$

where $t(z)$ is the propagation and return time for an echo from depth z after heating, $t_0(z)$ is the time before heating, and c is the SOS, which is a function of depth ζ and the temperature $\theta(\zeta)$ at that depth. The linear coefficient of thermal expansion, α, depends on the medium and is also a function of depth.[52] In this approach, variation in SOS with temperature is assumed to be linear up to about 45°C, but the method has proven useful in assessing high-temperature ablation.[62]

The echo shift occurring between two successive RF images is estimated using the speckle tracking technique described before has been most successful in the axial dimension, that is, along the propagation axis. Repeating this process along adjacent beams can generate a 2D or 3D map of the shifts in a region of interest.[60] Shifts in the lateral and elevation are usually smaller than those in the axial direction, and their assessment tends to more subject to noise. Temperature maps are typically generated in 1D, based on differentiation of shifts along the propagation

direction, because shifts in this direction are less influenced by noise, using the time-shift relation, $t_\delta(z)$, in Equation 13.1 given before. Its application is illustrated in the section on thermal strain imaging following.

13.3.2 Tracking and Compensating for Real and Apparent Motion

Estimation of tissue motion is a key step in many applications associated with ultrasonic imaging, such as elasticity imaging,[49,63-65] estimation of the velocity of blood flow,[66-68] and noninvasive temperature estimation, including our work.[43,69-71] These approaches have focused on 2D motion in ultrasound images.

13.3.2.1 Rigid Motion

Apparent tissue motion from rigid-motion, cross-correlation algorithms in both axial and lateral directions from eight overlapping regions in 10 by 40 mm images of four specimens of bovine liver, two of turkey breast, and one of pork-rib muscle had mean values within ±0.5 mm over the 37 to 50°C temperature range.[43] Tissue features appeared to move closer to the transducer in the turkey, pork, and two of the liver specimens, which is consistent with the increase in the SOS in the water path between the tissue and transducer. In the other two liver specimens, presumably nonuniform thermal effects in tissue were larger than the changes due to SOS changes in the water bath.

A nonuniform component of tissue motion in the seven specimens cited herein was indicated by the lateral motion. This tissue-dependent component also contributed to the axial motion, particularly at temperatures above 47°C. There were also differences in the apparent lateral motion in the two specimens of turkey breast. One exhibited nearly constant lateral motion. Its tissue fibers were parallel to the array of transducer elements. The other one showed a change of several tenths of a millimeter near 47°C. Its striations were perpendicular to the array.

13.3.2.2 Nonrigid Motion

Differences in apparent motion from varying fixed regions suggest that apparent motion during heating may not be rigid. If a rigid-motion assumption does not hold, the size of the tissue region over which rigid motion compensation will be successful is limited. To overcome this limitation and to compensate for motion over the whole tissue region of interest in a single operation, we developed nonrigid motion-compensation algorithms that can operate in 1D, 2D, or 3D. They are based on optimization of a function of the cross correlation of an image at a particular temperature with a reference image.[46, 73]

The cost function for our motion-compensation algorithms is the normalized cross-correlation function of two RF datasets. Let $I_r(\mathbf{x})$ and $I_t(\mathbf{x})$ be the image at the reference temperature r and a shifted image at temperature t, respectively, where

$$I_r(\mathbf{x}) = I_t(\mathbf{x} + \mathbf{x}) \quad (13.2)$$

where $\mathbf{x}$, $\Delta\mathbf{x} \in R^2$ or R^3 are the spatial coordinates and the motion in those coordinates. Our goal was to find an estimate of the motion, $\hat{\mathbf{x}}$, such that $I_t(\mathbf{x} + \hat{\mathbf{x}})$ is as close to $I_r(\mathbf{x})$ as possible. The similarity between the two images was measured by their correlation:

$$C[I_r(\mathbf{x}), I_t(\mathbf{x})] = \frac{\sum_{\mathbf{x}} I_r(\mathbf{x}) I_t(\mathbf{x})}{\sqrt{\left(\sum_{\mathbf{x}} I_r^2(\mathbf{x}) \sum_{\mathbf{x}} I_t^2(\mathbf{x})\right)}}. \quad (13.3)$$

For motion compensation prior to CBE computation, $\hat{\mathbf{x}}$ was modeled to vary linearly over the image region. Specifically, it was a linear function of the motion at control points, chosen as the corners of the 2D image or 3D image volume

$$\hat{\mathbf{x}} = g(\hat{\mathbf{x}}_1, \ldots, \hat{\mathbf{x}}_n) \quad (13.4)$$

where n is the number of control points. Our goal of finding $\hat{\mathbf{x}}$ is equivalent to searching for the estimate of $\Delta\mathbf{x}_1, \ldots, \Delta\mathbf{x}_n$ that maximizes the correlation function. This procedure is a multi-variable optimization problem with the correlation as the cost function

$$(\hat{\mathbf{x}}_1, \ldots, \hat{\mathbf{x}}_n) = argmax_{(\mathbf{x}_1, \ldots, \mathbf{x}_n)} C[I_r(\mathbf{x}), I_t(\mathbf{x} + \hat{\mathbf{x}})]. \quad (13.5)$$

We used the built-in functions in MATLAB® to solve this optimization problem. The motion field over a tissue volume was represented as a linear function of 3D motion vectors at eight reference points, the corners of the data volume. Examples of motion fields in 2D are shown in Figure 13.3 from images of a live nude mouse.[72] The direction of the arrows represents the direction of motion; the length of the arrows represents the magnitude of the motion. It is clear that the motion was nonrigid.

Our 3D motion-compensation algorithm has been used to correct for the motion in the images during 3D heating experiments.[73] In each experiment, a sequence of ultrasound RF images in 3D was obtained at increasing temperature. Motion between adjacent pairs of 3D image sets was estimated and accumulated relative to the reference image set. Figure 13.4 shows the 3D frame for one specimen of turkey breast muscle, along with the motion in the axial, lateral, and elevation directions with temperature at the center slice of the 3D volume.

Accumulated motion in all directions shown in Figure 13.4 was < 340 μm. On average, tissue movement was < 20 μm per 0.5°C step. This small change is consistent with visual observation of echo shift and apparent motion in images. That displacement is nearly an order of magnitude less than motion tracking and compensation methods based on correlation can handle in this application. Displacements for which we compensated took place over several minutes. Frame intervals for conventional ultrasound imaging systems can be orders of magnitude smaller. A typical frame interval is 30 msec. Thus motion between

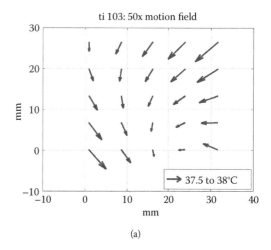

(a)

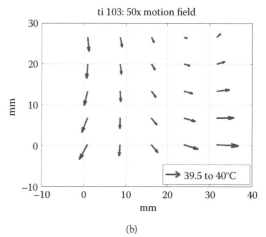

(b)

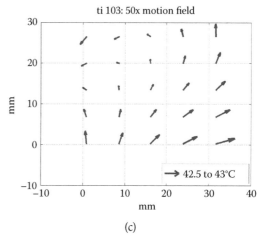

(c)

FIGURE 13.3 Nonrigid motion in 2D in a live nude mouse from (a) 37.5 to 38.0, (b) 39.5 to 40.0, and (c) 42.5 to 43.0°C. Motion is clearly nonrigid in all three frames. Arrow lengths are 50 times the actual motion field (similar to Figure 13.5 in Arthur et al.[72]).

frames encountered during clinical applications of ultrasonic thermometry can be much smaller than we encountered in our *in vitro* studies. Thus frame intervals for thermometry can be tailored to optimize the effectiveness and speed of motion-compensation algorithms.

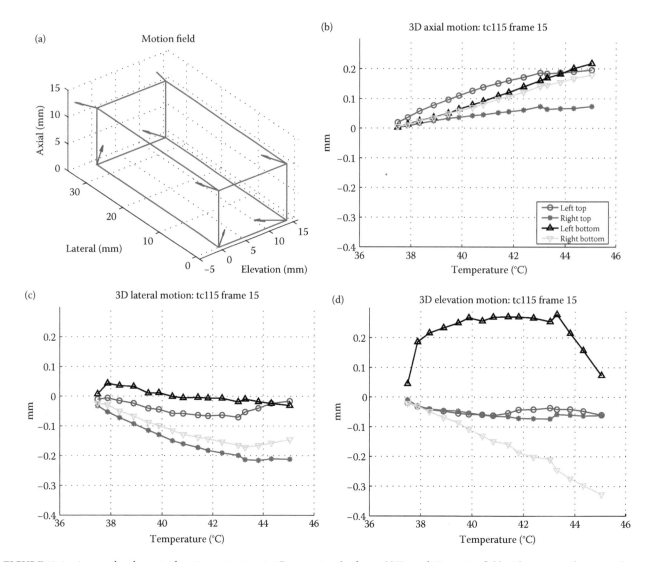

FIGURE 13.4 Accumulated nonrigid motion estimation in 3D over a 5 cm³ volume. (a) Typical 3D motion field with 5 × magnification at the corners of the tissue volume. (b) Axial, (c) lateral, and (d) elevation displacements for one specimen of turkey breast muscle at the center image (#15) of the 3D image set. Displacement values were found via trilinear interpolation of the estimated nonrigid 3D motion field over the image volume. The magnitude of the motion in a 0.5°C step was < 20μm (similar to Figure 13.3 in Arthur et al.[73]).

13.4 Ultrasonic Measurement of Temperature

Temperature dependence of ultrasonic tissue parameters has been reported extensively from *in vitro* analyses of properties that characterize tissue response to insonification.[74–84] These early investigations reported changes in tissue characteristics with temperature in order to evaluate thermal errors in tissue characterization. Some investigators did consider, however, the possibility of using temperature dependence of tissue properties as a means to track temperature changes.[76,77,85]

The primary ultrasonic parameter examined for its dependence on temperature in early work on measurement of temperature was SOS.[77,80,85] In these initial studies investigators tried to obtain SOS maps of the medium from which to infer temperature distributions. This approach, however, has never

been instituted clinically.[86] Perhaps this approach has not been implemented because in order to measure SOS, it is necessary to measure both distance and time to image an identifiable target from two directions, or to use a crossed-beam (multiple beams) method.[87] Such measurement is further complicated by the fact that ultrasonic windows do not always exist *in vivo* to allow insonification of a region of interest from two views. Another problem is that the temperature dependence of SOS differs depending on the tissue type (e.g., whether tissue has high water or fat content.)

More recently, the use of ultrasonic parameters as a guide for thermal therapy has been revisited with several different parameters being considered. In particular, papers by Sun and Ying;[35] Seip, Simon, Ebbini, and coworkers;[59,60,88] Maass-Moreno, Damianou, and coworkers;[58,89,90] and our group[42,70,91] have reported the changes in received ultrasonic signals due to changes in ultrasonic tissue characteristics with temperature. These changes have been investigated both theoretically and *in vitro* with an eye toward using these signals for noninvasive monitoring of thermal therapy. Most of these investigators have looked at the consequences of changes in SOS and thermal expansion with temperature that causes echo shifts or changes in the amplitudes of backscattered ultrasonic signals.

13.4.1 Experimental Concerns

Among the inherent advantages of using ultrasound for temperature imaging is that it is nonionizing, convenient, and relatively inexpensive. A-mode signals and phased-array images are routinely produced in real time. To take advantage of these attributes, effective temperature imaging should be expected to operate in real time with relatively simple additional signal processing requirements that employ existing ultrasonic equipment using ultrasonic properties extracted, if possible, from a single backscatter view.

To obtain sufficiently accurate measures of ultrasonic properties for temperature imaging, it may be necessary to compensate for known limitations of image generation in conventional imaging systems. These include effects of the measurement system itself, insertion loss, reflection and transmission losses, attenuation in tissue, and effects of beam diffraction.[92–99]

13.4.1.1 Temperature Standards

The conventional standard for temperature measurements in tissue is a reading from a thermocouple, which can fit into a hypodermic needle. Because of their invasive nature, however, thermocouple grids can only be used sparsely in tissue. For *in vivo* temperature imaging MRI is the standard modality during thermal therapy.[38–40,100,101] Nevertheless, MRI volumetric measurements themselves are corroborated with thermocouple readings.

Thermocouples are calibrated with a National Institute of Standards and Technology traceable thermometer. Typically, individual thermocouples are accurate to within $\pm 0.1°C$. To calibrate the thermocouples and the system that monitors them, they may be placed in a water bath with a heater. At equilibrium for a temperature of interest in the water bath, the NIST traceable standard reading and the thermocouple value are taken simultaneously. This process is repeated over the temperature range of interest. Typically, multiple calibration experiments must be conducted to assess the mean thermocouple errors and their standard deviations. The thermocouple monitoring system may employ an internal reference that varies each time the unit is used. This offset must be measured and used to correct the thermocouple reading during each experiment to calibrate thermocouple temperatures to within $0.2°C$.

13.4.1.2 Calibration of Tissue Properties during Uniform Heating

Measurements of temperature-dependent ultrasonic properties can be made by heating specimens in an insulated tank filled with deionized water, which had been degassed by vacuum pumping in an appropriate vessel. Degassing is necessary to prevent the formation of bubbles during heating. Gas bubbles scatter ultrasound and corrupt quantitative assessment of ultrasonic parameters.

Most ultrasonic imaging systems are based on notebook computers or other forms of computer-based systems. The computer in these imaging systems can be used to control tissue heating by setting the temperature of a circulating heater. The temperature in the tissue, monitored by a thermocouple, can be reported to a routine in the computer of the imaging system. When the desired temperature is reached the heater can be turned off and an image frame or image loop acquired and saved. The computer can also be used to automatically control the position of the transducer array position in the elevation direction via stepper motor to acquire 3D image sets. This process can be fully automated. For example, all keystrokes needed to switch between peripheral control and imaging can be administered using keystroke-emulation software, such as AutoIT (hiddensoft.com). An alternate for some imaging system is to control all of the operations via a proprietary software developers kit.

13.4.1.3 Temperature Imaging during Non-Uniform Heating

Heat sources in thermal therapies produce nonuniform temperature distributions because of the nature of the source itself or because of inhomogeneity in the tissue being heated. Hyperthermia heating is more nearly uniform than ablation therapy, which is concentrated to destroy a tumor or aberrant pathway in the heart, for example. The computer of the imaging system can be used to automate heating and image acquisition, as noted previously.

A nonuniform heat source may have many forms depending on the therapy. Heat sources include microwave antennas and low-frequency ultrasonic arrays for hyperthermia, and high-intensity ultrasound and RF electrodes for high-temperature ablation. These sources may be used in experiments to test ultrasonic thermometry for clinical settings, but other sources may be used to develop ultrasonic temperature imaging in the laboratory.

We created the fixture shown in Figure 13.5 to heat gelatin phantoms and tissue specimens[102] from a central hot-water source, while the phantom or tissue is surrounded by water at fixed reference temperature, usually $37°C$. This fixture uses a thermocouple grid to corroborate ultrasonic temperature images estimates. A 3D image set can be taken in which the tips of the thermocouples on each side of the specimen can be seen in the first and last images of the set, which are not used in generating temperature images, but can be used to place the thermocouple reading in the temperature images to assess their accuracy. Furthermore, with the metal of the thermocouples and their holders removed, the fixture can be used to perform MRI temperature imaging with the same hot-water heating source.

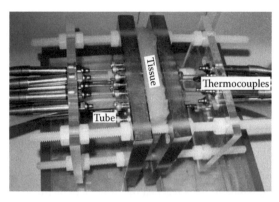

FIGURE 13.5 Configuration for nonuniform heating experiments. Tissue was imaged in a 37°C water bath with a 7.5 MHz linear array in the plane of the tissue. Heating was provided by 65°C water pumped through a 1 cm diameter silicon tube in the center of the tissue specimen. Thermocouple holders with fitted slots for the thermocouple shafts provided support to reduce movement of the thermocouple tips in the tissue.

13.4.2 Attenuation

The accepted method for measuring the attenuation coefficient is by use of a shadowed-reflector "substitution" technique. A reference RF trace (corresponding to the specular echo from a stainless-steel reflector through water only) and an RF trace corresponding to the reflected signal after it has passed through the tissue specimen are required. If the power spectra of the reference trace and the through-sample trace are found, then attenuation can be computed by subtracting the power spectrum through the sample from the spectrum of the reference in the logarithmic domain, correcting for interface insertion losses,

then normalizing the resulting spectrum by the sample thickness.[103] The frequency dependence of the attenuation coefficient is usually determined by a linear fit to the data over the useful bandwidth of the measurement system.[103–105]

Attenuation is more difficult to measure from single backscattered signals in tissue, that is, without a reference signal. Methods for setting bounds on the estimates of the attenuation of small tissue regions, particularly at temperatures below 50°C, may be useful in determining limits on accuracy and spatial resolution of attenuation-based temperatures.[106] The slope of attenuation found for a linear fit of attenuation shown in Figure 13.6 for four temperatures in dog myocardium is linear with temperature.[75] The attenuation curves at 2 MHz are not fit well, however, by a straight line, but are by a single-pole model[107] curve superimposed on the measurements in this figure. The pole locations shown in Figure 13.6 as a function of temperature may be a better choice for characterizing the temperature dependence of attenuation for temperature imaging, particularly at temperatures below 50°C.

Attenuation changes with temperature appear to be more pronounced at temperatures above 50°C than in the hyperthermia range. Several groups have investigated the temperature dependence above 50°C in tissue.[90,108,109] Damianou and coworkers investigated the temperature and the frequency dependence of ultrasonic attenuation and absorption in soft tissues.[90] They found that attenuation was highly dependent on temperature, but only at temperatures >50°C. Techavipoo and coworkers measured attenuation of canine tissue from 25 to 95°C with different tissue samples heated to different target temperatures to reduce cumulative tissue degradation. They found that attenuation at 3, 4, and 5 MHz was relatively unchanged from 40 to 60°C, but increased sharply above 60°C.[110] In measurements of insertion loss at room temperature before and after heating, increases in

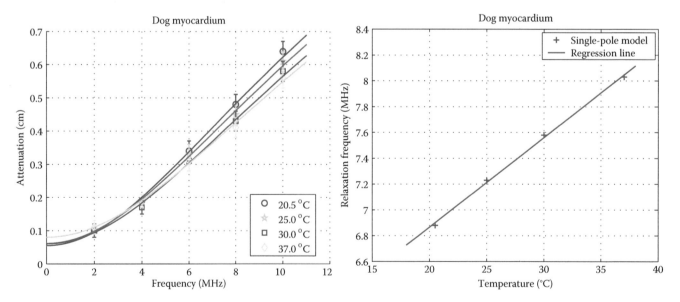

FIGURE 13.6 (Left) Attenuation of dog myocardium at four temperatures predicted by a single-pole tissue model. Data points were taken from the work of O'Donnell et al.[75] Temperatures indicated with each part are in degrees Celsius. Height of the vertical bar at each measurement is the standard error. (Right) Dependence of relaxation frequency of the single-pole model on the temperature of dog myocardium.[107] Transfer functions of the single-pole model were calculated from data shown in the left panel above.

attenuation of up to 2.4 dB/cm at 3.5 MHz were found in porcine liver after heating to 80°C in 300 sec.[109]

Ribault and coworkers also looked at the effect of temperature rise on frequency-dependent attenuation and found that tissue damage (lesion formation) caused a change in attenuation in porcine liver *in vitro*.[5] This effect was found by looking at the backscattered signal over the volume of the lesion and comparing the power received before and after high intensity focused ultrasound (HIFU). Other investigators have observed similar effects in the past.[111] Thus attenuation is of interest for thermometry and may have application with appropriate processing at temperatures below 50°C, but appears to be a parameter of interest at temperatures above 50°C, which may make it an attractive parameter for assessing high-temperature ablation.

13.4.3 Thermal Strain Using Echo Shift

Temperature maps based on thermal strain to find the change in temperature $\Delta T(z)$ along the propagation direction can be estimated by differentiating Equation 13.1, the expression for echo shift $t_\delta(z)$:[58]

$$T(z) = k \frac{\partial}{\partial z} t_\delta(z), \qquad (13.6)$$

where k is based on the change in SOS with temperature and the thermal expansion coefficient α of the medium of interest. For example, Maass-Moreno and coworkers investigated the ability to predict temperature in HIFU therapy from echo shifts in turkey breast muscle.[58,89] They found that results were consistent with their theoretical predictions. In investigations by Ebbini and coworkers, tracking echo shifts from scattering volumes was shown to be promising, as was the work of other investigators looking at echo shift for temperature estimation.[44,60,88]

Combining echo shift temperature estimation with elastography can improve the quality of monitoring of thermal lesions induced by focused ultrasound.[112] Furthermore, echo-shift estimation has been implemented with a zero-crossing method for tracking temporal positions of echo to determine the difference between A-mode signals gelatin phantoms before and after heating.[113] This technique yielded as much as a seven-fold improvement in computational efficiency compared to conventional cross-correlation methods with similar temperature estimation results.

Sun and Ying have also found some success in being able to predict temperatures using time-gated echo shifts, but they acknowledge the difficulty of using this method for general temperature monitoring because prior knowledge of both SOS and thermal expansion coefficients is necessary.[35] Obtaining *a priori* knowledge of both SOS and thermal expansion coefficients is a formidable problem for the *in vivo* case, as demonstrated by a quick look at the early tissue-characterization literature, which shows that SOS can vary greatly in different types of tissues. In fact the speed change due to temperature in lipid tissue is opposite in direction to the SOS change in aqueous tissue. These complications could cause difficulties in determining temperature in the complicated inhomogeneous tissues likely to be found in an *in vivo* situation.

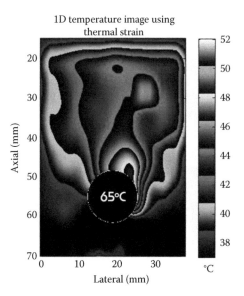

1D temperature image using thermal strain

FIGURE 13.7 (See color insert.) 1D thermal strain temperature image from axial echo shifts in turkey breast muscle during nonuniform heating using the fixture in Figure 13.5. Tissue was surrounded by water at 37°C. At time = 0 seconds 65°C water was pumped through the tissue center (black disk). Compare to the 3D CBE image in Figure 13.14 from a different specimen of turkey. Thermocouple readings in planes adjacent to the image were used to scale the strain image to give temperature estimates.

Varghese, Zagzebski, and coworkers investigated the spatial distribution of heating using echo shifts in studies that included *in vivo* measurements[61] and more recently *in vivo* temperature estimates during high-temperature ablation.[62] Temperature estimates were obtained using cross-correlation methods described previously and in Equation 13.1. Resulting temperature maps were used to display the initial temperature rise and to continuously update a thermal map of the treated region that was simultaneously monitored using thermosensors. Figure 13.7 shows a thermal strain image in a specimen of turkey breast muscle we generated from measurements taken with the fixture in Figure 13.5. The constant k in Equation 13.6 was estimated by comparison of the thermal strain image to thermocouple readings.

13.4.4 Change in Backscattered Energy

In a search for an ultrasonic parameter that changed monotonically with temperature, we modeled the backscattered energy from individual scatterers to an interrogating ultrasonic wave.[91] According to that model, the change in backscattered energy due to temperature was primarily dependent on the changes in SOS and density of the medium compared to their values in sub-wavelength inhomogeneities (scatterers) within the medium. Our predicted change in backscattered energy (CBE) at any temperature T with respect to its value at some reference temperature T_R is

$$CBE(T) = \frac{\alpha(T_R)}{\alpha(T)} \frac{\eta(T)}{\eta(T_R)} \frac{[1 - e^{-2\alpha(T)x}]}{[1 - e^{-2\alpha(T_R)x}]} \qquad (13.7)$$

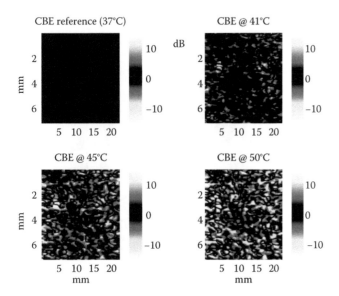

FIGURE 13.8 Change in backscattered energy in ultrasound images of bovine liver from 37 to 50°C after compensation for apparent motion. All images were referred, pixel-by-pixel, to the energy in the reference image at 37°C. Each color bar is in dB (similar to Figure 6, but from a different specimen and over a larger region of interest, in Arthur et al.[43]).

where, as functions of temperature, $\alpha(T)$ is the attenuation within the tissue volume and $\eta(T)$ is the backscatter coefficient of the tissue volume. Distance x is the path length in the tissue volume. We inferred the temperature dependence of the backscatter coefficient from the scattering cross section of a sub-wavelength scatterer.[114,115] Neglecting the effect of the small change in the wavenumber (<1.5%) with temperature, we assumed[91]

$$\frac{\eta(T)}{\eta(T_R)} = \frac{\left(\frac{\rho_m c(T)_m^2 - \rho_s c(T)_s^2}{\rho_s c(T)_s^2}\right)^2 + \frac{1}{3}\left(\frac{3\rho_s - 3\rho_m}{2\rho_s + \rho_m}\right)^2}{\left(\frac{\rho_m c(T_R)_m^2 - \rho_s c(T_R)_s^2}{\rho_s c(T_R)_s^2}\right)^2 + \frac{1}{3}\left(\frac{3\rho_s - 3\rho_m}{2\rho_s + \rho_m}\right)^2}, \quad (13.8)$$

where ρ and c are the density and speed of sound of the scatterer s and medium m. This expression applies to conditions where the wavelength λ is larger than $2\pi a$, where a is the radius of the scatterer. Assuming a speed of sound of 1.5 mm/μs and a frequency of 7.5 MHz, λ is 0.2 mm, which means that Equation 13.7 using Equation 13.8 applies to scatterers smaller than 30 μm.

Changes in backscattered energy were modeled assuming that the scattering potential of the volume was proportional to the scattering cross section of sub-wavelength scatterers. We predicted with this model that the change in backscattered energy could increase or decrease depending on what type of inhomogeneity caused the scattering. These calculations suggested that the change in backscattered energy could vary depending on the type of scatterers in a given tissue region.

In 1D studies, we showed that it is possible to isolate and measure backscattered energy from individual scattering regions, and that measured CBE was nearly monotonically dependent on

temperature.[42] In our studies of CBE in images, apparent motion of image features has been tracked and compensated for automatically as described previously, so that CBE can be measured at each pixel in motion-compensated images.[70] Figure 13.8 shows CBE images, that is, the energy changes relative to the reference image at 37°C. As predicted, the energy change is both positive (increasing) and negative (decreasing) with temperature. This approach allows use of the whole ultrasonic image rather than just the signals from selected scattering regions, but at the price of possibly increasing noise in estimates of CBE.

After compensating for apparent motion in images of bovine liver, turkey breast, and pork muscle, the mean change in backscattered energy at each pixel over eight image regions in all tissue specimens was calculated with respect to a reference temperature (37°C). As temperature increased, for some scattering regions the CBE was positive, for others it was negative as seen in Figure 13.9 for our initial predictions, from simulations of images of scatterer populations, and from *in vitro* measurements in different types of tissue. Because the means of CBE from pixels with positive and negative relative backscattered energy changed nearly monotonically, CBE is a suitable parameter for temperature estimation. From uniform-heating studies its accuracy and spatial resolution appear to be suitable for temperature imaging.[73]

13.4.4.1 Stochastic-Signal Framework

In our initial work, CBE from backscattered signals was computed as a ratio at each pixel in the envelope-detected images of energies at temperature T and T_R.[43] It was characterized by averaging ratios larger than and less than 1, denoted as positive CBE (PCBE) and negative CBE (NCBE), which describe the increase and decrease in the backscattered energy, respectively.

When i_{en} is represented by a random process, computation of the signal ratio can be modeled as a ratio between two random variables, y_T and y_R:[46]

$$z = \frac{y_T}{y_R} \quad (13.9)$$

where y_R and y_T are the random variables representing the B-scans at the reference and current temperatures. The ratio, z, is also a random variable whose probability density function (PDF), $f_Z(z)$, is determined by the joint distribution of (y_R, y_T):[116]

$$f_Z(z) = \int_{-\infty}^{\infty} |y_R| f_{Y_R Y_T}(y_R, y_R z) dy_R, \quad (13.10)$$

where y_R, y_T, and $z > 0$.

The computation of PCBE in our initial work can be written as

$$PCBE = \frac{1}{N_+} \sum_{k \in \{k | z_k > 1\}} z_k = \frac{\frac{1}{N} \sum_{k \in \{k | z_k > 1\}} z_k}{\frac{N_+}{N}},$$

where N is the number of pixels in image and N_+ is the number of ratio pixels with value larger than 1. Assuming z_k's are independent,

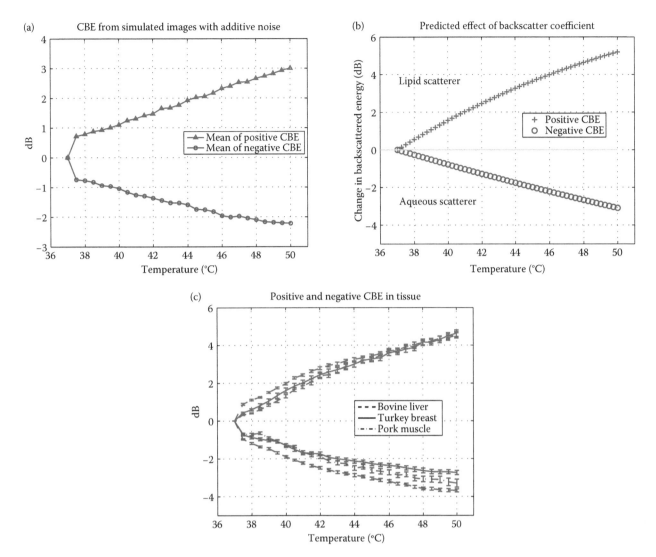

FIGURE 13.9 (a) Predicted CBE for single, sub-wavelength lipid and aqueous scatterers in an aqueous medium with thermal properties given by our initial theoretical study (similar to Figure 5 in Arthur et al.[91]). (b) CBE from simulated 1×1 cm B-mode images with collections of scatterers (2000 aqueous and lipid scatterers in a 2:1 ratio) in the presence of noise (similar to Figure 8 in Trobaugh et al.[57]). (c) Means of measured CBE in positive and negative regions of backscattered-energy images in four specimens of bovine liver, two of turkey breast, and one of pork muscle. The error bar is the standard error of the mean estimated from eight regions of interest in each of the tissue specimens (similar to Figure 8 in Arthur et al.[43]).

identically distributed random variables, the nominator approximates the integral $\int_1^\infty z f_Z(z)dz$. The denominator approximates the probability of z being larger than 1. Thus, PCBE is defined as the normalized mean of z over $z \in [1,\infty)$. Similarly, NCBE is defined as normalized mean of z over $z \in (0,1)$. Thus

$$PCBE = \frac{\int_1^\infty z f_Z(z)dz}{\int_1^\infty f_Z(z)dz}; \quad NCBE = \frac{\int_0^1 z f_Z(z)dz}{\int_0^1 f_Z(z)dz}. \quad (13.11)$$

From these definitions, PCBE and NCBE are in fact statistics of the signal ratio and are determined by the ratio distribution $f_Z(z)$.[46]

13.4.4.2 Effect of Signal-to-Noise Ratio, Temperature Range, and Insonification Frequency

In order to investigate CBE for populations of scatterers, we developed an ultrasonic image simulation model, including temperature dependence for individual scatterers based on predictions from our theoretical model.[57] CBE computed from images simulated for populations of randomly distributed scatterers behaves similarly to experimental results, with monotonic variation of CBE for both individual pixel measurements and over image regions. Effects on CBE of scatterer type and distribution, size of the image region, and signal-to-noise ratio affect temperature accuracy and spatial resolution. This simulation model also provides the basis for gaining a better understanding of the effects of

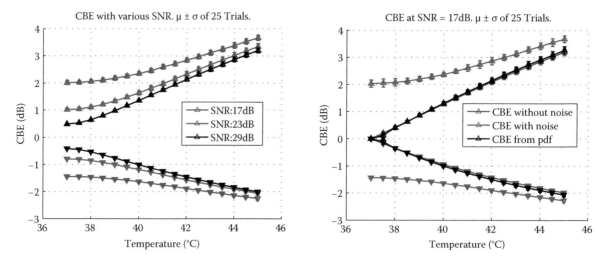

FIGURE 13.10 (Left) Mean ± standard deviation of positive (PCBE) and negative (NCBE) CBE from simulated images over the range of signal-to-noise ratios (SNRs) seen in our experiments. CBE was calculated from the means of the image ratios. Both the initial value of CBE (at 37°C) and the slope with temperature are affected by SNR.[57] (Right) Mean ± standard deviation of positive (PCBE) and negative (NCBE) CBE from simulated images with SNRs of 17 dB computed from the means of the image ratios. These curves are compared to CBE from images with infinite SNR and CBE computed using pdfs from the 17 dB SNR images. CBE from the pdfs in each case is close to CBE found without noise (infinite SNR). (From Guo, Y., *A Framework for Temperature Imaging using the Change in Backscattered Ultrasonic Signals*, PhD thesis, Washington University, St. Louis, MO, 2009. With permission.)

motion on CBE, limitations of motion-compensation techniques, and accuracy of temperature estimation, including tradeoffs between temperature accuracy and available spatial resolution.

Signal-to-noise ratio (SNR) in images from a given region size, with a particular scatterer type and population, affects the accuracy of the CBE temperature image from that tissue region. Figure 13.10 shows CBE from simulated images with various SNRs, typical of those seen in experiments. The initial (37°C) value of CBE and the slope with temperature are a function of SNR. Figure 13.10 also shows the effect on CBE of using the framework described in the previous section that allows for reduction of the noise effect by using the probability density function (pdf) from the low (17 dB) SNR images. Reduction of the noise effects on CBE improves temperature accuracy for a given spatial resolution. Noise effects can also be reduced by image averaging. Averaging 20 images results in about the same improvement in CBE shown in Figure 13.10 by using the pdfs from the images.

We studied effects of noise level on temperature imaging using Pennes's bioheat equation, which has had widespread application in the reconstruction of temperature fields in biological tissue.[117] For the *in vitro* case, in which perfusion and metabolism can be neglected, the heat flow equation at temperature T becomes

$$\rho C_p \frac{\partial T}{\partial t} = \nabla \cdot (k \nabla T) + Q, \qquad (13.12)$$

where ρ is density, C_p is specific heat, k is heat conductivity, and Q is the heat delivered to the specimen.

The bioheat equation (Equation 13.12) was implemented using finite-element software to simulate temperature distributions expected in the experimental fixture shown in Figure 13.5[102, 118]

Thermal parameters were taken from values for muscle given in the literature.[119] Temperature images with time are shown in the upper panel of Figure 13.11 for an initial temperature surrounding the medium of 37°C. At time zero, temperature on the inner surface (location of a heating tube in experiments) was raised to 65°C to match subsequent experiments.[102]

CBE with nonuniform heating was computed from simulated B-mode images with additive Gaussian random noise using the FEM temperature images. Figure 13.11 also shows temperature maps based on the computed CBE for an SNR of 24 dB, typical of that seen in an experiment. For noiseless B-mode images, the CBE temperature images followed the FEM simulated temperature maps closely. The error in estimation for the noiseless condition was 0.01 ± 0.2°C. The mean estimation error for 24 dB SNR was 0.4 ± 0.2°C. Note that the CBE temperature image was less affected by noise where the change in temperature was largest, that is, where CBE value was larger than the noise.

Another aspect of noise effects is seen in Figure 13.12, which shows CBE increasing to 60°C. This behavior may make CBE temperature imaging useful in high-temperature ablation border zones. The initial jump in CBE near 37°C indicates B-mode images that were noisier than usual for our imaging system. On the right in Figure 13.12, however, the jump at 37°C is small. The noise levels were reduced by synthetic-aperture imaging. In this case the completed dataset (64² signals) from a 64-element array was used to generate each pixel in each B-mode image, resulting in noise reduction due to spatial averaging compared to conventional phased-array imaging. Note too that the array operated at 2.25 Mhz, which shows that CBE can be effective for temperature imaging at both 2.25 and 7.5 MHz.

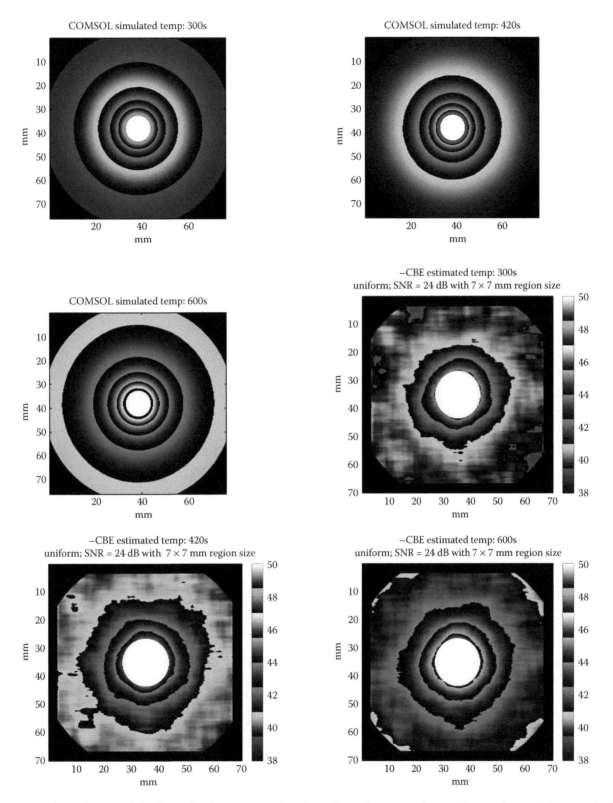

FIGURE 13.11 (**See color insert.**) (Left) Simulated temperatures based on a finite-element simulation of heating from a 65°C source in the center and a fixed 37°C surrounding temperature. (Right) Temperature images estimated with CBE from simulated images for a homogeneous scatterer distribution over the heated region with an SNR of 24dB. The mean estimation error for was 0.4 ± 0.2°C. (From Basu, D., *3D Temperature Imaging Using Ultrasonic Backscatter Energy during Non-Uniform Tissue Heating*, PhD thesis, Washington University, St. Louis, MO, 2010. With permission.)

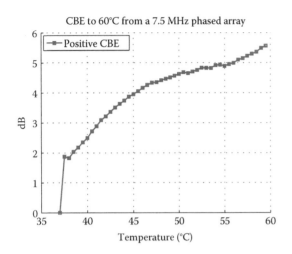

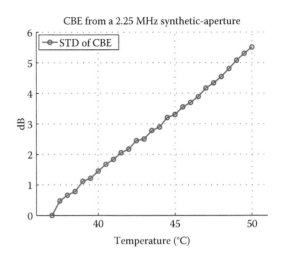

FIGURE 13.12   CBE Thermal Sensitivity from 37 to 60°C at 7.5 MHz and from 37 to 50°C at 2.25 MHz. (Left) Positive change in backscattered energy at 7.5 MHz from an *ex vivo* specimen of porcine muscle. (Right) CBE from synthetic-focus images at 2.25 MHz in a 10 × 30 mm region of bovine liver.

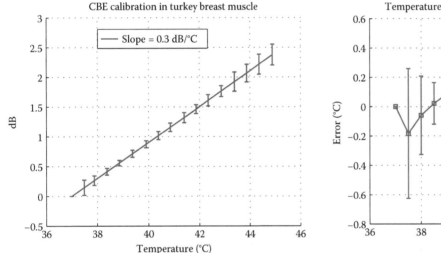

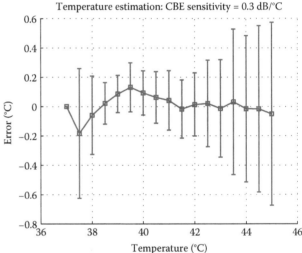

FIGURE 13.13   Calibration of CBE thermal sensitivity and temperature estimation in turkey breast muscle.[73] (Left) Mean ± standard deviation of the change in backscattered energy (CBE) in 20 volumes (1 cm³ each) from eight 4 × 6 × 2 cm specimens of turkey breast muscle. (Right) Mean estimation error ± standard deviation of error values for temperature estimation using CBE over all 20 tissue volumes with a fixed CBE sensitivity of 0.3 dB/°C.

13.4.4.3 3D Calibration and Temperature Estimation during Uniform Heating

To investigate the effect of temperature on backscattered energy in 3D using backscattered energy, we calibrated CBE in 1 cm³ volumes of turkey breast muscle during uniform heating. This volume size was selected to match our long-term goal to measure 3D temperature distributions to within 0.5°C in 1 cm³ volumes for monitoring hyperthermia treatments.[35, 120]

Temperature was estimated in separate 1 cm³ volumes of turkey.[73] Both calibration and estimation were done after compensation for motion in 3D. CBE as a function of temperature was measured in 20 1 cm³ volumes from eight specimens of turkey breast muscle using the standard deviation of the energy ratio in a given tissue volume.[43] The slope of each CBE curve was

well matched with a linear regression line (see Figure 13.13). Correlation coefficients ranged from 0.991 to 0.999. The error of the regression-line fit over all volumes and temperatures was 0 ± 0.07 dB. By definition, CBE at the reference temperature (37°C) is zero. The slope for the 20 cubic centimeter volumes was 0.300 ± 0.016 dB/°C.

Temperature estimation in 1 cm³ volumes of turkey breast muscle was based on calibration from a linear regression for measured CBE in a separate 1 cm³ volume of turkey breast muscle. Figure 13.13 shows an example of estimation of temperature based on calibration using a fixed CBE thermal sensitivity (0.3 dB/°C). Maximum positive and minimum negative errors were 0.53 and −0.39°C.

The mean of the maxima of the absolute value of the errors encountered in this *in vitro* study of CBE temperature estimation over all 1 cm³ volumes was < 0.5°C for temperatures from 38 to 42°C over all possible estimation tests using a single, separate cubic centimeter region for calibration and < 1°C for all temperatures. These results show that CBE, whose thermal behavior is monotonic, can serve as the basis for accurate volumetric temperature imaging.

13.4.4.4 Temperature Imaging during Nonuniform Heating

To generate and validate CBE temperature images during nonuniform heating, the fixture shown in Figure 13.5 was used to hold and heat abattoir specimens of turkey breast muscle.[102] The specimen was allowed to come to equilibrium at 37°C, then 65°C water was pumped through a silicone tube in the center of the specimen, using ultrasonic gel as a coupling agent between the tissue and the tube.

3D ultrasonic volumes were obtained at 7.5 MHz by moving a transducer array in the elevation direction under stepper motor control. The first and last images of the 3 mm thick 3D set captured the position of thermocouple tips for validating the CBE temperature images, as shown in Figure 13.14. The CBE temperature images were found after motion compensation using the sensitivity of CBE to temperature determined in the calibration studies, namely 0.3 dB/°C.

Figure 13.14 also shows the CBE-estimated and measured temperatures at the location of four thermocouples at three times after the start of heating. For this specimen, CBE temperatures errors were within about 1°C over the 900 s period of heating. Similar results were found over multiple specimens.[102] As expected, the estimated temperature patterns showed temperature increasing outwards from the heat source with consistent patterns over time, although not in a radial pattern as seen in homogeneous gelatin phantoms, perhaps due to inhomogeneous tissue structures.

13.5 Discussion

Ultrasonic thermometry is based on thermal effects in soft tissue that are manifest as changes in the speed of sound, attenuation, and backscatter of ultrasound. The use of echo shifts and changes in backscattered energy are the most promising techniques for estimating temperatures in the hyperthermia (41–45°C) range. Both methods are useful to temperatures above 60°C, which make them attractive for temperature imaging in the border zones of high-temperature ablations. Although attenuation is more difficult to measure than either echo shift or change in energy, its increased sensitivity to temperatures above 50°C also makes it of interest for assessing high-temperature ablation.

A key processing step common to all the methods just mentioned is the tracking of apparent motion of scattering regions *in vitro*. Motion detection takes on even more importance during *in vivo* studies because of the likely additional motion of the subject. Time shift as a function of depth is the basis for temperature

estimation using echo shifts. Motion must be tracked and correction applied before temperature can be estimated using changes in backscattered energy. Given the importance of determining tissue motion for estimating temperature with ultrasound, and that its determination is the most computationally intensive part of ultrasonic temperature imaging, it is likely that more efficient and sophisticated methods for motion tracking will be applied to this problem. If temperature estimation based on both echo shift and CBE continue to show promise, the accuracy and reliability of temperature estimation with ultrasound may be enhanced by a technique that combines both approaches.

Echo shift, attenuation, and CBE methods depend on being able to accurately calibrate, that is, find the thermal sensitivity to those parameters. The echo shift method, in addition to precise knowledge of the echo shift with tissue depth, requires a knowledge of α, the linear coefficient of thermal expansion, and β, a descriptor of the change in SOS with temperature. Similarly, calibration of the CBE method requires knowledge of how backscattered energy changes for a given type of tissue. The sensitivity to noise for these calibration data has yet to be fully determined for either method. That sensitivity will determine the temperature accuracy that is possible for a given spatial resolution for either method. A crucial step in identifying a viable ultrasonic approach to temperature estimation remains careful evaluation of its performance during *in vivo* application.

Volumetric temperature imaging is an important tool for guiding and assessing the effects of thermal therapy. Although MRI provides volumetric temperature information, it is expensive, adds increased complexity and duration to treatments, and may not be available in many hospitals. If ultrasound methods, which have been successful at imaging complex temperature patterns *in vitro*, can be shown to provide adequate thermal and spatial accuracy *in vivo*, they would provide important advantages over MRI temperature imaging. The significance of such systems would be high and could be an important element in increasing the utilization of thermal therapies.

13.6 Summary and Conclusions

Ultrasound is an attractive modality for volumetric temperature imaging to monitor thermal therapies because it is nonionizing, portable, convenient, inexpensive, and has relatively simple signal-processing requirements. This modality has proven useful for estimation of temperatures from the hyperthermia range (41–45°C) to border zones of regions of high-temperature ablation (>60°C).

The most prominent methods for exploiting ultrasound as a noninvasive thermometer rely on either (1) echo shifts due to changes in tissue thermal expansion and speed of sound (SOS), (2) variation in the attenuation coefficient, or (3) change in backscattered energy from tissue inhomogeneities. Each method has its strengths in terms of temperature range for which it yields a useful thermal signal and how well it can handle tradeoffs between temperature accuracy and spatial resolution.

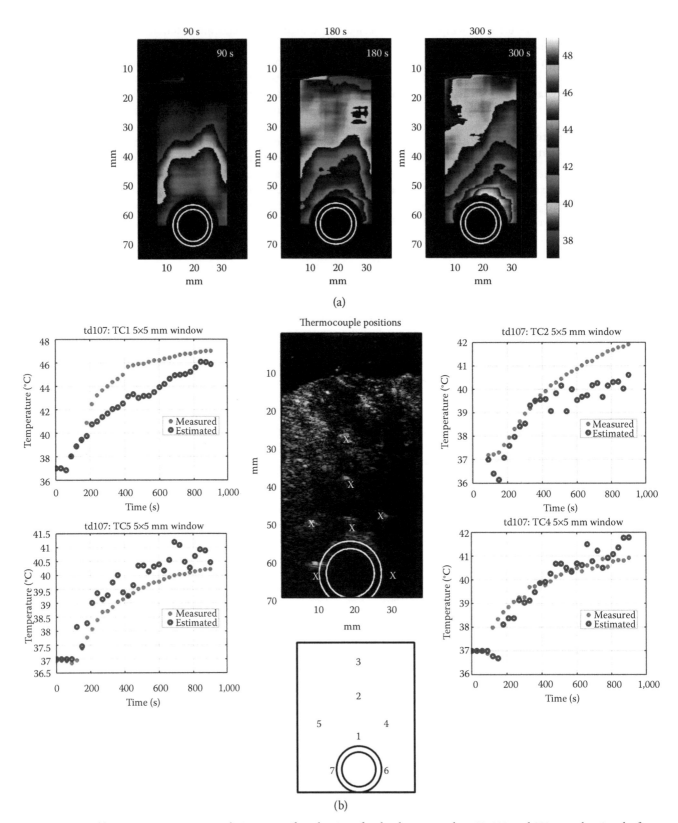

FIGURE 13.14 (a) CBE temperature images during nonuniform heating of turkey breast muscle at 90, 180, and 300 seconds using the fixture in Figure 13.5. Tissue was surrounded by water at 37°C. At time = 0 seconds 65°C water was pumped through the tissue center (double ring). (b) Thermocouple readings in planes adjacent to the 3D image set were used to verify CBE temperature estimates. CBE values were within about 1°C of the thermocouple readings. (From Basu, D., *3D Temperature Imaging Using Ultrasonic Backscatter Energy During Non-Uniform Tissue Heating*, PhD thesis, Washington University, St. Louis, MO, 2010. With permission.)

The use of echo shifts has received the most attention in the last decade. By tracking scattering volumes and measuring the time shift of received echoes, investigators have been able to estimate temperature with encouraging preliminary *in vivo* studies. Acoustic attenuation is dependent on temperature, but with significant changes occurring only at temperatures above 50°C. This property may lead to further development of its use in high-temperature thermal ablation therapy. Minimal change in attenuation, however, below this temperature range reduces its attractiveness for use in clinical hyperthermia.

The change in backscattered energy is scatterer dependent. Taking advantage of scatterer-dependent behavior enhances the thermal signal. This behavior has been matched with novel simulation methods for diverse scatterer populations and can be enhanced with stochastic signal processing methods. Monotonic thermal dependence of the change in backscattered energy has been measured to 60°C. Temperature maps with 1–2°C accuracy and 0.5 cm² spatial resolution can be produced routinely during nonuniform heating *in vitro*.

All of the ultrasonic thermometry methods, just like temperature imaging from MRI, must be able to cope with motion of the image features on which temperature estimates are based. Echo shift methods track and exploit that motion. Motion must be compensated in attenuation and CBE thermometry. Motion tracking and compensation are usually the most computationally intensive components of ultrasonic temperature imaging and limit frame rates for temperature imaging.

Thermal therapies are poised for rapid development and advancement due in part to a shift to volumetric temperature imaging from sparse invasive thermometry, offering improved monitoring as well as feedback for improved therapy control. Noninvasive temperature imaging with ultrasound could better (1) monitor and guide both hyperthermia treatment and high-temperature ablation and (2) deepen our understanding of tissue changes during hyperthermia and in ablation border zones now performed blindly, with limited invasive thermometry, or more expensive fixed-installation MRI thermometry that may limit options for heating sources. A crucial step in identifying a viable ultrasonic approach to temperature estimation is its performance during *in vivo* tests. The potential for significant clinical impact of ultrasonic thermometry is imminent with minimal addition, if any, to the existing hardware of ultrasonic imaging systems.

References

1. L Kreyberg, "Development of acute tissue damage due to cold," *Physiol Rev*, vol. 29, pp. 156–167, 1949.
2. LM Sutherland, JAR Williams, RTA Padbury, DC Gotley, B Stokes, and GJ Maddern, "Radiofrequency ablation of liver tumors," *Arch Surg*, vol. 141, pp. 181–190, 2006.
3. Xiaoming He and JC Bischof, "Quantification of temperature and injury response in thermal therapy and cryosurgery," *Critical Reviews In Biomedical Engineering*, vol. 31, no. 5-6, pp. 355–421, 2003.
4. K Hynynen, A Chung, M Buchanan, T Fjield, D Daum, V Colucci, P Lopath, and F Jolesz, "Feasibility of using ultrasound phased arrays for MRI monitored noninvasive surgery," *IEEE Trans on UFFC*, vol. 43, pp. 1043–1053, 1996.
5. M Ribault, J Chapelon, D Cathignol, and A Gelet, "Differential attenuation imaging for the characterization of high intensity focused ultrasound lesions," *Ultrasonic Imaging*, vol. 20, pp. 160–177, 1998.
6. GR ter Haar, "High intensity focused ultrasound for the treatment of tumors," *Echocardiography*, vol. 18, pp. 317–322, 2001.
7. MD Sherar, MR Gertner, CK Yue, ME O'Malley, A Toi, AS Gladman, SR Davidson, and J Trachtenberg, "Interstitial microwave thermal therapy for prostate cancer: Method of treatment," *J Urol*, vol. 166, pp. 1707–1714, 2001.
8. C McCann, JC Kumaradas, MR Gertner, SR Davidson, AM Dolan, and MD Sherar, "Feasibility of salvage interstitial microwave thermal therapy for prostate," *Phys Med Biol*, vol. 48, pp. 1041–1052, 2003.
9. CM Tempany, EA Stewart, N McDannold, BJ Quade, FA Jolesz, and K Hynynen, "MR imaging-guided focused ultrasound surgery of uterine leiomyomas: A feasibility study," *Radiology*, vol. 226, pp. 897–905, 2003.
10. A Anand and PJ Kaczkowski, "Monitoring formation of high intensity focused ultrasound (hifu) induced lesions using backscattered ultrasound," *Acoustics Research Letters Online*, vol. 5, no. 3, pp. 88–94, 2004.
11. GW Divkovic, M Liebler, K Braun, T Dreyer, PE Huber, and JW Jenne, "Thermal properties and changes of acoustic parameters in an egg white phantom during heating and coagulation by high intensity focused ultrasound," *J Ultrasound in Med & Biol*, vol. 33, no. 6, pp. 981–986, 2007.
12. J Overgaard, D Gonzales, M Hulshof, G Arcangeli, O Dahl, O Mella, and S Bentzen, "Hyperthermia as an adjuvant to radiation therapy of recurrent or metastatic malignant melanom: A multicenter randomized trial by the European Society for Hyperthermic Oncology," *Int J Hyperthermia*, vol. 12, pp. 3–20, 1996.
13. MW Dewhirst, L Prosnitz, D Thrall, D Prescott, S Cleff, C Charles, J Macfall, G Rosner, T Samulski, E Gillette, and S LaRue, "Hyperthermic treatment of malignant diseases: Current status and a view toward the future," *Seminars in Oncology*, vol. 24, pp. 616–625, 1997.
14. RJ Myerson, EG Moros, and JL Roti-Roti, "Hyperthermia," in *Principles and Practice of Radiation Oncology*, CA Perez and LW Brady, Eds., chapter 24, pp. 637–683. Lippincott-Raven, Philadelphia, third edition, 1998.
15. HK Lee, AG Antell, CA Perez, WL Straube, G Ramachandran, RJ Myerson, B Emami, EP Molmenti, A Buckner, and M A Lockett, "Specific absorption rate as a predictor of outcome in superficial tumors treated with hyperthermia and radiation therapy," *Int J Radiat Oncol Biol Phys*, vol. 40, pp. 365–375, 1998.

16. RJ Myerson, WL Straube, EG Moros, BN Emami, HK Lee, CA Perez, and ME Taylor, "Simultaneous superficial hyperthermia and external radiotherapy: Report of thermal dosimetry and tolerance to treatment," *Int J Hyperthermia*, vol. 15, pp. 251–266, 1999.

17. J van der Zee, D Gonzalez, G van Rhoon, J van Dijk, W van Putten, and A Hart, "Comparison of radiotherapy alone with radiotherapy plus hyperthermia in locally advanced pelvic tumours: A prospective, randomised, multicentre trial," *The Lancet*, vol. 355, pp. 1119–1125, 2000.

18. AM Westermann and et al., "First results of triple-modality treatment combining radiotherapy, chemotherapy, and hyperthermia for the treatment of patients with stage iib, iii and iva cervical carcinoma," *Cancer*, vol. 104, pp. 763–769, 2005.

19. JR MacFall and BJ Soher, "MR imaging in hyperthermia," *Radio Graphics*, vol. 27, pp. 1809–1818, 2007.

20. K Steinke, J King, DW Glenn, and DL Morris, "Percutaneous radiofrequency ablation of lung tumors with expandable needle electrodes: Tips from preliminary experience," *Am Roentgen Ray Society*, vol. 183, pp. 605–611, Sept 2004.

21. SN Goldberg, CJ Grassi, JF Cardella, JW Charboneau, and et al., "Image-guided tumor ablation: Standardization of terminology and reporting criteria," *Radiology*, vol. 235, pp. 728–739, 2005.

22. RO Illing, JE Kennedy, F Wu, GR ter Haar, AS Protheroe, PJ Friend, FV Gleeson, DW Cranston, RR Phillips, and MR Middleton, "The safety and feasibility of extracorporeal high-intensity focused ultrasound (hifu) for the treatment of liver and kidney tumours in a western population," *Br J Cancer*, vol. 93, no. 8, pp. 890–895, 2005.

23. H Higgins and DL Berger, "RFA for liver tumors: Does it really work?" *The Oncologist*, vol. 11, pp. 801–808, 2006.

24. S Solazzo, P Mertyna, H Peddi, M Ahmed, C Horkan, and SN Goldberg, "RF ablation with adjuvant therapy: Comparison of external beam radiation and liposomal doxorubicin on ablation efficacy in an animal tumor model," *Int J Hyperthermia*, vol. 24, pp. 560–567, 2008.

25. M Dewhirst, D Sim, S Sapareto, and W Coner, "Importance of minimum tumor temperature in determining early and long-term responses of spontaneous canine and feline tumors to heat and radiation," *Cancer Res*, vol. 44, pp. 43–50, 1984.

26. R Myerson, C Perez, B Emami, W Straube, R Kuske, L Leybovich, and J Von Gerichten, "Tumor control in long-term survivors following superficial hyperthermia," *Int J Radiat Oncol Biol Phys*, vol. 18, pp. 1123–1129, 1990.

27. K Leopold, M Dewhirst, T Samulski, R Dodge, S George, J Blivin, L Progsnitz, and J Oleson, "Relationships among tumor temperature, treatment time, and histopathological outcome using preoperative hyperthermia with radiation in soft tissue sarcomas," *Int J Radiat Oncol Biol Phys*, vol. 22, pp. 989–998, 1992.

28. J Hand, D Machin, C Vernon, and J Whaley, "Analysis of thermal parameters obtained during phase III trials of hyperthermia as an adjunct to radiotherapy in the treatment of breast carcinoma," *Int J of Hyperthermia*, vol. 13, pp. 343–364, 1997.

29. MW Dewhirst and PK Sneed, "Those in gene therapy should pay closer attention to lessons from hyperthermia," *Int J Radiat Oncol, Biol, Phys*, vol. 57, pp. 597–599, 2003, author reply, pp. 599–600.

30. HR Underwood, EC Burdette, KB Ocheltree KB, and RL Magin, "A multielement ultrasonic hyperthermia applicator with independent element control," *Int J Hyperthermia*, vol. 3, pp. 257–267, 1987.

31. TV Samulski, WJ Grant, JR Oleson, KA Leopold, MW Dewhirst, P Vallario, and JBlivin, "Clinical experience with a multi-element ultrasonic hyperthermia system: Analysis of treatment temperatures," *Int J Hyperthermia*, vol. 6, pp. 909–922, 1990.

32. P Stauffer, F Rossetto, M Leoncini, and G Gentilli, "Radiation patterns of dual concentric conductor microstrip antennas for superficial hyperthermia," *IEEE Trans Biomed Engr*, vol. 45, pp. 605–613, 1998.

33. EG Moros, X Fan, and WL Straube, "Experimental assessment of power and temperature penetration depth control with a dual frequency ultrasonic system," *Medical Physics*, vol. 26, pp. 810–817, 1999.

34. P Novak, EG Moros, WL Straube, and RJ Myerson, "SURLAS: A new clinical grade ultrasound system for sequential or concomitant thermoradiotherapy of superficial tumors: Applicator description," *Medical Physics*, vol. 32, no. 1, pp. 230–240, 2005.

35. Z Sun and H Ying, "A multi-gate time-of-flight technique for estimation of temperature distribution in heated tissue: Theory and computer simulation," *Ultrasonics*, vol. 37, pp. 107–122, 1999.

36. K Paulsen, M Moskowitz, T Ryan, S Mitchell, and P Hoopes, "Initial *in vivo* experience with EIT as a thermal estimator during hyperthermia," *Int J Hypertherm*, vol. 12, pp. 573–591, 1996.

37. P Meaney, K Paulsen, A Hartov, and R Crane, "Microwave imaging for tissue assessment: Initial evaluation in multi-target tissue-equivalent phantoms," *IEEE Trans Biomed Eng*, vol. 43, pp. 878–890, 1996.

38. D Carter, J MacFall, S Clegg S, X Wan, D Prescott, H Charles, and T Samulski, "Magnetic resonance thermometry during hyperthermia for human high-grade sarcoma," *Int J Radiat Oncol Biol Phys*, vol. 40, pp. 815–822, 1998.

39. J Nadobny, W Wlodarczyk, L Westhoff, J Gellermann, R Felix R, and P Wust, "A clinical water-coated antenna applicator for MR-controlled deep-body hyperthermia: A comparison of calculated and measured 3-D temperature data sets," *IEEE Trans Biomed Engineering*, vol. 52(3), pp. 505–519, 2005.

40. J Gellermann, W Wlodarczyk, H Ganter, J Nadobny, H Fahling, M Seebass, R Felix, and P Wust, "A practical approach to thermography in a hyperthermia/magnetic resonance hybrid system: Validation in a heterogeneous phantom," *Int J Radiat Oncol Biol Phys*, vol. 61, no. 1, pp. 267–277, 2005.

41. RM Arthur, WL Straube, JW Trobaugh, and EG Moros, "Noninvasive estimation of hyperthermia temperatures with ultrasound," *International J of Hyperthermia*, vol. 21, pp. 589–600, 2005.

42. RM Arthur, WL Straube, JD Starman, and EG Moros, "Noninvasive temperature estimation based on the energy of backscattered ultrasound," *Medical Physics*, vol. 30, pp. 1021–1029, 2003.

43. RM Arthur, JW Trobaugh, WL Straube, and EG Moros, "Temperature dependence of ultrasonic backscattered energy in motion-compensated images," *IEEE Trans on UFFC*, vol. 52, pp. 1644–1652, 2005.

44. M Pernot, M Tanter, J Bercoff, KR Waters, and M Fink, "Temperature estimation using ultrasonic spatial compound imaging," *IEEE Trans on UFFC*, vol. 51, no. 5, pp. 606–615, 2004.

45. KR Erikson, PL Carson, and HF Stewart, "Field evaluation of the AIUM standard 100 mm test object," in *Ultrasound in Medicine*, D White and R Barnes, Eds., vol. II, pp. 445–451. Plenum Press, New York, 1976.

46. Y Guo, *A Framework for Temperature Imaging Using the Change in Backscattered Ultrasonic Signals*, PhD thesis, Department of Electrical and Systems Engineering, Washington University, St. Louis, MO, USA, 2009.

47. A Goshtasby, "Image registration by local approximation methods," *Image and Vision Computing*, vol. 6, pp. 255–261, 1988.

48. GE Christensen, SC Joshi, and MI Miller, "Volumetric transformations of brain anatomy," *IEEE Trans on Med Imaging*, vol. 16, pp. 864–877, 1999.

49. MA Lubinski, SY Emelianov, and M O'Donnell, "Speckle tracking methods for ultrasonic elasticity imaging using short-time correlation," *IEEE Trans on UFFC*, vol. 46, pp. 82–96, 1999.

50. JM Fitzpatrick, DLG Hill, and CR Maurer, "Image registration," in *Handbook of Medical Imaging*, M Sonka and JM Fitzpatrick, eds., Bellingham, Washington, 2000, vol. 2 Medical Image Processing and Analysis, pp. 447–513, SPIE The International Society for Optical Engineering.

51. FV Viola and WF Walker, "A comparison of the performance of time-delay estimators in medical ultrasound," *IEEE Trans on UFFC*, vol. 50, pp. 392–401, 2003.

52. NR Miller, JC Bamber, and PM Meaney, "Fundamental limitations of noninvasive temperature imaging by means of ultrasound echo strain estimation," *J Ultrasound in Med & Biol*, vol. 28, no. 10, pp. 1319–1333, 2002.

53. FL Lizzi, R Muratore, CX Deng, JA Ketterling, SK Alam, S Mikaelian, and A Kalisz, "Radiation-force technique to monitor lesions during ultrasonic therapy," *J Ultrasound in Med & Biol*, vol. 29, no. 11, pp. 1593–1605, 2003.

54. W Liu, U Techavipoo, T Varghese, JA Zagzebski, Q Chen, and FT Lee, Jr, "Elastographic versus x-ray CT imaging of radio frequency ablation coagulations: An *in vitro* study," *Med Phys*, vol. 31, no. 6, pp. 1322–1332, 2004.

55. NR Miller, JC Bamber, and GR Ter Haar, "Imaging of temperature-induced echo strain: Preliminary *in vitro* study to assess feasibility for guiding focused ultrasound surgery," *J Ultrasound in Med & Biol*, vol. 30, no. 3, pp. 345–356, 2004.

56. TD Mast, DP Pucke, SE Subramanian, WJ Bowlus, SM Rudich, and JF Buell, "Ultrasound monitoring of *in vitro* radio frequency ablation by echo decorrelation imaging," *J Ultrasound Med*, vol. 27, pp. 1685–1697, 2008.

57. JW Trobaugh, RM Arthur, WL Straube, and EG Moros, "A simulation model for ultrasonic temperature imaging using change in backscattered energy," *J of Ultrasound in Med & Biol*, vol. 34, no. 2, pp. 289–298, Feb 2008, PMCID: PMC2269725.

58. R Maass-Moreno and CA Damianou, "Noninvasive temperature estimation in tissue via ultrasound echo-shifts: Part I Analytical model," *J Acoust Soc Am*, vol. 100, pp. 2514–2521, 1996.

59. R Seip, P VanBaren, C A Cain, and E S Ebbini, "Noninvasive real-time multipoint temperature control for ultrasound phased array treatments," *IEEE Trans on UFFC*, vol. 43, no. 6, pp. 1063–1073, 1996.

60. C Simon, P VanBaren, and E Ebbini, "Two-dimensional temperature estimation using diagnostic ultrasound," *IEEE Trans on UFFC*, vol. 45, pp. 1088–1099, 1998.

61. T Varghese, JA Zagzebski, Q Chen, U Techavipoo, G Frank, C Johnson, A Wright, and Jr. FT Lee, "Ultrasound monitoring of temperature change during radiofrequency ablation: Preliminary *in-vivo* results," *J Ultrasound in Med. & Biol.*, vol. 28, no. 3, pp. 321–329, 2002.

62. MJ Daniels and T Varghese, "Dynamic frame selection for *in vivo* ultrasound temperature estimation during radiofrequency ablation," *Phys Med Biol*, vol. 55, pp. 4735–4753, 2010.

63. M O'Donnell, AR Skovoroda, BM Shapo, and SY Emelianov, "Internal displacement and strain imaging using ultrasonic speckle tracking," *IEEE Trans on UFFC*, vol. 41, no. 3, pp. 314–325, 1994.

64. C Xunchang, MJ Zohdy, SY Emelianov, and M O'Donnell, "Lateral speckle tracking using synthetic lateral phase," *IEEE Trans on UFFC*, vol. 51, no. 5, pp. 540–550, 2004.

65. M Vogt and H Ermert, "Development and evaluation of a high-frequency ultrasound-based system for *in vivo* strain imaging of the skin," *IEEE Trans on UFFC*, vol. 52, no. 3, pp. 375–385, 2005.

66. SG Forster, PM Embree, and WD O'Brien, Jr, "Flow velocity profile via time-domain correlation: Error analysis and computer simulation," *IEEE Trans on UFFC*, vol. 37, no. 2, pp. 164–175, 1990.

67. LN Bohs and GE Trahey, "A novel method for angle independent ultrasonic imaging of blood flow and tissue motion," *IEEE Trans on Biomed Engr*, vol. 38, no. 3, pp. 280–286, 1991.

68. X Lai and H Torp, "Interpolation methods for time-delay estimation using cross-correlation method for blood velocity measurement," *IEEE Trans on UFFC*, vol. 46, no. 2, pp. 277–290, 1999.

69. IA Hein and WD O'Brien, Jr, "Current time-domain methods for assessing tissue motion by analysis from reflected ultrasound echoes," *IEEE Trans on UFFC*, vol. 40, no. 2, pp. 84–102, 1993.

70. RM Arthur, JW Trobaugh, WL Straube, EG Moros, and S Sangkatumvong, "Temperature dependence of ultrasonic backscattered energy in images compensated for tissue motion," in *Proceedings of the 2003 International IEEE Ultrasonics Symposium*, New York, 2003, vol. No 03CH37476C, pp. 990–993, IEEE Press.

71. ES Ebbini, "Phase-coupled two-dimensional speckle tracking algorithm," *IEEE Trans on UFFC*, vol. 53, no. 5, pp. 972–990, 2006.

72. RM Arthur, WL Straube, JW Trobaugh, and EG Moros, "*In vivo* change in ultrasonic backscattered energy with temperature in motion-compensated images," *Int J of Hyperthermia*, vol. 24, no. 5, pp. 389–398, Aug 2008, PMID: 18608589.

73. RM Arthur, D Basu, Y Guo, JW Trobaugh, and EG Moros, "3D *in vitro* estimation of temperature using the change in backscattered ultrasonic energy," *IEEE Transactions on UFFC*, vol. 57, no. 8, pp. 1724–1733, Aug 2010, PMID: 20679004.

74. F Jansson and E Sundmar, "Determination of the velocity of ultrasound in ocular tissues at different temperatures," *Acta Ophthalmologica*, vol. 39, pp. 899–910, 1961.

75. M O'Donnell, JW Mimbs, BE Sobel, and JG Miller, "Ultrasonic attenuation of myocardial tissue: Dependence on time after excision and on temperature," *J Acoust Soc Am*, vol. 62, pp. 1054–1057, Oct 1977.

76. JC Bamber and CR Hill, "Ultrasonic attenuation and propagation speed in mammalian tissues as a function of temperature," *J Ultrasound in Med & Biol*, vol. 5, pp. 149–157, 1979.

77. T Bowen, WG Connor, RL Nasoni, AE Pifer, and RR Sholes, "Measurement of the temperature dependence of the velocity of ultrasound in soft tissues," in *Ultrasonic Tissue Characterization II*, M Linzer, Ed. National Bureau of Standards, 1979, vol. NBS Spec Publ 525, pp. 57–61, US Government Printing Office.

78. PM Gammell, DH LeCroissette, and RC Heyser, "Temperature and frequency dependence of ultrasonic attenuation," *J Ultrasound in Med & Biol*, vol. 5, pp. 269–277, 1979.

79. RL Nasoni, T Bowen, WG Connor, and RR Sholes, "*In vivo* temperature dependence of ultrasound speed in tissue and its application to noninvasive temperature monitoring," *Ultrasonic Imaging*, vol. 1, pp. 34–43, 1979.

80. B Rajagopalan, JF Greenleaf, PJ Thomas, SA Johnson, and RC Bahn, "Variation of acoustic speed with temperature in various excised human tissues studied by ultrasound computerized tomography," in *Ultrasonic Tissue Characterization II*, M Linzer, ed., Washington, DC, 1979, National Bureau of Standards, vol. NBS Spec Publ 525, pp. 227–233, US Government Printing Office.

81. FW Kremkau, RW Barnes, and CP McGraw, "Ultrasonic attenuation and propagation speed in normal human brain," *J Acoust Soc Am*, vol. 70, pp. 29–38, 1981.

82. AC Lamont and BJ Cremin, "The effect of temperature on ultrasonic images in infant cadavers," *British Journal of Radiology*, vol. 59, pp. 271–272, 1986.

83. D Shore and CA Miles, "Attenuation of ultrasound in homogenates of bovine skeletal muscle and other tissues," *Ultrasonics*, vol. 26, pp. 218–222, 1988.

84. RN McCarthy, LB Jeffcott, and RN McCartney, "Ultrasound speed in equine cortical bone: Effects of orientation, density, porosity and temperature," *J Biomechanics*, vol. 23, pp. 1139–1143, 1990.

85. O Prakash, M Fabbri, M Drocourt, JM Escanye, C Marchal, ML Gaulard, and J Robert, "Hyperthermia induction and its measurement using ultrasound," in *Proceedings of IEEE Symposium on Ultrasonics*, New York, 1980, vol. 80CH1689-9, pp. 1063–1066, IEEE Press.

86. SA Johnson, DA Christensen, CC Johnson, JF Greenleaf, and B Rajagopalan, "Non-intrusive measurement of microwave and ultrasound-induced hyperthermia by acoustic temperature tomography," in *Proceedings of IEEE Symposium on Ultrasonics*, New York, 1977, vol. 77CH1264-1 SU, pp. 977–982, IEEE Press.

87. J Ophir, "Estimation of the speed of ultrasound propagation in biological tissues: A beam-tracking method," *IEEE Trans on UFFC*, vol. UFFC-33, no. 4, pp. 359–368, July 1986.

88. R Seip and ES Ebbini, "Noninvasive estimation of tissue temperature response to heating fields using diagnostic ultrasound," *IEEE Trans on Biomed Engr*, vol. 42, pp. 828–839, 1995.

89. Sanghvi NT Maass-Moreno R, Damianou CA, "Noninvasive temperature estimation in tissue via ultrasound echo-shifts: Part II *in vitro* study," *J Acoust Soc Am*, vol. 100, pp. 2522–2530, 1996.

90. CA Damianou, NT Sanghvi, FJ Fry, and R Maass-Moreno, "Dependence of ultrasonic attenuation and absorption in dog soft tissues on temperature and thermal dose," *J Acoust Soc Am*, vol. 102, pp. 628–634, 1997.

91. WL Straube and RM Arthur, "Theoretical estimation of the temperature dependence of backscattered ultrasonic power for noninvasive thermometry," *J Ultrasound in Med & Biol*, vol. 20, pp. 915–922, 1994.

92. RA Sigelmann and JM Reid, "Analysis and measurements of ultrasonic backscattering from an ensemble of scatterers excited by sine wave bursts," *J Acoust Soc of Am*, vol. 53, pp. 1351–1355, 1973.

93. M O'Donnell, JW Mimbs, and JG Miller, "Relationship between collagen and ultrasonic backscatter in myocardial tissue," *J Acoust Soc of Am*, vol. 69, pp. 580–588, 1981.

94. FL Lizzi, M Greenebaum, EJ Feleppa, M Elbaum, and DJ Coleman, "Theoretical framework for spectrum analysis in ultrasonic tissue characterization," *J Acoust Soc of America*, vol. 73, pp. 1366–1373, 1983.

95. EL Madsen, MF Insana, and JA Zagzebski, "Method of data reduction for accurate determination of acoustic backscatter coefficients," *J Acoust Soc Am*, vol. 76, pp. 913–923, 1984.

96. JG Mottley and JG Miller, "Anisotropy of the ultrasonic backscatter of myocardial tissue: I. Theory and measurements in vitro," *J Acoust Soc of America*, vol. 83, pp. 755–761, 1988.

97. KA Wear, MR Milunski, SA Wickline, JE Perez, BE Sobel, and JG Miller, "Differentiation between acutely ischemic myocardium and zones of completed infarction in dogs on the basis of frequency-dependent backscatter," *J Acoust Soc of Ama*, vol. 85, no. 6, pp. 2634–2641, 1989.

98. X Chen, D Phillips, KQ Schwarz, JG Mottley, and KJ Parker, "The measurement of backscatter coefficient from a broadband pulse-echo system: A new formulation," *IEEE Trans on UFFC*, vol. 44, pp. 515–525, 1997.

99. BK Hoffmeister, AK Wong, ED Verdonk, SA Wickline, and JG Miller, "Comparison of the anisotropy of apparent integrated ultrasonic backscatter from fixed human tendon and fixed human myocardium," *J Acoust Soc of America*, vol. 97, pp. 1307–1313, 1995.

100. BD de Senneville, B Quesson, and CTW Moonen, "Magnetic resonance temperature imaging," *Int J of Hyperthermia*, vol. 21, pp. 515–531, 2005.

101. J Gellermann, W Wlodarczyk, A Feussner, H Fahling, J Nadobny, B Hildebrandt, R Felix, and P Wust, "Methods and potentials of magnetic resonance imaging for monitoring radiofrequency hyperthermia in a hybrid system," *Int J of Hyperthermia*, vol. 21, pp. 497–513, 2005.

102. D Basu, *3D Temperature Imaging Using Ultrasonic Backscatter Energy During Non-Uniform Tissue Heating*, PhD thesis, Department of Electrical and Systems Engineering, Washington University, St. Louis, MO, USA, 2010.

103. ED Verdonk, BK Hoffmeister, SA Wickline, and JG Miller, "Anisotropy of the slope of ultrasonic attenuation in formalin fixed human myocardium," *J Acoust Soc of America*, vol. 99, pp. 3837–3843, 1996.

104. M O'Donnell, JW Mimbs, and JM Miller, "The relationship between collagen and ultrasonic attenuation in myocardial tissue," *J Acoust Soc Am*, vol. 65, pp. 512–517, 1979.

105. JG Mottley and JG Miller, "Anisotropy of the ultrasonic attenuation in soft tissues: Measurements *in vitro*," *J Acoust Soc of America*, vol. 88, pp. 1203–1210, 1990.

106. R Kuc, "Bounds on estimating the acoustic attenuation of small tissue regions from reflected ultrasound," *Proceedings of the IEEE*, vol. 73, pp. 1159, July 1985.

107. RM Arthur and KV Gurumurthy, "A single-pole model for the propagation of ultrasound in soft tissue," *J Acoust Soc Am*, vol. 77, pp. 1589–1597, April 1985.

108. AE Worthington, J Trachtenberg, and MD Sherar, "Ultrasound properties of human prostate tissue during heating," *J Ultrasound in Med & Biol*, vol. 28, pp. 1311–1318, 2002.

109. RL Clarke, NL Bush, and GR ter Haar, "The changes in acoustic attenuation due to *in vitro* heating," *J Ultrasound in Med & Biol*, vol. 29, pp. 127–135, 2003.

110. U Techavipoo, T Varghese, Q Chen, TA Stiles, JA Zagzebski, and GR Frank, "Temperature dependence of ultrasonic propagation speed and attenuation in excised canine liver tissue measured using transmitted and reflected pulses," *J Acoust Soc Am*, vol. 115, no. 5, pp. 2859–2865, 2004.

111. TC Robinson and PP Lele, "An analysis of lesion development in the brain and in plastics by high intensity focused ultrasound at low-megahertz frequencies," *J Acoust Soc Am*, vol. 5, pp. 1333–1351, 1972.

112. H-L Liu, M-L Li, P-H Tsui, M-S Lin, S-M Huang, and J Bai, "A unified approach to combine temperature estimation and elastography for thermal lesion determination in focused ultrasound thermal therapy," *Phys Med Biol*, vol. 56, pp. 169–186, 2011.

113. K-C Ju and H-L Liu, "Zero-crossing tracking technique for noninvasive ultrasonic temperature estimation," *J Ultrasound Med*, vol. 29, pp. 1607–1615, 2010.

114. KK Shung, MB Smith, and B Tsui, *Principles of Medical Imaging*, Academic Press, San Diego, CA, 1992, pp. 90–99.

115. PM Morse and KU Ingard, *Theoretical Acoustics*, McGraw-Hill, New York, 1968, p. 427.

116. V Krishnan, *Probability and Random Processes*, Wiley-Interscience, Hoboken, NJ, 2006.

117. HH Pennes, "Analysis of tissue and arterial temperatures in the resting human forearm," *J Applied Physiology*, vol. 1, pp. 93–122, 1948.

118. T Drizdal, M Vrba, M Cifra, P Togni, and J Vrba, "Feasibility study of superficial hyperthermia treatment planning using comsol multiphysics," *Microwave Techniques*, vol. 1, pp. 1–3, 2008.

119. Z Wang, JC Lin, W Mao, W LIu, MB Smith, and CM Collins, "SAR and temperature: Simulations and comparison to regulatory limits for MRI," *J Mag Res Imaging*, vol. 26, pp. 437–441, 2007.

120. EG Moros, PM Corry, and CG Orton, "Point/counterpoint: Thermoradiotherapy is underutilized for the treatment of cancer," *Med Phys*, vol. 34, no. 1, pp. 1–4, 2007.

Focused Ultrasound Applications for Brain Cancer

Meaghan A. O'Reilly
Sunnybrook Health Sciences Centre

Kullervo Hynynen
Sunnybrook Health Sciences Centre

14.1 Introduction

In 2007, the United States' National Cancer Institute reported that over 126,000 individuals in the United States alone were affected by primary brain or other central nervous system (CNS) cancers [Altekruse et al., 2010]. In addition to these numbers, metastatic brain cancer affects a significant percentage of those suffering from other primary cancers. Lung and breast cancer patients are particularly prone to developing brain metastases, with one study finding that as many as 19% of lung cancer patients will be diagnosed with brain metastases, and as many as 5% of breast cancer patients [Barnholtz-Sloan et al., 2004]. Old autopsy data from patients with breast carcinoma suggest the incidence may be higher, with as many as 30% developing CNS metastases [Tsukada et al., 1983]. Survival rates are poor once brain metastases develop, with a reported median survival of just seven months in patients with good prognosis [Gaspar et al, 1997]. Thorough reviews on both primary [Behin et al., 2003] and metastatic [Kamar and Posner, 2010] brain cancer exist that describe diagnosis and classification of the various tumor types.

Treatment practices vary depending on tumor type and location. Surgical intervention is an option for some tumors, but always results in some damage to surrounding structures. Surgical interventions are highly invasive, and many patients do not meet inclusion criteria for surgical removal. Chemotherapy or other anticancer agents can be delivered to the brain, however these agents have difficulty diffusing into tissue due to the tight junctions in the cerebral vasculature that form the blood-brain barrier (BBB). The BBB prevents molecules beyond a molecular weight of approximately 500 Da [Pardridge, 2005] from crossing into tissue. Unfortunately, this prevents most potentially useful therapeutic molecules from having a significant effect in the brain. In tumor tissue, the BBB can be leaky, but still prevents adequate drug delivery. One means of circumventing the BBB is to inject or implant the anticancer agents directly into the brain tissue and allow them to diffuse through the tissue. This approach, however, requires a surgical window in the skull. Radiotherapy or brachytherapy present other options for the treatment of brain tumors, however they require exposing the patient to ionizing radiation, which can harm the healthy tissues [DeAngelis, 1989; Leibel et al., 1989; Chang et al., 2009].

Research into focused ultrasound (FUS) in the brain began in the 1940s when Lynn et al. conducted the first focused ultrasound experiments in biological tissues [Lynn et al., 1942] and used FUS to produce neurological effects in laboratory animals [Lynn et al., 1942; Lynn and Putnam, 1944]. Since the first experiments, much research has been devoted to translating FUS into a clinical tool for brain therapy. FUS provides many advantages over traditional brain cancer treatments. A nonionizing energy source, FUS can be used to noninvasively produce a range of biological effects. Focused ultrasound has the potential to revolutionize cancer treatment, not only by noninvasively ablating tumors that are untreatable via traditional surgery but by providing new drug delivery techniques to improve delivery of therapeutic agents. This chapter will discuss the range of therapeutic effects that ultrasound can produce in the brain, the difficulties inherent to transcranial FUS, and how these problems have been addressed to date.

14.2 Focused Ultrasound in the Brain

14.2.1 Ultrasound and the Skull Bone

The greatest obstacle for the use of focused ultrasound in the brain is the skull bone. First, a large acoustic impedance mismatch occurs at bone/tissue interfaces, resulting in high reflection losses that increase with incidence angle (3 to 15 dB at normal incidence [Fry and Barger, 1978]). At low frequencies, below 500 kHz, reflection losses dominate overall signal loss through human skull bone, while at higher frequencies attenuation (absorption + scattering) losses also contribute significantly, resulting in very poor transmission at frequencies above 1 MHz [Fry and Barger, 1978]. Attenuation increases with increasing frequency as the wavelength of the ultrasound becomes small relative to the thickness of the bone. Skull bone has an outer dense cortical layer surrounding a layer of porous trabecular bone (Figure 14.1).

The skull has irregular surfaces, and varying thickness and density. The speed of sound in skull bone depends on its density [Clement et al., 2002] and thus varies across the bone thickness and from location to location, but is on average 2900 m/s [Fry and Barger, 1978; Pichardo et al., 2011], double that in water. The result is that sound passing through the cranium undergoes location-specific phase delays resulting in distortion of the focus.

At low frequencies the focal distortion is less severe due to the long wavelength, but increases greatly with frequency. In the late 1990s, Hynynen and Jolesz demonstrated that a sharp focus could be produced through human skull below 0.5 MHz, while above 1 MHz phase correction using a multi-element array was required to eliminate field distortions [Hynynen and Jolesz, 1998]. However, even at low frequencies, skull heating remains a problem, as skull bone absorbs energy at a higher rate than soft tissues. Therefore, the potential exists for the temperature rise in the skull to approach or surpass the temperature rise at the transducer focus [Connor and Hynynen, 2004].

Finally, the long sonations associated with therapeutic ultrasound and the presence of highly reflective bone surfaces can give rise to standing waves in the skull cavity [Azuma et al., 2004; Baron et al., 2009] and within the bone [Connor and Hynynen,

2004], causing undesired heating and unpredictable *in situ* pressures. Several techniques have been employed to overcome these challenges.

14.2.2 The Skull Window Approach

The first ultrasound treatments in the brain were performed through a craniotomy window. This approach allows a sharp focus to be achieved in the brain without concerns for skull heating, and was for many years the only method to conduct ultrasound brain therapy in clinical trials [Fry and Fry, 1960; Heimburger, 1985; Guthkelch et al., 1991]. The bone could be replaced by a material with better acoustic properties to best transmit the ultrasound energy [Tobias et al., 1987]. By removing the skull bone, focusing can be achieved with a single-element spherically focused transducer, minimizing hardware and software requirements. However, the skull window approach is invasive and limits treatment to the area exposed by the acoustic window. A superior approach is to focus through the skull bone.

14.2.3 Focusing through the Skull

Focusing through the skull can be achieved using multi-element arrays with applied phase delays and amplitude modulation to compensate for skull effects. There are several techniques for calculating the array element driving signals. The simplest method requires a hydrophone to be placed at desired focus to record the signals from each array element turned on in sequence [Smith et al., 1977; Thomas and Fink, 1996; Hynynen and Jolesz, 1998; Clement et al., 2000]. The deviations in phase and amplitude from the expected are measured by the hydrophone and can then be compensated for by adjusting the RF-driving signals of the transmit elements to negate the effects of the skull (Figure 14.2).

Similarly, a source can be placed at the focus and the emissions captured by the transmit elements. Using a time-reversal mirror, a sharp focus can be achieved and amplitude correction can also be applied [Fink, 1992; Thomas and Fink, 1996]. Time-reversal mirrors allow adjustment of amplitude and phase at the

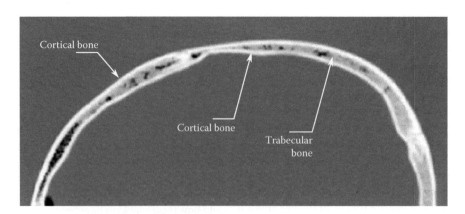

FIGURE 14.1 CT image of an *ex vivo* human cranium showing internal structures.

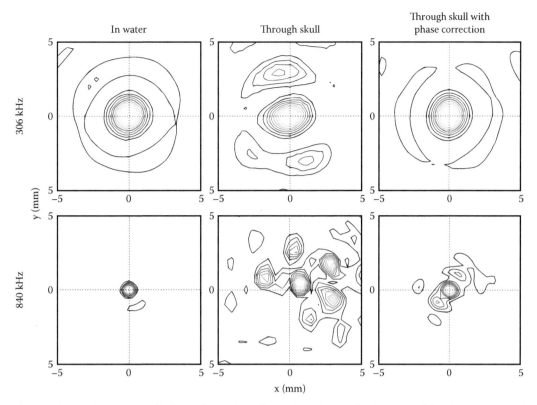

FIGURE 14.2 (Top row) Lateral pressure profile from a hemispherical transducer (f = 306 kHz) in water (left column), in water though a human skullcap (center column), and through a human skullcap with hydrophone-based phase correction (right column). (Bottom row) Lateral pressure profile from a hemispherical transducer (f = 830 kHz) in water (left column), in water though a human skullcap (center column), and through a human skullcap with hydrophone-based phase correction (right column). (Based on data from Song, J. and K. Hynynen, *IEEE Trans Biomed Eng* 57, 2010.)

focus, but a higher level of control can be achieved using a spatiotemporal inverse filter that allows constraints to be placed on the acoustic field around a focal point providing superior focusing capabilities [Tanter et al., 2001]. Placement of a hydrophone at the focus is not practical for *in vivo* work, and while time-reversal could be achieved by placing an ultrasound emitter in the brain during biopsy, it is still an invasive technique.

Hynynen and Sun [1999] proposed an entirely noninvasive method for correcting skull-induced distortions of the acoustic field. They proposed to use MRI information to model the shape of the skull bone and calculate the required element delays by simulating sound propagation from a virtual source. Their model consisted of three homogenous layers: water, bone, and brain tissue. This model was expanded to assign homogeneous material properties based on CT density information [Clement and Hynynen, 2002]. Aubry et al. [2003] further advanced this technique by using CT derived density information to produce a model with heterogeneous material properties. Simulation based focusing can be achieved using pre-procedure CT image information and registering it with MR images of the patient during the procedure.

Although simplified models can be used to perform focusing in real time for clinical treatments, more precise simulation-based focusing can be time consuming. To provide a quick and more precise, noninvasive focusing method, a further extension

of time-reversal focusing was proposed where cavitation bubbles act as the acoustic source [Kripfgans et al., 2000; Pernot et al., 2006]. Acoustic droplet vaporization could be used to generate bubbles at the focus [Kripfgans et al., 2000], or the cavitation bubbles can be induced at the focus in the absence of droplets using a very short pulse [Pernot et al., 2006]. The recorded acoustic signature is used to calculate time delays in the same manner as if a transducer were placed at the focus. The phase corrections are then applied to maximize focal efficiency. Recent work using this technique restored 97% of the focal pressures that can be achieved using hydrophone-based correction through a single *ex vivo* human skull in water, compared with 83% using only CT-based correction methods [Gateau et al., 2010]. This technique has produced good results, greatly reducing focusing time over full-scale simulation-based techniques, but its safety has yet to be examined *in vivo* where cavitation thresholds are highly variable.

In addition to phase correction techniques, White et al. [2005] examined two techniques for amplitude correction, which, combined with phase correction, can be used to compensate for losses through skull and achieve better focal reconstruction, but with the possibility of localized skull heating. Additionally, it was shown that a multifrequency approach can result in an increase in focal intensity over brief periods [White et al., 2006].

14.2.4 System Design for Transcranial Ultrasound Therapy

Transducer and system design are important elements of transcranial therapy. Compromises are necessary in order to produce a functional system. For example, low frequencies are desirable to minimize attenuation and distortion effects. However, low frequencies also increase the risk of standing waves [Azuma et al., 2004] and reduce the cavitation threshold in tissue [Hynynen, 1991], both of which have strong safety implications. Low frequencies also increase focal spot size, but above 1 MHz skull heating becomes an issue and prevents effective thermal ablation unless cavitation-enhanced methods are used [Sun and Hynynen, 1998]. At this time, the highest frequency prototype device for transcranial FUS operates at 1 MHz [Aubry et al., 2010].

Transducer geometry plays an important role in successful device design. Small aperture arrays suffer from weak focusing ability, especially in the axial direction. Not only does this increase focal spot size, but it increases the risk of standing wave formation in the skull cavity, especially at low frequencies (Figure 14.3). High focal number (FN) transducers produce almost planar ultrasound conditions in the pre- and post-focal regions, creating ideal conditions for standing waves at bone interfaces (Baron et al., 2009). Small aperture transducers also increase skull heating by concentrating the delivered energy through a small area of the skull. The adopted approach to minimize skull heating is to use a large aperture transducer, which can produce large focal gains [Sun and Hynynen, 1998]. Clement et al. [2000] conducted the first experiments using a hemispherical transducer for transcranial surgery (Figure 14.4). This first transducer consisted of 64 elements operating at 0.7 MHz and

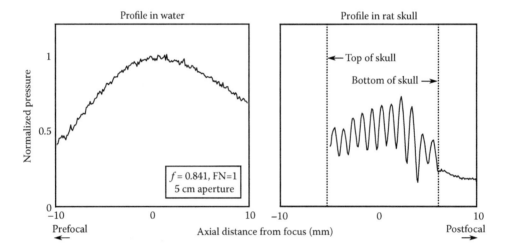

FIGURE 14.3 Axial pressure profile in an *ex vivo* rat skull of an FN 1 transducer operating at 0.841 MHz. (Based on data from O'Reilly, M.A., Y. Huang, and K. Hynynen, *Phys Med Biol* 55, 2010.)

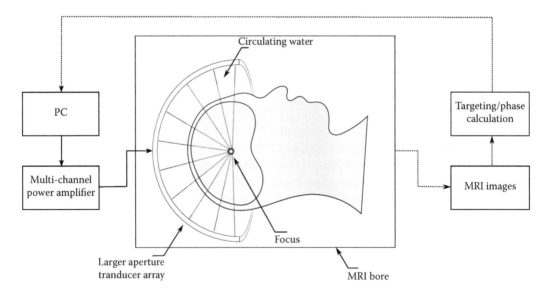

FIGURE 14.4 Illustration of transcranial therapy using a hemispherical array.

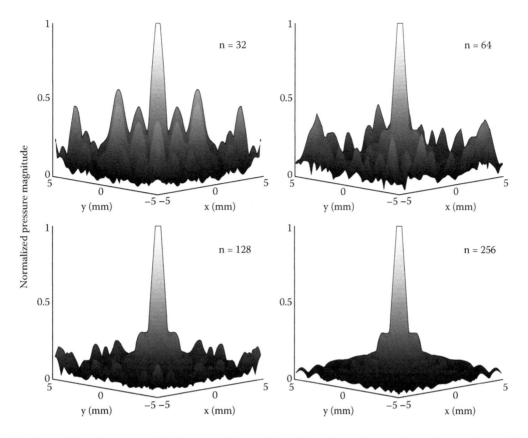

FIGURE 14.5 Simulation of lateral pressure profile for a 30 cm diameter hemispherical transducer with ring elements (external diameter 10 mm; internal diameter 7 mm) operating at 840 kHz with symmetrically arranged elements. Normalized profiled for (top left) 32 elements, (top right) 64 elements, (bottom left) 128 elements, (bottom right) 256 elements.

set within a 15 cm diameter hemispherical dome [Clement et al., 2000]. With this transducer, lesions were made in rabbit thigh through a human skull, but temperature rises in the *ex vivo* skull were approximately 12–18°C. The conclusion of the study was that transcranial FUS could be feasible using a large aperture array and active cooling of the head to reduce absolute temperatures at the skin-skull interface.

Large aperture, multi-element arrays are the best solution for transcranial FUS for thermal ablation of tumors. In addition to reducing skull heating, large aperture arrays are very tightly focused and therefore minimize standing wave effects. However, array performance improves with the number of elements in the array, often resulting in large hardware requirements. As the number of elements in an array increases, there is a reduction in side lobes (Figure 14.5).

A smaller number of elements can be used if a sparse random array is utilized. Randomization of the array elements minimizes side lobes by disrupting symmetric patterns of interference (Figure 14.6). Pernot et al. [2007] employed a 277-element random sparse array (Imasonic, Inc., Besançon, France) operating at 1 MHz to demonstrate noninvasive FUS in sheep brain. However, arrays with fewer elements can suffer from potential increased skull heating due to nonuniform energy transmission,

or poorer electronic steering capabilities if large elements are used. Electronic steering is desirable since the array can be placed once and need not be repositioned every time a new target is selected. This greatly reduces overall treatment times, since electronic steering can be used to quickly treat large volumes. The 200-element sparse random array presented by Pernot et al. [2003] provides ±15 mm electronic steering capability.

In 2010, Song and Hynynen presented a novel 1372 element hemispherical array operating at 306 kHz or 840 kHz [Song and Hynynen, 2010]. The array is made up of cylindrical elements that are laterally coupled to be driven in length mode. The novelty of the array is that the use of the lateral resonance mode reduces the electrical impedance of the elements. This eliminates the need for matching circuits to improve array efficiency and reduces overall hardware requirements. At 306 kHz, this dense array provides ±50 mm steerable range in the lateral dimension.

Two commercially available clinical prototype systems currently exist. Both are manufactured by Insightec (Exablate 4000, Insightec, Haifa, Israel) and have 1024 elements. The high frequency system operates at 660 kHz, and a low frequency system is offered that operates at 220 kHz. Both circulate water through the transducer dome to provide a coupling medium and cooling of the scalp.

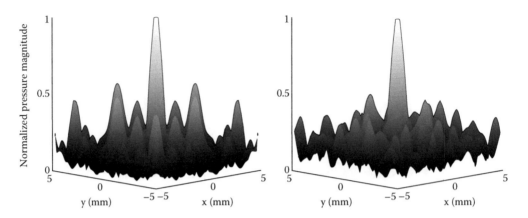

FIGURE 14.6 Simulation of lateral pressure profile for a 30 cm diameter hemispherical transducer ring element (external diameter 10 mm; internal diameter 7 mm) operating at 840 kHz with (left) 32 symmetrically arranged elements, (right) 32 randomly located elements.

Transducer design can improve treatment safety by minimizing heating and standing waves. Ultrasound pulse parameters are also very important in achieving desired results without unwanted effects. Thermal dose can be calculated from MR thermometry images, allowing sonication time and the cooling time between sonications to be selected. This allows control of bone heating and avoids under- or overtreating the target volume. In cavitation enhanced treatments, where microsecond length pulses are repeated at low duty cycle, the contrast agent dynamics must be considered. If the pulse repetition frequency (PRF) is too high, the bubble population in the vessels may not completely replenish between sonications [Goertz et al., 2010], which could reduce treatment efficacy [McDannold et al., 2008a]. If the PRF is too low, treatment efficiency may also be reduced [O'Reilly et al., 2011].

Considering safety, several methods have been proposed to minimize standing wave formation through pulse design. Sweep frequencies [Mitri et al., 2005] have been used to eliminate standing waves in vibroacoustography, while Tang and Clement used randomized phase shifts [Tang and Clement 2009] to disrupt the symmetry of the ultrasound and reduce standing wave effects. In a simulation study, Deffieux and Konofagou [2010] suppressed standing waves in human and primate skull models using periodic fast linear chirps. For preclinical BBB disruption (BBBD) research in small animal models, a modified pulse has been proposed that eliminates standing waves in the skull cavity [O'Reilly et al., 2010]. When large aperture transducers are used, standing waves are geometrically minimized since a sharp focus forms at the target, and at the skull base the acoustic field is widely distributed. Combining a large aperture transducer with a standing wave–suppressing pulse could improve procedure safety even further for close to the skull base sonications.

14.3 Therapeutic Effects of Ultrasound on the Brain

The therapeutic effects of ultrasound on the brain are varied, ranging from high intensity effects such as the thermal ablation

and tissue vaporization generated during focused ultrasound surgery, to low intensity effects such as blood-brain barrier (BBB) disruption.

14.3.1 Hyperthermia

Heat has long been known to have detrimental effects on tissues [Burger and Fuhrman, 1964a; Burger and Fuhrman, 1964b], and great potential exists for the use of moderate temperature increases for brain therapy. In hyperthermia, modest temperatures (42–48°C) are achieved and maintained over a period of minutes to hours. Total-body hyperthermia has been used to treat cancer with some response [Larkin, 1979]. However, given the noted detrimental effects of heat on mammalian brain [Harris, 1962; Burger and Fuhrman, 1964b, Lyons et al., 1986], a more localized delivery technique is desirable. Microwave hyperthermia combined with radiation therapy has been used to treat superficial brain tumors in the past [Engin, 1993]. However, the microwave energy is highly attenuated in tissues, limiting treatment to superficial tumors or use of interstitial applicators [Salcman and Samaras, 1983]. Ultrasound is able to better penetrate soft tissues than microwaves, allowing it to reach deeper seated tumors.

Lynn et al. first proposed the use of FUS in 1942, prior to which biological work involving ultrasound was limited to the use of plane waves. The use of FUS provides two main advantages over unfocused ultrasound. First, by focusing the ultrasound energy, higher intensities can be achieved at the transducer focus. Second, ultrasound can be delivered to targeted areas without affecting the surrounding tissue. These two factors make it an ideal modality for conducting targeted hyperthermia. Energy from the ultrasound beam passing through the tissue is deposited as heat. When short sonications at low PRFs are used, perfusion effects dominate and the local temperature rise is negligible. However, for continuous wave sonications or those using high PRFs, tissue heating can occur. The size and shape of the heated region can be controlled by scanning the transducer focus across the desired treatment volume (Figure 14.7).

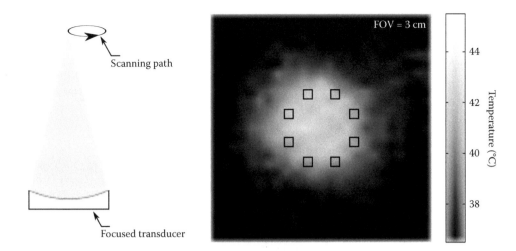

FIGURE 14.7 (Left) Illustration of mechanical circular scanning with a spherically focused transducer. (Right) Average spatial temperature distribution of during a 20-minute ultrasound hyperthermia exposure in a VX2 tumor in rabbit thigh using an FN = 2, 5 cm aperture transducer operating at 2.787 MHz. Eight control points are used to control temperature rise.

Ultrasound hyperthermia allows deep seated tumors to be treated since the sound energy can permeate farther into the tissue. Ultrasound hyperthermia was investigated in combination with radiation therapy starting in the 1970s [Marmor and Hahn, 1978], and much research into the field continued throughout the 1980s and early 1990s. Clinically, ultrasound-induced hyperthermia has been used with radiation to treat brain cancer [Shimm et al., 1988; Guthkelch et al., 1991] through a bone window. However, interest in mild hyperthermia as an adjuvant to radiation therapy in the brain has decreased since the 1990s, and the focus has shifted toward higher intensity thermal ablation procedures.

14.3.2 Thermal Ablation

In contrast to hyperthermia, in which modest temperature rises are sought, thermal ablation employs higher intensities to quickly induce high temperatures (55–60°C) and rapidly necrose a tissue volume. Early experiments with ultrasound exposures ranging in intensity and duration found that at lower intensities (100–1500 W/cm²), thermal effects dominate lesion formation, whereas at high intensities (>2000 W/cm²) [Fry et al., 1970] cavitation effects seem to be primarily responsible.

The first ultrasound brain ablations performed in humans were conducted in the 1950s and 1960s for a range of neurological disorders including Parkinson's disease and phantom limb pain [Fry and Fry, 1960]. These treatments were performed through a craniotomy window, requiring the patient to undergo traditional surgery to remove a portion of the skull. For this reason, thermal ablation in the brain did not advance as quickly as it might have. In fact, it was not until the late 1990s when ultrasound surgery through an intact skull was shown to be feasible [Hynynen and Jolesz, 1998] that interest in ultrasound-induced thermal ablation was revived. The introduction of MRI to guide and monitor treatments was another important development

in FUS research [Hynynen et al., 1993a; Hynynen et al., 1993b; Cline et al, 1993]. The safety and precision afforded by MRI-guidance and monitoring, combined with the noninvasiveness of through-skull treatments, has resulted in strong clinical interest in thermal ablations in the brain. MR-thermometry can be used to control thermal exposures real time during treatments [Vanne and Hynynen, 2003], and MRI can confirm lesion formation [Hynynen et al., 1994; Hynynen et al., 1997]. With specific application in the brain, MR-thermometry has been shown to be able to measure temperature rises below the thermal damage threshold [Hynynen et al., 1997], and the recorded temperature elevations can be related to the level of tissue damage [Vykhodtseva et al., 2000]. To further improve safety, it has been shown that the induced temperature elevations next to the skull can be accurately recorded using MR-thermometry during ultrasound heating of the brain [McDannold et al., 2004a]. The clinical prototype focused ultrasound brain system developed by InSightec (Haifa, Israel) described earlier has produced promising clinical ablation results in two centers [Martin et al., 2009; McDannold et al., 2010].

14.3.3 Cavitation Effects

The formation of vapor-filled cavities in a liquid was first described by Reynolds over a century ago [Reynolds, 1894]. The specific case of acoustic cavitation, vapor-filled cavities nucleated during the negative pressure phase of an ultrasonic wave, was described half a century later by Noltingk and Neppiras [Noltingk and Neppiras, 1950; Neppiras and Noltingk 1951]. Cavitation can be either stable or inertial, and can refer to either vapor- or gas-filled cavities [Neppiras 1980].

Inertial cavitation describes the formation and violent collapse of bubbles within a liquid, while stable cavitation refers to the oscillation of bubbles under a changing pressure field. The collapse of bubbles during inertial cavitation produces high

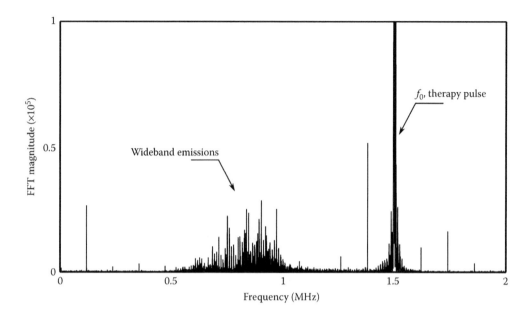

FIGURE 14.8 Wideband emissions detected around the center frequency (750 kHz) of a passive cavitation detector during focused ultrasound treatment in rat brain at 1.503 MHz.

local temperatures and can cause mechanical damage. This has the ability to enhance heating during FUS surgery [Hynynen, 1991; Holt and Roy, 2001; Sokka, 2003]. In the brain, however, early experiments showed that cavitation damage during high intensity exposures was not limited to the transducer focus, and more often occurred at boundaries between tissue and ventricles [Fry et al., 1970]. The lack of spatial control of cavitation in the brain, and the dangers associated with damage to that organ outside of the intended treatment volume, would suggest that bubble-enhanced thermal surgery in the brain at this stage requires more research to be properly utilized. One patient was treated in Boston at Brigham and Women's Hospital with a low frequency transducer (InSightec) intended for bubble-enhanced tumor surgery. The patient died several days after treatment due to bleeding that may or may not have been related to the procedure [Jolesz, 2009]. Cavitation monitoring during thermal surgery in the brain increases treatment safety by providing feedback to terminate treatment if broadband inertial cavitation signatures are detected (Figure 14.8).

Stable cavitation can also increase heating and energy deposition at the transducer focus, and has interesting potential applications in low intensity procedures. Power requirements for both inertial cavitation nucleation and for stable cavitation can be reduced through the use of preformed gas bubbles.

14.3.4 Microbubbles and Liquid Nanodroplets

Microbubbles used in ultrasound therapy are micron-sized encapsulated gas-filled bubbles. Used as ultrasound contrast agents due to their excellent scattering properties, microbubbles can enhance energy deposition at the focus. These microbubbles are typically on the order of a few micrometers in diameter,

filled with a perfluorocarbon gas and encapsulated with either an albumin shell (Optison, GE Healthcare) or phospholipid shell (Definity, Lantheus Medical Imaging; Sonovue, Bracco Diagnostics, Inc.). Microbubbles have been shown to nucleate inertial cavitation [Miller and Thomas, 1995] and can be manipulated with pressures far below the inertial cavitation threshold in brain tissue so bioeffects are likely to be limited to areas where the microbubbles interact with the acoustic field. However, while effects are mostly likely to be most prominent at the transducer focus, interactions can occur anywhere that there are bubbles in the sound field, leading to the possibility of effects occurring outside the focus [McDannold et al., 2006b].

Liquid nanodroplets provide what is perhaps a better way of localizing bubble mediated effects in the brain. Kripfgans et al. [2000] demonstrated that superheated dodecafluoropentan (DDFP) droplets could be manufactured and then vaporized, and later demonstrated that this could be achieved *in vivo* [Kripfgans et al., 2002]. DDFP droplets can be used over a wide range of sizes and frequencies where the vaporization threshold of the droplets is less than their inertial cavitation threshold [Schad and Hynynen, 2010]. Liquid nanodroplets do not interact with the ultrasound in the same way as microbubbles. By vaporizing the nanodroplets at the focus, bubbles can be formed with spatial precision, and effects outside the focus can be avoided. Both microbubbles and liquid nanodroplets can be loaded with drug payloads to be delivered when they are activated [Rapoport et al., 2009; Fabiilli et al., 2010]. This provides localized drug delivery. Both liquid nanodroplets and microbubbles have yet to be used clinically in the brain for therapeutic purposes, but their potential uses are being demonstrated in important preclinical research [Hynynen et al., 2001; McDannold et al., 2006a].

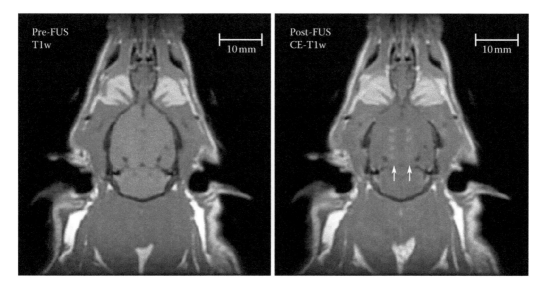

FIGURE 14.9 Pre-FUS baseline (left) and post-FUS contrast enhanced (right) T1w images of a Wistar rat brain showing enhancement at eight locations following sonication at 1.18 MHz. The arrow lines of four sonication points drawn in each hemisphere. The direction of the ultrasound travel is into the page. (Based on data from O'Reilly, M.A., A.C. Waspe, M. Ganguly et al., *Utrasound Med Biol*, 2011).

14.3.5 Blood–Brain Barrier Disruption

As mentioned before, brain tissue is protected by a series of tight junctions that prevent large molecules from passing from the vasculature into the tissue. While the purpose of the blood-brain barrier is to maintain the selective environment of this vital organ, it also prevents most useful therapeutic agents from reaching the tissue by limiting passage to molecules with good lipid solubility that are smaller than approximately 500 Da [Pardrige, 2005]. There exist pharmaceuticals that can disrupt the BBB, however they produce systemic, rather than localized disruption and expose the whole brain to both pathogens and the delivered drugs. Ultrasound disruption of the BBB provides a means for more localized BBBD.

Changes in the BBB under ultrasound exposure were first documented by Bakay et al. [1956] when they demonstrated that Trypan Blue dye leaked into the brain tissue around locations exposed to high intensity ultrasound. While high power BBBD can be achieved without neuronal damage [Vykhodtseva, 1995], it was not until 2001 that low power, microbubble mediated BBBD was demonstrated to produce consistent and repeatable opening without neuronal damage [Hynynen et al., 2001]. The documented disruption is localized, transient, and reversible. BBBD can be detected *in vivo* using contrast-enhanced T1w MRI as gadolidium does not permeate the intact BBB (Figure 14.9). The exact mechanism for the BBBD is unknown. McDannold et al. [2006a] correlated BBBD with an increase in harmonic acoustic emissions detected during treatment (Figure 14.10) and further established a frequency/pressure threshold relationship [McDannold et al., 2008b].

Microbubble-mediated, low power BBBD has been demonstrated in mice [Raymond et al., 2007; Choi et al., 2007], rats [Treat et al., 2007], rabbits [Hynynen et al., 2001], and

pigs [Xie et al., 2008], but has never been tested in humans. Promising results have been achieved in several studies delivering anticancer agents to the brain in preclinical models [Kinoshita et al., 2006; Treat et al., 2007], with enhanced animal survival after BBBD and doxorubicin infusion of rats with implanted 9L gliosarcoma tumors [Treat et al., 2009]. Delivery of 1,3-bis(2-chloroethyl)-1-nitrosourea (BCNU) through the BBB using FUS to treat glioblastoma in rats demonstrated enhanced BCNU uptake and good suppression of tumor growth over untreated controls [Liu et al., 2010]. However, there are several safety considerations that currently prevent clinical use of BBBD. The biggest limitation is the lack of a real-time monitoring technique to monitor and control the treatment, which will continue to keep BBBD from clinical implementation until a solution is presented.

Increased BBB permeability by mild hyperthermia has also been suggested. Cho et al. [2002] used mild hyperthermia to increase BBB permeability, however it has always been associated with brain tissue damage [McDannold et al., 2004b], and thus the most promising work to date has relied on the use of microbubbles.

14.4 Applications and Clinical Studies

Although there have been few clinical studies using focused ultrasound in the brain, there are an increasing number of large animal and cadaver feasibility studies, which suggests that clinical implementation is not too far off. The clinical study by Guthkelch et al. [1991] using ultrasound hyperthermia as an adjuvant to radiation therapy was the only phase I clinical trial to use ultrasound hyperthermia for brain cancer treatment. Current interests lie in the use of FUS for either thermal ablation in the brain or for BBBD.

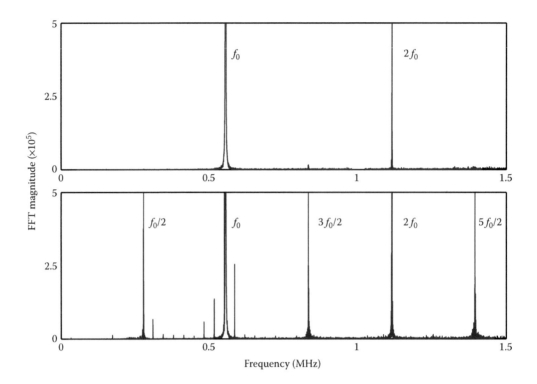

FIGURE 14.10 Example of acoustic emission spectra obtained during microbubble-mediated FUS-induced BBB disruption at 558 kHz. (Top) fundamental frequency and harmonic emissions, (bottom) fundamental frequency, harmonic emissions, and sub/ultraharmonic emissions. (Based on data from O'Reilly, M.A., and K. Hynynen, *IEEE Trans Biomed Eng* 57, 2010.)

In 1985, a small study of 14 patients was conducted to examine the feasibility of focused ultrasound surgery for the treatment of brain tumors [Heimburger, 1985]. Surgery was performed to create a craniotomy window. After the surgery, ultrasound lesions were induced in the brain through the replaced skin flap and the bone window. The study results were inconclusive.

In 2005, a Korean group presented at conference preliminary results of focused ultrasound surgery for the treatment of anaplastic glioma in a single patient [Park et al., 2006]. Although few details are available on the study, tumor shrinkage was observed, along with an improvement in the patient's symptoms.

In 2006, Ram et al. presented initial results from three patients with recurring gliomas treated using FUS ablation. The study used the ExAblate 2000 (InSightec Inc., Haifa, Israel) in-bed FUS system designed for treating uterine fibroids. The system was not designed for brain FUS, and the small aperture transducer necessitated the use of a craniotomy window in order to perform the ablation. Of the three patients in the study, one did not have any meaningful level of ultrasound energy deposition in the brain due to a technical malfunction, and one had an adverse event. In the patient with the adverse event, a secondary focus was formed in the mid-brain, possibly due to reflections, which caused hemiparesis. The remaining patient was treated without adverse event, however, only part of the tumor was treatable through the craniotomy window and the remaining section had to be surgically resected.

The preclinical feasibility study results in 10 pigs were reported after the clinical trial results had been published [Cohen et al., 2007]. In 10 pigs, the ExAblate 2000 system was used to ablate tissue through a craniotomy window without any adverse events. Despite good safety results in the feasibility study, the clinical trial encountered several problems with the system used. The secondary focus may have been avoided if tighter focusing from a larger aperture array had been used. McDannold et al. [2003] performed ablations in monkey brain using a small aperture transducer through a craniotomy window. When the focus of the transducer was near the bone, damage was seen in the surrounding tissue. Other large animal, *ex vivo* human skull, and clinical ablation studies have used larger transducers designed for transcranial brain therapy [Pernot et al., 2007; Marquet et al., 2009; McDannold et al., 2010; Martin et al., 2009].

Hynynen et al. [2006] performed the first noninvasive ablations in primates at 650 kHz using a system by InSightec and a CT-based correction algorithm. Marquet et al. [2006] also presented transcranial ablation results in primates at 900 kHz and CT-based aberration correction. More recently the same group has begun cadaver work validating a 512-element transcranial therapy array operating at 1 MHz [Aubry et al., 2010].

Results from two transcranial FUS clinical trials have been published to date. Both trials have used ExAblate transcranial ultrasound systems (InSightec). The first transcranial FUS trial was conducted in Boston using a 512-element Exablate system operating at 670 kHz on three patients with recurrent glioblastoma [McDannold et al., 2010]. In two patients, temperatures of 48°C and 51°C were achieved. One patient experienced sonication related pain, and temperatures to induce clear coagulation were not achieved. However, the results show promise for future use of transcranial FUS for tumor ablation. In the follow-up study in Zurich, nine patients participated in a study involving the use of transcranial FUS for the treatment of chronic pain [Martin et al., 2009]. This study used the InSightec system improved after the first three patients in Boston. This system had 1024 elements and operated at 670 kHz. Temperatures ranging between 51°C and 60°C were achieved, producing precise lesions (3–5 mm diameter) visible on MRI after 48 hours. Initial published results from the study were positive, showing good pain relief levels (30–100%). However, long-term follow-up data have yet to be published.

14.5 Potential Future Treatments

Transcranial FUS has great potential for future use in brain cancer treatments. Ideally, the coming years will see an increase in the number of tumor ablation studies performed worldwide. The advancement of BBB disruption into clinical practice has great potential to revolutionize the treatment of not only brain cancer but of a range of neurological disorders. Delivery of therapeutics would be more beneficial for the treatment of diffuse metastases than ablation, which is best suited for localized treatments. Additionally, BBBD procedures would be capable of targeting regions close to bony structures without the same concern for skull heating as in ablative procedures. A combination of treatments, as suggested by McDannold et al. [2006b] could provide the most powerful treatment tool. Tumor centers could be ablated using high intensities, while treating the periphery of the tumor using BBBD to deliver chemotherapy agents.

14.6 Needs for Advancement of Transcranial FUS into Routine Clinical Practice

The advancement of transcranial FUS into routine clinical practice currently faces several obstacles. Long-term follow-up to initial clinical trials is necessary to establish the safety of the procedures. Most importantly, transcranial FUS requires a real-time monitoring technique to provide information on treatment progress. MR-thermometry can provide information about thermal effects. However, monitoring of the acoustic field is necessary to ensure safety. The ExAblate 4000 allows a few elements to be used as passive cavitation detectors during therapy. Ideally, a larger receiver array would be integrated into the therapy array to allow passive imaging of cavitation events. This is especially important if BBBD is to be implemented clinically, as knowledge of the microbubble interactions with the acoustic field is necessary to ensure that cavitation events are not occurring outside the focus. Continued research interest in the field will help refine the technology and push transcranial FUS into the clinic.

Acknowledgments

The authors would like to thank Dr. Junho Song and Robert Staruch for their contributions to Figures 14.2 and 14.7. This work was supported by grant No R01 EB003268 from the National Health Institute and the Canada Research Chair program.

References

Altekruse, S.F., C.L. Kosary, M. Krapcho, N. Neyman, R. Aminou, W. Waldron et al. 2010. SEER Cancer Statistics Review, 1975–2007, National Cancer Institute. Bethesda MD, http://seer.cancer.gov/scr/1975-2007/, based on November 2009 Seer data submission, posted to the Seer Web site, 2010.

Aubry, J.F., L. Marsac, M. Pernot, B. Robert, A.-L. Boch, D. Chauvet et al. 2010. Ultrasons focalisés de forte intensité pour la thérapie transcrânienne du cerveau. *IRBM* 31:87–91.

Aubry, J.F., M. Tanter, M. Pernot, J.L. Thomas, and M. Fink. 2003. Experimental demonstration of noninvasive transskull adaptive focusing based on prior computed tomography scans. *J Acoust Soc Am* 113:84–93.

Azuma, T., K.-I. Kawabata, and S.-I. Umemura. 2004. Schlieren observation of therapeutic field in water surrouned by cranium radiated from 500 kHz ultrasonic sector transducer. *Proc. IEEE Ultrasonics Symp* 1001–1004.

Bakay, L., H. Ballantine, T. Hueter, and D. Sosa. 1956. Ultrasonically produced changes in the blood-brain barrier. *AMA Arch Neurol Psychiatry* 76:457–467.

Barnholtz-Sloan, J.S., A.E. Sloan, F.G. Davis, F.D. Vigneau, P. Lai, and R.E. Sawaya. 2004. Incidence proportions of brain metastases in patients diagnosed (1973 to 2001) in the Metropolitan Detroit Cancer Surveillance System. *J Clin Oncol* 22:2865–2872.

Baron, C., J.-F. Aubry, M. Tanter, S. Meairs, and M. Fink. 2009. Simulation of intracranial acoustic fields in clinical trials of sonothrombolysis. *Ultrasound Med Biol* 35:1148–1158.

Behin, A., K. Hoang-Xuan, A.F. Carpentier, and J.Y. Delattre. 2003. Primary brain tumours in adults. *Lancet* 361:323–331.

Burger, F.J. and F.A. Fuhrman. 1964a. Evidence of injury to tissues after hyperthermia. *Am J Physiol* 206:1062–1064.

Burger, F.J. and F.A. Furhman. 1964b. Evidence of injury by heat in mammalian tissues. *Am J Physiol* 206:1057–1061.

Chang, E.L., J.S. Wefel, K.R. Hess, P.K. Allen, F.F. Lang, D.G. Kornguth et al. 2009. Neurocognition in patients with brain metastases treated with radiosurgery or radiosurgery plus whole-brain irradiation: A randomised controlled trial. *Lancet Oncol* 10:1037–1044.

Cho, C.-W., Y. Liu, W.N. Cobb, T.K. Henthorn, K. Lillehei, U. Christians et al. 2002. Ultrasound-induced mild hyperthermia as a novel approach to increase drug uptake in brain microvessel endothelial cells. *Pharm Res* 19:1123–1129.

Choi, J.J., M. Pernot, S.A. Small, and E.E. Konofagou. 2007. Noninvasive, transcranial and localized opening of the blood-brain barrier using focused ultrasound in mice. *Ultrasound Med Biol* 33:95–104.

Clement, G.T. and K. Hynynen. 2002. A non-invasive method for focusing ultrasound through the human skull. *Phys Med Biol* 47:1219–1236.

Clement, G.T., J. Sun, T. Giesecke, and K. Hynynen. 2000. A hemisphere array for non-invasive ultrasound brain therapy and surgery. *Phys Med Biol* 45:3707–3719.

Cline, H.E., J.F. Schenck, R.D. Watkins, K. Hynynen, and F.A. Jolesz. 1993. Magnetic resonance-guided thermal surgery. *Magn Reson Med* 30:98–106.

Cohen, Z.R., J. Zaubermann, S. Harnof, Y. Mardor, D. Nass, E. Zadicario et al. 2007. Magnetic resonance imaging-guided focused ultrasound for thermal ablation in the brain: A feasibility study in a swine model. *Neurosurgery* 60:593–600.

Connor, C.W. and K. Hynynen. 2004. Patterns of thermal deposition in the skull during transcranial focused ultrasound surgery. *IEEE Trans Biomed Eng* 51:1693–1706.

DeAngelis, L.M., J.Y. Delattre, and J.B. Posner. 1989. Radiation-induced dementia in patients cured of brain metastases. *Neurology* 39:789–796.

Deffieux, T. and E.E. Konofagou. 2010. Numerical study of a simple transcranial focused ultrasound system applied to blood-brain barrier opening. *IEEE Trans Ultrason Ferroelectr Freq Control* 57: 2637–2653.

Engin, K., D.B. Leeper, L. Tupchong, F. M. Waterman, and C.M. Mansfield. 1993. Thermoradiation therapy for superficial malignant tumors. *Cancer* 72:287–296.

Fabiilli, M.L., K.J. Haworth, I.E. Sebastian, O.D. Kripfgans, P.L. Carson, and J.B. Fowlkes. 2010. Delivery of chlorambucil using an acoustically-triggered perfluoropentane emulsion. *Ultrasound Med Biol* 36:1364–1375.

Fink, M. 1992. Time reversal of ultrasonic fields. I. Basic principles. *IEEE Trans Ultrason Ferroelectr Freq Control* 39:555–566.

Fry, F.J. and J.E. Barger. 1978. Acoustical properties of the human skull. *J Acoust Soc Am* 63:1576–1590.

Fry, F.J., G. Kossoff, R.C. Eggleton, and F. Dunn. 1970. Threshold ultrasonic dosages for structural changes in the mammalian brain. *J Acoust Soc Am* 48:Suppl 2:1413.

Fry, W.J. and F.J. Fry. 1960. Fundamental neurological research and human neurosurgery using intense ultrasound. *IRE Trans Med Electron* ME-7:166–181.

Gaspar, L., C. Scott, M. Rotman, S. Asbell, T. Phillips, T. Wasserman et al. 1997. Recursive partitioning analysis (RPA) of prognostic factors in three Radiation Therapy Oncology Group (RTOG) brain metastases trials. *Int J Radiat Oncol Biol Phys* 37:745–751.

Gâteau, J., L. Marsac, M. Pernot, J.-F. Aubry, M. Tanter, and M. Fink. 2010. Transcranial ultrasonic therapy based on time reversal of acoustically induced cavitation bubble signature. *IEEE Trans Biomed Eng* 57:134–144.

Goertz, D.E., C. Wright, and K. Hynynen. 2010. Contrast agent kinetics in the rabbit brain during exposure to therapeutic ultrasound. *Ultrasound Med Biol* 36:916–924.

Guthkelch, A.N., L.P. Carter, J.R. Cassady, K.H. Hynynen, R.P. Iacono, P.C. Johnson et al. 1991. Treatment of malignant brain tumors with focused ultrasound hyperthermia and radiation: Results of a phase I trial. *J Neurooncol* 10:271–284.

Harris, A.B., L. Erickson, J.H. Kendig, S. Mingrino, and S. Goldring. 1962. Observations on selective brain heating in dogs. *J Neurosurg* 19:514–521.

Heimburger, R.F. 1985. Ultrasound augmentation of central nervous system tumor therapy. *Indiana Med* 78:469–476.

Holt, R.G. and R.A. Roy. 2001. Measurements of bubble-enhanced heating from focused, MHz-frequency ultrasound in a tissue-mimicking material. *Ultrasound Med Biol* 27:1399–1412.

Hynynen, K. 1991. The threshold for thermally significant cavitation in dog's thigh muscle in vivo. *Ultrasound Med Biol* 17:157–169.

Hynynen, K., A. Darkazanli, C.A. Damianou, E. Unger, and J.F. Schenck. 1994. The usefulness of a contrast agent and gradient-recalled acquisition in a steady-state imaging sequence for magnetic resonance imaging-guided noninvasive ultrasound surgery. *Investigative Radiology* 29:897–903.

Hynynen, K., A. Darkazanli, C.A. Damianou, E. Unger, and J.F. Schenck. 1993a. Tissue thermometry during ultrasound exposure. *Eur Urol* 23 Suppl 1:12–16.

Hynynen, K., A. Darkazanli, E. Unger, and J.F. Schenck. 1993b. MRI-guided noninvasive ultrasound surgery. *Med Phys* 20:107–115.

Hynynen, K. and F.A. Jolesz. 1998. Demonstration of potential noninvasive ultrasound brain therapy through an intact skull. *Ultrasound Med Biol* 24:275–283.

Hynynen, K., N. McDannold, G. Clement, F.A. Jolesz, E. Zadicario, R. Killiany et al. 2006. Pre-clinical testing of a phased array ultrasound system for MRI-guided noninvasive surgery of the brain–a primate study. *Eur J Radiol* 59:149–156.

Hynynen, K., N. McDannold, N. Vykhodtseva, and F. Jolesz. 2001. Noninvasive MR imaging-guided focal opening of the blood-brain barrier in rabbits. *Radiology* 220:640–646.

Hynynen, K. and J. Sun. 1999. Trans-skull ultrasound therapy: The feasibility of using image-derived skull thickness information to correct the phase distortion. *IEEE Transactions on Ultrasonics, Ferroelectrics, and Frequency Control* 46:752–755.

Hynynen, K., N.I. Vykhodtseva, A.H. Chung, V. Sorrentino, V. Colucci, and F.A. Jolesz. 1997. Thermal effects of focused ultrasound on the brain: Determination with MR imaging. *Radiology* 204:247–253.

Jolesz, F.A. 2009. Brain tumors. *Brain MRgFUS Workshop* 2009 ed. Washington, DC, March 23–24.

Kamar, F.G. and J.B. Posner. 2010. Brain metastases. *Semin Neurol* 30:217–235.

Kinoshita, M., N. McDannold, F. Jolesz, and K. Hynynen. 2006. Noninvasive localized delivery of Herceptin to the mouse brain by MRI-guided focused ultrasound-induced blood-brain barrier disruption. *Proc. Natl. Acad. Sci. U.S.A.* 103:11719–11723.

Kripfgans, O.D., J.B. Fowlkes, D.L. Miller, O.P. Eldevik, and P. L. Carson. 2000. Acoustic droplet vaporization for therapeutic and diagnostic applications. *Ultrasound Med Biol* 26:1177–1189.

Kripfgans, O.D., J.B. Fowlkes, M. Woydt, O.P. Eldevik, and P. L. Carson. 2002. *In vivo* droplet vaporization for occlusion therapy and phase aberration correction. *IEEE Trans Ultrason Ferroelectr Freq Control* 49:726–738.

Larkin, J.M. 1979. A clinical investigation of total-body hyperthermia as cancer therapy. *Cancer Res* 39:2252–2254.

Leibel, S.A., P.H. Gutin, W.M. Wara, P.S. Silver, D.A. Larson, M.S. Edwards et al. 1989. Survival and quality of life after interstitial implantation of removable high-activity iodine-125 sources for the treatment of patients with recurrent malignant gliomas. *Int J Radiat Oncol Biol Phys* 17:1129–1139.

Liu, H.-L., M.-Y. Hua, P.-Y. Chen, P.-C. Chu, C.-H. Pan, H.-W. Yang et al. 2010. Blood-brain barrier disruption with focused ultrasound enhances delivery of chemotherapeutic drugs for glioblastoma treatment. *Radiology* 255:415–425.

Lynn, J.G. and T.J. Putnam. 1944. Histology of cerebral lesions produced by focused ultrasound. *Am J Pathol* 20:637–649.

Lynn, J.G., R.L. Zwemer, A.J. Chick et al. 1942. A new method for the generation and use of focused ultrasound in experimental biology. *J Gen Physiol* 26:179–193.

Lyons, B.E., W.G. Obana, J.K. Borcich, and A.E. Miller. 1986. Chronic histological effects of ultrasonic hyperthermia on normal feline brain tissue. *Radiat Res* 106:234–251.

Marmor, J.B. and G.M. Hahn. 1978. Ultrasound heating in previously irradiated sites. *Int J Radiat Oncol Biol Phys* 4:1029–1032.

Marquet, F., M. Pernot, J.-F. Aubry, G. Montaldo, L. Marsac, M. Tanter et al., 2009. Non-invasive transcranial ultrasound therapy based on a 3D CT scan: Protocol validation and *in vitro* results. *Phys Med Biol* 54:2597–2613.

Marquet, F., M. Pernot, J.-F. Aubry, G. Montaldo, M. Tanter, and M. Fink. 2006. Non-invasive transcranial ultrasound therapy guided by CT-scans: An *in vivo* monkey study. *Conf Proc IEEE Eng Med Biol Soc* 1:683–687.

Martin, E., D. Jeanmonod, A. Morel, E. Zadicario, and B. Werner. 2009. High-intensity focused ultrasound for noninvasive functional neurosurgery. *Ann Neurol* 66:858–861.

McDannold, N., G.T. Clement, P. Black, F. Jolesz, and K. Hynynen. 2010. Transcranial magnetic resonance imaging-guided focused ultrasound surgery of brain tumors: Initial findings in 3 patients. *Neurosurgery* 66:323–332.

McDannold, N., R.L. King, and K. Hynynen. 2004a. MRI monitoring of heating produced by ultrasound absorption in the skull: *In vivo* study in pigs. *Magn Reson Med* 51:1061–1065.

McDannold, N., M. Moss, R. Killiany, D.L. Rosene, R.L. King, F.A. Jolesz et al. 2003. MRI-guided focused ultrasound surgery in the brain: Tests in a primate model. *Magn Reson Med* 49:1188–1191.

McDannold, N., N. Vykhodtseva, and K. Hynynen. 2008a. Effects of acoustic parameters and ultrasound contrast agent dose on focused-ultrasound induced blood-brain barrier disruption. *Ultrasound Med Biol* 34:930–937.

McDannold, N., N. Vykhodtseva, and K. Hynynen. 2008b. Blood-brain barrier disruption induced by focused ultrasound and circulating preformed microbubbles appears to be characterized by the mechanical index. *Ultrasound Med Biol* 34:834–840.

McDannold, N., N. Vykhodtseva, and K. Hynynen. 2006a. Targeted disruption of the blood-brain barrier with focused ultrasound: Association with cavitation activity. *Phys Med Biol* 51:793–807.

McDannold, N., N. Vykhodtseva, F.A. Jolesz, and K. Hynynen. 2004b. MRI investigation of the threshold for thermally induced blood-brain barrier disruption and brain tissue damage in the rabbit brain. *Magn Reson Med* 51:913–923.

McDannold, N.J., N.I. Vykhodtseva, and K. Hynynen. 2006b. Microbubble contrast agent with focused ultrasound to create brain lesions at low power levels: MR imaging and histologic study in rabbits. *Radiology* 241:95–106.

Miller, D.L. and R.M. Thomas. 1995. Ultrasound contrast agents nucleate inertial cavitation in vitro. *Ultrasound Med Biol* 21:1059–1065.

Mitri, F.G., J.F. Greenleaf, and M. Fatemi. 2005. Chirp imaging vibro-acoustography for removing the ultrasound standing wave artifact. *IEEE Trans Med Imaging* 24:1249–1255.

Neppiras, E.A. 1980. Acoustic cavitation. *Physics Reports (Review Section of Physics Letters)* 61:159–251.

Neppiras, E.A. and B.E. Noltingk. 1951. Cavitation produced by ultrasonics: Theoretical conditions for the onset of cavitation. *Proceedings of the Physical Society, Section B* 64:1032–1038.

Noltingk, B.E. and E.A. Neppiras. 1950. Cavitation produced by ultrasonics. *Proceedings of the Physical Society, Section B* 63:674–685.

O'Reilly, M.A., and K. Hynynen. 2010. A PVDF reciever for ultrasound monitoring of transcranial focused ultrasound therapy. *IEEE Trans Biomed Eng* 57:2286–2294.

O'Reilly, M.A., Y. Huang, and K. Hynynen. 2010. The impact of standing wave effects on transcranial focused ultrasound disruption of the blood-brain barrier in a rat model. *Phys Med Biol* 55:5251–5267.

O'Reilly, M.A., A.C. Waspe, M. Ganguly, and K. Hynynen. 2011. Focused-ultrasound disruption of the blood-brain barrier using closely-timed short pulses: Influence of sonication parameters and injection rate. *Ultrasound Med Biol* 37:587–594.

Pardridge, W.M. 2005. The blood-brain barrier: Bottleneck in brain drug development. *NeuroRx* 2:3–14.

Park, J-W., S. Jung, T-Y. Jung, and M.-C. Lee. 2006. Focused ultrasound surgery for the treatment of recurrent anaplastic astrocytoma: A preliminary report. In *Therapeutic Ultrasound: 5th International Symposium on Therapeutic Ultrasound*. eds. Clement, G.T., N.J. McDannold, and K. Hynynen. AIP conference proceedings, 238–240.

Pernot, M., J.F. Aubry, M. Tanter, J.L. Thomas, and M. Fink. 2003. High power transcranial beam steering for ultrasonic brain therapy. *Phys Med Biol* 48:2577–2589.

Pernot, M., J.-F. Aubry, M. Tanter, A.L. Boch, F. Marquet, M. Kujas et al. 2007. *In vivo* transcranial brain surgery with an ultrasonic time reversal mirror. *J Neurosurg* 106:1061–1066.

Pernot, M., G. Montaldo, M. Tanter, and M. Fink. 2006. "Ultrasonic stars" for time-reversal focusing using induced cavitation bubbles. *Applied Physics Letters* 88:034101–3.

Pichardo, S., V.W. Sin and K. Hynynen. 2011. Multi-frequency characterization of the speed of sound and attenuation coefficient for longitudinal transmission of freshly excised human skulls. *Phys Med Biol* 56:219–250.

Ram, Z., Z.R. Cohen, S. Harnof, S. Tal, M. Faibel, D. Nass et al. 2006. Magnetic resonance imaging-guided, high-intensity focused ultrasound for brain tumor therapy. *Neurosurgery* 59:949–955.

Rapoport, N.Y., A.M. Kennedy, J.E. Shea, C.L. Scaife, and K.-H. Nam. 2009. Controlled and targeted tumor chemotherapy by ultrasound-activated nanoemulsions/microbubbles. *J Control Release* 138:268–276.

Raymond, S.B., J. Skoch, K. Hynynen, and B. J. Bacskai. 2007. Multiphoton imaging of ultrasound/Optison mediated cerebrovascular effects in vivo. *J Cereb Blood Flow Metab* 27:393–403.

Reynolds, O. 1894. Experiments concerned with the boiling of water in an open tube at ordinary temperatures. *British Association, Section A.*

Salcman, M. and G.M. Samaras. 1983. Interstitial microwave hyperthermia for brain tumors. Results of a phase-1 clinical trial. *J Neurooncol* 1:225–236.

Schad, K.C. and K. Hynynen. 2010. *In vitro* characterization of perfluorocarbon droplets for focused ultrasound therapy. *Phys Med Biol* 55:4933–4947.

Shimm, D.S., K.H. Hynynen, D.P. Anhalt, R.B. Roemer, and J.R. Cassady. 1988. Scanned focussed ultrasound hyperthermia: Initial clinical results. *Int J Radiat Oncol Biol Phys* 15:1203–1208.

Smith, S.W., D.J. Phillips, O.T. von Ramm. 1977. Some advances in acoustic imaging through the skull. In *Symposium on Biological Effects and Characterizations of Ultrasound Sources*. eds. Hazzard, D.G. and M.L. Litz. 37–52., HEW, FDA.

Sokka, S.D., R. King, and K. Hynynen. 2003. MRI-guided gas bubble enhanced ultrasound heating in *in vivo* rabbit thigh. *Phys Med Biol* 48:223–241.

Song, J. and K. Hynynen. 2010. Feasibility of using lateral mode coupling method for a large scale ultrasound phased array for noninvasive transcranial therapy. *IEEE Trans Biomed Eng* 57:124–133.

Sun, J. and K. Hynynen. 1998. Focusing of therapeutic ultrasound through a human skull: A numerical study. *J. Acoust. Soc. Am.* 104:1705–1715.

Tang, S.C. and G.T. Clement. 2009. Acoustic standing wave suppression using randomized phase-shift-keying excitations. *J Acoust Soc Am* 126:1667–1670.

Tanter, M., J.-F. Aubry, J. Gerber, J.L. Thomas, and M. Fink. 2001. Optimal focusing by spatio-temporal inverse filter. I. Basic principles. *J. Acoust. Soc. Am.* 110:37–47.

Thomas, J.-L. and M. Fink. 1996. Ultrasonic beam focusing through tissue inhomogeneities with a time reversal mirror: Applications to transskull therapy. *IEEE Transactions on Ultrasonics, Ferroelectrics and Frequency Control* 43:1122–1129.

Tobias, J., K. Hynynen, R. Roemer, A.N. Guthkelch, A.S. Fleischer, and J. Shively. 1987. An ultrasound window to perform scanned, focused ultrasound hyperthermia treatments of brain tumors. *Med Phys* 14:228–234.

Treat, L.H., N. McDannold, N. Vykhodtseva,Y. Zhang, K. Tam, and K. Hynynen. 2007. Targeted delivery of doxorubicin to the rat brain at therapeutic levels using MRI-guided focused ultrasound. *Int J Cancer* 121:901–907.

Treat, L.H., Y. Zhang, N. McDannold, and K. Hynynen. 2009. MRI-guided focused ultrasound-enhanced chemotherapy of 9 L rat gliosarcoma: Survival study. *International Society of Magnetic Resonance in Medicine Annual Meeting* May 5–9, 2008 Toronto, Canada.

Tsukada, Y., A. Fouad, J.W. Pickren, and W.W. Lane. 1983. Central nervous system metastasis from breast carcinoma. Autopsy study. *Cancer* 52:2349–2354.

Vanne, A. and K. Hynynen. 2003. MRI feedback temperature control for focused ultrasound surgery. *Phys Med Biol* 48:31–43.

Vykhodtseva, N., V. Sorrentino, F.A. Jolesz, R.T. Bronson, and K. Hynynen. 2000. MRI detection of the thermal effects of focused ultrasound on the brain. *Ultrasound Med Biol* 26:871–880.

Vykhodtseva, N.I., K. Hynynen, and C. Damianou. 1995. Histologic effects of high intensity pulsed ultrasound exposure with subharmonic emission in rabbit brain in vivo. *Ultrasound Med Biol* 21:969–979.

White, J., G.T. Clement, and K. Hynynen. 2005. Transcranial ultrasound focus reconstruction with phase and amplitude correction. *IEEE Trans Ultrason Ferroelectr Freq Control* 52:1518–1522.

White, P.J., G.T. Clement, and K. Hynynen. 2006. Local frequency dependence in transcranial ultrasound transmission. *Phys Med Biol* 51:2293–2305.

Xie, F., M.D. Boska, J. Lof, J. Uberti, J. Tsutsui, and T. Porter. 2008. Effects of transcranial ultrasound and intravenous microbubbles on blood brain barrier permeability in a large animal model. *Ultrasound Med Biol* 34:2028–2034.

Extracorporeal Ultrasound-Guided High-Intensity Focused Ultrasound Ablation for Cancer Patients

Feng Wu
*Chongqing Medical University
and University of Oxford*

15.1 Clinical History of High-Intensity Focused Ultrasound Tumor Ablation

Using thermal energy to treat human neoplasms in clinics has a long history. It was reported that approximately five thousand years ago, physicians in Egypt used cautery with heated implements to destroy tumors [1]. In the past two decades, high-intensity focused ultrasound (HIFU) has been developed noninvasively to treat patients with solid tumors. It provides a thermal ablation for the precise and complete destruction of entire tumors in a three-dimension conformal fashion, with almost no limitation of tumor size and shape. This thermal therapy is attractive to both patients and physicians due to being less invasive with no incision, less scarring, cheap, less pain, and short recovery time. These technological advances result in an associated reduction in mortality, morbidity, hospital stay, and cost, and improved quality of life for cancer patients, initiating a change from open surgery toward less invasive techniques in the treatment of tumors [2–5].

The concept of using HIFU as a noninvasive therapy for destroying diseased tissues dates back about 70 years. In 1942, Lynn and colleagues reported for the first time that HIFU could cause tissue destruction with no damage to overlying and surrounding tissues [6]. In the 1950s and 1960s, William and Frank Fry at the University of Illinois in Champaign-Urbana, Illinois, did most of the early HIFU research work. They found that the lesion induced with HIFU exposure was well circumscribed. HIFU could successfully produce lesions deep in the brain of animals such as cat and monkey [7,8], and subsequently treated patients with Parkinson's disease and other neurological conditions after removing a piece of skull to create an "acoustic window" [9]. Early reports were encouraging in the treatment of Parkinson's disease. In the 1970s, Fred Lizzi and colleagues at the Riverside Research Institute in New York put considerable effort into applying HIFU in the field of ophthalmology. They constructed the first FDA-approved HIFU device (Sonocare CST-100) to investigate the possibility of using HIFU to treat glaucoma, choroidal melanomas, and capsular tears, and clinical results looked very exciting [10–13]. However, the advent of medical lasers for use in ophthalmology occurred simultaneously. Due to the ease of its use, the laser has superseded HIFU in most ophthalmological applications.

Based on their experience with extracorporeal shock wave lithotripsy, Guy Vallancien and colleagues at the institute Mutualiste Montsouris in Paris, France, constructed an extracorporeal HIFU device in the 1990s. His team used this ultrasound-guided pyrotherapy device to treat superficial bladder tumors in clinical trials. Five patients were enrolled in the phase I trial, and cystoscopy was performed before and after HIFU treatment. The disappearance of the tumor in two cases and coagulation necrosis in the remaining patients was noted [14]. In the phase II trial, a total of 25 patients with low-grade superficial bladder tumors were recruited. After treatment, 67% of the patients were tumor free at one year and no invasion or metastasis was detected with follow-up of 3–21 months [15]. However, when two patients with metastatic liver cancer were treated with the same device prior to surgical resection, the results looked unsatisfactory. There was no visible effect in one case, and in another there was extensive tissue laceration and patchy necrosis [16]. Gail ter Haar and colleagues at the Royal Marsden Hospital in London, United Kingdom, built a prototype of an ultrasound-guided HIFU (USgHIFU) device in the 1990s. This device employed a spherical ceramic transducer of 10 cm diameter and 15 cm focal length. It was driven at a frequency of 1.7 MHz and operated at free field spatial intensities between 1000 W cm^{-2} and 4660 W. cm^{-2} [17]. In the phase I trial, a total of 68 patients were treated with this device. The results demonstrated that HIFU treatment of liver cancer was well tolerated; some moderate local pain was observed, but only in a few patients [18].

My group at the Institute of Ultrasonic Engineering in Medicine, Chongqing Medical University in Chongqing, China, started USgHIFU research in 1988. Laboratory and animal studies including goat, pig, and monkey were mostly carried out from 1988 to 1997, and an extracorporeal HIFU prototype was designed and constructed in 1997 for clinical trials. It employed real-time ultrasound imaging to guide and monitor the procedure of HIFU ablation. On December 10, 1997, we used it to perform the first HIFU treatment in China for a boy with tibia osteosarcoma. The treatment was very successful and without any complications. After treating 1038 patients with solid tumors from 1997 to 2001 in 10 Chinese hospitals [19], the device (Model-JC HIFU system, Chongqing HAIFU, China) became the first HIFU system approved by the State Food and Drug Administration in China, an organization similar to the FDA in the United States. Solid malignancies treated with HIFU included primary and metastatic liver cancer, malignant bone tumor, breast cancer, soft tissue sarcoma, kidney cancer, pancreatic cancer, and advanced local tumors. Benign tumors, such as uterine fibroid, benign breast tumor, and hepatic hemangioma, were also treated. The same device was then introduced in the United Kingdom in 2002, and four clinical trials

were performed at the Churchill Hospital, University of Oxford, for the treatment of liver and kidney cancer. The clinical results were very promising, indicative that HIFU could be safe, feasible, and effective for the treatment of solid malignancies, leading to a CE approval in Europe for the device in 2005 [20].

Although the purpose of this chapter is to describe extracorporeal USgHIFU treatment, it is necessary to introduce some clinical experiences of using a transrectal USgHIFU device in the treatment of patients with prostate cancer. Up to now, two commercially available devices have been reported to treat prostate cancer in clinical practice. One transrectal device (Sonablate, Focused Surgery, United States) uses a 4 MHz PZT transducer for both imaging and treatment, and another (Ablatherm, EDAP, France) uses a 2.25–3.0 MHz rectangular transducer for treatment and a retractable 7.5 MHz probe for imaging guidance [21]. These devices have been widely used in the treatment of patients with prostate cancer, and clinical results are very promising [22]. In addition, Hynynen and colleagues at the Brigham and Women's Hospital in Boston, Massachusetts, did a lot of work incorporating HIFU into an MRI system and constructed an MRI-guided HIFU device in the 1990s [23]. With MRI thermometry techniques, the device can record focal temperature rises on the anatomical images during treatment procedure. This MRI-guided HIFU has been used clinically to treat uterine fibroids and breast neoplasms, and the results indicate successful ablation of targeted tumors [24, 25]. It has been approved by the FDA for the treatment of patients with uterine fibroids.

15.2 Physical Principles of HIFU Ablation

Ultrasound is a form of vibrational energy. It propagates as a mechanical wave by the motion of particles in the medium. The wave propagation leads to compressions and rarefactions of the particles, so that a pressure wave is transmitted along with the mechanical movement of the particles. As an ultrasound beam propagates through the body, it loses energy due to ultrasonic attenuation in tissue, which is caused by both scattering and absorption. The absorption of ultrasonic energy causes a local temperature rise in tissue if the rate of heating exceeds the rate of cooling. In HIFU, the absorption is greatest in the focus, where the acoustic intensity is at its highest. Additionally, during the rarefaction of the pressure wave, gas can be drawn out of the solution, and the subsequently formed bubbles may be acted on by the acoustic wave. When they reach the size of resonance, these bubbles suddenly collapse, causing mechanical stresses on surrounding tissues.

Two major effects are directly involved in the tissue damage induced by HIFU exposure. The first is a thermal effect from the conversion of mechanical energy into heat in the tissue, and the second is through cavitation. The thermal effect depends on the temperature achieved and the length of HIFU exposure. If the temperature rise is above a threshold of 56°C and the exposure time is 1 second [26], irreversible cell death will be induced through coagulation necrosis. In fact, the temperature at a focal volume may rise rapidly above 80°C during HIFU treatments [27]. A steep temperature gradient is detected between the focus and normal non-focal surrounding tissue, and therefore sharp demarcation between the treated and untreated tissue is demonstrated in histological examination.

The second mechanism is acoustic cavitation [28]. Acoustic cavitation can be defined as the interaction of a sound field with the microscopic gas bodies in a sonicated medium. The presence of small gaseous nuclei existing in subcellular organelles and fluid in tissue are the source of cavitation, which can expand and contract under influence of the acoustic pressure. During the collapse of bubbles, the acoustic pressure is more than several thousand pascals, and the temperatures reach several thousand degrees Celsius [29]. Therefore, it may cause tissue damage that is less predictable than the effect to tissue shape and position caused by heating [30]. However, recent experimental studies have been investigating the idea of promoting cavitation for enhancing the level of ablation and reducing required exposure times [31]. This is in contrast to previous approaches, where cavitation was viewed as an unpredictable damage mechanism that should be avoided [32].

15.3 Biological Effects of Thermal Ablation on Tumor

Thermal ablation can cause direct and indirect damage to a targeted tumor. Direct and indirect heat injury occurs during the period of heat deposition, and it is predominately determined by the total energy delivered to the targeted tumor [33]. Secondary injury usually occurs after thermal ablation, which produces a progression in tissue damage. It may involve a balance of several factors, including microvascular damage, cellular apoptosis, Kupffer cell activation, altered cytokine release, and antitumor immune response [34]. Direct injury is generally better defined than the secondary indirect effects.

15.3.1 Direct Thermal and Nonthermal Effects on Tumor

The effects of thermal ablation on a targeted tumor are determined by increased temperatures, thermal energy deposited, rate of removal of heat, and the specific thermal sensitivity of the tissue. As the tissue temperature rises, the time required to achieve irreversible cellular damage decreases exponentially. At temperatures between 50°C and 55°C, cellular death occurs instantaneously in cell culture [35]. Protein denaturation, membrane rupture, cell shrinkage, pyknosis, and hyperchromasia occur *ex vivo* between 60°C and 100°C, leading to immediate coagulation necrosis [36]. Tissue vaporization and boiling are superimposed on this process when the temperature is greater than 105°C. Carbonization, charring, and smoke generation occur while the temperature is over 300°C [37].

In addition, acoustic cavitation, one of mechanical effects induced by HIFU ablation, is the most important nonthermal mechanism for tissue disruption in the ultrasound field [28]. The presence of small gaseous nuclei existing in subcellular organelles and fluid in tissue are the source of cavitation, which can expand and contract under influence of the acoustic pressure. During the collapse of bubbles, the acoustic pressure is more than several thousand pascals, and the temperatures reach several thousand degrees Celsius, resulting in the local destruction of the tissue [29, 30].

15.3.2 Thermal Effects on Tumor Vasculature

Structural and functional changes are directly observed in tumor vasculature after thermal ablation. These changes are not as well described as thermal effects on the tissues, but they rely on varying temperatures. At temperatures between 40°C and 42°C, there is no significant change in tumor blood flow after 30–60 min exposure [38]. Beyond 42°C to 44°C, there is an irreversible decrease in tumor blood flow, with vascular stasis and thrombosis, resulting in heat trapping and progressive tissue damage [39]. When temperatures exceed 60°C, immediate destruction of tumor microvasculature occurs [40]. It cuts the blood supply to the tumor directly through the cauterization of the tumor feeder vessels, leading to deprivation of nutrition and oxygen. Thus, tissue destruction can be enhanced by the damage caused by thermal ablation to tumor blood vessels.

15.3.3 Secondary Effects on Tumor

Indirect injury is a secondary damage to tissue, which progresses after the cessation of thermal ablation stimulus [34]. It is based on histological evaluation of tissue damage at various time points after thermal ablation. The full extent of the secondary tissue damage becomes evident one to seven days after thermal ablation, depending on the model and energy source used [41, 42]. The exact mechanism of this process is still unknown. However, it may represent a balance of several promoting and inhibiting mechanisms, including induction of apoptosis, Kupffer cell activation, and cytokine release.

Cellular apoptosis may contribute to the progressive injury of tissue after thermal ablation. It is well established that apoptosis increases in a temperature-dependent manner, and temperatures between 40°C and 45°C cause inactivation of vital enzymes, thus initiating apoptosis of tumor cells [43, 44]. Most thermal ablation techniques create a temperature gradient that progressively decreases away from the site of probe insertion. The induction of apoptosis at a distance from the heat source may potentially contribute to the progression of injury. Increased rate of apoptosis is observed in the liver 24 hours after microwave ablation. The stimulation of apoptosis may be directly induced by temperature elevations, alterations in tissue microenvironment, and the release of various cytokines after thermal ablation.

Kupffer cell activity may be one of the major factors involved in the progressive injury after thermal ablation [34]. Heat induces Kupffer cells to secrete IL-1 [45] and tumor necrosis factor-α (TNF-α) [46], which are known to have *in vivo* antitumor activity [47] and to increase apoptosis in cancer cells [44]. Kupffer cells also induce the production of interferon that augments the liver-associated natural killer cell activity [48].

Thermal ablation may induce both regional and systemic production of cytokines through activation of inflammatory cells. Compared with controls, the circulating level of IFN-γ and vascular endothelial growth factor levels markedly increase after RFA [49,50]. The increased levels of IL-1 and TNF-α are also observed after RFA [51]. These cytokines may have direct cytotoxic effects such as inducing tumor endothelial injury and tumor cells more sensitive to heat-induced damage [52,53]. However, contrasting results were obtained for TNF-α level in two studies [50,54] and IL-1 level in one study [55], which remains unchanged after thermal ablation.

15.3.4 Antitumor Immune Response after HIFU Treatment

It has been noted that large amounts of tumor debris remain *in situ* after thermal ablation. As a normal process of healing response, the tumor debris is gradually reabsorbed by the individual patient, which takes a period ranging from months to few years. It is still unclear what kind of biological significance may exist during the absorption of the ablated tumor. However, some studies have shown that active immune response to the treated tumor could be developed after thermal ablation, and the host immune system could become more sensitive to the tumor cells [56–59]. This may lead to a potential procedure that reduces or perhaps eliminates metastases, and prevents local recurrence in cancer patients who have had original dysfunction of antitumor immunity before treatment.

Animal studies have suggested that HIFU may modulate host antitumor immunity. Yang and colleagues [60] used HIFU to treat C1300 neuroblastoma implanted in mouse flanks, followed by the rechallenge of the same tumor cells. A significantly slower growth of reimplanted tumors was observed in these mice while compared with the controls. After HIFU treatment, the cytotoxicity of peripheral blood T-lymphocytes was significantly increased in the H22 tumor–bearing mice treated with HIFU, and adoptive transfer of the activated lymphocytes could provide better long-term survival and lower metastatic rates in the mice rechallenged by the same tumor cells [56]. Similar results were confirmed in the mice implanted with MC-38 colon adenocarcinoma after HIFU ablation. HIFU treatment could also induce an enhanced CTLs activity *in vivo*, and thus provided protection against subsequent tumor rechallenge [61].

After HIFU ablation, large amounts of tumor debris remain *in situ*, and the host gradually reabsorbs them as the normal process of a healing response. Using a murine hepatocelluar carcinoma model, Zhang and colleagues [62] demonstrated that the remaining tumor debris induced by HIFU could be immunogenic as

an effective vaccine to elicit tumor-specific immune responses, including induction of CTL cytotoxic activity and protection against a lethal tumor challenge in naïve mice. When the tumor debris was loaded with immature DCs, it could significantly induce maturation of DCs, and increased cytotoxocity and TNF-α and IFN-γ secretion by CTL, thus initiating host-specific immune response after H22 challenge in the vaccinated mice [63]. Immediately after HIFU exposure to MC-38 colon adenocarcinoma cells *in vitro*, the release of endogenous danger signals including HSP60 was observed from the damaged cells. These signals could subsequently activate antigen presenting cells (APCs), leading to an increased expression of co-stimulatory molecules and enhanced secretion of IL-12 by the DCs, and elevated secretion of TNF-α by the macrophages [64]. In addition, HIFU could upregulate *in vitro* and *ex vitro* molecule expression of HSP70 [65,66], which are intracellular molecular chaperones that can enhance tumor cell immunogenicity, resulting in potent cellular immune responses.

The potency of APCs activation from mechanical lysis and a sparse-scan HIFU was much stronger than that from thermal necrosis and a dense-scan HIFU exposure, suggesting that optimization of HIFU ablation strategy may help to enhance immune response after treatment [67, 68]. Heat and acoustic cavitation are two major mechanisms involved in HIFU-induced tissue damage, and cavitation is a unique effect of HIFU when compared with other thermal ablation techniques. It causes membranous organelles to collapse, including mitochondria and endoplasmic reticulum, cell, and nuclear membrane. This breaks up tumor cells into small pieces, on which the tumor antigens may remain intact, or lead to the exposure of an immunogenic moiety that is normally hidden in tumor antigens [56]. Zhou and colleagues [68] used either heat- or HIFU-treated H22 tumor vaccine to inoculate naïve mice. The vaccination times were four sessions, once a week for four consecutive weeks, and each mouse was challenged with H22 tumor cells one week after the last vaccination. They found that the HIFU-treated tumor vaccine could significantly inhibit tumor growth and increase survival rates in the vaccinated mice, suggesting that acoustic cavitation could play an important role to stimulate host antitumor immune system.

Emerging clinical results revealed that systemic cellular immune response was observed in cancer patients after HIFU treatment. Rosberger and colleagues [69] reported five consecutive cases of posterior choroidal melanoma treated with HIFU. Three patients had abnormal, and two patients normal, CD4/CD8 ratios before treatment. One week after treatment, the ratio in two patients reverted to normal, while another was noted to have a 37% increase in his CD4T-cells relative to his CD8 cells. Wang and Sun [70] used multiple-session HIFU to treat 15 patients with late-stage pancreatic cancer. Although there was an increase in the average values of NK cell and T lymphocyte and subset in 10 patients after HIFU treatment, a significant statistical difference was observed in only NK cell activity before and after HIFU treatment (p < 0.05). Wu and colleagues [71] observed changes in circulating NK, T lymphocyte, and subsets in 16 patients with solid malignancy before and after HIFU treatment. The results showed a significant increase in the population of CD4+ lymphocytes (p < 0.01) and the ratio of CD4+/CD8+ (p < 0.05) after HIFU treatment. The abnormal levels of CD3+ lymphocytes returned to normal in two patients, CD4+/CD8+ ratio in three, CD19+ lymphocytes in one, and NK cell in one, respectively, in comparison to the values in the control group. In addition, serum levels of immunosuppressive cytokines including VEGF, TGF-β1, and TGF-β2 were significantly decreased in peripheral blood of cancer patients after HIFU treatment, indicating that HIFU may lessen tumor-induced immunosuppression and renew host antitumor immunity [72].

Clinical evidence suggests that HIFU treatment may also enhance local antitumor immunity in cancer patients. Kramer and colleagues [73] found that HIFU treatment could alter the presentation of tumor antigens in prostate cancer patients, which was most likely to be stimulatory. Histological examination showed significantly upregulated expression of HSP72, HSP73, glucose regulated protein (GRP) 75, and GRP78 at the border zone of HIFU treatment in prostate cancer. Heated prostatic cancer cells exhibited increased Th1-cytokine (IL-2, IFN-γ, TNF-α) release but decreased Th2-cytokine (IL-4, -5, -10) release of TILs. The upregulated expression of HSP70 was confirmed in the tumor debris of breast cancer after HIFU ablation [74], indicating that HIFU may modify tumor antigenicity to produce a host immune response. Xu and colleagues [75] found the number of tumor-infiltrating APCs, including DCs and macrophages, increased significantly along the margin of HIFU-treated human breast cancer, with an increased expression of HLA-DR, CD80, and CD86 molecules. Activated APCs may take up HSP-tumor peptide complex remaining in the tumor debris and present the chaperoned peptides directly to tumor-specific T lymphocytes with high efficiency, resulting in potent cellular immune responses against tumor cells after HIFU treatment. Furthermore, HIFU could induce significant infiltration of TILs in human breast cancer, including CD3, CD4, CD8, B lymphocytes, and NK cells. The numbers of the activated CTLs expressing FasL+ , granzyme+, and perforin+ significantly increased in the HIFU-treated tumor, suggesting that specific cellular antitumor immunity could be locally triggered after HIFU treatment [76].

15.4 HIFU Therapeutic Plan

The aim of HIFU treatment is to deliver extracorporeal focused ultrasound energy to a well-defined targeted volume at depth through the intact skin, and thereby induce coagulation of the tumor without causing damage to overlying or surrounding vital structures. Because tumor geometry is usually complicated, three-dimensional (3D) therapeutic planning based on images is essential to achieve a complete ablation. The spatial distribution of thermal ablation delivered depends on a well-thought out available plan, therapeutic device, and doctor expertise. The processes of treatment planning for extracorporeal USgHIFU system and technical considerations relative to this therapeutic planning will be described in Section 15.4.6.

15.4.1 Pre-HIFU Planning

The process of HIFU treatment for the individual patient with malignant tumor includes a number of steps, and each of these steps requires special knowledge and close collaboration with other members of the team to ensure accurate execution of a complicated set of procedures. It begins with the pathologic diagnosis of disease and accurate assessment of the tumor using TNM classification. As modalities for treating solid malignancy are usually multiple, combining local therapies such as surgery or radiation with systemic therapy such as chemotherapy, a decision as to whether the planned treatment is curative or palliative is made from a multidisciplinary clinical meeting held with surgeons, oncologists, radiologists, radiation oncologists, and pathologists. The role and scheduling of chemotherapy or vascular embolization of the tumor in relation to a session of HIFU are defined at the start. Subsequently, the location and extent of the tumor relative to overlying and adjacent critical normal tissue are determined by a variety of imaging modalities. However, sometimes pre-HIFU adjuvant treatments may influence the site and size of the targeted volume. For instance, patients with typical osteosarcoma are treated with several circles of neo-adjuvant chemotherapy, and obvious regression of the tumor is usually observed in these patients.

15.4.2 Imaging for HIFU Planning

Imaging aspects includes tumor size, shape, number, margin, and location within the organ relative to large blood vessel, nervous fiber, and vital structures that might be at risk of injury by the thermal ablation. Using contrast-enhanced technique, tumor vascularity is assessed to determine the acoustic energy distribution in the tumor.

Imaging taken for HIFU treatment planning is usually different from that taken for diagnostic use. Apart from imaging aspects of the tumor, adjacent organs close to the targeted volume require attention for the purpose of safety. Using available imaging techniques, the relationship between the tumor and its surrounding organs is evaluated. For instance, a hepatic tumor in the lower part of the left liver may be near to stomach, duodenum, and colon. Compared to magnetic resonance imaging (MRI) and computed tomography (CT), real-time ultrasound imaging is more important to assess the relationship between the tumor and gastrointestinal tracts.

MRI and CT examinations include non-enhanced and contrast-enhanced scans. Either MRI or CT is selected to perform for the acquisition of tumor data. The sequences used for imaging are detailed in Section 15.6.

Ultrasound imaging includes B-mode imaging, color Doppler and power ultrasound imaging, and contrast ultrasound examination. Attention must be given to the characteristics of the tumor gray-scale, margin, motion, and vascularity, particularly big blood vessels (more than 2 mm in diameter) in the tumor and/or surrounding the tumor. If ultrasound imaging is used for the guidance of ablative procedures, non-enhanced and contrast-enhanced MRI/CT will be performed as an essential imaging examination to compensate for some of the shortcomings of ultrasound imaging, such as less sensitivity in predicting the tumor margin, and occasional poor lesion detection as a result of overlying bone (e.g., ribs). It needs to be routine practice to compare the difference between the MRI/CT and ultrasound imaging appearances before planning HIFU treatment protocol.

15.4.3 Tumor Volume Localization

The targeted tumor volume is described as the macroscopic extent of the tumor, which is palpable, visible, and detectable by available radiological examinations. If the determination of the targeted volume depends on the diagnostic techniques including CT/MRI and ultrasound imaging, it is essential to indicate which methods have been used for its determination.

Tumor cells most likely extend beyond the actual tumor margins into normal tissue, and recently available image techniques may not exactly circumscribe acute tumor margins. The failure to achieve a satisfactory ablation margin will be accompanied by a high local recurrence ratio. However, available data for quantifying this margin of tumor volume are lacking. The definition of targeted volume is based on knowledge from surgical and postmortem specimens, and patterns of tumor recurrence, as well as from clinical experience. It may be the most difficult and subjective step in the HIFU planning process.

For curative treatment, the aim is to induce complete coagulation of a targeted tumor and the estimated extent of the microscopic spread surrounding the visible tumor. This is a principle that is routine in conventional surgery in order to ensure the removal of adjacent microsatellites, and to allow for the uncertainty that frequently exists concerning the exact location of definite tumor margins. A number of factors, such as the age of the patient and considerations of normal tissue tolerance, may affect the maximum volume considered to be appropriate for treatment. For instance, hepatocellular carcinoma is frequently seen in the setting of hepatic cirrhosis with partial liver dysfunction. Definition of targeted volume is cautiously determined for each patient. If the extent of ablative tissue is the same as that of the surgical resection, including the tumor and a 2 cm margin of obviously normal tissue surrounding the tumor, HIFU may offer its equivalent effectiveness and considerably decreased risk of morbidity and mortality in this high-risk patient population.

15.4.4 Acoustic Path for Ultrasound Energy Entry

Acoustic path is defined as a direct path between a HIFU transducer and a targeted volume at depth, where the focused ultrasound beam from the radiating surface can directly penetrate through the overlying tissue of the targeted volume and thereby destroy the tumor. This path is usually the shortest route of focused ultrasound beam from the skin to the targeted volume.

Because of the fact that the ultrasound energy is attenuated while it penetrates through the tissue, it is very important to achieve a large surface area for the acoustic path, and then the ultrasound beam can sufficiently be converged through the targeted volume. This geometrical gain of focusing is necessary to overcome the attenuation loss and to induce complete coagulation of the targeted volume. However, several factors are involved to influence the extent of the path, including the diameter of HIFU transducer, frequency, tumor depth and volume, and the structure of the overlying tissue, as well as the principal method that is used to make the transducer.

15.4.5 Ultrasonic Properties of Overlying Tissues

Although many factors can influence the energy deposition of focused ultrasound beam, the absorbed power density in the targeted volume can be estimated. One of the most important factors is attributed to the ultrasonic property of overlying tissues between the skin and the target. The beam is attenuated by absorption and scattering while it propagates through these overlying tissues. It is also reflected from the surfaces of various structures in the overlying tissues. Furthermore, the heterogeneity of the overlying tissues, such as ribs and muscular tendons, may change the beam propagation direction, focus location, and shape and size of a focal area. Thus, it is very important to know the ultrasonic properties of the overlying tissues, and a precise analysis of the acoustic properties of the overlying tissues during the treatment planning for an individual patient may avoid undesirable effects, and thereby achieve a successful HIFU ablation.

Tissue ultrasound amplitude attenuation is a sum of the losses from absorption and scattering of ultrasonic energy in tissue. Each tissue has an attenuation value of its own. For instance, the amplitude attenuation values in lung and bone tissue are approximately 430–480 Npm^{-1} MHz^{-1} and 150–350 Npm^{-1} MHz^{-1}, respectively, which are the highest values among the living tissues [77–79]. Fatty tissue has lower values, ranging from 5 to 9 Npm^{-1} MHz^{-1}, and the soft tissue values are generally about 10 Npm^{-1} MHz^{-1} [80–82]. Compared to normal tissues, the attenuation in the tumor is usually higher. The attenuation in brain tumor is higher than in the normal brain [83], and similar difference is also detected between breast cancer and normal breast [84]. However, the definition of the attenuation in the overlying tissue and the tumor is relative. While the deep part of a targeted tumor is treated with HIFU, it also includes the normal tissues between the skin and the superficial margin of the tumor and the neoplastic tissue in the front of the focus.

Images taken from the treatment planning can provide useful information about the overlying tissues. By using a HIFU device they are reassessed under the guidance of real-time ultrasound imaging when the patient lies on the treatment bed in a correct therapy position. The composition, interface, and thickness of the overlying tissue are carefully recorded. Then, ultrasound attenuation in the overlying tissue may be simply estimated and should be documented in writing with a diagram. These data may obviously vary due to the tumor location and patient position for treatment. Acoustic properties of the overlying tissues, the interfaces between the tissues, and incident angle of ultrasound beam can obviously affect the ultrasound attenuation. A large amount of energy reflection occurs when a focused ultrasound beam propagates from muscle layer to the bone surface such as ribs. This can cause the loss of energy deposition in the targeted volume, and may increase local temperature of muscle-bone interface because the amplitude attenuation coefficient of ultrasound is about 10–20 times higher in bone than in soft tissue [78,79]. Furthermore, attention must be given to the large variations in the overlying tissue thickness among individual patients (e.g., thin or overweight patient).

It is believed that there is only a small amount of energy reflection between the soft tissue interfaces because of small variations in the acoustic impedance of soft tissues. The largest one is the interface between fat and muscle tissues, but it causes only 1–2% of the beam intensity to be reflected [78,79]. It must be noted that some exceptional cases are observed in clinical application. For instance, the beam reflection is theoretically very low on the interface between subcutaneous tissue and rectus abdominis. As fibrous tissue forms a strong layer of connective tissue on the surface of the rectus abdominis, it can cause a higher reflection of ultrasound beam on this interface.

15.4.6 Therapeutic Planning Using US-Guided HIFU Device

Using an USgHIFU device the position of a patient is decided with the help of an imaging system equipped in the device. This position must be comfortable, reproducible, and suitable for acquisition of imaging for planning and subsequent treatment. It is technically ideal to achieve either the shortest way or enough surface extent of the acoustic path for the treatment of a targeted volume. The choice of position for treatment is dependent on the tumor location, size, and shape, which are shown on both MRI/CT and ultrasound imaging before the planning process, though palpation is important to the superficial tumors such as breast cancer. It is necessary that the patient be treated in one position during the whole process of the procedure. The change of patient position may result in alterations in internal and external anatomy and risk of mis-target. However, if the patients require a change of position for the purpose of treatment during the procedure, it must be technically designed in advance and carefully recorded, so that the over- or underdosage of the acoustic energy within the targeted volume can be avoided.

The position of the patient, all positioning aids, and anatomical measurements should be accurately documented in writing with a diagram and digital photos to ensure reproducibility through all stages of the planning process and subsequent treatment. However, sometimes the position may be constrained by equipment limitations. For instance, the arm position may be restricted by the limited size of the treatment bed. Attention should be given when the patient's position is selected.

Immobilization is required during the HIFU procedure. It varies according to the technique being used and the location of the tumors. Total physical restraint is essential to protect the patients who are receiving one kind of anesthesia method. Appropriate immobilization can be achieved with the various fixing aids such as hanging belts of sponge, pillows, leg restraints, and footrests. It must be ascertained that the respiration of patients is free while the fixation is performed, without any pressure effect on the chest and the abdomen.

Clinical studies have reported that there is organ motion occurring in the body. Either diaphragm movement during respiration or the pulsating of large arteries usually causes this motion. Furthermore, organ movement may not be uniform, but dependent on the individual patient. It is recommended that this variation be previously estimated, and then measured using real-time ultrasound imaging during the planning session for each patient. The movement of the diaphragm caused by respiration can obviously move some internal organs such as liver, pancreas, and kidney. It is very important to control the large movement of these organs during the HIFU procedure, particularly for the small volume tumor. An alternative technique of active breathing control involves immobilized breathing using general anesthesia; this method may be used for treatment of the liver and kidney, where excursions of up to 3 cm may occur.

Once the patient is correctly positioned and secured, the skin overlying the tumor is brought in contact with degassed water or a rubber bag filled with degassed water via acoustic gel, so that the ultrasound beam can transmit from the water into the tissue. The positions of the entire targeted tumor and surrounding vital structures are determined using the real-time diagnostic ultrasound imaging. With the movement of the diagnostic probe from one side of the tumor to the other, images of the targeted volume are achieved (Figure 15.1). Then, 3D ultrasound imaging of the tumor can be done by using sequential scans of the treatment volume.

For therapeutic purposes, the entire tumor is segmented into slices for ultrasound imaging; the spacing between slices depends on the size, shape, and location of the region of tumor and acoustic window. The extent of the separation is usually about 5–10 mm though it varies in different tumors. The targeted volume is defined as the volume of tissue that includes the tumor visualized on the ultrasound imaging and regions considered to be at risk for microscopic extension. If the margin of the tumor is not clear enough, MRI or CT scans are available to make it clearer. By using a tracker ball or mouse with an interactive software program, the contours of the targeted volume can be manually outlined at the planning computer on each ultrasound slice and then transferred to the planning system. Finally, the targeted volume that includes the margin with appropriate margins at all levels is built into the computer memory.

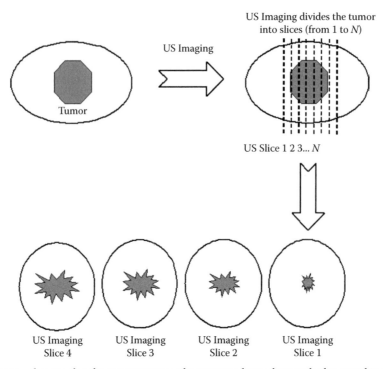

FIGURE 15.1 Schematic diagram showing that the entire tumor can be segmented into slices with ultrasound imaging.

15.5 HIFU 3-D Conformal Therapy

The volume of tissue ablation induced with one HIFU exposure can be regarded as a lesion. The focal volume of a HIFU transducer is usually ellipsoid or cigar shaped, with dimensions of 10–20 mm along the beam axis and 1–2 mm in the transverse direction. Therefore, the lesion induced with a single HIFU exposure is usually very small. But, while lesions are placed side by side, confluent volumes of ablation can be achieved.

To ablate clinically relevant volumes of tissue for the treatment of carcinomas, many of these small lesions should be positioned side by side systematically to "paint out" or cover a targeted volume, without any remaining live tissue between each lesion. If a suitable therapy plan is correctly performed, HIFU is able to ablate various shapes and sizes of solid tumors in a conformal fashion.

As a result, conformal HIFU therapy can be clinically defined as a precise procedure to ablate an entire tumor by moving high-energy concentrated focus side by side in a 3D fashion. At the beginning of a HIFU procedure, the targeted tumor is identified and divided into parallel slices of 5-mm separation by moving the diagnostic probe. Using HIFU exposure regimes, the tumor on each slice is completely ablated, and this process is repeated slice by slice to achieve complete coagulation of the targeted tumor, as shown as Figure 15.2. There are four exposure regimes that are usually used in HIFU procedure. They include single exposure for a cigar-shaped lesion, either multiple single exposures or linear scan exposures for a line-shaped lesion, and a convergent scan for the lesion of deep tumor.

A single exposure can be made to induce a cigar-shaped lesion when the location of the focal volume is immoveable. The exposure time for each pulse ranges from one second to several seconds. The single exposure can be repeated at predetermined intervals in the same position. Multiple single exposures can produce a line-shaped lesion by placing single-exposure lesions side by side with a present overlap and with a predetermined time interval between exposures. The multiple single exposures can be repeated in the perpendicular direction to form a slice-shaped lesion. A line-shaped lesion can also be achieved by a linear track exposure in which the activated transducer is moved at a constant speed over a line. This may be made by traversing one or more times in one direction only, or by scanning in both directions ("there and back") without pausing at the furthest extremity. Several of these tracks may be superimposed in one exposure period at chosen (preset) time intervals. Superimposition of tracks leads to an increase in the extent of ablation in the direction perpendicular to the direction of motion due to thermal conduction. As a result, a slice-shaped lesion can be induced in this way. From one slice-shaped lesion to the next, confluent volumes of ablation can be achieved.

The selection of the previous exposure regimes during HIFU procedure is very complicated in clinical practice. It depends on the component of overlying tissue structure, acoustic window, the depth of tumor from the skin, vital structures surrounding tumor, tumor vascularity, and size. For instance, single exposure and a linear scan exposure can be chosen for the treatment of a superficial, poorly vascularized tumor. However, multiple single exposures are usually used in the treatment of deep vascularized tumors. In clinics they can be separately used for an

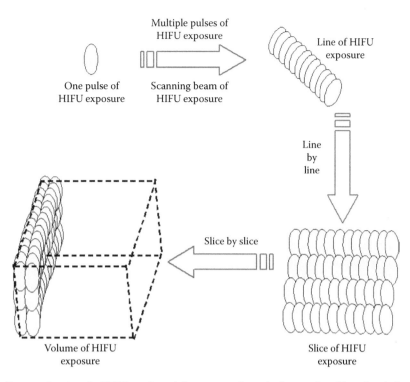

FIGURE 15.2 Schematic diagram showing the HIFU conformal therapeutic plan, which is used to ablate the whole volume of solid tumor.

individual patient simultaneously. Therefore, only doctors with a knowledge base from a specialized training course could perform this treatment. The experience that doctors have and reasonable judgment that they can make during HIFU procedures are important at the early stage of HIFU clinical application. As therapeutic data are extensively collected for each type of solid malignancy, improvement is expected in the near future.

15.6 Medical Imaging in HIFU Ablation

Recent developments of imaging modalities such as CT and MRI have dramatically led to significant advances in HIFU therapy. Accurate anatomical information is an essential precondition for planning and implementing HIFU procedure to the entire extent of the malignancy. In fact, radiological and medical images of various types are employed at each step of the HIFU treatment process, including diagnosis, assessment of the extent of disease, treatment planning, treatment and assessment of tumor response, follow-up, and outcome evaluation. Therefore, these images are of utmost importance in HIFU treatment. With the help of new imaging modalities, doctors are now able to define treatment volumes and critical structures with great precision, thus reducing marginal misses and energy exposure to normal tissues. Such capabilities may permit a higher rate of complete tumor ablation, potentially leading to improvement in local tumor treatment.

Medical images are classified into two types: anatomic and physiologic images. The former category usually includes CT, MRI, ultrasonography, and digital subtraction angiography (DSA); the latter is single photon emission computed tomography (SPECT), and positron emission tomography (PET). Anatomic images can provide accurate anatomical information about tumor extent, whereas physiologic images can provide physiologic information on the function of tumors. They are both acquired as digital data and can be mathematically processed.

15.6.1 Medical Images Used in HIFU

CT is an x-ray imaging technique used to visualize thin slices of the body. Early CT scanners employed a translate-rotate motion and required long times ranging from 20 seconds to 5 minutes to acquire a complete set of data for one image. Modern scanners are much faster, acquiring the data for one image in only 1 to 3 seconds. Faster and more accurate CT examination has recently become possible with a new device, the helical scanner. When contrast agents are used, enhanced CT can provide improved visualization of certain tumors. CT images can be used in various areas of HIFU treatment, including the delineation of the targeted volume, the determination of relative geometry of critical structures, and follow-up evaluation of treatment response.

MRI usually presents unique anatomical information and tumor detail. It can offer excellent discrimination of certain tumors with high contrast and good resolution, as well as the ability to select arbitrary planes such as transversal, sagittal, and coronal directions for imaging. When contrast agents such as gadolinium compounds are injected, the appearance of the resulting enhanced MRI image is much better than that of its non-enhanced counterpart. The utilization of MRI in HIFU treatment is the same as with CT. However, in thermometry techniques, MRI can be used as an imaging-guided device to record focal temperature rises on anatomic images during HIFU procedures [85–88].

Due to its extensive availability, real-time visualization, flexibility, nonionization, and low cost, ultrasonography is widely used in clinical practice. It can show tomographic views in almost any orientation. Compared to CT/MRI, ultrasonography produces images with poor resolution. Furthermore, as ultrasound beams cannot penetrate bone or gas-filled cavities such as the lung and gastrointestinal tract, the application is limited. Medical ultrasound imaging systems can be employed as a real-time imaging guided device to identify and target the tumor to be treated, and to monitor HIFU procedures [89–93]. They can also be used to assess the tissue response through grayscale changes caused by cavitation and tissue boiling in focal volume, which is an indication of ablation following each exposure [94].

DSA is a standard part of the evaluation of patients with a suspicious malignant tumor. It demonstrates the tumor vascularity and stain, the extent of tumor mass, and the anatomic variants of the regional arterial supply. The use of this technique yields satisfactory diagnostic and anatomic details in most cases of solid malignancies. Because it is an interventional examination, extensive applications of DSA are relatively limited in clinical practice.

SPECT is a radionuclide scanning technique that uses a radioisotope as a tracer to assess abnormal tumor function, rather than to provide simple anatomy. The radioisotopes, such as ^{99m}Tc sestamibi and ^{99m}Tc methylene diphosphate, usually have a high extraction rate at the malignant tumors. It can provide physiological imaging information through tomographic images of the radioactivity distribution within tumor tissue.

PET can determine the level of localized radioactivity by detecting a sufficient number of the two 180-degree opposed photos and using imaging reconstruction algorithms. It is another functional type of imaging that can provide localized physiological information on the presence of tumors.

15.6.2 Medical Imaging for HIFU Planning

Diagnostic images used in the preparation of HIFU ablation have been described in Section 15.4.2, imaging for HIFU planning. Most anatomic image modalities can be employed to provide anatomical information for determining the location, number, and size of tumors and surrounding vital structures. For instance, in the treatment of liver cancer it is very important preoperatively to know the relationship between the tumor and its surrounding structures, such as bile ducts, gallbladder, gastrointestinal tract, diaphragm, and large blood vessels. These pre-HIFU images are fundamental images on which the suitability of a patient for HIFU ablation is assessed. In addition, physiological imaging may be used as a pre-HIFU image for the purpose of assessing tumor response to the ablation.

15.6.3 Ultrasound Imaging for HIFU Procedure

To optimize the ablation of a tumor, it is fundamental to use precise and dependable imaging techniques for ascertaining the adequacy of the treatment. This imaging can be used as an imaging-guided device to identify and target the tumor to be treated, monitor the therapy procedure, and assess therapy response within targeted tissue. In particular, recent studies involved with the development of imaging procedures enable rapid assessment of the extent of targeted tissue destruction caused by HIFU ablation. As a result, either ultrasonography or MRI is employed to guide extracorporeal HIFU procedure. Individual preferences or personal research interests may dictate the selection of imaging technique.

There are advantages and disadvantages in both ultrasonography and MRI when each type of device is incorporated into a HIFU system. Clinical ultrasound imaging devices are inexpensive, extensively available, flexible, capable in real-time visualization, and valuable in the treatment of organs such as the liver and kidney, which are moved by respiration. The main disadvantage of ultrasonography is poorer imaging resolution than MRI, particularly in predicting tumor margins. It may provide occasional poor lesion detection through lack of inherent tissue contrast or because of overlying bone structures such as ribs. MRI can provide three-dimensional imaging with better resolution. Using indirect thermometry technique it can measure focal temperature rises following HIFU exposure. However, high cost, long treatment time, and problems in tracking a moving target such as the liver may limit extensive application of MRI-guided HIFU devices in clinics.

Because of ultrasonography's poor resolution, good-quality MRI is effectively used to compensate for some of the shortcomings of diagnostic ultrasound imaging, if an USgHIFU device has a problem in predicting the tumor margin. A combined utilization of MRI and ultrasonography before HIFU treatment can fully provide accurate imaging information for determining the number, sizes of tumors, and their relationship to surrounding vital structures. While the USgHIFU device is used in our clinical application, we feel that preoperative MR imaging is very helpful to establish the 3D coordinates of the targeted tumor in a HIFU planning session. We routinely compare the difference between the MRI and US image appearances in creating a HIFU treatment plan for the complete ablation of a tumor.

B-mode ultrasound imaging can provide significant changes in imaging information within the focal volume during a HIFU procedure. These changes are made evident by increased levels of tissue grayscale on ultrasound imaging immediately after HIFU exposure, and most of them become gradually less evident and sometimes disappear within several minute after the ablation. In our animal studies, both *in vivo* and *ex vivo*, we find that hyperechoic zones in the targeted tissue correspond mainly to the extent of the coagulation necrosis. There is a close relationship between the extent of necrosis as measured by gross examination and the hyperechoic extent measured immediately

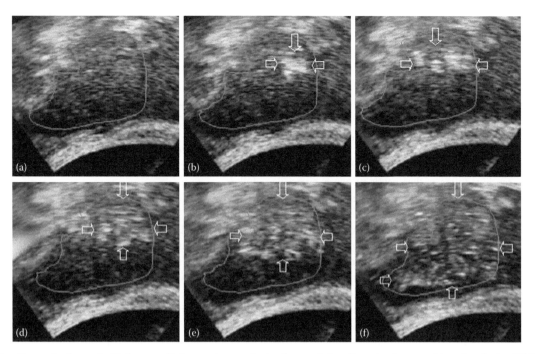

FIGURE 15.3 Grayscale change in big hepatocellular carcinoma obtained on real-time B-mode ultrasound imaging during HIFU procedure. Compared to tumor imaging (outlined) before HIFU (a), hyperechogenic region (arrows) is seen in targeted region (b–e), until complete ablation is achieved in this slice of the tumor (f).

after HIFU on the ultrasound image in treated tissues. Because HIFU exposures consist of single exposures, multiple single exposures, and scan track exposures, a cigar-shaped lesion and a line-shaped lesion, as well as a slice-shaped lesion, can be obviously detected on real-time ultrasound imaging following relative HIFU exposures.

These grayscale changes observed during HIFU procedures are of importance in the treatment of solid malignancy. They provide feedback control imaging information, monitor therapeutic effects, and control ultrasound energy deposition in the focal volume during the ablation procedure. Figure 15.3 shows the process of grayscale changes on real-time ultrasound imaging during the ablation for a patient with big hepatocellular carcinoma. While the focus of HIFU is moved along one slice of the tumor in a scanning fashion from deep to superficial part, a hyperechogenic region is clearly detected on ultrasound imaging in this slice of liver cancer.

Although the mechanism involved in tissue grayscale changes during HIFU procedure is not clear, to date it is generally considered that gas bubbles that originate from acoustic cavitation and tissue boiling (the vaporization of tissue water) are the main factors. The development of ultrasound contrast agents may improve the accuracy of ultrasound imaging for evaluating the final therapeutic effect of vascularized tumor immediately after the treatment is finished. To our knowledge, however, there is no previous experience of real-time evaluation of the therapeutic effects with ultrasound contrast agents during HIFU procedures.

15.6.4 Follow-Up Imaging for Assessment of HIFU Ablation

It is important to detect untreated regions of a targeted tumor that require immediate retreatment, and to demonstrate complete ablation of treated tumors as early as possible following HIFU thermal ablation. Because postprocedural core biopsy does not provide pathologic results for the entire treated tumor, follow-up imaging is essential to evaluate the short-term efficacy of *in situ* HIFU ablation. It is generally believed that, if untreated, residual viable tumor foci will regrow, resulting in therapeutic failure. The failure to detect focal areas of viable tumor in the treated regions early after thermal ablation can lead to local recurrence and to long distance metastases.

Prior to and following each treatment episode, contrast-enhanced CT/MRI, ultrasonography, DSA, SPECT, and PET can be employed to evaluate the response of targeted tumors after HIFU procedure is finished. These modalities, anatomical or physiological, can provide imaging information in determining the therapeutic effects on both tumor vascularity and on cellular functions immediately after HIFU, as well as changes in tumor size during long-term follow-up periods.

Using color and power Doppler methods, ultrasonography can assess changes in tumor vascularity after HIFU ablation. But, in some cases, especially in the tumors with poor blood supply, ultrasound imaging is not sufficiently sensitive for detection of damage to tumor blood vessels. Using ultrasound contrast agents, ultrasonography can provide imaging information to evaluate the therapeutic response of nonvascularized tumors after HIFU treatment.

Furthermore, DSA can clearly show variants of regional blood supply in the treated tumor after HIFU ablation. Because of interventional examination, applications of DSA are relatively limited. As a physiological imaging, SPECT can provide tomographic images of the radioactivity distribution in a small number of tumors such as osteosarcoma. However, it does not clearly identify the therapeutic response in most tumors, particularly in patients with liver cancer. Theoretically, PET can be employed to detect physiological function in almost solid malignancies.

Non-enhanced CT obtained 1–2 weeks after HIFU ablation is not sufficiently sensitive to assess therapeutic response, especially in encapsulated lesions such as those of hepatocellular carcinoma, and of osteosarcoma lesions located in the long bone marrow cavity. However, non-enhanced MRI may characteristically reveal varied signals of ablated tumors on both T1- and T2-weighted images after HIFU ablation. Contrast-enhanced CT is useful for demonstrating changes in tumor vascular perfusion, and for distinguishing the difference between nonviable and residual viable tumor directly in the treated regions. The most striking changes are seen in contrast-enhanced MRI, where it is common to observe the absence of contrast enhancement in the treated tumors that originate from liver, pancreas, kidney, breast, bone, soft tissue, and uterus. Areas of hypoattenuation, shown on MRI or CT, that did not enhance after contrast material administration can be considered to represent necrotic tissue. Still-enhancing areas are assumed to reveal residual viable tumor. In addition, a thin peripheral rim of enhancement may be detected surrounding the coagulation necrosis. This densely enhancing peripheral rim on delayed contrast imaging should not be misconstrued as residual tumor. In fact, this rim represents an inflammatory reaction to thermal ablation. However, a thick irregular rim at the edge of a treatment site is the most common feature of partial thermal ablation. In conclusion, among the follow-up images, enhanced CT or MRI should be standard imaging criteria for response assessment of treated tumors after HIFU ablation.

There is a lack of consensus on a standard regimen for follow-up imaging. Our clinical experience shows that diagnostic images, particularly enhanced MRI or CT, can be used to assess the efficacy of HIFU therapy 1–2 weeks afterwards. In addition to reporting tumor size and the diameter of the ablation volume, assessment of contrast enhancement or lack in the treated tumor can provide imaging information about tumor response and residual live tumor cells after HIFU ablation. Follow-up imaging at 3 months is necessary to detect new-growth tumors in untreated regions. Imaging at 6–12 months can show obvious regression of the lesion and the region of coagulation necrosis. Most frequently, the treated volume without blood supply shrinks by less than 20–50% in volume.

15.7 Clinical Applications of Extracorporeal US-Guided HIFU for Solid Malignancies

Since HIFU was first used in the 1950s for patients with neurological disease during surgical procedure, it has been studied in the laboratory to treat normal tissue and implanted tumors in animals. The recent resurgence of this noninvasive therapy is driven by the advances in the development of medical imaging techniques, which makes it possible to target HIFU extracorporeally and precisely ablate a tumor at depth. To date, two extracorporeal HIFU devices have been clinically used worldwide. They incorporate either MRI or B-mode ultrasonography to monitor the therapeutic procedure. The advantages and limitations of both imaging modalities used in HIFU have been described in Section 15.6, "Medical Imaging in HIFU Ablation." Clinical applications of the extracorporeal USgHIFU device follow.

Extracorporeal USgHIFU has been widely used for the treatment of solid tumor in China since 1999. The malignancies treated with HIFU include those of the liver, breast, kidney, pancreas, soft tissue, and bone; the benign tumors treated include uterine fibroids and breast fibroadenomas [93]. Most of these clinical applications present encouraging short-term outcomes; long-term follow-up survival data are now emerging for the treatment of hepatocellular carcinoma, breast cancer, and osteosarcoma. As large numbers of patients with solid malignancy are treated in China, the clinical experiences with extracorporeal USgHIFU are very important and should be shared worldwide. Much of the material of Section 15.7 is about the clinical knowledge and practice of HIFU in China; it is not intended to be exhaustive but to serve as an illustration of our clinical experience to date. Much of the clinical application is very recent, and detailed survival data will follow in subsequent publications.

15.7.1 Purposes of HIFU Therapy

In practice, cancer therapy usually needs to have multiple treatment methods for long-term survival benefit in addition to local therapy such as surgery. For instance, in the treatment of patients with breast cancer, surgery, chemotherapy, radiotherapy, and endocrine therapy must be provided in tandem because many clinical results indicate that the combination of these modalities can provide better survival benefit than one in isolation. HIFU is a local therapy for noninvasive destruction of the tumor, and it is essential to combine it with other therapies in clinical applications. However, some cancers, such as renal cell carcinoma and pancreatic cancer, are not sensitive to chemotherapy and radiotherapy. Similar to surgical operation, HIFU can become the only therapeutic option offered to the patients.

There are two goals of HIFU in the treatment of patients with solid malignancy. One goal of this ablation in patients with early stage cancer is to achieve a cure, and HIFU should be used as a local treatment to induce complete necrosis of the targeted tumor. Additional treatments, such as chemotherapy, radiotherapy, and endocrine therapy, are essential to patients with breast cancer for conservation of the diseased breast, if HIFU is used locally in patients with early stage breast cancer. In surgical oncology it is necessary to resect the entire tumor along with an adequate tumor-free margin to prevent local recurrence. In the same way, HIFU treatment should adopt a similar principle and aim to kill the entire malignant focus along with tumor-free margin of healthy tissue. However, the definition of an adequate tumor-free margin varies with the type of malignant tumor. For instance, the surgical margin in hepatic resection for colorectal metastasis was defined as being, preferably, 2 cm but no less than 1 cm of normal liver [95]. If the objective of HIFU is to replicate the success of liver resection, the same tumor-free margin should be treated.

The other goal of HIFU treatment is palliative for patients with advanced-stage cancer. They are usually those who have an unresectable tumor and for whom conventional tumor therapies, including chemotherapy and radiotherapy, have failed to control tumor growth. HIFU can be clinically used to impede tumor growth and improve the quality of life for such patients. Among those treated with HIFU in China, most are advanced-stage patients who are beyond the scope of conventional treatments. In these circumstances, HIFU can be successfully performed as a palliative method using partial or complete ablation. Symptoms such as pain caused by tumor disappear after HIFU, and survival time can be extended.

15.7.2 Anesthesia Selection for HIFU Therapy

Almost all patients have an uncomfortable sense of pain originating from the targeted tissue during HIFU procedures. Also, it is almost impossible for them to tolerate one fixed position without any motion for a long time. Therefore, either local or general anesthesia, as well as sedation, is essential to HIFU treatment. The selective standard of anesthesia is dependent on two factors. (1) The most important is the patient's general condition that decides which kind of anesthesia is suitable, and how far the targeted organ can move. (2) If clinical examinations show that the patient is able to receive any kind of anesthesia, the movement of the targeted organ becomes the dominant factor to influence the decision made by an anesthetist who is experienced in collaborating with HIFU doctors.

The anesthetist selects the anesthetic, primarily on the basis of targeted organ motion, ablation time, patient position, tumor location, and therapeutic ultrasound exposure. General anesthesia is usually used to ensure immobilization of targeted organs such as liver, kidney, and pancreas during the HIFU procedure. Endotracheal intubation and mechanical ventilation enables single lung ventilation on one side and, therefore, controls the movement of these organs, which is caused by respiration. Furthermore, it has the supplementary benefit of permitting temporary suspension of respiration with controlled pulmonary inflation, as necessary, to ablate a liver or kidney tumor behind the ribs.

15.7.3 HIFU Therapy for Liver Cancer

Because hepatocellular carcinoma (HCC) is frequently seen in the setting of hepatic cirrhosis, surgery can be performed in only 10% to 20% of patients with HCC [96–100]. Most HCC patients lose the chance to be treated with surgery. So the liver has been a major target of HIFU for a long time because of the noninvasive nature of HIFU, and there is an increasing body of work describing HIFU in the treatment of both primary and secondary liver cancer in human clinical trials.

In the early 1990s, Vallancien et al. [101] treated two patients with liver metastases prior to surgical resection; in one there was no visible effect, and in the other there was extensive tissue laceration and patchy necrosis. In 2001, we reported for the first time that HIFU could successfully treat patients with primary liver cancer, and a demarcated coagulation necrosis was clearly detected by pathological examinations in the HIFU-treated HCC [102]. From March 1998 to October 2001, a total of 474 patients with liver cancer, including primary and metastatic liver cancer, received HIFU treatment at ten hospitals in China [93]. Almost all patients had unresectable HCC ranging from 4 to 15 cm in diameter. Among them, most patients were advanced-stage patients with hepatic cirrhosis, and HIFU was used as a palliative therapy in clinical practice.

Wu et al. [103] reported a prospective, nonrandomized clinical trial in which 55 HCC patients with cirrhosis were treated with USgHIFU therapy in Chongqing, China. Of these, 51 patients had unresectable HCC, and tumor size ranged from 4 to 14 cm in diameter with mean diameter of 8.14 cm. The results showed that HIFU was safe, and no severe side effects were observed after the treatment. The overall survival rates at 6, 12, and 18 months were 86.1%, 61.5%, and 35.3%, respectively. The survival rates were significantly higher in patients in stage II than those in stage III_A ($P < 0.0132$) and in stage III_C ($P < 0.0265$). From November 1998 to May 2000, 50 consecutive patients with stage IV_A HCC were enrolled in a randomized, controlled clinical trial to assess the local therapeutic efficacy of USgHIFU therapy combined with transcatheter arterial chemoembolization (TACE) and TACE alone [104]. These patients were divided into two groups: TACE alone was performed in group 1 (n = 26), and HIFU combined with TACE was performed in group 2 (n = 24). Tumors ranged from 4–14 cm in diameter (mean 10.5 cm). Follow-up images showed absence or reduction of blood supply in the lesions after HIFU ablation when compared with those after TACE alone. The median survival times for patients were 11.3 months in the group 2 and 4 months in the group 1 ($P = 0.0042$). The 6-month survival rate of patients was 80.4–85.4% in group 2 and 13.2% in group 1 ($P = 0.0029$), and the 1-year survival rate was 42.9% and 0%, respectively ($P < 0.01$). Li et al. [105] also reported similar results of HIFU combined with TACE for 89 patients with large unresectable HCC. Follow-up results showed that 1–5 year overall survival rates were much higher in the HIFU combined TACE group than those in the TACE group. Recently, Zhang et al. [106] reported effects of HIFU ablation on 39 HCC patients whose lesions were close to the major hepatic blood vessels. Although the distance between tumor and main blood vessel was less than 1 cm, HIFU could achieve complete coagulation necrosis of HCC lesions, with no evidence of discernible damage to the major vessels in all patients. In addition, HIFU could be successfully used to treat needle-track seeding of HCC [107]. The results showed that during a mean follow-up of 10 months, complete ablation was persistently observed in eight patients, but one patient developed a recurrent lesion near the treated area, which was completely ablated by second HIFU ablation.

HIFU ablation has also been used for palliation in patients with advanced-stage liver cancer. Li et al. [108] reported a series of 100 patients with liver cancer who were treated with HIFU, including 62 patients with primary liver cancer and 38 with metastatic liver cancer. After the treatment, clinical symptoms, such as loss of appetite, weight loss, and discomfort or pain in the liver region, were obviously relieved in 87% patients. Compared to pre-HIFU images, follow-up MRI or CT examinations showed partial or complete coagulation necrosis of the targeted tumors. In another study, patients with unresectable HCC received either HIFU plus supportive treatment (HIFU group, $n = 151$) or supportive treatment only (control group, $n = 30$), according to their willingness [109]. A complete or a partial response was achieved in 28.5% ($n = 43$) or 60.3% ($n = 91$) of cases in the HIFU group. In contrast, the response rates were 0% and 16.7%, respectively, in the control group. In addition, the one- and two-year survival rates were 50.0% and 30.9% in the HIFU group, which were significantly higher than those in the control group (both $P < 0.01$).

In Oxford, United Kingdom, Illing et al. [110] reported interim results of two prospective, nonrandomized HIFU clinical trials for liver cancer in a Western population. Using either radiological images such as MRI and contrast ultrasound, or histological examination, the safety and effectiveness of HIFU were evaluated in the treatment of 22 patients with metastatic liver cancer. All patients who completed one treatment session with HIFU had some evidence of discrete ablation on both imaging and histology follow-up. These interim results showed that HIFU ablation resulted in discrete ablation zones of liver tumors in all evaluable patients (100%). The adverse event profile was favorable when compared to open or minimally invasive techniques. In the histological study, the predominant characteristic of HIFU-ablated tissue is coagulation necrosis, but heat fixation is evident in some areas [111]. Heat-fixed cells appear normal under haematoxylin and eosin staining, indicating that this is unreliable as an indicator of HIFU-induced cell death. The radiological study shows time to maximum enhancement maps and their 3D views in contrast-enhanced MRI can be used as a noninvasive tool to assess and potentially to quantify the success of HIFU ablation [112]. These results reveal that HIFU is capable of achieving selective ablation of predefined regions of liver tumor targets, and that MRI evidence of complete ablation of the target region can be taken to infer histological success.

In Yokohama, Japan, Numata et al. [113] reported the preliminary experiences of using both contrast-enhanced 3D ultrasound imaging and contrast-enhanced CT/MRI for the evaluation of HIFU effect on 21 HCC lesions. One month after HIFU ablation, contrast-enhanced CT/MRI showed complete ablation in 18 lesions and partial ablation in three lesions. There was a good correspondence between contrast-enhanced 3-D ultrasound imaging immediately after HIFU and contrast-enhanced CT/MRI one month after HIFU. Orsi et al. [114] treated 17 patients with 24 liver metastases at difficult locations in Milan, Italy. After one session of HIFU treatment, PET-CT and/or MDCT at day

one showed complete response in 22/24 liver metastases one day after HIFU. No side effects were observed during a median of 12 months of follow-up. However, in Seoul, South Korea, Park et al. [115] reported different results of using HIFU therapy for 13 patients with liver metastasis from colon and stomach cancer. Complete ablation was observed in eight patients, and partial ablation in five cases. Among the patients with complete ablation, three patients had no evidence of recurrence, two presented new foci of disease in the target organ or distant malignancy, and three underwent local tumor progression during follow-up period. From October 2006 to December 2008, Ng et al. [116] treated 49 patients with unresectable HCC with a curative intent in Hong Kong, China. The median size of the treated tumors was 2.2 cm ranging from 0.9 to 8 cm, and each patient underwent a single session of HIFU. The one- and three-year overall survival rates were 87.7% and 62.4%, respectively. Child-Pugh liver function grading was the significant prognostic factor influencing the overall survival rate.

15.7.4 HIFU Therapy for Breast Cancer

The first randomized controlled clinical trial was performed by our group to explore the possibility of HIFU for the treatment of patients with localized breast cancer in Chongqing, China [117]. In this study, patients were treated with either modified radical mastectomy ($n = 25$) or HIFU followed by modified radical mastectomy within one to two weeks ($n = 23$). The HIFU-treated area included the tumor and 1.5–2.0 cm of surrounding normal tissue. Short-term follow-up and pathological and immunohistochemical stains were performed to assess the therapeutic effects on the tumor and complications of HIFU. Using gross examination and triphenyltetrazolium chloride (TTC) staining, complete coagulation necrosis was confirmed in the treated tissue, which included the tumor and a wide margin of obviously normal breast tissue surrounding the tumor. Mean values of the largest parallel and perpendicular dimensions, and volume of HIFU lesions in excised breasts, were significantly larger than those of the targeted tumors, respectively [118]. Histological examinations showed that HIFU-treated tumor cells underwent typical characteristics of coagulation necrosis in the peripheral region of the ablated tumor in all patients. However, in 11 of 23 patients, hematoxylin and eosin staining showed normal cellular structure in the central ablated tumor. By using electronic microscopy and nicotinamide adenine dinucleotide-diaphorase stain, those who had normal-appearing cancer cells were not viable, indicating that HIFU could cause the heat fixation of ablated tumor through thermal effect [119]. Immunohistochemical staining showed that no expression of proliferating cell nuclear antigen (PCNA), matrix metalloproteinase-9 (MMP-9), and cell surface glycoprotein CD44v6 was detected within the treated tumor cells in the HIFU group, suggesting that the treated tumor cells lost the abilities of proliferation, invasion, and metastasis. [120].

In a nonrandomized prospective study, we evaluated the long-term clinical effects of HIFU on patients with breast cancer [121]. Twenty-two patients with biopsy-confirmed breast cancer were enrolled if they were deemed either unsuitable for surgery ($n = 6$) or had refused surgical resection ($n = 16$). Using the tumor node metastasis (TNM) staging, four patients were at stage I, nine at stage IIA, eight at stage IIB, and one at stage IV. The tumor and surrounding margins of 1.5–2.0 cm were treated. Axillary lymph node dissection was performed on patients with stage IIB disease 4–8 weeks following ablation (except in three patients who refused the surgery). All patients received six cycles of adjuvant chemotherapy and radiotherapy after HIFU ablation. On the completion of the chemotherapy, two years of hormone therapy (tamoxifen) followed. Outcome measures included radiological and pathologic assessment of the treated tumor, cosmesis, local recurrence, and cumulative survival rates.

After HIFU therapy, all patients experienced a palpable breast lump as anticipated, which extended to the whole treatment area (tumor and margin), and was therefore greater than the original tumor. Although patients were advised of this in advance, it did give rise to anxiety, and two of the 21 patients were elected to have mastectomy as a result. Radiological imaging showed absence of tumor blood flow in 19 of 22 patients, the regression of treated lesions in all patients, and disappearance in eight patients after HIFU treatment. Follow-up biopsy revealed coagulation necrosis of the targeted tumor and subsequent replacement by fibroblastic tissue.

After a median follow-up of 54.8 months, one patient died, one was lost to follow-up, and 20 were still alive. Two patients with stage IIB disease developed local recurrence at the 18th and 22nd months. They received modified radical mastectomy, followed by chemotherapy. Five-year disease-free survival and recurrence-free survival were reported as 95% and 85%, respectively. Cosmetic result was judged as good to excellent in 94% of patients. It is noted that in the Chinese patient population, breast cancer is often diagnosed at the advanced stage and that the average tumor size was larger than that which would typically be treated conservatively in the West (the typical size for conservative resection in China versus in the West is 4 cm versus 2 cm in diameter).

Currently, another clinical trial is underway at the European Institute of Oncology in Milan, Italy, wherein 12 patients with small breast cancers (<1.5 cm) have been treated with USgHIFU (personal communication). All tumors were removed for pathologic evaluation after HIFU. The pathologic results showed that all the tumors were completely ablated. Even though local edema was observed in 30% of the patients, the edema usually subsided in one day. No other side effects were observed.

15.7.5 HIFU Therapy for Bone Malignancy

An ultrasound beam is easily transmitted through soft tissues and parenchymatous organs *in vivo*, except air-containing organs such as lungs. Because of its high attenuation in osseous tissues, it is generally believed that US propagation through the bone is almost impossible in the diagnosis and treatment of bone disease. However, recent transcranial ultrasound techniques, like transcranial color-coded real-time sonography and

transcranial Doppler, have allowed noninvasive imaging of brain parenchyma and color flow imaging of intracranial vessels, and have become reliable methods for examination of patients with stroke [122–127]. These studies imply that despite the anatomical obstacle, US beams have been transmitted through bones.

In the field of HIFU, it has been discovered that focused US can cause thermal lesions in animal brains through the skull [128, 129]. As an aggressive malignant neoplasm, most osteosarcomas may be lytic regardless of the limited production of mineralized osteoid or bone tissue [130]. Slight to complete cortical destruction within tumor lesions makes possible ultrasound beam propagation through the damaged osteogenic structure into the medullary space in which osteosarcoma originates. Therefore, HIFU can be used as a noninvasive therapy to treat patients with osteosarcoma through the bone containing soft tissue of tumor and the weakened bone.

HIFU has been performed as a noninvasive approach in the treatment of patients with osteosarcoma. In a perspective clinical trial, we used USgHIFU combined with neoadjuvant chemotherapy to treat 34 patients with biopsy-proven malignant bone tumors (Enneking's Stage II$_b$) for conserving the diseased-limb, and HIFU was used alone for 10 patients with stage III$_b$ for palliative intent [131]. After a mean follow-up time of 17.6 months, the overall survival rate of 44 cases was 84.1%. For the 34 cases of stage IIb, 30 cases continued to survive disease-free, two died of lung and brain metastases, and the other two had local recurrence. Among 10 of the stage IIIb cases, five survived with tumor, one had local recurrence, and five died of lung metastases. The complication occurrence rate was 18.2%, including two cases with secondary infection, three with pathologic fracture, one with epiphysis separation, and two with common peroneal nerve injury. Enneking comprehensive function scoring of the 36 cases was not less than 15 points.

Li et al. [132] reported the preliminary experience of using USgHIFU for the treatment of 25 patients with primary and metastatic bone malignancy. Follow-up MRI/PET-CT showed that therapeutic response rate was 84.6% in 13 patients with primary bone tumors, including six patients (46.2%) with complete response and five (38.4%) with partial response. The therapeutic response rate was 75.0% in 12 patients with metastatic bone tumors, including five cases (41.7%) with complete response and four (33.3%) with partial response. The one-, two-, three-, and five-year survival rates were 100%, 85%, 69%, and 39%, respectively, in patients with primary bone tumors, and 83%, 17%, 0%, and 0%, respectively, in patients with metastatic bone tumors.

Recently, Chen et al. [133] evaluated long-term follow-up results of USgHIFU ablation for 80 patients with primary bone malignancies. From December 1997 to November 2004, 80 patients with primary bone malignancy were treated with USgHIFU, including 60 in stage IIb and 20 in stage III (Enneking staging). HIFU combined with chemotherapy was performed on 62 patients with osteosarcoma, one with periosteal osteosarcoma, and three with Ewing's sarcoma; the remaining 14 patients (with chondrosarcoma, malignant giant cell tumor of bone, sarcoma of the periosteum, or unknown histology) received HIFU alone.

MRI/CT and single photon emission CT (SPECT) were used to assess tumor response. Follow-up images demonstrated complete ablation of malignant bone tumors in 69 patients, with greater than 50% tumor ablation in the remaining 11 patients. The one-, two-, three-, four- and five-year overall survival rates were 93.3%, 82.4%, 75.0%, 63.7%, and 63.7%, respectively, in the patients with stage IIb cancer, and 79.2%, 42.2%, 21.1%, 15.8%, and 15.8%, respectively, in those with stage III disease. Among the patients with stage IIb disease, long-term survival rates were better in the 30 patients who received the full treatment (complete HIFU and full cycles of chemotherapy) than those in the 24 patients who did not finish the chemotherapy cycles and the six patients who underwent partial ablation only.

15.7.6 HIFU Therapy for Renal Cancer

Renal cell cancer (RCC) has the seventh highest incidence of all cancers in adults, accounting for approximately 3% of all solid malignancies. At the time of diagnosis, 75% of tumors are organ confined and as such should be amenable to a local treatment. RCC is both chemotherapy and radiotherapy resistant. As a result, the mainstay of treatment for tumors remains surgery, with five-year survival rates greater than 80% after resection [134].

In the early 1990s, Vallancien et al. [14] reported the first clinical feasibility study of using extracorporeal USgHIFU to treat RCC patients. They found evidence of ablation in the treated areas following excision of the kidney, but encountered a high rate of skin burns, with 10% of patients suffering this complication.

In a case report, Kohrmann et al. [135] reported treating a patient with three renal tumors using a hand-held HIFU device. Two of the tumors reduced in size following ablation, however one remained unaffected. Susani et al. [136] included two patients with renal tumors in a phase I trial. They claim accurate placement of lesions using the HIFU device, but detail is sparse. Marberger et al. [137] reported a series of 16 patients who had renal tumors treated with HIFU. In 14 patients a 10 mm^3 volume of renal tumor was treated with HIFU and this was followed by immediate surgical resection of the kidney. In nine patients, areas of acute tissue necrosis were seen, although the lesions only measured between 15% and 35% of the original targeted volume. Two patients were treated with curative intent; however, both had incomplete ablation with residual disease visible on follow-up MRI. Hacker et al. [138] described no major side effects following ablation of tissue in 43 kidneys, porcine and human, using an experimental handheld extracorporeal technology. However, technical success was mixed and the authors concluded that further work was required in the dosage and application of this system.

We also reported our preliminary clinical experience of using USgHIFU for the treatment of 12 patients with advanced-stage renal cell carcinoma and one patient with metastatic kidney cancer [139]. All patients received HIFU treatment safely, including 10 who had partial ablation for palliative intent and three who had complete tumor ablation for curative intent. After HIFU, hematuria disappeared in 7 of 8 patients and flank pain of presumed malignant origin disappeared in 9 of 10 patients. Postoperative images showed a decrease in or absence of tumor blood supply in the treated region and significant shrinkage of the ablated tumor. Of the 13 patients, seven died (median survival 14.1 months, range 2 to 27) and six were still alive with median follow-up of 18.5 months (range 10 to 27).

In Oxford, United Kingdom, Illing et al. [110] published interim results of a prospective, nonrandomized clinical trial to evaluate the safety and effectiveness of HIFU in the treatment of small kidney tumors. After a single HIFU session, the patients were evaluated with either radiological images, such as MRI and contrast ultrasound, or histological examinations. They found evidence of ablation in 5 of 10 patients with small RCC. While compared to open or minimally invasive techniques, the adverse event profile was favorable. Recently, Ritchie et al. [140] reported a three-year follow-up result of HIFU ablation for 15 patients with renal tumors. Among them, mean 30% decrease in tumor size was observed by contrast-enhanced MRI in 10 patients who remained on follow-up at a mean of 36 months (ranging from 14 to 55) after HIFU, with a central loss of contrast enhancement. Four patients had irregular enhancement on imaging and had alternative therapies. One patient had surgery due to persisting central enhancement. In addition, Chakera et al. [141] reported a case of a localized RCC of recipient origin that developed in the donor allograft, which was detected eight years after renal transplantation. Treatment with HIFU followed by partial nephrectomy was successful, averting the need for dialysis therapy.

15.7.7 HIFU Therapy for Pancreatic Cancer

Pancreatic carcinoma is the fourth leading cause of cancer-related deaths in the United States and the Western world. Nearly 80% of patients have unresectable disease on diagnosis. There remains no effective modality for the treatment of patients with locally advanced disease, and chemo-radiation is the current best practice although a great deal of debate remains on this issue. The median survival time is 6–10 months for patients with locally advanced pancreatic cancer and 3–6 months for patients with metastatic disease.

From December 2000 to September 2002, USgHIFU was performed to treat eight patients with advanced pancreatic cancer for palliative intent [142]. Of these, three patients had stage III disease, and five patients had stage IV disease. Five of the eight patients had liver metastases, and one had bone metastasis. All of the patients had constant localized pain. After HIFU, preexisting severe back pain of presumed malignant origin disappeared in each patient. Follow-up images showed reduction or absence of tumor blood supply in the treated region and significant shrinkage of the ablated tumor in all patients, ranging from 20% to 70%. Four of the eight patients died (median survival time 11.25 months, range 2–17 months), and the remaining four patients were still alive with median follow-up time of 11.5 months (range 9–16 months).

Wang et al. [70] reported on an extended population of 15 patients that included seven patients with lesions in the head of the pancreas. Thirteen of the 15 patients had pain associated with the cancer prior to treatment. Pain was fully alleviated in 11 patients and partly alleviated in the other two after HIFU treatment. Using a USgHIFU device, Orsi et al. [113] treated seven patients with unresectable pancreatic cancer. All seven patients were almost completely palliated in symptoms in 24 h after treatment. The median survival time was 11 months. MDCT or MRI performed 24 h after treatment did not detect any injury of the surrounding organs. Portal vein thrombosis was observed in one patient who was discharged 20 days later.

Recently, Wang et al. [143] published the clinical result of using HIFU for treating 40 patients with advanced pancreatic cancer, including 13 patients with stage III, and 27 patients with stage IV. Pain relief was achieved in 87.5% of the patients, and median pain relief time was 10 weeks. The median overall survival time was eight months in all patients, with 10 months in stage III patients and six months in stage IV patients. Six- and 12-months survival rates were 58.8% and 30.1%, respectively. HIFU was also combined with gemcitabine to treat 37 patients with locally advanced pancreatic cancer [144]. The median follow-up period was 16.5 months (range: 8.0–28.5 months). Abdominal pain was relieved in 22 patients (78.6%) after HIFU treatment. The overall therapeutic response rate was 43.6%, including two cases with complete response and 15 cases with partial response. The median time to progression and overall survival were 8.4 months and 12.6 months, respectively. The estimates of overall survival at 12 and 24 months were 50.6% and 17.1%, respectively.

15.7.8 HIFU Therapy for Soft Tissue Sarcoma

From December 1997 to October 2001, a total of 77 patients with soft tissue sarcoma received HIFU treatment in China [145]. Most of the patients had recurrent soft tissue sarcoma after surgery. Among them, 18 patients were treated with HIFU in Chongqing, China. Before treatment, pathological examination showed liposarcoma in six patients, synovial sarcoma in two, fibrosarcoma in two, malignant peripheral nerve sheath tumor in two, and other soft tissue sarcomas in five. The tumor size ranged from 5.5 to 16 cm in diameter (mean 8.6 cm), and follow-up time varied from 11 to 39 months (median 21 months). After HIFU treatment, contrast-enhanced MRI showed the complete ablation of the targeted tumor. Of the total, 16 patients are still alive (survival rate, 90%), and two patients died of metastasis after HIFU treatment. Three patients had local recurrence and then underwent a second HIFU treatment for the purpose of control.

15.8 Conclusions

HIFU ablation has been shown to be technically feasible and effective for the treatment of solid malignancies. It may offer complete ablation of cancer, with less morbidity, less damage, lower cost, and shorter hospital stay. There is clearly a place for

HIFU in the clinical management of solid malignancies. These common and serious problems affect many thousands of people every year and if HIFU can offer an option to even a small proportion of these patients then it is vital to continue pushing the technology forward. HIFU provides a noninvasive therapeutic option that once perfected will add a useful extra string to the clinician's bow.

However, until now, HIFU therapy for solid malignances has been mostly conducted in research settings for the assessment of technical safety, efficacy, and feasibility, and a few of those described herein have been used alone in clinical practice. Where clinically appropriate, HIFU should give at least the same results as surgical excision, with the extent of the negative surgical margins being determined by imaging. Although recent results have been very encouraging, multiple-central, long-term follow-up trials are essential to evaluate the long-term efficacy and cost-effectiveness of HIFU treatments in cancer. Not until these issues have been resolved, and the results from prospective, randomized clinical trials worldwide become available, can this noninvasive ablative technique be considered as a candidate for conventional therapy for widespread clinical applications.

Similar to surgical removal, the goal of thermal ablation is to eliminate a targeted cancer, which includes the cancer with a margin of normal tissue. As the targeted tissue is destroyed and left in place to be resorbed, using imaging techniques to identify the margin of a unifocal tumor and adjacent microsatellites is more important in thermal ablation than surgery. With no detailed pathology of the ablated tissue, the need for accurate pre-ablation assessment of the extent of cancer is essential, particularly in core biopsy for histological evaluation.

HIFU technology continues to develop. Much supplementary investigation is necessary to further evaluate the HIFU treatment plans, the relationship between HIFU dosage and the extent of coagulation necrosis, and factors that can influence focused ultrasound energy deposition in target tissue including tissue structure, movement, function, and perfusion. For instance, in our animal studies, it has been indicated that when mechanical and pharmacological means were used to manipulate tissue perfusion, perfusion-mediated tissue cooling could directly affect the shape and size of tissue necrosis induced by HIFU ablation. On the basis of these important findings, it seems that reduced tissue perfusion causes an increase in the volume of coagulation necrosis. Therefore, a better understanding of perfusion effects and a new method of controlling its cooling forces are essential to improve results.

Beyond optimization of technical and physiological parameters, it is clear that HIFU ablation should be undertaken when there is precise knowledge not only of the number and location of the lesions but also of the biological characteristics and natural history of the tumor. The goal of tumor therapy is that all cancer cells should be completely killed in the patient's body. For patients with cancer, the therapeutic strategy for the disease should be a multiple treatment plan, which includes local treatments such as surgery and radiotherapy, and systemic therapy such as chemotherapy and immunotherapy. A similar multidisciplinary approach including other modalities is important in the treatment of solid malignancies. HIFU is a local ablation, the same as surgery. Therefore, it is essential to combine HIFU with other therapies, including chemotherapy and radiotherapy. However, there are many questions needing to be answered, such as how to select the sequence of the combination and what types of standards are employed in the multidisciplinary therapies for each cancer, if HIFU is to be widely used in clinical practice. So, success achieved in the application of HIFU treatment is mainly dependent not only on the HIFU technique but also on better understanding of the natural characteristics of tumors.

References

1. Breasted J. H., ed., *The Edwin Smith surgical papyrus,* Chicago, Chicago University Press, 1930.
2. Timmerman R. D., Bizekis C. S., Pass H. I., Fong Y., Dupuy D. E., Dawson L. A., and Lu D. Local surgical, ablative, and radiation treatment of metastases. *CA Cancer* J Clin 2009; 59:145–170.
3. Liapi E., and Geschwind J.F. Transcatheter and ablative therapeutic approaches for solid malignancies. *J Clin Oncol* 2007; 25:978–986.
4. Hong K., Georgiades C.S., and Geschwind J.F. Technology insight: Image-guided therapies for hepatocellular carcinoma—intra-arterial and ablative techniques. *Nat Clin Pract Oncol* 2006; 3:315–324.
5. Hafron J., and Kaouk J. H. Ablative techniques for the management of kidney cancer. *Nat Clin Pract Urol* 2007; 4:261–-269.
6. Lynn J. G., Zwemer R. L., Chick A. J., and Miller A. G. A new method for the generation and use of focused US in experimental biology. *J Gen Physiol* 1942; 26:179–193.
7. Fry W. J., Mosberg W. H., Barnard J. W., and Fry F. J. Production of focal destructive lesions in the central nervous system with ultrasound. *J Neurosurg* 1954; 11:471–478.
8. Fry W. J., Barnard J. W., Fry F. J., Krumins R. F., and Brennan J. F. Ultrasonic lesions in the mammalian central nervous system. *Science* 1955; 122:517–518.
9. Fry F. J. Precision high intensity focusing ultrasonic machines for surgery. 1958; 37:152–156.
10. Lizzi F. L. High-precision thermotherapy for small lesions. *Eur Urol* 1993; 23 (Suppl 1):23–28.
11. Coleman D. J., Silverman R. H., Iwamoto T., Lizzi F. L., Rondeau M. J., Driller J., Rosado A., Abramson D. H., and Ellsworth R. M. Treatment of glaucoma with high-intensity focused ultrasound. *Ophthalmology* 1986; 93: 831–833.
12. Rosecan L. R., Iwamoto T., Rosado A., Lizzi F. L., and Coleman D. J. Therapeutic ultrasound in the treatment of retinal detachment: Clinical observations and light and electronmicroscopy. *Retina* 1985; 5:115–122.
13. Coleman D. J., Lizzi F. L., Burgess S. E., Silverman R. H., Smith M. E., Driller J., Rosado A., Ellsworth R. M., Haik B. G., and Abramson D. H. Ultrasonic hyperthermia and radiation in the management of intraocular malignant melanoma. *Am J Ophthalmol* 1986; 101:635–642.
14. Vallancien G., Chartier-Kastler E., Harouni M., Chopin D., and Bougaran J. Focused extracorporeal pyrotherapy: Experimental study and feasibility in man. *Semin Urol* 1993; 11:7–9.
15. Vallancien G., Harouni M., Guillonneau B., Veillon B., and Bougaran J. Ablation of superficial bladder tumors with focused extracorporeal pyrotherapy. *Urology* 1996; 47:204–207.
16. Vallancien G., Chartier-Kastler E., Bataille N., Chopin D., Harouni M., and Bougaran J. Focused extracorporeal pyrotherapy. *Eur Urol* 1993; 23 (Suppl 1):48–52.
17. ter Haar G. R., Clarke R. L., Vaughan M. G., and Hill C. R. Trackless surgery using focussed ultrasound: Technique and case report. *Min Inv Ther* 1991; 1: 13–19.
18. Visioli A. G., Rivens I. H., ter Haar G. R., Horwich A., Huddart R. A., Moskovic E., Padhani A., and Glees J. Preliminary results of a phase I dose escalation clinical trial using focused ultrasound in the treatment of localised tumours. *Eur J Ultrasound* 1999; 9:11–18.
19. Wu F., Wang Z. B., Chen W. Z., Wang W., Gui Y., Zhang M., Zheng G., Zhou Y., Xu G., Li M., Zhang C., Ye H., and Feng R. Extracorporeal high intensity focused ultrasound ablation in the treatment of 1038 patients with solid carcinomas in China: An overview. *Ultrason Sonochem* 2004; 11:149–154.
20. Illing R. O., Kennedy J. E., Wu F., ter Haar G. R., Protheroe A. S., Friend P. J., Gleeson F. V., Cranston D. W., Phillips R. R., and Middleton M. R. The safety and feasibility of extracorporeal high-intensity focused ultrasound (HIFU) for the treatment of liver and kidney tumours in a Western population. *Br J Cancer* 2005; 93:890–895.
21. Illing R., and Chapman A. The clinical applications of high intensity focused ultrasound in the prostate. *Int J Hyperthermia* 2007; 23:183–191.
22. Rebillard X., Soulié M., Chartier-Kastler E., Davin J. L., Mignard J. P., Moreau J. L., and Coulange C.; Association Francaise d'Urologie. High-intensity focused ultrasound in prostate cancer; a systematic literature review of the French Association of Urology. *BJU Int* 2008; 101:1205–1213.
23. Jolesz F. A., Hynynen K., McDannold N., and Tempany C. MR imaging-controlled focused ultrasound ablation: A noninvasive image-guided surgery. *Magn Reson Imaging Clin N Am* 2005; 13:545–560.
24. Gianfelice D., Khiat A., Amara M., Belblidia A., and Boulanger Y. MR imaging-guided focused US ablation of breast cancer: Histopathologic assessment of effectiveness—initial experience. *Radiology* 2003; 227:849–55.
25. Tempany C. M., Stewart E. A., McDannold N., Quade B. J., Jolesz F. A., and Hynynen K. MR imaging-guided focused ultrasound surgery of uterine leiomyomas: A feasibility study. *Radiology* 2003; 226:897–905.
26. Kennedy J. E., Ter Haar G. R., and Cranston D. High intensity focused ultrasound: Surgery of the future? *Br J Radiol* 2003; 76:590–599.

27. Hill C. R., and ter Haar G.R. Review article: High intensity focused ultrasound—potential for cancer treatment. *Br J Radiol* 1995; 68:1296–1303.

28. Balibar S., and Maris H. J. Negative pressures and cavitation in liquid helium. *Physics Today* 2000; 53, 29–34.

29. Coussios C. C., Farny C. H., Haar G. T, and Roy R. A. Role of acoustic cavitation in the delivery and monitoring of cancer treatment by high-intensity focused ultrasound (HIFU). *Int J Hyperthermia* 2007; 23:105–120.

30. Hill C. R., Rivens I., Vaughan M. G, and ter Haar G. R. Lesion development in focused ultrasound surgery: A general model. *Ultrasound Med Biol* 1994; 20:259–269.

31. Stride E. P., and Coussios C. C. Cavitation and contrast: The use of bubbles in ultrasound imaging and therapy. *Proc Inst Mech Eng H* 2010; 224:171–191.

32. Clement G. T. Perspectives in clinical uses of high-intensity focused ultrasound. *Ultrasonics* 2004; 42:1087–1093.

33. Overgaard J. The current and potential role of hyperthermia in radiotherapy. *Int J Radiat Oncol Biol Phys* 1989; 16:535–534.

34. Nikfarjam M., Malcontenti-Wilson C., and Christophi C. Focal hyperthermia produces progressive tumor necrosis in dependent of the initial thermal effects. *J Gastrointest Surg* 2005; 9:410–417.

35. Nikfarjam M., Muralidharan V., and Christophi C. Mechanisms of focal heat destruction of liver tumors. *J Surg Res* 2005; 127:208–223.

36. Wheatley D. N., Kerr C., and Gregory D. W. Heat-induced damage to HeLa-S3 cells: Correlation of viability, permeability, osmosensitivity, phase-contrast light-, scanning electron- and transmission electron-microscopical findings. *Int J Hyperthermia* 1989; 5:145–162.

37. Heisterkamp J., van Hillegersberg R., Sinofsky E., and IJzermans J. N. Heat-resistant cylindrical diffuser for interstitial laser coagulation: Comparison with the bare-tip fiber in a porcine liver model. *Lasers Surg Med* 1997; 20:304–309.

38. Germer C. T., Roggan A., Ritz J. P., Isbert C., Albrecht D., Müller G., and Buhr H. J. Optical properties of native and coagulated human liver tissue and liver metastases in the near infrared range. *Lasers Surg Med* 1998; 23:194–203.

39. Emami B., and Song C. W. Physiological mechanisms in hyperthermia: A review. *Int J Radiat Oncol Biol Phys* 1984; 10:289–295.

40. Tranberg K. G. Percutaneous ablation of liver tumours. *Best Pract Res Clin Gastroenterol* 2004; 18:125–145.

41. Muralidharan V., Nikfarjam M., Malcontenti-Wilson C., and Christophi C. Effect of interstitial laser hyperthermia in a murine model of colorectal liver metastases: Scanning electron microscopic study. *World J Surg* 2004; 28:33–37.

42. Matsumoto R., Selig A. M., Colucci V. M., and Jolesz F. A. Interstitial Nd:YAG laser ablation in normal rabbit liver: Trial to maximize the size of laser-induced lesions. *Lasers Surg Med* 1992; 12:650–658.

43. Wiersinga W. J., Jansen M.C., Straatsburg I. H., Davids P. H., Klaase J. M., Gouma D. J., and van Gulik T. M. Lesion progression with time and the effect of vascular occlusion following radiofrequency ablation of the liver. *Br J Surg* 2003; 90:306-312.

44. Benndorf R., and Bielka H. Cellular stress response: Stress proteins—physiology and implications for cancer. *Recent Results Cancer Res* 1997; 143:129–144.

45. Barry M. A., Behnke C. A., and Eastman A. Activation of programmed cell death (apoptosis) by cisplatin, other anti-cancer drugs, toxins and hyperthermia. *Biochem Pharmacol* 1990; 40:2353-2362.

46. Hori K., Mihich E., and Ehrke M. J. Role of tumor necrosis factor and interleukin 1 ingamma-interferon-promoted activation of mouse tumoricidal macrophages. *Cancer Res* 1989; 49:2606–2614.

47. Decker T., Lohmann-Matthes M. L., Karck U., Peters T., and Decker K. Comparative study of cytotoxicity, tumor necrosis factor, and prostaglandin release after stimulation of rat Kupffer cells, murine Kupffer cells, and murine inflammatory liver macrophages. *J Leukoc Biol* 1989; 45:139-146.

48. Adams D. O., and Hamilton T. A. The cell biology of macrophage activation. *Annu Rev Immunol* 1984; 2:283-318.

49. Kirn A., Bingen A., Steffan A. M., Wild M. T., Keller F., and Cinqualbre J. Endocytic capacities of Kupffer cells isolated from the human adult liver. *Hepatology* 1982; 2:216–222.

50. Napoletano C., Taurino F., Biffoni M., De Majo A., Coscarella G., Bellati F., Rahimi H., Pauselli S., Pellicciotta I., Burchell J. M., Gaspari L. A., Ercoli L., Rossi P., and Rughetti A. RFA strongly modulates the immune system and anti-tumor immune responses in metastatic liver patients. *Int J Oncol* 2008; 32:481–490.

51. Evrard S., Menetrier-Caux C., Biota C., Neaud V., Matholin-Pélissier S., Blay J. Y., and Rosenbaum J. Cytokines pattern after surgical radiofrequency ablation of liver colorectal metastases. *Gastroenterol Clin Biol* 2007; 31:141–145.

52. Ali M. Y., Grimm C. F., Ritter M., Mohr L., Allgaier H. P., Weth R., Bocher W. O., Endrulat K., Blum H. E., and Geissler M. Activation of dendritic cells by local ablation of hepatocellular carcinoma. *J Hepatol* 2005; 43:817--822.

53. Watanabe N., Niitsu Y., Umeno H., Kuriyama H., Neda H., Yamauchi N., Maeda M., and Urushizaki I. Toxic effect of tumor necrosis factor on tumor vasculature in mice. *Cancer Res* 1988; 48:2179-2183.

54. Isbert C., Ritz J. P., Roggan A., Schuppan D., Rühl M., Buhr H. J., and Germer C. T. Enhancement of the immune response to residual intrahepatic tumor tissue by laser-induced thermotherapy(LITT) compared to hepatic resection. *Lasers Surg Med* 2004; 35:284–292.

55. Schell S. R., Wessels F. J., Abouhamze A., Moldawer L. L., and Copeland E. M. 3rd. Pro- and antiinflammatory cytokine production after radiofrequency ablation of unresectable hepatic tumors. *J Am Coll Surg* 2002; 195:774–781.

56. Wu F., Zhou L., and Chen W. R. Host antitumour immune responses to HIFU ablation. *Int J Hyperthermia* 2007; 23:165–71.

57. Gravante G., Sconocchia G., Ong S. L, Dennison A. R., and Lloyd D. M. Immunoregulatory effects of liver ablation therapies for the treatment of primary and metastatic liver malignancies. *Liver Int* 2009;29:18–24.

58. Fagnoni F. F., Zerbini A., Pelosi G., and Missale G. Combination of radiofrequency ablation and immunotherapy. *Front Biosci* 2008; 13:369–81.

59. Sabel M. S. Cryo-immunology: A review of the literature and proposed mechanisms for stimulatory versus suppressive immune responses. *Cryobiology* 2009; 58:1–11.

60. Yang R., Reilly C. R., Rescorla F. J., Sanghvi N. T., Fry F. J., Franklin T. D. Jr., and Grosfeld J. L. Effects of high-intensity focused ultrasound in the treatment of experimental neuroblastoma. *J Pediatr Surg* 1992; 27:246–250.

61. Hu Z., Yang X. Y., Liu Y., Sankin G. N., Pua E. C., Morse M. A., Lyerly H. K., Clay T. M., and Zhong P. Investigation of HIFU-induced anti-tumor immunity in a murine tumor model. *J Transl Med* 2007; 5:34.

62. Zhang Y., Deng J., Feng J., and Wu F. Enhancement of antitumor vaccine in ablated hepatocellular carcinoma by high-intensity focused ultrasound: A preliminary report. *World J Gastroenterology* 2010; 16: in press.

63. Deng J., Zhang Y., Feng J., and Wu F. Dendritic cells loaded with ultrasound-ablated tumour induce *in vivo* specific antitumour immune responses. *Ultrasound Med Biol* 2010; 36:441–448.

64. Hu Z., Yang X. Y., Liu Y., Morse M. A., Lyerly H. K., Clay T. M, and Zhong P. Release of endogenous danger signals from HIFU-treated tumor cells and their stimulatory effects on APCs. *Biochem Biophys Res Commun* 2005; 335:124–131.

65. Kruse D. E., Mackanos M. A., O'Connell-Rodwell C. E., ContagC.H., and FerraraK.W. Short-duration-focused ultrasound stimulation of Hsp70 expression *in vivo*. *Phys Med Biol* 2008; 53:3641-3660.

66. Hundt W., O'Connell-Rodwell C. E., Bednarski M. D., Steinbach S., and Guccione S. *In vitro* effect of focused ultrasound or thermal stress on HSP70 expression and cell viability in three tumor cell lines. *Acad Radiol* 2007; 14:859–870.

67. Liu F., Hu Z., Qiu L., Hui C., Li C., Zhong P., and Zhang J. Boosting high-intensity focused ultrasound-induced antitumor immunity using a sparse-scan strategy that can more effectively promote dendritic cell maturation. *J Transl Med* 2010; 8:7.

68. Zhou P., Fu M., Bai J., Wang Z., and Wu F. Immune response after high-intensity focused ultrasound ablation for H22 tumor. *J Clin Oncol* 2007; 25 (S18):21169.

69. Rosberger D. F., Coleman D. J., Silverman R., Woods S., Rondeau M., and Cunningham-Rundles S. Immunomodulation in choroidal melanoma: Reversal of inverted CD4/CD8 ratios following treatment with ultrasonic hyperthermia. *Biotechnol Ther* 1994; 5:59–68.

70. Wang X., and Sun J. High-intensity focused ultrasound in patients with late-stage pancreatic carcinoma. *Chin Med J (Engl.)* 2002; 115:1332–1335.

71. Wu F., Wang Z. B., Lu P., Xu Z. L., Chen W. Z., Zhu H., and Jin C. B. Activated anti-tumor immunity in cancer patients after high intensity focused ultrasound ablation. *Ultrasound Med Biol* 2004; 30:1217–1222.

72. Zhou Q., Zhu X. Q., Zhang J., Xu Z. L., Lu P., and Wu F. Changes in circulating immunosuppressive cytokine levels of cancer patients after high intensity focused ultrasound treatment. *Ultrasound Med Bio* 2008; 34:81–88.

73. Kramer G., Steiner G. E., Grobl M., Hrachowitz K., Reithmayr F., Paucz L., Newman M., Madersbacher S., Gruber D., Susani M., and Marberger M. Response to sublethal heat treatment of prostatic tumor cells and of prostatic tumor infiltrating T-cells. *Prostate* 2004; 58:109–120.

74. Wu F., Wang Z. B., Cao Y. D., Zhou Q., Zhang Y., Xu Z. L., and Zhu X. Q. Expression of tumor antigens and heat-shock protein 70 in breast cancer cells after high-intensity focused ultrasound ablation. *Ann Surg Oncol* 2007; 14:1237–1242.

75. Xu Z. L., Zhu X. Q., Lu P., Zhou Q., Zhang J., and Wu F. Activation of tumor-infiltrating antigen presenting cells by high intensity focused ultrasound ablation of human breast cancer. *Ultrasound Med Biol* 2009; 35: 50–57.

76. Lu P., Zhu X. Q., Xu Z. L., Zhou Q., Zhang J., and Wu F. Increased infiltration of activated tumor-infiltrating lymphocytes after high intensity focused ultrasound ablation of human breast cancer. *Surgery* 2009; 145:286–93.

77. Dunn F. Attenuation and speed of ultrasound in lung. *J Acoust Soc Am* 1974; 56: 1638–1639.

78. Wells P. N. T., Ed., *Biomedical ultrasonics,* Academic Press, Inc. London (1977).

79. Hill C. R., Bamber J. C., and ter Haar G. R., Eds., *Physical principles of medical ultrasound,* 2nd ed. John Wiley & Sons Ltd., Chichester (2004).

80. Goss S. A., Frizzell L. A., and Dunn F. Ultrasonic absorption and attenuation in mammalian tissues. *Ultrasound Med Biol* 1979; 5:181–186.

81. Goss S. A., Johnson R. L., and Dunn F. Comprehensive compilation of empirical ultrasonic properties of mammalian tissues. *J Acoust Soc Am* 1978; 64:423–457.

82. Goss S. A., Johnson R. L., and Dunn F. Compilation of empirical ultrasonic properties of mammalian tissues. *J Acoust Soc Am* 1980; 68:93–108.

83. Kikuchi Y., Uchida R., Tanaka K., and Wagai T. Early cancer diagnosis through ultrasonics. *J Acoust Soc Am* 1957; 29:824–833.

84. Calderon C., Vilkomerson D., Mezrich R., Etzold K. F., Kingsley B., and Haskin M. Differences in the attenuation of ultrasound by normal, benign, and malignant breast tissue. *J Clin Ultrasound* 1976; 4: 249–254.

85. Cline H. E., Schenek J. F., Hynynen K., Watkins R. D., Schenek J. F., and Jolesz F. A. MR-guided focused ultrasound surgery. *J Comput Assist Tomogr* 1992; 16:956–965.

86. Cline H. E., Hynynen K., Hardy C. J., Watkins R. D., Schenck J. F., Jolesz F. A. MR temperature mapping of focused ultrasound surgery. *Magn Reson Med* 1994; 31:628–636.

87. Hynynen K., Darkazanli A., Unger E., and Schenck J. F. MRI-guided noninvasive ultrasound surgery. *Med Phys* 1993; 20:107–115.

88. Hardy C. J., Cline H. E., and Watkins R. D. One-dimensional NMR thermal mapping of focused ultrasound surgery. *J Comput Assist Tomogr* 1994; 18:476–483.

89. Gelet A., Chapelon J. Y., Bouvier R., Souchon R., Pangaud C., Abdelrahim A. F., Cathignol D., and Dubernard J. M. Treatment of prostate cancer with transrectal focused ultrasound: Early clinical experience. *Eur Urol* 1996; 29: 174–183.

90. Foster R. S., Bihrle R., Sanghvi N. T., Fry F. J., and Donohue J. P. High-intensity focused ultrasound in the treatment of prostatic disease. *Eur Urol* 1993; 23 (supple 1): 29–33.

91. Coleman D. J., Lizzi F. L., Driller J., Rosado A. L., Chang S., Iwamoto T., and Rosenthal D. Therapeutic ultrasound in the treatment of glaucoma. I. Experimental model. *Ophthalmology* 1985; 92:339–346.

92. Visioli A. G., Rivens I. H., terHaar G. R., Horwich A., Huddart R. A., Moskovic E., Padhani A., and Glees J. Preliminary results of a phase I dose escalation clinical trial using focused ultrasound in the treatment of localised tumours. *EurJUltrasound* 1999; 9:11–18.

93. Wu F., Wang Z. B., Chen W. Z., Wang W., Gui Y., Zhang M., Zheng G., Zhou Y., Xu G., Li M., Zhang C., Ye H., and Feng R. Extracorporeal high intensity focused ultrasound ablation in the treatment of 1038 patients with solid carcinomas in China: An overview. *Ultrason Sonochem* 2004; 11:149–154.

94. Wu F., Wang Z. B., and Wang Z. L. Changes in ultrasonic image of tissue damaged by high intensity focused ultrasound *in vivo*. *J Acoustic Soc Am* 1998; 103:2869.

95. Cady B., Jenkins R. L., Steele G. D., Lewis W. D., Stone M. D., McDermott W. V., Jessup J. M., Bothe A., Lalor P., Lovett E. J., Lavin P., and Linehan D. C. Surgical margin in hepatic resection for colorectal metastasis: A critical and improvable determinant of outcome. *Ann Surg* 1998; 227:566–571.

96. Farmer D. G., Rosove M. H., Shaked A., and Busuttil R. W. Current treatment modalities for hepatocellular carcinoma. *Ann Surg* 1994; 219:236–247.

97. Lin D., Lin S. M., and Liaw Y. F. Non-surgical treatment of hepatocellular carcinoma. *J Gastroenterol Hepa* 1997; 12:S319–S328.

98. Lin T. Y., Lee C. S., Chen K. M., and Chen C. C. Role of surgery in the treatment of primary carcinoma of the liver: A 31-year experience. *Br J Surg* 1987; 74:839–842.

99. Zibari G. B., Riche A., Zizzi H. C., McMillan R.W., Aultman D.F., Boykin K.N., Gonzalez E., Nandy I., DiesD. F., Gholson C.F., Holcombe R.F., and McDonald J.C. Surgical and nonsurgical management of primary and metastatic liver tumors. *Am Surg* 1998; 64:211–221.

100. Goldberg S. N., Gszelle G. S., and Mueller P. R. Thermal ablation therapy for focal malignancy: A unified approach to underlying principles, techniques, and diagnostic imaging guidance. *Am J Roentgenol* 2000; 174:323–331.

101. Vallancien G., Harouni M., Veillon B., Mombet A., Prapotnich D., Brisset J. M., and Bougaran J. Focused extracorporeal pyrotherapy: Feasibility study in man. *J Endourol* 1992; 6:173–181.

102. Wu F., Chen W. Z., Bai J., Zou J. Z., Wang Z. L., Zhu H., and Wang Z. B. Pathological changes in human malignant carcinoma treated with high-intensity focused ultrasound. *Ultrasound Med Biol* 2001; 27: 1099–1106.

103. Wu F., Wang Z. B., Chen W. Z., Zhu H., Bai J., Zou J. Z., Li K. Q., Jin C. B., Xie F. L., and Su H. B. Extracorporeal high intensity focused ultrasound ablation in the treatment of patients with large hepatocellular carcinoma. *Ann Surg Oncol* 2004; 11:1061–1069.

104. Wu F., Wang Z. B., Chen W. Z., Zhu H., Bai J., Zou J. Z., Li K. Q., Jin C. B., Xie F. L., and Su H. B. Advanced hepatocellular carcinoma: Treatment with high-intensity focused ultrasound ablation combined with transcatheter arterial embolization. *Radiology* 2005; 235: 659–667.

105. Li C., Zhang W., Zhang R., Zhang L., Wu P., and Zhang F. Therapeutic effects and prognostic factors in high-intensity focused ultrasound combined with chemoembolisation for larger hepatocellular carcinoma. *Eur J Cancer* 2010; 46:2513–2521.

106. Zhang L., Zhu H., Jin C., Zhou K., Li K., Su H., Chen W., Bai J., and Wang Z. High-intensity focused ultrasound (HIFU): Effective and safe therapy for hepatocellular carcinoma adjacent to major hepatic veins. *Eur Radiol* 2009; 19:437–445.

107. Wang Y., Wang W., Wang Y., and Tang J. Ultrasound-guided high-intensity focused ultrasound treatment for needle-track seeding of hepatocellular carcinoma: Preliminary results. *Int J Hyperthermia* 2010; 26:441–447.

108. Li C. X., Xu G. L., Jiang Z. Y., Li J. J., Luo G. Y., Shan H. B., Zhang R., and Li Y. Analysis of clinical effect of high-intensity focused ultrasound on liver cancer. *World J Gastroenterol* 2004; 10:2201–2204.

109. Li Y. Y., Sha W. H., Zhou Y. J., and Nie Y. Q. Short and long term efficacy of high intensity focused ultrasound therapy for advanced hepatocellular carcinoma. *J Gastroenterol Hepatol* 2007; 22: 2148–2154.

110. Illing R. O., Kennedy J. E., Wu F., Ter Haar G. R., Protheroe A. S., Friend P. J., Gleeson F. V., Cranston D. W., Phillips R. R., and Middleton M. R. The safety and feasibility of extracorporeal high-intensity focused ultrasound (HIFU) for the treatment of liver and kidney tumours in a Western population. *Br J Cancer* 2005; 93:890–895.

111. Leslie T. A., Kennedy J. E., Illing R. O., Ter Haar G. R., Wu F., Phillips R. R., Friend P. J., Roberts I. S., Cranston D. W., and Middleton M. R. High-intensity focused ultrasound ablation of liver tumours: Can radiological assessment predict the histological response? *Br J Radiol* 2008; 81:564–571.

112. Noterdaeme O., Leslie T. A., Kennedy J. E., Phillips R. R., and Brady M. The use of time to maximum enhancement to indicate areas of ablation following the treatment of liver tumours with high-intensity focused ultrasound. *Br J Radiol* 2009; 82:412–420.

113. Orsi F., Zhang L., Arnone P., Orgera G., Bonomo G., Vigna P. D., Monfardini L., Zhou K., Chen W., Wang Z., and Veronesi U. High-intensity focused ultrasound ablation: Effective and safe therapy for solid tumors in difficult locations. *Am J Roentgenol* 2010; 195:W245–252.

114. Numata K., Fukuda H., Ohto M., Itou R., Nozaki A., Kondou M., Morimoto M., Karasawa E., and Tanaka K. Evaluation of the therapeutic efficacy of high-intensity focused ultrasound ablation of hepatocellular carcinoma by three-dimensional sonography with a perflubutane-based contrastagent. *Eur J Radiol* 2010; 75:e67–75.

115. Park M. Y., Jung S. E., Cho S. H., Piao X. H., Hahn S. T., Han J. Y., and Woo I. S. Preliminary experience using high intensity focused ultrasound for treating liver metastasis from colon and stomach cancer. *Int J Hyperthermia* 2009; 25:180–188.

116. Ng K. K., Poon R. T., Chan S. C., Chok K. S., Cheung T. T., Tung H., Chu F., Tso W. K., Yu W. C., Lo C. M., and Fan S. T. High-intensity focused ultrasound for hepatocellular carcinoma: A single-center experience. *Ann Surg* 2011; 253:981–257.

117. Wu F., Wang Z. B., Cao Y. D., Chen W. Z., Bai J., Zou J. Z., and Zhu H. A randomised clinical trial of high-intensity focused ultrasound ablation for the treatment of patients with localised breast cancer. *Br J Cancer* 2003; 89: 2227–2233.

118. Wu F., Wang Z. B., Cao Y. D., Zhu X. Q., Zhu H., Chen W. Z., and Zou J. Z. "Wide local ablation" of localized breast cancer using high intensity focused ultrasound. *J Surg Oncol* 2007; 96:130–136.

119. Wu F., Wang Z. B., Cao Y. D., Xu Z. L., Zhou Q., Zhu H., and Chen W. Z. Heat fixation of cancer cells ablated with high-intensity-focused ultrasound in patients with breast cancer. *Am J Surg* 2006; 192(2): 179–184.

120. Wu F., Wang Z. B., Chen W. Z., Zhu H., Bai J., Zou J. Z., Li K. Q., Jin C. B., Xie F. L., Su H. B., and Gao G. W. Changes in biologic characteristics of breast cancer treated with high-intensity focused ultrasound. *Ultrasound Med Biol* 2003; 29: 1487–1492.

121. Wu F., Wang Z. B., Chen W. Z., Zhu H., Bai J., Zou J. Z., Li K. Q., Jin C. B., Xie F. L., and Su H. B. Extracorporeal high intensity focused ultrasound treatment for patients with breast cancer. *Breast Cancer Res Treat* 2005; 92: 51–60.

122. Becker G., Winkler J., Hofmann E., and Bogdahn U. Differentiation between ischemic and hemorrhagic stroke by transcranial color-coded real-time sonography. *J Neuroimaging* 1993; 3:41–47.

123. Bogdahn U., Becker G., Winkler J., Greiner K., Perez J., and Meurers B. Transcranial color-coded real-time sonography in adults. *Stroke* 1990; 21:1680–1688.

124. Kimura K., Hashimoto Y., Hirano T., Uchino M., and Ando M. Diagnosis of middle cerebral artery occlusion with transcranial color-coded real-time sonography. *Am J Neuroradiol* 1996; 17:895–899.

125. Kushner M. J., Zanette E. M., Bastianello S., Mancini G., Sacchetti M. L., Carolei A., and Bozzao L. Transcranial Doppler in acute hemispheric brain infarction. *Neurology* 1991; 41:109–113.

126. Martin P. J., Pye I. F., Abbott R. J., and Naylor A. R. Color-coded ultrasound diagnosis of vascular occlusion in acute ischemic stroke. *J Neuroimag* 1995; 5:152–156.

127. Smith S. W., Trahey G. E., and von Ramm O. T. Phased array ultrasound imaging through planar tissue layers. *Ultrasound Med Biol* 1986; 12:229–243.

128. Fry F. J. Transkull transmission of an intense focused ultrasonic beam. *Ultrasound Med Biol* 1977; 3:179–184.

129. Fry F. J., and Barger J. E. Acoustical properties of the human skull. *J Acoust Soc Am* 1978; 63:1576–1590.

130. Weis L., in *Surgery for bone and soft-tissue tumors*, eds. M. A. Simon and D. Springfield, Lippincott-Raven, Philadelphia, 1998.

131. Li C., Zhang W., Fan W., Huang J., Zhang F., and Wu P. Noninvasive treatment of malignant bone tumors using high-intensity focused ultrasound. *Cancer* 2010; 116:3934–3942.

132. Chen W., and Zhou K. High-intensity focused ultrasound ablation: A new strategy to manage primary bone tumors. *Curr Opin Orthop* 2006; 16: 494–500.

133. Chen W., Zhu H., Zhang L., Li K., Su H., Jin C., Zhou K., Bai J., Wu F., and Wang Z. Primary bone malignancy: Effective treatment with high intensity focused ultrasound ablation. *Radiology* 2010; 255: 968–978.

134. Reddan D. N., Raj G. V., and Polascik T. J. Management of small renal tumors: An overview. *Am J Med* 2001; 110:558–562.

135. Kohrmann K. U., Michel M. S., Gaa J., Marlinghaus E., and Alken P. High intensity focused ultrasound as non-invasive therapy for multilocal renal cell carcinoma: Case study and review of the literature. *J Urol* 2002; 167:2397–2403.

136. Susani M., Madersbacher S., Kratzik C., Vingers L., and Marberger M. Morphology of tissue destruction induced by focused ultrasound. *Eur Urol* 1993; 23(Suppl 1):34–38.

137. Marberger M., Schatzl G., Cranston D., and Kennedy J. E. Extracorporeal ablation of renal tumours with high-intensity focused ultrasound. *BJU Int* 2005; 95(Suppl 2):52–55.

138. Hacker A., Michel M. S., Marlinghaus E., Kohrmann K. U., and Alken P. Extracorporeally induced ablation of renal tissue by high-intensity focused ultrasound. *BJU Int* 2006; 97: 779–785.

139. Wu F., Wang Z. B., Chen W. Z., Bai J., Zhu H., and Qiao T. Y. Preliminary experience using high intensity focused ultrasound for the treatment of patients with advanced stage renal malignancy. *J Urol* 2003; 170:2237–2240.

140. Ritchie R. W., Leslie T., Phillips R., Wu F., Illing R., ter Haar G., Protheroe A., and Cranston D. Extracorporeal high intensity focused ultrasound for renal tumours: A 3-year follow-up. *BJU Int* 2010; 106:1004–1009.

141. Chakera A., Leslie T., Roberts I., O'Callaghan C. A., and Cranston D. A lucky fall? Case report. *Transplant Proc* 2010; 42:3883–3886.

142. Wu F., Wang Z. B., Zhu H., Chen W. Z., Zou J. Z., Bai J., Li K. Q., Jin C. B., Xie F. L., and Su H. B. Feasibility of US-guided high-intensity focused ultrasound treatment in patients with advanced pancreatic cancer: Initial experience. *Radiology* 2005; 236: 1034–1040.

143. Wang K., Chen Z., Meng Z., Lin J., Zhou Z., Wang P., Chen L., and Liu L. An algesic effect of high intensity focused ultrasound therapy for unresectable pancreatic cancer. *Int J Hyperthermia* 2011; 27:101–107.

144. Zhao H., Yang G., Wang D., Yu X., Zhang Y., Zhu J., Ji Y., Zhong B., Zhao W., Yang Z., and Aziz F. Concurrent gemcitabine and high-intensity focused ultrasound therapy in patients with locally advanced pancreatic cancer. *Anticancer Drugs* 2010; 21:447–452.

145. Wu F., Wang Z. B., Zhu H., Chen W. Z., Zou J. Z., Bai J., Li K. Q., Jin C. B., Xie F. L., Su H. B., and Gao G. W. Extracorporeal focused ultrasound surgery for treatment of human solid carcinomas: Early Chinese clinical experience. *Ultrasound Med Biol* 2004; 30:245–260.

Using Hyperthermia to Augment Drug Delivery

Mark W. Dewhirst
Duke University Medical Center

16.1 Introduction

This chapter will provide an overview of the use of hyperthermia to augment drug delivery and antitumor effects of drug-carrying liposomes. The focus of this chapter will be on applications as opposed to design of liposomal formulations. Interested readers are encouraged to seek out several in-depth reviews on nuances of liposome formulations that have been published previously (Kong and Dewhirst 1999, Koning et al. 2010, Landon et al. 2011, Lindner and Hossann 2010, Lindner et al. 2005, Tashjian et al. 2008).

16.2 Mild Hyperthermia Is Ideal for Augmenting Drug Delivery and Efficacy in Tumors

We will focus on the effects of hyperthermia treatment in the temperature range of 39–42°C, because it is in this temperature range that profound physiologic effects occur in tumors that can mediate improvements in drug delivery (Song et al. 2001) (Figure 16.1). There are literally dozens of reports showing that temperatures in this range increase tumor perfusion, oxygenation, and vascular permeability (Dewhirst et al. 2005,

Song 1984). The efficacy of many drugs is limited by hypoxia, so any treatment that improves tumor oxygenation is also likely to improve the efficacy of chemotherapy (Brown and Wilson 2004). When temperatures are elevated further, however, hyperthermia can cause vascular damage, which would impede drug delivery (Song et al. 2001). There are circumstances where thermal ablative therapy has been combined with liposomal drugs, which will be discussed later in more detail. However, even in this situation, the addition of liposomal drugs is designed to take advantage of vascular changes at the edge of the ablation zone, which again is in the mild hyperthermia temperature range (Ahmed and Goldberg 2004, Mostafa et al. 2008, Poon and Borys 2009).

16.3 Fundamental Limitations for Effective Drug Delivery and Antitumor Activity

One of the basic limitations of traditional drug delivery following intravenous free drug administration is that drugs do not reach the target volume in sufficient quantities to be efficacious. Part of the limitation is due to the fact that there is indiscriminate distribution throughout the body. As a result,

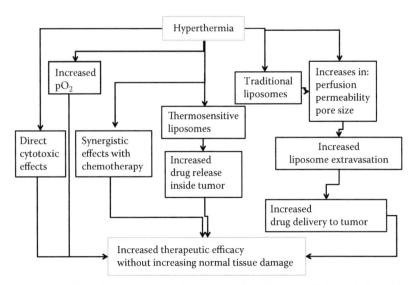

FIGURE 16.1 Summary of beneficial effects of hyperthermia, when combined with liposomal drugs. (Adapted from Kong, G., and Dewhirst, M. W. *International Journal of Hyperthermia*, 15, 5, 1999.)

normal tissue toxicity limits the total amount of drug that can be delivered. Additionally, even when drug reaches the site of the tumor, it is extremely difficult to reach all tumor cells as a result of aberrant properties of tumor vasculature, which include large intravascular distances and arteriovenous shunting (Minchinton and Tannock 2006, Primeau et al. 2005). If it were possible to improve tumor perfusion, then one could, in principle, deliver drug more effectively to tumors. It has been shown that hyperthermia can mildly augment the delivery of free drug to tumors, for example (Kong et al. 2000). Part of the enhancement in delivery could be the result of heat-mediated increases in perfusion, coupled with the fact that hyperthermia can augment the cellular uptake of a number of different drugs and enhance their cytotoxicity by inhibiting DNA damage repair mechanisms (Bates and Mackillop 1986). Thermally mediated increases in vascular permeability could also play a role (Friedl et al. 2003, Umeno et al. 1994), at least for drugs that are protein bound. Many chemotherapeutic agents exhibit protein binding (Demant and Sehested 1993, Khan et al. 2008, Li et al. 2007, Ramanathan-Girish and Boroujerdi 2001, Zemlickis et al. 1994). It has been shown that heating in the range of 42°C for an hour can increase extravasation of albumin in tumors by 25% over what occurs under normothermic conditions (Kong et al. 2001). There are many preclinical reports demonstrating that the combination of intravenous free drugs with hyperthermia yields improvements in drug delivery and antitumor effects (Herman and Teicher 1994). A recently reported phase III trial from Europe showed improvements in local tumor control and progression free survival in patients with locally advanced soft tissue sarcomas, randomized to standard of care chemotherapy alone versus chemotherapy combined with hyperthermia (Issels et al. 2010). Presumably, the effects discussed here played a role in the positive effects seen with this trial, the largest ever conducted in this patient group.

16.4 Effects of Hyperthermia on Liposome Uptake in Tissues

A number of studies have evaluated the effects of hyperthermia on liposomal drug accumulation in tumors, at the preclinical and clinical level. Kong and Dewhirst examined these data extensively in a review paper published in 1999 (Kong and Dewhirst 1999). Hyperthermia increased liposomal drug accumulation in tumors by factors of 2–8, compared with free drug plus hyperthermia, depending upon the drug, tumor model, and temperatures used. One of the challenges in the historical data was that there were a variety of target temperatures used, as well as a range of different liposomal formulations and sequences between timing of liposome administration and treatment with hyperthermia. Thus it was difficult to conclude the optimal conditions to maximize liposomal drug delivery to tumors using hyperthermia. To address this issue, Kong systematically examined a number of parameters that could influence liposome extravasation, using a single tumor model. He first investigated the effects of 42°C hyperthermia on liposomal extravasation, using a range of different-sized liposomes (Kong et al. 2000). The largest thermally mediated enhancement in liposomal extravasation occurred with 100 nm liposomes. Augmentation of extravasation was also seen with 200 and 400 nm liposomes, but not to the same extent. Thermal dependence of extravasation was specific for tumor tissue, as hyperthermia did not appear to enhance liposomal extravasation in normal tissues. These results clearly showed that hyperthermia was not increasing liposomal extravasation in tumor microvessels by increasing general permeability. Alternatively, it appeared to be capable of opening pores in the vascular endothelial lining of tumor microvessels that were large enough to enhance 100 nm–400 nm liposomal drug delivery (Figure 16.2a). Kong went on to examine the temperature dependence of liposomal extravasation, finding a threshold for enhancement of extravasation in tumors between 39 and 40°C (Kong et al. 2001). The rate of extravasation doubled for each 1°C temperature rise, up to a maximum temperature of 42°C (Figure 16.2b). Above this temperature, hemorrhage obscured the ability to observe further thermal effects on extravasation in the models that he used.

Kong also found evidence for enlarged pores up to 4 hr post heating, but by 6 hr, they had closed to pre-heating levels. When he attempted to reopen them by a second heat, delivered 8 hr after the first heating, they did not reopen. This provided evidence that the closing of the pores was likely associated with the heat shock response and in effect was a manifestation of thermotolerance (Kong et al. 2001).

To examine whether hyperthermia would augment liposomal accumulation in a spontaneous tumor, Matteucci et al. radiolabelled a long-circulating liposomal drug formulation and examined tumor uptake of these liposomes with a gamma camera in pet cats with spontaneous soft tissue sarcomas. Baseline scans were performed without heating and then followed with heating several days later in the same animals. Similar to what was seen in rodent models, hyperthermia enhanced the uptake of liposomes by factors between 2 and 13 fold (Matteucci et al. 2000). Kleiter later determined that the relative enhancement of liposomal uptake after hyperthermia treatment, as observed by gamma camera in rats with transplanted mammary carcinomas, was linearly proportional to doxorubicin uptake for sterically stabilized Doxil™ liposomes (Figure 16.3) (Kleiter et al. 2006).

Goldberg examined the effects of thermal ablation on accumulation of the pegylated liposome Doxil™, finding that RF ablation administered prior to Doxil™ administration led to greater than a five-fold increase in drug accumulation, with the greatest concentrations being seen at the periphery of the ablation zone. It is likely that the enhanced liposome accumulation observed with this combination therapy is the result of the combination

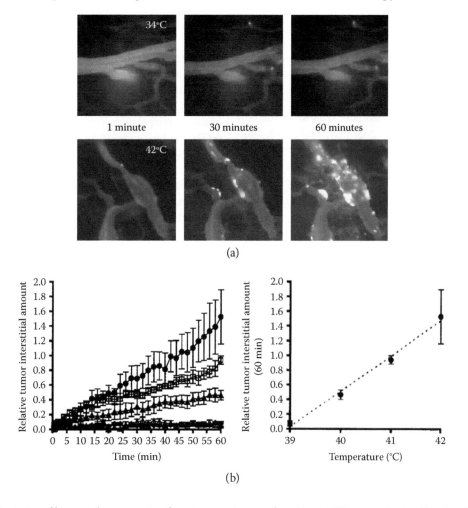

FIGURE 16.2 (a) Depiction of liposomal extravasation from tumor microvessels at 34 vs. 42°C, up to 60 min of heating. Data acquired using intravital microscopy of SKOV3 tumor microvessels in skin-fold window chambers. The liposomes were fluorescently labeled, which permitted visualization of the tumor microvessels as well as the extravasation of the liposomes, which can be visualized as bright perivascular spots of accumulation. (From Kong, G., Braun, R. D., Dewhirst, M. W., *Cancer Res*, 60, 2000. With permission.) (b) (left panel). Rate of liposomal extravasation from SKOV3 tumor microvasculature, as a function of temperature. ● = 42°C, ∗ = 41°C, π = 40°C, lowest sets of curves represent 34 and 39°C, respectively. (right panel). Relative amount of interstitial liposomes as a function of temperature. The slope of this curve predicts that the accumulation of liposomes doubles for each 1°C temperature rise. (From Kong, G. et al., *Cancer Res*, 60, 2000. With permission.)

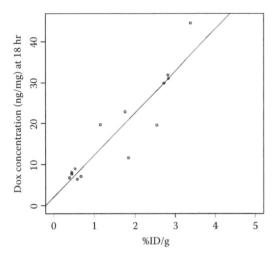

FIGURE 16.3 Data showing that measurement of radiolabelled liposome accumulation in tumors after heating, using a gamma camera, can be used to accurately determine the accumulation of doxorubicin-containing liposomes after hyperthermia treatment. These studies were conducted in rats with R3230Ac mammary carcinomas that were grown in the flank. (From Kleiter, M. M. et al., *Clin Cancer Res*, 12, 2006. With permission.)

of vascular destruction and enlarged pore sizes at the margin of the ablation zone (Monsky et al. 2002). Thermal ablation did not affect doxorubicin accumulation when it was administered as free drug in this study.

16.5 Development of Thermally Sensitive Liposomes

The Doxil™ liposome (a long circulating pegylated doxorubicin-containing liposome) was originally approved for clinical use because it exhibits antitumor activity comparable to that of free drug (O'Brien et al. 2004), but with less cardiac toxicity (Gabizon et al. 2003, Grenader et al. 2010), which is the main dose limiting target organ for free doxorubicin. The limitation of the Doxil™ type formulation relates to the polyethylene glycol that is added to the liposome surface to enhance the circulation time. Whereas this maximizes passive accumulation to tumor via the enhanced permeability and retention (EPR) effect, this same polymer coating reduces the rate at which drug is released once the liposome reaches the target volume (Landon et al. 2011). The slow drug release rate greatly reduces the antitumor effect. A number of laboratories have investigated means to trigger content release from liposomes in tumors as a means to overcome this limitation. Examples of triggers that take advantage of inherent features of the tumor microenvironment include sensitivity to acid pH (tumor pH values are often lower than normal tissue) (Hong and Kim 2011) and relying on intrinsically enhanced extracellular enzyme activity (such as metalloproteases) in tumors (Andresen et al. 2010).

In 1978, Yatvin published the first report of a solid phase liposome that underwent a phase transition at 44°C, which led to enhanced drug release rates (Yatvin et al. 1978). The lipid composition of the liposomes that lead to thermal sensitivity for drug release have been reviewed elsewhere and will not be discussed here (Hossann et al. 2007, Lindner et al. 2004, Needham and Ponce 2007). Yatvin speculated that such liposomes should greatly increase antitumor effect compared with free drug, if heating methods were directed specifically toward the tumor. He also suggested that thermosensitive liposomes might release drug intravascularly, thereby driving drug down its concentration gradient into the tumor (Yatvin et al. 1978). Although proof of principle was shown by these authors, the short circulation time of these original nonsterically stabilized formulations limited their general applicability. Gaber and Papahadjopoulos were the first to formulate a sterically stabilized thermosensitive liposome (Gaber et al. 1995) and demonstrated enhanced drug release upon reaching the transition temperature *in vitro* and *in vivo* (Gaber et al. 1996, Wu et al. 1997). These liposomes and those of Yatvin had two additional limitations, however. The release temperature was too high (43–45°C), and the rate of release was too slow (30 min)(Gaber et al. 1996). Dewhirst recognized the limitations of these formulations in discussions with David Needham in the early 1990s. The performance characteristics defined by Dewhirst and Needham were used to develop a thermosensitive liposome that (1) exhibited release in a more realistic temperature range that could be achieved clinically (39–42°C) and (2) showed rapid drug release upon reaching the transition temperature. This latter design characteristic was based on Dr. Dewhirst's a priori knowledge of typical microvascular flow velocities in tumors (mm/sec) (Dewhirst et al. 1989), which clearly dictated that to achieve maximal drug deposition in tumors, release should occur in seconds, as opposed to 30 min, the release time associated with the earlier formulations. A rapid release time would ensure that most drug would be deposited in the tumor before intravascular liposomes could exit from the tumor.

Needham subsequently reported on a novel pegylated liposome design that incorporated a single chain fatty acid into the liposome bilayer (Needham et al. 2000, Needham and Dewhirst 2001). This represented a significant departure from earlier formulations that exclusively utilized mixtures of dual chain fatty acids of different chain lengths to achieve thermal sensitivity. This liposome, which has been coined the low temperature sensitive liposome, or LTSL, exhibited exactly the design characteristics that were originally stipulated. Lindner and coworkers have reported on a similar formulation that utilizes synthetic fatty acids to achieve thermally triggered drug release (Lindner et al. 2004). This group has also optimized the pegylation of such formulations to yield longer circulation times than of the Needham formulation (Hossann et al. 2007, Li et al. 2010). It remains to be seen whether a long circulating LTSL is advantageous. The rationale would be to take advantage of the enhanced vascular permeability brought on by hyperthermia, which has been reported to last up to 4 hr in one tumor model (Kong et al. 2001). Further studies are needed to ascertain the likelihood that this will perform better than the Needham formulation, which exhibits a 2 hr half-life in humans (Poon and Borys 2009). Given that a typical hyperthermia treatment lasts 30–60 min, the shorter half-life may be adequate.

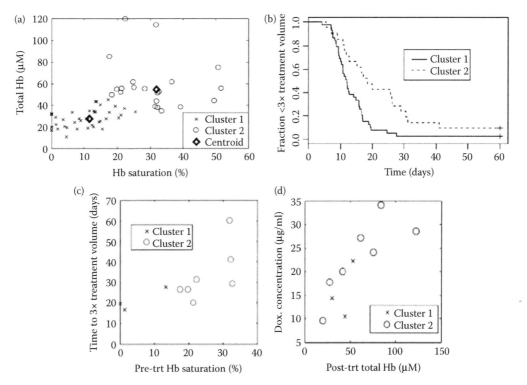

FIGURE 16.4 Effects of doxorubicin concentration, perfusion, and oxygenation on antitumor effect of LTSL-Dox + hyperthermia. (a) Results of a k-means cluster analysis of hemoglobin saturation and total hemoglobin, measured at baseline for individual animals bearing SKOV3 ovarian carcinoma xenografts that were treated with LTSL-Dox or free doxorubicin ± hyperthermia. Two clusters were identified when comparing total Hb vs. Hb saturation, prior to treatment. Cluster 1 had relatively low Hb saturation and low hemoglobin concentration, compared with cluster 2. The time to reach 3 × treatment volume was significantly shorter for cluster 1, which tended to more hypoxic and have lower blood volume, compared with cluster 2 ($p = 0.003$; b). (c) and (d) show highly significant correlations between pretreatment Hb saturation and time to 3 × pretreatment volume, and the posttreatment total Hb and doxorubicin concentration ($r = 0.8$, $p = 0.01$, and $r = 0.89$, $p = 0.001$, respectively, by Spearman rank correlation). These data show that even for LTSL-Dox with hyperthermia, individual variations in the amount of drug delivered and the oxygenation state have a strong influence on growth delay time. (Palmer, G. et al., *J Control Release*, 142, 2010. With permission.)

The effects of hyperthermia on nanoparticle extravasation from tumor microvessels and on thermosensitive liposome content release are summarized in Figure 16.4. Under normothermic conditions, a few liposomes extravasate via the EPR effect to heterogeneously collect in perivascular regions. Hyperthermia at 40–42°C increases pore sizes between endothelial cells, permitting greatly enhanced extravasation of liposomes into the interstitial space. Thermosensitive liposomes rapidly release their contents upon reaching their transition temperature, which bathes the heated zone in free drug, which can diffuse down its concentration gradient to reach tumor cells.

16.6 Factors That Influence Effectiveness of LTSL-Dox When Combined with Hyperthermia

16.6.1 Drug Delivery and Oxygenation

The primary drug that has been studied with this formulation is doxorubicin, because it is easily loaded into liposomes, using a pH gradient loading method (Needham and Dewhirst 2001).

Kong and Needham were the first to directly compare the antitumor efficacy of hyperthermia combined with LTSL loaded with doxorubicin (LTSL-Dox) to that of a Doxil™-type formulation and free doxorubicin (Kong et al. 2000, Needham et al. 2000). Using the FaDu human squamous cell carcinoma line, they were able to show that hyperthermia + LTSL-Dox yielded 25x and 4x higher drug concentrations in tumor, compared with hyperthermia combined with free drug or the Doxil™-type formulation, respectively. This increase in drug concentration was correlated with enhanced antitumor effect, yielding long-term tumor control in several subjects in replicate experiments, whereas free doxorubicin exhibited virtually no antitumor effect. They further showed that the amount of doxorubicin bound to DNA was higher in the LTSL-Dox plus hyperthermia group than other groups (Kong et al. 2000). Since DNA is one of the putative targets of doxorubicin, this result was consistent with the improved antitumor effect seen with this drug. These results strongly suggest that the increase in drug concentration achieved with LTSL-Dox with hyperthermia is key to its improved antitumor effect.

In a recent report, Palmer et al. showed that the extent of growth delay achieved in SKOV3 ovarian carcinoma xenografts following LTSL-Dox + hyperthermia was related to doxorubicin

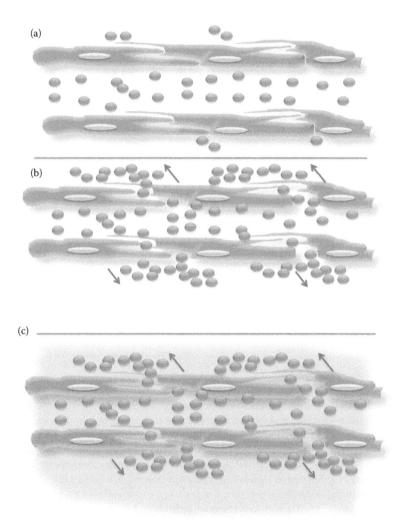

FIGURE 16.5 Effects of hyperthermia on nanoparticle extravasation from tumor microvessels. (a) Under normothermic conditions, a few liposomes extravasate via the EPR effect, to heterogeneously collect in perivascular regions. (b) 40–42°C hyperthermia increases pore sizes between endothelial cells, permitting greatly enhanced extravasation of liposomes into the interstitial space. The rate of extravasation doubles for every °C between 40°C and 42°C. This effect does not occur in normal tissues, thereby providing potential for a therapeutic gain. (c) Thermosensitive liposomes rapidly release content upon reaching their transition temperature, which bathes the heated zone in free drug, which can diffuse down its concentration gradient to reach tumor cells. Elongated structures with interdigitating protrusions depict endothelial cells, circular structures represent liposomes. Ellipsoidal structures depict endothelial cell nuclei. Size of liposomes is not drawn to scale. They are typically 100 microns or 1/100 the size of a red blood cell.

concentration achieved within individual tumors, as measured by a noninvasive fluorescence spectroscopic method (Palmer et al. 2010). This work also demonstrated the importance of tumor oxygenation and perfusion in affecting tumor response to LTSL-Dox + hyperthermia, as longer growth delays were seen in tumors with higher hemoglobin content (which reflects blood volume/perfusion) and hemoglobin saturation (which is related to overall oxygenation status) (Figure 16.5).

Yarmolenko et al. compared the antitumor efficacies of hyperthermia + LTSL-Dox across several tumor lines (Yarmolenko et al. 2010). LTSL-Dox with 42°C hyperthermia enhanced the antitumor effect of all these tumor lines, compared with LTSL-Dox alone, heat alone, or untreated controls, but the extent of enhancement varied between tumor lines. In an attempt to understand the reasons for variation, the authors investigated

the *in vitro* sensitivity to doxorubicin (as assessed by clonogenic assay), the *in vitro* doubling time, intratumoral drug concentrations, and intratumoral pH. The only parameter that was consistently associated with the relative improvement in growth delay was doubling time. Tumors with longer *in vitro* doubling times tended to respond better and have longer response durations to this treatment. The authors concluded that *a priori* knowledge of tumor doubling time might prove useful in designing treatment schedules for individual patients with this drug formulation.

16.6.2 Vascular Targeting and HIF-1

Chen et al. reported that the efficacy of LTSL-Dox with hyperthermia may be related to its vascular targeting effects. Using intravital microscopy, they showed that LTSL-Dox with hyperthermia leads to relatively rapid vascular shutdown (within 6 hr) that persists out to at least 24 hr after treatment (Chen et al. 2004). The extent of vascular shutdown was greater in the FaDu tumor line than in the 4T07 tumor, which was found to be relatively resistant to the vascular damaging effects of this formulation (Chen et al. 2008). Chen compared the *in vitro* sensitivity of human umbilical vein endothelial cells (HUVEC), 4T07 and FaDu, to doxorubicin, using a colorimetric method. They concluded that the sensitivities of the two tumor lines and HUVEC to LTSL-Dox were equivalent, so the difference in antivascular effects could not be attributed to variations in tumor or endothelial cell sensitivity (Chen et al. 2008). However, it should be noted that the clonogenic assay studies performed by Yarmolenko clearly demonstrated that 4T07 is the most resistant of all tumor lines studied thus far, including the FaDu tumor line (Yarmolenko et al. 2010). Thus, one cannot rule out that variations in the vascular targeting effects of LTSL-Dox are not attributable to variations in tumor cellular sensitivity to this combination treatment. The link between vascular sensitivity and cell sensitivity to killing could be attributable to differences in VEGF levels, since this pro-angiogenic cytokine, which is produced by tumor cells, is known to promote survival of endothelial cells in response to cytotoxic therapy (Gupta et al. 2002). If more tumor cells are killed, VEGF levels could drop, thereby enhancing endothelial cell sensitivity to this formulation. However, the issue is likely more complex than this.

It has been reported recently that hyperthermia can increase levels of the hypoxia-inducible transcription factor, HIF-1, in tumors as a result of NADPH oxidase activation in tumor cells (Moon et al. 2010). The increase in HIF-1 after hyperthermia is associated with promotion of angiogenesis in 4T1 tumors. Similarly, we have found that doxorubicin can increase HIF-1 levels in aerobic 4T1 tumor cells via a mechanism involving reactive species formation (unpublished observations). Thus, the difference in vascular sensitivity between tumor lines may likely be a complex story involving the balance between how many cells are killed versus whether or not HIF-1 and its downstream cytoprotective protein targets are upregulated in surviving tumor cells. Clearly, more work is needed to probe this question further, because in doing so, additional strategies may be elucidated that could enhance the efficacy of the LTSL-Dox formulation.

Chen went on to demonstrate that the antivascular efficacy of LTSL-Dox with hyperthermia could be enhanced if vascular permeability was enhanced using platelet activating factor (Figure 16.6) (Chen et al. 2008). The mechanism underlying the enhanced vascular targeting effects of this drug being associated with enhanced permeability has not been elucidated, although the authors speculated that it might be related to enhanced drug concentrations achieved in vascular endothelium. One could also speculate that endothelial cells that are subjected to platelet activating factor may be more vulnerable to toxicity as a result of decreased cell-to-cell contact.

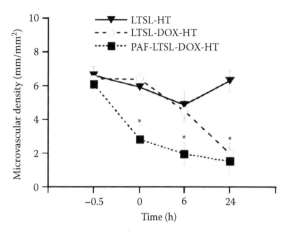

FIGURE 16.6 Skin-fold window chamber data showing changes in microvascular density in 4T07 tumors treated with LTSL-Dox with hyperthermia, empty LTSL with hyperthermia, and the combination of platelet activating factor (PAF), LTSL-Dox and hyperthermia. $^*p < 0.05$, when comparing PAF + LTSL-Dox and hyperthermia to the other two groups at 0 and 6 hr. At 24 hr, the extent of vascular shutdown between empty LTSL and the other two groups was significant at 24 hr ($p < 0.05$). Error bars – standard error of the mean (SEM). (From Chen, Q. et al., *International Journal of Hyperthermia*, 24, 2008. With permission.)

16.7 Use of Imaging to Quantify Drug Delivery from LTSL-Dox and Establishment of Treatment Optimization Strategies

One of the methods used to load doxorubicin into liposomes involves $MnSO_4$. When this salt is loaded into the interior of a liposome, doxorubicin enters into the liposome down its concentration gradient and is trapped in a complex with manganese (Chiu et al. 2005). Viglianti utilized the paramagnetic properties of Mn^{++} to evaluate Mn^{++} and doxorubicin release from LTSL-Dox-Mn, using MRI (Viglianti et al. 2004). He first demonstrated that it was possible to measure content release *in vitro*, because the relaxivity of the Mn^{++} increases when it is released from LTSL. This occurs because the interaction of Mn^{++} with bulk water increases, after release. The relaxivity of Mn^{++} in bulk water is also temperature dependent, so this needs to be kept in mind when interpreting imaging results that incorporate this type of contrast agent. In a subsequent paper, Viglianti demonstrated that the extent of T1 shortening associated with Mn^{++} release was linearly correlated with doxorubicin concentration in tumors that were subjected to heating during MR acquisition of T1 data (Viglianti et al. 2006).

Ponce evaluated how patterns of drug delivery by hyperthermia combined with LTSL-Dox-Mn could influence the efficacy of this formulation (Ponce et al. 2007). The rapid drug release properties of the LTSL provided a perfect foil for establishing a rationale for drug dose painting as a means to enhance the antitumor effects of LTSL-Dox. The method that she used involved real-time imaging of Mn^{++} delivery using MRI. Using the calibrations

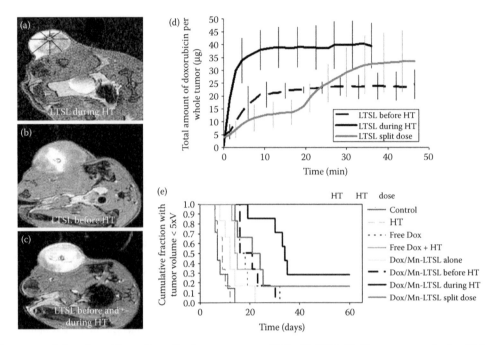

FIGURE 16.7 Intratumoral drug deposition patterns for three sequences of HT with LTSL-MnDox, as assessed using T1-based MRI. (a) Drug delivery pattern when tumor is heated before and during liposomes administration. (b) Drug delivery pattern when drug is given prior to heating. (c) Drug delivery pattern is more uniform if half is given prior to heating and the other half after heating has started. (d) Drug concentration in tumor over time, as measured using MRI. More drug is delivered if the tumor is heated first. The highest overall drug concentration and antitumor effect (e) was observed for the sequence that used HT first, followed by LTSL-Dox (a). (Reproduced from Ponce, A. M. et al., *Journal of the National Cancer Institute*, 99, 2007. With permission.)

provided previously by Viglianti, she was able to measure drug concentration distributions in these tumors, based on the change in T1 relaxation in each voxel of the tumor. The heating device consisted of a catheter placed centrally along the axis of the tumor, through which hot water was circulated. This device was MR compatible, which permitted visualization of drug deposition in real-time during heating. Three scenarios were followed, which altered the drug deposition pattern: (1) If the tumor was preheated before drug was administered, the drug deposition pattern was primarily peripheral. This occurred because the temperature of the periphery was above the transition temperature for the liposome and the inflow of blood to the tumor started at the periphery (Figure 16.7). (2) If the liposomes were administered first, followed by the onset of heating, the drug deposition pattern was centrally located and spread outward as heat spread radially from the heating catheter. (3) A more uniform drug delivery pattern was achieved by delivering half the dose and starting to heat, followed by injection of the second half of the dose after the tumor achieved thermal steady state. Interestingly, the greatest antitumor effect was observed for case #1. The reason for the greater antitumor effect was surmised to be the result of an antivascular effect, where the greatest concentration of drug was deposited in the region of the feeding tumor vasculature.

Investigators examining high intensity focused ultrasound have followed the lead of Ponce to publish other papers using this same type of liposome, in combination with HIFU, to dose paint drug into target tumors (Mylonopoulou et al. 2010, Negussie et al. 2011, Staruch et al. 2011).

16.8 Clinical Applications of LTSL-Dox

16.8.1 Canine Trial

The first phase I study of LTSL-Dox was conducted by Hauck et al. in dogs with spontaneous canine tumors (Hauck et al. 2006). Privately owned dogs with solid sarcomas or carcinomas were enrolled. The tumors were required to be in a location that was heatable using a 433 MHz microwave heating system. Escalating doses of LTSL-Dox from 0.7 to 1.0 mg/kg were administered at each of three courses, scheduled three weeks apart. Pharmacokinetics were evaluated during the first treatment cycle. A total of 21 patients were enrolled. The maximum tolerated dose (MTD) was 0.93 mg/kg, which is approximately 10% lower than the reported MTD of free drug for the dog. The first two dogs enrolled in this trial experienced anaphylactoid reactions, characterized by a sudden transitory drop in blood pressure, an increase in end inspiratory pressure, and in one case extensive facial edema. Subsequent studies performed in normal dogs identified this as being related to a profound histaminemia. Subsequent animals were premedicated with steroids and antihistamines, which minimized these toxicities. These toxicities influenced the pretreatment regimens now employed in human trials with this drug. The primary toxicity encountered was neutropenia and renal toxicity (two with fatal isothenuria), with dose-limiting toxicities (DLTs) observed at 0.93 mg/kg. Grade 2 cardiac toxicities were observed at the lower two dose levels.

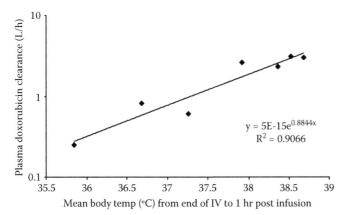

FIGURE 16.8 Doxorubicin plasma clearance vs. mean deep rectal (systemic) temperature at end of heating period in tumor-bearing dogs treated in a phase I trial of LTSL-Dox + HT. The range of temperatures is below that required for LTSL drug release. Thus, the change in clearance rate is likely a reflection of the treatment volume vs. body weight size of the animals being treated. When the treatment volume is large relative to body size, then more drug is released in the heated volume, leading to an elevated clearance rate. It is possible that free drug is being released intravascularly in the heated volume as well, which would increase the clearance rate. (Reproduced from Hauck, M. L. et al., *Clin Cancer Res*, 12, 2006. With permission.)

The terminal half-life of doxorubicin was 1.6 hr in this study. Importantly, the doxorubicin clearance (L/hr) was found to be linearly related to the rectal temperature at the end of the heating period in these animals (Figure 16.8). The rectal temperatures were below that required for triggered drug release. This result was likely associated with treatment volume. Larger tumor volumes were more likely to result in relatively more systemic drug release as a result of the local heating. It has been theoretically estimated recently that the transit time of an LTSL through a 3 cm tumor may be <5 s (Gasselhuber et al. 2010). If this is the case, some liposomes may escape the heated tumor region, since the drug release time of LTSL is on the order of 10–20 s. If they remain intravascular, they may dump drug outside the heated volume. If this is the case, then one might expect to see free drug in plasma during and after heating at a level higher than would be expected after LTSL-Dox administration without heating. The formulation does exhibit slow release at 37°C. As will be shown later, free drug is measurable in the plasma of humans treated with LTSL-Dox in the context of thermal ablation for liver cancer.

The tumors treated in this canine trial were relatively large (median of 91 cm³), but of the 20 evaluable animals that received at least two courses of chemotherapy, two experienced progression, 12 had stable disease, and the remaining six demonstrated partial responses to the treatment.

16.8.2 Human Trials

Phase I trials have been completed in patients with chest wall recurrences of breast cancer and in patients with either primary or metastatic liver cancer. The former were treated with superficial hyperthermia devices, whereas the liver cancer patients were treated with thermal ablation. Results of the chest wall recurrence trial have not been published, nor has a full report been published for the liver cancer trials. One limited report has been published from the phase I liver trial, however (Poon and Borys 2009). A description of the clinical rationale and limited results for the liver trial follow.

16.8.2.1 Primary Hepatocellular Carcinoma

Primary liver cancer is the third leading cause of cancer deaths worldwide. Statistics from 2007 showed that there were over 700,000 cases and nearly as many deaths (680,000) (Sung and Thung 2010). There is high incidence in Eastern Asia, such as China and Japan, and the African Congo, but relatively low incidence in the Western world. It is associated with transmission of hepatitis B and C viruses. However, the most common risk factor is cirrhosis, where it is thought that inflammation and proliferation likely play a role in its development. The difference between the incidence and death rate clearly indicates that there is currently not an effective therapy for this disease. Less than 20% of patients are suitable candidates for surgical resection, but when successful, the five year survival rate is over 75%. Hepatic reserve is an important consideration in the treatment of this disease, which has led to the concept of localized therapies, such as ethanol ablation, cryoblation, chemoembolization, and RF ablation.

LTSL-Dox has been tested as an adjuvant to thermal ablation for medium (3.1–5 cm) and large (>5 cm) tumors in seriously impaired livers. The rationale for this approach is to widen the margin of the ablation zone, to kill tumor cells that have invaded adjacent normal liver, and/or to increase the treatment margin in regions of thermally significant vessels where temperatures fall below that required for thermal ablation. RFA has been a successful treatment for small lesions in this disease. Local recurrences have been reported to be less than 20% for lesions <3 cm, but the local failure rate is higher for larger lesions.

A phase I trial was conducted in 24 patients who received RF ablation for hepatocellular carcinoma or metastatic liver cancer (Poon and Borys 2009). The MTD was 50 mg/M². The pharmacokinetics of the drug are similar to what was observed in the canine trial, with most of the doxorubicin exposure occurring in the 6 hr after the end of drug infusion (Figure 16.9). The dose-limiting toxicities were grade 3 alanine aminotransferase increase and grade 4 neutropenia, which were observed at a dose of 60mg/M². Half of the tumors treated in this trial were between 3.8 and 6.5 cm in diameter. There was a statistically significant dose effect. The median time to failure for patients who received less than the MTD was 80 days, whereas for those who received the MTD dose, it was 374 days ($p = 0.0380$). These results strongly suggest that LTSL-Dox has the capacity to augment RF ablation. Based on this result, a phase III multi-institutional randomized trial was opened by Celsion Corporation in 2008 (Phase 3 Study of ThermoDox With Radiofrequency Ablation (RFA) in Treatment of Hepatocellular Carcinoma (HCC); NCT00617981). The trial is currently open,

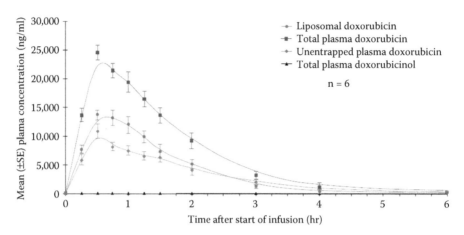

FIGURE 16.9 Pharmacokinetics of liposomal and free doxorubicin in patients with liver tumors, treated with LTSL-Dox and thermal ablation at the MTD dose of 50 mg/M^2. (Reproduced from Poon, R. T., and Borys, N., *Expert Opin Pharmacother*, 10, 2009. With permission.)

with a planned primary completion date of 2011 and study completion date of 2013.

16.8.2.2 Chest Wall Recurrences of Breast Cancer

Radiotherapy administered after surgical resection has impacted local recurrences of breast cancer and has impacted survival (Zagar et al. 2010). It is currently estimated that between 5 and 15% of breast cancer patients who receive radiotherapy will experience a chest wall recurrence. The symptoms of such recurrences have substantial influence on the quality of life for these patients. Symptoms can include pain, ulceration, bleeding, and the distress of having a visible local failure (Zagar et al. 2010). Hyperthermia has been used in combination with radiotherapy for this disease and has been shown to be superior to radiotherapy alone in randomized studies (Jones et al. 2005, Vernon et al. 1996). This combination therapy has been approved by the FDA for at least two decades. In spite of its relative success, local failures are still common, which has led to the rationale to add LTSL-Dox to hyperthermia as an alternate to thermoradiotherapy. A phase I trial was completed at Duke last year using this combination and results will be submitted for review soon.

16.9 Future Directions

The choice of therapeutic payloads to consider adding to the LTSL platform is dependent upon choosing diseases that might benefit and then discerning whether such diseases are heatable to the temperature range needed for drug release. Preferably, one would want to use drugs for which there is thermal augmentation of cell killing as well, although that is not an absolute requirement. The success with LTSL-Dox that has been seen thus far is below the temperature needed to see synergistic cell killing with hyperthermia (Hahn et al. 1975), for example. Choice of drugs to load within liposomes is also dependent upon lipid solubility. Drugs that are highly lipid soluble will be incorporated into the lipid bilayer as opposed to being in the aqueous center, and the performance characteristics of drug release may

be dependent upon the type of drug that is loaded. The science of drug loading into these nanoparticles is extensive, and each type of drug formulation has to be carefully characterized and optimized (Landon et al. 2011).

16.9.1 Additional Applications for LTSL-Dox

The performance of LTSL-Dox provides strong rationale for investigating its potential use in other diseases:

Liver metastases. If the phase III trial for hepatocellular carcinoma is positive, this would provide rationale for expanding its applications for treatment of liver metastases.

Locally advanced breast cancer. Locally advanced breast cancer is another potential target. This disease can be heated to the proper temperature range, and there is experience in using nonthermally sensitive doxorubicin containing liposomes in this target population (Vujaskovic et al. 2010).

16.9.1.1 LTSL-Cisplatin and LTSL-Mitomycin C

As an example of the principles outlined before, we discuss, following, the rationale for clinical development of two LTSL formulations that are targeted for treatment of bladder cancer. The bladder is heatable, using the BSD 2000 family of RF-phased array devices, so it is a good candidate site.

We have recently developed a cisplatin-containing LTSL, and the first clinical application with this formulation will be for muscle-invasive bladder cancer. Bladder cancer has the fourth highest prevalence and fifth highest incidence of all solid tumors (Altecruse et al. 2010, http://seer.cancer.gov/csr/1975_2007/, based on November 2009 SEER data submission, posted to the SEER Web site, 2010, Canadian Cancer Society/National Cancer Institute of Canada 2005). When the cost of treating a single patient from diagnosis to death is considered, it also ranks as the single most expensive cancer to diagnose and treat (Botteman et al. 2003, Riley and Lubitz 1989, Riley et al. 1995). Approximately 75% of new bladder cancers are nonmuscle invasive (NMIBC), and such tumors are characterized by high local recurrence rates.

Intravesical therapies, such as bacillus Calmette-Guérin (BCG) immunotherapy and mitomycin C (MMC) chemotherapy, can prevent these local recurrences and are therefore recommended as adjuvant therapy for high-risk patients after transurethral resection of the bladder tumor (TURBT) (Herr et al. 1988, Herr et al. 1995, Malmstrom et al. 2009, Shelley et al. 2003, Shelley et al. 2000, Sylvester et al. 2005). Unfortunately, despite receiving intravesical therapy many patients (>50%) will experience a local recurrence of cancer. More importantly, 15–30% will progress to muscle-invasive bladder cancer (MIBC), a disease that carries a five-year mortality rate of ~50% (Addeo et al. 2010). Improving bladder cancer patient survival and quality of life requires three things: (1) a truly effective treatment that can prevent local recurrences of NMIBC, (2) thwart the progression of NIMBC to MIBC, and (3) a therapy that can eradicate MIBC once it has arisen and thereby prevent its spread to distant organ sites. Our current development includes a mitomycin C containing LTSL, which we intend to take forward for treatment of nonmuscle invasive bladder cancer. The rationale for this will be to use thermal enhancement of cell killing that is achieved with mitomycin C, combined with an approach that we believe will enhance penetration depth of the drug into the bladder wall, thereby reducing likelihood for local recurrence. A randomized phase III trial has been previously reported indicating that intravesicular mitomycin C with local hyperthermia can prolong progression free survival in patients with nonmuscle invasive bladder cancer (Colombo et al. 2011). We believe that this success can be further built upon by using a liposomal formulation of mitomycin C.

The neoadjuvant combination of cisplatin and gemcitabine, followed by cystectomy, is the current standard of care for treatment of muscle invasive bladder cancer. For MIBC, current neoadjuvant chemotherapy regimens only increase five-year survival by about 5% over surgery alone (Shariat et al. 2006). Thus, there is a need to improve survival. There is evidence that increasing the pathologic complete response (CR) rate has a positive influence on progression free survival (Sawhney et al. 2006). The rationale for using the LTSL-cisplatin in this setting will be to increase the pathologic CR rate.

References

Addeo, R., Caraglia, M., Bellini, S., Abbruzzese, A., Vincenzi, B., Montella, L. et al. 2010 Randomized phase III trial on gemcitabine versus mytomicin in recurrent superficial bladder cancer: Evaluation of efficacy and tolerance. *J Clin Oncol*, 28: 543–8.

Ahmed, M., Goldberg, S. N. 2004 Combination radiofrequency thermal ablation and adjuvant IV liposomal doxorubicin increases tissue coagulation and intratumoural drug accumulation. *Int J Hyperthermia*, 20: 781–802.

Altekruse, S., Kosary, C., Krapcho, M., Neyman, N., Aminou, R., Waldron, W. et al.: *SEER Cancer Statistics Review, 1975–2007*. Bethesda, MD: National Cancer Institute, 2010. http://seer.cancer.gov/csr/1975_2007

Andresen, T. L., Thompson, D. H., Kaasgaard, T. 2010 Enzyme-triggered nanomedicine: Drug release strategies in cancer therapy (Invited Review). *Molecular Membrane Biology*, 27: 353–63.

Bates, D. A., Mackillop, W. J. 1986 Hyperthermia, adriamycin transport, and cytotoxicity in drug-sensitive and -resistant Chinese hamster ovary cells. *Cancer Res*, 46: 5477–81.

Botteman, M. F., Pashos, C. L., Redaelli, A., Laskin, B., Hauser, R. 2003 The health economics of bladder cancer: A comprehensive review of the published literature. *Pharmacoeconomics*, 21: 1315–30.

Brown, J. M., Wilson, W. R. 2004 Exploiting tumour hypoxia in cancer treatment. *Nat Rev Cancer*, 4: 437–47.

Canadian Cancer Society/National Cancer Institute of Canada: Toronto, C., 2005 Canadian Cancer Statistics 2005. 0835-2976, April 2005.

Chen, Q., Krol, A., Wright, A., Needham, D., Dewhirst, M. W., Yuan, F. 2008 Tumor microvascular permeability is a key determinant for antivascular effects of doxorubicin encapsulated in a temperature sensitive liposome. *International Journal of Hyperthermia*, 24: 475–82.

Chen, Q., Tong, S., Dewhirst, M. W., Yuan, F. 2004 Targeting tumor microvessels using doxorubicin encapsulated in a novel thermosensitive liposome. *Molecular Cancer Therapeutics*, 3: 1311–17.

Chiu, G. N. C., Abraham, S. A., Ickenstein, L. M., Ng, R., Karlsson, G., Edwards, K. et al. 2005 Encapsulation of doxorubicin into thermosensitive liposomes via complexation with the transition metal manganese. *Journal of Controlled Release*, 104: 271–88.

Colombo, R., Salonia, A., Leib, Z., Pavone-Macaluso, M., Engelstein, D. 2011 Long-term outcomes of a randomized controlled trial comparing thermochemotherapy with mitomycin-C alone as adjuvant treatment for non-muscle-invasive bladder cancer (NMIBC). *BJU Int*, 107: 912–8.

Demant, E. J. F., Sehested, M. 1993 Recognition of anthracycline binding domains in bovine serum-albumin and design of a free fatty-acid sensor protein. *Biochimica Et Biophysica Acta*, 1156: 151–60.

Dewhirst, M. W., Tso, C. Y., Oliver, R., Gustafson, C. S., Secomb, T. W., Gross, J. F. 1989 Morphologic and hemodynamic comparison of tumor and healing normal tissue microvasculature. *Int J Radiat Oncol Biol Phys*, 17: 91–9.

Dewhirst, M. W., Vujaskovic, Z., Jones, E., Thrall, D. 2005 Re-setting the biologic rationale for thermal therapy. *Int J Hyperthermia*, 21: 779–90.

Friedl, J., Turner, E., Alexander, H. R. 2003 Augmentation of endothelial cell monolayer permeability by hyperthermia but not tumor necrosis factor: Evidence for disruption of vascular integrity via VE-cadherin down-regulation. *International Journal of Oncology*, 23: 611–16.

Gaber, M. H., Hong, K., Huang, S. K., Papahadjopoulos, D. 1995 Thermosensitive sterically stabilized liposomes: Formulation and *in vitro* studies on mechanism of doxorubicin release by bovine serum and human plasma. *Pharm Res*, 12: 1407–16.

Gaber, M. H., Wu, N. Z., Hong, K., Huang, S. K., Dewhirst, M. W., Papahadjopoulos, D. 1996 Thermosensitive liposomes: Extravasation and release of contents in tumor microvascular networks. *Int J Radiat Oncol Biol Phys*, 36: 1177–87.

Gabizon, A., Shmeeda, H., Barenholz, Y. 2003 Pharmacokinetics of pegylated liposomal doxorubicin: Review of animal and human studies. *Clinical Pharmacokinetics*, 42: 419–36.

Gasselhuber, A., Dreher, M. R., Negussie, A., Wood, B. J., Rattay, F., Haemmerich, D. 2010 Mathematical spatio-temporal model of drug delivery from low temperature sensitive liposomes during radiofrequency tumour ablation. *Int J Hyperthermia*, 26: 499–513.

Grenader, T., Goldberg, A., Gabizon, A. 2010 Monitoring long-term treatment with pegylated liposomal doxorubicin: How important is intensive cardiac follow-up? *Anti-Cancer Drugs*, 21: 868–71.

Gupta, V. K., Jaskowiak, N. T., Beckett, M. A., Mauceri, H. J., Grunstein, J., Johnson, R. S. et al. 2002 Vascular endothelial growth factor enhances endothelial cell survival and tumor radioresistance. *Cancer J*, 8: 47–54.

Hahn, G. M., Braun, J., Har-Kedar, I. 1975 Thermochemotherapy: Synergism between hyperthermia (42–43 degrees) and adriamycin (of bleomycin) in mammalian cell inactivation. *Proc Natl Acad Sci USA*, 72: 937–40.

Hauck, M. L., LaRue, S. M., Petros, W. P., Poulson, J. M., Yu, D., Spasojevic, I. et al. 2006 Phase I trial of doxorubicin-containing low temperature sensitive liposomes in spontaneous canine tumors. *Clin Cancer Res*, 12: 4004–10.

Herman, T. S., Teicher, B. A. 1994 Summary of studies adding systemic chemotherapy to local hyperthermia and radiation. *Int J Hyperthermia*, 10: 443–9.

Herr, H. W., Laudone, V. P., Badalament, R. A., Oettgen, H. F., Sogani, P. C., Freedman, B. D. et al. 1988 Bacillus Calmette-Guerin therapy alters the progression of superficial bladder cancer. *J Clin Oncol*, 6: 1450–5.

Herr, H. W., Schwalb, D. M., Zhang, Z. F., Sogani, P. C., Fair, W. R., Whitmore, W. F. et al. 1995 Intravesical bacillus Calmette-Guerin therapy prevents tumor progression and death from superficial bladder cancer: Ten-year follow-up of a prospective randomized trial. *J Clin Oncol*, 13: 1404–8.

Hong, Y. J., Kim, J. C. 2011 Egg phosphatidylcholine liposomes incorporating hydrophobically modified chitosan: pH-Sensitive release. *Journal of Nanoscience and Nanotechnology*, 11: 204–09.

Hossann, M., Wiggenhorn, M., Schwerdt, A., Wachholz, K., Teichert, N., Eibl, H. et al. 2007 *In vitro* stability and content release properties of phosphatidylglyceroglycerol containing thermosensitive liposomes. *Biochim Biophys Acta*, 1768: 2491–9.

Issels, R. D., Lindner, L. H., Verweij, J., Wust, P., Reichardt, P., Schem, B. C. et al. 2010 Neo-adjuvant chemotherapy alone or with regional hyperthermia for localised high-risk soft-tissue sarcoma: A randomised phase 3 multicentre study. *Lancet Oncol*, 11: 561–70.

Jones, E. L., Oleson, J. R., Prosnitz, L. R., Samulski, T. V., Vujaskovic, Z., Yu, D. et al. 2005 Randomized trial of hyperthermia and radiation for superficial tumors. *J Clin Oncol*, 23: 3079–85.

Khan, S. N., Islam, B., Yennamalli, R., Zia, Q., Subbarao, N., Khan, A. U. 2008 Characterization of doxorubicin binding site and drug induced alteration in the functionally important structural state of oxyhemoglobin. *Journal of Pharmaceutical and Biomedical Analysis*, 48: 1096–104.

Kleiter, M. M., Yu, D., Mohammadian, L. A., Niehaus, N., Spasojevic, I., Sanders, L. et al. 2006 A tracer dose of technetium-99m-labeled liposomes can estimate the effect of hyperthermia on intratumoral doxil extravasation. *Clin Cancer Res*, 12: 6800–7.

Kong, G., Anyarambhatla, G., Petros, W. P., Braun, R. D., Colvin, O. M., Needham, D. et al. 2000 Efficacy of liposomes and hyperthermia in a human tumor xenograft model: Importance of triggered drug release. *Cancer Res*, 60: 6950–7.

Kong, G., Braun, R., Dewhirst, M. 2001 Characterization of the effect of hyperthermia on nanoparticle extravasation from tumor vasculature. *Cancer Res*, 61: 3027–32.

Kong, G., Braun, R. D., Dewhirst, M. W. 2000 Hyperthermia enables tumor-specific nanoparticle delivery: Effect of particle size. *Cancer Res*, 60: 4440–5.

Kong, G., Dewhirst, M. W. 1999 Hyperthermia and liposomes. *Int J Hyperthermia*, 15: 345–70.

Koning, G. A., Eggermont, A. M., Lindner, L. H., ten Hagen, T. L. 2010 Hyperthermia and thermosensitive liposomes for improved delivery of chemotherapeutic drugs to solid tumors. *Pharm Res*, 27: 1750–4.

Landon, C., Park, J. Y., Needham, D., Dewhirst, M. W. 2011 Hyperthermia and nanoscale drug delivery: The materials design and preclinical and clinical testing of low temperature-sensitive liposomes used in combination with mild hyperthermia in the treatment of local cancer. *Open Nanomedicine Journal*, 3: 38–64.

Li, L., ten Hagen, T. L., Schipper, D., Wijnberg, T. M., van Rhoon, G. C., Eggermont, A. M. et al. 2010 Triggered content release from optimized stealth thermosensitive liposomes using mild hyperthermia. *J Control Release*, 143: 274–9.

Li, L. W., Wang, D. D., Sun, D. Z., Liu, M., Qu, X. K. 2007 Thermodynamic study on interaction between anti-tumor drug 5-fluorouracil and human serum albumin. *Acta Chimica Sinica*, 65: 2853–57.

Lindner, L. H., Eichhorn, M. E., Eibl, H., , Eibl, H., Teichert, N., Schmitt-Sody, M. et al. 2004 Novel temperature-sensitive liposomes with prolonged circulation time. *Clin Cancer Res*, 10: 2168–78.

Lindner, L. H., Hossann, M. 2010 Factors affecting drug release from liposomes. *Curr Opin Drug Discov Devel*, 13: 111–23.

Lindner, L. H., Reinl, H. M., Schlemmer, M., Stahl, R., Peller, M. 2005 Paramagnetic thermosensitive liposomes for MR-thermometry. *Int J Hyperthermia*, 21: 575–88.

Malmstrom, P. U., Sylvester, R. J., Crawford, D. E. , Friedrich, M., Krege, S., Rintala, E. et al. 2009 An individual patient data meta-analysis of the long-term outcome of randomised studies comparing intravesical mitomycin C versus Bacillus Calmette-Guerin for non-muscle-invasive bladder cancer. *Eur Urol*, 56(2): 247–56.

Matteucci, M. L., Anyarambhatla, G., Rosner, G., Azuma, C., Fisher, P. E., Dewhirst, M. W. et al. 2000 Hyperthermia increases accumulation of technetium-99m-labeled liposomes in feline sarcomas. *Clin Cancer Res*, 6: 3748–55.

Minchinton, A. I., Tannock, I. F. 2006 Drug penetration in solid tumours. *Nat Rev Cancer*, 6: 583–92.

Monsky, W. L., Kruskal, J. B., Lukyanov, A. N. et al. 2002 Radiofrequency ablation increases intratumoral liposomal doxorubicin accumulation in a rat breast tumor model. *Radiology*, 224: 823–9.

Moon, E. J., Sonveaux, P., Porporato, P. E., Danhier, P., Gallez, B., Batinic-Haberle, I. et al. 2010 NADPH oxidase-mediated reactive oxygen species production activates hypoxia-inducible factor-1 (HIF-1) via the ERK pathway after hyperthermia treatment. *Proc Natl Acad Sci USA*, 107: 20477–82.

Mostafa, E. M., Ganguli, S., Faintuch, S., Mertyna, P., Goldberg, S. N. 2008 Optimal strategies for combining transcatheter arterial chemoembolization and radiofrequency ablation in rabbit VX2 hepatic tumors. *J Vasc Interv Radiol*, 19: 1740–8.

Mylonopoulou, E., Arvanitis, C. D., Bazan-Peregrino, M., Arora, M., Coussios, C. C. 2010. Ultrasonic activation of thermally sensitive liposomes. In: *9th International Symposium on Therapeutic Ultrasound*. Edited by K. Hynynen and J. Souquet. Melville: Amer Inst Physics, vol. 1215, pp. 83–87.

Needham, D., Anyarambhatla, G., Kong, G., Dewhirst, M. W. 2000 A new temperature-sensitive liposome for use with mild hyperthermia: Characterization and testing in a human tumor xenograft model. *Cancer Res*, 60: 1197–201.

Needham, D., Dewhirst, M. W. 2001 The development and testing of a new temperature-sensitive drug delivery system for the treatment of solid tumors. *Adv Drug Deliv Rev*, 53: 285–305.

Needham, D., Ponce, A. M. 2007 Nanoscale drug delivery vehicles for solid tumors: A new paradigm for localized drug delivery using temperature sensitive liposomes. In: *Nanotechnology for Cancer Therapy*. Edited by M. A. Mansoor. Boca Raton: CRC Press, pp. 678–719.

Negussie, A. H., Yarmolenko, P. S., Partanen, A., Ranjan, A., Jacobs, G., Woods, D. et al. 2011 Formulation and characterisation of magnetic resonance imageable thermally sensitive liposomes for use with magnetic resonance-guided high intensity focused ultrasound. *Int J Hyperthermia*, 27: 140–55.

O'Brien, M. E. R., Wigler, N., Inbar, M., Rosso, R., Grischke, E., Santoro, A. et al. 2004 Reduced cardiotoxicity and comparable efficacy in a phase III trial of pegylated liposomal doxorubicin HCl (CAELYX (TM)/Doxil (R)) versus conventional doxorubicin for first-line treatment of metastatic breast cancer. *Annals of Oncology*, 15: 440–49.

Palmer, G., Boruta, R., Viglianti, B., Lan, L., Spasojevic, I., Dewhirst, M. 2010 Non-invasive monitoring of intra-tumor drug concentration and therapeutic response using optical spectroscopy. *J Control Release*, 142: 457–64.

Palmer, G. M., Boruta, R. J., Viglianti, B. L., Lan, L., Spasojevic, I., Dewhirst, M. W. 2010 Non-invasive monitoring of intra-tumor drug concentration and therapeutic response using optical spectroscopy. *Journal of Controlled Release*, 142: 457–64.

Ponce, A. M., Viglianti, B. L., Yu, D. H., Yarmolenko, P. S., Michelich, C. R., Woo, J. et al. 2007 Magnetic resonance imaging of temperature-sensitive liposome release: Drug dose painting and antitumor effects. *Journal of the National Cancer Institute*, 99: 53–63.

Poon, R. T., Borys, N. 2009 Lyso-thermosensitive liposomal doxorubicin: A novel approach to enhance efficacy of thermal ablation of liver cancer. *Expert Opin Pharmacother*, 10: 333–43.

Primeau, A. J., Rendon, A., Hedley, D., Lilge, L., Tannock, I. F. 2005 The distribution of the anticancer drug doxorubicin in relation to blood vessels in solid tumors. *Clin Cancer Res*, 11: 8782–8.

Ramanathan-Girish, S., Boroujerdi, M. 2001 Contradistinction between doxorubicin and epirubicin: In-vitro interaction with blood components. *Journal of Pharmacy and Pharmacology*, 53: 815–21.

Riley, G. F., Lubitz, J. D.: Longitudinal patterns in medicare costs for cancer decedents. In: *Cancer Care and Cost: DRGs and Beyond*. Edited by R. M. Scheffler and N. C. Andrews. Ann Arbor: Health Administration Press Perspectives, vol. 1, pp. 89–106, 1989

Riley, G. F., Potosky, A. L., Lubitz, J. D., Kessler, L. G. 1995 Medicare payments from diagnosis to death for elderly cancer patients by stage at diagnosis. *Med Care*, 33: 828–41.

Sawhney, R., Bourgeois, D., Chaudhary, U. B. 2006 Neo-adjuvant chemotherapy for muscle-invasive bladder cancer: A look ahead. *Ann Oncol*, 17: 1360–9.

Shariat, S. F., Karakiewicz, P. I., Palapattu, G. S., Lotan, Y., Rogers, C. G., Amiel, G. E. et al. 2006 Outcomes of radical cystectomy for transitional cell carcinoma of the bladder: A contemporary series from the Bladder Cancer Research Consortium. *J Urol*, 176: 2414–22; discussion 22.

Shelley, M. D., Court, J. B., Kynaston, H., Wilt, T. J., Coles, B., Mason, M. 2003 Intravesical bacillus Calmette-Guerin versus mitomycin C for Ta and T1 bladder cancer. *Cochrane Database Syst Rev* CD003231.

Shelley, M. D., Court, J. B., Kynaston, H., Wilt, T. J., Fish, R. G., Mason, M. 2000 Intravesical Bacillus Calmette-Guerin in Ta and T1 Bladder Cancer. *Cochrane Database Syst Rev* CD001986.

Song, C. W. 1984 Effect of local hyperthermia on blood flow and microenvironment: A review. *Cancer Res*, 44: 4721s–30s.

Song, C. W., Park, H., Griffin, R. J. 2001 Improvement of tumor oxygenation by mild hyperthermia. *Radiat Res*, 155: 515–28.

Staruch, R., Chopra, R., Hynynen, K. 2011 Localised drug release using MRI-controlled focused ultrasound hyperthermia. *Int J Hyperthermia*, 27: 156–71.

Sung, M., Thung, S. N. 2010. Primary neoplasms of the liver. In: *Cancer Medicine*, 8th ed. Edited by W. K. Hong, R. C. Bast, W. N. Hait et al. Shelton, CT: People's Medical Publishing House, pp. 1124–31.

Sylvester, R. J., van der Meijden, A. P., Witjes, J. A., Kurth, K. 2005 Bacillus calmette-guerin versus chemotherapy for the intravesical treatment of patients with carcinoma *in situ* of the bladder: A meta-analysis of the published results of randomized clinical trials. *J Urol*, 174: 86-91; discussion 91–2.

Tashjian, J. A., Dewhirst, M. W., Needham, D., Viglianti, B. L. 2008 Rationale for and measurement of liposomal drug delivery with hyperthermia using non-invasive imaging techniques. *International Journal of Hyperthermia*, 24: 79–90.

Umeno, H., Watanabe, N., Yamauchi, N., Tsuji, N., Okamoto, T., Niitsu, Y. 1994 Enhancement of blood stasis and vascular-permeability in Meth-A tumors by administration of hyperthermia in combination with tumor-necrosis-factor. *Japanese Journal of Cancer Research*, 85: 325–30.

Vernon, C. C., Hand, J. W., Field, S. B., Machin, D., Whaley, J. B., van der Zee, J. et al. 1996 Radiotherapy with or without hyperthermia in the treatment of superficial localized breast cancer: Results from five randomized controlled trials. International Collaborative Hyperthermia Group. *Int J Radiat Oncol Biol Phys*, 35: 731–44.

Viglianti, B. L., Abraham, S. A., Michelich, C. R., Yarmolenko, P. S., MacFall, J. R., Bally, M. B. et al. 2004 *in vivo* monitoring of tissue pharmacokinetics of liposome/drug using MRI: Illustration of targeted delivery. *Magnetic Resonance in Medicine*, 51: 1153–62.

Viglianti, B. L., Ponce, A. M., Michelich, C. R., Yu, D., Abraham, S. A., Sanders, L. et al. 2006 Chemodosimetry of *in vivo* tumor liposomal drug concentration using MRI. *Magn Reson Med*, 56: 1011–8.

Vujaskovic, Z., Kim, D. W., Jones, E., Lan, L., McCall, L., Dewhirst, M. W. et al. 2010 A phase I/II study of neoadjuvant liposomal doxorubicin, paclitaxel, and hyperthermia in locally advanced breast cancer. *Int J Hyperthermia*, 26: 514–21.

Wu, N. Z., Braun, R. D., Gaber, M. H., Lin, G. M., Ong, E. T., Shan, S. et al. 1997 Simultaneous measurement of liposome extravasation and content release in tumors. *Microcirculation*, 4: 83–101.

Yarmolenko, P. S., Zhao, Y., Landon, C., Spasojevic, I., Yuan, F., Needham, D. et al. 2010 Comparative effects of thermosensitive doxorubicin-containing liposomes and hyperthermia in human and murine tumours. *Int J Hyperthermia*, 26: 485–98.

Yatvin, M. B., Weinstein, J. N., Dennis, W. H., Blumenthal, R. 1978 Design of liposomes for enhanced local release of drugs by hyperthermia. *Science*, 202: 1290–3.

Zagar, T. M., Oleson, J. R., Vujaskovic, Z., Dewhirst, M. W., Craciunescu, O. I., Blackwell, K. L. et al. 2010 Hyperthermia combined with radiation therapy for superficial breast cancer and chest wall recurrence: A review of the randomised data. *Int J Hyperthermia*, 26: 612–7.

Zemlickis, D., Klein, J., Moselhy, G., Koren, G. 1994 Cisplatin protein-binding in pregnancy and the neonatal-period. *Medical and Pediatric Oncology*, 23: 476–79.

17

Magnetic Nanoparticles for Cancer Therapy

Michael L. Etheridge
University of Minnesota

John C. Bischof
University of Minnesota

Andreas Jordan
Charité-University Medicine

17.1 Introduction

Electromagnetic field–based thermal therapies have demonstrated the capability to selectively deposit large amounts of energy in tissue, resulting in localized temperature increases capable of hyperthermia and thermoablation. However, current clinical approaches have met with considerable limitations. Microwaves, radiofrequency (RF) waves, and lasers exhibit significant absorption at interfaces with differing electrical properties (Wust et al. 1991a). This results in attenuation at surfaces, issues with focusing energy, and unintended hot spots, leading to difficulty in treating deep-seated tumors. In addition, the geometry of the treated region is limited by the shape of the probe or array (VanSonnenberg, McMullen, and Solbiati 2005), requiring overtreatment of the surrounding areas or skill- and time-intensive repositioning to ensure complete treatment of complex tumors.

It has been well characterized that low-frequency, alternating magnetic fields show very little attenuation in biological tissues, and it was determined that implanted, energy-absorbing materials provide a means of targeting heat into deep-seated tissues. Early studies on magnetic field–based therapies by Oleson et al. (Oleson, Cetas, and Corry 1983; Oleson, Heusinkveld, and Manning 1983), Brezovich et al. (Brezovich, Atkinson, and Lilly 1984; Brezovich 1988), and Stauffer et al. (Stauffer et al. 1984; Stauffer, Cetas, and Jones 2007) utilized implanted ferromagnetic thermal seeds (on the order of millimeters), but the application was limited by the need to surgically implant each seed, and because efficacy of heat generation was critically dependent

on the correct orientation of the seeds within the applied field. However, this early work provided important observations and equations for understanding magnetic field and tissue interactions, which have been crucial in the development of the next phase of magnetic field–based therapies—magnetic fluid hyperthermia (MFH). Magnetic fluids are aqueous dispersions of nano- or microscale particles that are excited by alternating magnetic fields to produce localized heat and do not demonstrate the same critical alignment problems as thermoseeds. In addition, MFH offers significant advantages over traditional electromagnetic-based therapies, including:

1. The potential for completely noninvasive treatment. Nanoparticles injected intravenously could preferentially collect in the tumor tissue through the enhanced permeability and retention (EPR) effect (Iyer et al. 2006) and tumor-specific targeting (Byrne, Betancourt, and Brannon-Peppas 2008). These deep-seated deposits can then be excited by an external field, with no need for surgical intervention. Current MFH techniques require minimally invasive, interstitial injection to attain adequate concentrations for treatment, but truly noninvasive procedures are the long-term ambition.

2. The potential capability to treat complex tumor geometries, while minimizing effects to surrounding tissue. If the tumor is preferentially loaded with nanoparticles (either through interstitial delivery or targeting), heating can be better confined to the region of interest.

3. Iron oxide nanoparticles have been shown to form stable deposits in treated tissue (Johannsen et al. 2010), providing the capability for repeated, cyclic treatments after a single administration.

The dominant physical mechanisms behind heating in macroscale thermoseeds and magnetic nanoparticles differ and have been widely discussed in literature (Jordan 2009; Hergt et al. 2002; Hergt et al. 2004; Hergt et al. 2005; Hergt, Dutz, and Röder 2008; Popplewell, Rosensweig, and Johnston 2002; Rosensweig 2002; Barry 2008; Jones et al. 1992; Moroz, Jones, and Gray 2002). Thermoseeds mainly take advantage of resistive heating induced by eddy currents, whereas heating in magnetic nanoparticles occurs through hysteresis or superparamagnetic relaxation mechanisms. Early investigation demonstrated that nanoscale, superparamagnetic particles were superior to microscale, multi-domain particles in terms of specific absorption rate (SAR) due to these varying mechanisms (Jordan et al. 1993; Jordan 2009). This relaxation-based heating shows strong dependence on applied field strength and frequency, nanoparticle magnetic properties, and nanoparticle size distribution. The dependence on nanoparticle properties and size highlights the importance of well-controlled methods of synthesis. Wet-phase chemistry approaches, including coprecipitation and thermal decomposition, are the most common methods, generally producing colloidal iron oxide (magnetite-Fe_3O_4 and maghemite-Fe_2O_3) particles, but multifunctional core-shell structures and surface functionalized particles (with drugs, isotopes, and biologics) are currently an area of heavy research (Lu, Salabas, and Schüth 2007; Gupta and Gupta 2005; Krishnan 2010).

A number of different groups have demonstrated the efficacy of magnetic nanoparticle-based heating in vivo, utilizing a variety of nanoparticles, field applicators, field parameters, and thermometry methods (Gilchrist et al. 1957; Medal et al. 1959; Chan et al. 1993; Gordon, Hines, and Gordon 1979; Lerch and Pizzarello 1986; Rand, Snow, and Brown 1981a; Sato et al. 1990; Mitsumori et al. 1994; Luderer et al. 1983; Borrelli, Luderer, and Panzarino 1984). This extensive preclinical work led to the initiation of several clinical studies utilizing interstitially injected aminosilane-coated magnetite nanoparticles activated under a specially designed clinical field applicator. The results of the most advanced clinical studies, utilizing thermal therapy in combination with conventional irradiation, have demonstrated a clear clinical benefit in terms of survival for recurrent glioblastoma multiforme patients. Despite these promising initial indications, research continues in an attempt to (1) improve the efficiency of heat generation, reducing the required dosages, (2) optimize the field delivery to better focus energy deposition (Wust et al. 2006), and (3) develop biological targeting, allowing for systemic delivery and a truly noninvasive procedure (Stelter et al. 2009; Hoopes et al. 2009).

17.2 Scientific Background

17.2.1 Physical Principles

Although magnetic fields demonstrate minimal tissue interactions compared to other forms of electromagnetic radiation, alternating magnetic fields will induce eddy current losses in any conductive medium, and this limits usable fields for biological applications. Treatments utilizing magnetic nanoparticles attempt to minimize this field-tissue interaction while maximizing interactions between the field and the energy-absorbing nanoparticle deposits. Energy conversion in the particles occurs through hysteresis losses in multidomain particles or through relaxation losses (Brownian and Néelian) in superparamagnetic, single-domain particles. Domain and superparamagnetic behavior is determined by the magnetic material and particle size, with the latter mode generally appearing below about 20 nm (for iron oxide). Heating efficiency is mainly determined by the magnetic material properties, applied field parameters, and the nanoparticle size distribution. The following discussion will describe these physical mechanisms in detail.

17.2.2 Effects of AC Magnetic Fields in Human Application: Calculations and Clinical Experience

Many forms of electromagnetic radiation exhibit strong interactions with tissue, and this allows direct application in thermal therapies, such as microwave, RF, and laser ablation. However, all these modalities exhibit significant attenuation in surface layers, complicating potential treatment of deep-seated tissues. In contrast, alternating magnetic fields with frequencies up to 10 MHz have demonstrated essentially no attenuation in tissue equivalents with radii equal to that of a human torso (Young, Wang, and Brezovich 2007), offering a platform for uniformly penetrating deep tissue areas.

The components of human tissue are largely diamagnetic and, in general, magnetic effects are negligible. However, application of an alternating electromagnetic induction field will produce eddy currents in any conducting media, including biological tissue (Atkinson, Brezovich, and Chakraborty 2007), and like all currents, are subject to losses. These eddy currents increase radially, so in the human body, maximum losses will be expected in regions with the greatest cross-sectional area (such as the torso). Assuming a uniform field and treating the torso as a cylinder, the volumetric power generation (P) can be estimated by integrating the time-averaged current density over the cross-sectional area, giving:

$$P = \sigma \, (\pi \, \mu_0 \, f \, H_a)^2 \, r^2 \tag{17.1}$$

where σ is the bulk tissue conductivity, μ_0 is the permeability of free space, f is the applied frequency, H_a is the applied field strength, and r is the effective torso radius. The eddy current losses demonstrate three quadratic dependencies, with frequency, field strength, and radius. Thus, losses will increase significantly with increases in field strength, and frequency and will be most prominent near the exterior of large cross-sections of tissue.

Atkinson et al. performed a series of clinical studies to determine the range of tolerable parameters for alternating magnetic field-based treatments (Atkinson, Brezovich, and Chakraborty 2007). Results indicated that field tolerance could be roughly estimated as a limit to the product of frequency and field strength,

$(fH_a) < 4.85 \times 10^8$ A/m-s. These approximate limits were verified by Wust et al. through additional clinical study (Wust et al. 1991a; Wust et al. 1991b) and were matched with early experimental ferrofluid heating data to demonstrate the promise of the field (Jordan et al. 1993). These results are illustrated in Figure 17.1, which shows expected power absorption for an experimentally characterized ferrofluid, maintaining a 25 mW/ml inductive heating limit for a representative torso (a) and cranium (b). This clinically determined limit correlates well with that of Atkinson et al. Field combinations above the curve will likely result in significant patient discomfort, while operating below the limits should provide safe treatment. Since tissue heating is proportional to the square of radius, it is quite apparent that higher field limits are achievable for the case of the cranium (noting that the ferrite concentration is 5 mg/ml for the torso and only 1 mg/ml for the cranium). In both cases, significantly higher ferrofluid heating can be achieved for higher field

strengths and frequencies less than approximately 500 kHz, with optimal performance at around 100 kHz (Etheridge and Bischof 2012a) and so all subsequent discussions will focus on the physical phenomena occurring in this lower frequency range.

17.2.3 Activation of Iron Oxide Nanoparticles in Alternating Magnetic Fields

Detailed descriptions on the principles of magnetics and magnetic materials can be found in textbooks from Cullity (Cullity and Graham 2009) and O'Handley (O'Handley 2000), and Gubin has recently published a textbook focusing specifically on magnetism in nanoparticles (Gubin 2009). Briefly, however, magnetism arises at the atomic level from unpaired electron spins, which behave like atomic dipole moments. Ferromagnetism is the strongest form of magnetism and is due to strong exchange

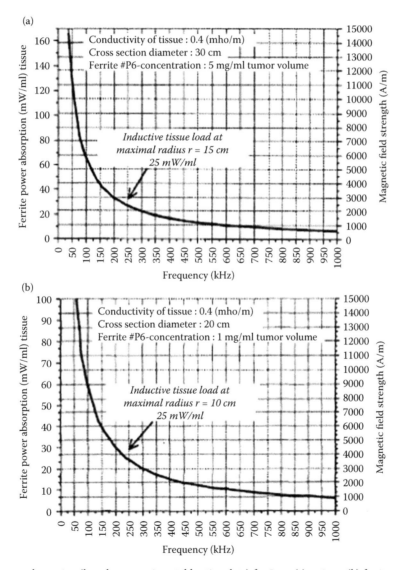

FIGURE 17.1 Expected power absorption (based on experimental heating data) for 5 mg (a) or 1 mg (b) ferrite per ml tumor volume with an inductive tissue load of 25 mW/ml at a maximum radius of 15 cm (a) or 10 cm (b), as a function of field frequency and magnetic field strength. (From Wust, P. et al. *Strahlentherapie Und Onkologie,* 166, 12, 1991a; Wust, P., *Strahlentherapie Und Onkologie,* 167, 3, 1991b.)

interactions between atomic moments in metals, most commonly, Fe, Ni, and Co. Ferrimagnetism is similar to ferromagnetism, but results from exchange interactions in ionic solids, such as metallic oxides. Both ferro- and ferrimagnetic materials demonstrate strong enough interactions to maintain a magnetic field in the absence of an applied field, but when a strong external field is applied, the atomic moments will align in the applied field direction. Diamagnetism, paramagnetism, and antiferromagnetism are additional forms of magnetic behavior, but will not be discussed in any detail here. Superparamagnetism is a unique form of magnetic behavior that arises in nanoscale particles and will be described in more detail below.

Like many physicochemical properties, a material's magnetic behavior can change as its characteristic dimensions approach the nanoscale, and this affects the loss mechanisms in an alternating field (Jordan et al. 1993; Hergt et al. 2002; Lu, Salabas, and Schüth 2007). Heat generation in magnetic materials under alternating magnetic fields can be generally considered in three regimes: eddy current heat generation (bulk materials), hysteresis heating in multidomain structures (nanoscale and larger), and relaxation losses in single-domain, superparamagnetic nanoparticles (Hergt et al. 2002). A summary of size-dependent magnetic behavior and heating mechanisms is included in Figure 17.2 and will be discussed in more detail later. Eddy currents have already been described with regard to bulk heating in tissue and are a significant source of heat generation in the use of magnetic seeds for hyperthermia (Atkinson, Brezovich, and Chakraborty 2007). However, eddy current effects are insignificant in the heating of nanoparticles, due to their small dimensions and the low conductivity of the iron oxides commonly used (Lu, Salabas, and Schüth 2007), so the subsequent discussion will focus on hysteresis and relaxation losses in magnetic nanoparticles, which have both been shown to produce clinically relevant levels of heating (Etheridge and Bishcof 2012a).

Typical magnetic materials demonstrate unique domains of magnetism (parallel magnetic moments), separated by narrow zones of magnetic, directional transition termed domain walls. Domains form to minimize the overall magnetostatic energy of the material, but as dimensions approach the nanoscale, the energy reduction provided by multiple domains is overcome by the energy cost of maintaining the domain walls, and it becomes energetically favorable to form a single magnetic domain. A number of methods for estimating the critical radius for single-domain behavior have been proposed (Lu, Salabas, and Schüth 2007; Gubin 2009), and the results can vary notably depending on the approach. Some estimated values from literature have been included in Figure 17.2, with typical diameters on the order of tens of nanometers.

When an external field is applied to a magnetic material, the potential energy of the magnetic moments is minimized by aligning with the external field, but energy is also required to rotate the moments. In a multidomain material, the domains that are aligned with the external field expand at the expense of the surrounding domains. This motion of the domain walls is associated with thermal energy losses. The strength of the external field determines the extent of domain wall motion, until the material reaches magnetic saturation and is maximally aligned with the external field. Upon field reversal, the reverse process occurs, but in moving back through a zero field, the domain walls do not return all the way to their original position and there is a remnant magnetization (M_r). Thus, under an alternating field, the material's magnetization creates a hysteresis loop. The coercivity (H_c) is the field required to reduce the magnetization back to zero. Comparisons of several example hysteresis curves adapted from data in Hergt et al. are included in Figure 17.3 (Hergt et al. 2002). Loop 1 demonstrates lower saturation magnetization than Loop 2, but a much higher coercivity. The power loss can be approximated by integrating within the hysteresis B-H loop for each cycle, and so higher heating rates would be expected for Loop 1. Estimating the expected losses requires measurement of the hysteresis behavior at the fields of interest. Significant losses can be obtained for materials with high magnetic saturation and coercivity. However, magnetic saturation generally requires relatively high fields, and hysteresis loops produced under clinically relevant fields can shrink significantly, as shown in Figure 17.3.

Despite the absence of domain walls, hysteresis behavior can still occur in single-domain particles, but involves more complicated processes for reversal, such as buckling and fanning. Classical physical treatments of these reversal losses (Stoner-Wohlfarth model) have fallen short in explaining experimentally measured losses in this single-domain hysteresis range, but phenomenological modeling has demonstrated potential as a predictive tool. Hergt et al. utilized experimental data on various magnetic particles, ranging in size from 30 to 100 nm, to produce expressions that closely predicted losses based on the applied field parameters and particle size distributions (Hergt, Silvio Dutz, and Röder 2008). The experimental values and theoretical predictions offered heating rates comparable to those of superparamagnetic nanoparticles.

At even smaller dimensions, magnetic nanoparticles exhibit another type of unique behavior, superparamagnetism, in which thermal motion causes the magnetic moments to randomly flip directions, eliminating any remnant magnetization. Thus, a normally ferro- or ferrimagnetic material will only exhibit magnetism under an applied field. This behavior arises because below a critical volume, the anisotropic energy barrier ($K_u V_m$) of the magnetic particle is reduced to the point where it can be overcome by the energy of thermal motion ($k_B T$). The definition of superparamagnetism is somewhat arbitrary, in that it relies on the choice of a measurement time (τ_m), for which the behavior is observed and is generally taken to be 100 seconds. The approximate critical diameter (d_c) for superparamagnetic behavior can be determined by assuming a spherical geometry and modifying the equation describing the probability of relaxation (O'Handley 2000):

$$\frac{\tau_m}{\tau_0} = \exp\left(\frac{K_u V_m}{k_B T}\right) \times d_c = \left[\frac{6}{\pi} \, ln\left(\frac{\tau_m}{\tau_0}\right) \frac{k_B T}{K_u}\right]^{\frac{1}{3}} \qquad (17.2)$$

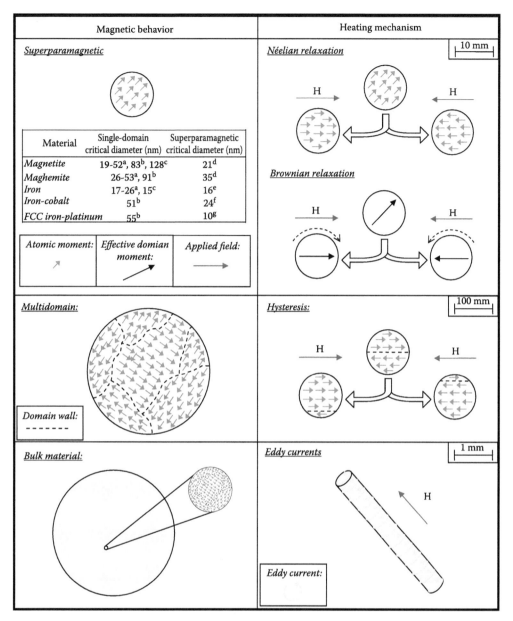

FIGURE 17.2 Regimes of magnetic behavior and mechanisms of heat generation. Values for single-domain critical diameters taken from: [a]Gubin 2009, [b]Krishnan 2010, and [c]Lu, Salabas, and Schüth 2007. Values for superparamagnetic critical diameters calculated from properties in: [d]Rosensweig 2002, [e]O'Handley 2000, [f]Kline et al. 2009, and [g]Maenosono and Saita 2006.

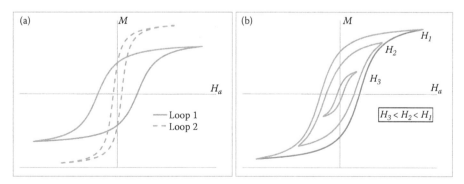

FIGURE 17.3 Example hysteresis loops adapted from data for magnetite powder in Hergt et al. (2002), demonstrating differences in loops based on material (a) and applied field strength (b).

where τ_0 is the attempt time (generally taken to be 10^{-9} seconds), V_m is the volume of magnetic material, k_B is Boltzmann's constant, and T is the absolute temperature. The approximate critical diameters for superparamagnetism for several common magnetic nanoparticle materials are included in Figure 17.2. Although remnant magnetization and hysteresis behavior are eliminated in superparamagnetic particles, significant losses can still occur through moment relaxation mechanisms.

The following description and equations are based on reviews provided by Rosensweig (Rosensweig 2002) and Hergt et al. (Hergt et al. 2002). Relaxation of superparamagnetic particles is described by two mechanisms: Brownian and Néelian relaxation, illustrated in Figure 17.2. Brownian relaxation involves the physical rotation as the overall particle moment aligns with the external field, creating frictional losses. Néelian relaxation involves the rotation and losses of the individual moments within the particle. Characteristic time constants can be used to describe each process:

Brownian:
$$\tau_B = \frac{3\eta V_H}{k_B T} \tag{17.3}$$

Néelian:
$$\tau_N = \frac{\sqrt{\pi}}{2} \tau_0 \frac{e^\Gamma}{\sqrt{\Gamma}}, \quad \Gamma = \frac{K_u V_M}{k_B T} \tag{17.4}$$

where η is the viscosity of the suspending medium, V_H is the hydrodynamic volume of the particle (including coatings), and Γ is used to represent the ratio of anisotropic to thermal energies. These two processes occur simultaneously, governing behavior much like two resistors in parallel. The shorter time constant will thus have a tendency to dominate, and the effective relaxation time (τ) can be found by:

$$\tau = \frac{\tau_B \tau_N}{\tau_B + \tau_N}. \tag{17.5}$$

Magnetic work is traditionally expressed as the product of the field strength and the change in magnetic induction. The work performed by the external field is going to result in a change in internal energy (U). Taking the fundamental relationship between induction (B), magnetization (M), and applied field:

$$B = \mu_0(H_a + M) \tag{17.6}$$

and integrating by parts, the cyclic increase in internal energy can be found by:

$$U = -\mu_0 \oint M \, dH. \tag{17.7}$$

For a sinusoidal alternating magnetic field, the time-dependent field and magnetization can be expressed in terms of the field strength and frequency:

$$H(t) = H_a \cos(2\pi f t) \tag{17.8}$$

$$M(t) = H_a(\chi' \cos(2\pi f t) + \chi'' \sin(2\pi f t)) \tag{17.9}$$

where χ' and χ'' are the in-phase and out-of-phase components of the ferrofluid magnetic susceptibility, respectively. Substitution and integration of Equation 17.7 yields the cyclic change in internal energy, which can then be multiplied by the frequency to give the volumetric power generation:

$$P = \chi''(\mu_0 \pi f H_0^2). \tag{17.10}$$

Thus the rate of heating is dependent on the out-of-phase component of susceptibility (lagging the applied field) and the incident power density, which is the term in parentheses. This expression is equivalent to SAR in watts per cubic meter of fluid (or tissue). This can be easily converted into more standard units of cubic centimeters or grams tissue. In addition, absorption for magnetic nanoparticles is often expressed in terms of watts per mass iron, which can also be obtained through simple conversions. This value is often termed SAR_{Fe} or specific loss power (SLP). Both SAR and SLP will be used throughout the remainder of the chapter, and it is important to keep the distinction straight.

The ferrofluid susceptibility is dependent on both nanoparticle and field properties, so it is helpful to express this term through more fundamental parameters. Frequency dependence can be given by:

$$\chi'' = \frac{2\pi f \tau}{1 + (2\pi f \tau)^2} \chi_0 \tag{17.11}$$

where χ_0 is the equilibrium susceptibility, which can be conservatively estimated by the chord susceptibility, following the Langevin equation (Rosensweig 2002):

$$\chi_0 = \chi_i \frac{3}{\xi}\left[\coth(\xi) - \frac{1}{\xi}\right], \xi = \frac{\mu_0 M_s H_a V_m}{k_B T} \tag{17.12}$$

where χ_i is the initial susceptibility and ξ is the Langevin parameter. The initial susceptibility is determined by differentiating the Langevin relationship:

$$\chi_i = \frac{\mu_0 \phi M_s^2 V_m}{3 k_B T}. \tag{17.13}$$

Equations 17.10 through 17.13 then provide the capability to predict SAR based on nanoparticle and field parameters, which is often expressed in the simplified form, which follows. In this form, SAR depends on nanoparticle concentration through the volume fraction in Equation 17.12. However, if it is converted to watts per gram magnetic material, the SLP will be constant for a given frequency and field strength (i.e., no concentration dependence).

$$SAR = \mu_0 \pi \chi_0 f H_0^2 \frac{2\pi f \tau}{1 + (2\pi f \tau)^2} \sim \left[\frac{W}{m^3}\right]. \tag{17.14}$$

Relaxation behavior depends strongly on nanoparticle size, and heating demonstrates a peak efficiency at a specific radius, depending on the magnetic material. Bulk values for saturation magnetization (M_s) and anisotropy (K_u) for some relevant materials are

TABLE 17.1   Properties for Some Potential Magnetic Nanoparticle Materials

Material	M_S (kA/m)	K_u (kJ/m³)	c_p (J/kg-K)	ρ (kg/m³)
Magnetite[a]	446	23	670	5180
Maghemite[a]	414	4.7	746	4600
Iron[b]	1707	48	450	7870
Iron-Cobalt[c]	1815	15	172	8031
FCC Iron-Platinum[d]	1140	206	327	15,200

[a]Values taken from Rosensweig (2002)
[b]Values taken from O'Handley (2000)
[c]Values taken from Kline et al. (2009)
[d]Values taken from Maenosono and Saita (2006)

included in Table 17.1. Bulk heat capacity (c_p) and density (ρ) have also been included, as these are important for subsequent heating calculations. Characteristic size-dependent heating curves for iron oxide (magnetite and maghemite) are also illustrated in Figure 17.4. Location of the peak is largely dependent on the material anisotropy, but will also vary slightly based on the frequency, viscosity, and temperature. Amplitude of the peak depends strongly on the material magnetization, and so larger moment materials will produce higher rates of heating (all other factors equal). Puri et al. completed a theoretical analysis, comparing the heating capabilities of some potential magnetic materials in a spherical, perfused tissue system (Kappiyoor et al. 2010). The results suggested that barium-ferrite and cobalt-ferrite would not be able to produce sufficient heating at physiologically relevant concentrations and field parameters, but magnetite, maghemite, and FCC iron-platinum demonstrated adequate SARs for treatment. The authors also suggested the rate of heating for iron-cobalt was too high for safe application. However, treatment could be applied at lower nanoparticle concentrations or lower fields, and it is more likely that the potential cytoxic effects of cobalt will be a larger hurdle to safe application in vivo.

In addition, the strong size dependence indicates that polydispersity will be an important consideration. Most magnetic nanoparticle populations demonstrate significant polydispersity

and often follow log-normal distributions with (r_0, σ). The effective polydisperse SAR can be solved for by integrating across the probability distribution function, as shown in Equation 17.15. The general effects of polydispersity are illustrated for magnetite in Figure 17.4. It is clear that polydispersity significantly flattens the peak, but this broadening will also reduce the sensitivity to small shifts in the mean size.

$$SAR = \int_0^\infty SAR(r)\,g(r)dr, \quad g(R) = \frac{1}{\sigma\,R\,\sqrt{2\pi}}\exp\left[-\frac{\ln\left(\dfrac{R}{r_0}\right)^2}{2\,\sigma^2}\right].$$

(17.15)

The magnetic field parameters are another important factor in determining heating effects. Referring back to Equation 17.10, volumetric power generation depends directly on the applied frequency and square of the applied field strength. However, field effects on susceptibility also need to be taken into account. Figure 17.5 illustrates the effects of applied frequency and field strength on susceptibility and SAR for magnetite. Susceptibility (χ'') reaches a slight peak for a frequency $f = 250$ kHz and demonstrates a linear decrease for increasing field strength. The power generation is roughly linearly dependent on both frequency and field strength, within the usable field parameters. Frequency effects will approach a plateau for frequencies on the order of several MHz, but again, other unintended heating effects will dominate in these ranges, limiting clinical use for MFH, so this is generally of no consequence.

Brownian heating will also increase significantly in cases of very low fluid viscosity. However, this is not generally relevant in biological cases, where the particle is in aqueous suspension, but could have an impact for loosely attached coatings or particles in a multifunctional polymer matrix. In contrast, Brownian heating can be effectively eliminated if particles are bound to biological structures or form aggregates with large hydrodynamic diameters. Temperature also appears in a number of the equations and will have minor effects on both magnetic properties

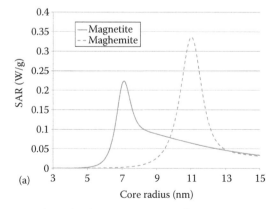

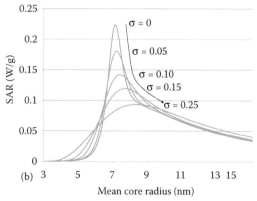

FIGURE 17.4   Size-dependent heating comparison for magnetite/maghemite (a) and polydisperse magnetite (b) in water. Field at 10 kA/m and 250 kHz.

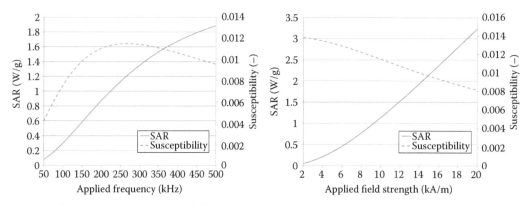

FIGURE 17.5 Effects of frequency (a) and field strength (b) on magnetic fluid susceptibility and SAR. Field fixed at 10 kA/m (a) or 250 kHz (b).

and relaxation behavior, but the impact is not significant within the temperature ranges relevant to biological heating.

The previous description assumes uniformly dispersed, non-interacting particles. Interparticle interactions will considerably increase the complexity of the problem, but can be represented as a system of interacting dipoles. This problem has not been thoroughly addressed in literature, but is very relevant, as particles often will form tight aggregates in vivo, affecting their collective magnetic behavior. Contradicting effects have been shown in literature, with aggregation leading to both reported increases (Dennis et al. 2009) and decreases (Jordan et al. 1996) in heating effects for different nanoparticle systems. Thus interparticle interactions may provide another potential dimension of engineering analysis for optimizing heating.

17.2.4 Alternating Magnetic Field Generation

There are many different means of creating magnetic fields, but a great majority of developmental work is performed with fields created by inductive coils, due to the ease of application and high field uniformity within the coil. Basic characterization of heating for ferrofluid samples is often performed in experimental setups similar to that illustrated in Figure 17.6. The strength of the uniform field within the coil can be theoretically estimated by (O'Handley 2000):

$$H_a = \frac{NI}{L} \tag{17.16}$$

where N is the number of turns in the coil, I is the coil current, and L is the coil height, but this will typically overestimate the

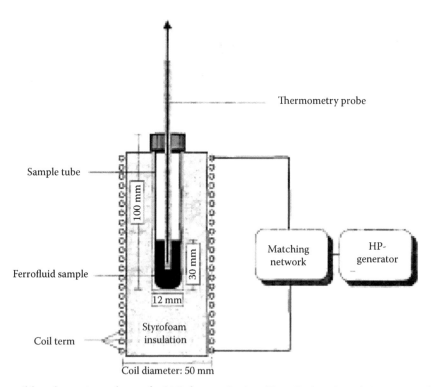

FIGURE 17.6 Inductive coil-based experimental setup for SAR characterization. (From Jordan, A. et al., *International Journal of Hyperthermia* 25, 7, 1993.)

actual field strength and so numerical techniques or direct measurement should be used to characterize the field distribution in a coil. The alternating field is applied and the temperature of the sample can be monitored through a temperature probe. The SAR can then be estimated through the rate of temperature rise method (Chou 1990). Another type of field generator more relevant to the clinical setting will be discussed in Section 17.5.1.4.

17.3 Synthesis and Modification of Iron Oxide Nanoparticles

Elucidating the theory behind magnetic nanoparticle heating highlights the factors that are important for engineering optimization of nanoparticle systems, but it is the methods of synthesis and resulting nanoparticle constructions that ultimately determine the physicochemical and physiological behavior. Bare particles do not generally demonstrate stability or physiological compatibility without surface modification. Core-shell structures are critical for providing viable in vivo application. Considerations regarding different core-shell structures and methods of synthesis will be discussed in the following section. More comprehensive reviews of magnetic nanoparticle synthesis and surface modification have been provided by Gupta et al. (Gupta and Gupta 2005), Lu et al. (Lu, Salabas, and Schüth 2007), and Krishnan (Krishnan 2010).

17.3.1 Synthesis and Core-Shell Structures

Synthesis of magnetic nanoparticles for in vivo biomedical applications requires well-controlled processes that can reliably provide particles with well-defined size distributions, consistent magnetic properties, good structural and chemical stability under physiological conditions, and high biocompatibility. Although the particles' magnetic behavior is largely determined by the metallic or metallic-oxide core, this core must be functionalized with a coating (and/or shell) that determines subsequent interactions in solution and in biological systems. Chemical synthesis methods have been the dominant

route for producing magnetic nanoparticle structures, and iron oxide has been the magnetic material of choice for in vivo applications, due to its well-documented biocompatibility and metabolic pathways (Weissleder et al. 1989).

Coprecipitation and thermal decomposition are the preferred methods for synthesizing iron oxide nanoparticles, owing largely to good control over nanoparticle size, a large literature base supporting process development, and high economic viability (Lu, Salabas, and Schüth 2007; Krishnan 2010). Coprecipitation involves an aqueous solution reaction between an Fe^{2+}/Fe^{3+} salt and a base under an inert atmosphere. Reactions can take place at room or elevated temperatures. Coprecipitation methods can produce large quantities of nanoparticles with highly reproducible quality once the kinetic synthesis parameters have been set, including ionic ratios, reaction temperatures, and solution pH. Thermal decomposition involves decomposition of organometallic compounds in high temperature organic solvents, containing stabilizing surfactants. Common surfactants include oleic acid, fatty acids, and hexadecylamine. Resulting nanoparticle properties, size, and polydispersity are largely determined by the ratios of reactants and surfactants, reaction temperature, reaction time, and aging period.

Although coprecipitation and thermal decomposition methods can be used to synthesize highly reproducible nanoparticle populations, these populations often demonstrate notable polydispersity, and as discussed in previous sections, this can have a significant impact on relaxation behavior and heating. Some common methods exist for reducing polydispersity based purely on size and density, but a preferred method for magnetic fluids is magnetic fractionation (Jordan et al. 2003). In magnetic fractionation, the aqueous nanoparticle solution is poured through a column under a high, static magnetic flux and washed with deionized water until the washout is clear. The magnetic flux can then be decreased in a stepwise manner down to zero, performing a similar washout at each increment, ideally producing nanoparticle fractions with increasing magnetism. This technique was demonstrated on dextran coated, superparamagnetic iron oxide in a field decreasing from 1100 mT, with results included in Figure 17.7. Two fractions were taken at the highest field. A clear increase in the specific saturation

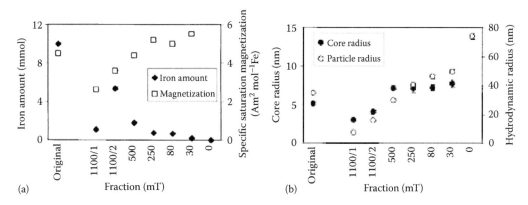

FIGURE 17.7 Iron concentration (a) and radius distributions (b) created by magnetic fractionation technique. (From Jordan, A. et al., *Journal of Nanoparticle Research* 5, 5, 2003.)

magnetization is demonstrated, with a correlated increase in particle size. SAR was also characterized in the study and followed the expected trends based on the measured core radii.

Equally important to performance is the particles' coating. The coating not only stabilizes the core structurally and in solution (preventing aggregation and settling), but also determines the biological interactions and pathways. Biological applications generally require water soluble particles, but the surfactants used in most chemical methods are hydrophobic and can be cytotoxic (Krishnan 2010). Therefore, aqueous stability and biocompatibility requires additional modification, most commonly a polymer surface coating consisting of dextran or polyethylene glycol (PEG) (Jordan et al. 1999). However, bioinert materials, such as silica and gold, are also under serious investigation as coatings (Krishnan 2010). Both shell materials provide excellent aqueous stability and facilitate surface modification. Gold, in particular, has been extensively characterized for biomolecular surface modification. However, one critical, and as of yet poorly understood, consideration in magnetic nanoparticle coating is the effects on magnetic behavior. Different surface coatings have been shown to lead to either decreases (magnetic "dead layer") or increases in the magnetic moment and anisotropy of core structures, with no clear, general correlations determined to date (Lu, Salabas, and Schüth 2007). This becomes an even more significant consideration in vivo, as biological interactions can modify coatings, which can then subsequently affect magnetization. Most significantly, nanoparticles can be internalized into lysosomes (Chou, Ming, and Chan 2010) and subjected to "cellular digestion" through intravesicle pH down to 4. Coatings that are not able to withstand these harsh conditions are broken down, leading to particle aggregation and other changes.

Beyond purely chemically driven modifications, the particle surface can also be functionalized with biomolecular targeting agents. These ligands can be generally classified into proteins (antibodies and fragments), nucleic acids (aptamers, etc.), and other ligands (peptides, vitamins, carbohydrates), with complementary receptors that are overexpressed in certain forms of cancer (Chou, Ming, and Chan 2010). These ligands can mediate cell-specific delivery and uptake. Aminosilane coated superparamagnetic magnetite particles with HIV-1 tat targeting

peptides have been successfully synthesized and demonstrated improved uptake in vivo (Stelter et al. 2009).

Combinations of various modes of synthesis and surface modification also provide the capability for multifunctional nanoparticle platforms. The ability to synthesize organic interlayer stabilized magnetite-gold (Smolensky et al. 2011), silica-magnetite-gold (Lu, Salabas, and Schüth 2007), iron-cobalt-gold (Kline et al. 2009), and iron-iron-oxide (Zeng et al. 2007) nanoparticle core-shell structures offers the potential capability for multimodal platforms for diagnosis, imaging, and treatment. Plasma-reactor-based synthesis methods have also demonstrated the feasibility of producing such core-shell structures in one continuous, in-line process (Kline et al. 2009; Zhang et al. 2008), which may offer benefits over the serial reactions required in many wet chemistry methods. In addition, iron oxide nanoparticles have been encapsulated in biodegradable, thermoresponsive polymer shells with the capability for drug-loading and stimulated release (Zhang, Srivastava, and Misra 2007). This provides a highly targeted mode for delivery of potential combinatorial therapies.

17.3.2 Characterization

The importance of the nanoparticle physical and magnetic properties has been highlighted, and so adequate characterization of these properties is another key to understanding performance. Nanoparticle size is generally characterized through standard techniques, including transmission electron microscopy (TEM), X-ray diffraction (XRD), and dynamic light scattering (DLS). Standard magnetic measurements techniques have also proved capable, with vibrating sample magnetometers (VSMs) or superconducting quantum interference devices (SQUIDs) providing key magnetic performance data. Additionally, as discussed briefly in Section 17.2.4, SAR can be measured readily in small samples subjected to an alternating magnetic field through the rate of temperature rise method (Chou 1990). This method is applied frequently throughout the literature, and despite a very wide range of reported SLPs, results are often in reasonable agreement with that predicted by theory (Zhang, Gu, and Wang 2007; Qin, Etheridge, and Bischof 2011; Etheridge et al. 2012b). Measured SLP for a number of in vitro studies utilizing superparamagnetic nanoparticles is included

TABLE 17.2 Specific Loss Power (in Watts per Gram Ferrite) for a Number of In Vitro Heating Characterization Studies

Group	Core Material	Size (nm)	Coating	Medium	H (kA/m)	f (kHz)	SLP (W/g)
Jordan et al. 1993	MnZnFeO	7.6	Dextran	Water	0.5	200–1000	0.05–0.5
	Iron Oxide	3.1	Dextran	Water	0.5	200–1000	0.15–0.8
			Dextran	Dextran	0.2–13.2	520	10–235
Hergt et al. 1998	Magnetite	10	N/A	Kerosene	6.5	300	45
		10	N/A	Ether			29
		8	N/A	Water			21
		6	Dextran	Water			<0.1
Hilger et al. 2002	Magnetite	8	400	Water	6.5	400	84
		3–10					56
		3–10					31
		3–10					54

in Table 17.2, but there are many other studies available, for a wide range of magnetic nanoparticle systems.

17.4 Biological Effects

A nanoparticle's surface coating is a major determinant for biological interactions and pathways. Upon administration into the body, the surface chemistry, in combination with size and geometry, determines which biological proteins adsorb to the particle surface, which in turn largely determines subsequent biological processing (Aggarwal et al. 2009). On the cellular level, the nanoparticle coating also determines the mechanism of cellular uptake. Nanoparticles are generally internalized through direct interaction with membrane-embedded receptors or indirectly through association with the membrane lipid bilayer (Chou, Ming, and Chan 2010). Both processes result in some form of endocytosis, in which the nanoparticles are internalized into a membrane-bound vesicle. Specialized cells, including macrophages, monocytes, and neutrophils, can also internalize particles through phagocytosis.

17.4.1 Effects of Nanoparticle Surface Coating

Nanoparticle coating and cell type have been shown to have a major impact on uptake of iron oxide nanoparticles in vitro (Jordan et al. 1999). Jordan et al. demonstrated differential endocytosis of dextran- and silane-coated magnetite nanoparticles in vitro with normal human fibroblasts, colonic adenocarcinoma (WiDr), malignant human glioma (RuSi-RS1), and normal human cerebral cortical neuronal (HCN-2) cells. Cells were grown in medium containing one of two types of magnetite nanoparticles, both at a concentration of 0.6 mg/ml. The first (#P6) consisted of 3.3 nm diameter cores with a dextran coating for a total hydrodynamic diameter of 50 to 70 nm with a negative surface charge. The second (#BU48) featured 13.1 nm diameter cores coated with aminosilane, for a total hydrodynamic diameter of 17 nm with a positive surface charge. Biocompatibility had been previously demonstrated for both particles. Uptake concentrations and cellular distribution were determined after 0, 6, 24, 48, 72, 144, 168, and 192 hours. Intracellular iron concentration was characterized by magnetophoresis and a colorimetric iron assay. Intracellular uptake was characterized by TEM, and surface attachment was characterized by SEM. Time-dependent uptake for two of the cell lines is summarized in Figure 17.8. The wide variety of uptake trends is apparent. The fibroblast cells demonstrate the most significant uptake. However, the uptake profiles vary significantly between the two nanoparticles. The #P6 nanoparticles exhibit high initial uptake followed by more gradual uptake through 192 hours. The #BU48 nanoparticles show low initial uptake, with a sharp peak in intracellular concentration around 168 hours, followed by a steep decline (which was attributed to exocytosis). All the cell lines demonstrated measurable iron uptake.

TEM indicated intracellular nanoparticles were contained in phagosomes or lysosomes. The #P6 particles often occurred as aggregates, which was due to the loss of their dextran coating in

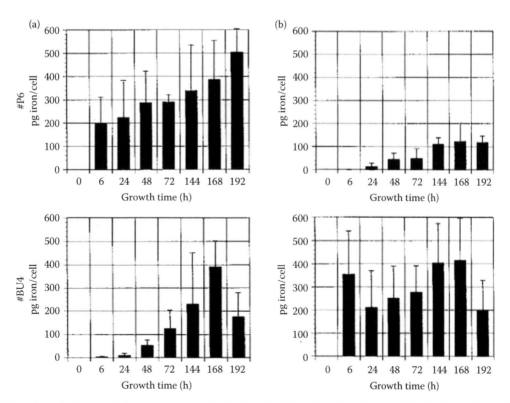

FIGURE 17.8 Time-dependent intracellular iron concentration for fibroblast (a) and malignant glioma (b) cells. (From Jordan, A. et al., *Journal of Magnetism and Magnetic Materials* 194, 1, 1999.)

the low pH of the lysosomes. However, analysis of the diffraction patterns suggested that the exposed cores retained their magnetite structure. Nearly all the #BU48 particles appeared to remain clearly separated. SEM showed wide variance in the surface attachment behavior between the cell lines and nanoparticle types, which is likely a large determinant in the observed differences in uptake behavior.

Similar results were obtained by Kalambur et al. in human prostate tumor cells (LNCaP) (Kalambur, Longmire, and Bischof 2007). Two types of nanoparticles were studied, both with 10 nm magnetite cores. One had an anionic surfactant coating and the other a neutral dextran coating. The cells were incubated in medium containing concentrations of nanoparticles at 0.05, 0.1, 0.5, and 1 mg Fe per ml, and uptake was measured by magnetophoresis and colorimetric iron assays at 1, 6, 24, 48, and 72 hours. The resulting uptake kinetics are included in Figure 17.9. The surfactant-coated particles demonstrated much higher uptake than the dextran-coated particles, with apparent saturation behavior for both concentration and time. It was suggested that this difference in kinetics was due to differences in the uptake mechanisms. The saturation behavior would suggest an adsorptive endocytotic pathway for the surfactant-coated particles versus a fluid-phase pinocytotic pathway for the dextran-coated particles, suggested by the linear increase with external concentration. The adsorptive process could potentially be further enhanced by specifically targeting the nanoparticles to receptors on the cell membranes or perhaps through cationic surfactants.

Natarajan et al. characterized the in vivo uptake of PEG-coated iron-oxide nanoparticles, tagged with a breast cell targeting monoclonal antibody, ChL6 (Natarajan et al. 2008). Biodistributions for targeted and untargeted nanoparticles with mean core diameters of 20, 30, and 100 nm were characterized in mice bearing human breast cancer HBT 3477. The nanoparticles were intravenously injected, and blood and tissue data were collected at 4, 24, and 48 hours. The tumor uptake for the targeted particles after two days was between 4% and 9% of the injected dose, which was substantially higher than uptake for the untargeted particles, at less than 0.5% of the injected dose. This suggests significant interaction between the antibodies and cancer cell receptors, which is a promising result for prospective, clinical modes of systemic delivery.

17.4.2 Thermal Dose In Vitro

A large number of in vitro studies have demonstrated the ability of magnetic nanoparticles to heat cells to a cytotoxic level in the presence of an alternating magnetic field. Jordan et al. expanded the uptake studies described in Section 17.4.1 to include heating in a magnetic field at 520 kHz and 4 to 12.5 kA/m, for times between 5 and 120 minutes (Jordan et al. 1999). Survival rates were compared with those of a treatment in a constant temperature water bath at 43°C and 45°C. One of the most prominent results was the inability of the dextran-coated particles to heat after intracellular uptake. It was expected that degradation of the dextran coating in the lysosomes led to tight particle aggregates, producing high interparticle interaction and eliminating the superparamagnetic heating behavior. However, extracellular #P6 and intra- and extracellular #BU48 produced significant heating in the alternating magnetic field, demonstrating cellular deactivation at levels (at least) equivalent to the water bath hyperthermia. In addition, the iron oxide treatments also appeared to result in some sort of sensitization effect over the first 60 minutes, in which decreased survival occurred. It was speculated that this might be a result of membrane disruption or organelle-specific damage, due to localized heating of the nanoparticles. However, the ability of magnetic nanoparticles to produce heating rates high enough to create localized, intracellular temperature increases has been questioned from an analytical perspective (Rabin 2002; Keblinski et al. 2006), and thermal effects are more likely confined to macroscopic, bulk heating (Qin, Etheridge, and Bischof 2011; Etheridge and Bischof 2012a).

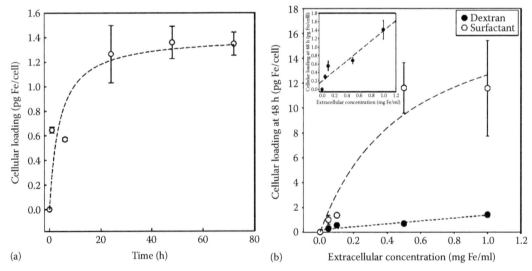

FIGURE 17.9 (a) Time and (b) concentration dependent uptake kinetics for surfactant and dextran coated nanoparticles in LNCaP. (From Kalambur, V. S, E. K. Longmire, and J. C. Bischof, *Langmuir* 23, 24, 2007.)

TABLE 17.3 Prominent Preclinical Magnetic Nanoparticle Heating Studies

Group	Year	Particle Structure	Magnetic Material	Magnetic Core Dimension (nm)	Animal Model	Tumor Model	Successful Indications
Gilchrist et al.	1957	Nanoparticle	Maghemite	20–100	Dog	Lymph Node	Δ14°C after 3 minutes of heating in lymph node.
Gordon et al.	1979	Dextran-Coated Nanoparticle	Magnetite	< 6	Rat	Breast Carcinoma	Systemic injection with Δ8°C and tumor necrosis.
Rand et al.	1981	Neede-Shaped Microparticle	Ferromagnetic	100–1000	Rabbits	Renal Carcinoma	Tumor surface temp of 50°C with complete cancer cell necrosis.
Jordan et al.	1997, 2005	Coated Nanoparticle	Magnetite	3, 15	Mouse Rat	Breast Carcinoma, Malignant Glioma	Δ12°C resulted in 4.5 increase in survival over control.
Hilger et al.	2002	Coated Nanoparticle	Magnetite	10, 200	Mouse	Breast Carcinoma	Δ12°C and Δ73°C with tumor coagulation and necrosis.
Johannsen et al.	2005, 2006	Aminosilane-Coated Nanoparticle	Magnetite	15	Rat	Prostate Carcinoma	Combination, low-dose radiotherapy with 88% tumor regression.
Ohno et al.	2002	Stick-Type Carboxymethyl-cellulose	Magnetite	10	Rat	Glioma	Δ5°C with significant 30-day increase in survival over control.
Yanase et al.	1998, 1999	Magnetic Cationic Liposome (MCL)	Magnetite	10	Rat	Glioma	Δ6–8°C with tumor regression in 90% of treated rats.
Le et al.	2001	Magnetic Cationic Liposome (MCL)	Magnetite	10	Mouse	Glioma	Anitbody targeting resulted in 60% retention of interstitial injection.
Ito et al.	2003	Magnetic Cationic Liposome (MCL)	Magnetite	10	Mouse	Melanoma	Δ6-8°C with combined IL-2 or GM-CSF results in complete regression in 40 to 75% of mice.
Tanaka et al.	2005	Magnetic Cationic Liposome (MCL)	Magnetite	10	Mouse, Rat	Melanoma	Combined dendritic cell immunotherapy.
Kawai et al.	2005	Magnetic Cationic Liposome (MCL)	Magnetite	10	Rat	Prostate Carcinoma	Δ7°C with demonstrated tumor regression.
Motoyama et al.	2008	Magnetic Cationic Liposome (MCL)	Magnetite	10	Rat	Breast Carcinoma	Δ7°C with tumor regression in 75% of the treated rats.
Natarajan et al.	2008	PEG-Coated Nanoparticle	Iron Oxide	20, 30, 100	Mouse	Breast Carcinoma	Antibody targeting achieved 4–9% deposition of IV injected dose.
Dennis et al.	2009	Dextran-Coated Nanoparticle	Magnetite	44	Mouse	Breast Carcinoma	Up to 52°C with nearly complete tumor regression at highest dose.

A more detailed discussion of scaling between nano-, micro-, and macroscale heating effects is included in Chapter 19, this book. An alternative explanation suggests that ferric ions produced by the nanoparticles result in additional oxidative stress. In a separate study, it was found that iron concentrations of 1 mM produce no cytotoxic effects at 37°C, but became toxic at temperatures of 43°C (Freeman, Spitz, and Meredith 1990).

17.4.3 Thermal Dose In Vivo

Encouraged by the promising results demonstrated in vitro a number of groups have also pursued in vivo animal work, with successful results spanning over five decades. A summary of some of the most prominent studies is included in Table 17.3. In addition to the successful indications, the exclusive use of iron oxide (mainly magnetite) as the magnetic core material should be noted. This again is due to the demonstrated ability to produce heating and well-characterized biocompatibility. Additional detail on several studies will follow.

Jordan et al. used intratumoral injection to administer superparamagnetic, dextran coated magnetite particles in intramuscularly implanted mammary carcinoma of the mouse (Jordan et al. 1997). Upon ferrofluid injection, a bandage securing six magnets was fixed on the tumor site. This static field gradient was intended

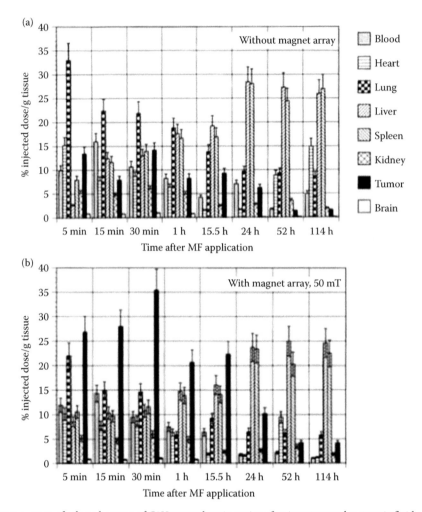

FIGURE 17.10 Tissue iron content of selected organs of C3H tumor bearing mice after intratumoral magnetic fluid administration without (a) and with (b) an external magnet array. (From Jordan, A. et al., *International Journal of Hyperthermia* 13, 6, 1997.)

to increase retention in the tumor. The resulting time-dependent biodistribution is included in Figure 17.10. Retention in the tumor was increased significantly through the use of the magnetic array, as well as observed decreases in iron deposition for bystander organs. Intratumoral steadystate temperatures of 47 ± 1°C were maintained for 30 minutes with a whole body field at 520 kHz and 6 to 12.5 kA/m. No significant heating was observed in other tissues. The histological results were fairly heterogeneous and likely reflected inhomogeneity of nanoparticle deposition. Some of the tumors showed no evidence of regrowth after 50 days, while others grew readily after treatment.

Magnetic fluid hyperthermia was also used for treatment in rat tumor models with glioblastoma multiforme and prostate carcinoma (Jordan et al. 2006; Johannsen et al. 2005). Intratumoral injection of aminosilane-coated particles produced stable nanoparticle deposits, capable of multiple treatments under alternating magnetic fields variable from 0 to 18 kA/m at 100 kHz. Intratumoral temperatures in the glioblastoma model were measured at 43°C to 47°C (held for 30 minutes), and resulted in an increased survival rate of 1.7- to 4.5-fold over the control.

Maximum intratumoral temperatures of up to 70°C were measured in the prostate carcinoma model, with mean maximum and mean minimum temperatures of 54.8°C and 41.2°C, respectively. Treatment resulted in a 44% to 51% inhibition of tumor growth over the control. Post-mortem histological analysis showed mean iron biodistribution at 82.5% in the tumor, 5.3% in the liver, 1.0% in the lung, and 0.5% in the spleen.

An additional in vivo study demonstrated the potential for magnetic fluid hyperthermia as a combinatorial therapy, with adjunct radiotherapy treatment in a rat tumor model with prostate carcinoma (Johannsen et al. 2006). Aminosilane-coated nanoparticles were injected intratumorally, with two subsequent thermal treatments or two radiation doses ranging from 2 × 10 Gy to 2 × 30 Gy. Thermal therapy was also combined with the lowest radiation dosage. Mean maximum and mean minimum intratumoral temperatures were measured at 57.1°C and 42.5°C, respectively. The combined low-dose radiation thermal treatment matched the effectiveness of the high-dose radiation therapies, with a reduction in tumor growth of 87.5% to 89.2% over controls.

17.5 Clinical Application in Cancer Therapy

17.5.1 Components of a Clinical MFH Thermotherapy System

Following is a description of the components necessary for successful clinical application of magnetic nanoparticle-based thermal therapy treatment. Reference is made to the MagForce NanoTherapy (MagForce Nanotechnology AG, Berlin, Germany) system, as this is the only clinical magnetic fluid hyperthermia system in use at this time. However, equivalent or similar system components will be required for any clinical MFH-based treatment. A high-level treatment flow chart has been included in Figure 17.11, for illustrative purposes. Imaging is not explicitly described as a discrete system component in the subsequent discussion, but is an integral tool utilized throughout the process.

17.5.1.1 Magnetic Nanoparticles

Heating dependence on magnetic nanoparticle properties and potential core-shell structures have already been discussed. However, keys to an effective clinical application include high SAR, high biocompatibility (acute and long-term), and stability under physiological conditions. NanoTherm® (MagForce Nanotechnologies AG, Berlin, Germany) is an aqueous dispersion of 15 nm iron oxide cores coated with aminosilane, for a total hydrodynamic diameter of 100 nm (Dudeck et al. 2006). The solution has an iron concentration of 112 mg/ml and is directly injected as a number of 0.5 ml deposits.

17.5.1.2 Pretreatment Planning and Nanoparticle Imaging

Understanding the physical mechanisms behind magnetic nanoparticle heating and characterization of their performance in vitro and in preclinical study has provided an extensive knowledge base for predicting performance in vivo, but convenient tools are still required to effectively translate this understanding to the clinical setting. Robust and simple methods are necessary to help clinicians determine the required nanoparticle loading and expected treatment efficacy, without involved calculation.

The Pennes bioheat equation is a well-accepted method for solving biological heat transfer problems (Pennes 1948):

$$\rho c_{cp} \frac{dT}{dt} = k \nabla T + (\rho c_p)_{blood} \omega_b (T - T_a) + q_{SAR} \qquad (17.17)$$

where ρ and c_p are the density and specific heat of the tissue and blood, T is temperature, T_a is the arterial blood temperature, t is time, k is the tissue thermal conductivity, ω_b is blood perfusion, and q_{SAR} has been added for nanoparticle heat generation (i.e., SAR from Equations 17.14 and 17.15). Contributions of metabolic heat generation will be insignificant compared to nanoparticle SAR and has been neglected. This equation is often solved for the steady-state treatment temperature, in which case the left

Pre-operative planning:

Imaging: MRI, CT, and US for 3D imaging of tumor geometry and location.

Planning software: Utilize 3D tumor images to plan implantation trajectories and deposits.

MF implantation:

Imaging: Minimally invasive implantation under CT, TRUS with fluoroscopy, or stereotactic guidance.

Pre-treatment planning:

Imaging: CT measurement of magnetic nanoparticle distribution.

Planning software: Estimate treatment temperature distribution based on nanoparticle distribution and estimated perfusion.

Thermometry: In vivo temperature measurement to estimate mean perfusion rate.

Treatment planning:

Field applicator: NanoActivator @ 100 kHz:
Pelvic: 3–5 kA/m
Thoracic/neck: <10 kA/m
Cranial: 12–15 kA/m
→ Target: 41–43°C for 60 min

Thermometry: Intratumoral temperature measurement to calculate thermal equivalent dose.

Adjunct therapy: Combined radio- and chemotherapy before, during, and/or after thermal treatment.

Follow up:

Imaging: PET or SPECT to monitor tumor progression. MRI cannot be used due to susceptibility artifacts caused by enduring nanoparticle deposits.

FIGURE 17.11 Components involved in clinical application of MFH.

side can be set to zero. Perfusion can be estimated from values in literature or various imaging techniques—MRI (Williams et al. 1992), CT (Eastwood et al. 2002), or US (Schrope and Newhouse 1993)). Calculation and measurement of SAR (see Sections 17.2.3 and 17.3.2) has already been discussed, so a solution for either nanoparticle density or treatment temperature can be found, given the other value. This can be accomplished analytically for simple geometries or with finite element methods in more realistic cases. This relation then provides a tool for calculating the required amounts of nanoparticles or expected temperature profiles.

Although magnetic nanoparticles are used as contrast agents to enhance MR imaging techniques, the high concentrations required for hyperthermia applications result in localized susceptibility artifacts, which eliminate the possibility for quantification. However, the ability to measure nanoparticle concentrations based on differences in density has been demonstrated with computed tomography (CT) imaging (Johannsen, Gneveckow, Taymoorian, Cho et al. 2007). A comparison between signal intensities from MRI and CT is included in Figure 17.12. CT signal intensity is expressed in units of Hounsfield Units (HU). Previous studies have demonstrated the capability to differentiate nanoparticle deposits down to a concentration of approximately

4 mg Fe/ml, correlating to a signal intensity change of about 20 HU per mg Fe/ml (Wust et al. 2006). Thus CT imaging can be used to measure nanoparticle distribution after injection and subsequently determine appropriate field parameters and treatment time. In addition, while traditional MR techniques are limited in quantifying high concentrations of iron oxide nanoparticles, a new approach, sweep imaging with Fourier transform (SWIFT) MR, has demonstrated preliminary capabilities to image at greater than 1 mg Fe/ml (Hoopes et al. 2012).

The NanoPlan module (MagForce Nanotechnologies AG, Berlin, Germany) puts these methods to commercial use. The software package allows physicians to calculate needle trajectories and deposit locations for nanoparticle implantation, based on 3D reconstructions obtained from previous imaging. Homogeneous

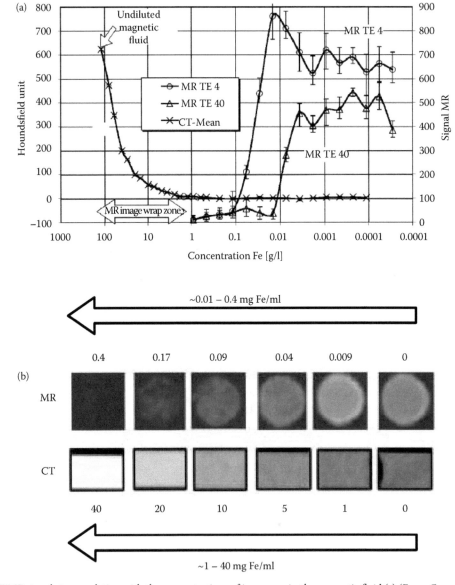

FIGURE 17.12 CT/MR signals in correlation with the concentrations of iron mass in the magnetic fluid (a) (From Gneveckow, U. et al., *Medical Physics* 31, 2004); and CT/MR contrast images for Petri dishes containing varying concentrations of iron mass (b) (From Kalambur, V. S, S. Hui, and J. C Bischof, *Proceedings of SPIE*, 6440, 2007.)

deposition is attempted by planning trajectories that provide interdeposit spacing between 8 and 10 mm throughout the tumor (Thiesen and Jordan 2009). In addition, the software can be used in conjunction with postimplantation CT nanoparticle concentration data, to estimate field-dependent temperature distributions in the tissue, using perfusion estimates based on minimally invasive thermometry. Preliminary experience with these methods and software has yielded reasonable agreement of intratumoral temperatures, but considerable deviations are found outside of the tumor.

17.5.1.3 Implantation Procedure

Effective thermal treatment of a tumor requires that all areas of the tumor are heated to a therapeutic level, and this is best achieved through a homogenous distribution of nanoparticles. This, in turn, requires an effective method for implantation. Some success has been demonstrated for image-guided interstitial injection, while local arterial infusion indicates potential promise for some future applications.

Interstitial injection is most effective under some form of image guidance. CT guidance has been used to control nanoparticle injection in the cranium (stereotactically administered), cervical area, and other soft tissue sites, while transrectal ultrasound (TRUS) with X-fluoroscopy guidance has been used in the prostate. While these techniques have generally met with success, some significant mechanical resistance has been encountered in tissues that received prior radiotherapy (Wust et al. 2006; Johannsen et al. 2010). This provided difficulty in following the planned trajectories and raises concern over the subsequent ability of the nanoparticles to diffuse through the tissue. Overall, however, the interstitial injection techniques have proved clinically viable. Computed images of planned and actual nanoparticle distributions are included in Figure 17.13, along with the resulting temperature maps during treatment. Intraoperative injection under direct visual and endoscopic guidance has also been attempted, with mixed results (Wust et al. 2006; Steinbach et al. [in draft]). Transarterial injection has demonstrated some promise in preferential deposition of nanoparticles for treatment of liver carcinoma (Dudeck et al. 2006), which would offer major advantages over minimally invasive methods. This procedure will be discussed in more detail in Section 17.5.2.5.

Nanoparticle retention is another critical factor for successful implantation. Significant diffusion from the injection site can result in unwanted heating of surrounding tissues, and stable deposits provide the capability for repeated treatments over time with a single injection. No significant nanoparticle deposits have been identified outside of the injection site throughout the clinical imaging completed to date, and stable deposits in the prostate have been detectable a year after implantation (Johannsen et al. 2010). This provides for safe, repeatable treatment.

17.5.1.4 Magnetic Field Applicator

A means of safely and effectively applying an external magnetic field is necessary to activate the implanted nanoparticles. Major considerations in design of such an applicator include uniformity of field, patient comfort, and capability for treatment throughout the body. Although multiturn inductive coils provide an adequate platform for small animal preclinical studies, a more robust system is required for clinical applications. Currently, the only applicator under clinical investigation is the NanoActivator (MagForce Nanotechnologies AG, Berlin, Germany), illustrated in Figure 17.14. Patients are placed horizontally on the bed and slid in the y-direction into the applicator. A ferrite yoke with pole shoes above and below the patient is coupled with a resonant circuit that creates an alternating magnetic field at 100 kHz. A roughly cylindrical field with 20 cm diameter is created between the two pole shoes. Field variability is shown in Figure 17.14. The magnetic field strength depends linearly on the coil current and is adjustable from 2 to 18 kA/m. An aperture can also control the gap between the pole shoes, adjustable from 210 to 320 mm, with some decrease in field strength for wider gaps. However, the field is relatively homogeneous with very little radial variance.

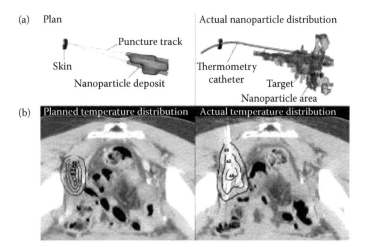

FIGURE 17.13 Comparison of planned and actual nanoparticle distributions for treatment of cervical cancer (a), with resulting temperature distributions (b). (From Wust, P. et al., *International Journal of Hyperthermia* 22, 8, 2006.)

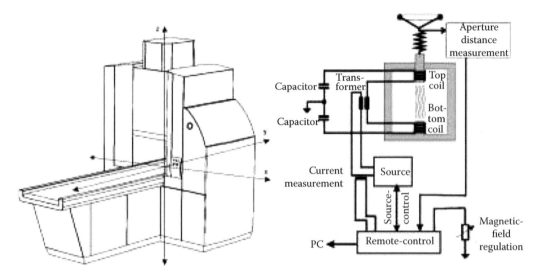

FIGURE 17.14   NanoActivator™ with internal schematic. (From Gneveckow, U. et al., *Medical physics* 31, 2004.)

Although the system has the capability to deliver fields up to 18 kA/m, practical field strengths have been limited in clinical application. Significant patient discomfort resulting from hot spots and subjective feelings of pain result in varying levels of tolerance for different regions of the body. Tolerated fields have typically been 3 to 5 kA/m for the pelvic region, 8.5 kA/m for the upper thoracic region, and >10 kA/m for the head (Wust et al. 2006). Hot spots occur at skin folds, where the induced current densities are highest, and bone interfaces, where it is expected a phenomenon is occurring similar to RF heating at bounda-rie, due to mismatches in electrical properties (Johannsen, Gneveckow, Taymoorian, Thiesen et al. 2007).

17.5.1.5 Real-Time Thermometry and Dosimetry

Real-time thermometry is critical for validation of predicted tem-perature distributions and determining the treatment effects in terms of the thermal equivalent dose. Unfortunately, compatible, noninvasive thermography techniques are not available for use with MFH. MR thermometry has demonstrated some promise for noninvasive thermal measurement (Poorter et al. 1995), but is ineffective in MFH due to the susceptibility artifacts created by the high concentration, magnetic nanoparticle deposits (Wust et al. 2006). However, some potential for noninvasive temperature measurement has been proposed with US (Amini, Ebbini, and Georgiou 2005) and CT (Fallone, Moran, and Podgorsak 1982) thermography and could be developed for compatible application with MFH. An additional technique uses the fifth and third har-monics response of magnetic nanoparticles in a sinusoidal field to estimate the bulk temperature, with an accuracy of 0.3°C demon-strated in vitro (Weaver, Rauwerdink, and Hansen 2009).

Currently, though, minimally invasive, probe-based ther-mometry is the only method of temperature measurement during MFH. Temperature mapping is accomplished through the insertion of single or multiple 1 mm catheters contain-ing fiber-optic temperature probes. Catheters are positioned pretreatment and spatial temperature distributions are taken by sliding the probe longitudinally along the catheter's axis (Maier-Hauff et al. 2011).

A common and well-accepted method for describing thermal dosimetry was proposed by Sapareto and Dewey, in which the measured temperature-time curve is normalized to an equivalent, cumulative time at 43°C (Sapareto and Dewey 1984). A represen-tative temperature T_{90}, can be taken as the temperature exceeded by 90% of the treated tumor volume, and can be used to calculate the cumulative equivalent minutes at 43°C (CEM43), based on the following:

$$CEM43 = t\ R^{43-T_{90}} \qquad (17.18)$$

where t is the time spent at the temperature T_{90} and R is either 0.25 for T < 43°C or 0.5 for T > 43°C. This relation can be used to calculate the thermal dose at a constant treatment temperature or integrated across a temperature curve.

17.5.2 Clinical Results

Seven clinical trials have been completed as of 2011, with two additional studies in progress. A summary of the trials is included in Table 17.4. Phase I trials are aimed to investigate feasibility, toxicity, and tolerability of MFH treatments. Demonstration of feasibility generally included homogeneous implantation of the magnetic fluid, the capability to maintain therapeutic tem-peratures in the treatment area, and validation of the calculated temperature distributions. Phase II study is intended to demon-strate efficacy and further evaluate safety. More specific results and outcomes will be discussed in the subsequent sections.

TABLE 17.4 Summary of MagForce NanoTherm Therapy Clinical Trials Completed as of 2011

Start Year	Phase	Indication	Patients	Adjunct Therapy	Implantation Guidance	Tolerable Field (kA/m)	Median T90 (°C)	Median CEM (min)
2003	Phase I	Recurrent Glioblastoma Multiforme	14	Radiotherapy	StealthStation®	3.8–13.5	40.5	7.7
2004	Phase I	Recurrent Prostate Carcinoma	10	—	TRUS, X-fluoroscopy	3–5	40.1	7.8
2004	Phase I	Recurrent Prostate Carcinoma	8	Brachytherapy	TRUS, X-fluoroscopy	3–5	39.9	5.8
2004	Phase I	Rectal and Cervical Carcinoma, Sarcoma Prostate Carcinoma Cervical Carcinoma	6 8 8	Chemoradiotherapy Brachytherapy Chemoradiotherapy	CT TRUS, X-fluoroscopy Intraoperative	Pelvic: 3–5 Thoracic: <7.5	NA	NA
2004	Phase I	Hepatocellular Carcinoma	13	Chemotherapy	Transarterial Infusion		NA	
2005	Phase II	Recurrent Glioblastoma Multiforme	66	Radiotherapy	StealthStation®	NA	NA	NA
2006	Phase I	Incurable Esophageal Carcinoma	2 9	— Chemoradiotherapy	Endoscopic	~3.6	NA	NA

NA = data not yet available

Sources: Dudeck, O., K. et al., *Investigative Radiology* 41, 6, 2006; Wust, P. U. et al., *International Journal of Hyperthermia* 22, 8, 2006; Johannsen, M. B. et al., *International Journal of Hyperthermia*, 0, 2006; Steinbach, M. et al., Nanotherapy induced hyperthermia in advanced esophageal cancer: Results of a Phase I Study, in draft; Maier-Hauff, K. et al., *Journal of Neuro-Oncology*, 2011; Maier-Hauff, K. et al., *Journal of Neuro-Oncology* 81, 1, 2007.

Although, the results have generally been promising, some side effects were encountered. In many cases, unwanted heating occurred at the skin or bone interfaces, generally resulting in discomfort, but occasionally causing superficial burns. Other side effects have included tachycardia, headaches, elevated blood pressure, focal convulsions (Maier-Hauff et al. 2011), and acute urinary retention (Johannsen, Gneveckow, Taymoorian, Thiesen et al. 2007). In addition, in current practice, the concentrations of iron oxide required to create therapeutic temperatures is much higher than those optimally predicted by theory, so there is significant opportunity for increasing the heating efficiency of the nanoparticles used.

17.5.2.1 Glioblastoma Multiforme

Phase I and phase II trials have been completed investigating MFH for treatment of glioblastoma multiforme in combination with fractionated radiotherapy (Thiesen and Jordan 2009; Maier-Hauff et al. 2007). The phase I trial included 14 patients with locally recurrent or nonresectable tumors. Patients received between 4 and 10 biweekly thermal treatments, depending on the total weeks of irradiation. Single radiotherapy fractions of 2 Gy were administered, for complete dosage between 16 to 70 Gy. Nanoparticle injection was preplanned and administered under stereoscopic guidance with StealthStation® (Medtronic, Minneapolis, Minnesota). During treatment, field strength was increased until the patient experienced subjective feelings of discomfort, the field was reduced, and the temperatures were maintained for 60 minutes. Field strengths from 3.8 to 13.5 kA/m were well tolerated. Invasive thermometry was used to monitor intratumoral temperature during treatment. Median

maximum intratumoral temperature was 44.6°C and T_{90} ranged from 39.3 to 45.5°C, with a median of 40.5°C. Median calculated CEM43 was 7.7 minutes. There was no measurable increase in skin temperature, but body temperature increased by 1.0 to 1.5°C on average.

Implantation and treatment were well tolerated, with no signs of systemic toxicity. Median patient survival was 14.5 months, which was promising compared to survival prognoses ranging from 2.7 to 11.5 months. With appropriate consent, histology was performed on sections of the treated tissue after death (van Landeghem et al. 2009). Multifocal deposits were found in the necrotic regions of the treated tissue. There was significant uptake by macrophages, and the aggregates were partially surrounded by rings of macrophages. The glioblastoma cells demonstrated uptake to a lesser extent, with only about 5% of cells containing nanoparticles. Hemorrhage was also found along the canals of instilled nanoparticles. One patient did not receive thermal treatment due to health complications after implantation, and the postmortem histology showed significantly decreased phagocytotic activity, compared to patients receiving hyperthermia treatment.

Phase II study has also been completed on 66 patients with glioblastoma multiforme (Maier-Hauff et al. 2011). The primary endpoint was survival following diagnosis of first tumor recurrence (OS-2), with a secondary endpoint of survival after primary diagnosis (OS-1). The same methods and procedures were utilized as in phase I study. Patients received six biweekly thermal treatments with fractionated radiotherapy occurring directly before or after, with median overall dose of 30 Gy. Median peak treatment temperature was 51.2°C, with an overall maximum of 82°C. Direct comparative analysis of endpoints is difficult, but results were promising with a median OS-2 of 13.4 months (as compared to 5.8 months for chemotherapy) and median OS-1 of 23.2 months (as compared to 14.6 months for the reference group). In addition, levels of key metabolites of iron were tested before and after injection, with no indication of iron release from the nanoparticle deposits.

17.5.2.2 Prostate Carcinoma

Two separate phase I trials have been completed investigating MFH treatment of locally recurrent prostate cancer as a monotherapy and with adjunctive permanent seed brachytherapy (Johannsen, Gneveckow, Taymoorian, Cho et al. 2007; Johannsen et al. 2010; Wust et al. 2006; Johannsen, Gneveckow, Taymoorian, Thiesen et al. 2007). Ten patients received six weekly thermotherapy sessions and for another eight patients, 125-Iodine seeds were also implanted at the time of nanoparticle injection. Injections were made transperineally with the aid of a template and under TRUS with X-fluoroscopy guidance. Deposit trajectories were preplanned, but significant mechanical resistance was encountered in the pre-irradiated tumors, and it was difficult to achieve a homogenous distribution. Treatments began at a field strength of 2.5 kA/m, which was gradually increased to the threshold before the patient experienced significant discomfort, and temperatures were held for 60 minutes. Hot spots generally occurred at skin folds of the scrotum or rectum and could be somewhat alleviated with cooling and by keeping the skin dry, but tolerated fields only reached 3 to 5 kA/m. Skin temperatures up to 44°C were measured. Minimally invasive temperature mapping was conducted during the first and last sessions with four catheter-based probes, and intraluminal temperatures were monitored in the urethra and rectum throughout all treatments. Temperature mapping indicated fairly heterogeneous temperature distributions throughout the tumor. Treatment of the prostate is often complicated due to high perfusion rates of the surrounding structures, but this was also likely exacerbated by heterogeneous nanoparticle distributions. Intratumoral temperatures agreed reasonably well with the predicted temperatures, but the measured urethral temperature was on average 1.1°C lower than the calculated values.

For the monotherapy group, a maximum temperature of 55°C was achieved, with a median T_{90} of 40.1°C, equating to a median CEM43 of 7.8 minutes (mean 20.9 minutes). The large variability in these values again demonstrates the heterogeneity of the temperatures achieved, which could be a significant obstacle for application as a viable treatment. Although it was not a specific end point, prostate-specific antigen (PSA) levels were also measured; 8 of the 10 patients demonstrated PSA decreases, ranging from <25% to 70%, with some apparent correlation to the homogeneity of the nanoparticle deposits and temperatures reached. In the combined therapy group, median T_{90} of 39.3°C and CEM43 of 5.8 minutes were achieved. These were adequate thermal doses for combination treatment, but fell short as a monotherapy (Johannsen et al. 2006).

17.5.2.3 Esophageal Carcinoma

Phase I investigation of MFH treatment in 11 patients with incurable esophageal carcinoma is in progress (Steinbach et al. [draft]). Two patients received thermal monotherapy and nine patients received combination thermal, chemoradiotherapy. Nanoparticle injection was performed endoscopically, under camera guidance. Preplanning of nanoparticle deposits was not possible, due to the flexible geometry of the esophagus. Significant mechanical resistance was encountered in pre-irradiated tissue, and some leakage occurred after injection, but was remedied by sealing with fibrin glue. Imaging of the nanoparticle deposits indicated that most of the injected load was focused in the center mass of the tumor. Insufficient loading occurred in two patients, and injection had to be repeated. Upon treatment, all patients complained of pain arising from hot spots (generally at bone interfaces) and tolerated an average field of 3.6 kA/m. Combination therapy involved simultaneous infusion of cisplatin, with adjunct radiotherapy at an average dose of 45 Gy. Treatment had to be stopped in four patients due to pneumonia and decreases in general health conditions. Perforation of the esophageal wall occurred during injection in one patient. Results of the study are still under analysis and not yet available.

17.5.2.4 Local Recurrences of Different Origins

Phase I study was conducted on 22 patients with various forms of locally recurrent cancer, aimed mainly at investigating feasibility for several modes of implantation and validity of calculated temperatures at various sites (Wust et al. 2006). Groups included: implantation under CT guidance (sarcoma, cervical and ovarian carcinoma, rectal cancer), implantation under TRUS with X-fluoroscopy guidance (prostate), and intraoperative implantation under visual guidance (treatment of residual tumors after cervical resection). Implantation was preplanned for the image-guided injection methods, but intraoperative injection relied on the surgeon's discretion and the magnetic fluid was administered liberally. However, no nanoparticle deposits were detected in two of the intraoperative implantation patients. It is suspected that the nanoparticle solution drained through tissue gaps and was suctioned during surgery. One to 15 thermal treatments were administered once to twice a week for 60 minutes. Tolerable field strengths varied between different regions of the body. Generally, field strengths of 3 kA/m to 6 kA/m were used in the pelvic regions and up to 7.5 kA/m in the thoracic and neck regions. All patients also received combination radio- and/or chemotherapy. Results of the study are still under analysis and are not yet available.

17.5.2.5 Liver Carcinoma

Phase I study was completed on 13 patients, intended to investigate the feasibility of transarterial injection of magnetic fluid for potential treatment of hepatocellular carcinoma (HCC) (Dudeck et al. 2006). The patients underwent transarterial chemostabilization (TACE) in combination with the infusion of iron oxide nanoparticles. HCC draws its blood supply mainly from the hepatic arterial system, while normal hepatic parenchyma (NHP) are perfused by the portal venous system. Administration into the arterial system should then result in preferential delivery to HCC. Six patients received NanoTherm injections, while six patients received injections of ferucarbotran (Resovist, Schering, Berlin, Germany), with a 62 nm hydrodynamic diameter. In each case, patients received a total of 3.92 mg of iron, which is a consistent dosage administered intravenously for MRI diagnostics. Chemotherapy agents included doxorubicin and cisplatin. After nanoparticle injection, the artery was infused with biodegradable starch microspheres until the tumor was occluded, to prevent excessive washout. Baseline and follow-up was conducted with 3.0 T MRI. The MagForce particles demonstrated significantly higher MR signal intensity than Resovist, and it was speculated that this was due to mechanisms related to the differences in the particles' coatings. The intratumoral concentration of nanoparticles was estimated at 0.032 to 0.01 mg/ml, and the surrounding NHP were largely clear of nanoparticles. One patient received a liver transplant 28 days after injection, and histological analysis was performed on the resected liver. No nanoparticles were found in the tumor vessels, but numerous macrophages in the interstitial tumor capsule demonstrated particle uptake. The nanoparticle concentrations achieved in this study were much too low for hyperthermic treatments, but some feasibility was demonstrated for arterial embolization hyperthermia (Moroz et al. 2002).

17.6 What Comes Next?

Decades of research have clearly demonstrated the promise of magnetic fluid hyperthermia, which is now being realized in clinical application. However, technology development is a long and arduous process, and despite years of critical advances, the field is still in its adolescent phase. Table 17.5 highlights some of the current technological strengths and areas for potential improvement. Moving beyond these current limitations is going to involve continued research in three key areas: furthering basic understanding, optimizing the heating efficiency, and improving nanoparticle delivery.

TABLE 17.5 Strengths and Weaknesses of Current Technology

Strengths	Areas for Improvement
• Biocompatibility of iron oxide nanoparticles	• Current capabilities as a monotherapy
• Demonstrated SAR capable of therapeutic temperatures	• Heterogeneity of nanoparticle distributions and treatment
• Uniform fields capable of penetrating deep tissues, with minimal adverse effects	• High nanoparticle concentrations required for treatment
• Stable nanoparticle deposits capable of repeated treatments over time	• Minimally invasive techniques required for implantation and thermometry
• Successful preliminary indications in combinatorial treatments	• Uncontrolled "hot spots" limit applied fields

Although a fairly detailed knowledge base on the mechanisms behind MFH has been developed, specific areas of basic understanding still require some significant development. Magnetic fluids have been heavily characterized as dispersed solutions but exhibit significant interactions in biological systems. Theoretical modeling generally assumes noninteracting particles, but in vitro and in vivo study have demonstrated that particles will form aggregates and that this can have significant effects on magnetic behavior and heating. Current models need to be modified to factor in these interactions. In addition, although, thermodynamic scaling arguments suggest that localized, intracellular heating is not occurring, enhanced sensitization and/or cytotoxic effects have been demonstrated with MFH (Jordan et al. 1999; Ogden et al. 2009; Creixell et al. 2011; Rodríguez-Luccioni et al. 2011). The mechanisms behind these enhanced effects need to be elucidated to fully realize the advantages of MFH. Another tool that will significantly advance basic understanding in the field is noninvasive thermometry. Currently, minimally invasive probe techniques offer few data points and introduce unneeded complications to experiments and procedures. Significant development will be required, but compatible MR, CT, or US thermometry techniques could produce three-dimensional temperature maps that could be closely matched with histology and modeling, providing an in-depth knowledge of in vivo heating and treatment efficacy. This level of information will be critical in developing the capabilities to provide well-controlled treatment to complex tumor geometries.

While current iron oxide nanoparticles provide adequate SAR to produce therapeutic temperatures, there is certainly room for improvement. Higher magnetic moment nanoparticles are an area of intense research, including iron-iron-oxide composites (Zeng et al. 2007), iron-cobalt core structures (Kline et al. 2009), and magnetosomes (Hergt et al. 2005). Improved methods for synthesizing well-controlled, monodisperse, iron oxide nanoparticles at the optimal heating radii (refer back to Figure 17.4) would also significantly increase the SAR achieved. Beyond the nanoparticle construct, modifying the applied field provides another means of improving heat generation. Current field application has been largely limited by unwanted hot spots produced away from the treatment zone. The field geometry could be modified to better focus the fields in different regions of the body (Wust et al. 2006). In addition, optimizing the alternating wave form (for example, square waves (Morgan and Victora 2010) or pulsed waves (Foner and Kolm 2009)) could provide higher power input.

Enhanced nanoparticle delivery is another major area of development. Uniform nanoparticle distributions confined to the tumor will provide more complete treatment and better selectivity in focusing treatment to the tumor. Enhanced delivery methods could employ biomolecular targeting to preferentially deposit nanoparticles in the tumor region after intravenous injection (Stelter et al. 2009; Hoopes et al. 2009), but such methods have yet to demonstrate adequate accumulation for heating. Magnetic nanoparticles incorporated into multifunctional platforms may also provide enhanced therapeutic, imaging, and diagnostic capabilities, in addition to improved biodistribution profiles. Thermally responsive polymers (Zhang, Srivastava, and Misra 2007) and liposomes (Babincová et al. 2002) combined with iron oxide nanoparticles have already demonstrated the great potential for single-platform adjunctive therapies, and it is likely that the future of cancer medicine will be built on the promise of such synergistic approaches.

References

Aggarwal, P., J. B Hall, C. B McLeland, M. A Dobrovolskaia, and S. E McNeil. 2009. Nanoparticle interaction with plasma proteins as it relates to particle biodistribution, biocompatibility and therapeutic efficacy. *Advanced Drug Delivery Reviews* 61, no. 6: 428–437.

Amini, A. N, E. S Ebbini, and T. T Georgiou. 2005. Noninvasive estimation of tissue temperature via high-resolution spectral analysis techniques. *Biomedical Engineering, IEEE Transactions on* 52, no. 2: 221–228.

Atkinson, W. J, I. A Brezovich, and D. P Chakraborty. 2007. Usable frequencies in hyperthermia with thermal seeds. *Biomedical Engineering, IEEE Transactions on*, no. 1: 70–75.

Babincová, M., P. Cicmanec, V. Altanerová, C. Altaner, and P. Babinec. 2002. AC-magnetic field controlled drug release from magnetoliposomes: Design of a method for site-specific chemotherapy. *Bioelectrochemistry* 55, no. 1 (January): 17–19.

Barry, S. E. 2008. Challenges in the development of magnetic particles for therapeutic applications. *International Journal of Hyperthermia* 24, no. 6: 451–466.

Borrelli, N. F., A. A. Luderer, and J. N. Panzarino. 1984. Hysteresis heating for the treatment of tumours. *Physics in Medicine and Biology* 29: 487.

Brezovich, I. A, W. J Atkinson, and M. B Lilly. 1984. Local hyperthermia with interstitial techniques. *Cancer Research* 44, no. 10: 4752s.

Brezovich, I. A. 1988. Low frequency hyperthermia: Capacitive and ferromagnetic thermoseed methods. *Medical Physics Monograph, Medical Physics*, New York.

Byrne, J. D., T. Betancourt, and L. Brannon-Peppas. 2008. Active targeting schemes for nanoparticle systems in cancer therapeutics. *Advanced Drug Delivery Reviews* 60, no. 15 (December 14): 1615–1626.

Chan, D.C.F., D.B. Kirpotina, and P.A. Bunn. 1993. Synthesis and evaluation of colloidal magnetic iron oxides for the site-specific radiofrequency-induced hyperthermia of cancer. *Journal of Magnetism and Magnetic Materials* 122, no. 1: 374–378.

Chou, C. K. 1990. Use of heating rate and specific absorption rate in the hyperthermia clinic. *International Journal of Hyperthermia* 6, no. 2: 367–370.

Chou, L. Y. T, K. Ming, and W. C. W Chan. 2010. Strategies for the intracellular delivery of nanoparticles. *Chemical Society Reviews* 40: 233–245.

Creixell, M., A. C. Bohórquez, M. Torres-Lugo, and C. Rinaldi. 2011. EGFR-targeted magnetic nanoparticle heaters kill cancer cells without a perceptible temperature rise. *ACS Nano* 5, no. 9: 7124–7129.

Cullity, B. D., and C. D. Graham. 2009. *Introduction to magnetic materials*. Wiley-IEEE (Hoboken, NJ).

Dennis, C. L, A. J. Jackson, J. A. Borchers, P. J. Hoopes, R. Strawbridge, A. R. Foreman, J. van Lierop, C. Grüttner, and R. Ivkov. 2009. Nearly complete regression of tumors via collective behavior of magnetic nanoparticles in hyperthermia. *Nanotechnology* 20, no. 39 (9): 395103.

Dudeck, O., K. Bogusiewicz, J. Pinkernelle, G. Gaffke, M. Pech, G. Wieners, H. Bruhn, A. Jordan, and J. Ricke. 2006. Local arterial infusion of superparamagnetic iron oxide particles in hepatocellular carcinoma. *Investigative Radiology* 41, no. 6: 527–535.

Eastwood, J. D, M. H Lev, T. Azhari, T. Y Lee, D. P Barboriak, D. M Delong, C. Fitzek et al. 2002. CT Perfusion scanning with deconvolution analysis: Pilot study in patients with acute middle cerebral artery stroke. *Radiology* 222, no. 1: 227.

Etheridge, M. L. and J. C. Bischof. 2012a. Optimizing magnetic nanoparticle based thermal therapies within the physical limits of heating. *Annals of Biomedical Engineering* (In Press).

Etheridge, M. L., N. Manuchehrabadi, R. Franklin, and J. C. Bischof. 2012b. Superparamagnetic iron oxide nanoparticle heating: A basic tutorial. In *Nanoparticle Heat Transfer and Fluid Flow*, W. J. Minkowycz, E. M. Sparrow, and J. P. Abraham (eds.), 97–121. CRC Press (New York).

Fallone, B. G., P. R. Moran, and E. B. Podgorsak. 1982. Noninvasive thermometry with a clinical x-ray CT scanner. *Medical Physics* 9: 715.

Foner, S., and H. H Kolm. 2009. Coils for the production of high-intensity pulsed magnetic fields. *Review of Scientific Instruments* 28, no. 10: 799–807.

Freeman, M. L, D. R Spitz, and M. J Meredith. 1990. Does heat shock enhance oxidative stress? Studies with ferrous and ferric iron. *Radiation Research* 124, no. 3: 288–293.

Gilchrist, R. K., R. Medal, W. D. Shorey, R. C. Hanselman, J. C. Parrott, and C. B. Taylor. 1957a. Selective inductive heating of lymph nodes. *Annals of Surgery* 146, no. 4: 596.

Gneveckow, U., A. Jordan, R. Scholz, V. Brüss, N. Waldöfner, J. Ricke, A. Feussner, B. Hildebrandt, B. Rau, and P. Wust. 2004. Description and characterization of the novel hyperthermia- and thermoablation-system MFH300F for clinical magnetic fluid hyperthermia. *Medical Physics* 31: 1444.

Gordon, R. T., J. R. Hines, and D. Gordon. 1979. Intracellular hyperthermia a biophysical approach to cancer treatment via intracellular temperature and biophysical alterations. *Medical Hypotheses* 5, no. 1: 83–102.

Gubin, S. P. 2009. *Magnetic nanoparticles*. Wiley-VCH (Weinheim).

Gupta, A. K, and M. Gupta. 2005. Synthesis and surface engineering of iron oxide nanoparticles for biomedical applications. *Biomaterials* 26, no. 18: 3995–4021.

Hergt, R., W. Andra, C. G. d'Ambly, I. Hilger, W. A. Kaiser, U. Richter, and H. G Schmidt. 2002. Physical limits of hyperthermia using magnetite fine particles. *Magnetics, IEEE Transactions on* 34, no. 5: 3745–3754.

Hergt, R., S. Dutz, and M. Röder. 2008. Effects of size distribution on hysteresis losses of magnetic nanoparticles for hyperthermia. *Journal of Physics: Condensed Matter* 20: 385214.

Hergt, R., R. Hiergeist, I. Hilger, W. A. Kaiser, Y. Lapatnikov, S. Margel, and U. Richter. 2004. Maghemite nanoparticles with very high AC-losses for application in RF-magnetic hyperthermia. *Journal of Magnetism and Magnetic Materials* 270, no. 3: 345–357.

Hergt, R., R. Hiergeist, M. Zeisberger, D. Schüler, U. Heyen, I. Hilger, and W. A Kaiser. 2005. Magnetic properties of bacterial magnetosomes as potential diagnostic and therapeutic tools. *Journal of Magnetism and Magnetic Materials* 293, no. 1: 80–86.

Hergt, R., S. Dutz, and M. Röder. 2008. Effects of size distribution on hysteresis losses of magnetic nanoparticles for hyperthermia. *Journal of Physics: Condensed Matter* 20, no. 38 (9): 385214.

Hergt, R., R. Hiergeist, M. Zeisberger, D. Schüler, U. Heyen, I. Hilger, and W. A. Kaiser. 2005. Magnetic properties of bacterial magnetosomes as potential diagnostic and therapeutic tools. *Journal of Magnetism and Magnetic Materials* 293, no. 1 (May): 80–86.

Hilger, I., K., W. Frühauf, R. Andrä, R. Hiergeist, R. Hergt, and W. A. Kaiser. 2002. Heating potential of iron oxides for therapeutic purposes in interventional radiology. *Academic Radiology* 9, no. 2: 198–202.

Hilger, I., R. Hiergeist, R. Hergt, K. Winnefeld, H. Schubert, and W. A Kaiser. 2002. Thermal ablation of tumors using magnetic nanoparticles: An in vivo feasibility study. *Investigative Radiology* 37, no. 10: 580.

Hoopes, P. J., J. A. Tate, J. A. Ogden, R. R. Strawbridge, S. N. Fiering, A. A. Petryk, S. M. Cassim et al. 2009. Assessment of intratumor non-antibody directed iron oxide nanoparticle hyperthermia cancer therapy and antibody directed IONP uptake in murine and human cells. In *Proceedings of SPIE*, 7181:71810P.

Hoopes, P. J., A. A. Petryk, B. Gimi, A. J. Giustini, J. B. Weaver, J. C. Bischof et al. 2012. In vivo imaging and quantification of iron oxide nanoparticle uptake and biodistribution. In *Proceedings of SPIE*, 8317: 83170R–83170R–9.

Ito, A., K. Tanaka, K. Kondo, M. Shinkai, H. Honda, K. Matsumoto, T. Saida, and T. Kobayashi. 2003. Tumor regression by combined immunotherapy and hyperthermia using magnetic nanoparticles in an experimental subcutaneous murine melanoma. *Cancer Science* 94, no. 3: 308–313.

Iyer, A. K, G. Khaled, J. Fang, and H. Maeda. 2006. Exploiting the enhanced permeability and retention effect for tumor targeting. *Drug Discovery Today* 11, no. 17: 812–818.

Johannsen, M., U. Gneveckow, K. Taymoorian, C. Hee Cho, B. Thiesen, R. Scholz, N. Waldöfner, S. A Loening, P. Wust, and A. Jordan. 2007. Thermal therapy of prostate cancer using magnetic nanoparticles. *Actas Urológicas Espanolas* 31: 660–667.

Johannsen, M., U. Gneveckow, K. Taymoorian, B. Thiesen, N. Waldöfner, R. Scholz, K. Jung, A. Jordan, P. Wust, and S. A. Loening. 2007. Morbidity and quality of life during thermotherapy using magnetic nanoparticles in locally recurrent prostate cancer: Results of a prospective phase I trial. *International Journal of Hyperthermia* 23, no. 3: 315–323.

Johannsen, M., B. Thiesen, U. Gneveckow, K. Taymoorian, N. Waldöfner, R. Scholz, S. Deger, K. Jung, S. A Loening, and A. Jordan. 2006. Thermotherapy using magnetic nanoparticles combined with external radiation in an orthotopic rat model of prostate cancer. *The Prostate* 66, no. 1: 97–104.

Johannsen, M., B. Thiesen, A. Jordan, K. Taymoorian, U. Gneveckow, N. Waldöfner, R. Scholz et al. 2005. Magnetic fluid hyperthermia (MFH) reduces prostate cancer growth in the orthotopic Dunning R3327 rat model. *The Prostate* 64, no. 3: 283–292.

Johannsen, M., B. Thiesen, P. Wust, and A. Jordan. 2010. Magnetic nanoparticle hyperthermia for prostate cancer. *International Journal of Hyperthermia*, no. 0: 1–6.

Johannsen, M., U. Gneveckow, B. Thiesen, K. Taymoorian, C. H. Cho, N. Waldoefner, R. Scholz, A. Jordan, S. Loening, and P. Wust. 2006. Thermotherapy of prostate cancer using magnetic nanoparticles: Feasibility, imaging, and three-dimensional temperature distribution. *European Urology* 52, no. 6: 1653–1662.

Jones, S. K., B. N. Gray, M. A. Burton, J. P. Codde, and R. Street. 1992. Evaluation of ferromagnetic materials for low-frequency hysteresis heating of tumours. *Physics in Medicine and Biology* 37: 293.

Jordan, A., T. Rheinländer, N. Waldöfner, and R. Scholz. 2003. Increase of the specific absorption rate (SAR) by magnetic fractionation of magnetic fluids. *Journal of Nanoparticle Research* 5, no. 5: 597–600.

Jordan, A., R. Scholz, K. Maier-Hauff, F. K. H. van Landeghem, N. Waldoefner, U. Teichgraeber, J. Pinkernelle et al. 2006. The effect of thermotherapy using magnetic nanoparticles on rat malignant glioma. *Journal of Neuro-Oncology* 78, no. 1: 7–14.

Jordan, A., R. Scholz, P. Wust, H. Fähling, J. Krause, W. Wlodarczyk, B. Sander, T. Vogl, and R. Felix. 1997. Effects of magnetic fluid hyperthermia (MFH) on C3H mammary carcinoma in vivo. *International Journal of Hyperthermia* 13, no. 6: 587–605.

Jordan, A., R. Scholz, P. Wust, H. Schirra, and others. 1999. Endocytosis of dextran and silan-coated magnetite nanoparticles and the effect of intracellular hyperthermia on human mammary carcinoma cells in vitro. *Journal of Magnetism and Magnetic Materials* 194, no. 1: 185–196.

Jordan, A., P. Wust, H. Fähling, W. John, A. Hinz, and R. Felix. 1993. Inductive heating of ferrimagnetic particles and magnetic fluids: Physical evaluation of their potential for hyperthermia. *International Journal of Hyperthermia* 25, no. 7: 499–511.

Jordan, A., P. Wust, R. Scholz, B. Tesche, H. Fähling, T. Mitrovics, T. Vogl, J. Cervos-Navarro, and R. Felix. 1996. Cellular uptake of magnetic fluid particles and their effects on human adenocarcinoma cells exposed to AC magnetic fields in vitro. *International Journal of Hyperthermia* 12, no. 6: 705–722.

Jordan, A. 2009. Hyperthermia classic commentary: "Inductive heating of ferrimagnetic particles and magnetic fluids: Physical evaluation of their potential for hyperthermia" by Andreas Jordan et al., *International Journal of Hyperthermia*, 1993; 9:51–68. *International Journal of Hyperthermia* 25, no. 7 (November): 512–516.

Kalambur, V. S, S. Hui, and J. C Bischof. 2007. Multifunctional magnetic nanoparticles for biomedical applications. In *Proceedings of SPIE*, 6440: 64400V.

Kalambur, V. S, E. K. Longmire, and J. C. Bischof. 2007. Cellular level loading and heating of superparamagnetic iron oxide nanoparticles. *Langmuir* 23, no. 24: 12329–12336.

Kappiyoor, R., M. Liangruksa, R. Ganguly, and I. K. Puri. 2010. The effects of magnetic nanoparticle properties on magnetic fluid hyperthermia. *Journal of Applied Physics* 108, no. 9: 094702.

Kawai, N., A. Ito, Y. Nakahara, M. Futakuchi, T. Shirai, H. Honda, T. Kobayashi, and K. Kohri. 2005. Anticancer effect of hyperthermia on prostate cancer mediated by magnetite cationic liposomes and immune-response induction in transplanted syngeneic rats. *The Prostate* 64, no. 4: 373–381.

Keblinski, P., D. G. Cahill, A. Bodapati, C. R. Sullivan, and T. A. Taton. 2006. Limits of localized heating by electromagnetically excited nanoparticles. *Journal of Applied Physics* 100: 054305.

Kline, T. L, Y. H Xu, Y. Jing, and J. P Wang. 2009. Biocompatible high-moment FeCo-Au magnetic nanoparticles for magnetic hyperthermia treatment optimization. *Journal of Magnetism and Magnetic Materials* 321, no. 10: 1525–1528.

Krishnan, K. M. 2010. Biomedical nanomagnetics: A spin through possibilities in imaging, diagnostics, and therapy. *IEEE Transactions on Magnetics* 46, no. 7: 2523.

Le, B., M. Shinkai, T. Kitade, H. Honda, J. Yoshida, T. Wakabayashi, and T. Kobayashi. 2001. Preparation of tumor-specific magnetoliposomes and their application for hyperthermia. *Journal of Chemical Engineering of Japan* 34, no. 1: 66–72.

Lerch, I. A., and D. J. Pizzarello. 1986. The physics and biology of tumor specific, particle-induction hyperthermia. *Med Phys* 13: 786.

Lu, A. H, E. L. Salabas, and F. Schüth. 2007. Magnetic nanoparticles: Synthesis, protection, functionalization, and application. *Angewandte Chemie International Edition* 46, no. 8: 1222–1244.

Luderer, A. A, N. F. Borrelli, J. N. Panzarino, G. R. Mansfield, D. M. Hess, J. L. Brown, E. H. Barnett, and E. W. Hahn. 1983. Glass-ceramic-mediated, magnetic-field-induced localized hyperthermia: Response of a murine mammary carcinoma. *Radiation Research* 94, no. 1: 190–198.

Maenosono, S., and S. Saita. 2006. Theoretical assessment of FePt nanoparticles as heating elements for magnetic hyperthermia. *IEEE Transactions on Magnetics* 42, no. 6 (6): 1638–1642.

Maier-Hauff, K., R. Rothe, R. Scholz, U. Gneveckow, P. Wust, B. Thiesen, A. Feussner et al. 2007. Intracranial thermotherapy using magnetic nanoparticles combined with external beam radiotherapy: Results of a feasibility study on patients with glioblastoma multiforme. *Journal of Neuro-Oncology* 81, no. 1: 53–60.

Maier-Hauff, K., F. Ulrich, D. Nestler, H. Niehoff, P. Wust, B. Thiesen, H. Orawa, V. Budach, and A. Jordan. 2011. Efficacy and safety of intratumoral thermotherapy using magnetic iron-oxide nanoparticles combined with external beam radiotherapy on patients with recurrent glioblastoma multiforme. *Journal of Neuro-Oncology* 103, no. 2: 317–324.

Medal, R., W. Shorey, R. K. Gilchrist, W. Barker, and R. Hanselman. 1959. Controlled radio-frequency generator for production of localized heat in intact animal: Mechanism and construction. *Archives of Surgery* 79, no. 3: 427.

Mitsumori, M., M. Hiraoka, T. Shibata, Y. Okuno, S. Masunaga, M. Koishi, K. Okajima et al. 1994. Development of intra-arterial hyperthermia using a dextran-magnetite complex. *International Journal of Hyperthermia* 10, no. 6: 785–793.

Morgan, S. M, and R. H. Victora. 2010. Use of square waves incident on magnetic nanoparticles to induce magnetic hyperthermia for therapeutic cancer treatment. *Applied Physics Letters* 97: 093705.

Moroz, P., S. K. Jones, and B. N. Gray. 2002. Magnetically mediated hyperthermia: Current status and future directions. *International Journal of Hyperthermia* 18, no. 4: 267–284.

Motoyama, J., N. Yamashita, T. Morino, M. Tanaka, T. Kobayashi, and H. Honda. 2008. Hyperthermic treatment of DMBA-induced rat mammary cancer using magnetic nanoparticles. *BioMagnetic Research and Technology* 6, no. 1: 2.

Natarajan, A., C. Gruettner, R. Ivkov, G. L. DeNardo, G. Mirick, A. Yuan, A. Foreman, and S. J. DeNardo. 2008. NanoFerrite particle based radioimmunonanoparticles: Binding affinity and in vivo pharmacokinetics. *Bioconjugate Chemistry* 19, no. 6 (6): 1211–1218.

Ogden, J. A., J. A. Tate, R. R. Strawbridge, R. Ivkov, and P. J. Hoopes. 2009. Comparison of iron oxide nanoparticle and waterbath hyperthermia cytotoxicity. In *Proceedings of SPIE* 7181: 71810K.

O'Handley, R. C. 2000. *Modern magnetic materials: Principles and applications*. Wiley (New York).

Ohno, T., T. Wakabayashi, A. Takemura, J. Yoshida, A. Ito, M. Shinkai, H. Honda, and T. Kobayashi. 2002. Effective solitary hyperthermia treatment of malignant glioma using stick type CMC-magnetite. In vivo study. *Journal of Neuro-Oncology* 56, no. 3: 233–239.

Oleson, J. R., T. C. Cetas, and P. M. Corry. 1983. Hyperthermia by magnetic induction: Experimental and theoretical results for coaxial coil pairs. *Radiation Research* 95, no. 1: 175–186.

Oleson, J. R., R. S. Heusinkveld, and M. R. Manning. 1983. Hyperthermia by magnetic induction: II. Clinical experience with concentric electrodes. *International Journal of Radiation Oncology, Biology, Physics* 9, no. 4: 549–556.

Pennes, H. H. 1948. Analysis of tissue and arterial blood temperatures in the resting human forearm. *Journal of Applied Physiology* 1, no. 2: 93.

Poorter, J. D., C. D. Wagter, Y. D. Deene, C. Thomsen, F. Staahlberg, and E. Achten. 1995. Noninvasive MRI thermometry with the proton resonance frequency (PRF) method: In vivo results in human muscle. *Magnetic Resonance in Medicine* 33, no. 1: 74–81.

Popplewell, J., R. E. Rosensweig, and R. J. Johnston. 2002. Magnetic field induced rotations in ferrofluids. *Magnetics, IEEE Transactions on* 26, no. 5: 1852–1854.

Qin, Z., M. Etheridge, and J. Bischof. Nanoparticle heating: Nanoscale to bulk effects of electromagnetically heated iron oxide and gold for biomedical applications. In *Proceedings of SPIE*, 7901: 79010-C–79010C-15.

Rabin, Y. 2002. Is intracellular hyperthermia superior to extracellular hyperthermia in the thermal sense? *International Journal of Hyperthermia* 18, no. 3: 194–202.

Rand, R. W., H. D. Snow, and W. J. Brown. 1981. Thermomagnetic surgery for cancer. *Applied Biochemistry and Biotechnology* 6, no. 4: 265–272.

Rodríguez-Luccioni, H. L., M. Latorre-Esteves, J. Méndez-Vega, O. Soto, A. R. Rodríguez, C. Rinaldi, and M. Torres-Lugo. 2011. Enhanced reduction in cell viability by hyperthermia induced by magnetic nanoparticles. *International Journal of Nanomedicine* 6: 373–380.

Rosensweig, R. E. 2002. Heating magnetic fluid with alternating magnetic field. *Journal of Magnetism and Magnetic Materials* 252: 370–374.

Sapareto, S. A., and W. C. Dewey. 1984. Thermal dose determination in cancer therapy. *International Journal of Radiation Oncology, Biology, Physics* 10, no. 6: 787–800.

Sato, M., G. Nakajima, T. Namikawa, and Y. Yamazaki. 1990. Magnetic properties and microstructures of Fe sub (3) O sub (4)- gamma Fe sub (2) O sub (3) intermediate state. *IEEE Transactions on Magnetics* 26, no. 5: 1825–1827.

Schrope, B. A., and V. L. Newhouse. 1993. Second harmonic ultrasonic blood perfusion measurement. *Ultrasound in Medicine & Biology* 19, no. 7: 567–579.

Smolensky, Eric D., Michelle C. Neary, Yue Zhou, Thelma S. Berquo, and Valérie C. Pierre. 2011. Fe3O4@organic@Au: Core–shell nanocomposites with high saturation magnetisation as magnetoplasmonic MRI contrast agents. *Chemical Communications* 47, no. 7: 2149.

Stauffer, P. R., T. C. Cetas, and R. C. Jones. 2007. Magnetic induction heating of ferromagnetic implants for inducing localized hyperthermia in deep-seated tumors. *Biomedical Engineering, IEEE Transactions on*, no. 2: 235–251.

Stauffer, P. R., T. C. Cetas, A. M. Fletcher, D. W. DeYoung, M. W. Dewhirst, J. R. Oleson, and R. B. Roemer. 1984. Observations on the use of ferromagnetic implants for inducing hyperthermia. *IEEE Transactions on Bio-Medical Engineering* 31, no. 1: 76.

Steinbach, M., S. Koswig, S. Dresel, M. Huenerbein, P. Wust, J. Ollek, A. Jordan, and B. Rau. Nanotherapy induced hyperthermia in advanced esophageal cancer: Results of a phase I study. *TBD* (In Draft).

Stelter, L., J. G. Pinkernelle, R. Michel, R. Schwartländer, N. Raschzok, M. H. Morgul, M. Koch et al. 2009. Modification of aminosilanized superparamagnetic nanoparticles: Feasibility of multimodal detection using 3T MRI, small animal PET, and fluorescence imaging. *Molecular Imaging and Biology* 12, no. 1 (7): 25–34.

Tanaka, K., A. Ito, T. Kobayashi, T. Kawamura, S. Shimada, K. Matsumoto, T. Saida, and H. Honda. 2005. Intratumoral injection of immature dendritic cells enhances antitumor effect of hyperthermia using magnetic nanoparticles. *International Journal of Cancer* 116, no. 4: 624–633.

Thiesen, B., and A. Jordan. 2009. Clinical applications of magnetic nanoparticles for hyperthermia. *International Journal of Hyperthermia* 24, no. 6: 467–474.

van Landeghem, F. K. H, K. Maier-Hauff, A. Jordan, K. T. Hoffmann, U. Gneveckow, R. Scholz, B. Thiesen, W. Brück, and A. von Deimling. 2009. Post-mortem studies in glioblastoma patients treated with thermotherapy using magnetic nanoparticles. *Biomaterials* 30, no. 1: 52–57.

Van Sonnenberg, E., W. McMullen, and L. Solbiati. 2005. *Tumor ablation: Principles and practice*. Springer (New York).

Weaver, John B., Adam M. Rauwerdink, and Eric W. Hansen. 2009. Magnetic nanoparticle temperature estimation. *Medical Physics* 36, no. 5: 1822.

Weissleder, R., D. D. Stark, B. L. Engelstad, B. R. Bacon, C. C. Compton, D. L. White, P. Jacobs, and J. Lewis. 1989. Superparamagnetic iron oxide: Pharmacokinetics and toxicity. *American Journal of Roentgenology* 152, no. 1: 167.

Williams, D. S., J. A. Detre, J. S. Leigh, and A. P. Koretsky. 1992. Magnetic resonance imaging of perfusion using spin inversion of arterial water. *Proceedings of the National Academy of Sciences of the United States of America* 89, no. 1: 212.

Wust, P., U. Gneveckow, M. Johannsen, D. Böhmer, T. Henkel, F. Kahmann, J. Sehouli, R. Felix, J. Ricke, and A. Jordan. 2006. Magnetic nanoparticles for interstitial thermotherapy– feasibility, tolerance and achieved temperatures. *International Journal of Hyperthermia* 22, no. 8: 673–685.

Wust, P., J. Nadobny, H. Fähling, H. Riess, K. Koch, W. John, and R. Felix. 1991a. Determinant factors and disturbances in controlling power distribution patterns by the hyperthermia-ring system BSD-2000. 1. *Clinical Obervations and Phantom Measurements. Strahlentherapie Und Onkologie* 166, no. 12: 822–830.

Wust, P., J. Nadobny, H. Fähling, H. Riess, K. Koch, W. John, and R. Felix. 1991b. Determinant factors and disturbances in controlling power distribution patterns by the hyperthermia-ring system BSD-2000. 2. Measuring techniques and analysis. *Strahlentherapie Und Onkologie* 167, no. 3: 172.

Yanase, M., M. Shinkai, H. Honda, T. Wakabayashi, J. Yoshida, and T. Kobayashi. 1998. Antitumor immunity induction by intracellular hyperthermia using magnetite cationic liposomes. *Cancer Science* 89, no. 7: 775–782.

Young, J. H., M. T Wang, and I. A. Brezovich. 2007. Frequency/depth-penetration considerations in hyperthermia by magnetically induced currents. *Electronics Letters* 16, no. 10: 358–359.

Zeng, Q., I. Baker, J. A. Loudis, Y. F. Liao, and P. J. Hoopes. 2007. Synthesis and heating effect of iron/iron oxide composite and iron oxide nanoparticles. In *Proceedings of SPIE*, 64400H-64400H-11. San Jose, California.

Zhang, B., Y. C. Liao, S. L. Girshick, and J. T. Roberts. 2008. Growth of coatings on nanoparticles by photoinduced chemical vapor deposition. *Journal of Nanoparticle Research* 10, no. 1: 173–178.

Zhang, J. L., R. S. Srivastava, and R. D. K. Misra. 2007. Core-shell magnetite nanoparticles surface encapsulated with smart stimuli-responsive polymer: Synthesis, characterization, and LCST of viable drug-targeting delivery system. *Langmuir* 23, no. 11: 6342–6351.

Zhang, L. Y., H. C. Gu, and X. M. Wang. 2007. Magnetite ferrofluid with high specific absorption rate for application in hyperthermia. *Journal of Magnetism and Magnetic Materials* 311, no. 1 (April): 228–233.

18

Application of Gold Nanoparticles (GNP) in Laser Thermal Therapy

Zhenpeng Qin
University of Minnesota

John C. Bischof
University of Minnesota

18.1 Introduction

A compelling vision for nanoparticles (NPs) in medicine involves their use to selectively diagnose, image, and destroy disease by throwing a noninvasive "switch." This chapter explores the ability of laser light to be a switch that activates NPs accumulated in tissue to heat and destroy disease. NPs can be locally or systemically delivered such that they accumulate within the cellular and extracellular spaces of diseased tissue such as tumors. How NPs distribute throughout the body and tumors will be briefly discussed, and comprehensive reviews will be noted. When present, the *scattering* of light by the NP can yield contrast for imaging, while light *absorption* by the NP can be used for therapeutic heating and photothermal imaging. While the advantages of using NPs for image contrast have been extensively reviewed by others (Qian 2008, Hu 2006, Zharov 2007), this chapter will focus on the ability and effects of NP laser heat generation at multiple scales. The potential advantages over traditional thermal therapies in terms of tumor specific destruction will be discussed. While various NPs are under investigation and some are in clinical trials, gold nanoparticles (GNP) are among the most mature for this technique due to the fact that they are among the strongest NP absorbers of laser light and are already being used clinically for photothermal treatment of cancer.

The Pennes bioheat equation is widely used for the prediction of temperature change during thermal therapies (Pennes 1948). Although more sophisticated models incorporating the vasculature of the tissue are available, they require more parameters that are usually not readily available (Baish 2000). The Pennes bioheat equation can be written as

$$\rho C \frac{\partial T}{\partial t} = \nabla \cdot (k \nabla T) + (\rho C)_b \omega (T_b - T) + q_m + \text{SAR}. \quad (18.1)$$

The terms on the right hand side (W/m³) are heat diffusion, blood perfusion (denoted by subscript b), metabolic heat generation (denoted by subscript m), and external heat source (specific absorption rate – SAR), respectively.

The presence of GNP in the tumor can increase the selective absorption of energy (i.e., laser) and therefore heating within the tumor compared with traditional thermal therapies. The temperature increase of laser photothermal therapy using GNPs comes from the heat generated by a large amount of individual GNPs. The heat generation of single GNP (Q_{nano}) under laser light can be written as the product of absorption cross section area (C_{abs}, m^2) and laser fluence (I, W/m²) in Equation 18.2. The absorption cross-section area is the characteristic area of the particle that absorbs the energy passing through.

$$Q_{nano} = C_{abs} \cdot I \quad \text{(W)}. \tag{18.2}$$

With GNP of a certain concentration ($N - \#$ NPs/ml), SAR is the sum of heat generated by all NPs,

$$\text{SAR} = N \cdot Q_{nano} = N \cdot C_{abs} \cdot I = \alpha \cdot I \quad \text{(W/m}^3). \tag{18.3}$$

This equation illustrates the important fundamental terms necessary for laser photothermal therapy with GNPs. First, the laser fluence ($I = \text{W/m}^2$) should be known in order to estimate the heat generation and temperature change. Second, the optical properties (C_{abs}) of GNP (and tissue as well) should be well characterized. Third, and finally, a sufficient number of GNPs (N) have to be delivered to the tumor.

This chapter discusses the important terms in photothermal GNP therapy and their application. First, laser-tissue interactions (Section 18.2) and laser fluence estimation in the tissue with and without GNPs are discussed (Section 18.3) (I). Then the optical properties of GNP (Section 18.4) are reviewed to provide an understanding of absorption cross section (C_{abs}). With the heat generated by GNP, the thermal response of GNP laser heating (Section 18.5) will then be analyzed at the single NP (nano-), cellular (micro-), and tissue (macro- or bulk) levels. The physical response around single GNP and biological responses of *in vitro* and *in vivo* systems (Section 18.6) are then reviewed with brief discussion on GNP biodistribution (N in Equation 18.3). Finally, existing clinical trials on GNP for photothermal therapy are reviewed (Section 18.7), and this chapter will end with several suggestions for future studies (Section 18.8).

18.2 Overview of Laser-Tissue Interactions

18.2.1 Tissue Optical Properties and the Therapeutic Window

Tissue is comprised of various structures that combine to create a complicated optical property landscape. For instance, the presence of multiple compartments such as cells, vasculature, and interstitial space can each affect the optical properties of tissue and make the overall absorption and scattering far different from other bulk materials. Absorbers (i.e., chromophores) and scatterers in the tissue structure determine the light interaction with the tissue. It is often convenient to treat tissue as an absorbing matrix with randomly distributed scatterers and then assess bulk optical properties (Welch 1995). Experimental characterization of the tissue then yields a bulk absorption coefficient, scattering coefficient, and scattering phase function. The common chromophores that absorb in the tissue are water, hemoglobin,

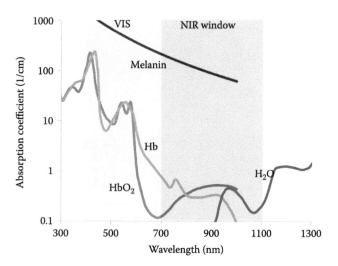

FIGURE 18.1 Optical absorption of tissue components including deoxy-hemoglobin (Hb), oxy-hemoglobin (HbO$_2$), water (H$_2$O), and melanin. Visible (VIS) and near infrared (NIR) wavelength regimes are noted. (Data source: the absorption coefficients of hemoglobin, melanin, and water are plotted based on the data compiled by Scott Prahl at Oregon Medical Laser Center [http://omlc.ogi.edu/spectra/].)

and melanin. The absorption of each component is dependent on the wavelength, and the overall absorption depends on the tissue composition. For instance, hemoglobin and melanin strongly absorb in the visible region, while water absorbs more in the infrared, as shown in Figure 18.1. The wavelength window between visible and infrared, which is around 700~1100 nm, is often referred to as the therapeutic window (Weissleder 2001, Jöbsis-vanderVliet 1999), or the near infrared (NIR) window. The NIR window is the gap between the absorbers in the visible and infrared regions, and as a result these wavelengths allow deeper light penetration. In addition to absorption, tissue is also highly scattering, with an average path length of 0.05~0.2 mm between two scattering events (Welch 1995). Both the absorption and scattering determine how tissue interacts with light. For more detailed discussion of tissue optical properties, their wavelength dependence, and the measurement techniques, one can refer to previous literature (Welch 1995, Patterson 1989, Farrell 1992, Cheong 1990).

18.2.2 Biomedical Lasers and Laser Tissue Interactions with and without GNPs

Lasers that can be used to either image or heat tissue can be classified into different categories depending on the lasing medium within the laser (solid, gas, dye, etc.), operating mode (pulsed or continuous wave [CW]), and wavelength (visible, IR, etc.). Table 18.1 lists several common biomedical lasers and their operating parameters (including lasing wavelength and corresponding tissue absorber, power level, and operating mode). The lasing medium and operating mode tend to determine the spectrum and stability of the laser. Further, the irradiance (W/m^2) and interaction time define an area upon which mechanisms of laser tissue interactions spread

TABLE 18.1 Common Biomedical Lasers Listed by Wavelength from Short to Long

Laser type	Wavelength	Tissue Absorber	Power[c]	CW/Pulsed
Excimer lasers[a]	UV, 126~350nm	Hb, Melanin	~0.2J/pulse	Pulsed
Ion lasers (Ar⁺ and Kr⁺)	350~1100nm	Hb, Melanin, NIR window	mW to 40W	CW
He-Cd laser	325nm, 442nm	Hb, Melanin	~100mW	CW
Semiconductor (Diode) laser	400 nm~microns	Hb, Melanin H_2O	several W	CW
Cu vapor	512nm	Hb, Melanin	1–10 W	Pulsed
KTP[b]	532nm	Hb, Melanin	100 W	Pulsed
Nd:YAG laser	532nm 1064nm	Hb, HbO_2, NIR window	up to 100MW	CW or pulsed
Ti-sapphire laser	650nm~1100nm	Hb, HbO_2, NIR window	0.5~1.5W	CW or pulsed
Ruby laser	694nm	Melanin	high	Pulsed
Alexandrite laser	800nm	NIR window	up to 100W	Pulsed
Dye lasers	UV-Vis-IR, tunable	Varies	up to kW	CW or pulsed
CO_2 laser	Far IR,~10 μm	H_2O	mW to kW	CW or pulsed

Note: The corresponding tissue absorbers, laser power level, and operating model (CW or pulsed) for each laser are listed.

[a] The wavelength is excimer dependent, such as XeF, XeCl, KrF, ArF, and KrCl.

[b] KTP laser is pumped by Nd:YAG laser.

[c] The power for pulsed laser is the average over time when expressed in W.

out in the laser-tissue interaction map (Figure 18.2a). For instance, while CW laser delivers energy continuously, pulsed laser is able to confine the energy into short periods of time called pulses (with duration as short as femto-seconds, 10^{-15} s). Depending on a subset of laser parameters (power and pulse duration), the mechanism of tissue interaction can change (Judy 2000, Niemz 2004). Examples include: photochemical (laser-triggered chemical reaction or decomposition (Srinivasan 1986)); photothermal (heat generation and temperature increase (Welch 1995)); photo ablative (intensive temperature increase and tissue vaporization (Vogel 2003)); and photomechanical (generation of tensile stress due to explosive ablation (Paltauf 2003, Jacques 1993)). Photochemical and photothermal interactions usually require continuous low power laser, while photo ablative and photomechanical interactions need pulsed high power laser. However, these effects can overlap, for example, photo ablation also has thermal and mechanical effects. Some examples of laser treatment include photodynamic therapy, which is photochemical in nature (Niemz 2004), laser interstitial thermotherapy (LITT), which is photothermal (Müller 1995), and photo-ablation of benign prostatic hyperplasia (BPH) (Kuntz 2007).

The introduction of GNPs increases the absorption of laser energy (discussed in detail later in optical properties of GNPs), thereby reducing the laser energy needed to achieve the same effects, such as photothermal therapy, bubble formation, and plasma formation. This can be easily visualized in Figure 18.2b where a comparison is made with laser parameters from the literature that were used to achieve the effects described herein, in systems with and without GNPs. The exposure time decreases as

one move toward the left (x-axis), and the laser power decreases as one moves down (y-axis). Thus, the presence of GNPs is shown to reduce the power and exposure time necessary to achieve the same outcome. It is worth noting that the figure is plotted on a logarithmic scale, and as a result noticeable shifts in the figure represent order of magnitude change.

18.2.3 Applying Laser to the Tumor

Depending on the position of the tumor to be treated, appropriate methods of applying laser energy should be adopted. In general, there are three ways, including: (1) direct illumination if close to the surface, (2) interstitial laser fiber, and (3) endoscopes or catheters. The penetration depth of all laser approaches is wavelength dependent and in the best case can reach up to ~1 cm (Peng 2008, Niemz 2004). Most preclinical studies use rodent systems, and the tumors are grown near the surface (Hirsch 2003, von Maltzahn 2009). In these studies, direct laser illumination is applied to the tumor. The laser attenuates when penetrating in the tissue, and the collimated intensity decreases exponentially according to Beer's law:

$$I(z) = I_0(1 - r_{ref}).e^{-(\mu_a + \mu_s)^*z} \quad (W/m^2) \qquad (18.4)$$

where I is the laser intensity from the source [W/m²], r_{ref} is the portion that is reflected at the air-tissue interface, μ_a is the absorption coefficient [m⁻¹], μ_s is the scattering coefficient [m⁻¹],

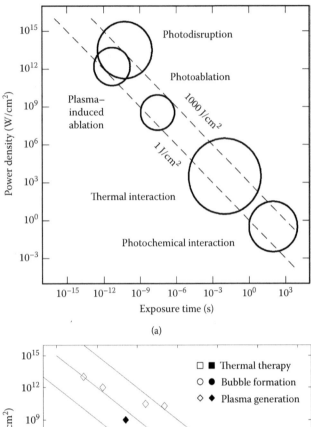

(a)

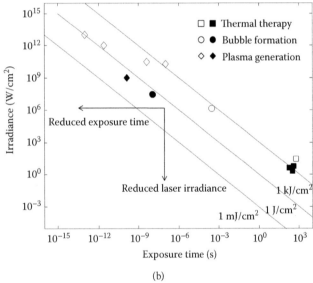

(b)

FIGURE 18.2 (a) Map of laser-tissue interactions including photochemical, photothermal, photo ablative, plasma induced ablative, and photodisrutptive modes. Note that this map shows the approximate position of the interaction modes. (Reproduced from Niemz, M., *Laser-Tissue Interactions: Fundamentals and Applications*, Springer Verlag, 2004. With kind permission from Springer Science + Business Media B.V.) (b) The thresholds of laser parameters for thermal therapy, bubble formation (photo ablative), and plasma generation. Open symbols are without GNP, and filled symbols are with GNP. The dashed lines in the figure indicate the same amount of energy input in terms of J/cm². (From Müller, G. and Roggan, A., Laser-induced interstitial thermotherapy. SPIE-International Society for Optical Engineering, 1995; Hirsch, L. et al., *Proceedings of the National Academy of Sciences of the United States of America*, 100, 2003; Gobin, A. et al., *Nano Lett*, 7, 2007; von Maltzahn et al., *Cancer Res*, 69, 2009; Vogel, A. and Venugopalan, V., *Chem. Rev*, 103, 2003; Takeda, Y., Kondow, T. and Mafuné, F., *J. Phys. Chem. B*, 110, 2006.)

and z is the depth. Direct illumination is easy to apply but is limited by the penetration depth of the laser.

For deep tissue, interstitial laser fibers can be used, for example, to treat brain tumors (Schwartz 2009). The laser attenuates more rapidly than the case given by Equation 18.4 because of the radial geometry. Usually a water-cooled applicator is applied to reduce the tissue temperature near the fiber and avoid vaporization and carbonization of the tissue. Endoscopes and catheters are typically used for vascular disease, for example, in cardiology.

In summary, for tumor photothermal therapy, the main methods of applying laser are either direct illumination for surface tumors or minimally invasive interstitial fiber for deep tumors. To more quantitatively obtain the laser fluence in the tissue with or without the presence of GNPs, light transport models are needed and reviewed next.

18.3 Laser Fluence Estimation in Tissues

The introduction of NPs alters the optical properties of the tissue and hence the thermal response. In this section, the methodologies commonly used in modeling the light transport and recent applications for NP laden tissue are reviewed. The light transport is described by the radiative transport equation (RTE), given by (Welch 1995, Modest 2003):

$$\frac{dL(r,\hat{s})}{ds} = -\mu_a L(r,\hat{s}) - \mu_s L(r,\hat{s}) + \mu_s \int_{4\pi} p(\hat{s},\hat{s}')L(r,\hat{s})d\omega' + S(r,\hat{s})$$

(18.5)

where $L(r,\hat{s})$ is the radiance in position r and $\hat{s}$ direction with unit [W/m²/sr] and the integration of $L(r,\hat{s})$ over all directions is the fluence (I), μ_a is the absorption coefficient [1/m], μ_s is the scattering coefficient, $p(\hat{s},\hat{s}')$ is the scattering phase function [1/sr] describing the contribution of scattering from $\hat{s}'$ direction to $\hat{s}$ direction, and $S(r,\hat{s})$ is the source term. It states that the change of light fluence is determined by losses from the absorption (first term) and scattering (second term), and gain from scattering (third term) and energy source (last term). RTE is an integro-differential equation and is difficult to solve directly. Several approximations for solving the RTE equation have been used for laser GNP heating and in some cases can be compared directly to experimental results.

By employing the P1 approximation (first approximation to spherical harmonics) to RTE, Bayazitoglu and coworkers (Tjahjono 2008, Vera 2009) computationally studied the effect of NP scattering and absorption on SAR. It was shown that GNPs with high scattering (for example, gold nanoshell with high C_{sca}/C_{ext} in Table 18.2) increase the internal diffuse radiation and create a more even SAR, while GNPs with high absorption produce a large amount of SAR at the entry region (i.e., the region near the laser source). However, this approach is difficult to apply for complex geometries.

Another simplifying assumption to solve RTE is the diffusion approximation (DA), which is applicable for highly scattering

TABLE 18.2 Representative GNP Modifications (Size and Shape) and Their Optical Properties Including Surface Plasmon Resonance Wavelength (λ_{SPR}), Absorption Cross Section (C_{abs}), and the Ratio of Absorption over Extinction (C_{abs}/C_{ext})

Particle Type	Schematic	Size Info	λ_{SPR} (nm)	C_{abs} (nm2)	C_{abs}/C_{ext}	Ref.
Sphere		D = 40 nm	528	2,927	0.94	(Jain 2006)
Si core/Au shell		R_1 = 60 nm R_2 = 75 nm	800	45,769	0.53	(Day 2009)
Au₂S core/Au shell		R_1 = 21 nm R_2 = 25 nm	~780	2,750	>0.9	(Cole 2009)
Au rod		L_1 = 12 nm[a] L_2 = 36 nm	797	5,674	0.93	(Jain 2006)
Au cage		L = 36.7 nm t = 3.3 nm[b]	800	7,260	0.90	(Chen 2005)
Au star		100 nm	770	–	–	(Nehl 2006)

[a] L_1 and L_2 are the width and length of the rod.
[b] Au cage L is the outer length of the cubic and t is the thickness of the cage.

medium (Welch 1995). The advantage of the diffusion approximation is its relative simplicity to solve numerically. In DA, the absorption (μ_a) and reduced scattering ($\mu_s' = \mu_s'(1-g)$, where g is the anisotropy) coefficients are lumped into the optical diffusion coefficient (D):

$$D = \frac{1}{3(\mu_s' + \mu_a)}. \tag{18.6}$$

Elliott et al. (Elliott 2007) used DA to model the laser fluence in phantoms containing different concentrations of NPs and calculated the temperature distribution within the phantoms with the finite element method. The calculated temperature was compared with measurements by magnetic resonance temperature imaging (MRTI), and showed reasonable agreement at low gold nanoshell concentrations (1.19×10^9 NPs/ml). A follow-up study by Elliott et al. (Elliott 2009) showed that with higher gold nanoshell concentration (up to 2.5×10^9 NPs/ml), the predictions from DA give unsatisfactory results. Instead, the delta P1 approximation (treat forward-directed and scattered light separately) gives a better prediction for both the lower and higher gold nanoshell concentrations investigated. Modifications to speed the calculation and increase accuracy of the RTE equation have also been proposed (Xu 2010). Note that all of these approximations should be used with care, and special attention should be paid to the applicable conditions and desirable accuracy, for example, the failure of DA for high gold nanoshell concentration discussed here (Elliott 2009).

Another approach to evaluate light transport or fluence involves Monte Carlo ray tracing to directly simulate photon transport in tissues. Monte Carlo is computationally expensive as a sufficient number of photons need to be launched and traced for the technique to be accurate. This method has been used to estimate laser fluence within the gold nanoshell targeted tumor (Feng 2009). In addition, an optimization algorithm was used

to calibrate the model parameters that were changing during the laser photothermal treatment by Feng et al. (Feng 2009). With the use of supercomputers, real-time predictions are possible when averaging over 10 runs with 1 million photons in each run (Feng 2009). In another example, Lin et al. (Lin 2005) used Monte Carlo to explore the use of light interaction with gold nanoshells for early cancer detection. Specifically, a small number of gold nanoshells (<0.05% volume fraction) can induce measurable change in tissue diffuse reflectance.

To model light transport with the RTE and Monte Carlo method described here, the optical properties of the NP-laden tissue are fundamental inputs and need to be described accurately. Under idealized conditions, for example, the phantom gel used by Elliott et al. (Elliott 2007), the gold nanoshells are assumed to be uniformly distributed. In this case, the optical properties of NP-laden medium are obtained by the superposition of the NP and medium properties linearly (Lin 2005, Kirillin 2009). However, GNPs are known to interact with biological systems, such as cells and tissue, leading to non-ideal conditions. These include: (1) inhomogeneous distribution of NPs such as accumulation around blood vessels, as shown in Perrault et al. (Perrault 2009); and (2) aggregation of NPs in cells upon internalization (Chithrani 2007). Further studies to address these non-idealities and possible nonlinear effects on optical properties are still necessary. Next, the optical properties of GNPs under idealized conditions are discussed.

18.4 Optical Properties of GNPs

18.4.1 What Is the Ideal Absorber?

Most laser-tissue interactions rely upon absorption of laser energy within the tissue. Thus, being able to modify and control absorption is important and can be approached through the addition of exogenous absorbers. Ideally, the absorbing agent should be biocompatible, stable, easily functionalized using surface chemistry, and, of course, highly absorbing. Among the various materials studied (Burda 2005), gold can satisfy all of these criteria, particularly in NP form. The optical properties of GNP, especially absorption that is related to heat generation for thermal therapy, will be reviewed next.

18.4.2 Optical Properties of GNPs

When light interacts with metallic particles, the electromagnetic field of light drives the oscillation of free electrons in the metal. When the frequency of light is approaching that of electron oscillations, the amplitude of electron oscillation reaches the maximum value, and this is called surface plasmon resonance (SPR) (Jain 2007b). The energy of light is transferred to the metal particle and then is either: (1) absorbed by the particle and dissipated as heat; or (2) reemitted by the particle at the same frequency (Rayleigh scattering) or at a shifted frequency (Raman scattering) (Modest 2003). Both the absorption and scattering is much stronger compared with traditional dyes because of the SPR.

The GNP size and shape determine its optical properties (Kelly 2003, Link 1999, Link 2000). Figure 18.3 shows the structure and the absorbance for three types of selected GNPs: gold nanosphere, gold nanoshell, and gold nanorod. As one can see, the SPR of the gold nanosphere is in the visible range (514–532 nm), while the SPR of the gold nanoshell and nanorod (longitudinal) are tuned to the near infrared regime (700–900 nm), which lies in the NIR window of tissue (Figure 18.1).

Often cross-sectional areas are used to describe NP optical properties, and the amount of energy absorbed or scattered is equal to the absorption or scattering cross section multiplied by the laser fluence ($Q_{nano} = C_{abs} I$, $E_{sca} = C_{sca} I$). Extinction consists of absorption and scattering. Table 18.2 gives a nonexclusive list of recently developed GNPs with different shapes and optical properties including the surface plasmon resonance (SPR) wavelength (λ_{SPR}), absorption cross sections (C_{abs}), and the ratio of absorption over total extinction (C_{abs}/C_{ext}). The listed NPs all have SPR in NIR region except spherical GNP. The Au/Si shell has the highest absorption cross section, but the size is significantly larger than other particles. Gold nanorod and cage have good absorption cross sections considering the relatively small size. Comparing the ratio of absorption over extinction (C_{abs}/C_{ext}), the low value for Au/Si shell indicates significant scattering. In general, scattering increases as the GNP size increases (Jain 2006, Bohren 1983). Comparing the absorption cross section of these GNPs (10^3~10^4 nm^2) shows 5~6 orders of magnitude increase from traditional dyes like indocyanine green with $C_{abs} = 1.66 \times 10^{-2}$ nm^2 at light wavelength of ~800 nm (Hirsch 2006, Jain 2007a). As discussed in the introduction, the absorption cross section is an important parameter for heat generation, and the high value for GNP means much more heat can be generated under the same laser fluence.

Due to the importance of GNP optics for photothermal therapy, accurate methods to theoretically predict and experimentally measure the optical properties of GNPs are needed. An important tool for prediction of the optical properties of a given GNP is Maxwell's equation. While the general solution of Maxwell's equation can be difficult, there are analytical solutions for simple geometries such as sphere and shell (Bohren 1983, Link 2000). Note that size-dependent material properties need to be considered for particles less than 10 nm (Kreibig 1995). For more complex-shaped NPs, numerical methods including boundary element method (BEM), discrete-dipole approximation (DDA), and finite difference in the time domain

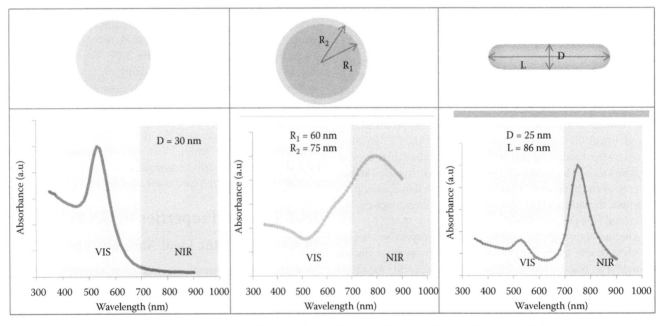

FIGURE 18.3 Optical properties of selected GNPs. Gold nanosphere, gold nanoshell, and gold nanorod with specific dimensions are shown.

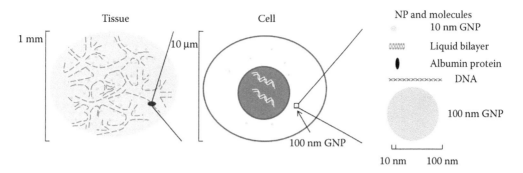

FIGURE 18.4 Relative dimensions of interacting components, including tissue, cell, NP, and molecules. The size of NPs are comparable with some macromolecules (lipid bilayer, albumin protein, DNA) but are much smaller than the cell and tissue.

(FDTD) (Myroshnychenko 2008, Schatz 2007) can be used to estimate GNP optical properties. Experimental measurement of NP-system optical properties rely on UV-vis-IR spectrophotometry for optical extinction of ensemble NPs and electron microscopy (EM) for the geometry of single NPs. Combining theoretical and experimental characterization methods, the absorption and scattering properties are then obtained (Ungureanu 2010). While one can use the traditional combinatorial approach, new developments using dark field scattering and photothermal imaging to probe both the scattering and absorption directly at the single NP level have been recently published (Tcherniak 2010).

For more detailed discussion of GNP optical properties, readers are directed to the classic book by Bohren and Huffman (Bohren 1983) and several more recent reviews (Link 2000, Myroshnychenko 2008). The chemistry and synthesis of GNP is also discussed elsewhere and will not be discussed further here (Burda 2005, Daniel 2004, Huang 2009, Yu 1997, Sau 2004). As reviewed previously, the absorption cross section is a representative property that characterizes the light-to-heat conversion within a GNP. The ensuing thermal response of NP-laden systems will be discussed next.

18.5 Laser GNP Effects I—Thermal Response at Multiple Scales

Having characterized the optical properties and laser fluence along with the number of GNPs present, one can estimate the SAR and temperature change by using the tools discussed earlier. Alternatively, the temperature change can be measured directly to estimate SAR. In either case the temperature rise in the system is the desired effect. In this section, scaling of temperature rise during GNP photothermal therapy at the single NP, single cell, and tissue level are presented and discussed. The concept of *thermal confinement*, important in SAR estimation and treatment planning, is then introduced.

18.5.1 Heat Generation and Scaling

The light to heat conversion in a NP-impregnated system and the resulting bulk temperature increase determines the outcome of photothermal therapy. To analyze this process thermally, it

is important to have a sense of the length scales for detailed heat transfer analysis. An NP is typically 10 to 100 nm in diameter. When compared with typical biological systems as shown in Figure 18.4, NP size is comparable to macromolecules, such as proteins and DNA, enabling a number of interesting nano-bio interfacing applications (Nel 2009). On the other hand, an NP is more than 100 times smaller than a human cell (~10 μm), and over five orders of magnitude smaller than a tumor (cm). Because of the large differences in length scales, the thermal responses are quite different in terms of magnitude and time scale.

To illustrate the importance of scaling, we analyze the temperature response of three biologically relevant situations, also shown in Figure 18.4: (1) the heating of a single NP and its immediate surrounding; (2) the heating of a single cell loaded with GNPs; and (3) the heating of a tumor loaded with GNPs. The question of how a single NP heats and affects its immediate surrounding is the most fundamental question, with impact in all three situations. In addition, single NP heating has some interesting and unique applications in its own right, such as nano- or molecular surgery (Csaki 2007) and photothermal imaging of single GNPs beyond the diffraction limit (Boyer 2002). The heating of a single cell loaded with GNPs is important for selective cell ablation, for example, detecting and treating circulating tumor cells (Galanzha 2009). Finally, the heating of the entire tumor loaded with GNPs (i.e., GNP photothermal surgery) leads to thermal injury and subsequent tumor necrosis and regression, a topic of increasing clinical interest.

18.5.2 Thermal Response of GNP Heating

18.5.2.1 Nanoscale Heating (T_nano)

The heating of a single GNP and the interaction with its immediate surroundings can be treated as the heat dissipation from a sphere to its surroundings medium (i.e., water in a biological environment as shown in Figure 18.5a) (Goldenberg 1952). To justify the use of continuum theory for the analysis, the mean free path for gold and water and the Knudsen numbers for different length scales are listed in Tables 18.3 and 18.4 along with other thermal properties. The mean free path of water is about 0.2 nm, which is two orders of magnitude smaller than the NP size (10~100 nm), so it can be safely

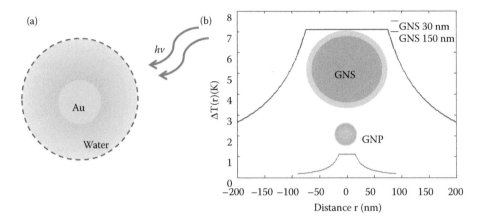

FIGURE 18.5 GNP heats up the immediate surrounding (assume one GNP in infinite medium) (a) Schematic of gold nanoparticle and its immediate surrounding; (b) The steady state temperature profile around the nanoparticle for two NPs, 30 nm spherical GNP and 150 nm gold nanoshell with 120 nm Si core, laser fluence 10^4 W/cm².

treated as a continuum. However, the mean free path in gold (31 nm) is comparable to the size of a GNP, and one cannot use the heat equation to describe the heat transfer within the particle. Here GNP with radius r_p is treated as a lumped system with the same temperature, and the heat equation for the surrounding medium ($r > r_p$) is given by:

$$\rho C \frac{\partial T}{\partial t} = \frac{1}{r^2} \frac{\partial}{\partial r} \left(r^2 k \frac{\partial T}{\partial r} \right). \qquad (18.7)$$

The complete solution to the transient heating is available elsewhere (Goldenberg 1952, Keblinski 2006). Here the nomenclature of Keblinski et al. (Keblinski 2006) is used, and the characteristic time of the transient heating process is described by:

$$\tau_{nano} = \frac{r_p^2}{\alpha} \qquad (18.8)$$

where α is the thermal diffusivity. This characteristic time, t_{nano}, is a measure of the time for the temperatures of the GNP surface and surrounding medium to reach steady state given a uniform internal heating of the GNP. Assuming water as the surrounding medium, and the particle sizes of 10 nm and 100 nm, the characteristic time τ_{nano} is on the order of 1 ns and 100 ns, respectively. This is also related to the *thermal confinement* concept introduced later and is schematically shown in Figure 18.6. Note that the characteristic time is only a rough estimate of the time scale, and one has to solve the governing equation to obtain more accurate estimations. Experimental measurements of heat dissipation from GNPs in aqueous solution show the same trend (i.e., the thermal relaxation time is proportional to the square of the particle radius (Hu 2002)).

After a short time (1–100 ns calculated before), the temperature distribution around a single continuously heated GNP in an infinite medium reaches steady state and can be easily obtained. Applying the constant heat flux boundary condition at the GNP surface ($r = r_p$) and negligible temperature change far away from the particle $T(r \to \infty) = T_\infty$, the surrounding medium temperature due to the heated GNP is given by

$$T(r) - T_\infty = \frac{\dot{Q}_{nano}}{4\pi k r}, \quad r \geq r_p \qquad (18.9)$$

TABLE 18.3 Diffusion Length and Knudsen Number Scaling

	Diffusion Length		Knudsen Number	
Time Scale (s)	H₂O	Au	H₂O	Au
10^{-15} (femto second)	0.012 nm	0.36 nm	16.67	86.11
10^{-12} (pico second)	0.38 nm	11 nm	0.53	2.82
10^{-9} (nano second)	12 nm	0.36 μm	0.02	0.09
10^{-6} (micro second)	0.38 μm	11 μm	5.26×10^{-4}	2.82×10^{-3}
10^{-3} (milli second)	12 μm	360 μm	1.67×10^{-5}	8.61×10^{-5}
1 (second)	380 μm	11 mm	5.26×10^{-7}	2.82×10^{-6}
60 (minute)	3 mm	87 mm	6.67×10^{-8}	3.56×10^{-7}
3600 (hour)	23 mm	68 cm	8.70×10^{-9}	4.56×10^{-8}

Note: (1) diffusion length = square root of (diffusivity × time); (2) the characteristic length for Knudsen number is the diffusion length (L); (3) continuum assumption holds for Kn << 1, i.e., $t \geq 10^{-9}$ s (nano second). Mean free paths (λ) for water and gold are listed in Table 18.4.

TABLE 18.4 Thermal Properties of Water and Gold Including the Thermal Conductivity, Density, Specific Heat, Thermal Diffusivity, and Mean Free Path

Properties	Liquid Water (H$_2$O)	Solid Gold (Au)	Tissue[a]
Thermal conductivity (k, W/m·K)	0.613	317	0.567
Density (ρ, kg/m³)	0.997×10^3	19.3×10^3	1.05×10^3
Specific heat (c_p, kJ/kg·K)	4.179	0.129	1.5
Thermal diffusivity (α, m²/s)	1.47×10^{-7}	1.27×10^{-4}	1.5×10^{-7}
Mean free path (λ, nm)	<1 nm, 0.2 nm	31	[b]

Note: All properties are at room temperature except noted.
[a] Human liver, data from Cooper et al. (1972)
[b] Tissue mean free path should be similar to water

where $\dot{Q}_{nano}$ is the heat generated by or in the GNP, and k is the thermal conductivity of the medium. The maximum temperature occurs at the NP surface, given by

$$T_{nano} = \frac{\dot{Q}_{nano}}{4\pi k r_p}. \tag{18.10}$$

One can use the typical laser parameters for GNP photothermal therapy to calculate this nanoscale temperature increase, T_{nano}. For example, Hirsch et al. (Hirsch 2003) designed a gold nanoshell with $r_p = 65 nm$ absorption cross section $C_{abs} = 3.8 \times 10^{-14} m^2$ and CW laser fluence $I = 4W/cm^2$. This yields $T_{nano} = 0.003k$ in water using properties in Table 18.4. The reason for this extremely small temperature increase is because of the large surface-to-volume ratio of the NP, which yields rapid heat dissipation. While a single GNP doesn't provide enough temperature increase for photothermal therapy at the nanoscale, the macroscopic temperature increase is significant as shown later, due to the collective heating effect of

many GNPs (Keblinski 2006, Richardson 2009). Another possibility of generating larger temperature increase is to increase the laser fluence with CW or pulsed laser. Figure 18.5b shows several degrees of temperature increase for spherical GNP and gold nanoshell with a laser fluence of 10^4 W/cm². Note that this is three orders of magnitude higher than the laser fluence used for *in vivo* photothermal therapy as shown and discussed later in Table 18.7.

18.5.2.2 Microscale Heating (T$_{micro}$)

The same methodology presented for single NP heating can be applied to the heating of a cell. The time for a heated cell to reach equilibrium with the surrounding medium can be written as

$$\tau_{micro} = \frac{r_{cell}^2}{\alpha}. \tag{18.11}$$

Considering a cell with radius $r_{cell} = 5\,\mu m$ and using the diffusivity of water, the characteristic time is around 0.7 ms. To

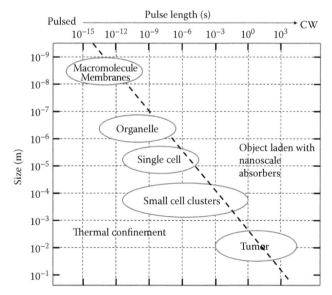

FIGURE 18.6 The concept of thermal confinement. To confine the heat within the object shown, one needs to work with the parameters at the lower left regime below the dashed line. (Adapted from Huettman, G. et al. *Proceedings of SPIE*, 5142, 2003.)

estimate the temperature increase of a single cell loaded with GNPs, one can treat the cell as being loaded with a uniform distribution of NPs, with a given number density, N. The temperature of the cell comes from the superposition of heating in all GNPs (Keblinski 2006):

$$T_{micro} = \frac{N r_{cell}^2 \dot{Q}_{nano}}{2k}.$$ (18.12)

For example, if one assumes there are 5000 gold nanoshells in a cell ($r_{cell} = 10$ μm) and irradiation is with a 35 W/cm² laser as described in Hirsch et al. (Hirsch 2003), the temperature increase is only about 0.09°C using the previous equation. This is because we assume that only one cell is present in an infinite medium. The ability of gold nanoshells to kill cells shown in this work and many other *in vitro* studies is due to the collective effect of many GNPs in the cell medium, causing the macroscopic heating described next. To selectively kill a single cancer cell with GNPs, one needs at least 2~3 orders of magnitude higher laser irradiance to provide over 10°C of temperature increase within a cell. This high laser irradiance can be provided by pulsed lasers as shown in the literature (Zharov 2005b, Kalambur 2007). However, pulsed laser will induce both thermal and mechanical effects as discussed later in Section 18.6.2.

18.5.2.3 Bulk Tissue Heating (T$_{macro}$)

As the length scale increases for bulk systems so does the diffusion time. The macroscopic diffusion time for a tumor can be written as

$$\tau_{macro} = \frac{r_{tumor}^2}{\alpha}.$$ (18.13)

This results in a diffusion time τ_{macro} of 7 s and 12 min for a 1 mm and 1 cm diameter tumor, respectively. Using a modification to the analysis, the temperature increase at the center of the tumor can be estimated as (Keblinski 2006, Rabin 2002)

$$\tau_{macro} = \frac{N R^2 \dot{Q}_{nano}}{2k}.$$ (18.14)

Again using the typical values for the photothermal therapy from Hirsch et al., consider tumor size $R = 5$ mm and gold nanoshell number density $N = 10^{15} \text{m}^{-3}$ (10^9/ml), the corresponding temperature increase is $T_{macro} = 31K$, which is sufficient for thermal therapy. One can perform this scaling easily given the NP's absorption cross section, concentration, and applied laser intensity as shown in Table 18.7.

18.5.3 The Concept of Thermal Confinement

In the discussion of heat diffusion and temperature scaling, an important underlying concept is *thermal confinement*

(Huettmann 2003, Anderson 1983, Vogel 2003). To treat a small target area with minimal injury to the surrounding tissue, the majority of the heat generated needs to be confined in the target area; this is usually called selective photothermolysis (Anderson 1983). There are usually two ways to achieve this: (1) to use a laser "microbeam" to achieve confined thermal damage; and (2) to use a broad beam but treat a pigmented or absorber-loaded target. For both methods, thermal confinement is essential for selective thermal damage. To confine the thermal energy, one has to induce the damage before the heat diffuses to the surrounding medium. The diffusion time or the thermal relaxation time for a sphere is given by

$$\tau_r = \frac{d^2}{27\alpha}.$$ (18.15)

The thermal relaxation time scales with the square of the target size. So the diffusion time scales down rapidly as the size of the target decreases, shown in Figure 18.6. For example, to conduct "nanosurgery" to macromolecules with several K rise at the nanoparticle surface, nano- to femto-second pulsed lasers have to be used (Csaki 2007, Huettmann 2003). Note that with the use of nano absorbers, for example a GNP, "nanosurgery" is possible and can operate beyond the diffraction limit (i.e., <250 nm) while not otherwise possible with laser "microbeam." This possibility raises the question of whether thermal injury kinetics at high temperatures and short times will scale from lower temperatures and longer times. This area is not well understood although some work is beginning to address this interesting question (Yan 2010).

An important application of thermal confinement is SAR estimation, as discussed next. This is assuming that the heat diffusion during the time of estimation is negligible.

18.5.4 SAR Estimation

In cases when not all the parameters in Equation 18.3 can be obtained, one can measure SAR experimentally. In principle, SAR can be measured by two separate methods: (1) characterizing the laser fluence and absorption; or (2) directly measuring the temperature change within laser NP-heated systems. By obtaining the optical absorption and local fluence rate, the SAR can be calculated by the product of the two parameters as shown in Equation 18.3. This requires the measurement of α, the absorption coefficient of the tissue (laden with NPs), and local fluence rate. The most direct way to measure local fluence is by invasive measurement with an optical fiber, although this can pose some challenges (Welch 1995).

The second approach to measuring SAR takes advantage of thermal confinement discussed earlier and uses local temperature change during short time periods before thermal diffusion dominates. This approach has the advantage that measuring

TABLE 18.5 Noninvasive Thermal Imaging Techniques for Thermal Therapy Monitoring

Method	Mapping	Spatial Resolution	Sensitivity	Speed	Ref.
IR[a]	Surface	~18 μm	~0.1°C	60Hz	(Childs 2000)
MRTI[b]	Yes	~0.16mm	-0.01 ppm/°C	0.2Hz	(Hirsch 2003, Schwartz 2009)
Photoacoustic	Yes	~50 μm	~0.16°C[c]	>10Hz	(Shah 2008b)
Ultrasound	Yes	~0.3mm	~0.05°C	>10Hz	(Shah 2008a, Shah 2008b, Liu 2010)

Note: The spatial resolution, temperature sensitivity, and speed of each technique are estimated from the references listed.

[a] Infrared thermometry

[b] Magnetic resonance temperature imaging

[c] The resolution of photoacoustic temperature imaging depends on the stability of pulsed laser used.

temperature is much easier than measuring the fluence rate distribution. The SAR can then be obtained by calculating the initial slope of the local temperature change, usually over a matter of seconds:

$$SAR = \rho C \left(\frac{dT}{dt} \right)_{initial} \quad (W/cm^3). \quad (18.16)$$

Recently, noninvasive or minimally invasive methods, such as infrared thermography, magnetic resonance temperature imaging (MRTI), ultrasound thermometry, and photoacoustic thermometry, have been proposed to guide and monitor thermal therapy. Typical properties of these methods, including the 3D mapping capability, spatial resolution, temperature sensitivity, and speed, are listed in Table 18.5. Among these methods, Shah et al. (Shah 2008b) showed that simultaneous ultrasound and photoacoustic temperature measurements give results within less than 0.5°C difference, with photoacoustic thermal imaging showing higher signal to noise than ultrasound. Magnetic resonance temperature imaging (MRTI) is also a promising technique that is undergoing several clinical trials to monitor laser thermal treatment (Clinicaltrials.gov study number NCT00392119, NCT00720837, and NCT00787982). These techniques offer exciting opportunities for *in vivo* treatment monitoring, SAR, and injury assessment based on delivered GNPs.

18.6 Laser GNP Effects II—Physical and Biological Responses at Multiple Scales

The heating of GNP with laser can induce some physical effects, for example, phase change of the surrounding medium, and biological responses when present in cellular and tissue systems. In this section, these effects are reviewed at the nanoscale, cellular, and tissue levels similar to the previous section. For the *in vitro* (cellular) and *in vivo* (tissue) effects, fundamental questions, including GNP biodistribution, mechanism of injury, and treatment outcome, are discussed.

18.6.1 Nanoscale Effects

High intensity CW and pulsed laser can transiently heat up a GNP at the nanoscale (Zharov 2006, Keblinski 2006, Carlson 2011). Figure 18.5b shows several degrees of temperature increase for a gold nanoshell and sphere under laser fluence of 10^4 W/cm². With even higher power pulsed lasers, the temperature of the GNP can increase over thousands of degrees within a very short period of time (1–100 ns scaled above). Depending on the energy input, different responses in the surrounding medium and GNP occur. These include the phase change of the medium (melting ice and polymer) (Richardson 2006, Govorov 2006), selective protein denaturation around the NP (Huettmann 2003, Pitsillides 2003), acoustic wave formation because of the particle expansion (Zharov 2007), vaporization of the water around the particle (Zharov 2005b), melting of the particle itself, vaporization of GNP (Letfullin 2006), optical breakdown and plasma formation (Takeda 2006), and eventually particle fragmentation and degradation (Letfullin 2006). These processes have been recently summarized by Pustovalov et al. (Pustovalov 2008), and are shown schematically in Figure 18.7, along with the phase diagrams of water and gold (refer to phase change properties in Table 18.6). Both theoretical (Pustovalov 2008, Merabia 2009) and experimental efforts have been undertaken to understand these processes, however, the complexity of the problem and difficulty of measurements at small time and length scales have limited our complete understanding. Each of these processes have important applications, for example, acoustic wave formation for photoacoustic imaging (Shah 2008b), and bubble formation around the GNP and/or GNP fragmentation for enhanced tumor treatment (Zharov 2005a). While bubble generation is promising for both diagnostic and therapeutic purposes, the fragmentation of GNPs may be difficult to control and requires high laser energy input (Lapotko 2009).

18.6.2 *In Vitro* Cellular Level Effects

The introduction of GNP to the cell increases the absorption and thus heat generation within the cell. Selective destruction can be achieved by targeting GNPs to the cancer cells and then applying laser irradiation. However, successful thermal destruction is only possible with specific conditions (NP and laser parameters).

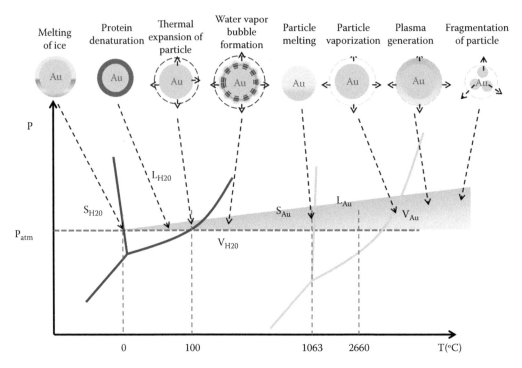

FIGURE 18.7 The responses and accompanied phenomena of GNP heating under laser activation. From left to right, as the laser power increases, the GNP temperature increases and eventually leads to fragmentation of GNP. (Modified from Pustovalov, V. K., Smetannikov, A. S., and Zharov, V. P., *Laser Physics Letters*, 5, 2008.)

As estimated in the previous section, the characteristic time for thermal relaxation of single cells is sub-millisecond, and temperature elevation is small for CW laser. Thus, a high power pulsed laser is usually required to induce significant temperature increase and cell death for single cells. The mechanisms of cell injury with the presence of GNP under laser irradiation can be different depending on the localization of GNP within cells and laser parameters used. Next, the biodistribution of GNPs in cells, and the thermal and nonthermal injuries (mechanical and chemical effects) after laser irradiation are reviewed.

18.6.2.1 Cellular GNP Biodistribution

Understanding the interactions of GNPs with cells is necessary to effectively target contrast agent and therapeutics within the cell. The size, shape, and surface chemistry of NPs affect the cellular uptake amount and intracellular pathways, and may eventually lead

TABLE 18.6 Thermodynamic Properties for Gold and Water Phase Changes

Substance	Phase Transition	Temperature (K)	Temperature (°C)	Heat of Phase Change/Pressure
Au	Melting	1336.15	1063.00	64.5 kJ/kg
	Vaporization	2933	2660	1578 kJ/kg
	Triple point[a]	1336	1063	0.025 Pa
	Critical point[b]	7400	7127	5300×10^5 Pa
H_2O	Melting	273.15	0.00	334 kJ/kg
	Vaporization	373.15	100.00	2256 kJ/kg
	Triple point	273.16	0.01	0.0061×10^5 Pa
	Critical point	647	374	220.6×10^5 Pa

Note: The melting, vaporization, triple, and critical points are listed with corresponding temperature (in Kelvin and Celsius) and latent heat. For triple point and critical point, the pressure is listed instead of latent heat. All the data are taken from NIST standard reference data, the NIST chemistry WebBook (http://webbook.nist. gov/chemistry/) unless noted.

[a] The triple point of Au was calculated from thermodynamic relations (http://oregonstate.edu/instruct/ me581/Homework/F08/ME581Hmwk3.html).

[b] Boboridis et al. (1999)

to different intracellular fates and biological responses (Chou 2011). The most common pathway is receptor mediated endocytosis. It has been shown that the cellular uptake of GNP is size and shape dependent, and an optimum size exists (~50nm) for maximum GNP cellular uptake (Chithrani 2006). For detailed discussion of the intracellular delivery of NPs, one can refer to the extensive review of Chou et al. (Chou 2011) and the references therein.

18.6.2.2 Cellular Thermal Injury (Kinetics)

Cell injury after heating is considered a kinetic process depending on both the temperature and time (Sapareto 1978) and protein denaturation is highly correlated to thermally induced cell death (Lepock 1993, He 2004). However, the mechanism and kinetics of thermally induced cell death and protein denaturation at high temperatures above 50°C and especially near boiling, such as that achievable under pulsed laser irradiation, is an area of active investigation (He 2003, He 2009, Yan 2010). According to the Arrhenius kinetic model, the heat induced cell injury (Ω) or survival fraction (S) can be described as

$$\Omega = \ln\left(\frac{1}{S}\right) = \int_0^\tau A e^{\frac{-E}{RT(t)}} dt \qquad (18.17)$$

where A is the frequency factor (s^{-1}), ΔE is the activation energy (J mol^{-1}), R is the universal gas constant, $T(t)$ is the temperature history the cell experiences, t is the time, and τ is the duration of the heating. The kinetic parameters, A and ΔE, are cell type (and thus tissue) and assay dependent (Bhowmick 2000, He 2009, He 2004). The thermal injury of cell (or tissue) accumulates faster at higher temperatures. According to the clinically used thermal dose model (TID), a derivative of the Arrhenius model, it takes roughly half the time to induce the same injury for every degree increase in temperature. Not surprisingly, the TID model and the Arrhenius model predict similar injury in various urologic cell and tissue systems (i.e., kidney, BPH, and prostate tumor) between 43.5 and 50°C. However, the TID model is less accurate than the Arrhenius model at higher temperatures (He 2009). For instance, TID predicts a longer time to accumulate injury at temperatures above 50°C than the Arrhenius model (i.e., 25% longer for kidney and BPH). Computational and experimental studies to explain how thermal injury kinetics scale at higher temperatures (i.e., near boiling) is of continuing interest in the field of thermal injury and directly relevant to laser tissue and laser GNP therapies (Yan 2010).

Using CW laser, it has been shown that laser treatment is selective for cells that have taken up GNPs versus cells without GNPs (Huang 2006, El-Sayed 2006, Hirsch 2003, Loo 2005). For example, using antibody-conjugated gold nanorods, Huang et al. showed that the treatment of malignant oral epithelial cells after incubation with gold nanorods requires about half the energy versus cells that did not take up nanorods (Huang 2006). This CW laser treatment usually takes several minutes, which corresponds to a steady state macroscopic temperature increase instead of just local subcellular heating. The spatiotemporal changes of

temperature have been modeled or characterized for a laser beam irradiating the monolayer cells with GNR (Huang 2010). The cell viabilities at different positions were then predicted and compared with experimental measurement. It is still unclear whether the presence of GNPs in cells affects the thermal injury kinetics.

On the other hand, if a pulsed laser is used with a pulse duration that is shorter than or comparable with the thermal equilibrium time of the GNP or cell, local injury can be achieved (selective nanophotothermolysis) (Zharov 2005a, Zharov 2006, Anderson 1983). For example, in a mixed-cell suspension Kalambur et al. showed that a nanosecond pulsed laser can selectively destroy malignant tumor cells loaded with NPs while leaving normal cells intact (without NPs) (Kalambur 2007). This implies highly localized effects such as bubble formation around the NPs, which are suggested to form and then collapse thereby destroying cell membranes by cavitation as discussed next (Zharov 2005a, Zharov 2006, Anderson 1983).

18.6.2.3 Nonthermal Injury (Mechanical and Chemical)

If the absorbed energy is high enough, phase change (i.e., boiling) occurs. The ensuing bubble generation and cavitation can induce mechanical stress and lead to selective killing of cancer cells. A schematic of the pressure-temperature water phase diagram showing the relative position of these and other heating events is shown in Figure 18.7.

Lin and coworkers showed that the immuno-targeted GNPs (30 nm) induce selective cell injury under pulsed laser irradiation (Pitsillides 2003). Detailed heat transfer analysis showed that bubble generation and cavitation damage was responsible for cell killing. Recently, Lapotko and coworkers showed that the bubbles generated by GNPs can be fine-tuned for imaging and therapeutic purposes (Lapotko 2009, Lukianova-Hleb 2010). Small photothermal bubbles serve as sensitive imaging contrast agents for target cells without significant damage to the cell, while bigger bubbles lead to disruption of cellular structure and cell death.

The mechanical effects on the cell strongly depend on the localization of GNPs in the cell. If GNPs are targeted and bound to the cell membrane, increased cell membrane permeability can be achieved, and this could be used to deliver foreign molecules, such as drugs, proteins, and genes, to the selectively targeted cell (Pitsillides 2003). Further increasing the laser energy will compromise the integrity of the cell membrane and lead to cell death. Tong et al. showed that it requires much less energy to lyse the cell with membrane-bound gold nanorods than when these nanorods are in the cytoplasm (Tong 2007). Zharov et al. showed that the microbubbles overlapping on the cell membrane synergistically enhance the photothermolysis of targeted cancer cells (Zharov 2005b). These studies speak to the ability of laser heating to selectively heat GNPs at the subcellular level (i.e., no bulk heating) with consequent cell destruction.

Recently, it was shown that the presence of GNP under low power laser irradiation also increases the intracellular reactive oxygen species (ROS) level, which causes damage to the endosomal membranes (Krpetie 2010) and mitochondria (Tong 2009), and leads to apoptosis of the cell. The detailed mechanism of the ROS generation with GNPs is not clear and requires further studies.

18.6.3 *In Vivo* Effects

The use of GNP for photothermal therapy has been studied with *in vivo* tumor model systems (such as rodent and canine) and is currently in a clinical trial (discussed in Section 18.7). This section will focus on the issues of implementing GNP for photothermal therapy for tumors first by delivering the GNPs to the tumor (i.e., biodistribution) followed by laser heating.

18.6.3.1 Overview of GNP Biodistribution

The successful intravenous delivery of NPs to the tumor *in vivo* relies on the enhanced permeability and retention (EPR) effects due to the leaky vasculature in the tumor tissue. As the NPs are circulating in the blood stream, they penetrate the leaky vessels and accumulate in the tumor. Biodistribution studies show that the reticuloendothelial system (RES) organs, such as liver and spleen, take up the majority of the NPs injected (Perrault 2009, Goel 2009). It is critical to increase the blood circulation time of GNPs to enhance the NP delivery to the tumor. Some controversy remains as to whether active delivery with tumor targeting ligands will increase the total NP accumulation in the tumor (Nie 2010). The tumor uptake of GNPs for several representative

studies is listed in Table 18.7. One can see that the accumulation of GNPs in the tumor is quite different between different studies. After penetrating the leaky vessels, NPs will diffuse away from the microvasculature. This process is size and charge dependent. Larger NPs diffuse slower and tend to stay near the microvasculature, while smaller NPs diffuse faster and distribute more uniformly throughout the tumor tissue over time (Perrault 2009). The surface charge of NPs also affects the transport of NP in the tissue. Positively charged NPs are more quickly internalized in cancer cells, and negatively charged NPs diffuse deeper into the tissue (Kim 2010). These fundamental studies can help the engineering design of NPs for different purposes. For example, small negatively charged NPs would be favorable for delivery of attached drugs or contrast agents into deep tissue. The shape of the NPs may also affect the transport within the tissue, but this effect requires further study.

Comprehensive reviews on biodistribution (Khlebtsov 2010), toxicity (Khlebtsov 2010, Fischer 2007), and immune response of GNPs (Dobrovolskaia 2009, Dobrovolskaia 2007), are available for the interested reader. Once the GNP is delivered to the tumor and the amount of GNP is sufficient for heating (based on the scaling of Equation 18.14), the laser can be applied for photothermal therapy as discussed next.

18.6.3.2 Thermal Injury (Kinetics)

So far the thermal injury kinetics for *in vivo* laser GNP photothermal therapy has not been carefully studied. Tissue responses

TABLE 18.7 Representative Studies on *In Vivo* Laser and Nanoparticle Doses for Cancer Photothermal Therapy and Scaling of Temperature Change

NP type	NP Info.	NP Dose[a]	NP Conc. at Tumor[b]	Laser Dose	ΔTscaling (K)	ΔTexp (K)	Ref.
Au nanoshell	110 nm Si core/10 nm Au shell, PEG[c]	20~50 μL interstitial injection, 1.5×10^{10} NPs/ml	5–13 μg/g [d]	820 nm, 4 W/cm², 5 mm[f], <6 min	20 ~ 50	28~60	(Hirsch 2003)
	110 nm Si core/10 nm Au shell, PEG	100 μL, 2.4×10^{11} NPs/ml	NA	808 nm, 4 W/cm², 5 mm, 3 min	NA	20[g]	(O'Neal 2004)
	119 nm Si core/12 nm Au shell, PEG	150 μL, 1.5×10^{11} NPs/ml	12.5 μg/g at 20 hr	808 nm, 4 W/cm², 5 mm, 3 min	30	NA	(Gobin 2007)
	35/55 nm Au/Au₂S	75 μL, 7.7×10^{11} NPs/ml	40 μg/g at 24 hr	808 nm, 4 W/cm², 5 mm, 3 min	50	16~30[g]	(Gobin 2010)
Au nanorod	14 × 47 nm, PEG	20 mg Au/kg	31 μg/g at 72 hr [e]	810 nm, 2 W/cm², 5 mm, 5 min	20	40	(von Maltzahn 2009)
	14 × 45 nm, PEG	6 ml/kg at 2.5×10^{12} NPs/ml;	NA	808 nm, 3.5 W, 3 min, diffuse fiber 1 cm tip	–	30	(Goodrich 2010)

Note: The information include the NP and its dose, the measured NP concentration at the tumor (if available), laser dose, temperature scaling, and reported temperature change. Note that the temperature obtained through scaling is just a rough estimate of the magnitude of temperature increase.

[a] The administration of NP is systemic unless otherwise noted.
[b] The unit μg/g is similar to μg/mL as the tissue density is ~1g/mL.
[c] PEG means that the particle is coated with polyethylene glycol (PEG)
[d] Assumptions include the tumor volume of 0.5 cm³ and uniform GNP distribution
[e] Animal weight of 22 g was used (von Maltzahn 2009).
[f] Laser beam size
[g] The mouse skin temperature of 30°C is assumed.

under heating have been extensively studied for traditional thermal therapies such as radiofrequency (RF), microwave (MW), and laser therapy (van Sonnenberg 2005). In addition to the heating effects for the cells and tissues under *in vitro* conditions, *in vivo* conditions often show enhanced injury due to vascular effects as previously shown (Bhowmick 2002, He 2004). These enhancements can be demonstrated by the same Arrhenius approach discussed previously for cellular injury, but using new parameters that govern vascular response (Bhowmick 2002, He 2004). Normal tissues respond to heat by increasing blood perfusion, sometimes up to an order of magnitude from nominal control conditions. When heating finally overstresses the tissue, blood perfusion will drop and ultimately vascular stasis occurs. Interestingly for tumor tissues, the initial increase of blood perfusion prior to vascular shutdown is small, predominately due to the incomplete and leaky vasculature of the tumor (Song 1984). Similar effects with GNP photothermal injury demonstrating a vascular effect are expected, although this has not been studied to our knowledge.

18.6.3.3 GNP Photothermal Therapy Doses and Outcome

In order to design and translate photothermal therapy with GNPs, it is important to design and control the doses of GNP and laser to obtain the desired and optimal SAR and corresponding temperature increase within the tumor for thermal injury. Typical parameters of GNP and laser fluence for small animal models are shown in Table 18.7. The doses of GNP to the animal and the resultant GNP delivery to the tumor are listed if available. One can see that the accumulation of GNP in the tumor covers a wide range from several µg/g to as high as 40 µg/g. The laser dose variation is smaller, with laser intensity between 2–4 W/cm² and duration 3–6 min. The corresponding temperature increase with the laser irradiation is also included and most of these studied reached temperatures over 50°C based on surface measurement by infrared thermography (von Maltzahn 2009, Gobin 2010) or MRTI (Hirsch 2003, Schwartz 2009). Within several minutes of heating, the tumor necrosis occurs according to the thermal injury kinetics from traditional thermal therapy.

In vivo studies using GNP for photothermal therapy showed larger temperature increase and improved treatment outcome over laser only. Comparing the temperature increase of GNP loaded versus control tumors shows over a 10°C increase with GNP over control under the same laser irradiation (Hirsch 2003, Goodrich 2010, von Maltzahn 2009). This led to significant tumor regression in GNP loaded versus control tumors after the treatment (von Maltzahn 2009). Other studies report an increased survival in laser-irradiated mice with GNP loaded versus control tumors reported on the Kaplan-Meier survival curve (O'Neal 2004, Gobin 2007, Gobin 2010, Goodrich 2010). Further, by comparing laser thermal therapy with and without the application of GNPs, it can be shown that GNP presence significantly reduces the laser power needed for cancer destruction because of enhanced absorption. For example, using the same interstitial laser fiber, only 3.5 W was required for 3 min to treat a tumor after systematic injection of gold nanoshells (Schwartz

2009), while over 10 W for 10 min was required without GNP (Müller 1995). This leads to an order of magnitude reduction in the laser energy needed to treat a tumor due to enhanced absorption of GNPs, as shown in Figure 18.7.

18.7 Clinical Studies

Previous clinical studies for rheumatoid arthritis (RA) and drug delivery confirm the biocompatibility of gold and help demonstrate its safety for photothermal therapy trials. For example, since the 1920s, gold in compound form (injectable gold) (Rau 2005) has been shown to be well tolerated and used for the treatment of RA. Further, the nanoparticle form of gold has successfully completed a phase I clinical trial (Clinicaltrials. gov study number NCT00356980) as a drug delivery agent "Aurimmune" by CytImmune Sciences Inc. (TNF bound to the surface of GNP) (Libutti 2010). The clinical trial with GNP photothermal therapy has focused on head and neck tumors, where the survival rate has improved little during the past 50 years (El-Sayed 2010). In this case, "AuroLase" therapy (Nanospectra Biosciences Inc.) with gold nanoshells is used for refractory and recurrent head and neck cancer (Clinicaltrials.gov study number NCT00848042). The gold nanoshells coated with polyethylene glycol (PEG) are infused into the blood stream and then accumulate passively in the solid tumor. No evidence of particle-induced systemic toxicity was found in animal studies (Payne 2010). A diffuse fiber is then inserted into the solid tumor and NIR laser light is applied to activate the particles.

Further clinical studies on combinatorial therapies are expected. For example, the presence of GNPs has recently been shown to enhance radiotherapy because of the enhanced x-ray absorption (Hainfeld 2004). Further, GNPs with TNF have been shown to sensitize tumors to thermal therapies (Shenoi 2011). It is thus expected that combinatorial drug or radiation treatments with GNP photothermal therapy may be attempted in the near future.

18.8 Conclusion and Future Studies

The area of laser photothermal therapy for cancer treatment has attracted intensive research efforts recently, and the number of publications in this area has risen dramatically in the past five years, from 13 in 2005 to over 150 in 2010 (search term: gold nanoparticle AND cancer, ISI Web of Knowledge; http://www. isiknowledge.com). With the increased absorption with the GNPs, tumor destruction can be more selective compared with traditional thermal therapies, and the laser exposure can be significantly reduced as discussed in Figure 18.2b. Many of the interesting properties of GNPs have been reviewed in this chapter, however, new multifunctional NPs that have both diagnostic and therapeutic functions are also being developed (Sanvicens 2008). Effective translation of both the traditional and new multifunctional NPs to clinical use in photothermal therapy may be hampered by safety concerns and a lack of techniques to monitor the location and heating patterns of the NPs within tissue.

A nonexhaustive list of future work expected to contribute to the further understanding and use of GNPs in photothermal therapy follows:

- Improved understanding of laser light movement in tissues with and without GNP loading
- Improved understanding and measurement of optical properties of GNP laden tissue
- Improved (noninvasive and accurate) assessment of GNP intra-organ biodistribution (tumor and RES organs)
- Enhanced tumor uptake of NP and reduced RES organ capture through NP design
- Real-time thermometry to monitor the laser nanoparticle treatment and dose estimation
- Efficient and accurate pretreatment planning and intraoperative tools incorporating the GNP biodistribution and thermometry
- Demonstration and use of enhanced cell injury mechanisms (thermal and nonthermal) from GNPs during CW and pulsed laser treatment
- Development of adjuvants for GNP photothermal therapy that allow improved cancer destruction.

In summary, NP laser photothermal therapy is an exciting area made possible by the unique optical and transport properties of GNPs and tissue. Continued work on biodistribution, targeting, and absorption properties of GNPs and in image guidance, thermometry, and adjuvant use will continue to improve this important new minimally invasive cancer treatment in the years to come.

Acknowledgments

The authors thank bio-heat mass transfer (BHMT) lab members (Michael Etheridge and Neha Shah) for the careful reading and comments of the manuscript. This work is supported partially by Minnesota Futures grant from the University of Minnesota.

References

Anderson, R. and Parrish, J. (1983) Selective photothermolysis: Precise microsurgery by selective absorption of pulsed radiation. *Science,* 220, 524.

Baish, J. W. (2000) Microvascular heat transfer. In Bronzino, J. (ed.) *The biomedical engineering handbook.* CRC Pr I Llc.

Bhowmick, S., Hoffmann, N. E., and Bischof, J. C. (2002) Thermal therapy of prostate tumor tissue in dorsal skin flap chamber. *Microvscular Research,* 64, 170–173.

Bhowmick, S., Swanlund, D. J. & Bischof, J. C. (2000) Supraphysiological thermal injury in Dunning AT-1 prostate tumor cells. *Journal of Biomechanical Engineering,* 122, 51–59.

Boboridis, K., Pottlacher, G. and Jäger, H. (1999) Determination of the critical point of gold. *International Journal of Thermophysics,* 20, 1289–1297.

Bohren, C. and Huffman, D. (1983) *Absorption and scattering of light by small particles.* New York: Wiley.

Boulnois, J.-L. (1986) Photophysical processes in recent medical laser developments: A review. *Lasers in Medical Science,* 1, 47–66.

Boyer, D., Tamarat, P., Maali, A. et al. (2002) Photothermal imaging of nanometer-sized metal particles among scatterers. *Science,* 297, 1160.

Burda, C., Chen, X., Narayanan, R. et al. (2005) Chemistry and properties of nanocrystals of different shapes. *Chem. Rev,* 105, 1025–1102.

Carlson, M. T., Khan, A. and Richardson, H. H. (2011) Local temperature determination of optically excited nanoparticles and nanodots. *Nano Letters,* 11, 1061–1069.

Chen, J., Saeki, F., Wiley, B. J. et al. (2005) Gold nanocages: Bioconjugation and their potential use as optical imaging contrast agents. *Nano Letters,* 5, 473–477.

Cheong, W., Prahl, S. and Welch, A. (1990) A review of the optical properties of biological tissues. *IEEE Journal of Quantum Electronics,* 26, 2166–2185.

Childs, P., Greenwood, J. and Long, C. (2000) Review of temperature measurement. *Review of Scientific Instruments,* 71, 2959.

Chithrani, B. and Chan, W. (2007) Elucidating the mechanism of cellular uptake and removal of protein-coated gold nanoparticles of different sizes and shapes. *Nano Lett,* 7, 1542–1550.

Chithrani, B. D., Ghazani, A. A. and Chan, W. C. W. (2006) Determining the size and shape dependence of gold nanoparticle uptake into mammalian cells. *Nano Letters,* 6, 662–668.

Chou, L. Y. T., Ming, K. and Chan, W. C. W. (2011) Strategies for the intracellular delivery of nanoparticles. *Chemical Society Reviews,* 40, 233–245.

Clinicaltrials.Gov. Study Number Nct00392119, Nct00787982, Nct00356980, and Nct00848042.

Cole, J. R., Mirin, N. A., Knight, M. W. et al. (2009) Photothermal efficiencies of nanoshells and nanorods for clinical therapeutic applications. *Journal of Physical Chemistry C,* 113, 12090–12094.

Cooper, T. and Trezek, G. (1972) A probe technique for determining the thermal conductivity of tissue. *Journal of Heat Transfer,* 94, 133.

Csaki, A., Garwe, F., Steinbruck, A. et al. (2007) A parallel approach for subwavelength molecular surgery using gene-specific positioned metal nanoparticles as laser light antennas. *Nano Lett,* 7, 247–53.

Daniel, M. and Astruc, D. (2004) Gold nanoparticles: Assembly, supramolecular chemistry, quantum-size-related properties, and applications toward biology, catalysis, and nanotechnology. *Chemical Reviews,* 104, 293–346.

Day, E. S., Morton, J. G. and West, J. L. (2009) Nanoparticles for thermal cancer therapy. *Journal of Biomechanical Engineering,* 131, 074001–5.

Dobrovolskaia, M., Germolec, D. and Weaver, J. (2009) Evaluation of nanoparticle immunotoxicity. *Nature Nanotechnology,* 4, 411–414.

Dobrovolskaia, M. and McNeil, S. (2007) Immunological properties of engineered nanomaterials. *Nature Nanotechnology,* 2, 469–478.

El-Sayed, I. (2010) Nanotechnology in head and neck cancer: The race is on. *Current Oncology Reports,* 12, 121–128.

El-Sayed, I. H., Huang, X. and El-Sayed, M. A. (2006) Selective laser photo-thermal therapy of epithelial carcinoma using anti-EGFR antibody conjugated gold nanoparticles. *Cancer Letters,* 239, 129–135.

Elliott, A., Schwartz, J., Wang, J. et al. (2009) Quantitative comparison of delta P1 versus optical diffusion approximations for modeling near-infrared gold nanoshell heating. *Medical Physics,* 36, 1351.

Elliott, A. M., Stafford, R. J., Schwartz, J. et al. (2007) Laser-induced thermal response and characterization of nanoparticles for cancer treatment using magnetic resonance thermal imaging. *Medical Physics,* 34, 3102–3108.

Farrell, T., Patterson, M. and Wilson, B. (1992) A diffusion theory model of spatially resolved, steady-state diffuse reflectance for the noninvasive determination of tissue optical properties *in vivo. Med. Phys,* 19, 879–888.

Feng, Y., Fuentes, D., Hawkins, A. et al. (2009) Nanoshell-mediated laser surgery simulation for prostate cancer treatment. *Engineering with Computers,* 25, 3–13.

Fischer, H. and Chan, W. (2007) Nanotoxicity: The growing need for *in vivo* study. *Current Opinion in Biotechnology,* 18, 565–571.

Galanzha, E., Shashkov, E., Kelly, T. et al. (2009) *In vivo* magnetic enrichment and multiplex photoacoustic detection of circulating tumour cells. *Nature Nanotechnology,* 4, 855–860.

Gobin, A., Lee, M., Halas, N. et al. (2007) Near-infrared resonant nanoshells for combined optical imaging and photothermal cancer therapy. *Nano Lett,* 7, 1929–1934.

Gobin, A. M., Watkins, E. M., Quevedo, E. et al. (2010) Near infrared resonant gold/gold sulfide nanoparticles as a photothermal cancer therapeutic agent. *Small,* 6, 745–752.

Goel, R., Shah, N., Visaria, R. et al. (2009) Biodistribution of TNF-alpha-coated gold nanoparticles in an *in vivo* model system. *Nanomedicine (Lond),* 4, 401–10.

Goldenberg, H. and Tranter, C. J. (1952) Heat flow in an infinite medium heated by a sphere. *British Journal of Applied Physics,* 296.

Goodrich, G. P., Bao, L. L., Gill-Sharp, K. et al. (2010) Photothermal therapy in a murine colon cancer model using near-infrared absorbing gold nanorods. *Journal of Biomedical Optics,* 15, 018001.

Govorov, A., Zhang, W., Skeini, T. et al. (2006) Gold nanoparticle ensembles as heaters and actuators: Melting and collective plasmon resonances. *Nanoscale Research Letters,* 1, 84–90.

Hainfeld, J., Slatkin, D. and Smilowitz, H. (2004) The use of gold nanoparticles to enhance radiotherapy in mice. *Physics in Medicine and Biology,* 49, N309.

He, X., Bhowmick, S. and Bischof, J. C. (2009) Thermal therapy in urologic systems: A comparison of arrhenius and thermal isoeffective dose models in predicting hyperthermic injury. *J Biomech Eng,* 131, 074507.

He, X. and Bischof, J. (2003) Quantification of temperature and injury response in thermal therapy and cryosurgery. *Critical Reviews in Biomedical Engineering,* 31, 355–421.

He, X., Wolkers, W. F., Crowe, J. H. et al. (2004) *In situ* thermal denaturation of proteins in dunning AT-1 prostate cancer cells: Implication for hyperthermic cell injury. *Ann Biomed Eng,* 32, 1384–98.

Hirsch, L., Stafford, R., Bankson, J. et al. (2003) Nanoshell-mediated near-infrared thermal therapy of tumors under magnetic resonance guidance. *Proceedings of the National Academy of Sciences of the United States of America,* 100, 13549.

Hirsch, L. R., Gobin, A. M., Lowery, A. R. et al. (2006) Metal nanoshells. *Annals of Biomedical Engineering,* 34, 15–22.

Hu, M., Chen, J., Li, Z. et al. (2006) Gold nanostructures: Engineering their plasmonic properties for biomedical applications. *Chemical Society Reviews,* 35, 1084–1094.

Hu, M. and Hartland, G. (2002) Heat dissipation for Au particles in aqueous solution: Relaxation time versus size. *J. Phys. Chem. B,* 106, 7029–7033.

Huang, H.-C., Rege, K. and Heys, J. J. (2010) Spatiotemporal temperature distribution and cancer cell death in response to extracellular hyperthermia induced by gold nanorods. *ACS Nano,* 4, 2892–2900.

Huang, X., El-Sayed, I. H., Qian, W. et al. (2006) Cancer cell imaging and photothermal therapy in the near-infrared region by using gold nanorods. *Journal of the American Chemical Society,* 128, 2115–2120.

Huang, X., Neretina, S. and El-Sayed, M. (2009) Gold nanorods: From synthesis and properties to biological and biomedical applications. *Advanced Materials,* 21, 4880–4910.

Huettmann, G., Radt, B., Serbin, J. et al. (2003) Inactivation of proteins by irradiation of gold nanoparticles with nano-and picosecond laser pulses. *Proceedings of SPIE,* 5142, 88–95.

Jacques, S. (1993) Role of tissue optics and pulse duration on tissue effects during high-power laser irradiation. *Applied Optics,* 32, 2447–2454.

Jain, P., Lee, K., El-Sayed, I. et al. (2006) Calculated absorption and scattering properties of gold nanoparticles of different size, shape, and composition: Applications in biological imaging and biomedicine. *J. Phys. Chem. B,* 110, 7238–7248.

Jain, P. K., El-Sayed, I. H. and El-Sayed, M. A. (2007a) Au nanoparticles target cancer. *Nano Today,* 2, 18–29.

Jain, P. K., Huang, X., El-Sayed, I. H. et al. (2007b) Review of some interesting surface plasmon resonance-enhanced properties of noble metal nanoparticles and their applications to biosystems. *Plasmonics,* 2, 107–118.

Jöbsis-Vandervliet, F. (1999) Discovery of the near-infrared window into the body and the early development of near-infrared spectroscopy. *Journal of Biomedical Optics,* 4, 392.

Judy, M. M. (2000) Biomedical lasers. In Bronzino, J. D. (ed.) *The biomedical engineering handbook.* CRC Pr I Llc.

Kalambur, V. S., Longmire, E. K. and Bischof, J. C. (2007) Cellular level loading and heating of superparamagnetic iron oxide nanoparticles. *Langmuir,* 23, 12329–36.

Keblinski, P., Cahill, D. G., Bodapati, A. et al. (2006) Limits of localized heating by electromagnetically excited nanoparticles. *Journal of Applied Physics,* 100, 054305.

Kelly, K., Coronado, E., Zhao, L. et al. (2003) The optical properties of metal nanoparticles: The influence of size, shape, and dielectric environment. *J. Phys. Chem. B,* 107, 668–677.

Khlebtsov, N. and Dykman, L. (2010) Biodistribution and toxicity of engineered gold nanoparticles: A review of *in vitro* and *in vivo* studies. *Chemical Society Reviews,* 40, 1647–1671.

Kim, B., Han, G., Toley, B. J. et al. (2010) Tuning payload delivery in tumour cylindroids using gold nanoparticles. *Nat Nano,* 5, 465–472.

Kirillin, M., Shirmanova, M., Sirotkina, M. et al. (2009) Contrasting properties of gold nanoshells and titanium dioxide nanoparticles for optical coherence tomography imaging of skin: Monte Carlo simulations and *in vivo* study. *Journal of Biomedical Optics,* 14, 021017.

Kreibig, U. and Vollmer, M. (1995) *Optical properties of metal clusters,* Springer Verlag.

Krpetie, Z., Nativo, P., See, V. et al. (2010) Inflicting controlled nonthermal damage to subcellular structures by laser-activated gold nanoparticles. *Nano Letters,* 10, 4549–4554.

Kuntz, R. M. (2007) Laser treatment of benign prostatic hyperplasia. *World Journal of Urology,* 25, 241–247.

Lapotko, D. (2009) Plasmonic nanoparticle-generated photothermal bubbles and their biomedical applications. *Nanomedicine,* 4, 813–845.

Lepock, J. R., Frey, H. E. and Ritchie, K. P. (1993) Protein denaturation in intact hepatocytes and isolated cellular organelles during heat shock. *The Journal of Cell Biology,* 122, 1267–1276.

Letfullin, R. R., Joenathan, C., George, T. F. et al. (2006) Laser-induced explosion of gold nanoparticles: Potential role for nanophotothermolysis of cancer. *Nanomed,* 1, 473–80.

Libutti, S., Paciotti, G., Byrnes, A. et al. (2010) Phase I and pharmacokinetic studies of CYT-6091, a novel PEGylated colloidal gold-rhTNF nanomedicine. *Clinical Cancer Research,* 16, 6139–6149.

Lin, A., Lewinski, N., West, J. et al. (2005) Optically tunable nanoparticle contrast agents for early cancer detection: Model-based analysis of gold nanoshells. *Journal of Biomedical Optics,* 10, 064035.

Link, S. and El-Sayed, M. (1999) Size and temperature dependence of the plasmon absorption of colloidal gold nanoparticles. *J. Phys. Chem. B,* 103, 4212–4217.

Link, S. and El-Sayed, M. A. (2000) Shape and size dependence of radiative, non-radiative and photothermal properties of gold nanocrystals. *International Reviews in Physical Chemistry,* 19, 409–453.

Liu, D. and Ebbini, E. S. (2010) Real-time 2-D temperature imaging using ultrasound. *Biomedical Engineering, IEEE Transactions on,* 57, 12–16.

Loo, C., Lowery, A., Halas, N. et al. (2005) Immunotargeted nanoshells for integrated cancer imaging and therapy. *Nano Letters,* 5, 709–711.

Lukianova-Hleb, E., Hanna, E., Hafner, J. et al. (2010) Tunable plasmonic nanobubbles for cell theranostics. *Nanotechnology,* 21, 085102.

Merabia, S., Shenogin, S., Joly, L. et al. (2009) Heat transfer from nanoparticles: A corresponding state analysis. *Proceedings of the National Academy of Sciences,* 106, 15113–15118.

Modest, M. (2003) *Radiative heat transfer.* San Diego: Academic Press.

Müller, G. and Roggan, A. (1995) Laser-induced interstitial thermotherapy. SPIE-International Society for Optical Engineering, Bellingham.

Myroshnychenko, V., Rodriguez-Fernandez, J., Pastoriza-Santos, I. et al. (2008) Modelling the optical response of gold nanoparticles. *Chemical Society Reviews,* 37, 1792–1805.

Nehl, C. L., Liao, H. and Hafner, J. H. (2006) Optical properties of star-shaped gold nanoparticles. *Nano Letters,* 6, 683–688.

Nel, A., Madler, L., Velegol, D. et al. (2009) Understanding biophysicochemical interactions at the nano-bio interface. *Nat Mater,* 8, 543–557.

Nie, S. (2010) Understanding and overcoming major barriers in cancer nanomedicine. *Nanomedicine,* 5, 523–528.

Niemz, M. (2004) *Laser-tissue interactions: Fundamentals and applications,* Springer Verlag.

O'Neal, D. P., Hirsch, L. R., Halas, N. J. et al. (2004) Photo-thermal tumor ablation in mice using near infrared-absorbing nanoparticles. *Cancer Letters,* 209, 171–176.

Paltauf, G. and Dyer, P. (2003) Photomechanical processes and effects in ablation. *Chem. Rev,* 103, 487–518.

Patterson, M., Chance, B. and Wilson, B. (1989) Time resolved reflectance and transmittance for the noninvasive measurement of tissue optical properties. *Appl. Opt,* 28, 2331–2336.

Payne, D. (2010) Personal communication of unpublished data (Nanospectra Bioscience Inc.) at Thermal Therapy short course (University of Minnesota, May 2010).

Peng, Q., Juzeniene, A., Chen, J. et al. (2008) Lasers in medicine. *Reports on Progress in Physics,* 71, 056701.

Pennes, H. (1948) Analysis of tissue and arterial blood temperatures in the resting human forearm. *Journal of Applied Physiology,* 1, 93.

Perrault, S., Walkey, C., Jennings, T. et al. (2009) Mediating tumor targeting efficiency of nanoparticles through design. *Nano Letters,* 9, 1909–1915.

Pitsillides, C. M., Joe, E. K., Wei, X. et al. (2003) Selective cell targeting with light-absorbing microparticles and nanoparticles. *Biophys J,* 84, 4023–32.

Pustovalov, V. K., Smetannikov, A. S. and Zharov, V. P. (2008) Photothermal and accompanied phenomena of selective nanophotothermolysis with gold nanoparticles and laser pulses. *Laser Physics Letters,* 5, 775–792.

Qian, X. M. and Nie, S. M. (2008) Single-molecule and single-nanoparticle SERS: From fundamental mechanisms to biomedical applications. *Chemical Society Reviews*, 37, 912–920.

Rabin, Y. (2002) Is intracellular hyperthermia superior to extra-cellular hyperthermia in the thermal sense? *International Journal of Hyperthermia*, 18, 194–202.

Rau, R. (2005) Have traditional DMARDs had their day? *Clinical Rheumatology*, 24, 189–202.

Richardson, H. H., Carlson, M. T., Tandler, P. J. et al. (2009) Experimental and theoretical studies of light-to-heat conversion and collective heating effects in metal nanoparticle solutions. *Nano Letters*, 9, 1139–1146.

Richardson, H. H., Hickman, Z. N., Govorov, A. O. et al. (2006) Thermooptical properties of gold nanoparticles embedded in ice: Characterization of heat generation and melting. *Nano Letters*, 6, 783–788.

Sanvicens, N. and Marco, M. P. (2008) Multifunctional nanoparticles—properties and prospects for their use in human medicine. *Trends in Biotechnology*, 26, 425–433.

Sapareto, S. A., Hopwood, L. E., Dewey, W. C. et al. (1978) Effects of hyperthermia on survival and progression of Chinese hamster ovary cells. *Cancer Research*, 38, 393–400.

Sau, T. K. and Murphy, C. J. (2004) Room temperature, high-yield synthesis of multiple shapes of gold nanoparticles in aqueous solution. *Journal of the American Chemical Society*, 126, 8648–8649.

Schatz, G. C. (2007) Using theory and computation to model nanoscale properties. *Proceedings of the National Academy of Sciences*, 104, 6885–6892.

Schwartz, J., Shetty, A., Price, R. et al. (2009) Feasibility study of particle-assisted laser ablation of brain tumors in ortho-topic canine model. *Cancer Research*, 69, 1659.

Shah, J., Aglyamov, S. R., Sokolov, K. et al. (2008a) Ultrasound imaging to monitor photothermal therapy—feasibility study. *Opt Express*, 16, 3776–85.

Shah, J., Park, S., Aglyamov, S. et al. (2008b) Photoacoustic imaging and temperature measurement for photothermal cancer therapy. *Journal of Biomedical Optics*, 13, 034024.

Shenoi, M., Shah, N., Griffin, G. et al. (2011) Nanoparticle pre-conditioning for enhanced thermal therapies in cancer. *Nanomedicine*, in press.

Song, C. (1984) Effect of local hyperthermia on blood flow and microenvironment: A review. *Cancer Research*, 44, 4721s.

Srinivasan, R. (1986) Ablation of polymers and biological tissue by ultraviolet lasers. *Science*, 234, 559.

Takeda, Y., Kondow, T. and Mafuné, F. (2006) Degradation of protein in nanoplasma generated around gold nanoparticles in solution by laser irradiation. *J. Phys. Chem. B*, 110, 2393–2397.

Tcherniak, A., Ha, J. W., Dominguez-Medina, S. et al. (2010) Probing a century old prediction one plasmonic particle at a time. *Nano Letters*, 10, 1398–1404.

Tjahjono, I. K. and Bayazitoglu, Y. (2008) Near-infrared light heating of a slab by embedded nanoparticles. *International Journal of Heat and Mass Transfer*, 51, 1505–1515.

Tong, L. and Cheng, J. (2009) Gold nanorod-mediated photo-thermolysis induces apoptosis of macrophages via damage of mitochondria. *Nanomedicine*, 4, 265–276.

Tong, L., Zhao, Y., Huff T. et al. (2007) Gold nanorods mediate tumor cell death by compromising membrane integrity. *Advanced Materials*, 19, 3136–3141.

Ungureanu, C., Amelink, A., Rayavarapu, R. G. et al. (2010) Differential pathlength spectroscopy for the quantitation of optical properties of gold nanoparticles. *ACS Nano*, 4, 4081–4089.

van sonnenberg, E., Mcmullen, W. N. and Solbiati, L. (2005) *Tumor ablation: Principles and practice*, New York, New York : Springer.

Vera, J. and Bayazitoglu, Y. (2009) Gold nanoshell density variation with laser power for induced hyperthermia. *International Journal of Heat and Mass Transfer*, 52, 564–573.

Vogel, A. and Venugopalan, V. (2003) Mechanisms of pulsed laser ablation of biological tissues. *Chem. Rev*, 103, 577–644.

von Maltzahn, G., Park, J. H., Agrawal, A. et al. (2009) Computationally guided photothermal tumor therapy using long-circulating gold nanorod antennas. *Cancer Res*, 69, 3892–900.

Weissleder, R. (2001) A clearer vision for *in vivo* imaging. *Nature Biotechnology*, 19, 316–316.

Welch, A. and Van Gemert, M. (1995) *Optical-thermal response of laser-irradiated tissue*, Plenum Press New York.

Xu, X., Meade, A. and Bayazitoglu, Y. (2010) Fluence rate distribution in laser-induced interstitial thermotherapy by mesh free collocation. *International Journal of Heat and Mass Transfer*, 53, 4017–4022.

Yan, C., Pattani, V., Tunnell, J. W. et al. (2010) Temperature-induced unfolding of epidermal growth factor (EGF): Insight from molecular dynamics simulation. *Journal of Molecular Graphics and Modelling*, 29, 2–12.

Yu, Chang, S.-S., Lee, C.-L. et al. (1997) Gold nanorods: Electrochemical synthesis and optical properties. *Journal of Physical Chemistry B*, 101, 6661–6664.

Zharov, V. P., Galanzha, E. I., Shashkov, E. V. et al. (2007) Photoacoustic flow cytometry: Principle and application for real-time detection of circulating single nanoparticles, pathogens, and contrast dyes *in vivo*. *Journal of Biomedical Optics*, 12, 051503–14.

Zharov, V. P., Galitovskaya, E. N., Johnson, C. et al. (2005a) Synergistic enhancement of selective nanophotothermoly-sis with gold nanoclusters: Potential for cancer therapy. *Lasers in Surgery and Medicine*, 37, 219–226.

Zharov, V. P., Letfullin, R. R. and Galitovskaya, E. N. (2005b) Microbubbles-overlapping mode for laser killing of cancer cells with absorbing nanoparticle clusters. *Journal of Physics D-Applied Physics*, 38, 2571–2581.

Zharov, V. P., Mercer, K. E., Galitovskaya, E. N. et al. (2006) Photothermal nanotherapeutics and nanodiagnostics for selective killing of bacteria targeted with gold nanoparticles. *Biophys J*, 90, 619–27.

19

Thermochemical Ablation

Erik N. K. Cressman
University of Minnesota
Medical Center

The origins of thermochemical ablation stem from a clinical problem with many solutions of which none is ideal. The problem is that to be given a diagnosis of hepatocellular carcinoma (primary liver cancer, hepatoma, HCC) in many parts of the world is to be given a death sentence. The question foremost in the mind of most of these patients is in the absence of a cure, how long they have left to live. The answer is usually measured in months. This reflects a combination of factors that include the insidious yet aggressive nature of the disease, the cost and relative sensitivities of any available screening programs, the fact that many people do not know they are at risk and should be screened, and the limited availability of effective treatments. Worldwide, HCC is one of the leading causes of cancer death, and unlike more treatable malignancies, most people with HCC will die from HCC. Making a bad situation worse is the fact that the incidence of HCC is increasing.[1-3]

Cirrhosis from hepatitis B (HBV) is less common than cirrhosis from other forms of hepatitis, but due to the number of people infected with HBV, it is thought to be the most common cause of HCC. Vaccination programs for hepatitis B as a long-term solution are yielding very positive results, but there are already many developing cases of liver cancer from this single source. There is as yet no vaccination for hepatitis C (HCV), alcoholic liver disease is very common, and the incidence of fatty liver disease that leads to cirrhosis is arguably overtaking all other causes for this cancer. Etiologies are varied and are changing with the population and risk factors. With the rapid pace of advancements in molecular biology, researchers and clinicians now understand how diverse their target in fact is. This makes individualized treatment both critical and extremely challenging from a pharmacologic point of view. The situation as outlined herein has thus led to an enormous effort from many different angles for treatment.

Surgical interventions, such as liver transplantation or partial hepatectomy, can be potentially curative, and transplantation has the added advantage of treating the underlying cirrhosis. Unfortunately, there are numerous difficulties with this approach. The large majority of patients are not surgical candidates at presentation, and regardless of cost, the number of transplants performed is dwarfed by the demand. Another issue is that the nature of the disease is such that it will often recur elsewhere in the liver within a relatively short time.

Chemotherapy would seem to be a reasonable option for the treatment of liver cancer as it is for so many other forms of cancer. HCC has proven to be a difficult challenge, however, with well over 100 clinical trials of various regimens showing no survival benefit. Indeed, even achieving stable disease or a partial response has been uncommon. The advent of newer, targeted therapies such as sorafenib has been heralded as a turning point in the battle as a survival benefit was shown for the first time.[4, 5] The results must be viewed with some caution, however. The duration of the benefit was on average slightly less than three additional months. Furthermore, since the initial trials were run, there have been many papers published on how toxicity has resulted in a great many patients either lowering the dose of the drug or stopping it altogether.[6] Since the survival benefit was observed at the full dose and many patients cannot tolerate the full dose, it is not clear that the drug provides the anticipated benefit for a large portion of the population.

With surgery often excluded, and limited benefit from drug therapies, medicine has turned toward ablative therapies to treat HCC. The goal in the best cases would be to treat as effectively as possible while at the same time sparing as much non-tumor tissue as possible. There is an important semantic issue to make at this point, which is that in clinical practice it is inaccurate and in fact dangerously ignorant to refer to tumor versus normal tissue. This is critical because, as noted before, HCC most commonly occurs in the setting of cirrhosis. The cirrhotic liver is by no means "normal," and this has implications for how aggressive a physician can be in treating the disease. One must keep in mind that the underlying survival curve for decompensated liver disease is an intrinsic limiting factor in the maximal benefit that can be attained by any therapy. Lacking a replacement liver, there is a survival ceiling, as it were, beyond which there is no additional benefit.

To place thermochemical ablation in context, it is helpful to briefly review a number of existing or developing options. These include chemical ablation,[7, 8] thermal ablation (hot or cold),[9–11] external beam radiation, and, most recently, high intensity focused ultrasound (HIFU, in simplest terms another form of thermal ablation),[12–14] and irreversible electroporation. The transarterial therapies such as bland embolization, chemoembolization, and radioembolization are outside the scope of this discussion. It is worth noting, however, that there is a recognized survival benefit in appropriately chosen patients for chemoembolization on the order of 1–2 additional years of life and in many cases more.

Historically, chemical ablation is the oldest method for practical purposes and is performed by instillation of 95–100% ethyl alcohol into tumors under ultrasound or CT guidance using a sidehole needle.[15] The tumoricidal effect of this treatment has been attributed to dehydration and denaturation of the cells and proteins. Mechanistically, these are ideas are intertwined and to some extent redundant as protein structure is intimately related to the hydration state. The other chemical agent that has been used, although less widely accepted, is 50% acetic acid.[16, 17] It is thought to work in a similar manner, but adding a pH change that is assumed to play a role in the denaturation. Since it penetrates and solvates collagen better than ethanol, it is thought to do a better job than ethanol diffusing throughout tumor septae. This theoretically should allow better treatment of small satellite nodules that are not grossly apparent on imaging, leading to less local recurrence.

There are two main issues with these agents. One, inherent to all injectable therapies, is distribution of these agents that tracks along paths of least resistance. This can lead to injury to nearby structures and incomplete treatment, even in encapsulated tumors. The other is the intrinsic systemic toxicity resulting from exposure to either agent. Ethanol toxicity can manifest as CNS depression, respiratory depression, pulmonary hypertension, and cardiovascular collapse.[18–21] Less is known about acetic acid, but renal toxicity from hemolysis has been reported.[22] In either case, multiple sessions treating with small volumes has been the rule. To address the distribution issue, a multitine, multi-sidehole needle has been developed. Published reports show promising results, but it has not been widely adopted.[23–25]

Both of these issues are addressed by the thermal methods, but they too have limitations. Cryoablation has been reported but in the liver has not been as widely accepted as the hyperthermic methods, mainly radiofrequency (RF) and microwave (MW) ablation. Mechanistically, again hyperthermic protein denaturation has been invoked as the predominant mechanism of action and is described in detail elsewhere in this volume. The predictability of shape of the resulting coagulum or devitalized zone and injury to nearby tissues are the main problems for RF and MW ablation. The presence of larger blood vessels acting as heat sinks and even the perfusion-mediated cooling of tissues at the capillary level can lead to residual viable tumor and local recurrence. Assuming the area is completely treated, an advantage over chemical ablation is that lesions can generally be treated in one session. Commonly, these procedures are done under general anesthesia, which adds to both the risk and the cost. Economic situations are highly variable, but in general capital budgeting is required for power generators for RF and MW technology, and the single-use probes are somewhat costly.

The situation as outlined herein presented an opportunity to ask if it was possible to improve on chemical ablation in some way. Since the dose-limiting issue is systemic toxicity, this was a logical avenue to pursue. Initially, efforts were focused on decreasing the acid load by neutralization after the fact, which is to say injection of a base to neutralize the acid within the tissues. This poses at least two problems, however. One is how to ensure that the base would come in contact with the acid, and the other is what would happen if base does not come in contact with acid. This incomplete reaction would lead to treating other areas with base, systemic exposure to excess base, or most likely some degree of both problems could occur. It would be possible, if base were to react with the acid, though, for heat to be evolved. Thus was born the initial concept for thermochemical ablation. This was subsequently shown in our lab (unpublished results) and others to occur in tissues.[26] However, for the reasons outlined here, it did not appear to be a viable strategy. The concept then evolved, and the question of how to proceed became essentially inverted. That is to say, if the sole emphasis was on releasing heat in the tissues, how much would be available?

To answer this question, at least three general variables must be considered. First, what kind of tissue is under consideration? In other words, what is the specific heat of the target tissue? Second, how high of a temperature elevation is targeted and for how long? This equates to the traditional question of the target thermal dose. Finally, what kind of chemistry might be applied to the problem? Using an exotherm as the sole criterion leaves open far too many possibilities.

The first question can be dealt with in a straightforward fashion. For liver cancer, liver tissue, and liver tumor a specific heat of 3.6 kJ/kg-K has some basis in the literature.[27–29] This allows one to actually calculate requirements if a certain size tumor, 3 cm for example, is heated to 55°C, a temperature at or near which devitalization is nearly instantaneous. Allowing for a 1 cm margin, and for the moment ignoring any perfusion-mediated cooling factors, a sphere of 5 cm in diameter would have a volume of approximately 67 mL. Thus to heat such a volume of tissue from a body temperature of 37°C to 55°C, the change in temperature is 18°C. Assuming a time frame ranging from 1 sec to 100 sec, the energy input requirement to raise this volume of tissue by 18°C is 4–400 W. This calculation provided a target range for feasibility considerations for any particular chemistry.

The reaction of an acid and base was mentioned previously as an example of exothermic chemistry. The molar heat of formation for water, which is one of the products of acid-base neutralization, is 55 kJ/mole with a common example shown in Figure 19.1.

$$HCl + NaOH \longrightarrow NaCl + H_2O + Heat$$

FIGURE 19.1 Neutralization reaction of hydrochloric acid with sodium hydroxide to produce common table salt and water. The reaction also releases a substantial amount of energy on a per mole basis. Note that energy is not produced per se.

Is this sufficient for a thermal ablation? To answer that question we must consider several factors. Concentration and volume clearly play a role in the amount of energy released from any exothermic chemical reaction, so these must be factored into any range calculations. We chose to assess concentration ranges of 1–10 mole/liter and volumes of 1–10 mL. These concentrations span a wide practical range, and the volumes are in the range of what is already clinically used for chemical ablation. These conditions would provide a range of energy release spanning 0.5–5000 W depending on the time over which the reaction is carried out. We concluded that acid-base neutralization reactions do have the potential to release enough energy to raise the desired volume of tissue by the desired amount. Indeed, even with such small quantities this analysis highlights why general chemistry laboratory students are required for lab safety to store acids in one cupboard and bases in another.

Thus far, we have concerned ourselves with simple acids. Additional energy could be obtained from polyprotic acids such as sulfuric (H_2SO_4) and phosphoric acid (H_3PO_4). With these reagents, there are two acidic protons. Translated into practical terms, this means that more heat would be released if a second equivalent of base were added to the reaction to react with the additional available proton. The third proton of phosphoric acid is not sufficiently acidic (indeed, it is far on the alkaline side of the spectrum) to yield any useful amount of energy.

The energy release from some other familiar biochemical reactions is helpful in placing this kind of chemistry in context, as shown in Figure 19.2.

Perhaps the best known is the ubiquitous molecule we know as ATP, which releases 31 kJ/mole when broken down into ADP and P_i. This equates to a K_{eq} of 10^5, which is irreversible. Hydrolysis of this compound provides the energy for many of the essential functions of life and is also the reason our bodies stay warmer than our surroundings. The highest energy compound familiar in biochemistry is PEP, or phosphoenolpyruvate. Hydrolysis of PEP releases 62 kJ/mole, so it is apparent that formation of water, then, releases nearly as much energy as the most exothermic reaction that commonly occurs in the body.

Neutralization chemistry is only one category of exothermic chemistry. A brief examination of several other kinds of reactions (listed in Table 19.1) is useful in order to appreciate the full potential of potential inherent in thermochemical ablation. Some, for example, combustion of fuels such as hydrocarbons, rocket fuel, and explosives, clearly are not useful despite the enormous energy release possible. This is due in part to the fact that delivery of adequate amounts of oxygen to combine with fuel is challenging, but also to the fact that such reactions generate copious amounts of exhaust gases in situ that would have to be safely and quickly removed.

Hydrolysis reactions are another category, exemplified by the breakdown of ATP and PEP just described. However, this kind of chemistry can be further extended through the application of other electrophilic species. One example is acetyl chloride, the hydrolysis of which has been measured in the range of 90 kJ/mole,[30] or nearly half again as energetic as PEP and approaching twice that of the formation of water and illustrated in Figure 19.3.

The product of this reaction is acetic acid, which as noted earlier on is itself a chemical ablation agent. However, in addition to the equivalent of acetic acid, the byproduct of the hydrolysis reaction is itself an acid. This product, hydrochloric acid, is even stronger than acetic acid. Thus, not only does the reaction release a substantial amount of heat energy, it produces two equivalents of acid. Strong electrophiles may therefore find utility, particularly where small volumes yet higher ablation efficiency is useful.

Oxidation-reduction or redox chemistry is another category with potential. For example, thermite reactions can be an order of magnitude more energetic than acid-base chemistry, with exotherms on the order of 800 kJ/mole or even more. At first glance, a more-is-better approach might seem to be the best of all possible worlds. However, the extreme conditions required to initiate these reactions are usually beyond practical reach (often exceeding 400°C), and they can be very difficult or impossible to control once started. Translated into practical terms, this means that a molten bolus could easily burn a hole through a patient, the table, and potentially even the floor of the IR suite much like flowing lava. This kind of reaction is also not readily extinguished and must burn itself out. Clearly, there are some practical upper limits to how much energy is released and how quickly.

A more controlled example of redox chemistry is the permanganate oxidation of carbohydrates and similar structures.[31] Here the oxidizing agent is soluble in water, and concentration and substrate both can play a role in the amount of energy released. Relatively simple carbohydrates and related alcohols, such as ethylene glycol, glycerol, and simple sugars such as glucose, fructose, and sucrose provide an easily accessible fuel source. More complex carbohydrates and other fuel substrates such as glycogen and some polymers are not as accessible to the oxidizer and do no react as readily, and neither does tissue. The resulting exotherms are progressively smaller as the rate of reaction is slowed down.

Heat of solvation is an abundant source of energy. This area revolves around the greater stability (lower energy state) of hydrated reagents and the heat evolved as this occurs. Examples would include the hydration of sulfuric acid or sodium hydroxide. The dilution of concentrated sulfuric acid releases approximately 95 kJ/mole of energy. Another solvation process that has particular appeal would be in situ hydration of quicklime, also known as slaking of lime. Calcium oxide hydrates and becomes calcium hydroxide in a very vigorous reaction that releases approximately 64 kJ/mole of energy. In either case, the pH would be drastically altered as well. Indeed, in the unfortunate situation of an industrial accident with an acid burn, both effects, heat from solvation and drastic pH change, are in full operation with a resulting bad outcome. Handling of quicklime is always done with protective equipment for similar reasons.

A final source of heat is release of energy that is intrinsic to a molecular structure itself rather than bond energies. An example is ring-strain energy found in three- and four-membered ring compounds such epoxides, cylcopropanes, aziridines, and the like.

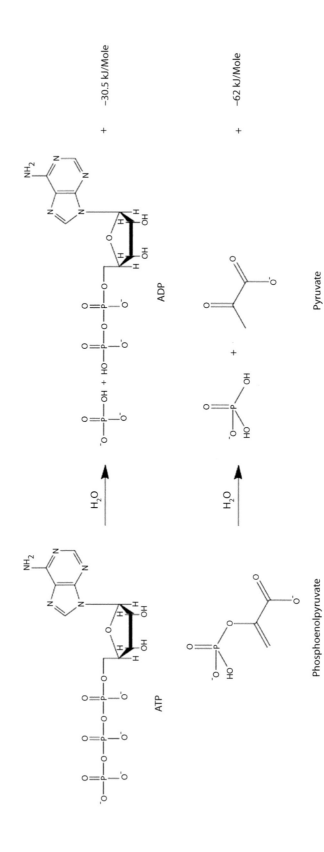

FIGURE 19.2 Two familiar biochemical substrates used in eukaryotic cells to release energy. Adenosine triphosphate is the common currency for many pathways and is kinetically more stable than the high energy phosphoenolpyruvate.

TABLE 19.1   Broad Categories of Exothermic Chemical Transformations That Could Be Exploited for Ablation of Tissues

Sources of Chemical Energy
Acid-base neutralization (formation of water)
Hydrolysis
Oxidation-reduction reactions (redox chemistry)
Heat of solvation
Ring strain energy

TABLE 19.2   Criteria for Selection of Reagents Potentially Useful in Thermochemical Ablation

Some Selection Criteria
Liquid to facilitate injection
Easy to handle
Safe to handle
Low cost
Readily available
Adequate release of energy

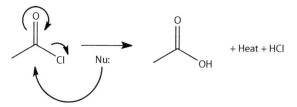

FIGURE 19.3   Hydrolysis of acetyl chloride by water. This reactive compound is converted to acetic acid in a vigorous reaction that simultaneously generates an equivalent of an even stronger acid, hydrochloric acid.

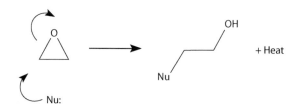

FIGURE 19.4   The ring opening of a compound with strain energy such as an epoxide can produce an exotherm that is extremely energetic (127 kJ/Mole).

Opening the ring of an epoxide in a reaction with a nucleophilic species evolves approximately 127 kJ/mole, shown in Figure 19.4.

This may be attractive from one point of view, but these compounds are typically alkylating agents and therefore many are carcinogenic.

19.1 Practical Considerations

There are several constraints to bear in mind for exploiting any chemistry for a percutaneous therapeutic application. Several criteria are listed in Table 19.2.

In general, liquids are preferable to solids because of the ease of injection into tissues and the ability to spread from the injection site. The reagents should be easy and safe to handle, with minimal or very low risk to both patient and personnel should any spillage occur. Low cost and ready availability both would be desirable, and the amount of energy released would need to be commensurate with the task at hand. Applying these criteria as filters for the categories discussed herein leads naturally to acid-base neutralization. Provided the pK_a values are in a

useful range to release enough energy, and that the salts produced are safe at the doses necessary, it should be possible to coagulate tissues by this method. Details about precise reagent choice and regarding delivery therefore need to be investigated and addressed. Chemistry will be discussed in more detail first, followed by device development and applications.

In order to assess a number of potential combinations, development of a phantom was necessary. Criteria for a suitable phantom were identified and are listed in Table 19.3.

Ultimately, a commonly available household product called baby oil gel found in pharmacies and large discount stores was drafted into service as a simple calorimeter. Using this approach, we characterized the relationships between reagent strength and concentration.[32] The chemistry was remarkably consistent under these conditions, showing clearly that for a given concentration, the stronger reagents evolved larger amounts of heat energy. It was also clear that higher concentrations of reagents released more energy.[33] This makes it possible to compare a wide variety of reagents and conditions.

Much attention is given to the ability to monitor treatment progression in procedures such as ablations. With chemical ablation, some have added iodinated contrast material to the reagent, such as was reported using acetic acid doped with a small amount of contrast.[34] The difficulty with having separate imaging and therapeutic agents is that it is based on the assumption that the materials interact with and spread through tissues equally. A more rigorous approach would consider the heterogeneous nature of the tissues to be analogous to the packing material in column chromatography. In such a case, the injection point would be considered to correspond to the top of a column, and the materials in the mixture would be expected to pass through at different rates. It would be no surprise to find

TABLE 19.3   Criteria for a Phantom (Calorimeter) for In Vitro Studies of Candidate Reagents for Thermochemical Ablation

Requirements for a Useful Phantom
Clear
Colorless
Medium viscosity
Neutral density
Nontoxic
Low cost
Chemically stable to conditions
Nonvolatile

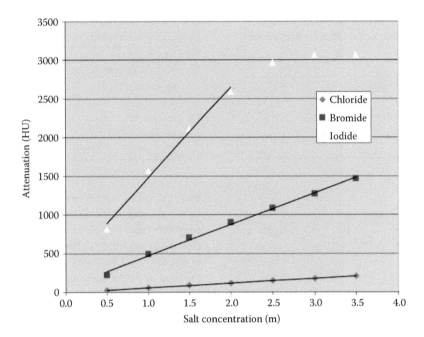

FIGURE 19.5 Attenuation increases associated with increasing iodine concentration. In general, a change of 50–100 HU (Hounsfield Units) is more than adequate to perceive a difference in contrast. It is apparent that even low concentrations of iodide are readily detected and higher concentrations effectively saturate the system.

that contrast, reagents, and salt products would migrate through tissues at different rates.

To address this we considered that it should be possible to use an acid with the anion positioned lower on the periodic table with a higher atomic number and atomic mass. Two obvious choices were HBr and HI. Both of these are strong acids, and when tested in the phantom, performed equally as well as their less dense counterpart, HCl, at the maximum concentrations available. In particular, HI is attractive because it has a dramatic effect on attenuation at CT even at relatively low concentrations, as shown in Figure 19.5, where the relative attenuation increases with the increase in iodine content.

Thus, it is possible to use a portion of HI mixed with HCl to obtain a suitably dense mixture for thermochemical ablation. Unlike adding contrast, which has a very different structure and physical properties than the acid, it seems safe to assume that where the density from iodide is present, there too is acid.

The next logical step was in vitro testing in tissues. In order to do this, though, it was necessary to invent a device that would channel two reagents into the tissues but only allow reaction at or very near the tip. The resources for a custom device were not available, and furthermore we considered it desirable for others to be able to repeat the results elsewhere without a large investment. A survey of existing components readily available in the interventional radiology suite led to a combination shown in Figure 19.6.

The components are a coaxial biopsy cannula/trocar system, a rotary hemostatic valve attached to the hub, and a smaller coaxial needle. The needle must be of sufficient length to pass through the valve such that the tip extends nearly to the tip of the cannula. There are then two injection ports, one through the

inner needle, and one via the side arm of the hemostatic valve. It is relatively straightforward to connect these via extension tubing to syringes loaded with reagents, and a syringe pump can be incorporated to ensure consistent injection rates.

Use of such a device then allowed us to test thermochemical ablation in tissues such as ex vivo liver. We used a combination of thermocouple data and infrared imaging for the initial set of experiments.[35] Although infrared imaging is not capable of penetrating deeply into tissues and is not a volumetric modality, it did provide the first level of data upon which to build. Sectioned tissues are shown in Figure 19.7 to demonstrate the shape of the heated area at least on the exposed surfaces, in which the specimens were bivalved as soon as possible after the completion of injections to visualize the heated area. Here again, as in the in vitro phantom experiments, the chemistry was consistent.

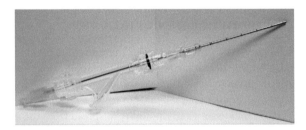

FIGURE 19.6 Components of a prototype thermochemical ablation device. A rotary hemostatic valve is connected to a biopsy outer cannula, and through this assembly a long needle is inserted coaxially such that the tips are at nearly the same distance. When the two injection ports are connected to appropriate solutions via tubing, the reagents will only mix and react near the tip, exiting as a hot, hyperosmolar solution.

FIGURE 19.7 Infrared image of bivalved ex vivo pig liver specimen taken as soon as practical after completion of injection using 0.5 mL each of HCl and NaOH at 7 mol/L. The area of the excursion is relatively rounded and homogeneous within the camera's resolution. Note that the central area is saturated, indicating a temperature above the range setting (20°C–45°C) on the camera.

Higher temperatures were produced when higher concentrations were used, and larger injection volumes produced larger areas of coagulation.

The next step was to compare two injection methods, simultaneous and sequential, to positive and negative controls. We chose to look at peak temperatures, volume of tissue coagulated, and the shape of lesions (the sphericity or roundness coefficient). In addition, we were interested in any changes in heart rate or blood pressure in response to the injections. Based on previous work, we used 10 M solutions of reagents, which therefore would produce 5 M salt solutions if both starting materials were to react completely.

Experiments were performed in healthy swine with surgical exposure and intraoperative ultrasound guidance for device placement. This was necessary in order to avoid positioning in or directly adjacent to large vessels, which are ubiquitous in a healthy liver. A thermocouple was positioned within a few mm at the same depth as the device. Contrast was added to the acid in order to aid in localization of the lesions by CT of the explanted specimens. Each animal was treated in the same manner with four injections done in the same order. First, a saline sham with contrast was performed. Following this, acid and then the simultaneous injections were performed. Finally, the sequential injections were done. It quickly became apparent that sequential injections were not tolerated as well, particularly given that one of the animals went into cardiac arrest after a sequential injection and required resuscitation to finish the experiment. Fortunately, according to the protocol, the sequential injection was the final injection and we needed only to restore circulation long enough to be satisfied that the acute crisis had passed. This situation was not seen with any of the other injections. Upon completion of the experiments, the animals were sacrificed and the livers were explanted. The organ was then scanned with CT imaging, and the results were correlated with the findings at histopathology.

The results were interesting from several angles. One was that as we had hoped, the simultaneous injection produced the highest peak temperatures, depicted in Figure 19.8. This implies that in the sequential injections not all of the acid and base reacted together. We believe that unreacted reagents, especially the base since it is the second reagent, likely escape into the vasculature with sequential injection.

Another interesting result was that again, as we had hoped, the volumes of coagulation were larger than when acid was injected alone, shown in Figure 19.9. At first glance this might seem to be an unrealistic comparison. After all, the volume injected in the acid alone was only half that used with either of the other two methods since the base was also injected. However, one of the points in this research was to show that using the same amount of acid we could obtain a larger volume of injection with a different method and conditions.

It might be argued that additional experiments should be done using the same total volume for the acid alone, but this raises two problems. One is that by diluting with water to keep the total amount of acid the same would reduce the concentration of the acid and therefore not be an accurate representation of the current practice. The other would be to keep the concentration the same and just use twice the volume. This also would not be a suitable comparison, as we would then be using twice as much acid in the acid alone experiments.

Predictability of lesion shape is key in ablation therapies. With injected therapies, the amount that is used, the viscosity, the hydraulic conductivity of the tissue, and the rate of injection all affect the distribution. To analyze the results, we performed a sphericity analysis. The coefficient reflects the degree of deviation away

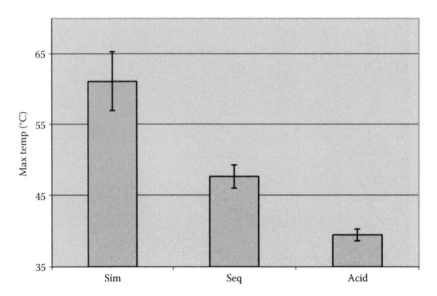

FIGURE 19.8 Average temperature graphed as a function of method of injection. Simultaneous injections produced the highest average and peak temperatures compared to sequential injection. Reagents were used at a concentration of 10 mol/L.

from a sphere. To our surprise, the ratios of the different methods were not significantly different and were reasonably high.

Finally, there is the question of total volume. Here there was no question that the volume of coagulation obtained with sequential injection was clearly the greatest among all the conditions tested. However, as noted before, this was not tolerated nearly as well physiologically. In both methods, the volume was a large improvement over the use of acid alone.

The CT imaging of the explants proved disappointing, with a relatively poor correlation between distribution on imaging and what was found on sectioning (Figure 19.10a and b). The distribution of contrast within the tissues was uneven and did not track closely enough to the pathology to be useful. The

correlation was closer with acid alone, as mentioned earlier. Histologically, coagulative necrosis was quite apparent, illustrated in Figure 19.11.

Given that simultaneous injection seemed to be tolerated well and improved upon use of acid alone, we then investigated the relationships between both temperature and final volume of coagulation depending on the concentration of the reagents and the volume injected. Animals were treated according to a protocol in which four identical injections were administered. Each animal was assigned to a position on a grid using total injection volumes of 1, 2, and 4 cc. The concentrations used were 5, 10, and 15 M. For statistical considerations, we used four animals in the center of the grid.

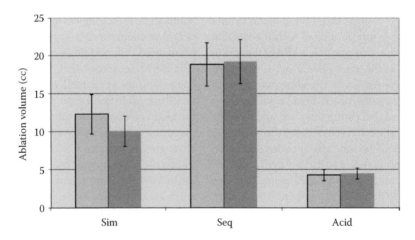

FIGURE 19.9 Comparison of volumes for zones of coagulation among different conditions for acetic acid by itself, simultaneous injection with sodium hydroxide, and acid injected first followed by sodium hydroxide solution. Acid alone produced the smallest volume, and the volumes obtained from either of the other methods were larger.

(a)

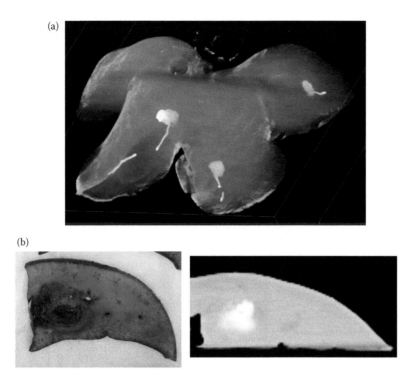

(b)

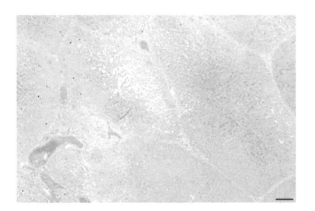

FIGURE 19.10 (a) CT reconstruction of the explanted liver. The saline sham was not visualized but contrast persisted in the treated areas with each tract marked by a piece of pasta dipped in India ink. (b) A representative CT cross section and the corresponding section at gross pathology. The attenuation due to contrast is not homogenous and is likely therefore unreliable for use as a real-time marker for an ablation.

All animals tolerated the injections well. Even four such injections at the highest volume and the highest concentrations produced no appreciable changes in the physiology. Gross pathology of formalin-fixed specimens is shown in Figure 19.12. Appearance of the tissue resembles charring from hyperthermia in many cases, but based on the temperatures seen, this was a possibility in only a rather small subset of cases and would have only been likely right near the center of an injection. These observations suggest that effects other than temperature are also a factor in the outcomes. Peak temperature observed at the highest concentration (15 M) was 105°C, and the maximum average peak temperature was 91°C. The correlation between concentration and peak temperature was apparent, as depicted in Figure 19.13.

Volume relationships, however, were not so simple and are still not completely understood. This is illustrated in Figure 19.14. The largest coagulum was not obtained using the most extreme or energetic conditions (largest volume at highest concentration). At this stage, it is not clear whether this result is an artifact or if it would prove a durable result. The nature of liver tissue is such that it is rather porous, rich with vasculature, and very little is yet known about the mechanism. Furthermore, there is likely a maximum attainable volume from a point injection before

FIGURE 19.11 Histologic section of thermochemical ablation using 10 mol/L reagents. Coagulative necrosis is evident with ruptured cell membranes, pale cytoplasm, shrunken and irregular nuclei, and hemorrhage (40X, H&E stain, bar equals 100 microns).

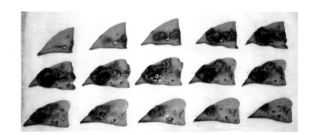

FIGURE 19.12 Typical formalin-fixed specimen following thermochemical ablation, sectioned at 3 mm for volume and sphericity calculations. Simultaneous injection, 10 mol/L with 2cc each of acetic acid and sodium hydroxide.

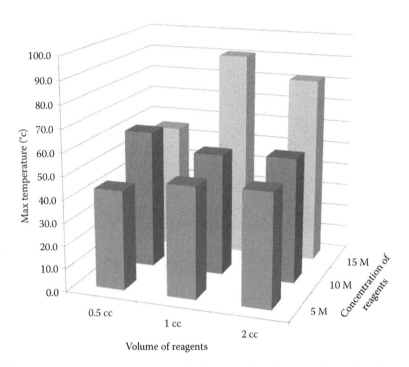

FIGURE 19.13 Relationships between recorded temperatures at the thermocouple and the conditions for each experiment. The highest concentration (15 mol/L) gave rise to the highest peak and average temperatures.

intravascular or other undesired spillage or tracking would dominate an injection. This becomes an engineering challenge to develop a device similar in some respects to an existing multi-tined needle currently used for single-agent chemical ablation.

The temperature excursion, as high as it is in these injections, is relatively brief. Furthermore, with the data obtained thus far, nothing is known about the temperatures extending outward to the edge of the ablation zone. Three-dimensional imaging such

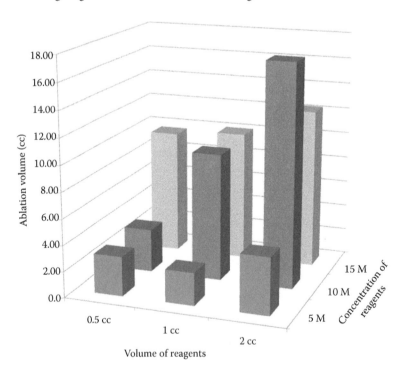

FIGURE 19.14 Relationships between volumes of the coagulum obtained and the conditions of each experiment. The highest volume of coagulum is not seen with the most concentrated reagents and highest volume. For both the lowest and highest concentrations, the coagulum volume does not vary greatly with change in injection volume.

as MR thermometry will be necessary to document the degree of the change at the periphery, the gradient that must exist from the center of the injection outward, and the time course of the decay back to baseline.

Given that the peak temperature starts to decline as soon as the input stops (analogous to running out of fuel), and that it is almost certainly not as hot at the edge as measured near the tip, other mechanisms must also be active. Two likely candidates are the hyperosmolarity of the reaction products and Hofmeister effects. To understand osmolarity better, it is helpful to have some context within the human body. Most cells live in a milieu in the range of 270–300 mOsm. The exception is in the distal convoluted tubule of the kidney, where an environment of up to 1200 mOsm has been observed in a normal kidney as it concentrates urine. This is an adapted, chronic condition at body temperature. In contrast, with a 15 M acid and base, the osmotic strength of the salt solution would be 15 kOsm. This approaches two orders of magnitude higher in concentration than almost anywhere in the body other than the specialized area in the kidney noted previously, and in an ablation it is an abrupt change. Furthermore, to varying degrees, the change in temperature will likely have some effects that are cooperative. A final note on this topic is that in a number of reagent choices, it is entirely possible to generate a salt at a concentration much higher than the solubility even at elevated temperatures. Thus, in the local environment it would be possible in theory to essentially inject a salt crystal deposit.

A detailed explanation of Hofmeister effects is outside the scope of this discussion, but a minimal understanding may shed some additional light on another aspect of how thermochemical ablation may work. Briefly, salts vary in their ability to either stabilize or destabilize (i.e., denature) proteins.[36–38] The eminent protein chemist Franz Hofmeister late in the nineteenth century formalized this in the series that now bears his name and is shown in Figure 19.15.

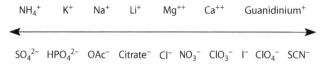

FIGURE 19.15 The Hofmeister series of anions and cations arranged according to denaturing ability. The more chaotropic (denaturing, from chaos or disorder) species are on the right side of the spectrum, and the more kosmotropic (favoring order) species are on the left.

He ranked various salts according to their effects on proteins and noted that anions generally have more effect than cations. It is useful to note where in the Hofmeister series the current salt, NaOAc, is positioned. As is apparent in the figure, this salt is at best intermediate in its intrinsic ability to denature proteins. Theoretically, then, the choice of the salt could be altered more in the direction of a denaturant. Few things are as simple as they seem, however, and thermochemical ablation using acid-base chemistry definitely entails other considerations. In an effort to shed more light on this, we studied the effects of exposure to numerous salts at 37°C at various concentrations using the MTT viability assay. This allowed us to separate the effects of salt and concentration from a temperature excursion; an example assay is shown in Figure 19.16 with sodium acetate as the salt. We used HuH7 cells as they represent a human hepatoma lineage.

Although little toxicity is observed at six hours with 200–400 mM sodium acetate, it is clear that few cells survive an extended exposure at this concentration. How much of this result is simply a hyperosmolarity effect rather than anything more complex is yet unresolved.

From the preceding discussion, we must keep in mind which reagents would be used to generate the salts, and determine how much of an energy release could be expected. This is predicated

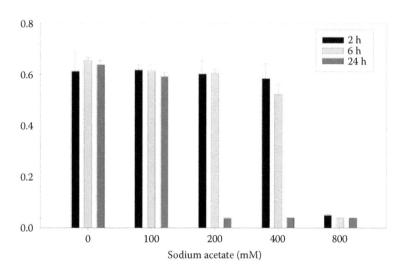

FIGURE 19.16 MTT viability assays of sodium acetate exposure on HuH7 cells with respect to duration and concentration. The survival drops dramatically at 24 h of exposure even at lower concentrations. These conditions might be observed at the periphery of a treated area, and the concentrations would be much higher closer to the center.

on the assumption, likely although yet unproven, that thermal effects are a significant contributor to the overall outcome. Systemic toxicity for a salt must also be considered. The initial choice of acetic acid and sodium hydroxide to produce sodium acetate was at best an educated guess. Our reasoning was that there is prior clinical experience with acetic acid alone as noted earlier, the amount of energy released although intermediate appeared adequate, and the salt is readily converted in the body to bicarbonate.[39] Indeed, it is actually difficult to measure serum acetate concentrations for this reason.

19.2 Future Directions

There are many avenues that could provide fruitful opportunity to explore in the future in the realm of thermochemical ablation. As a thermal method, a natural question concerns the thermal dose and the distribution of that dose. Thus, various methods of thermometry and thermography will play an important role in delineating what is occurring within a treated area over time. Another obvious one is that to date this new method has not been tested against tumors in vivo. With regard to performance, differences in hydraulic conductivity between tumor and surrounding tissues can be substantial. Certain typically encapsulated tumors, such as HCC in a cirrhotic liver, may respond well, but a role for an injectable therapy in metastatic disease surrounded by normal, nonfibrotic tissue is less clear. Perhaps a single-agent thermochemical ablation such as with an energetic electrophile distributed in small volumes via a multi-tined needle will prove beneficial. No doubt device development will be essential in achieving better results than those obtained with relatively crude prototypes. Studies to date have only been conducted in the acute setting. Much work remains to truly understand the physiology and safety profile for any particular reaction in survival settings ranging from acute to chronic. Determination of the mechanisms of action will be key to understanding this technique. These will be different depending on the conditions chosen, but some common themes will likely emerge related to temperature, concentration, and Hofmeister effects to begin with. Imageable agents have not yet been shown to perform satisfactorily in vivo, though ex vivo results are encouraging and the chemistry appears to be consistent. Lastly, it is also possible that other applications not directly related to tumor therapy could be considered. Virtually anywhere that hyperthermia is used for therapeutic purposes could provide a starting point for further research. Given the simplicity, ready availability, and relative safety of this concept, it appears that there is much opportunity in the field of thermochemical ablation for future research.

References

1. Kuo YH, Lu SN, Chen CL et al. Hepatocellular carcinoma surveillance and appropriate treatment options improve survival for patients with liver cirrhosis. *Eur J Cancer.* 2010;46(4):744–751.

2. Schumacher PA, Powell JJ, MacNeill AJ et al. Multimodal therapy for hepatocellular carcinoma: A complementary approach to liver transplantation. *Ann Hepatol.* 2010;9(1):23–32.

3. Cabibbo G, Enea M, Attanasio M, Bruix J, Craxi A, Camma C. A meta-analysis of survival rates of untreated patients in randomized clinical trials of hepatocellular carcinoma. *Hepatology.* 2010;51(4):1274–1283.

4. Rimassa L, Santoro A. Sorafenib therapy in advanced hepatocellular carcinoma: The SHARP trial. *Expert Rev Anticancer Ther.* 2009;9(6):739–745.

5. Llovet JM, Ricci S, Mazzaferro V et al. Sorafenib in advanced hepatocellular carcinoma. *N Engl J Med.* 2008;359(4):378–390. 10.1056/NEJMoa0708857.

6. Lee WJ, Lee JL, Chang SE et al. Cutaneous adverse effects in patients treated with the multitargeted kinase inhibitors sorafenib and sunitinib. *Br J Dermatol.* 2009;161(5):1045–1051.

7. Clark TW. Chemical ablation of liver cancer. *Tech Vasc Interv Radiol.* 2007;10(1):58–63.

8. Clark TW, Soulen MC. Chemical ablation of hepatocellular carcinoma. *J Vasc Interv Radiol.* 2002;13(9 Pt 2):S245–52.

9. Bhardwaj N, Strickland AD, Ahmad F, Dennison AR, Lloyd DM. Liver ablation techniques: A review. *Surg Endosc.* 2010; 24(2):254–265.

10. Boutros C, Somasundar P, Garrean S, Saied A, Espat NJ. Microwave coagulation therapy for hepatic tumors: Review of the literature and critical analysis. *Surg Oncol.* 2010;19(1):e22–32. 10.1016/j.suronc.2009.02.001.

11. Kudo M. Radiofrequency ablation for hepatocellular carcinoma: Updated review in 2010. *Oncology.* 2010;78 Suppl 1:113–124. 10.1159/000315239.

12. Chaussy C, Thuroff S, Rebillard X, Gelet A. Technology insight: High-intensity focused ultrasound for urologic cancers. *Nat Clin Pract Urol.* 2005;2(4):191–198.

13. Caballero JM, Borrat P, Paraira M, Marti L, Ristol J. Extracorporeal high-intensity focused ultrasound: Therapeutic alternative for renal tumors. *Actas Urol Esp.* 2010;34(5):403–411.

14. Bradley WG, Jr. MR-guided focused ultrasound: A potentially disruptive technology. *J Am Coll Radiol.* 2009;6(7):510–513.

15. Livraghi T, Benedini V, Lazzaroni S, Meloni F, Torzilli G, Vettori C. Long term results of single session percutaneous ethanol injection in patients with large hepatocellular carcinoma. *Cancer.* 1998;83(1):48–57.

16. Ohnishi K, Ohyama N, Ito S, Fujiwara K. Small hepatocellular carcinoma: Treatment with US-guided intratumoral injection of acetic acid. *Radiology.* 1994;193(3):747–752.

17. Ohnishi K, Yoshioka H, Ito S, Fujiwara K. Treatment of nodular hepatocellular carcinoma larger than 3 cm with ultrasound-guided percutaneous acetic acid injection. *Hepatology.* 1996;24(6):1379–1385.

18. Arnulf F, Monika S, Herwig S et al. Atropine for prevention of cardiac dysrhythmias in patients with hepatocellular carcinoma undergoing percutaneous ethanol instillation: A randomized, placebo-controlled, double-blind trial. *Liver Int.* 2009;29(5):715–720.

19. Burton KR, O'Dwyer H, Scudamore C. Percutaneous ethanol ablation of hepatocellular carcinoma: Periprocedural onset alcohol toxicity and pancreatitis following conventional percutaneous ethanol ablation treatment. *Can J Gastroenterol.* 2009;23(8):554–556.

20. Sidi A, Naik B, Urdaneta F, Muehlschlegel JD, Kirby DS, Lobato EB. Treatment of ethanol-induced acute pulmonary hypertension and right ventricular dysfunction in pigs, by sildenafil analogue (UK343-664) or nitroglycerin. *Ann Card Anaesth.* 2008;11(2):97–104.

21. Sidi A, Naik B, Muehlschlegel JD, Kirby DS, Lobato EB. Ethanol-induced acute pulmonary hypertension and right ventricular dysfunction in pigs. *Br J Anaesth.* 2008;100(4):568–569.

22. Van Hoof M, Joris JP, Horsmans Y, Geubel A. Acute renal failure requiring haemodialysis after high doses percutaneous acetic acid injection for hepatocellular carcinoma. *Acta Gastroenterol Belg.* 1999;62(1):49–51.

23. Lencioni R, Crocetti L, Cioni D et al. Single-session percutaneous ethanol ablation of early-stage hepatocellular carcinoma with a multipronged injection needle: Results of a pilot clinical study. *J Vasc Interv Radiol.* 2010;21(10):1533–1538. 10.1016/j.jvir.2010.06.019.

24. Kuang M, Lu MD, Xie XY et al. Ethanol ablation of hepatocellular carcinoma up to 5.0 cm by using a multipronged injection needle with high-dose strategy. *Radiology.* 2009; 253(2):552–561.

25. Ho CS, Kachura JR, Gallinger S et al. Percutaneous ethanol injection of unresectable medium-to-large-sized hepatomas using a multipronged needle: efficacy and safety. *Cardiovasc Intervent Radiol.* 2007;30(2):241–247.

26. Deng ZS, Liu J. Minimally invasive thermotherapy method for tumor treatment based on an exothermic chemical reaction. *Minim Invasive Ther Allied Technol.* 2007;16(6):341–346.

27. Cooper TE, Trezek GJ. Correlation of thermal properties of some human tissue with water content. *Aerosp Med.* 1971;42(1):24–27.

28. Haemmerich D, Schutt DJ, dos Santos I, Webster JG, Mahvi DM. Measurement of temperature-dependent specific heat of biological tissues. *Physiol Meas.* 2005;26(1):59–67.

29. Haemmerich D, dos Santos I, Schutt DJ, Webster JG, Mahvi DM. In vitro measurements of temperature-dependent specific heat of liver tissue. *Med Eng Phys.* 2006;28(2):194–197. 10.1016/j.medengphy.2005.04.020.

30. Devore JA, O'Neal HE. Heats of formation of the acetyl halides and of the acetyl radical. *Journal of Physical Chemistry.* 1969;73(8):2644–2648.

31. Cressman EN, Tseng HJ, Talaie R, Henderson BM. A new heat source for thermochemical ablation based on redox chemistry: Initial studies using permanganate. *Int J Hyperthermia.* 2010;26(4):327–337.

32. Misselt AJ, Edelman TL, Choi JH, Bischof JC, Cressman EN. A hydrophobic gel phantom for study of thermochemical ablation: Initial results using a weak acid and weak base. *J Vasc Interv Radiol.* 2009;20(10):1352–1358.

33. Freeman LA, Anwer B, Brady RP et al. In vitro thermal profile suitability assessment of acids and bases for thermochemical ablation: Underlying principles. *J Vasc Interv Radiol.* 2010;21(3):381–385.

34. Arrive L, Rosmorduc O, Dahan H et al. Percutaneous acetic acid injection for hepatocellular carcinoma: Using CT fluoroscopy to evaluate distribution of acetic acid mixed with an iodinated contrast agent. *AJR Am J Roentgenol.* 2003;180(1):159–162.

35. Farnam JL, Smith BC, Johnson BR et al. Thermochemical ablation in an ex-vivo porcine liver model using acetic acid and sodium hydroxide: Proof of concept. *J Vasc Interv Radiol.* 2010;21(10):1573–1578. 10.1016/j.jvir.2010.06.012.

36. Shimizu S, McLaren WM, Matubayasi N. The Hofmeister series and protein-salt interactions. *J Chem Phys.* 2006;124:234905.

37. Broering JM, Bommarius AS. Evaluation of Hofmeister effects on the kinetic stability of proteins. *J Phys Chem B.* 2005;109(43):20612–20619.

38. Tadeo X, Pons M, Millet O. Influence of the Hofmeister anions on protein stability as studied by thermal denaturation and chemical shift perturbation. *Biochemistry (N Y).* 2007;46(3):917–923.

39. Tsai IC, Huang JW, Chu TS, Wu KD, Tsai TJ. Factors associated with metabolic acidosis in patients receiving parenteral nutrition. *Nephrology (Carlton).* 2007;12(1):3–7.

Index

T - #0247 - 111024 - C0 - 280/210/18 - PB - 9780367576639 - Gloss Lamination